PRINCIPLES OF NONLINEAR OPTICAL SPECTROSCOPY

OXFORD SERIES IN OPTICAL AND IMAGING SCIENCES

1. D.M. Lubman (ed.)
   Lasers and Mass Spectrometry
2. D. Sarid
   Scanning Force Microscopy With Applications to Electric, Magnetic, and Atomic Forces
3. A.B. Schvartsburg
   Non-linear Pulses in Integrated and Waveguide Optics
4. C.J. Chen
   Introduction to Scanning Tunneling Microscopy
5. D. Sarid
   Scanning Force Microscopy With Applications to Electric, Magnetic, and Atomic Forces, revised edition
6. S. Mukamel
   Principles of Nonlinear Optical Spectroscopy

PRINCIPLES OF NONLINEAR OPTICAL SPECTROSCOPY

Shaul Mukamel

University of Rochester
Rochester, New York

New York Oxford
OXFORD UNIVERSITY PRESS

Oxford University Press

Oxford New York
Athens Auckland Bangkok Bogotá Buenos Aires Calcutta
Cape Town Chennai Dar es Salaam Delhi Florence Hong Kong Istanbul
Karachi Kuala Lumpur Madrid Melbourne Mexico City Mumbai
Nairobi Paris São Paulo Singapore Taipei Tokyo Toronto Warsaw

and associated companies in
Berlin Ibadan

Published by Oxford University Press, Inc.,
198 Madison Avenue, New York, New York 10016
http://www.oup-usa.org

Library of Congress Cataloging-in-Publication Data

Mukamel, S (Shaul), 1948–
Principles of nonlinear optical spectroscopy / Shaul Mukamel.
p. cm.—(Oxford series in optical and imaging sciences ; 6)
Includes bibliographical references and index.
ISBN-13 978-0-19-513291-5

1. Spectrum analysis. 2. Density operators.
3. Nonlinear optics. 4. Femtosecond pulses.
I. Title. II. Series.
QC454.M32M85 1995 543'.0858—dc20
94-10702

Printed in the United States of America
on acid-free paper

To my parents Marcel and Meir Mukamel,
To Eran, Ronen, and Dana.

Preface to the Paperback Edition

It gives me great pleasure to thank the numerous colleagues and students from all over the world who have used the hardcover edition in graduate level classes and had made extremely useful comments and suggestions for improvements. Special thanks go to V. Chernyak, K. Okumura, M. Schultz, M. Toutounji and G. van der Zwan for their tremendous help. Major revisions and updates will have to wait for the future. The revisions made in this paperback edition were limited to the corrections of typos and to some clarifications of the presentation. The Guggeheim Fellowship and the Alexander von Humboldt Senior Scientist Award are gratefully acknowledged. The support and stimulating discussions with members of the Rochester Theory Center for Optical Science and Engineering are particularly appreciated.

Rochester, N.Y. S.M.
January 1999

Preface

Nonlinear optical interactions of laser fields with matter provide powerful spectroscopic tools for the understanding of microscopic interactions and dynamic processes. The ability to control pulse durations (to a few femtoseconds), bandwidths (up to a 1 Hz resolution), peak intensities (up to 10^{19} W/cm^2), and repetition rates provides novel probes of elementary dynamic events. The advent of laser pulse sequences with controlled shapes and phases have opened up new directions with exciting possibilities. Recent investigations of polyatomic molecules, for example, have closely followed the dynamics of vibrational motions, photodissociation, curve crossing, or isomerization. These studies apply to molecules in the gas phase, in clusters, in a matrix (solvent, polymers, glass, crystals, etc.), in restricted geometries (monolayers, multilayers, and zeolites), and to isolated ultracold molecules in supersonic beams. Similar progress has taken place in the studies of semiconductor nanostructures (quantum wells, dots, and wires). In addition, the ability to control nonlinear optical properties (e.g., nonlinear susceptibilities) is an important ingredient in the design and fabrication of new optical materials and devices.

One of the major obstacles facing researchers in this field is the flood of experimental techniques and terminologies that create a serious language barrier. Every new peak or minor modification in experimental set-up is given a name (usually with a four-letter acronym) and it is often a hopeless task to try to compare the information content of various techniques. In addition, the absence of a common language prohibits effective communication among various disciplines (e.g., semiconductor and organic materials) resulting often in parallel and redundant duplication of efforts and developments. A general microscopic theory of the nonlinear optical response is essential for a global interpretation of experimental observables.

This book attempts to provide a systematic and unifying viewpoint for a wide class of nonlinear spectroscopic techniques, in time domain and frequency domain. It grew out of a graduate-level course taught by the author at the University of Rochester over a period of 10 years, attended by physics, optics, chemistry, and engineering graduate students. It is directed toward researchers as well as graduate students who enter this complex and rapidly developing field. Current active research areas such as ultrafast time-domain techniques, the interplay of phase coherence in the matter and the coherence properties of the radiation field, experiments involving phase-controlled pulse sequences, and cooperative effects in nanostructures will be emphasized.

The approach adopted in this book is based on formulating the nonlinear response by representing the state of matter by the density operator and following its evolution in Liouville space. The density operator is commonly used in the theoretical formulation of spectroscopic measurements. The optical Bloch equations, for example, are based on a density operator, but treat only the few degrees of freedom directly coupled to the radiation field. Other degrees of freedom are incorporated phenomenologically as a homogeneous or an inhomogeneous broadening. Therefore, these equations, although very powerful, are restricted to simple models of the material system. Here, on the other hand, we keep the other degrees of freedom "alive" in the theoretical description treating them also by a density operator. In this way we are able to express the optical response through correlation functions without alluding to a specific model for the material system. We further establish numerous fundamental relationships among the various spectroscopic techniques.

The references at the end of each chapter should provide convenient leads for further study of recent developments. In the interest of clarity, and considering the enormous scope of this field, a systematic historical account was not always given. Recent review articles are cited when possible. The selection of citations is naturally subjective and I apologize for the many authors whose works are not explicitly referenced.

The outline of this book is as follows: following the introduction (Chapter 1), which presents a brief summary of experimental techniques and theoretical problems, we proceed in Chapter 2 to survey a few elementary tools of quantum mechanics relevant for calculating the time evolution of the wavefunction in Hilbert space. We further introduce notation and Green-function techniques that are repeatedly used throughout the book. The main formal tools are developed in Chapter 3, where we discuss the density operator evolution using the quantum Liouville equation and introduce the notion of Liouville space. The strength of this elegant formulation is that it is completely analogous to the conventional quantum dynamics surveyed in Chapter 2, provided the wavefunction is replaced by the density operator and the Hamiltonian is replaced by the Liouville operator. We simply work in a linear vector (Hilbert) space of higher dimensionality. For an N level system the wavefunction ψ is an N component vector whereas any dynamic variable such as the Hamiltonian H is an $N \times N$ matrix. In Liouville space the state of the system is represented by the density operator ρ, which is an N^2 component vector, and the Liouville operator L is an $N^2 \times N^2$ matrix.

After we pay the price of working in a space with higher dimensionality, we can immediately use all the elementary tools (Dirac's bra-ket notation, scalar products, propagators, Green functions, etc.) introduced in Chapter 2 in this new space. This approach and language are common in the studies of nuclear magnetic resonance and can be immediately applied to complex many-level systems. Converting it to a semiclassical formulation in the coordinate representation, and incorporating relevant models for nuclear degrees of freedom such as the Brownian oscillator and rotor and their coupling with electronic transitions is less straightforward. These aspects will be explored as well. Based on the author's experience, Chapter 3 should be the "potential barrier" for most readers.

it requires the adoption of a new and somewhat unconventional notation and getting used to thinking and "living" in Liouville space. I hope that the book will show how this investment pays off immediately and handsomely. In Chapter 4 we survey some elementary concepts of quantum electrodynamics related to the radiation–matter interaction. Optical polarization is the only material quantity that enters into the Maxwell equations. It therefore provides the complete material information necessary for the calculation of any optical measurement. We provide a microscopic rigorous definition for the optical polarization without invoking the dipole approximation. We further show how elementary measurements such as absorption and multiwave mixing can be related to the polarization. In Chapter 5 we develop the time-ordered perturbation theory for the density operator that allows the introduction of the material (linear and nonlinear) response functions and optical susceptibilities. Exact formal expressions for these quantities are derived, and the rewards of working in Liouville space are summarized.

In Chapter 6 we consider the linear and nonlinear optical response of a multilevel system. Various methods for maintaining the bookkeeping of interactions with the radiation field and introducing relaxation effects are discussed and compared. These include the density operator, the wavefunction, and the Heisenberg equation approaches. Expressions for the susceptibilities $\chi^{(1)}$, $\chi^{(2)}$, and $\chi^{(3)}$ are derived, graphic representation in terms of Liouville space pathways or double sided Feynman diagrams is introduced, and the connection to the optical Bloch equations is made. The harmonic oscillator (Drude) model for the linear response, and its possible extension to optical nonlinearities by adding anharmonities, is discussed. We further give a microscopic definition of homogeneous and inhomogeneous broadening. Semiclassical methods for simulating the response functions using classical trajectories and the phase averaging procedure are discussed in Chapter 7. In Chapter 8 we solve the response functions using the second-order cumulant expansion. Applications of this solution include the Multimode Brownian Oscillator model for nuclear dynamics, which can represent a wide range of physical situations ranging from coherent high-frequency vibrational modes to overdamped solvation modes that enter into the time-dependent Stokes shift. The mapping of classical simulations onto this model is discussed. Further applications include pressure broadening in gases as well as dielectric continuum and stochastic models of line-broadening. Closed expressions for $\chi^{(3)}$ for a system interacting with a bath with an arbitrary time scale (e.g., a polyatomic chromophore in a solvent) are derived. Chapters 9 through 15 are devoted to applications of the nonlinear response formalism to the analysis of various spectroscopic techniques. An application of the third-order response function to frequency-domain techniques such as fluorescence, spontaneous Raman, and coherent Raman spectroscopy is made in Chapter 9. Two important classes of techniques related to the third-order nonlinear response are presented in the following two chapters; in Chapter 10 we consider photon echoes, and in Chapter 11 we treat resonant gratings, impulsive pump-probe, hole burning, and fluorescence spectroscopy. A semiclassical picture and approximations for the nonlinear response function can be obtained by using the Wigner phase-space representation for the density operator. This results in an extremely powerful wavepacket representation of the material dynamics underlying the nonlinear response function. This picture is developed in Chapter 12. Applications of this approach to nonimpulsive measurements involving light pulses with finite duration such as pump-probe and hole-burning spectroscopy are made in Chapter 13, and off resonance Raman spectroscopy is treated in Chapter 14. In Chapter 15 we consider polarization (birefringence and dichroism) spectroscopies. The tensor nature of the nonlinear response function is discussed and a rotational diffusion model is applied for its evaluation. The complete wavepacket expression of the nonlinear response function, which holds for resonant as well as off-resonant detunings, is derived. The last two chapters treat the more complex problem of interacting chromophores. The simplest approach for optically dense media with a finite density of chromophores, the local-field approximation, is introduced in Chapter 16. In Chapter 17 we discuss the many-body aspects and cooperative effects in the nonlinear response, which are missed by the local-field approximation. The theoretical and experimental study of these many-body problems is still in its infancy compared with the well-established formulation of single-absorber problems. The quasiparticle vs. the molecular eigenstates approaches are compared. The chapter summarizes the open problems and challenges in this area.

The formal expressions for optical susceptibilities and response functions developed in Chapter 5 apply for a very general model of the matter. The response function is nonlocal in time as well as in space, reflecting spatial coherences induced in the system. In macroscopic systems the spatial information is contained in the wavevector dependence of susceptibilities (spatial dispersion). In addition, susceptibilities are high rank tensors that depend on the polarization directions of the various fields. However, all the applications given in Chapters 6 through 14 are limited to the simplest case of a medium consisting of non-interacting small particles. The common Bloch-Maxwell formulation is a special case of this model. The tensor notation is avoided whenever possible and the tensor aspects of the response are discussed only in Chapter 15. The more complex problem of many-interacting particles is treated in Chapters 16 and 17, where we also analyze the spatial nonlocality of the response. The wavevector dependence of optical susceptibilities can be used to probe directly transport processes of elementary excitations by using, e.g., transient grating spectroscopy. Cooperative effects, as well as spontaneous emission, are also introduced only in these chapters. Perhaps a more systematic approach would have been to start the applications in this book with the most general expression for the optical response including many-absorber models and spatial dispersion. We could then show how the various models, approximation schemes, and experimental techniques could be derived starting with a single unified framework. However, I believe that this order of presentation makes the book more readily accessible to a broader audience since the reader does not have to face the full complexity of the problem from the start. This also makes it easier to skip the more complex chapters and focus on the first parts. Chapters 2 and 3 contain a summary of mathematical techniques and approximations used in quantum dynamics, and Chapter 4 contains a rigorous treatment of the electromagnetic field and its interaction with the matter. When this book is used for a course, it is advisable to skip much of these chapters initially and to introduce the various techniques and concepts when applicable.

I wish to thank my colleagues and students at the University of Rochester who provided a stimulating research and teaching environment in the areas of nonlinear spectroscopy and optics. The useful contributions, suggestions, and assistance of W. B. Bosma, J. T. Buontempo, V. Chernyak, M. Cho, L. E.

Fried, J. K. Jenkins, A. E. Johnson, V. Khidekel, O. Kühn, J. A. Leegwater, R. F. Loring, P. Rott, M. A. Sepulveda, F. C. Spano, Y. Tanimura, M. Van Leeuwen, T. Wagersreiter and N. Wang made this book possible. Stimulating discussions with D. S. Chemla, K. Duppen, M. D. Fayer, G. R. Fleming, W. H. Knox, J. Krause, D. McMorrow, R. J. D. Miller, A. Muenter, A. B. Myers, K. A. Nelson, C. V. Shank, G. J. Small, G. I. Stegeman, D. A. Wiersma, K. R. Wilson, and A. H. Zewail were instrumental in making the connection with current experiments. Special thanks go to A. Ben-Reuven for introducing me to the intricacies of Liouville space and to Y. J. Yan, J. Knoester, and U. Fano for their critical comments and invaluable suggestions. The generous support of the Air Force Office of Scientific Research, the National Science Foundation and the Center for Photoinduced Charge Transfer, the Teacher Scholar award of the Camille and Henry Dreyfus Foundation, the Alfred P. Sloan Foundation, and the Petroleum Research Fund, administered by the American Chemical Society, is gratefully acknowledged. Finally, I wish to thank Ms. Julie Nowak for the careful typing and patience through the numerous revisions of the book.

Rochester, N.Y. S. M.
April 1994

Contents

PRINCIPLES OF NONLINEAR OPTICAL SPECTROSCOPY

CHAPTER 1

Introduction

The structure and dynamic processes of matter may be probed directly by its interactions with electromagnetic fields. It is hard to imagine our microscopic understanding of nature without the invaluable tools provided by nuclear magnetic resonance, microwave, infrared, optical, ultraviolet, and X-ray spectroscopies. There are many different reasons for the fundamental interest in optical techniques. Spectroscopists use the radiation field as a convenient probe for material systems. Optical engineers are more interested in the electromagnetic field itself (its intensity, coherence, and statistical properties), or in the design of new optical materials with specified characteristics (fast switching, large susceptibilities). This diversity of motivations and viewpoints has created a terminology barrier that the present book attempts to overcome.

Optical processes may be classified by a variety of criteria such as the order of the response with respect to the applied fields, the type of resonances, and the distinction between time- and frequency-domain techniques and between resonant and off-resonant techniques. These classifications will be outlined below.

LINEAR VERSUS NONLINEAR SPECTROSCOPY

Optical measurements can be naturally classified according to their power-law dependence on the external electric field. The *optical polarization*, which will be precisely defined and discussed in depth in Chapter 4, is the relevant material quantity that couples with the radiation field. The polarization component to nth order in the field is denoted $P^{(n)}$, [1]

$$P = P^{(1)} + P_{\mathrm{NL}}$$

$$P_{\mathrm{NL}} = P^{(2)} + P^{(3)} + \cdots$$

$P^{(1)}$ is the *linear polarization* that controls the linear optical response. Processes such as absorption, light propagation, reflection, and refraction involving a weak incoming field are all related to $P^{(1)}$.

We shall now make a brief survey of various low-order nonlinear techniques which are connected with the *nonlinear polarization* P_{NL}. The simplest and most common techniques are related to multiwave mixing. An n wave mixing process

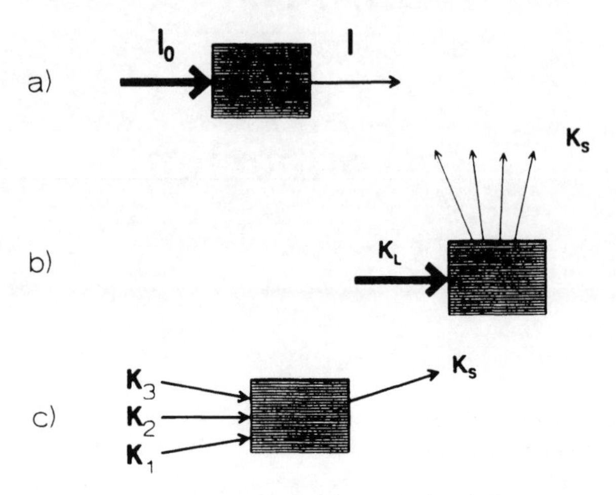

FIG. 1.1 Schematic representation of typical optical measurements. (a) Absorption spectroscopy. Attenuation of an incident beam by the medium. (b) Spontaneous light emission. An incident beam $\mathbf{k}_L$ induces spontaneous emission $\mathbf{k}_S$. The emission is commonly classified as direct scattering (Raman) and incoherent scattering (fluorescence). (c) Four-wave mixing. Three incident beams $\mathbf{k}_1$, $\mathbf{k}_2$, and $\mathbf{k}_3$ generate a coherent signal $\mathbf{k}_S$ in one of the phase matched directions $\mathbf{k}_S = \mathbf{k}_1 + \mathbf{k}_2 + \mathbf{k}_3$.

involves the interaction of n laser fields with wave vectors $\mathbf{k}_1, \mathbf{k}_2, \ldots$ and $\mathbf{k}_n$ and frequencies $\omega_1, \omega_2, \ldots$ and ω_n, respectively, with the material system. A coherently generated signal with wave vector $\mathbf{k}_s$ and frequency ω_s is then detected (Figure 1.1), where

$$\mathbf{k}_s = \pm\mathbf{k}_1 \pm \mathbf{k}_2 \pm \mathbf{k}_3 \cdots \pm \mathbf{k}_n \tag{1.1a}$$

and

$$\omega_s = \pm\omega_1 \pm \omega_2 \pm \omega_3 \cdots \pm \omega_n. \tag{1.1b}$$

Equations (1.1) imply that $\mathbf{k}_s$ and ω_s are given by any linear combination of the applied wave vectors and frequencies. Various processes differ in the choice of $\mathbf{k}_s$ and ω_s [i.e., the particular choice of signs in Eqs. (1.1)]. For $n = 2$ we have three wave mixing processes related to $P^{(2)}$. Examples are second harmonic generation ($\mathbf{k}_s = 2\mathbf{k}_1, \omega_s = 2\omega_1$) or more generally sum frequency generation ($\mathbf{k}_s = \mathbf{k}_1 + \mathbf{k}_2, \omega_s = \omega_1 + \omega_2$) and difference frequency generation ($\mathbf{k}_s = \mathbf{k}_1 - \mathbf{k}_2, \omega_s = \omega_1 - \omega_2$). $P^{(2)}$, which describes second-order nonlinearities, vanishes for random isotropic media with inversion symmetry, so that the lowest order optical nonlinearity is often related to the third-order polarization $P^{(3)}$.

Four-wave mixing (4WM) processes related to third-order nonlinearities ($n = 3$) play an important role in current studies of nonlinear optical phenomena. There are numerous spectroscopic techniques related to $P^{(3)}$; these include third harmonic generation, photon echo, transient grating, and coherent anti-Stokes Raman (CARS). In these applications the signal field generated is in a new direction ($\mathbf{k}_s$ is different from any of the incoming waves $\mathbf{k}_j$). It is possible to interpret the signal in terms of a grating formed by two beams and a third beam that undergoes a Bragg diffraction from that grating. In other applications such as hole-burning and pump-probe spectroscopy the signal is generated in the direction of one of the incoming beams and we measure its effect on that beam (self action).

TIME- VERSUS FREQUENCY-DOMAIN TECHNIQUES

Optical techniques also differ by the temporal profiles of the applied fields. In one limit the applied fields (and the signal) are stationary. In the opposite limit the applied fields are very short pulses, i.e., much shorter than any material timescale, except for the optical period (optical pulses should necessarily last at least a few optical cycles and they cannot be shorter than an optical period). These are ideal time-domain techniques [2]. Realistic experiments involving pulses with finite duration are characterized by a finite spectral and temporal resolution and are intermediate between these two ideal frequency and time-domain limits. In principle, time-domain and frequency-domain observables are related by multiple Fourier transforms and carry the same information. In practice, one of these approaches could be advantageous for a particular application.

The progress in the development of ultrashort pulses at various wavelengths is illustrated in Figure 1.2 and the exponential increase in time resolution is shown in Figure 1.3. Figure 1.4 shows a trace of a 6 fs pulse and Figure 1.5 summarizes the advances in the development of high power lasers.

High-resolution linear optical spectra of simple systems such as atomic vapor at low pressure or ultracold small molecules in supersonic beams are traditionally interpreted in terms of level positions and transition dipole elements. In the time domain, pulsed experiments carried out on systems with a sparse level structure may prepare a coherent superposition of these levels and could show coherent oscillations (quantum beats). In complex systems involving many degrees of freedom, the spectra contain a large (often macroscopic) number of highly

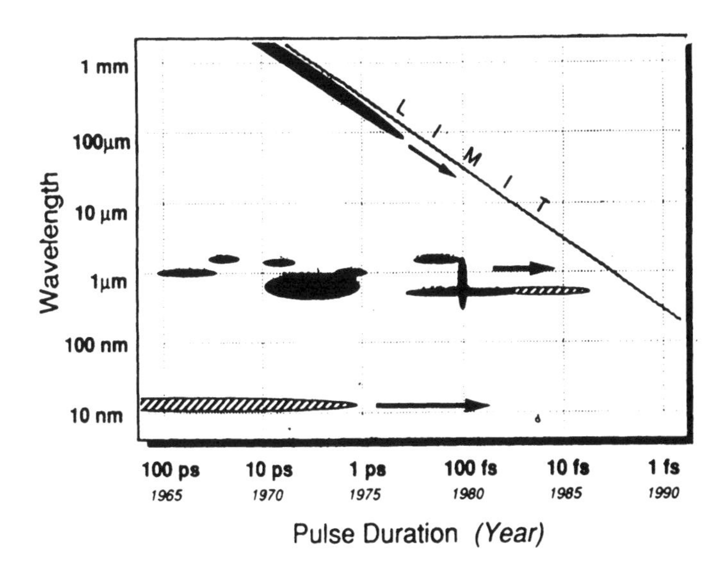

FIG. 1.2 Progress in the development of ultrafast sources (courtsy of G. A. Mourou).

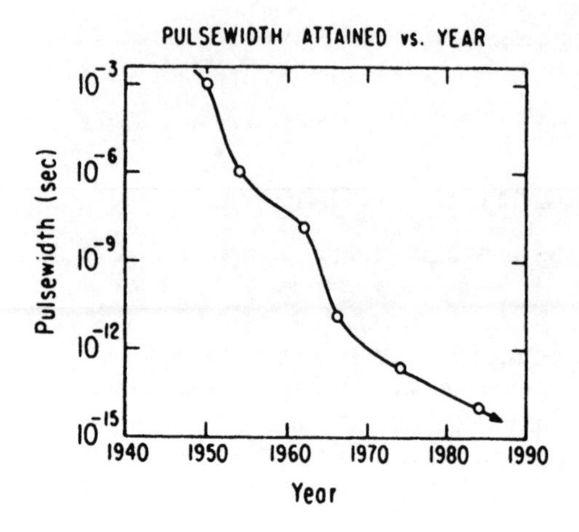

FIG. 1.3 Development of ultrashort pulse capability for time-resolved spectroscopy. [From G. R. Fleming, *Chemical Applications of Ultrafast Spectroscopy* (Oxford, London) (1986).].

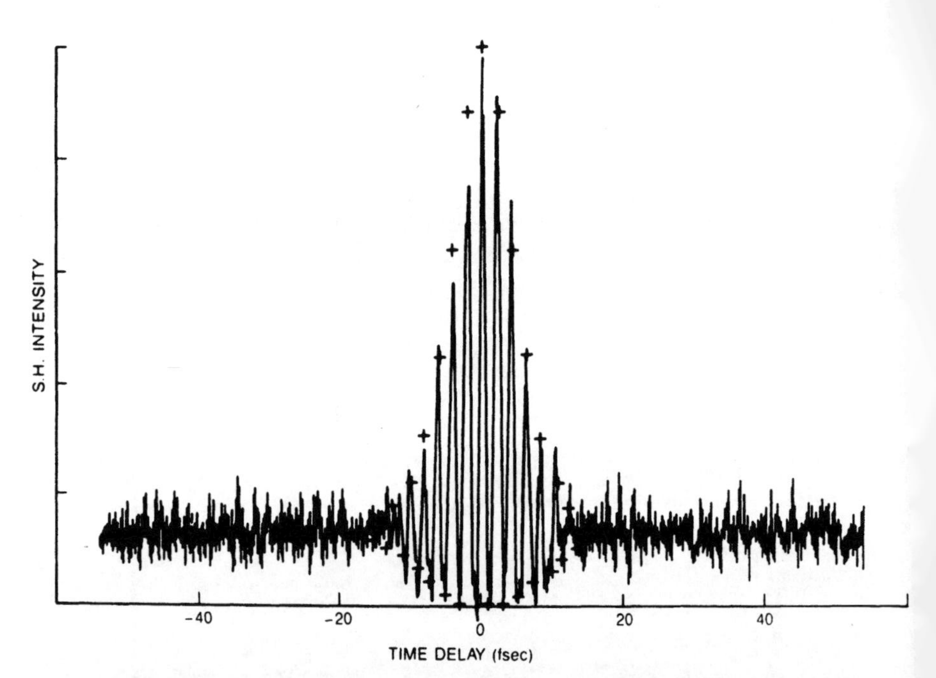

FIG. 1.4 Interferometric autocorrelation measurement of a 6 fs optical pulse (~3 optical cycles at 620 nm). The maxima and minima calculated for a 6 fs sech^2 pulse are indicated by crosses. [R. L. Fork, C. H. Brito Cruz, P. C. Becker, and C. V. Shank, "Compression of optical pulses to six femtoseconds by using cubic phase compensation," *Opt. Lett.* **12**, 483 (1987).]

congested lines that are not resolved individually. The spectra show line broadening in which information on individual eigenstates is highly averaged. The spectroscopic information is then contained in the envelopes (i.e., "lineshapes") of the spectral features. The time-domain signature of this situation is as follows: As the density of states is increased, the beat pattern becomes more complex and eventually turns into an apparent irreversible decay that can be interpreted in

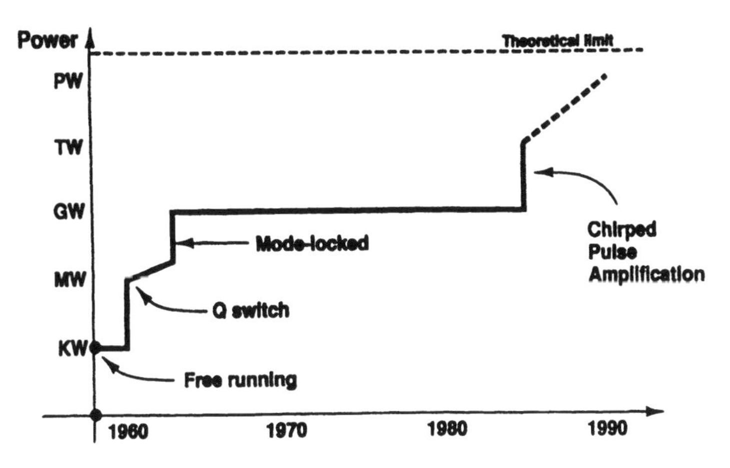

FIG. 1.5 Power of table top laser systems (courtesy of G. A. Mourou).

terms of relaxation processes. A frequency-domain description is therefore most appropriate for small and simple systems whereas a time-domain description is more adequate for complex and large systems.

A femtosecond pump-probe measurement is displayed in Figure 1.6. High resolution frequency-domain spectra are shown in a semiconductor (Figure 1.7), in an isolated polyatomic molecule (Figure 1.8), and in a dye-doped polymer (Figure 1.9).

Optical spectra in condensed phases often probe both high-frequency motions and low-frequency damped motions. The spectra thus consist of well-resolved and broadened spectral features, and a mixed frequency and time-domain description is most appropriate. The line positions bear on eigenstates in the frequency domain, but the lineshapes are analyzed through relaxation in the time domain. The theoretical description of nonlinear spectroscopy should depend on the temporal and spectral resolution. Generally a frequency-domain representation is adequate for long pulses whereas a time-domain description will work better for shorter pulses. We should emphasize that it is always possible to express the optical response of the system (be it linear or nonlinear) in terms of multiple summations over eigenstates. It is a matter of choice to switch to the time-domain representation in highly congested spectral regions where frequency-domain summations become tedious, but the time-domain quantities become smooth, simple, and more insightful.

Finally, we note that the ability to simultaneously control several time and/or frequency variables results in a multidimensional spectroscopy. Two-dimensional and multidimensional spectroscopic techniques are well established in nuclear magnetic resonance.

RESONANT VERSUS OFF-RESONANT RESPONSE

Spectroscopic applications of nonlinear optics are often carried out in a resonant mode whereby one field frequency or a combination of field frequencies is equal to

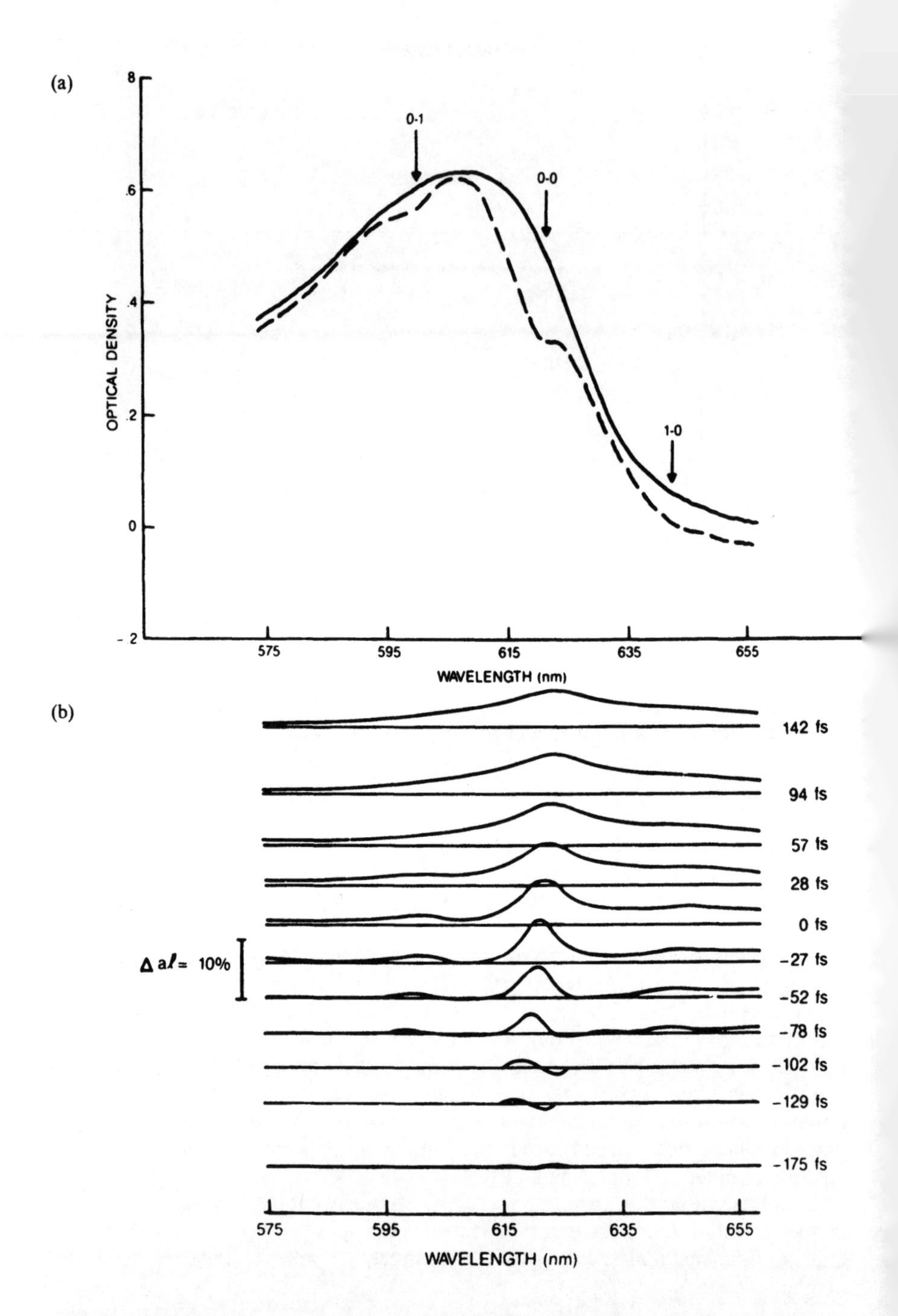

FIG. 1.6 (a) Pump-probe absorption spectrum of the molecule cresyl violet near zero time delay before (solid line) and after excitation (dashed line) with a 60 fs optical pulse. (b) Differential absorbence spectra plotted as a function of relative time delay following excitation with a 60 fs optical pulse at 618 nm for the molecule cresyl violet. [C. H. Brito Cruz, R. L. Fork, W. H. Knox, and C. V. Shank, "Spectral hole burning in large molecules probed with 10 fs optical pulses," *Chem. Phys. Lett.* **132**, 341 (1986).]

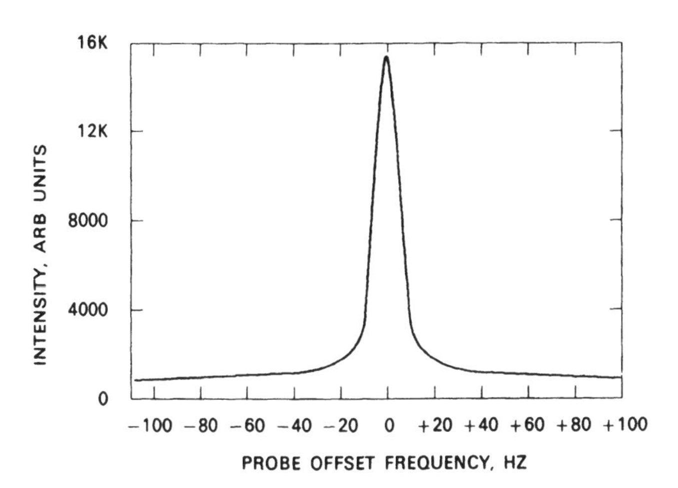

FIG. 1.7 Nearly degenerate four wave mixing signal in a semiconductor as a function of pump-probe detuning. [D. G. Steel and S. C. Rand, "Ultranarrow nonlinear optical resonances in solids," *Phys. Rev. Lett.* **55**, 2285 (1985).]

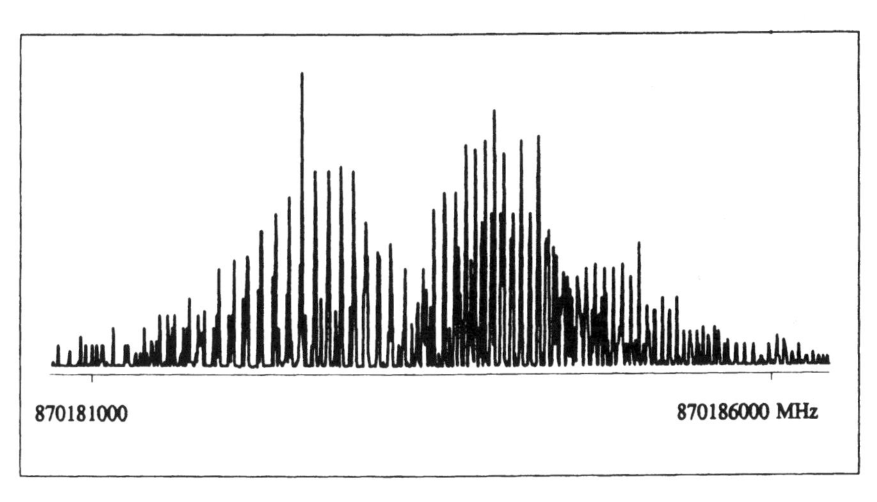

FIG. 1.8 High resolution molecular spectrum. The fluorescence excitation spectrum for all-*trans*-octatetraene in a molecular beam. The band shown in the first allowed vibronic band in the one-photon forbidden $2^1A_g \leftarrow 1^1A_g$ transition, at 29 024.9 cm^{-1} (0^0_0 + 76.2 cm^{-1}). Absolute frequency markers are in MHZ. [Courtesy of B. A. Tounge, R. L. Christensen, J. F. Pfanstiel, and D. W. Pratt.]

characteristic frequencies of matter. Resonant techniques provide a direct probe for specific eigenstates and their dynamic behavior. They are also sensitive to relaxation processes including spontaneous emission. Some examples of femtosecond resonant measurements are shown in Figures 1.10 and 1.11. Figure 1.10 shows pump-probe spectroscopy of the dynamics of curve crossing between the ionic and covalent states of NaI, and Figure 1.11 displays intramolecular and solvent dynamics of a large dye molecule in alcohol, as probed by a three pulse photon echo.

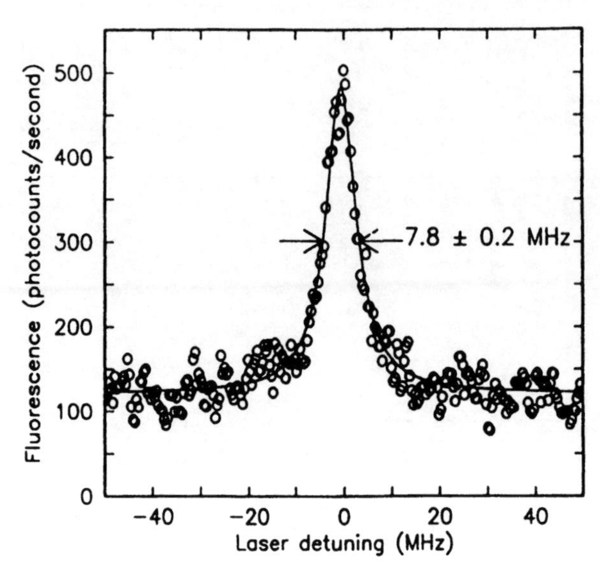

FIG. 1.9 Low-power fluorescence excitation spectrum for a single pentacene molecule in a sublimed crystal of p-terphenyl at 1.5 K. 0 MHz detuning ≡ 592.407 nm, which is in the wing of the O_1 site inhomogeneous line. The solid line is a Lorentzian fit to the data. [From W. P. Ambrose, Th. Basché, and W. E. Moerner, "Detection and spectroscopy of single pentacene molecules in a p-terphenyl crystal by means of fluorescence excitation," *J. Chem. Phys.* **95**, 7150 (1991).]

Off-resonant techniques which use laser frequencies that are far detuned from any resonant frequency (be it electronic or nuclear) are also commonly used [3]. Raman spectroscopy [4] and wave mixing techniques may be carried out under off-resonant conditions. This also is the case for many practical applications of optical materials, where it is important to avoid absorptive losses and other competing processes, and to obtain faster switching timescales (resonant techniques usually have slower response times related to the creation of excited state populations or charge carriers). We shall explore the nonlinear response in both regimes using a unified approach. The ability to predict the behavior in both regimes is a crucial test for the validity and the approximations involved in theoretical modeling.

The microscopic calculation of nonlinear susceptibilities in condensed phases is a fundamental problem that will be addressed in this book. Considerable effort has been devoted to calculating these quantities in terms of material properties and interactions. Much of the theory and the current level of understanding of nonlinear optical processes is based on the analysis of simple models for the material system. Most models lack a firm microscopic basis or their validity is limited to a certain class of systems. Historically, because of the close analogy and the formal connection with nuclear magnetic resonance (NMR) [5], many optical experiments were interpreted by mapping them into corresponding NMR models (e.g., the vector model and the Bloch sphere [6]). The celebrated Bloch equations [7], first proposed for the interpretation of nuclear magnetic resonance experiments, describe the evolution of a nuclear spin 1/2 system

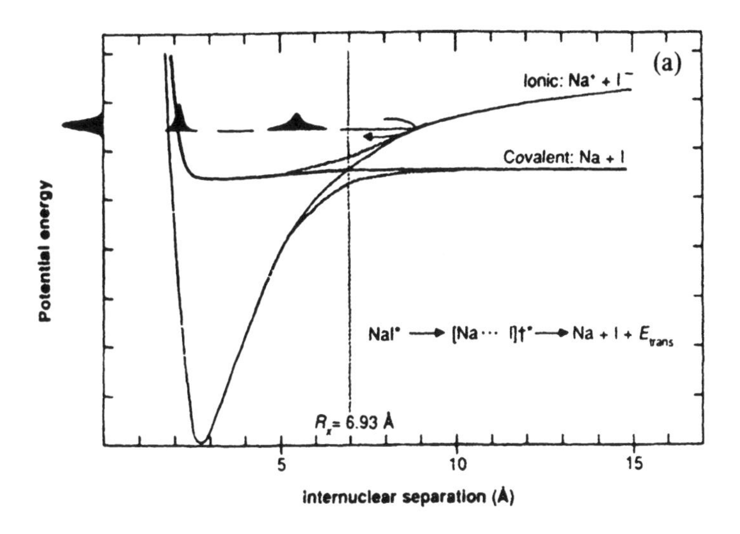

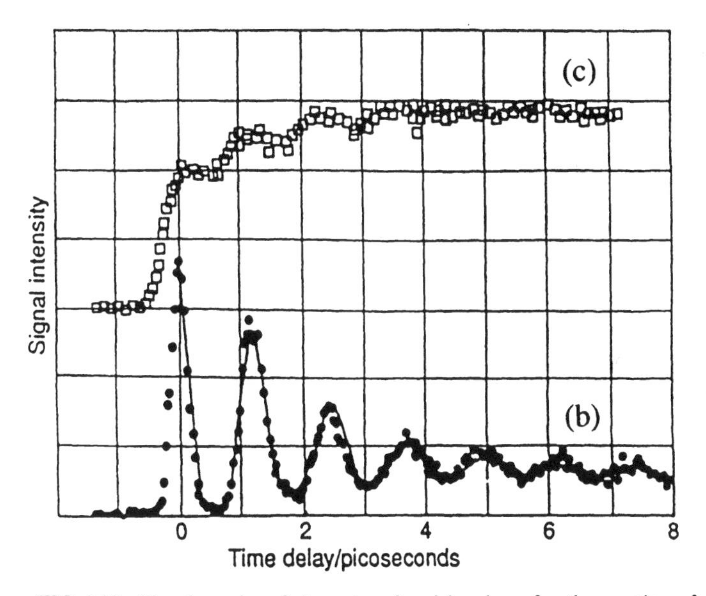

FIG. 1.10 The dynamics of elementary bond breakage for the reaction of NaI to yield Na and I, obtained by pump-probe spectroscopy. (a) the potential energy curves displaying the ionic and covalent states of the reaction. The pump pulse, indicated by the bell-shape curve on the vertical axis, takes the salt Na^+I^- to the covalent surface NaI*. The probe monitors the evolution of the wave packet (indicated by the bell-shaped curve moving along R) of dissociating molecules. Trace (b) shows the probe absorption to a third potential surface (not shown) tuned to probe the transition region. It shows the oscillatory behavior of the wave packet as it moves back and forth between the covalent and ionic curves. Each time the packet comes near the crossing point there is a probability of escape ($\sim$0.1), which is observed as a damping of the oscillations. Trace (c) corresponds to detecting on-resonance with the transition of the free Na atom (i.e. large separation) and shows a stepwise increase each time the wavepacket passes through the curve crossing region. [From A. H. Zewail, "Laser Femtochemistry," *Science* **242**, 1645 (1988).]

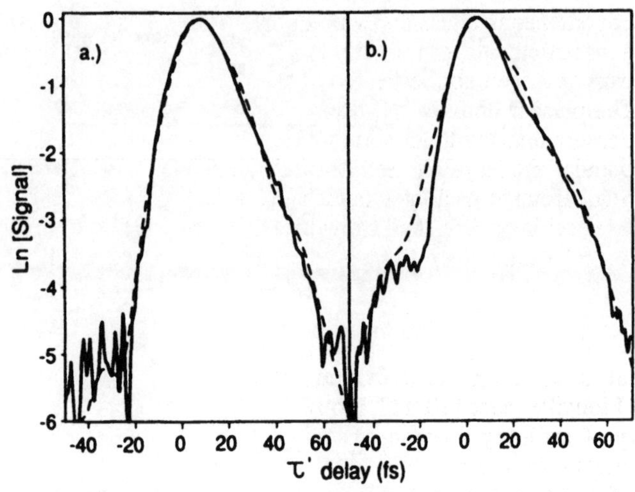

FIG. 1.11 Experimental (solid) and calculated (dashed) three pulse photon echo signals, scanning τ', the delay of the second pulse, with the delay of the third pulse $\tau = 120$ fs. (In the notation of Chapter 10.) The system is the laser dye LD690 (Oxazine 4) dissolved in (a) methanol and (b) 1-pentanol. The calculated signals were generated using the multimode Brownian oscillator model with 18 intramolecular modes seen in resonance Raman and 2 overdamped solvent modes, one fast and one slow. The solvent mode parameters were varied to model the signals. [From C. J. Bardeen and C. V. Shank, "Ultrafast dynamics of the solvent-solute interaction measured by femtosecond four-wave mixing: LD 690 in n-alcohols," *Chem. Phys. Lett.* **226**, 310 (1994).]

interacting with a classical radiation field, and with a thermal bath. The effects of the bath are incorporated via relaxation parameters representing level relaxation (T_1) and coherence dephasing (T_2). Feynman, Vernon, and Hellwarth [8] have established the exact isomorphism between these equations and the evolution of two levels interacting with an optical field, known as the optical Bloch equations. The Bloch equations and their limitations will be discussed in Chapter 6. The Bloch equations coupled with the Maxwell equations often provide a convenient methodology for the calculation of nonlinear susceptibilities and transient experiments. The origin of the nonlinear response, saturation, coherent transients, homogeneous, and inhomogeneous line broadening are all well understood and analyzed using the Bloch equations or their straightforward extensions to "few levels" (three, four, etc.). When faced with more complex situations, attempts are often made to "fix" the Bloch equations by using, e.g., a memory kernel ("non-Markovian" effects). Such attempts are of limited success, since a few-level model cannot be easily applied to complex situations such as a dye molecule in a solvent, which require a more microscopic and detailed treatment. There are many fundamentals differences between NMR and optical spectroscopy. For one, spin systems are described by simple universal Hamiltonians which only depend on a few coupling parameters. The level structure and interactions in the optical regime are much more complex. In magnetic resonance studies the samples are typically much smaller than the wavelength ($\sim$1 m), whereas in optics the reverse is true. This gives rise to wavevector selection (phase matching), grating configurations, and spatial resolution that do not normally exst in NMR. Spontaneous emission, which plays an important role in many resonant optical measurements, can usually be ignored in NMR.* For these reasons, optical spectroscopy need not be treated as the poor cousin of nuclear magnetic resonance! The present book is "NMR free" and develops the proper language for optical measurements without alluding to NMR concepts.

Another popular model often used in the calculation of optical nonlinearities is the driven anharmonic oscillator model [1], which is a generalization of the classical Drude oscillator. We shall show that this model is of very limited value for simple few-level systems but has tremendous advantages in the analysis of complex many-body systems.

In this book we present a microscopic framework for the description and the analysis of nonlinear optical spectroscopies of complex systems and optical materials that is based on the calculation of the density operator [9] and its evolution in Liouville space [10, 11]. Liouville space descriptions are widely used in the theory of ordinary linear optical lineshapes as well as in modeling the nonlinear optical response of few-level systems. We shall use the Liouville space approach to formulate the problem in terms of multitime correlation functions, i.e., the nonlinear response functions, which contain all the microscopic information relevant for any type of nonlinear optical measurement. Correlation function methods, which are based on a reduced description, are commonly used in the calculation of spectral line shapes in macroscopic systems, e.g., pressure broadening in the gas phase and lineshapes in liquids and solid matrices. One never attempts to calculate the exact eigenstates of the macroscopic system. The reason is twofold: (1) such calculation is extremely difficult due to the enormous number of degrees of freedom involved, and (2) the available broadened line shapes contain highly averaged information and do not reveal properties of individual eigenstates. The calculation of global eigenstates of macroscopic systems is therefore neither feasible nor desirable.

Correlation functions can be formally defined without alluding to a particular technique or a specific model for the matter, and, consequently, they provide a natural link between theory and experiment [12], and clarify the interrelationships among various techniques; we no longer need a new theory for each technique. A time-domain theoretical framework provides a better intuitive picture for the connection between spectroscopic measurements and microscopic dynamics in terms of time evolving wavepackets, which can be calculated semiclassically. The two ideal limiting cases of time-domain and frequency-domain experiments are derived from the same unified expression. The nonlinear susceptibility that is commonly used in the frequency domain is also obtained in terms of the response function. We show which time-domain and frequency-domain techniques are related by simple Fourier transform relationships and are therefore equivalent. We shall focus the applications primarily on some of the most widely used four-wave mixing techniques. Specific models for the matter will be introduced and the response function will be calculated. In most of this book we shall consider the simplest model system, namely a single absorber in a bath. Multiple absorber systems that require addressing the local-field problem as well as many-body and cooperative effects will be analyzed in the last two chapters.

---

* The role of cooperative radiative decay was discussed in W. S. Warren, W. Richter, A. H. Andreotti, and B. T. Farmer II, *Science* **262**, 2005 (1993).

NOTES

1. N. Bloembergen, *Nonlinear Optics* (Benjamin, New York, 1965).
2. G. A. Mourou, A. H. Zewail, P. F. Barbara, and W. H. Knox, Eds., *Ultrafast Phenomena IX* (Springer-Verlag, Berlin, 1994). (See also proceedings of previous conferences in the series.)
3. R. W. Hellwarth, *Progr. Quant. Electron.* **5**, 1 (1977).
4. N.-T. Yu and X.-Y. Li, Eds., *XIVth International Conference on Raman Spectroscopy* (Wiley, New York, 1994).
5. A. Abragam, *The Principles of Nuclear Magnetism* (Oxford, London, 1961).
6. Y. R. Shen, *The Principles of Nonlinear Optics* (Wiley, New York, 1984).
7. F. Bloch, *Phys. Rev.* **70**, 460 (1946).
8. R. P. Feynman, F. L. Vernon, and R. W. Hellwarth, *J. Appl. Phys.* **28**, 49 (1957).
9. J. Von Neumann, *Mathematische Grundlagen der Quantenmechanik* (Springer-Verlag, Berlin, 1932); English trans. by R. T. Beyer, *Mathematical Foundations of Quantum Mechanics* (Princeton University Press, Princeton, New Jersey, 1955).
10. U. Fano, *Rev. Mod. Phys.* **29**, 74 (1957).
11. R. Zwanzig, *Lect. Theoret. Phys.* **3**, 106 (1961); *Physica* **30**, 1109 (1964).
12. P. C. Martin, *Measurements and Correlation Functions* (Gordon and Breach, New York, 1968).

BIBLIOGRAPHY

M. Born and E. Wolf, *Principles of Optics*, 6th ed. (Pergamon, Oxford, 1980).
R. W. Boyd, *Nonlinear Optics* (Academic Press, Boston, 1992).
P. N. Butcher and D. Cotter, *The Element of Nonlinear Optics* (Cambridge University Press, Cambridge, 1990).
G. R. Fleming, *Chemical Applications of Ultrafast Spectroscopy* (Oxford, London, 1986).
C. Flytzanis, *Quantum Electronics*, Vol. **1**, 1, H. Rabin and C. L. Tang, Eds. (Academic Press, New York, 1975).
M. D. Levenson and S. S. Kano, *Introduction to Nonlinear Laser Spectroscopy* (Academic Press, New York, 1988).
M. D. Levenson, E. Mazur, P. S. Pershan, and Y. R. Shen, Eds., *Resonances.* A Volume in Honor of N. Bloembergen. (World Scientific, Singapore, 1990).
M. Schubert and B. Wilhelmi, *Nonlinear Optics and Quantum Electronics* (Wiley, New York, 1986).
A. Yariv, *Optical Electronics*, 3rd ed. (Holt Saunders, New York, 1985).

CHAPTER 2

Quantum Dynamics in Hilbert Space

In this chapter we introduce some useful tools of quantum dynamics including Green function and operator techniques that are commonly used in calculating the time evolution of the wavefunction in Hilbert space. In the next chapter we shall show how all of these tools can be extended for describing the evolution of the density operator in Liouville space. These techniques will be used throughout this book.

TIME-EVOLUTION OPERATOR WITH TIME-INDEPENDENT HAMILTONIANS

In the Schrödinger picture, the time evolution of a quantum mechanical system may be described by following its time-dependent state, which is represented by a Dirac ket $|\psi(t)\rangle$. The wavefunction represents the state in the coordinate representation and is given by

$$\psi(\mathbf{x}, t) \equiv \langle \mathbf{x}|\psi(t)\rangle$$

where $\mathbf{x}$ is a complete set of coordinates. We further define an *addition* (superposition) of two states

$$|\psi\rangle = \lambda_1|\psi_1\rangle + \lambda_2|\psi_2\rangle, \tag{2.1a}$$

where λ_1 and λ_2 are complex numbers. The *scalar product* of two states is defined by

$$\langle\psi(t)|\phi(t)\rangle \equiv \int \psi^*(\mathbf{x}, t)\phi(\mathbf{x}, t)\, d\mathbf{x}. \tag{2.1b}$$

The collection of all possible states, together with the addition and the scalar product operation, forms a linear vector space called the *Hilbert space*. In the Appendix we list the precise conditions that have to be met by the states together with the addition and the scalar product operations in order to form a Hilbert space.

The expectation value of any dynamic variable A is given by

$$\langle A(t) \rangle \equiv \langle \psi(t)|A|\psi(t) \rangle \equiv \int \psi^*(\mathbf{x}, t) A \psi(\mathbf{x}, t)\, d\mathbf{x}. \tag{2.2}$$

$|\psi(t)\rangle$ may be calculated by solving the time-dependent Schrödinger equation

$$\frac{\partial |\psi(t)\rangle}{\partial t} = -\frac{i}{\hbar} H |\psi(t)\rangle, \tag{2.3}$$

H being the total Hamiltonian of the system and $\hbar$ is Planck's constant.

The solution of Eq. (2.3) may be conveniently represented using the eigenvalues E_n and eigenvectors $|\varphi_n\rangle$ of the time-independent (stationary) Schrödinger equation

$$H|\varphi_n\rangle = E_n |\varphi_n\rangle. \tag{2.4}$$

$|\varphi_n\rangle$ form a complete and orthonormal basis set in Hilbert space. They therefore satisfy the completeness condition

$$\sum_n |\varphi_n\rangle\langle\varphi_n| = 1, \tag{2.5}$$

and the orthonormality

$$\langle \varphi_n | \varphi_{n'} \rangle \equiv \int \varphi_n^*(\mathbf{x}) \varphi_{n'}(\mathbf{x})\, d\mathbf{x} = \delta_{nn'}. \tag{2.6}$$

We next expand $|\psi(t)\rangle$ in the $|\varphi_n\rangle$ basis set

$$|\psi(t)\rangle = \sum_n |\varphi_n\rangle\langle\varphi_n|\psi(t)\rangle. \tag{2.7}$$

Once we choose a basis set, Eq. (2.3) can be represented using matrix notation. In the present basis set $|\psi(t)\rangle$ is a vector with components $\langle\varphi_n|\psi(t)\rangle$, and the Hamiltonian is represented by the diagonal matrix

$$H_{nm} \equiv \langle \varphi_n | H | \varphi_m \rangle \equiv \int d\mathbf{x}\, \varphi_n^*(\mathbf{x}) H \varphi_m(\mathbf{x}) = E_n \delta_{nm}. \tag{2.8}$$

When Eq. (2.7) is substituted in Eq. (2.3), and we multiply by $\langle\varphi_n|$ from the left, we get

$$\frac{d}{dt}\langle\varphi_n|\psi(t)\rangle = -\frac{i}{\hbar} E_n \langle\varphi_n|\psi(t)\rangle, \tag{2.9}$$

whose solution is

$$\langle\varphi_n|\psi(t)\rangle = \exp\left[-\frac{i}{\hbar} E_n (t - t_0)\right]\langle\varphi_n|\psi(t_0)\rangle, \tag{2.10}$$

where $\langle\varphi_n|\psi(t_0)\rangle$ are the initial expansion coefficients of the wavefunction, which are assumed to be known. We then have

$$|\psi(t)\rangle = \sum_n \exp\left[-\frac{i}{\hbar} E_n (t - t_0)\right]|\varphi_n\rangle\langle\varphi_n|\psi(t_0)\rangle. \tag{2.11}$$

At this point we introduce the *time evolution operator* defined as

$$|\psi(t)\rangle \equiv U(t, t_0)|\psi(t_0)\rangle. \tag{2.12}$$

When $U(t, t_0)$ acts on the wavefunction at time t_0, it transforms it to the wavefunction at time t. From this definition it immediately follows that

$$U(t_0, t_0) = 1. \tag{2.13}$$

The introduction of $U(t, t_0)$ allows us to solve the time evolution in general, without using specific initial conditions. Once $U(t, t_0)$ is obtained, it can act on any initial state $|\psi(t_0)\rangle$ to get the state at time t. Without U, on the other hand, we need to solve the Schrödinger equation over and over again when the initial conditions are varied. In addition, the time evolution operator opens up many new approximation schemes for quantum dynamics, as will be shown later in this chapter. Comparison of Eqs. (2.11) and (2.12) shows that

$$U(t, t_0) = \sum_n |\varphi_n\rangle \exp\left[-\frac{i}{\hbar} E_n(t - t_0)\right]\langle\varphi_n|. \tag{2.14}$$

Equation (2.14) gives the time evolution operator in a specific representation, i.e., the eigenstates of the Hamiltonian H. This representation is useful only for simple systems (e.g., the harmonic oscillator, the rigid rotor or the hydrogen atom) where the complete set of eigenstates is readily available. We would like therefore to rewrite this operator in a more general form that is not restricted to a particular representation. This will provide a more formal flexibility and will allow the development of other, e.g. semiclassical, approximations. To that end we need to introduce the notion of a *function of an operator*. Consider an operator A, and an ordinary function $f(x)$. What do we mean by $f(A)$, the function of the A operator? Assuming that $f(x)$ is well behaved and can be expanded in a Taylor series around $x = 0$, we can write

$$f(x) = \sum_{p=0}^{\infty} \frac{f^{(p)}(0)}{p!} x^p, \tag{2.15a}$$

$f^{(p)}(0)$ being the pth derivative of $f(x)$ at $x = 0$, and $f^{(0)}(0) \equiv f(0)$. A reasonable definition of $f(A)$ will then be in terms of the Taylor series

$$f(A) \equiv \sum_{p=0}^{\infty} \frac{f^{(p)}(0)}{p!} A^p. \tag{2.15b}$$

A more compact (and general) definition of $f(A)$ may be obtained as follows: Consider the eigenvalues (a_j) and eigenvectors $|\alpha_j\rangle$ of A, i.e.

$$A|\alpha_j\rangle = a_j|\alpha_j\rangle. \tag{2.16a}$$

Using Eq. (2.15) we then have

$$f(A)|\alpha_j\rangle = \sum_{p=0}^{\infty} \frac{f^{(p)}(0)}{p!} A_j^p|\alpha_j\rangle \equiv f(a_j)|\alpha_j\rangle. \tag{2.16b}$$

Assuming that $|\alpha_j\rangle$ form a complete and orthonormal set in our Hilbert space, we can then write

$$f(A) = \sum_j |\alpha_j\rangle f(a_j)\langle\alpha_j|. \tag{2.17}$$

Equation (2.17) will serve as our new definition of a function of an operator. It naturally coincides with the previous definition [Eq. (2.15)] when the function can be expanded in a Taylor series. Equation (2.17) implies that $f(A)$ is an operator that is represented by a diagonal matrix in the $|\alpha_j\rangle$ representation, i.e.,

$$A = \begin{pmatrix} a_1 & & 0 \\ & a_2 & \\ 0 & & \ddots \end{pmatrix} \Rightarrow f(A) = \begin{pmatrix} f(a_1) & & 0 \\ & f(a_2) & \\ 0 & & \ddots \end{pmatrix}. \tag{2.18}$$

Once $f(A)$ is defined in a given representation, we can always transform it to any other representation. Consider a new basis set

$$|\beta_k\rangle = \sum_j S_{kj}^* |\alpha_j\rangle, \tag{2.19}$$

where $S_{jk} \equiv \langle\beta_k|\alpha_j\rangle$ are transformation coefficients between the old and the new basis set. By multiplying Eq. (2.17) with the unit operator expanded in the new basis set we have

$$\begin{aligned} f(A) &= \sum_{kk'j} |\beta_k\rangle\langle\beta_k|\alpha_j\rangle f(a_j)\langle\alpha_j|\beta_{k'}\rangle\langle\beta_{k'}| \\ &\equiv \sum_{kk'} |\beta_k\rangle f'_{kk'}\langle\beta_{k'}|. \end{aligned} \tag{2.20}$$

$f'_{kk'}$ is the matrix representing $f(A)$ in the new, $|\beta\rangle$, basis set. Equation (2.20) implies that

$$f'_{kk'} = \sum_j S_{kj} f(a_j) S_{k'j}^*. \tag{2.21}$$

As an example for the use of the above two definitions of a function of an operator let us prove that the inverse of the operator $T = \exp(\lambda A)$ is $T' = \exp(-\lambda A)$ where A is any operator and λ is a complex number. Using the Taylor series we have

$$\exp(\lambda A) = 1 + \lambda A + \tfrac{1}{2}(\lambda A)^2 + \cdots \tag{2.22a}$$

$$\exp(-\lambda A) = 1 - \lambda A + \tfrac{1}{2}(\lambda A)^2 - \cdots \tag{2.22b}$$

Multiplying these equations and collecting terms order by order in powers of λ, we have

$$\exp(\lambda A)\cdot\exp(-\lambda A) = [1 + \lambda A + \tfrac{1}{2}(\lambda A)^2 + \cdots][1 - \lambda A + \tfrac{1}{2}(\lambda A)^2 - \cdots] = 1.$$

Alternatively we may use the definition [Eqs. (2.17)], and get

$$T = \sum_j |\alpha_j\rangle \exp(\lambda a_j)\langle\alpha_j|,$$

$$T' = \sum_{j'} |\alpha_{j'}\rangle \exp(-\lambda a_{j'})\langle\alpha_{j'}|,$$

$$TT' = \sum_{j,j'} |\alpha_j\rangle \exp(\lambda a_j)\langle\alpha_j|\alpha_{j'}\rangle \exp(-\lambda a_{j'})\langle\alpha_{j'}| = 1. \tag{2.23}$$

The last equality follows from the orthonormality $\langle\alpha_j|\alpha_j'\rangle = \delta_{jj'}$.

Using Eqs. (2.14) and (2.17), we can recast the time evolution operator in the form

$$U(t, t_0) = \exp\left[-\frac{i}{\hbar} H(t - t_0)\right]. \tag{2.24}$$

As an example, we shall now calculate the time evolution operator of a coupled two-level system ($|\phi_a\rangle$ and $|\phi_b\rangle$), with energies ε_a and ε_b, and a coupling V_{ab}, represented by the Hamiltonian

$$H = \begin{pmatrix} \varepsilon_a & V_{ab} \\ V_{ba} & \varepsilon_b \end{pmatrix}. \tag{2.25}$$

We shall denote

$$V_{ab} = V_{ba}^* = |V_{ab}| \exp(-i\chi), \qquad 0 < \chi < 2\pi$$

where $|V_{ab}|$ is the modulus of V_{ab} and χ is its phase. The eigenvalues of this Hamiltonian are

$$\varepsilon_\pm = \tfrac{1}{2}(\varepsilon_a + \varepsilon_b) \pm \tfrac{1}{2}\sqrt{(\varepsilon_a - \varepsilon_b)^2 + 4|V_{ab}|^2}, \tag{2.26}$$

and the corresponding eigenvectors*

$$|\psi_+\rangle = \cos\theta \exp(-i\chi/2)|\phi_a\rangle + \sin\theta \exp(i\chi/2)|\phi_b\rangle, \tag{2.27a}$$

$$|\psi_-\rangle = -\exp(-i\chi/2)\sin\theta|\phi_a\rangle + \exp(i\chi/2)\cos\theta|\phi_b\rangle, \tag{2.27b}$$

where θ is a transformation angle defined by

$$\tan 2\theta \equiv \frac{2|V_{ab}|}{\varepsilon_a - \varepsilon_b}, \qquad 0 < \theta < \pi/2.$$

The time evolution operator is thus given by

$$U(t, t_0) = |\psi_+\rangle \exp\left[-\frac{i}{\hbar}\varepsilon_+(t - t_0)\right]\langle\psi_+| + |\psi_-\rangle \exp\left[-\frac{i}{\hbar}\varepsilon_-(t - t_0)\right]\langle\psi_-|. \tag{2.28}$$

* Note that the phase χ of the coupling affects the eigenstates and consequently, the time evolution operator. When V_{ab} represents the coupling with the radiation field (see, e.g., Appendix 6B), χ is associated with the phase of the radiation field.

Since

$$|\psi_+\rangle\langle\psi_+| = \begin{pmatrix} \exp(-i\chi/2)\cos\theta \\ \exp(i\chi/2)\sin\theta \end{pmatrix} (\exp(i\chi/2)\cos\theta, \quad \exp(-i\chi/2)\sin\theta)$$

$$= \begin{pmatrix} \cos^2\theta & \exp(-i\chi)\sin\theta\cos\theta \\ \exp(i\chi)\sin\theta\cos\theta & \sin^2\theta \end{pmatrix},$$

$$|\psi_-\rangle\langle\psi_-| = \begin{pmatrix} -\exp(-i\chi/2)\sin\theta \\ \exp(i\chi/2)\cos\theta \end{pmatrix} (-\exp(i\chi/2)\sin\theta, \quad \exp(-i\chi/2)\cos\theta)$$

$$= \begin{pmatrix} \sin^2\theta & -\exp(-i\chi)\sin\theta\cos\theta \\ -\exp(i\chi)\sin\theta\cos\theta & \cos^2\theta \end{pmatrix},$$

We finally have the time evolution operator in the $|\phi_a\rangle$, $|\phi_b\rangle$ basis set

$$U(t, t_0) = \begin{pmatrix} \cos^2\theta & \exp(-i\chi)\sin\theta\cos\theta \\ \exp(i\chi)\sin\theta\cos\theta & \sin^2\theta \end{pmatrix} \exp\left[-\frac{i}{\hbar}\varepsilon_+(t - t_0)\right]$$
$$+ \begin{pmatrix} \sin^2\theta & -\exp(-i\chi)\sin\theta\cos\theta \\ -\exp(i\chi)\sin\theta\cos\theta & \cos^2\theta \end{pmatrix} \exp\left[-\frac{i}{\hbar}\varepsilon_-(t - t_0)\right]. \quad (2.29)$$

Suppose the system is initially (at time $t_0 = 0$) in the $|\phi_a\rangle$ state, i.e., $|\psi(0)\rangle = |\phi_a\rangle$, and we shall be interested in calculating the probability of the system to be found in the $|\phi_b\rangle$ state at time t.

$$P_{ba}(t) \equiv |\langle\phi_b|\psi(t)\rangle|^2 = |\langle\phi_b|U(t, 0)|\phi_a\rangle|^2. \quad (2.30)$$

In a matrix representation we have

$$\langle\phi_b|U(t, 0)|\phi_a\rangle = (0, 1)\begin{pmatrix} U_{aa}(t) & U_{ab}(t) \\ U_{ba}(t) & U_{bb}(t) \end{pmatrix}\begin{pmatrix} 1 \\ 0 \end{pmatrix}. \quad (2.31)$$

Upon the substitution of Eq. (2.29) and (2.31) in (2.30) we get [1]

$$P_{ba}(t) = \frac{4|V_{ab}|^2}{4|V_{ab}|^2 + (\varepsilon_a - \varepsilon_b)^2} \sin^2 \sqrt{4|V_{ab}|^2 + (\varepsilon_a - \varepsilon_b)^2}\, \frac{t}{2\hbar}. \quad (2.32)$$

This is known as the Rabi formula, and

$$\Omega_R \equiv \frac{1}{\hbar}\sqrt{4|V_{ab}|^2 + (\varepsilon_a - \varepsilon_b)^2},$$

is the Rabi frequency. Equation (2.32) is displayed in Figure 2.1.

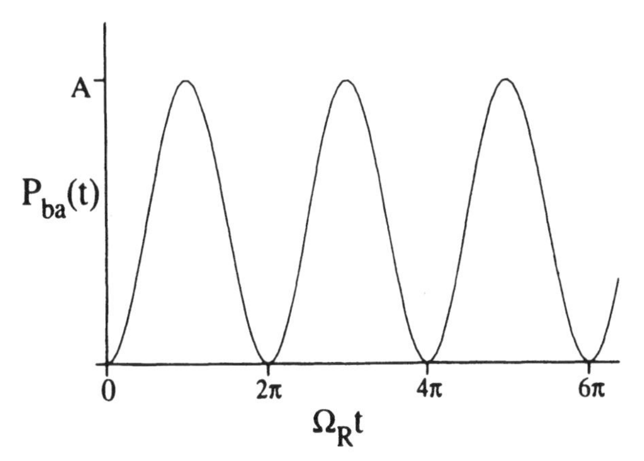

FIG. 2.1 Rabi oscillation [Eq. (2.32)]. Shown is the probability of the two-level system to be found in the state $|b\rangle$ at time t, given that it started at state $|a\rangle$ at $t = 0$. The oscillation amplitude is $A = 4|V_{ab}|^2/[4|V_{ab}|^2 + (\varepsilon_a - \varepsilon_b)^2]$.

We shall now introduce a few additional properties of operators. Consider the matrix element of an operator A

$$A_{ij} = \langle \phi_i | A | \phi_j \rangle.$$

Normally the operator acts "to the right," i.e., we define a new ket

$$|\phi_j'\rangle \equiv A|\phi_j\rangle$$

and the matrix element is evaluated by the scalar product of the resulting ket with $|\phi_j\rangle$, i.e.,

$$A_{ij} = \langle \phi_i | \phi_j' \rangle.$$

Alternatively, in some cases it may be convenient to act to the left, e.g., when ϕ_i is a simpler function than ϕ_j. To that end we introduce the *Hermitian conjugate* (adjoint) $A^\dagger$ of the operator A, defined by the relation

$$A_{ij} = \langle A^\dagger \phi_i | \phi_j \rangle.$$

Using the properties of a scalar product we then have

$$A_{ji}^\dagger = A_{ij}^*. \tag{2.33}$$

Hermitian conjugation is useful in representing transformations among basis sets. For example, Eq. (2.21) may be simply represented using a matrix notation. Denoting the diagonal matrix **f** [Eq. (2.18)], and the transformation matrix **S**, with elements S_{jk} we have

$$\mathbf{f}' = \mathbf{S}\mathbf{f}\mathbf{S}^\dagger. \tag{2.34}$$

An operator that is equal to its Hermitian conjugate $A = A^\dagger$ so that $A_{ji}^* = A_{ij}$ is denoted *hermitian* (*or self-adjoint*). The eigenvalues of Hermitian operators are real, and physical observables such as the Hamiltonian, the momentum, the angular momentum, etc. are represented by Hermitian operators.

A *unitary operator* has the property that its Hermitian conjugate is equal to its inverse, i.e.,

$$A^\dagger A = AA^\dagger = 1. \tag{2.35}$$

It follows directly from these definitions that the Hermitian conjugate of a product of operators is equal to the product of conjugates in reverse order

$$(AB)^\dagger = B^\dagger A^\dagger. \tag{2.36}$$

This can be easily verified by taking the ij matrix element of both sides. Using this relation we immediately see that the product of two unitary operators is unitary. Note, however, that the product of two Hermitian operators is not necessarily Hermitian.

The introduction of a Hermitian conjugate provides an important flexibility in the evaluation of matrix elements. Consider the matrix element of a product of two operators A and B,

$$S \equiv \langle\psi|AB|\phi\rangle. \tag{2.37a}$$

Naturally, the operators act "to the right," i.e., we define a new ket

$$|\phi'\rangle \equiv AB|\phi\rangle,$$

and the matrix element is calculated as

$$S = \langle\psi\,|\,\phi'\rangle. \tag{2.37b}$$

Alternatively, we can act to the right and define

$$|\psi'\rangle \equiv B^\dagger A^\dagger|\psi\rangle,$$

so that

$$S = \langle\psi'\,|\,\phi\rangle. \tag{2.37c}$$

We can also adopt an intermediate procedure by acting from the right and the left. We define

$$|\phi''\rangle \equiv B|\phi\rangle,$$

and

$$|\psi''\rangle \equiv A^\dagger|\psi\rangle,$$

so that

$$S = \langle\psi''\,|\,\phi''\rangle. \tag{2.37d}$$

All of these expressions are formally identical. In particular applications one form could be easier to implement.

For an operator of the form $T = \exp(iA)$ we have $T^\dagger = \exp(-iA^\dagger)$. Using Eq. (2.23) we can now prove that if the operator A is Hermitian then $T = \exp(iA)$ is unitary, since $T^\dagger T = \exp(-iA)\exp(iA) = 1$. This shows that the time evolution operator [Eq. (2.24)] is unitary.

PROPAGATION WITH TIME-DEPENDENT HAMILTONIANS: TIME ORDERING

When all the degrees of freedom of a dynamic system are included in the Hamiltonian, the Hamiltonian is time independent. In many cases it is useful to eliminate some degrees of freedom and treat them as "external" forces, with a predetermined, known, time evolution, and include explicitly in the Hamiltonian only the remaining coordinates. In that case the Hamiltonian becomes time dependent. The price that we have to pay is that calculating the time evolution then becomes a much more complex problem. An important example for spectroscopy is a material system that interacts with the radiation field. It is possible to adopt a time-independent Hamiltonian that includes the radiation field degrees of freedom (see Chapter 4). However, the classical approximation for the field has many advantages and often provides an adequate level of description. The choice that we face is between working with a time-independent Hamiltonian in a larger phase space that includes the matter and the radiation field or using a time-dependent Hamiltonian in a smaller phase space of the matter alone.

We shall now generalize the definition of the time-evolution operator to the case of a time-dependent Hamiltonian. When Eq. (2.12) is substituted in the Schrödinger equation (2.3) we get

$$\frac{\partial}{\partial t} U(t, t_0)|\psi(t_0)\rangle = -\frac{i}{\hbar} H(t)U(t, t_0)|\psi(t_0)\rangle.$$

Since this equation must hold for any initial vector $|\psi(t_0)\rangle$, then the operator U must satisfy the same equation

$$\frac{\partial}{\partial t} U(t, t_0) = -\frac{i}{\hbar} H(t)U(t, t_0). \tag{2.38}$$

Upon integrating both sides from time t_0 to time t and using Eq. (2.13), we can recast Eq. (2.38) in an integral form.

$$U(t, t_0) = 1 - \frac{i}{\hbar}\int_{t_0}^{t} d\tau H(\tau)U(\tau, t_0). \tag{2.39}$$

Equation (2.39) can be solved iteratively by plugging it into itself. Iterating once, we get

$$U(t, t_0) = 1 - \frac{i}{\hbar}\int_{t_0}^{t} d\tau\, H(\tau) + \left(-\frac{i}{\hbar}\right)^2 \int_{t_0}^{t} d\tau \int_{t_0}^{\tau} d\tau'\, H(\tau)H(\tau')U(\tau', t_0).$$

When the substitution is repeated we obtain the expansion (see Figure 2.2)

$$U(t, t_0) = 1 + \sum_{n=1}^{\infty}\left(-\frac{i}{\hbar}\right)^n \int_{t_0}^{t} d\tau_n \int_{t_0}^{\tau_n} d\tau_{n-1}\cdots\int_{t_0}^{\tau_2} d\tau_1\, H(\tau_n)H(\tau_{n-1})\cdots H(\tau_1). \tag{2.40}$$

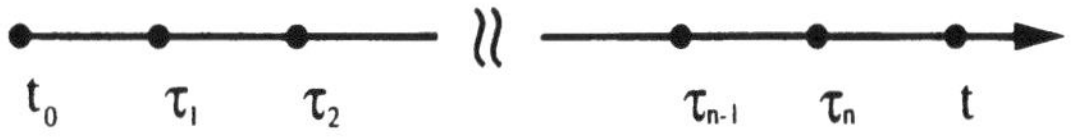

$t_0 \quad \tau_1 \quad \tau_2 \quad \cdots \quad \tau_{n-1} \quad \tau_n \quad t$

FIG. 2.2 Time variables in the time-ordered expansion [Eq. (2.40)].

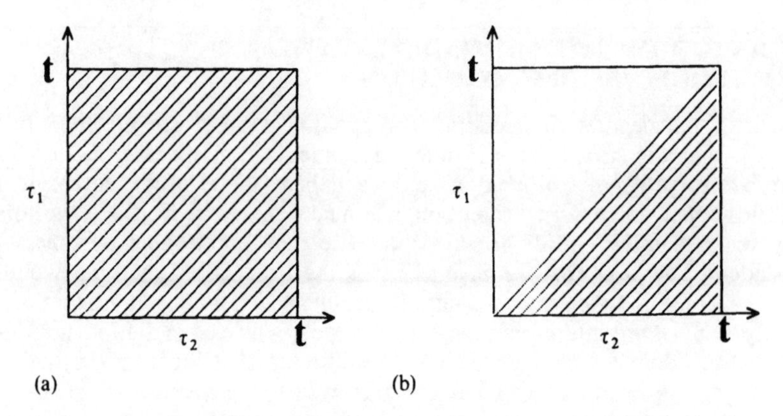

FIG. 2.3 (a) The integration domain for the second order ($n = 2$) term in the (non-time-ordered) expansion of an exponential [Eq. (2.41)]. (b) The same term in a time-ordered exponential [Eq. (2.40)].

Had we ignored the fact that $H(\tau)$ is an operator and treated it as a number, we could have solved Eq. (2.38) trivially resulting in

$$U(t, t_0) \stackrel{?}{=} \exp\left[-\frac{i}{\hbar}\int_{t_0}^{t} d\tau\, H(\tau)\right],$$

which would imply that

$$U(t, t_0) \stackrel{?}{=} 1 + \sum_{n=1}^{\infty} \frac{1}{n!}\left(-\frac{i}{\hbar}\right)^n \int_{t_0}^{t} d\tau_n \cdots \int_{t_0}^{t} d\tau_1\, H(\tau_n)H(\tau_{n-1})\cdots H(\tau_1). \quad (2.41)$$

Equation (2.41) is very similar to Eq. (2.40) but is incorrect. In Eq. (2.40) the τ_j time variables are *fully ordered* $t \geqslant \tau_n \geqslant \cdots \geqslant \tau_1 \geqslant t_0$ whereas in Eq. (2.41) all time orderings contribute. The difference is illustrated in Figure 2.3. If the Hamiltonians at different times were to commute, then time ordering will not be of a concern and the two expressions will be identical [note that the integration volume over $\tau_1 \cdots \tau_n$ in Eq. (2.41) is $n!$ times larger than in Eq. (2.40) and the division by $n!$ makes them equal]. It is therefore common to call Eq. (2.40) a *positive time ordered exponential* and denote it as follows [2]:

$$U(t, t_0) = \exp_+\left[-\frac{i}{\hbar}\int_{t_0}^{t} d\tau\, H(\tau)\right]. \quad (2.42)$$

It should be emphasized that this definition is nothing but an abbreviated notation for the expansion (2.40), which still needs to be calculated order by order. Such an expansion may be valid at short times but will always break down at longer times. The reason is that we expand perturbatively in the entire Hamiltonian. Several approximate methods have been developed to overcome this problem. In the following we shall outline the interaction picture and the Magnus expansion. Before doing that, however, we first note that in some cases it is desirable to perform the time evolution by acting to the left. Taking the Hermitian conjugate of Eqs. (2.12) we get

$$\langle\psi(t)| = \langle\psi(t_0)|U^\dagger(t, t_0). \tag{2.43}$$

When calculating the Hermitian conjugate of Eq. (2.42) we need to change $i \to -i$ and change the ordering of all operators. We thus have

$$U^\dagger(t, t_0) = \exp_-\left[\frac{i}{\hbar}\int_{t_0}^{t} d\tau\, H(\tau)\right], \tag{2.44}$$

with the *negative time ordered exponential* defined as

$$\exp_-\left[i\int_{t_0}^{t} d\tau\, H(\tau)\right] \equiv 1 + \sum_{n=1}^{\infty}\left(\frac{i}{\hbar}\right)^n \int_{t_0}^{t} d\tau_n \int_{t_0}^{\tau_n} d\tau_{n-1} \cdots \times \int_{t_0}^{\tau_2} d\tau_1\, H(\tau_1)H(\tau_2)\cdots H(\tau_n). \tag{2.45}$$

In the derivation of Eq. (2.44) we have used the hermiticity of the Hamiltonian $H^\dagger(\tau) = H(\tau)$.

THE TIME EVOLUTION OPERATOR REVISITED

The formal expression for the time evolution operator can be alternatively derived as follows: from the definition of the time evolution operator it follows that

$$|\psi(t)\rangle = U(t, t')U(t', t'')|\psi(t'')\rangle$$

so that

$$U(t, t'') = U(t, t')U(t', t''). \tag{2.46a}$$

More generally, we can break it up as many times as we like:

$$U(t, t_0) = U(t, t_n)U(t_n, t_{n-1})\cdots U(t_1, t_0). \tag{2.46b}$$

To proceed further we shall divide the $[t_0, t]$ time interval into a large number (N) of equal time segments Δt with $N\Delta t = t - t_0$. If Δt is very small we can write using Eq. (2.39)

$$U(t_n, t_{n-1}) \approx 1 - \frac{i}{\hbar} H(t_n)\Delta t. \tag{2.47}$$

This is the *infinitesimal evolution operator*. The following relations can now be shown:

1. Since H is Hermitian, the *infinitesimal evolution operator* is unitary. This can be verified as follows: The Hermitian conjugate of Eq. (2.47) is

$$U^\dagger(t_n, t_{n-1}) \approx 1 + \frac{i}{\hbar} H(t_n)\Delta t.$$

When U is multiplied by $U^{\dagger}$, the terms linear in Δt vanish and we have $UU^{\dagger} = 1 + 0(\Delta t^2)$. Since $\Delta t \to 0$, the infinitesimal evolution operator is unitary.

2. The total evolution operator can be represented as a product of infinitesimal operators and is therefore also unitary. We now have

$$U(t, t_0) = \left[1 - \frac{i}{\hbar} H(t_{n-1})\Delta t\right]\left[1 - \frac{i}{\hbar} H(t_{n-2})\Delta t\right] \cdots \left[1 - \frac{i}{\hbar} H(t_0)\Delta t\right]. \quad (2.48)$$

As a check let us consider again the case of a time-independent Hamiltonian. Equation (2.48) then becomes

$$U(t, t_0) = \left[1 - \frac{i}{\hbar} H\Delta t\right]^N,$$

or

$$U(t, t_0) = \left[1 - \frac{i}{N\hbar} H(t - t_0)\right]^N.$$

We now recall the definition of the number e:

$$\lim_{N \to \infty} (1 + a/N)^N = \exp(a).$$

Choosing $a = (-i/\hbar)H(t - t_0)$, we thus get

$$U(t, t_0) = \exp\left[-\frac{i}{\hbar} H(t - t_0)\right], \quad (2.49)$$

which is identical to Eq. (2.24). Of course, if $|\psi\rangle$ and H were ordinary numbers, Eq. (2.38) could be solved directly to yield Eq. (2.49). Since $|\psi\rangle$ is a vector and H is a matrix we had to proceed with more caution in the derivation.

Turning back to the time-dependent Hamiltonian, we can start collecting terms in the expansion of Eq. (2.48) in powers of Δt. To lowest order we have

$$U(t, t_0) \cong 1 - \frac{i}{\hbar} \sum_{j=0}^{n} H(t_j)\Delta t + \cdots$$

as $\Delta t \to 0$ we get

$$U(t, t_0) \cong 1 - \frac{i}{\hbar} \int_{t_0}^{t} d\tau\, H(\tau).$$

Collecting terms to higher orders in Δt, we then recover the entire expansion Eq. (2.40) order by order.

THE INTERACTION PICTURE

The expansion (2.40) is not very useful. Since it treats the entire Hamiltonian perturbatively, it usually applies only for very short times and it breaks down at longer times. In the following we shall recast the time evolution operator in the *interaction picture*, which allows us to treat part of the Hamiltonian (H_0) exactly and expand perturbatively only in the remainder of the Hamiltonian (H'). By a careful choice of H_0 and H' this expansion can be made to hold for long times, even when truncated at a low order. This forms the basis for the perturbation theory, which will be repeatedly used in this book.

In the interaction picture we first partition the Hamiltonian into two parts.

$$H(t) = H_0(t) + H'(t). \tag{2.50}$$

H_0 is usually a simpler Hamiltonian, which results in a time evolution that can be calculated exactly whereas $H'(t)$ is a more complicated part, which will be treated perturbatively. We shall denote the time evolution operator with respect to H_0, by $U_0(t, t_0)$. It satisfies the equation

$$\frac{\partial}{\partial t} U_0(t, t_0) = -\frac{i}{\hbar} H_0(t) U_0(t, t_0) \tag{2.51}$$

whose solution is

$$U_0(t, t_0) = \exp_+\left[-\frac{i}{\hbar}\int_{t_0}^{t} d\tau\, H_0(\tau)\right].$$

We next define the wavefunction in the interaction picture $|\psi_I(t)\rangle$ as follows:

$$|\psi_S(t)\rangle \equiv U_0(t, t_0)|\psi_I(t)\rangle, \tag{2.52}$$

where $|\psi_S(t)\rangle$ is the previously defined wavefunction (hereafter denoted the Schrödinger picture). The subscript I stands for the "interaction representation." On substituting Eq. (2.52) in Eq. (2.3), and recalling that U_0 is unitary, i.e., $U_0^\dagger(t, t_0) = U_0^{-1}(t, t_0)$ we get

$$\frac{\partial|\psi_I\rangle}{\partial t} = -\frac{i}{\hbar} H_I'(t)|\psi_I(t)\rangle, \tag{2.53a}$$

where

$$H_I'(t) \equiv U_0^\dagger(t, t_0) H'(t) U_0(t, t_0). \tag{2.53b}$$

Note that the time dependence of $H_I'(t)$ comes from two sources: the original explicit time dependence of H' and the transformation U_0 to the interaction picture.

We next introduce the time evolution operator in the interaction representation

$$|\psi_I(t)\rangle \equiv U_I(t, t_0)|\psi_I(t_0)\rangle. \tag{2.54}$$

Upon substituting Eq. (2.54) in Eq. (2.53a) we get

$$\frac{\partial}{\partial t} U_I(t, t_0) = -\frac{i}{\hbar} H_I'(t) U_I(t, t_0), \tag{2.55}$$

whose solution is

$$U_I(t, t_0) = \exp_+\left[-\frac{i}{\hbar}\int_{t_0}^{t} d\tau\, H_I'(\tau)\right]. \tag{2.56}$$

Combining these definitions we finally have

$$\begin{aligned}|\psi_S(t)\rangle &= U_0(t, t_0)|\psi_I(t)\rangle \\ &= U_0(t, t_0)U_I(t, t_0)|\psi_I(t_0)\rangle \\ &= U_0(t, t_0)U_I(t, t_0)|\psi_S(t_0)\rangle.\end{aligned}$$

Since by definition, $|\psi_I(t_0)\rangle = |\psi_S(t_0)\rangle$, we obtain

$$U(t, t_0) = U_0(t, t_0)U_I(t, t_0) = U_0(t, t_0)\exp_+\left[-\frac{i}{\hbar}\int_{t_0}^{t} d\tau\, H_I'(\tau)\right]. \tag{2.57}$$

Our final expression for the time evolution operator now becomes

$$U(t, t_0) = U_0(t, t_0) + \sum_{n=1}^{\infty}\left(-\frac{i}{\hbar}\right)^n \int_{t_0}^{t} d\tau_n \int_{t_0}^{\tau_n} d\tau_{n-1} \cdots \int_{t_0}^{\tau_2} d\tau_1$$
$$U_0(t, \tau_n)H'(\tau_n)U_0(\tau_n, \tau_{n-1})H'(\tau_{n-1}) \cdots U_0(\tau_2, \tau_1)H'(\tau_1)U_0(\tau_1, t_0). \tag{2.58}$$

If the time evolution operator with respect to H_0, $U_0(t, t_0)$, can be calculated exactly, then Eq. (2.58) provides a means for expanding the propagation in powers of $H'(\tau)$ alone [rather than powers of the entire Hamiltonian $H(t)$, as was done in Eq. (2.40)]. This result can be interpreted as follows: In nth order the system interacts n times with H', at time $\tau_1, \ldots, \tau_n$. The system first propagates freely from time t_0 to τ_1, $[U_0(\tau_1, t_0)]$, then interacts with $H'(\tau_1)$, then propagates to time τ_2 $[U_0(\tau_2, \tau_1)]$, then interacts at $H'(\tau_2)$, and so forth, until the last interaction at τ_n, followed by a free propagation $U_0(t, \tau_n)$. The time arguments $t \geq \tau_n \geq \cdots \geq \tau_1 \geq t_0$ are fully ordered, but apart from this constraint can assume any value. The n fold integration runs over all possible times. This expression is very useful, and with the proper partitioning of the Hamiltonian it may be truncated at a low order and still hold even for long times.

By writing Eq. (2.55) in an integral form analogous to Eq. (2.39), and combining with Eq. (2.57), we can recast Eq. (2.58) as

$$U(t, t_0) = U_0(t, t_0) - \frac{i}{\hbar}\int_{t_0}^{t} d\tau\, U_0(t, \tau)H'(\tau)U(\tau, t_0), \tag{2.59a}$$

or when acting on $|\psi_s(t_0)\rangle$

$$|\psi(t)\rangle = |\psi_0(t)\rangle - \frac{i}{\hbar}\int_{t_0}^{t} d\tau\, U_0(t, \tau)H'(\tau)|\psi(\tau)\rangle, \tag{2.59b}$$

where

$$|\psi_0(t)\rangle \equiv U_0(t, t_0)|\psi_s(t_0)\rangle.$$

Taking the Hermitian conjugate of Eq. (2.57) and recalling the definition of the negative time ordered exponential [Eq. (2.45)] we get

$$U^{\dagger}(t, t_0) = \exp_{-}\left[\frac{i}{\hbar}\int_{t_0}^{t} d\tau\, H_{\mathrm{I}}'(\tau)\right]U_0^{\dagger}(t, t_0), \tag{2.60a}$$

with

$$U_0^{\dagger}(t, t_0) = \exp_{-}\left[\frac{i}{\hbar}\int_{t_0}^{t} d\tau\, H_0(\tau)\right]. \tag{2.60b}$$

A particularly useful case is when H_0 is time independent. This is often the case in optical problems involving weak radiation fields, where we choose H_0 to represent the material Hamiltonian, and the time-dependent radiation–matter interaction is treated perturbatively. In this case, U_0 is simply given by $U_0(t, t_0) = \exp[-(i/\hbar)H_0(t - t_0)]$.

So far we used the interaction picture soley as an intermediate step in the derivation of an expression for the time evolution operator. This picture has, however, a broader and deeper significance. To see that let us consider now the expectation value of an arbitrary operator. Using Eq. (2.2) we have

$$\begin{aligned}\langle A(t)\rangle &= \langle\psi(t)|A|\psi(t)\rangle\\ &= \langle\psi(0)|U^{\dagger}(t, t_0)AU(t, t_0)|\psi(0)\rangle.\end{aligned}$$

We next introduce the operator A in the interaction picture

$$A_{\mathrm{I}}(t) \equiv U_0^{\dagger}(t, t_0)A_{\mathrm{S}}U_0(t, t_0). \tag{2.61}$$

Here A_{S} is the previously defined operator A in the Schrödinger picture. Using Eqs. (2.53a) and (2.57), and noting that $U^{\dagger}(t, t_0) = U_{\mathrm{I}}^{\dagger}(t, t_0)U_0^{\dagger}(t, t_0)$, the time evolution in the interaction picture is finally given by

$$\langle A(t)\rangle = \langle\psi_{\mathrm{I}}(t)|A_{\mathrm{I}}(t)|\psi_{\mathrm{I}}(t)\rangle, \tag{2.62a}$$

$$\dot{\psi}_{\mathrm{I}} = -\frac{i}{\hbar}H_{\mathrm{I}}'(t)\psi_{\mathrm{I}}, \tag{2.62b}$$

$$\dot{A}_{\mathrm{I}} = \frac{i}{\hbar}[H_0, A_{\mathrm{I}}], \tag{2.62c}$$

$$H_{\mathrm{I}}'(t) = U_0^{\dagger}(t, t_0)H'(t)U_0(t, t_0). \tag{2.62d}$$

Equation (2.62c) can be verified using the definition of A_{I} and Eq. (2.51). Equations (2.62) are valid for an arbitrary partitioning of the Hamiltonian. They offer a new viewpoint (hence the name "picture") on quantum evolution. In this picture we include part of the time evolution in the wavefunction and part in the operators. Choosing $H_0 = 0$, then $H' = H$ and we recover the previous *Schrödinger picture* where the time evolution is entirely with the wavefunction and the operators are time independent

$$\dot{\psi}_S = -\frac{i}{\hbar} H_s \psi_S, \tag{2.63a}$$

$$\dot{A}_S = 0. \tag{2.63b}$$

Making the opposite choice $H_0 = H$, then $H' = 0$ we obtain the *Heisenberg picture* whereby ψ is time independent and the entire evolution is with the operators:

$$\dot{\psi}_H = 0, \tag{2.63c}$$

$$\dot{A}_H = \frac{i}{\hbar}[H, A_H]. \tag{2.63d}$$

The interaction picture thus includes the Schrödinger and the Heisenberg pictures as special cases.

THE MAGNUS EXPANSION

We start by introducing the *cumulant expansion* [3–5]. Suppose we evaluate a certain quantity such as the time evolution operator, the wavefunction, etc. and expand it perturbatively in some small parameter λ. We thus have

$$A = A_0(1 + \lambda A_1 + \lambda^2 A_2 + \cdots). \tag{2.64}$$

We next make the ansatz

$$A \equiv A_0 \exp(F)$$

where F is also expanded in a Taylor series

$$F = \lambda F_1 + \lambda^2 F_2 + \cdots.$$

When this expansion is substituted in the above ansatz, we can expand the exponential and collect terms to a given order in λ. We can then express F_n in terms of $A_1, A_2 \cdots A_n$, and evaluate all F_n successively. To second order we have

$$A = A_0 \exp[\lambda A_1 + \lambda^2(A_2 - \tfrac{1}{2}A_1^2) + \cdots]. \tag{2.65}$$

If we calculate A_1 and A_2 and truncate the expansion (2.64), we are assuming (implicitly) that $A_3 = A_4 = \cdots = 0$. This is of course not generally true. We were just too lazy to calculate these higher order terms! If we use the same pieces of information (A_1 and A_2) to construct the expansion of F and truncate it to second order, we obtain an expansion of A [Eq. (2.65)], which contains terms of infinite order. Instead of neglecting higher order terms, we make an educated guess and express them (approximately) in terms of the lower order quantities. The truncation of F thus provides a *partial resummation* of the perturbation series for A.*

* This form of resummation is not unique. If instead of an exponential form, we represent A as a ratio two polynomials we obtain the Padé resummation. [C. M. Bender and S. A. Orszag, *Advanced Mathematical Methods for Scientists and Engineers*, McGraw-Hill, New York (1978).].

When the cumulant expansion is applied to a time-ordered exponential operator, such as the propagator, we obtain the *Magnus expansion*:

$$\exp_{+}\left[\lambda\int_{t_0}^{t} d\tau\, A(\tau)\right] \equiv \exp\left[\sum_{n=1}^{\infty}\frac{1}{n!}\lambda^n F_n(t, t_0)\right], \tag{2.65a}$$

$$F_1(t, t_0) = \int_{t_0}^{t} d\tau\, A(\tau), \tag{2.65b}$$

$$F_2(t, t_0) = \int_{t_0}^{t} d\tau_2 \int_{t_0}^{\tau_2} d\tau_1 [A(\tau_2), A(\tau_1)], \tag{2.65c}$$

$$F_3(t, t_0) = \int_{t_0}^{t} d\tau_3 \int_{t_0}^{\tau_3} d\tau_2 \int_{t_0}^{\tau_2} d\tau_1 \{[A(\tau_3), [A(\tau_2), A(\tau_1)]] + [[A(\tau_3), A(\tau_2)], A(\tau_1)]\}. \tag{2.65d}$$

When this expansion is applied to the time evolution operator in the interaction picture [Eq. (2.56)] we have

$$U_1(t, t_0) = \exp_{+}\left[-\frac{i}{\hbar}\int_{t_0}^{t} H_1'(\tau)\, d\tau\right] \equiv \exp\left[\sum_{n=1}^{\infty}\frac{1}{n!}\left(-\frac{i}{\hbar}\right)^n \mathscr{H}_n(t, t_0)\right], \tag{2.66}$$

where

$$\mathscr{H}_1 = \int_{t_0}^{t} H_1'(\tau_1)\, d\tau_1, \tag{2.67a}$$

$$\mathscr{H}_2 = \int_{t_0}^{t} d\tau_2 \int_{t_0}^{\tau_2} d\tau_1 [H_1'(\tau_2), H_1'(\tau_1)], \tag{2.67b}$$

$$\mathscr{H}_3 = \int_{t_0}^{t} d\tau_3 \int_{t_0}^{\tau_3} d\tau_2 \int_{t_0}^{\tau_2} d\tau_1 \{[H_1'(\tau_3), [H_1'(\tau_2), H_1'(\tau_1)]] + [[H_1'(\tau_3), H_1'(\tau_2)], H_1'(\tau_1)]\}. \tag{2.67c}$$

Here we have expressed the time-ordered exponential of H_1' as an ordinary (not time-ordered) exponential of a different, time-dependent operator, given by an infinite series. The nth term in the series $\mathscr{H}_n$ is given by a combination of nth order commutators of the original Hamiltonian H_1'. This expansion has been used in magnetic resonance spectroscopy to develop the average Hamiltonian theory [5]. It is particularly useful for periodic Hamiltonians as is the case, e.g., when using repetitive pulse sequences, where we need to evaluate the time evolution operator only for a single period.

Equations (2.66) can also be used for the ordinary expansion of $U(t, t_0)$ [Eq. (2.40)], by simply replacing all $H_1'(t)$ with $H(t)$.

GREEN FUNCTIONS AND CAUSALITY

For many applications it is more convenient to work in the frequency domain rather than in the time domain, and calculate the Fourier transform of the time evolution operator. This is done by introducing the *Green function* [6–8]. (It is almost a law of nature that time domain observables are more conveniently calculated in the frequency domain and vice versa.)

To transform $U(t, t_0)$ to the frequency domain, we need to treat separately the case $t > t_0$ (forward propagation) and the case $t < t_0$ (backward propagation). We further restrict the following discussion to time-independent Hamiltonians.

We first consider the forward propagation and define

$$G(t - t_0) \equiv \theta(t - t_0)U(t, t_0)$$
$$= \theta(t - t_0) \exp\left[-\frac{i}{\hbar} H(t - t_0)\right], \tag{2.68}$$

where $\theta(t - t_0)$ is the Heavyside step function

$$\theta(t - t_0) \equiv \begin{cases} 1 & t > t_0 \\ 0 & t < t_0. \end{cases}$$

We shall use the following integrals

$$-\frac{1}{2\pi i}\int_{-\infty}^{\infty} dE \frac{1}{E - E_n + i\varepsilon} \exp\left[-\frac{i}{\hbar} E(t - t_0)\right] = \begin{cases} 0 & t < t_0 \\ \exp\left[-\frac{i}{\hbar} E_n(t - t_0)\right] & t > t_0 \end{cases} \tag{2.69a}$$

and

$$\frac{1}{2\pi i}\int_{-\infty}^{\infty} dE \frac{1}{E - E_n - i\varepsilon} \exp\left[-\frac{i}{\hbar} E(t - t_0)\right] = \begin{cases} \exp\left[-\frac{i}{\hbar} E_n(t - t_0)\right] & t < t_0 \\ 0 & t > t_0. \end{cases} \tag{2.69b}$$

These integrals can be easily verified using contour integration (see Figure 2.4). ε is a positive number that is set to $\varepsilon = 0$ at the end. It is essential to introduce it, though, since the integral without ε is ill defined, and the sign of ε makes a big difference. On substituting Eq. (2.69a) in (2.14) and rearranging the terms, we get

$$\theta(t - t_0)U(t, t_0) = -\frac{1}{2\pi i}\int_{-\infty}^{\infty} dE \exp\left[-\frac{i}{\hbar} E(t - t_0)\right] G(E),$$

where

$$G(E) = \sum_n \frac{|\varphi_n\rangle\langle\varphi_n|}{E - E_n + i\varepsilon}.$$

Using our definition of a function of an operator we can recast this in the compact form

$$G(E) = \frac{1}{E - H + i\varepsilon}. \tag{2.70}$$

(a)

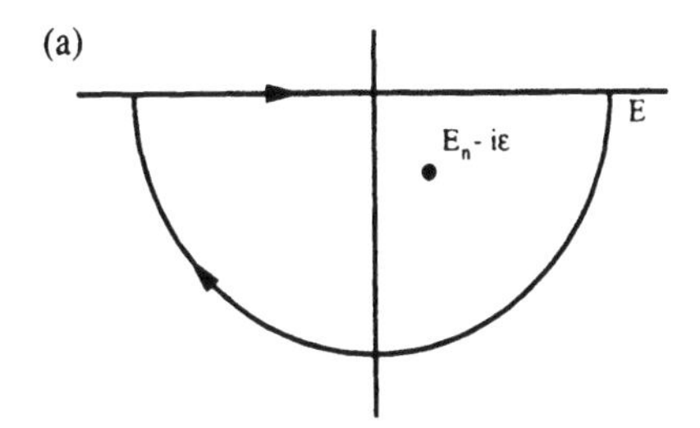

(b)

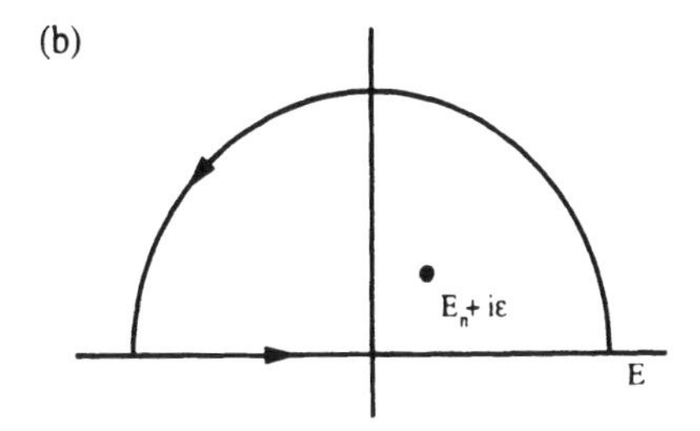

FIG. 2.4 (a) Contour integration for the retarded [Eq. (2.71a)] Green function in the complex E phase. The pole is in the lower half plane and the Green function is finite only for $t > 0$. (b) Contour integration for the advanced Green function. The pole is in the upper half plane and the Green function is finite for $t < 0$.

In summary we have

$$G(t - t_0) = -\frac{1}{2\pi i}\int_{-\infty}^{\infty} dE \exp\left[-\frac{i}{\hbar}E(t - t_0)\right]G(E), \qquad (2.71a)$$

$$G(E) = -\frac{i}{\hbar}\int_{-\infty}^{\infty} dt\, G(t - t_0)\exp\left[\frac{i}{\hbar}E(t - t_0)\right]. \qquad (2.71b)$$

Equation (2.71b) is the inverse transform of Eq. (2.71a). $G(t - t_0)$ [Eq. (2.68)] or its transform $G(E)$ [Eq. (2.70)], which represent the propagation forward in time, are called the retarded Green function. We note that the Green function is not a function but is rather an operator.

For backward propagation ($t < t_0$) we define the advanced Green function

$$G^-(t_0 - t) \equiv \theta(t_0 - t)U(t, t_0). \qquad (2.72)$$

Making use of Eq. (2.69b) we then have

$$\theta(t_0 - t)U(t, t_0) = \frac{1}{2\pi i}\int_{-\infty}^{\infty} dE \exp\left[-\frac{i}{\hbar}E(t - t_0)\right]G^-(E), \qquad (2.73)$$

where the advanced Green function in the frequency domain is given by

$$G^-(E) = \frac{1}{E - H - i\varepsilon}.$$

We finally obtain

$$G^-(t - t_0) = \frac{1}{2\pi i} \int_{-\infty}^{\infty} dE \exp\left[-\frac{i}{\hbar} E(t - t_0)\right] G^-(E), \quad (2.74a)$$

$$G^-(E) = \frac{i}{\hbar} \int_{-\infty}^{\infty} G^-(t - t_0) \exp\left[\frac{i}{\hbar} E(t - t_0)\right]. \quad (2.74b)$$

The retarded Green function is sometimes denoted G^+. For brevity, we shall delete the $+$ superscript as well as the $+i\varepsilon$. All Green functions are to be understood as "retarded" and the argument E understood as $E + i\varepsilon$, unless stated otherwise.

Perturbative expansions can be very conveniently carried out using the Green function. An important equation that the Green function satisfies is the Dyson equation. We introduce a zero order Green function

$$G_0(E) \equiv \frac{1}{E - H_0 + i\varepsilon}, \quad (2.75)$$

where

$$H = H_0 + V. \quad (2.76)$$

Using the identity $A^{-1} = B^{-1} + B^{-1}(B - A)A^{-1}$, where A and B are two operators, we can then write

$$G(E) = G_0(E) + G_0(E)VG(E), \quad (2.77)$$

or

$$G(E) = G_0(E) + G(E)VG_0(E). \quad (2.78)$$

Upon iteration of either Eq. (2.77) or (2.78) we get

$$G(E) = G_0(E) + G_0(E)VG_0(E) + G_0(E)VG_0(E)VG_0(E) + \cdots. \quad (2.79)$$

Equation (2.79) can be obtained by a Fourier transformation of Eq. (2.58), provided we take H to be time independent and set $H'(t) = V$.

PROJECTION OPERATORS, REDUCED EQUATIONS OF MOTION, AND EFFECTIVE HAMILTONIANS

Projection operators and self-energies can be used to perform a partial resummation of perturbation series such as Eq. (2.79). Projection operators are defined as follows: we consider a basis set in our Hilbert space $|\phi_i\rangle$. We then partition the space into two parts, P and Q, and introduce the operators [8, 9]:

$$P = \sum_{i=1}^{n} |\phi_i\rangle\langle\phi_i|, \qquad Q = \sum_{i=n+1}^{\infty} |\phi_i\rangle\langle\phi_i| \quad (2.80)$$

where n is some chosen cutoff that defines the partitioning. From these definitions it immediately follows that

$$P + Q = 1 \tag{2.81a}$$

$$P^2 = P \qquad Q^2 = Q \tag{2.81b}$$

$$PQ = QP = 0. \tag{2.81c}$$

Operators that satisfy Eqs. (2.81) are called *projection operators*. Equation (2.81b) implies that projection operators are *idempotent*.

We now return to the Schrödinger equation. Inserting the partitioning of the unit operators (Eq. (2.81a)) between the H and the ψ, and multiplying both sides from the left by either P or Q,

$$\underset{\substack{\uparrow \\ P,\,Q}}{\dot{\psi}} = -\frac{i}{\hbar}\underset{\substack{\uparrow \\ P+Q}}{H}\psi,$$

we obtain two coupled equations for $P\psi$ and $Q\psi$

$$P\dot{\psi} = -\frac{i}{\hbar}(PHP)(P\psi) - \frac{i}{\hbar}PHQ(Q\psi),$$

$$Q\dot{\psi} = -\frac{i}{\hbar}(QHP)(P\psi) - \frac{i}{\hbar}QHQ(Q\psi).$$

In matrix form these equations read

$$\begin{pmatrix} P\dot{\psi} \\ Q\dot{\psi} \end{pmatrix} = -\frac{i}{\hbar}\left(\begin{array}{c|c} PHP & PHQ \\ \hline QHP & QHQ \end{array}\right)\begin{pmatrix} P\psi \\ Q\psi \end{pmatrix}. \tag{2.82}$$

Solving for $Q\psi$ we have

$$Q\psi(t) = \exp\left(-\frac{i}{\hbar}QHQt\right)Q\psi(0) - \frac{i}{\hbar}\int_0^t d\tau \exp\left[-\frac{i}{\hbar}QHQ(t-\tau)\right]QHP[P\psi(\tau)].$$

Substituting into the equation for $P\psi$, and assuming that the system is initially in the P space so that $Q\psi(0) = 0$ we finally obtain

$$P\dot{\psi} = -\frac{i}{\hbar}PHP(P\psi) + \left(\frac{i}{\hbar}\right)^2 \int_0^t d\tau\, PHQ \exp\left[-\frac{i}{\hbar}QHQ(t-\tau)\right]QHP[P\psi(\tau)]. \tag{2.83}$$

We next introduce a similar partitioning of the Green function

$$\begin{pmatrix} P\psi(t) \\ Q\psi(t) \end{pmatrix} = \left(\begin{array}{c|c} PG(t)P & PG(t)Q \\ \hline QG(t)P & QG(t)Q \end{array}\right)\begin{pmatrix} P\psi(0) \\ Q\psi(0) \end{pmatrix}. \tag{2.84}$$

$$P\psi(t) = PG(t)P\psi(0), \tag{2.85}$$

where

$$PG(t)P = -\frac{1}{2\pi i}\int_{-\infty}^{\infty} dE \exp\left(-\frac{i}{\hbar}Et\right)PG(E)P. \tag{2.86}$$

By performing a Fourier transform on Eq. (2.83) we have

$$PG(E)P = \frac{1}{E - PH_{\text{eff}}(E)P}, \tag{2.87}$$

with the effective Hamiltonian

$$H_{\text{eff}}(E) = H + HQ\tilde{G}(E)QH,$$

and

$$\tilde{G}(E) = \frac{1}{E - QHQ}. \tag{2.88a}$$

We next partition the Hamiltonian using Eq. (2.76) and assume that the zero-order Hamiltonian H_0 commutes with the P projection-operator so that

$$PH_0Q = QH_0P = 0.$$

The effective Hamiltonian thus assumes the form

$$PH_{\text{eff}}(E)P = PH_0P + PR(E)P, \tag{2.88b}$$

with the self energy operator

$$PR(E)P = PVP + PVQ\tilde{G}(E)QVP. \tag{2.88c}$$

Equation (2.83) or (2.87) allows us to consider only partial information and solve for $P\psi$. The effect of $Q\psi$ is rigorously incorporated in the effective Hamiltonian H_{eff} through the self-energy operator $R(E)$. The other projections of the Green function can be obtained in a similar way. In summary, the complete set of projected Green functions are [9].

$$PG(E)P = P\frac{1}{E - H_0 - PR(E)P}P, \tag{2.89a}$$

$$QG(E)P = \tilde{G}(E)QVPG(E)P, \tag{2.89b}$$

$$PG(E)Q = PG(E)PVQ\tilde{G}(E), \tag{2.89c}$$

$$QG(E)Q = \tilde{G}(E) + \tilde{G}(E)QVPG(E)PVQ\tilde{G}(E). \tag{2.89d}$$

As an example of an application for the projected Green function, consider a model system of a single state $|s\rangle$ coupled to a continuum $\{|l\rangle\}$ (Figure 2.5):

$$H = |s\rangle E_s\langle s| + \sum_{l=1}^{N} |l\rangle E_l\langle l| + \sum_{l=1}^{N} (V_{sl}|s\rangle\langle l| + V_{ls}|l\rangle\langle s|). \tag{2.90}$$

This model proposed by Wigner and Weisskopf [10] is a prototype for irreversible decay. It can represent radiative-damping, autoionization in atoms,

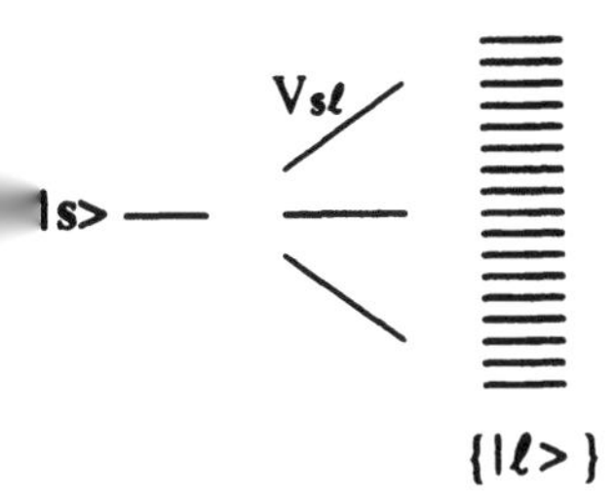

FIG. 2.5 Level coupling scheme for the Wigner–Weiskopf model.

intramolecular relaxation, coupling to a phonon bath, etc. We shall be interested in calculating the probability of the system to be found in the $|s\rangle$ state at time t, given that it was prepared in the same state as $t = 0$. To that end we need to calculate the matrix element $G_{ss}(E)$. The straightforward (and most tedious) way to do that is by calculating the eigenstates and eigenvalues of this Hamiltonian, which requires the diagonalization of an $(N + 1) \times (N + 1)$ matrix. We then have

$$H|j\rangle = E_j|j\rangle,$$

and

$$G_{ss}(E) = \sum_j \frac{|\langle s|j\rangle|^2}{E - E_j + i\varepsilon},$$

so that

$$P_{ss}(t) \equiv |G_{ss}(t)|^2 = \left|\sum_j |\langle s|j\rangle|^2 \exp\left(-\frac{i}{\hbar} E_j t\right)\right|^2.$$

Instead of calculating these quantities we shall introduce the projections

$$P = |s\rangle\langle s|; \qquad Q = \sum_l |l\rangle\langle l|.$$

We note that QHQ is diagonal in this example. In matrix form, the partitioned Hamiltonian for the Wigner–Weisskopf problem is given below

$$
\begin{array}{c}
\begin{array}{cc} \quad P & \qquad\qquad Q \end{array} \\
\left(\begin{array}{c|cccc}
E_s & V_{sl_1} & V_{sl_2} & V_{sl_3} & \cdots \\
\hline
V_{l_1s} & E_{l_1} & & 0 & \\
V_{l_2s} & & E_{l_2} & & \\
V_{l_3s} & & & E_{l_3} & \\
\vdots & & 0 & & \ddots
\end{array}\right)
\end{array}
$$

We can then immediately write

$$G_{ss}(E) = \frac{1}{E - E_s - \Delta_s(E) + \frac{i}{2}\hbar\Gamma_s(E)}, \tag{2.91}$$

with

$$R_{ss}(E) = \sum_l \frac{|V_{sl}|^2}{E - E_l + i\varepsilon} \equiv \Delta_s(E) - \frac{i}{2}\hbar\Gamma_s(E). \tag{2.92}$$

To proceed further, we assume that the $\{|l\rangle\}$ manifold is sufficiently dense to allow switching from a summation to an integration. We thus write

$$R_{ss}(E) = \int dE_l \frac{|V_{sl}|^2 \rho(E_l)}{E - E_l + i\varepsilon},$$

where $\rho(E_l)$ is the density of states in the manifold. Integrals of this type can be evaluated as follows. Consider the integral

$$\int dE' \frac{f(E')}{E' - E + i\varepsilon} = \int dE' f(E') \frac{E' - E}{(E' - E)^2 + \varepsilon^2} - i \int dE' \frac{f(E')\varepsilon}{(E' - E)^2 + \varepsilon^2}.$$

As $\varepsilon \to 0$ we have

$$\frac{\varepsilon}{(E' - E)^2 + \varepsilon^2} \to \pi\delta(E - E')$$

we thus get

$$\int dE' \frac{f(E')}{E' - E + i\varepsilon} \equiv \mathscr{P}\mathscr{P} \int dE' \frac{f(E')}{E' - E} - i\pi f(E),$$

where

$$\mathscr{P}\mathscr{P} \int dE' \frac{f(E')}{E' - E} \equiv \lim_{\varepsilon \to 0} \left[\int_{-\infty}^{-\varepsilon} dE' \frac{f(E')}{E' - E} + \int_{\varepsilon}^{\infty} dE' \frac{f(E')}{E' - E} \right].$$

$\mathscr{P}\mathscr{P}$ is called the *principal part* of the integral. We now have

$$\Delta_s(E) = \mathscr{P}\mathscr{P} \int \frac{|V_{sl}|^2 \rho(E_l)}{E - E_l} dE_l, \tag{2.93a}$$

$$\hbar\Gamma_s(E) = 2\pi \int |V_{sl}|^2 \delta(E - E_l) \rho(E_l) \, dE_l. \tag{2.93b}$$

If the continuum is dense, and we can ignore the smooth and slow energy dependence of Δ_s and Γ_s, we then have the single pole approximation

$$\Delta_s(E) \cong \Delta_s(E_s), \qquad \Gamma_s(E) \cong \Gamma_s(E_s), \tag{2.94}$$

which immediately yields

$$G_{ss}(t) = \theta(t) \, exp\left[-\frac{i}{\hbar}(E_s + \Delta_s)t - (\Gamma_s/2)t \right]. \tag{2.95a}$$

The probability of the system to be in the $|s\rangle$ state, if it is initially prepared in that state is then

$$P_{ss}(t) \equiv |G_{ss}(t)|^2 = \exp(-\Gamma_s t). \tag{2.95b}$$

The real part (Δ_s) of R is a level shift whereas the imaginary part (Γ_s) represents a relaxation rate (inverse lifetime).

This example illustrates an important general behavior: When the Q space represents a smooth continuum we can neglect the dependence of $H_{\text{eff}}(E)$ on E, denoting it simply H_{eff}. Equation (2.83) then reads

$$P\dot{\psi} = -\frac{i}{\hbar} PH_{\text{eff}}P(P\psi). \tag{2.96}$$

The effective Hamiltonian is particularly useful in this limit since the evolution can be simply described using its eigenstates whose number is much smaller than the eigenstates of the full Hamiltonian.

Finally, we note that Eq. (2.91) can be derived directly from the Dyson equation, without using projection operators. Taking matrix elements of the Dyson equation we get

$$G_{ss}(E) = \frac{1}{E - E_s} + \frac{1}{E - E_s} \sum_l V_{sl} G_{ls}(E),$$

$$G_{ls}(E) = \frac{1}{E - E_l} V_{ls} G_{ss}(E).$$

By solving these coupled algebraic equations we immediately obtain Eq. (2.91).

APPENDIX 2A

Linear Vector Spaces and Operators

Consider a collection (manifold) of objects (vectors) $R = \{|\mathbf{a}\rangle, |\mathbf{b}\rangle, |\mathbf{c}\rangle \cdots\}$. We shall denote a complex number by λ. We next define a multiplication operation of a vector by the complex number λ, which satisfies the following requirements.[11]

MULTIPLICATION RULES

(i) $\lambda|\mathbf{a}\rangle = |\mathbf{b}\rangle \in R$	(closure)
(ii) $(\lambda_1 + \lambda_2)|\mathbf{a}\rangle = \lambda_1|\mathbf{a}\rangle + \lambda_2|\mathbf{a}\rangle$	(distributivity)
(iii) $(\lambda_1\lambda_2)|\mathbf{a}\rangle = \lambda_1(\lambda_2|\mathbf{a}\rangle)$	(associativity)
(iv) $1 \cdot |\mathbf{a}\rangle = |\mathbf{a}\rangle$	(multiplication by unity)
(v) $0 \cdot |\mathbf{a}\rangle = |\mathbf{0}\rangle$	(existence of a "zero" vector)

The symbol $|\mathbf{b}\rangle \in R$ implies that the vector $|\mathbf{b}\rangle$ belongs to the set R. Condition (v) implies the existence of a zero vector in the manifold, such that the multiplication of any vector by the number 0 gives this zero vector.

We next introduce a second operation; the *addition* of vectors. This operation has the following properties:

ADDITION RULES

(vi) $|\mathbf{a}\rangle + |\mathbf{b}\rangle = |\mathbf{c}\rangle \in R$	(closure)
(vii) $|\mathbf{a}\rangle + |\mathbf{b}\rangle = |\mathbf{b}\rangle + |\mathbf{a}\rangle$	(commutativity)
(viii) $\lambda(|\mathbf{a}\rangle + |\mathbf{b}\rangle) = \lambda|\mathbf{a}\rangle + \lambda|\mathbf{b}\rangle$	(distributivity)
(ix) $(|\mathbf{a}\rangle + |\mathbf{b}\rangle) + |\mathbf{c}\rangle = |\mathbf{a}\rangle + (|\mathbf{b}\rangle + |\mathbf{c}\rangle)$	(associativity)
(x) $(|\mathbf{a}\rangle + |\mathbf{b}\rangle) = |0\rangle$, if and only if $|\mathbf{b}\rangle \equiv -|\mathbf{a}\rangle$	(the zero vector is unique)

We now introduce another operation: A *scalar product* of two vectors, which associates a complex number λ to any pair of vectors, and has the following properties.

SCALAR PRODUCT RULES

(xi) $\langle \mathbf{a} | \mathbf{b} \rangle = \lambda$

(xii) $\langle \mathbf{a} | \mathbf{b} \rangle = \langle \mathbf{b} | \mathbf{a} \rangle^*$

(xiii) $\langle \mathbf{a} | \mathbf{b} + \mathbf{c} \rangle = \langle \mathbf{a} | \mathbf{b} \rangle + \langle \mathbf{a} | \mathbf{c} \rangle$ (distributivity)

(xiv) $\langle \mathbf{a} | \lambda \mathbf{b} \rangle = \lambda \langle \mathbf{a} | \mathbf{b} \rangle$

The following properties are also satisfied. They result from the previous ones:

(1) $\langle \mathbf{a} + \mathbf{b} | \mathbf{c} \rangle = \langle \mathbf{a} | \mathbf{c} \rangle + \langle \mathbf{b} | \mathbf{c} \rangle$

(2) $\langle \lambda \mathbf{a} | \mathbf{b} \rangle = \lambda^* \langle \mathbf{a} | \mathbf{b} \rangle$

We next consider the scalar product of a vector with itself $\langle \mathbf{a} | \mathbf{a} \rangle \equiv |\mathbf{a}|^2$. Following (xii) it has to be real. Note that it is not necessarily positive. We shall be interested in spaces with nonnegative (positive definite) metric that satisfy $|\mathbf{a}|^2 \geq 0$, and the equality holds only for the zero vector $\mathbf{a} = 0$. The square root of that scalar product $\sqrt{\langle \mathbf{a} | \mathbf{a} \rangle}$ is called the norm (length) of the vector. A vector with norm equal to 1 is called normalized. If $|\mathbf{a}\rangle \neq |0\rangle$ and $|\mathbf{b}\rangle \neq |0\rangle$ but $\langle \mathbf{a} | \mathbf{b} \rangle = 0$ we say that $|\mathbf{a}\rangle$ and $|\mathbf{b}\rangle$ are orthogonal.

Vector spaces with nonnegative metric satisfy the *Schwartz inequality*

$$\langle \mathbf{a} | \mathbf{a} \rangle \langle \mathbf{b} | \mathbf{b} \rangle \geq \langle \mathbf{a} | \mathbf{b} \rangle \langle \mathbf{b} | \mathbf{a} \rangle$$

and the equality holds only if $|\mathbf{a}\rangle = \lambda |\mathbf{b}\rangle$.

A set of n vectors $\{\mathbf{a}_j\}$ is said to be *linearly independent* if and only if

$$\sum_{j=1}^{n} \lambda_j \mathbf{a}_j \neq 0,$$

except when $\lambda_1 = \lambda_2 = \cdots = \lambda_n = 0$.

When all properties (i)–(xiv) are satisfied these vectors form a linear vector space. The maximum number n of linear independent vectors that can be found in a given space is called the dimensionality of the space. Such a set forms a *basis set* and all other vectors can be represented as linear combinations of the basis set. As an example, the three unit vectors $\hat{x}, \hat{y}, \hat{z}$ from a basis set for vectors in three-dimensional space.

A *Hilbert space* is a vector space with infinite dimensionality.

An operator A is a mathematical operation that associates with every vector in the space another vector

$$A|\mathbf{a}\rangle = |\mathbf{b}\rangle.$$

The operator is linear if it satisfies

$$A(\lambda_1|\mathbf{a}\rangle + \lambda_2|\mathbf{b}\rangle) = \lambda_1 A|\mathbf{a}\rangle + \lambda_2 A|\mathbf{b}\rangle.$$

The object $|\mathbf{a}\rangle\langle\mathbf{b}|$ is denoted an *outer product* of two vectors (as opposed to the scalar product $\langle \mathbf{a} | \mathbf{b} \rangle$, which is denoted *inner product*). $|\mathbf{a}\rangle\langle\mathbf{b}|$ is a special example of a linear operator. Using a basis set $|\mathbf{a}_n\rangle$ we have

$$\sum_n |\mathbf{a}_n\rangle\langle\mathbf{a}_n| = 1 \qquad \text{(Completeness)}$$

$$\langle\mathbf{a}_n|\mathbf{a}_m\rangle = \delta_{nm} \qquad \text{(Orthogonality)}$$

By multiplying any vector by the completeness relation we have

$$|\mathbf{c}\rangle = \sum_n |\mathbf{a}_n\rangle\langle\mathbf{a}_n|\mathbf{c}\rangle.$$

This is an expansion of $|\mathbf{c}\rangle$ in the basis set $\{|\mathbf{a}_n\rangle\}$. The scalar products $\langle\mathbf{a}_n|\mathbf{c}\rangle$ represent the expansion coefficients. Once a basis set is specified, we can define a vector by simply giving the coefficients $\langle\mathbf{a}_n|\mathbf{c}\rangle$. These represent the vector $|\mathbf{c}\rangle$ in the $\{|\mathbf{a}_n\rangle\}$ representation. A general linear operator can be written as the superposition of all outer products of a basis set

$$A = \sum_{n,m} A_{nm}|\mathbf{a}_n\rangle\langle\mathbf{a}_m|.$$

The operator can thus be thought of as a matrix in our linear space.

Finally, the scalar product can be defined as the trace of the corresponding outer product (the trace of a matrix is the sum of its diagonal elements)

$$\mathrm{Tr}\,|\mathbf{b}\rangle\langle\mathbf{a}| \equiv \sum_n \langle\mathbf{a}_n|\mathbf{b}\rangle\langle\mathbf{a}|\mathbf{a}_n\rangle = \sum_n \langle\mathbf{a}|\mathbf{a}_n\rangle\langle\mathbf{a}_n|\mathbf{b}\rangle = \langle\mathbf{a}|\mathbf{b}\rangle.$$

This property will be particularly useful in the definition of Liouville space to be introduced in Chapter 3.

NOTES

1. C. Cohen-Tannoudji, B. Diu, and F. Laloe, *Quantum Mechanics Volumes I, II* (Wiley, New York, 1977).
2. R. P. Feynman and A. R. Hibbs, *Quantum Mechanics and Path Integrals* (McGraw-Hill, New York, 1965).
3. W. Magnus, *Commun. Pure Appl. Math.* **7**, 649 (1954).
4. R. M. Wilcox, *J. Math. Phys.* **8**, 962 (1967).
5. U. Haeberlen, *Advances in Magnetic Resonance*, Suppl. I (Academic Press, New York, 1976).
6. S. Doniach and E. H. Sondheimer, *Green's Functions for Solid State Physicists* (Benjamin, London, 1974).
7. E. N. Economou, *Green's Functions in Quantum Physics*, 2nd ed. (Springer-Verlag, Berlin, 1983).
8. M. L. Goldberger and R. M. Watson, *Collision Theory* (Wiley, New York, 1964).
9. L. Mower, *Phys. Rev.* **142**, 799 (1966).
10. G. Källen, *Quantum Electrodynamics* (Springer-Verlag, New York, 1972).
11. P. Roman, *Advanced Quantum Theory* (Addison-Wesley, New York, 1965, Appendix I).

BIBLIOGRAPHY

A. D. Davydov, *Quantum Mechanics* (Pergamon Press, Oxford, 1965).
P. A. M. Dirac, *The Principles of Quantum Mechanics*, 4th ed. (Oxford, London, 1976).
E. Merzbacher, *Quantum Mechanics* (Wiley, New York, 1970).
P. M. Morse and H. Feshbach, *Methods of Math Physics*, Vols. I, II (McGraw-Hill, New York, 1953).
L. I. Schiff, *Quantum Mechanics*, 3rd ed. (McGraw-Hill, New York, 1968).

CHAPTER 3

The Density Operator and Quantum Dynamics in Liouville Space

BASIC DEFINITIONS: PURE AND MIXED STATES

Consider a quantum system whose state is represented by a wavefunction $|\psi(t)\rangle$. The expectation value of any operator A is given by the matrix element

$$\langle A(t)\rangle = \langle\psi(t)|A|\psi(t)\rangle. \tag{3.1}$$

Using an arbitrary basis set, $\{|n\rangle\}$, the wavefunction can be expanded as

$$|\psi(t)\rangle = \sum_n |n\rangle c_n(t). \tag{3.2a}$$

Taking the Hermitian conjugate of Eq. (3.2a) we have

$$\langle\psi(t)| = \sum_m c_m^*(t)\langle m|. \tag{3.2b}$$

When Eqs. (3.2) are substituted in Eq. (3.1) we have

$$\langle A(t)\rangle = \sum_{n,m} c_n(t)c_m^*(t)\langle m|A|n\rangle \equiv \sum_{n,m} c_n(t)c_m^*(t)A_{mn}. \tag{3.3}$$

At this point we introduce the *density operator* (also known as the density matrix) [1, 2]

$$\rho(t) \equiv |\psi(t)\rangle\langle\psi(t)|. \tag{3.4}$$

When expanded in the basis set we have

$$\rho(t) = \sum_{n,m} c_n(t)c_m^*(t)|n\rangle\langle m| \equiv \sum_{n,m} \rho_{nm}(t)|n\rangle\langle m|, \tag{3.5}$$

where the density operator matrix element $\rho_{nm}(t) \equiv c_n(t)c_m^*(t)$. Equation (3.3) thus assumes the form

$$\langle A(t)\rangle = \sum_{n,m} A_{mn}\rho_{nm}(t) \equiv \mathrm{Tr}[A\rho(t)], \tag{3.6}$$

where Tr stands for the trace of an operator, i.e., the sum of its diagonal elements

$$\mathrm{Tr}\, B \equiv \sum_n B_{nn}. \tag{3.7}$$

The trace has several important properties, which follow immediately from its definition.

1. The trace of a product of operators is invariant to a cyclic permutation

$$\mathrm{Tr}(ABC) = \mathrm{Tr}(CAB) = \mathrm{Tr}(BCA). \tag{3.8}$$

Note that the trace of an operator does not always exist (e.g., the trace of the position x or the momentum operator p is infinite). Equation (3.8) thus holds only if one of the operators has a bounded spectrum so that the trace of the product exists.

2. The trace of a commutator vanishes,

$$\mathrm{Tr}[A, B] = 0. \tag{3.9}$$

This follows immediately from Eq. (3.8) and the same restrictions apply, i.e., that Tr AB exists. An obvious example when Eq. (3.9) does not hold is the position and the momentum $[x, p] = i\hbar$ and the trace of the commutator is infinite!

3. The trace is invariant to a unitary transformation. If S is a unitary operator (i.e., $S^\dagger = S^{-1}$) then

$$\mathrm{Tr}(S^\dagger AS) = \mathrm{Tr}\, A. \tag{3.10}$$

This follows directly from Eq. (3.8). We can also prove this relation by expanding it in a basis set

$$\mathrm{Tr}\, S^\dagger AS = \sum_{n,m,j} S^\dagger_{nm} A_{mj} S_{jn} = \sum_{j,m}\left(\sum_n S_{jn} S^\dagger_{nm}\right) A_{ml} = \sum_{j,m} \delta_{jm} A_{mj} = \mathrm{Tr}\, A.$$

Using vector and matrix notation, we have for the wavefunction and for the density matrix

$$|\psi(t)\rangle = \begin{pmatrix} c_1 \\ \vdots \\ c_N \end{pmatrix}, \qquad \rho(t) = \begin{pmatrix} c_1c_1^* & c_1c_2^* & \cdots \\ c_2c_1^* & c_2c_2^* & \\ \vdots & & \ddots \; c_Nc_N^* \end{pmatrix}.$$

In ordinary quantum mechanics the state of the system is represented by a vector and an operator is represented by a matrix. The expectation value of an operator [Eq. (3.1)] is given by multiplying the matrix by a column vector from the right and by a row vector from the left. Using the density operator we represent the state of the system by a matrix (rather than a vector). The expectation value of an operator [Eq. (3.6)] is then given by multiplying the A and the $\rho(t)$ matrices and summing over all diagonal elements of the product (the trace).

A system which can be characterized by a wavefunction is said to be in a *pure state.* According to the postulates of quantum mechanics, the system is then completely defined. However, a general state of a quantum system may or may not be pure. Consider an ensemble of systems with a given probability distribution to be in various quantum states. Denoting the probability of the system to be in the state $|\psi_k(t)\rangle$ by P_k, then the corresponding density operator is defined by

$$\rho(t) \equiv \sum_k P_k |\psi_k(t)\rangle\langle\psi_k(t)|, \tag{3.11}$$

where P_k are nonnegative

$$P_k \geq 0, \tag{3.12a}$$

and normalized

$$\sum_k P_k = 1. \tag{3.12b}$$

When all $P_k = 0$ except for one of them which is $P_k = 1$ then the system is in a pure state and can be represented by a wavefunction, as discussed earlier. Otherwise Eq. (3.11) represents a system whose state is incompletely defined. This is commonly denoted a *statistical mixture* or a *mixed state.* By the definition of a mixed state we have

$$\langle A(t)\rangle = \sum_k P_k \langle\psi_k(t)|A|\psi_k(t)\rangle = \mathrm{Tr}[A\rho(t)]. \tag{3.13}$$

Equation (3.6) thus holds for a mixed state as well. The distinction between pure and mixed states is fundamental and is not simply a matter of representation. A mixed state cannot be assigned a single wavefunction. This will be shown below.

Using an arbitrary basis set [Eq. (3.2)] we have the matrix elements

$$\rho_{nm} = \sum_k P_k \langle n|\psi_k\rangle\langle\psi_k|m\rangle.$$

We shall now list a few important properties of the density operator.

1. $\rho(t)$ is Hermitian, i.e.,

$$\rho^*_{nm}(t) = \rho_{mn}(t). \tag{3.14}$$

2. The diagonal elements of $\rho(t)$ (in any representation) are nonnegative

$$\rho_{nn} = \sum_k P_k \langle n|\psi_k\rangle\langle\psi_k|n\rangle = \sum_k P_k |\langle n|\psi_k\rangle|^2 \geq 0. \tag{3.15}$$

ρ_{nn} can thus be viewed as the probability of the system to be found in the state $|n\rangle$ by detection through an analyzer that rejects states orthogonal to $|n\rangle$. The off diagonal elements $\rho_{nm}(t)$ with $n \neq m$ are generally complex numbers and may contain a phase. Their modulus is $|\rho_{nm}|$ and phase $(1/2i)\ \ln(\rho_{nm}/\rho^*_{nm})$. We denote these two types of elements as *populations* and *coherences*, respectively.

3. $\mathrm{Tr}\,\rho(t) = 1.$ (3.16)

This is the normalization condition.

4. $\mathrm{Tr}\,\rho^2(t) \leq 1.$ (3.17)

For a pure state $\mathrm{Tr}\,\rho^2 = 1$ whereas for a mixed state $\mathrm{Tr}\,\rho^2 < 1$. This can be shown as follows: Since ρ is Hermitian, it can always be diagonalized:

$$S\rho S^\dagger = \rho' = \begin{pmatrix} P_1 & & 0 \\ & P_2 & \\ 0 & & \ddots \end{pmatrix}.$$

The invariance of the trace to a cyclic permutation yields

$$\mathrm{Tr}\,\rho = \mathrm{Tr}\,\rho' = \sum_k P_k = 1$$

$$\mathrm{Tr}\,\rho^2 = \mathrm{Tr}\,\rho'^2 = \sum_k P_k^2 \leq \left(\sum_k P_k\right)^2 = 1.$$

This implies the inequality (3.17), where the equality holds only for a pure state. Consequently, no wavefunction exists such that $\rho(t) = |\psi(t)\rangle\langle\psi(t)|$ when $\mathrm{Tr}\,\rho^2(t) < 1$.

5. Another general property of the density operator is that the magnitude of each off diagonal element is smaller than or equal to the geometric mean of the corresponding diagonal elements, i.e.,

$$\rho_{nn}\rho_{mm} \geq |\rho_{nm}|^2. \tag{3.18}$$

This can be shown as follows: Using Eq. (3.11) we have

$$\rho_{nn} = \sum_k P_k|\langle\psi_k|n\rangle|^2; \qquad \rho_{mm} = \sum_k P_k|\langle\psi_k|m\rangle|^2; \qquad \rho_{nm} = \sum_k P_k\langle n|\psi_k\rangle\langle\psi_k|m\rangle.$$

We now define the following vectors:

$$\alpha_n = \begin{pmatrix} \sqrt{P_1} & \langle\psi_1|n\rangle \\ \sqrt{P_2} & \langle\psi_2|n\rangle \\ \vdots & \vdots \end{pmatrix}, \qquad \alpha_m = \begin{pmatrix} \sqrt{P_1} & \langle\psi_1|m\rangle \\ \sqrt{P_2} & \langle\psi_2|m\rangle \\ \vdots & \vdots \end{pmatrix}.$$

Using this notation we can write the density operator elements as scalar products and we have $\rho_{nn} = \langle\alpha_n|\alpha_n\rangle$, $\rho_{mm} = \langle\alpha_m|\alpha_m\rangle$, and $\rho_{nm} = \langle\alpha_n|\alpha_m\rangle$. Equation (3.18) then directly follows from the Schwartz inequality: $\langle\alpha_n|\alpha_n\rangle\langle\alpha_m|\alpha_m\rangle \geq |\langle\alpha_n|\alpha_m\rangle|^2$, which holds for any vector space (see Appendix 2A).

The main properties of the density operator are summarized in Table 3.1.

TABLE 3.1 Properties of the Density Operator

Property	Pure state	Mixed state
Hermiticity	$\rho^*_{nm}(t) = \rho_{mn}(t)$	$\rho^*_{nm}(t) = \rho_{mn}(t)$
Nonnegative diagonal elements	$\rho_{nn} \geq 0$	$\rho_{nn} \geq 0$
Normalization	$\mathrm{Tr}\, \rho = 1$	$Tr\, \rho = 1$
Trace of ρ^2	$\mathrm{Tr}\, \rho^2 = 1$	$Tr\, \rho^2 < 1$
Schwartz inequality	$\rho_{nn}\rho_{mm} = \lvert\rho_{nm}\rvert^2$	$\rho_{nn}\rho_{mm} > \lvert\rho_{nm}\rvert^2$

THE REDUCED DENSITY OPERATOR: THE TRUTH BUT NOT THE WHOLE TRUTH

One important advantage of the density operator is its ability to represent an ensemble of systems in a compact way through a mixed state. Another useful property is that it offers the possibility to describe only a part of the system of interest, i.e., to provide a *reduced description.* Suppose the total Hamiltonian of a coupled many-body system is partitioned as follows:

$$H = H_S(\mathbf{q}_S) + H_B(\mathbf{q}_B) + H'(\mathbf{q}_S, \mathbf{q}_B). \tag{3.19}$$

H_S, H_B, and H' represent the system, a bath, and their interaction. Typically, the system contains a few degrees of freedom of interest $\mathbf{q}_S$, whereas the bath consists of many degrees of freedom $\mathbf{q}_B$. The total density operator in the joint system and bath phase space is too complex and cannot be calculated explicitly.

We shall be interested in following the dynamics of a *system operator*, which depends only on the system degrees of freedom, i.e.,

$$\langle A(\mathbf{q}_S)\rangle = \mathrm{Tr}[\rho(t)A(\mathbf{q}_S)]. \tag{3.20}$$

We shall now demonstrate that this does not require calculating the eigenstates of the entire system and bath. To that end we introduce the system and the bath eigenstates as

$$H_S|a\rangle = E_a|a\rangle$$

and

$$H_B|\alpha\rangle = E_\alpha|\alpha\rangle.$$

The direct product states $|a\alpha\rangle$ form a complete basis set in the joint system and bath space,

$$\sum_{a,\alpha} |a\alpha\rangle\langle a\alpha| = 1.$$

Note that due to H', these states are not the eigenstates of the total Hamiltonian; nevertheless, they form a complete basis set, which can be used to calculate the trace (we recall that the trace is invariant to the basis set). Using this basis set we have

$$\mathrm{Tr}\, \rho(t)A(\mathbf{q}_S) = \sum_{\substack{a,b \\ \alpha,\beta}} \langle a\alpha|\rho(t)|b\beta\rangle\langle b\beta|A(\mathbf{q}_S)|a\alpha\rangle.$$

Since the operator A does not depend on the bath degrees of freedom, its matrix elements can be factorized as follows:

$$\langle b\beta | A(q_S) | a\alpha \rangle = \langle b|A|a\rangle\langle \beta \mid \alpha\rangle,$$

where the first matrix element is calculated in the system subspace and the second in the bath subspace. The orthonormality of the bath eigenstates $\langle \beta \mid \alpha \rangle = \delta_{\alpha,\beta}$ then yields

$$\langle A \rangle = \sum_{\substack{a,b \\ \alpha}} \langle a\alpha | \rho(t) | b\alpha \rangle \langle b|A|a\rangle. \tag{3.21}$$

We can now define *the reduced system density operator* [3–5]:

$$\sigma_{ab}(t) \equiv \sum_{\alpha} \langle a\alpha | \rho(t) | b\alpha \rangle = \mathrm{Tr_B} \langle a | \rho(t) | b \rangle. \tag{3.22}$$

Equations (3.21) and (3.22) result in

$$\langle A(\mathbf{q}_S) \rangle = \sum_{a,b} \sigma_{ab}(t) A_{ba} \equiv \mathrm{Tr_S}[\sigma(t)A]. \tag{3.23}$$

Here $\mathrm{Tr_B}$ and $\mathrm{Tr_S}$ denote a partial trace over the bath or system degrees of freedom, respectively. Obviously $\mathrm{Tr} = \mathrm{Tr_B}\mathrm{Tr_S}$. All we need, therefore, is a reduced density operator defined in the subspace of the system's degrees of freedom! If we can find some way to calculate σ directly without worrying about the entire ρ, then we have simplified the problem considerably. This may be achieved using reduced equations of motion for σ [see Eq. (3.82)]. It should be noted that the present procedure can work the other way too. The expectation value of any operator that depends only on the bath degrees of freedom can be calculated using the reduced density operator of the bath calculated by tracing over the system degrees of freedom. This simplicity arises because we are interested in operators that depend only on either system or bath variables. Expectation values of operators that depend on both degrees of freedom require the entire density operator in the joint system and bath space. The important information, that is difficult to keep track of, and that is neglected in the reduced density operator is the *correlation* between the system and the bath, i.e., the fact that the total density operator is not a simple direct product of a system part and a bath part. This correlation is irrelevant for the expectation values of system operators. By calculating σ we therefore keep only the necessary information about the system.

TIME EVOLUTION OF THE DENSITY OPERATOR

We shall now consider the time evolution of the density operator and derive its equation of motion. Let us start with a pure state. The time derivative of the density operator consists of a sum of contributions from the bra and from the ket, i.e.,

$$\frac{\partial \rho}{\partial t} = \left(\frac{\partial}{\partial t}|\psi(t)\rangle\right)\langle\psi(t)| + |\psi(t)\rangle\left(\frac{\partial}{\partial t}\langle\psi(t)|\right).$$

Using the Schrödinger equation

$$\frac{\partial|\psi\rangle}{\partial t} = -\frac{i}{\hbar} H|\psi\rangle,$$

and its Hermitian conjugate

$$\frac{\partial\langle\psi|}{\partial t} = \frac{i}{\hbar}\langle\psi|H,$$

we get

$$\frac{\partial\rho}{\partial t} = -\frac{i}{\hbar} H|\psi\rangle\langle\psi| + \frac{i}{\hbar}|\psi\rangle\langle\psi|H$$

$$= -\frac{i}{\hbar}(H\rho - \rho H),$$

which finally yields *the quantum Liouville equation* (also known as the Liouville–Von Neumann equation)

$$\frac{\partial\rho}{\partial t} = -\frac{i}{\hbar}[H, \rho]. \tag{3.24}$$

Since the transformation from a pure to a mixed state in Liouville space, Eq. (3.11), is linear (i.e., the density operator for a mixed state is a superposition of pure state density operators), Eq. (3.24) holds for any density operator regardless on whether it represents a pure or a mixed state.

The formal solution of Eq. (3.24), as can be verified by substitution is

$$\rho(t) = U(t, t_0)\rho(t_0)U^\dagger(t, t_0). \tag{3.25}$$

where the U operator acting from the left propagates the ket [Eq. (2.42)] and $U^\dagger$ acting from the right propagates the bra.

An interesting general conclusion that follows directly from the Liouville equation is that if ρ is any function of the Hamiltonian H, $\rho = F(H)$, then $[H, \rho] = 0$ so that $\dot{\rho} = 0$. This is a trivial consequence of the fact that H commutes with any function of H. Such density operator is therefore stationary (time-independent). An important example is the equilibrium canonical density operator

$$\rho = Z^{-1}\exp(-\beta H), \tag{3.26a}$$

with the canonical partition function

$$Z \equiv \mathrm{Tr}\exp(-\beta H). \tag{3.26b}$$

Here $\beta \equiv (k_B T)^{-1}$, where k_B is the Boltzmann constant and T is the temperature. These expressions can be explicitly evaluated using the general definition of a function of an operator introduced in Chapter 2. Choosing the basis set of eigenfunctions of the Hamiltonian [Eq. (2.4)], we then have

$$Z = \sum_n \langle\varphi_n|\exp(-\beta H)|\varphi_n\rangle = \sum_n \exp(-\beta E_n)$$

and

$$\rho_{nm} = Z^{-1} \exp(-\beta E_n)\delta_{nm}.$$

The equilibrium canonical density operator is thus diagonal in this basis set

$$\rho = \frac{1}{Z}\begin{bmatrix} \exp(-\beta E_1) & & 0 \\ & \exp(-\beta E_2) & \\ 0 & & \ddots \end{bmatrix}. \tag{3.26c}$$

To demonstrate the difference between pure and mixed states, consider the two-level system $|\phi_a\rangle$ and $|\phi_b\rangle$ [Eq. (2.25)], and assume that its density operator is prepared initially in one of the following states:

(i) $\rho(t_0) = |\psi\rangle\langle\psi|$ with $|\psi\rangle = \dfrac{1}{\sqrt{2}}(|\phi_a\rangle + |\phi_b\rangle)$,

(ii) $\rho(t_0) = |\psi\rangle\langle\psi|$ with $|\psi\rangle = \dfrac{1}{\sqrt{2}}(|\phi_a\rangle - |\phi_b\rangle)$,

(iii) $\rho(t_0) = \frac{1}{2}|\phi_a\rangle\langle\phi_a| + \frac{1}{2}|\phi_b\rangle\langle\phi_b|$.

Using the $|\phi_a\rangle, |\phi_b\rangle$ basis set, these density operators can be written in matrix form

$$\text{(i)}\begin{pmatrix} \frac{1}{2} & \frac{1}{2} \\ \frac{1}{2} & \frac{1}{2} \end{pmatrix} \qquad \text{(ii)}\begin{pmatrix} \frac{1}{2} & -\frac{1}{2} \\ -\frac{1}{2} & \frac{1}{2} \end{pmatrix} \qquad \text{(iii)}\begin{pmatrix} \frac{1}{2} & 0 \\ 0 & \frac{1}{2} \end{pmatrix}$$

In all cases the diagonal elements are the same. (i) and (ii) are pure states whereas (iii) is a mixed state. This can be seen by, e.g., checking Tr ρ^2. Tr $\rho^2 = 1$, 1, 1/2 for (i), (ii), and (iii). These density matrices differ in the coherences and thus will have a different time evolution and may differ also in other properties such as the dipole moment. In Figure 3.1 we show the time–dependent

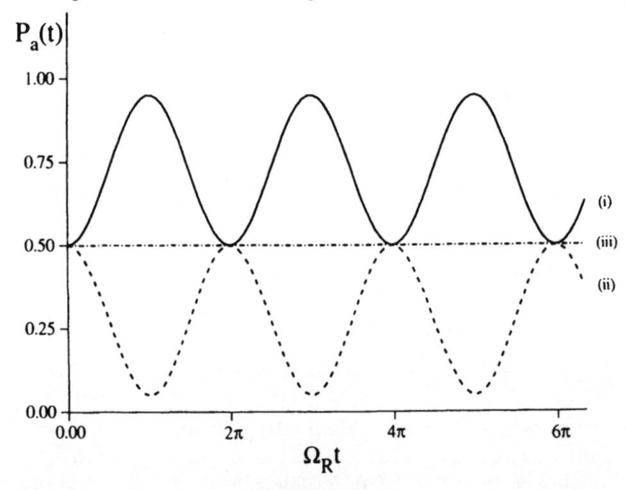

FIG. 3.1 The time-dependent probability of a two-level system to be found in state $|\phi_a\rangle$ at time t for three choices of the initial density matrix, as indicated. The Rabi frequency Ω_R was defined in Eq. (2.32) and $\theta = \pi/6$.

population of state $|\phi_a\rangle$ given that the initial density operator of the system $\rho(t_0)$ is either (i) or (ii) or (iii). The time evolution was calculated using Eq. (3.25) together with Eq. (2.29).

$$P_a(t) \equiv (1, 0)U(t, t_0)\rho(t_0)U^\dagger(t, t_0)\begin{pmatrix}1\\0\end{pmatrix}.$$

LIOUVILLE SPACE AND TETRADIC NOTATION

Having established the definition and the main properties of the density operator, we shall now introduce an alternative way of thinking about it, which makes its evaluation formally identical to the calculation of a wavefunction. This is the *Liouville space* dynamics that forms the backbone for the theoretical approach presented in this book. The term "Liouville space" was first coined by Fano [6]. Important contributions, particularly to the quantum theory of relaxation were made by Zwanzig and Redfield [3–10].

To introduce Liouville space let us start with an example: Consider the two-level system with the Hamiltonian [Eq. (2.25)]. Its density operator has four elements

$$\rho(t) = \begin{pmatrix}\rho_{aa}(t) & \rho_{ab}(t)\\ \rho_{ba}(t) & \rho_{bb}(t)\end{pmatrix}.$$

Let us take the *aa* element of both sides of the Liouville equation

$$\dot{\rho} = -\frac{i}{\hbar}[H, \rho]. \tag{3.27}$$

We then get

$$\dot{\rho}_{aa} = -\frac{i}{\hbar}(H_{aa}\rho_{aa} + H_{ab}\rho_{ba}) + \frac{i}{\hbar}(\rho_{aa}H_{aa} + \rho_{ab}H_{ba}).$$

We can proceed further and take all other matrix elements of the Liouville equation. This will result in the four coupled equations:

$$\dot{\rho}_{aa} = -\frac{i}{\hbar}(V_{ab}\rho_{ba} - V_{ba}\rho_{ab}),$$

$$\dot{\rho}_{bb} = -\frac{i}{\hbar}(V_{ba}\rho_{ab} - V_{ab}\rho_{ba}),$$

$$\dot{\rho}_{ab} = -\frac{i}{\hbar}(\varepsilon_a - \varepsilon_b)\rho_{ab} - \frac{i}{\hbar}V_{ab}(\rho_{bb} - \rho_{aa}),$$

$$\dot{\rho}_{ba} = -\frac{i}{\hbar}(\varepsilon_b - \varepsilon_a)\rho_{ba} - \frac{i}{\hbar}V_{ba}(\rho_{aa} - \rho_{bb}).$$

We can rewrite these equations in a matrix form as

$$\frac{d}{dt}\begin{pmatrix}\rho_{aa}\\ \rho_{bb}\\ \rho_{ab}\\ \rho_{ba}\end{pmatrix} = \frac{-i}{\hbar}\begin{bmatrix} 0 & 0 & -V_{ba} & V_{ab}\\ 0 & 0 & V_{ba} & -V_{ab}\\ -V_{ab} & V_{ab} & (\varepsilon_a-\varepsilon_b) & 0\\ V_{ba} & -V_{ba} & 0 & (\varepsilon_b-\varepsilon_a)\end{bmatrix}\begin{pmatrix}\rho_{aa}\\ \rho_{bb}\\ \rho_{ab}\\ \rho_{ba}\end{pmatrix}. \quad (3.28)$$

The important point about Eq. (3.28) is that here we consider the density operator to be a vector with four components. It satisfies an equation of motion which contains a 4×4 matrix. The celebrated Bloch equations are based on this type of representation. This form is not limited to the present two-level model but is very general. For an arbitrary system characterized by a Hamiltonian H, we introduce a complete basis set of functions in our Hilbert space $|j\rangle$, $|k\rangle$, $|m\rangle$, $|n\rangle\cdots$ and take the jk matrix element of both sides of Eq. (3.27). We then get

$$\dot{\rho}_{jk} = -\frac{i}{\hbar}[(H\rho)_{jk} - (\rho H)_{jk}], \qquad j, k = 1, 2, \ldots, N.$$

For an N level system, the density operator has N^2 matrix elements ρ_{jk}. The Liouville equation is a matrix equation which provides N^2 equations for our N^2 unknowns. (Each matrix element of the right-hand side is equal to its counterpart in the left-hand side.) When these matrix elements are evaluated we get

$$\frac{d\rho_{jk}}{dt} = -\frac{i}{\hbar}\sum_m [H_{jm}\rho_{mk} - \rho_{jm}H_{mk}]. \quad (3.29)$$

This can be rearranged in the following form:

$$\frac{d\rho_{jk}}{dt} = -\frac{i}{\hbar}\sum_{m,n} \mathscr{L}_{jk,mn}\rho_{mn}, \quad (3.30a)$$

which is completely analogous to the Schrödinger equation with the expansion (3.2a)

$$\frac{dc_j}{dt} = -\frac{i}{\hbar}\sum_m H_{jm}c_m. \quad (3.30b)$$

Upon the comparison of Eqs. (3.29) and (3.30a) we get

$$\mathscr{L}_{jk,mn} = H_{jm}\delta_{kn} - H^*_{kn}\delta_{jm} \quad (3.31)$$

Equation (3.30a) may be understood as follows: given a basis set with N basis functions we may think of ρ_{mn} as a column with N^2 elements $n, m = 1, \ldots, N$. (The precise order of elements in the column is arbitrary, but once chosen it needs to be kept.) The Liouville operator $\mathscr{L}$ is an $N^2 \times N^2$ matrix, and since each element ρ_{mn} is labeled by two indices, the matrix elements of $\mathscr{L}$ are labeled by four indices, $\mathscr{L}$ is therefore a *tetradic matrix or a superoperator*. In Eq. (3.29), H operates from the right *and* from the left. However, the Liouville operator acts on ρ only from the left, and Eqs. (3.30a) and (3.30b) are isomorphic. This space where the density operator is a vector is called the *Liouville space*. The dynamics of the density operator are more conveniently described in this space.

We shall now introduce a few additional definitions related to the tetradic notation, which will allow us to establish a complete isomorphism between the time evolution of the wavefunction in ordinary Hilbert space and the evolution of the density operator. It then becomes possible to use all the powerful methods developed for the wavefunction in Chapter 2, and apply them directly toward calculating the density operator.

In Hilbert space, we expand the wavefunction in a complete basis set of vectors $\{|j\rangle\}$ [Eq. (3.2a)]. In this basis $|j\rangle$ is represented by a unit vector whose jth element is 1, and all other elements vanish

$$|j\rangle \leftrightarrow \begin{pmatrix} 0 \\ \vdots \\ 1 \\ \vdots \\ 0 \end{pmatrix} j\text{th component.} \tag{3.32a}$$

Similarly, let us consider the expression

$$\rho = \sum_{j,k} \rho_{jk} |j\rangle\langle k|.$$

$|j\rangle\langle k|$ is the matrix whose element in the jth row and kth column is 1 and all other elements vanish

$$|j\rangle\langle k| \leftrightarrow j \begin{pmatrix} 0 & \overset{k}{\vdots} & 0 \\ \dots & 1 & \dots \\ 0 & \vdots & 0 \end{pmatrix}. \tag{3.32b}$$

We may think of the family of N^2 operators $|j\rangle\langle k|$, $j, k = 1, 2, \dots, N$ as a complete set of matrices and recast the density operator in the form

$$|\rho\rangle\rangle = \sum_{j,k} \rho_{jk} |jk\rangle\rangle, \tag{3.33a}$$

where we adopt a double bracket notation analogous to the brackets used in Hilbert space, and the ket $|jk\rangle\rangle$ denotes the Liouville space vector representing the Hilbert space operator $|j\rangle\langle k|$. Equation (3.33a) is of course analogous to

$$|\psi\rangle = \sum_{j} c_j |j\rangle. \tag{3.33b}$$

We next introduce a bra $\langle\langle jk|$ as the Hermitian conjugate to $|jk\rangle\rangle$,

$$\langle\langle jk| \equiv (|jk\rangle\rangle)^\dagger.$$

In Liouville space, any ordinary operator A such as the Hamiltonian, the momentum etc. can be thought of as a vector and denoted $|A\rangle\rangle$. By that we mean that we can expand it in our "basis set"

$$A = \sum_{j,k} A_{jk} |j\rangle\langle k| \tag{3.34a}$$

i.e.,

$$|A\rangle\rangle = \sum_{j,k} |jk\rangle\rangle A_{jk}. \tag{3.34b}$$

We further define a "bra" vector

$$\langle\langle B| \leftrightarrow B^{\dagger}, \tag{3.35}$$

and a scalar product of two operators

$$\langle\langle B | A\rangle\rangle \equiv \mathrm{Tr}(B^{\dagger}A). \tag{3.36}$$

In particular, we have an orthonormality

$$\langle\langle jk | mn\rangle\rangle \equiv \mathrm{Tr}[|k\rangle\langle j | m\rangle\langle n|] = \delta_{kn}\delta_{jm}, \tag{3.37a}$$

which is analogous to

$$\langle j | m\rangle = \delta_{jm}. \tag{3.37b}$$

Adopting the terminology of Appendix 2A, the outer products of vectors in Hilbert space constitute the vectors in the higher, Louville space. Once we accept that, then Eq. (3.36) simply follows naturally: it implies that the scalar product is the trace of the corresponding outer product.

Consider the following scalar product:

$$\langle\langle jk | A\rangle\rangle \equiv \mathrm{Tr}[|k\rangle\langle j|A] = \sum_{\alpha} \langle\alpha|k\rangle\langle j|A|\alpha\rangle = \langle j|A|k\rangle \equiv A_{jk}.$$

Upon the substitution in (3.34b) we have

$$|A\rangle\rangle = \sum_{jk} |jk\rangle\rangle\langle\langle jk | A\rangle\rangle, \tag{3.38}$$

which implies that the completeness condition (resolution of the unit operator) in Liouville space is

$$\sum_{j,k} |jk\rangle\rangle\langle\langle jk| = 1. \tag{3.39a}$$

This is the direct analog of the completeness condition in Hilbert space,

$$\sum_{j} |j\rangle\langle j| = 1. \tag{3.39b}$$

A collection of objects "vectors" $|A\rangle\rangle$ together with an addition operation $|\lambda_1 A + \lambda_2 B\rangle\rangle \equiv \lambda_1|A\rangle\rangle + \lambda_2|B\rangle\rangle$ where λ_1 and λ_2 are complex numbers, and the scalar product [Eq. (3.36)] form a linear vector space, provided they satisfy the 14 conditions listed in Appendix A of Chapter 2. It can be easily verified from the present definitions that all of these conditions are indeed satisfied. The *Liouville space* is therefore a linear vector space in which ρ (and any other operator such as the Hamiltonian etc.) is a vector.

We can now proceed one step further and introduce an outer product of the Liouville space vectors, which will allow us to define a linear operator in Liouville space. An operator in Liouville space is defined by

$$\mathscr{F} = \sum_{\substack{j,k \\ m,n}} |jk\rangle\rangle\langle\langle jk|\mathscr{F}|mn\rangle\rangle\langle\langle mn|, \tag{3.40a}$$

with Liouville space "matrix elements"

$$\mathscr{F}_{jk,mn} \equiv \langle\langle jk|\mathscr{F}|mn\rangle\rangle. \tag{3.40b}$$

To distinguish them from ordinary (Hilbert space) operators, operators in Liouville space are also called *superoperators* or *tetradic operators* (which signifies that their matrix elements have four indices). As an illustration let us calculate the matrix elements of the Liouville operator using the double bracket notation

$$\langle\langle B|\mathscr{L}|A\rangle\rangle \equiv \mathrm{Tr}\, B^\dagger \mathscr{L} A \equiv \mathrm{Tr}\, B^\dagger [H, A].$$

Choosing $|A\rangle\rangle = |mn\rangle\rangle$ and $|B\rangle\rangle = |jk\rangle\rangle$ we have

$$\begin{aligned}
\mathscr{L}_{jk,mn} &\equiv \langle\langle jk|\mathscr{L}|mn\rangle\rangle = \mathrm{Tr}\{|k\rangle\langle j|\mathscr{L}|m\rangle\langle n|\} \\
&= \mathrm{Tr}\{|k\rangle\langle j|H|m\rangle\langle n| - |k\rangle\langle j|m\rangle\langle n|H\} \\
&= \sum_i [\langle i|k\rangle\langle j|H|m\rangle\langle n|i\rangle - \langle i|k\rangle\langle j|m\rangle\langle n|H|i\rangle] \\
&= H_{jm}\delta_{kn} - H^*_{kn}\delta_{jm}.
\end{aligned}$$

This agrees with Eq. (3.31).

The importance of the Liouville space notation for quantum dynamics is that the Liouville equation,

$$\frac{\partial \rho}{\partial t} = -\frac{i}{\hbar}\mathscr{L}\rho, \tag{3.41a}$$

is now formally isomorphous to the Schrödinger equation,

$$\frac{\partial \psi}{\partial t} = -\frac{i}{\hbar}H\psi. \tag{3.41b}$$

TABLE 3.2 Comparison of the Liouville and the Hilbert spaces

Object	Hilbert space	Liouville space
State of system (ket)	$\|\psi\rangle$	$\|\rho\rangle\rangle$
Hermitian conjugate state (bra)	$\langle\psi\| \equiv (\|\psi\rangle)^\dagger$	$\langle\langle\rho\| \equiv (\|\rho\rangle\rangle)^\dagger$
Basis set	$\|n\rangle$	$\|nm\rangle\rangle (\leftrightarrow \|n\rangle\langle m\|)$
Orthogonality	$\langle n \| m\rangle = \delta_{nm}$	$\langle\langle jk \| mn\rangle\rangle = \delta_{kn}\delta_{jm}$
Unit operator	$1 = \sum_n \|n\rangle\langle n\|$	$I = \sum_{n,m} \|nm\rangle\rangle\langle\langle nm\|$
Scalar product	$\langle\psi \| \phi\rangle$	$\langle\langle A \| B\rangle\rangle$
Expansion of a vector	$\|\psi\rangle = \sum_n \|n\rangle\langle n \| \psi\rangle$	$\|A\rangle\rangle = \sum_{n,m} \|nm\rangle\rangle\langle\langle nm \| A\rangle\rangle$
Linear equation	$\sum_m A_{nm}\psi_m = \phi_n$	$\sum_{jk} \mathscr{L}_{nm,jk} A_{jk} = B_{nm}$
Equation of motion	$\|\dot{\psi}\rangle = -\frac{i}{\hbar} H\|\psi\rangle$	$\|\dot{\rho}\rangle\rangle = -\frac{i}{\hbar}\mathscr{L}\|\rho\rangle\rangle$
Matrix element	H_{nm}	$\mathscr{L}_{nm,n'm'} = H_{nn'}\delta_{mm'} - H^*_{mm'}\delta_{nn'}$
The time evolution operator	$\|\psi(t)\rangle\rangle = U(t, t_0)\|\psi(t_0)\rangle\rangle$ $\dot{U}(t, t_0) = -\frac{i}{\hbar} HU(t, t_0)$	$\|\rho(t)\rangle\rangle = \mathscr{U}(t, t_0)\|\rho(t_0)\rangle\rangle$ $\dot{\mathscr{U}}(t, t_0) = -\frac{i}{\hbar}\mathscr{L}\mathscr{U}(t, t_0)$
Green functions (time domain)	$G(t - t_0) = \theta(t - t_0)\exp\left[-\frac{i}{\hbar}H(t - t_0)\right]$	$\mathscr{G}(t - t_0) = \theta(t - t_0)\exp\left[-\frac{i}{\hbar}\mathscr{L}(t - t_0)\right]$
Green functions (frequency domain)	$G(E) \equiv (E - H + i\varepsilon)^{-1}$	$\mathscr{G}(\omega) \equiv (\omega - \mathbf{L} + i\varepsilon)^{-1}$; $\mathbf{L} \equiv \mathscr{L}/\hbar$

The time derivative of a vector (be it ψ or ρ) is given by a matrix (H or $\mathscr{L}$) times that vector. The only difference is the vector size (N vs. N^2). The Dirac-type double-bracket notation highlights this analogy and implies that the density operator is a vector $|\rho\rangle\rangle$ analogous to $|\psi\rangle$. In Table 3.2 we compare the Liouville space and the Hilbert space notation.*

This connection will allow us to apply all the results and techniques developed in Chapter 2 to Liouville space by replacing ψ with ρ and H with $\mathscr{L}$. We can define Liouville space Green functions, an interaction picture, a Magnus expansion etc. These relations, which need no further proof, since they result directly from the complete formal analogy between the two spaces will now be specified.

* Formally there is a homomorphism between Hilbert and Liouville space, related to a mathematical construct denoted Fiber bundles [N. E. Steenrod, *The Topology of Fiber Bundles*, Princeton University Press, Princeton, N.J. (1965)].

TIME EVOLUTION OPERATOR IN LIOUVILLE SPACE

Equation (3.25) is the general formal solution of the quantum Liouville equation. This can be verified by noting that

$$|\psi_k(t)\rangle = U(t, t_0)|\psi_k(t_0)\rangle \tag{3.42a}$$

and its Hermitian conjugate

$$\langle\psi_k(t)| = \langle\psi_k(t_0)|U^\dagger(t, t_0). \tag{3.42b}$$

When these expressions are substituted in Eq. (3.11) we recover Eq. (3.25), which is not written in Liouville space, since it treats the density operator as a matrix and the right-hand side is simply a product of three matrices. Alternatively, we can use the Liouville space form Eq. (3.41a) and proceed in a completely analogous manner to Chapter 2, replacing everywhere ψ by ρ and H by $\mathscr{L}$. We can then introduce the *Liouville space propagator.*

$$\rho(t) \equiv \mathscr{U}(t, t_0)\rho(t_0). \tag{3.43}$$

Upon the substitution of Eq. (3.43) in Eq. (3.41a) we can show that $\mathscr{U}$ satisfies the quantum Liouville equation

$$\frac{\partial \mathscr{U}(t, t_0)}{\partial t} = -\frac{i}{\hbar}\mathscr{L}(t)\mathscr{U}(t, t_0), \tag{3.44a}$$

with the initial condition

$$\mathscr{U}(t_0, t_0) = 1. \tag{3.44b}$$

For a time-independent Hamiltonian (and hence also time-independent Liouville Operator) we simply have

$$\mathscr{U}(t, t_0) \equiv \exp\left[-\frac{i}{\hbar}\mathscr{L}(t - t_0)\right]. \tag{3.45}$$

Combining Eqs. (3.25) and (3.43) we have

$$\mathscr{U}(t, t_0)\rho(t_0) \Leftrightarrow U(t, t_0)\rho(t_0)U^\dagger(t, t_0), \tag{3.46a}$$

and in particular for a time independent Hamiltonian (setting $t_0 = 0$)

$$\exp\left(-\frac{i}{\hbar}\mathscr{L}t\right)\rho(0) \Leftrightarrow \exp\left(-\frac{i}{\hbar}Ht\right)\rho(0)\exp\left(\frac{i}{\hbar}Ht\right). \tag{3.46b}$$

We cannot write an equality here because in the left-hand side ρ is considered a vector whereas in the right-hand side ρ is a matrix. However, the important point is that the two entities in the right-hand side and left-hand side represent the same object (although, in a different space).

The tetradic time evolution operator is analogous to its Hilbert space counterpart [Eq. (2.24)]. In Chapter 2 we expanded U in the eigenstates of the Hamiltonian [Eq. (2.4)]. Let us consider the action of $\mathscr{L}$ on $|\varphi_n\rangle\langle\varphi_m|$

$$\mathscr{L}|\varphi_n\rangle\langle\varphi_m| = H|\varphi_n\rangle\langle\varphi_m| - |\varphi_n\rangle\langle\varphi_m|H = (E_n - E_m)|\varphi_n\rangle\langle\varphi_m| = \hbar\omega_{nm}|\varphi_n\rangle\langle\varphi_m|, \tag{3.46c}$$

where we have introduced the transition frequency between two states $\hbar\omega_{nm} \equiv E_n - E_m$. In Liouville space notation we then have

$$\mathscr{L}|\varphi_n\varphi_m\rangle\rangle = \hbar\omega_{nm}|\varphi_n\varphi_m\rangle\rangle.$$

We thus see that if $|\varphi_n\rangle$ and $|\varphi_m\rangle$ are two eigenstates of H with eigenvalues E_n and E_m, then $|\varphi_n\varphi_m\rangle\rangle$ is an eigenvector of $\mathscr{L}$ with the eigenvalue $\hbar\omega_{nm}$.

Multiplying $\mathscr{U}(t, t_0)$ by the unit operator $\sum_{nm} |\varphi_n\varphi_m\rangle\rangle\langle\langle\varphi_n\varphi_m|$, and making use of the last equation, we obtain

$$\mathscr{U}(t, t_0) = \sum_{n,m} |\varphi_n\varphi_m\rangle\rangle \exp[-i\omega_{nm}(t - t_0)]\langle\langle\varphi_n\varphi_m|, \tag{3.47a}$$

which is analogous to

$$U(t, t_0) = \sum_n |\varphi_n\rangle \exp\left[-\frac{i}{\hbar} E_n(t - t_0)\right]\langle\varphi_n|. \tag{3.47b}$$

Once we have these operators expressed in a specific basis set (the eigenstates of H), we can immediately transform to an arbitrary basis $\{|j\rangle\}$ by multiplying from the left and from the right by the unit operator $\sum_j |j\rangle\langle j|$ or $\sum_{jk} |jk\rangle\rangle\langle\langle jk|$. We then have

$$\mathscr{U}(t, t_0) = \sum_{\substack{j,k \\ j',k'}} \sum_{n,m} |jk\rangle\rangle\langle\langle jk|\varphi_n\varphi_m\rangle\rangle\langle\langle\varphi_n\varphi_m|j'k'\rangle\rangle\langle\langle j'k'| \exp[-i\omega_{nm}(t - t_0)],$$

which yields

$$\mathscr{U}(t, t_0) = \sum_{\substack{j,k \\ j',k'}} \mathscr{U}_{jk,j'k'}(t, t_0)|jk\rangle\rangle\langle\langle j'k'|, \tag{3.48a}$$

with

$$\mathscr{U}_{jk,j'k'}(t, t_0) = \sum_{n,m} \langle j|\varphi_n\rangle\langle\varphi_m|k\rangle\langle\varphi_n|j'\rangle\langle k'|\varphi_m\rangle \exp[-i\omega_{nm}(t - t_0)]. \tag{3.48b}$$

This is analogous to

$$U(t, t_0) = \sum_{j,j'} U_{jj'}(t, t_0)|j\rangle\langle j'|, \tag{3.48c}$$

with

$$U_{jj'}(t, t_0) = \sum_n \langle j|\varphi_n\rangle\langle\varphi_n|j'\rangle \exp\left[-\frac{i}{\hbar} E_n(t - t_0)\right]. \tag{3.48d}$$

When the Hamiltonian is time dependent, we have the formal expression for the time evolution operator

$$\mathscr{U}(t, t_0) = \exp_+\left[-\frac{i}{\hbar}\int_{t_0}^{t} d\tau\, \mathscr{L}(\tau)\right], \tag{3.49}$$

where the time-ordered exponential was defined in Eqs. (2.40) and (2.42). Equation (3.49) is an abbreviated notation for

$$\mathscr{U}(t, t_0) = 1 + \sum_{n=1}^{\infty} \left(\frac{-i}{\hbar}\right)^n \int_{t_0}^{t} d\tau_n \int_{t_0}^{\tau_n} d\tau_{n-1} \cdots \int_{t_0}^{\tau_2} d\tau_1\, \mathscr{L}(\tau_n)\mathscr{L}(\tau_{n-1}) \cdots \mathscr{L}(\tau_1). \tag{3.50}$$

When acting on $\rho(t_0)$ we have

$$\mathscr{U}(t, t_0)\rho(t_0) = \rho(t_0) + \sum_{n=1}^{\infty} \left(\frac{-i}{\hbar}\right)^n \int_{t_0}^{t} d\tau_n \int_{t_0}^{\tau_n} d\tau_{n-1} \cdots \int_{t_0}^{\tau_2} d\tau_1$$
$$[H(\tau_n), \ldots [H(\tau_2), [H(\tau_1), \rho(t_0)]] \cdots], \tag{3.51a}$$

where we have used the relation

$$\mathscr{L}(\tau)A = [H(\tau), A]. \tag{3.51b}$$

By using the identity (Eq. (3.46a)), and taking matrix elements of both sides, we can show that

$$\mathscr{U}_{jk,j'k'}(t, t_0) = U_{jj'}(t, t_0)U^{\dagger}_{k'k}(t, t_0). \tag{3.52}$$

The Hermitian conjugate of Eq. (3.43) reads

$$\rho(t) = \rho(t_0)\mathscr{U}^{\dagger}(t, t_0), \tag{3.53a}$$

where $\mathscr{U}^{\dagger}$ is given by the negative time ordered exponential

$$\mathscr{U}^{\dagger}(t, t_0) = 1 + \sum_{n=1}^{\infty} \left(\frac{i}{\hbar}\right)^n \int_{t_0}^{t} d\tau_n \int_{t_0}^{\tau_n} d\tau_{n-1} \cdots \int_{t_0}^{\tau_2} d\tau_1\, \mathscr{L}(\tau_1)\mathscr{L}(\tau_2) \cdots \mathscr{L}(\tau_n). \tag{3.53b}$$

We thus have

$$\rho(t_0)\mathscr{U}^{\dagger}(t, t_0) = \rho(t_0) + \sum_{n=1}^{\infty} \left(\frac{i}{\hbar}\right)^n \int_{t_0}^{t} d\tau_n \int_{t_0}^{\tau_n} d\tau_{n-1} \cdots \int_{t_0}^{\tau_2} d\tau_1$$
$$[\ldots [[\rho(t_o), H(\tau_1)], H(\tau_2)] \ldots, H(\tau_n)], \tag{3.54}$$

which coincides with Eq.(3.51a).

THE INTERACTION PICTURE

We can also introduce an interaction picture in Liouville space. The partitioning of the Hamiltonian

$$H = H_0(t) + H'(t), \tag{3.55a}$$

implies a similar partitioning of the Liouville operator

$$\mathscr{L} = \mathscr{L}_0(t) + \mathscr{L}'(t), \tag{3.55b}$$

where

$$\mathscr{L}_0(t)A \equiv [H_0(t), A],$$

$$\mathscr{L}'(t)A \equiv [H'(t), A].$$

We then get

$$\mathscr{U}(t, t_0) = \mathscr{U}_0(t, t_0)\mathscr{U}_1(t, t_0), \tag{3.56a}$$

with

$$\mathscr{U}_0(t, t_0) = \exp_+\left[-\frac{i}{\hbar}\int_{t_0}^{t} d\tau\, \mathscr{L}_0(\tau)\right], \tag{3.56b}$$

and

$$\mathscr{U}_1(t, t_0) = \exp_+\left[-\frac{i}{\hbar}\int_{t_0}^{t} d\tau\, \mathscr{L}_1'(\tau)\right]. \tag{3.56c}$$

Here

$$\mathscr{L}_1'(\tau) \equiv \mathscr{U}_0^\dagger(\tau, t_0)\mathscr{L}'(\tau)\mathscr{U}_0(\tau, t_0). \tag{3.56d}$$

The Liouville space analogue of the time evolution operator [Eq. (2.57)] thus reads

$$\mathscr{U}(t, t_0) = \mathscr{U}_0(t, t_0)\exp_+\left[-\frac{i}{\hbar}\int_{t_0}^{t} d\tau\, \mathscr{L}_1'(\tau)\right], \tag{3.57}$$

with the time-ordered expansion

$$\mathscr{U}(t, t_0) = \mathscr{U}_0(t, t_0) + \sum_{n=1}^{\infty}\left(-\frac{i}{\hbar}\right)^n \int_{t_0}^{t} d\tau_n \int_{t_0}^{\tau_n} d\tau_{n-1}\cdots\int_{t_0}^{\tau_2} d\tau_1$$
$$\mathscr{U}_0(t, \tau_n)\mathscr{L}'(\tau_n)\mathscr{U}_0(\tau_n, \tau_{n-1})\mathscr{L}'(\tau_{n-1})\cdots\mathscr{U}_0(\tau_2, \tau_1)\mathscr{L}'(\tau_1)\mathscr{U}_0(\tau_1, t_0). \tag{3.58}$$

This is analogous to Eq. (2.58). Similar to Eq. (2.59) we can recast this in an integral form

$$\mathscr{U}(t, t_0) = \mathscr{U}_0(t, t_0) - \frac{i}{\hbar}\int_{t_0}^{t} d\tau\, \mathscr{U}_0(t, \tau)\mathscr{L}'(\tau)\mathscr{U}(\tau, t_0), \tag{3.59a}$$

which when acting on the equilibrium density operator yields

$$|\rho(t)\rangle\rangle = |\rho(-\infty)\rangle\rangle - \frac{i}{\hbar}\int_{t_0}^{t} d\tau\, \mathscr{U}_0(t, \tau)\mathscr{L}'(\tau)|\rho(\tau)\rangle\rangle. \tag{3.59b}$$

We can further take the Hermitian conjugate of Eq. (3.57) and calculate the time evolution by action from the right [Eq. (3.53a)]

$$\mathscr{U}^\dagger(t, t_0) = \exp_-\left[\frac{i}{\hbar}\int_{t_0}^{t} d\tau\, \mathscr{L}_1'(\tau)\right]\exp_-\left[\frac{i}{\hbar}\int_{t_0}^{t} d\tau\, \mathscr{L}_0(\tau)\right].$$

When the zero-order Hamiltonian H_0 is time independent, we simply have

$$\mathscr{U}_0(t, t_0) = \exp\left[-\frac{i}{\hbar}\mathscr{L}_0(t - t_0)\right].$$

To summarize, the interaction picture (Eqs. (2.62)), in Liouville space notation, reads

$$\langle A(t)\rangle = \mathrm{Tr}[A_{\mathrm{I}}(t)\rho_{\mathrm{I}}(t)], \tag{3.60a}$$

with

$$\dot{\rho}_{\mathrm{I}} = -\frac{i}{\hbar}\mathscr{L}'_{\mathrm{I}}(t)\rho_{\mathrm{I}}, \tag{3.60b}$$

$$\dot{A}_{\mathrm{I}} = \frac{i}{\hbar}\mathscr{L}_0 A_{\mathrm{I}}, \tag{3.60c}$$

$$\mathscr{L}'_{\mathrm{I}}(t) \equiv \mathscr{U}_0^{\dagger}(t, t_0)\mathscr{L}'(t)\mathscr{U}_0(t, t_0). \tag{3.60d}$$

Note the change in sign in Eqs. (3.60b) and (3.60c).

Returning to Hilbert space notation we have

$$\mathscr{L}'_{\mathrm{I}}\rho_{\mathrm{I}} = [H'_{\mathrm{I}}(t), \rho_{\mathrm{I}}]$$

and

$$\mathscr{L}_0 A_{\mathrm{I}} = [H_0, A_{\mathrm{I}}].$$

As included in our Hilbert space discussion, the interaction picture reduces to the Schrödinger and the Heisenberg pictures by a proper choice of $H_0(\mathscr{L}_0)$. In the Schrödinger picture we have

$$\dot{\rho}_{\mathrm{S}} = -\frac{i}{\hbar}\mathscr{L}\rho_{\mathrm{S}} \equiv -\frac{i}{\hbar}[H, \rho_{\mathrm{S}}], \tag{3.61a}$$

$$\dot{A}_{\mathrm{S}} = 0, \tag{3.61b}$$

whereas in the Heisenberg picture we get

$$\dot{A}_{\mathrm{H}} = \frac{i}{\hbar}\mathscr{L}A_{\mathrm{H}} \equiv \frac{i}{\hbar}[H, A_{\mathrm{H}}], \tag{3.61c}$$

$$\dot{\rho}_{\mathrm{H}} = 0. \tag{3.61d}$$

THE MAGNUS EXPANSION

In complete analogy to Eqs. (2.66) and (2.67), we can introduce the Magnus expansion in Liouville space and replace the time ordered exponential by an ordinary exponential

$$\mathscr{U}_1(t, t_0) = \exp\left[\sum_{n=1}^{\infty} \frac{1}{n!}\left(-\frac{i}{\hbar}\right)^n \mathscr{L}_n(t, t_0)\right], \tag{3.62}$$

with

$$\mathscr{L}_1 = \int_{t_0}^{t} \mathscr{L}_I'(\tau_1)\, d\tau_1, \tag{3.63a}$$

$$\mathscr{L}_2 = \int_{t_0}^{t} d\tau_2 \int_{t_0}^{\tau_2} d\tau_1 [\mathscr{L}_I'(\tau_2), \mathscr{L}_I'(\tau_1)], \tag{3.63b}$$

$$\mathscr{L}_3 = \int_{t_0}^{t} d\tau_3 \int_{t_0}^{\tau_3} d\tau_2 \int_{t_0}^{\tau_2} d\tau_1$$

$$\{[\mathscr{L}_I'(\tau_3), [\mathscr{L}_I'(\tau_2), \mathscr{L}_I'(\tau_1)]] + [[\mathscr{L}_I'(\tau_3), \mathscr{L}_I'(\tau_2)], \mathscr{L}_I'(\tau_1)]\}. \tag{3.63c}$$

LIOUVILLE SPACE GREEN FUNCTIONS

In analogy to Eq. (2.68) we introduce the Liouville space *retarded Green function* for the case of time-independent Hamiltonian

$$\mathscr{G}(t - t_0) \equiv \theta(t - t_0)\mathscr{U}(t, t_0)$$

$$= \theta(t - t_0) \exp\left[-\frac{i}{\hbar}\mathscr{L}(t - t_0)\right], \tag{3.64}$$

and in the frequency domain

$$\mathscr{G}(\omega) = \frac{1}{\omega - \mathbf{L} + i\varepsilon}, \tag{3.65}$$

where $\mathbf{L} \equiv \mathscr{L}/\hbar$. The two are related as follows [see Eqs. (2.71)]:

$$\mathscr{G}(t - t_0) = -\frac{1}{2\pi i}\int_{-\infty}^{\infty} d\omega \exp[-i\omega(t - t_0)]\mathscr{G}(\omega), \tag{3.66a}$$

$$\mathscr{G}(\omega) = -i\int_{-\infty}^{\infty} dt\, \mathscr{G}(t - t_0) \exp[i\omega(t - t_0)]. \tag{3.66b}$$

Similarly, we can introduce the *advanced Green functions*, which are responsible for backward propagation

$$\mathscr{G}^-(t - t_0) \equiv \theta(t_0 - t)\mathscr{U}(t, t_0) \tag{3.67}$$

and

$$\mathscr{G}^-(\omega) \equiv \frac{1}{\omega - \mathbf{L} - i\varepsilon}. \tag{3.68}$$

We then have [see Eqs. (2.74)]

$$\mathscr{G}^{-}(t-t_0)=\frac{1}{2\pi i}\int_{-\infty}^{\infty}d\omega\exp[-i\omega(t-t_0)]\mathscr{G}^{-}(\omega), \qquad (3.69a)$$

$$\mathscr{G}^{-}(\omega)=i\int_{-\infty}^{\infty}dt\exp[i\omega(t-t_0)]\mathscr{G}^{-}(t-t_0). \qquad (3.69b)$$

The Green function can be easily expanded in the eigenstates of the Hamiltonian. To that end we multiply it by the unit operator:

$$\mathscr{G}(\omega)=\frac{1}{\omega-\mathbf{L}+i\varepsilon}\sum_{nm}|\varphi_n\varphi_m\rangle\rangle\langle\langle\varphi_n\varphi_m|.$$

Making use of Eq. (3.46c), we then get

$$\mathscr{G}(\omega)=\sum_{nm}\frac{|\varphi_n\varphi_m\rangle\rangle\langle\langle\varphi_n\varphi_m|}{\omega-\omega_{nm}+i\varepsilon} \qquad (3.70a)$$

and

$$\mathscr{G}(t-t_0)=\theta(t-t_0)\sum_{nm}|\varphi_n\varphi_m\rangle\rangle\exp[-i\omega_{nm}(t-t_0)]\langle\langle\varphi_n\varphi_m|. \qquad (3.70b)$$

By performing a Fourier transform of Eq. (3.52) we can establish a connection between the Liouville space and the Hilbert space Green functions [9]

$$\mathscr{G}_{jk,j'k'}(\omega)=\frac{\hbar}{2\pi i}\int dE\,G_{jj'}(\hbar\omega+E)G^{*}_{kk'}(E). \qquad (3.71)$$

It is interesting to note that the poles of the ordinary Green function lie near an energy equal to an eigenvalue $E=E_n$ whereas the poles of the tetradic Green function are when ω is equal to a transition frequency ω_{nm}. The tetradic Green function is thus more physical since the absolute energy depends on a choice of an arbitrary zero reference, whereas the transition frequencies are physical observables.

We can further define a Liouville space Dyson equation, in complete analogy with Eq. (2.78). Consider a time-independent Liouvillian that is partitioned as follows

$$\mathscr{L}=\mathscr{L}_0+\mathscr{V}.$$

We now define the zero order Green function

$$\mathscr{G}_0(\omega)\equiv\frac{1}{\omega-\mathbf{L}_0+i\varepsilon},$$

with $\mathbf{L}_0\equiv\mathscr{L}_0/\hbar$. The Liouville space Dyson equation then reads

$$\mathscr{G}(\omega)=\mathscr{G}_0(\omega)+\mathscr{G}_0(\omega)\mathscr{V}\mathscr{G}(\omega), \qquad (3.72a)$$

or

$$\mathscr{G}(\omega)=\mathscr{G}_0(\omega)+\mathscr{G}(\omega)\mathscr{V}\mathscr{G}_0(\omega). \qquad (3.62b)$$

We can also define hermiticity in Liouville space. The Hermitian conjugate of an arbitrary Liouville space operator $\mathscr{F}$ is defined by its action to the left

$$\langle\langle B|\mathscr{F}|A\rangle\rangle = \langle\langle \mathscr{F}^\dagger B|A\rangle\rangle = \langle\langle A|\mathscr{F}^\dagger B\rangle\rangle^*, \tag{3.73a}$$

which implies [see Eq. (2.33)]

$$\mathscr{F}^\dagger_{ij,kl} = \mathscr{F}^*_{kl,ij}. \tag{3.73b}$$

The Hermitian conjugate of a product of Liouville space operators $\mathscr{F}$ and $\mathscr{S}$ is equal to the product of the conjugates in reverse order [see Eq. (2.36)],

$$(\mathscr{F}\mathscr{S})^\dagger_{ij,kl} = (\mathscr{F}\mathscr{S})^*_{kl,ij} = \sum_{m,n} \mathscr{F}^*_{kl,mn}\mathscr{S}^*_{mn,ij},$$

$$(\mathscr{S}^\dagger\mathscr{F}^\dagger)_{ij,kl} = \sum_{m,n} \mathscr{S}^\dagger_{ij,mn}\mathscr{F}^\dagger_{mn,kl} \equiv \mathscr{F}^*_{kl,mn}\mathscr{S}^*_{mn,ij},$$

i.e.,

$$(\mathscr{F}\mathscr{S})^\dagger = \mathscr{S}^\dagger\mathscr{F}^\dagger. \tag{3.74}$$

In Chapter 2 we considered a few equivalent possibilities of action of an operator from the left and the right [Eq. (2.37)]. This flexibility is very useful. We can immediately write the Liouville space analogues. Consider a Liouville space matrix element

$$S = \langle\langle A|\mathscr{F}_1\mathscr{F}_2|B\rangle\rangle, \tag{3.75a}$$

where $\mathscr{F}_1$ and $\mathscr{F}_2$ are any Liouville space operators. We then define

$$|B'\rangle\rangle \equiv \mathscr{F}_1\mathscr{F}_2|B\rangle\rangle,$$

so that

$$S = \langle\langle A|B'\rangle\rangle. \tag{3.75b}$$

Alternatively, we can define

$$|A'\rangle\rangle \equiv \mathscr{F}_2^\dagger\mathscr{F}_1^\dagger|A\rangle\rangle,$$

and write

$$S = \langle\langle A'|B\rangle\rangle. \tag{3.75c}$$

Finally we can have a "mixed" action

$$|A''\rangle\rangle \equiv \mathscr{F}_1^\dagger|A\rangle\rangle,$$

$$|B''\rangle\rangle \equiv \mathscr{F}_2|B\rangle\rangle,$$

$$S \equiv \langle\langle A''|B''\rangle\rangle. \tag{3.75d}$$

All three expressions are identical. In a particular application, one of these forms may be preferable. We can now extend Eqs. (3.75) for a product of an arbitrary number of operators,

$$\langle\langle A|\mathscr{F}_1\mathscr{F}_2\cdots\mathscr{F}_n|B\rangle\rangle = \langle\langle B|\mathscr{F}_n^\dagger\cdots\mathscr{F}_2^\dagger\mathscr{F}_1^\dagger|A\rangle\rangle^*. \tag{3.76}$$

The Liouville operator is Hermitian (i.e., equal to its Hermitian conjugate). This can be easily verified by taking the matrix elements:

$$\mathscr{L}_{ij,kl} = H_{ik}\delta_{jl} - H^*_{jl}\delta_{ik},$$

$$\mathscr{L}_{kl,ij} = H_{ki}\delta_{lj} - H^*_{lj}\delta_{ki},$$

which imply that

$$\mathscr{L}_{kl,ij} = \mathscr{L}^*_{ij,kl}. \tag{3.77}$$

It immediately follows that the Liouville space time evolution operator is unitary

$$\left[\exp\left(-\frac{i}{\hbar}\mathscr{L}t\right)\right]^\dagger = \exp\left(\frac{i}{\hbar}\mathscr{L}t\right). \tag{3.78a}$$

Similarly, we have for the Green function

$$[\mathscr{G}(\omega)]^\dagger = \mathscr{G}^-(\omega). \tag{3.78b}$$

PROJECTION OPERATORS AND THE REDUCED DENSITY OPERATOR: EFFECTIVE LIOUVILLE OPERATORS

In complete analogy with Hilbert space, we can partition Liouville space using projection operators P and Q

$$P = \sum_{n,m}{}' |nm\rangle\rangle\langle\langle nm|, \tag{3.79a}$$

$$Q = 1 - P, \tag{3.79b}$$

where the prime denotes a partial sum over a selected set of nm vectors. Q contains the remaining vectors. Liouville space is richer than Hilbert space and it is possible to define other types of projections. An important example involves creating the reduced density operator σ by tracing over a bath

$$\rho = P\rho + Q\rho, \tag{3.80}$$

$$P\rho = \rho_B^0 \operatorname{Tr}_B \rho \equiv \rho_B^0 \sigma, \tag{3.81}$$

where ρ_B^0 is some reference density operator of the bath, which is usually taken to be the equilibrium bath density operator. It can be easily shown that P is indeed a projection operator since it satisfies Eqs. (2.81).

We can further write a reduced equation of motion for the projected density operator $P\rho$ [no derivation is necessary, since we use the analogy with Eq. (2.83)], [3, 11]

$$P\dot{\rho} = -\frac{i}{\hbar}P\mathscr{L}P(P\rho) + \left(\frac{i}{\hbar}\right)^2 \int_0^t d\tau\, P\mathscr{L}Q \exp\left[-\frac{i}{\hbar}Q\mathscr{L}Q(t-\tau)\right]Q\mathscr{L}P[P\rho(\tau)]. \tag{3.82}$$

We can also write for the projected Green function [see Eq. (2.87)]

$$P\mathscr{G}(\omega)P = \frac{1}{\omega - P\mathbf{L}_{\text{eff}}(\omega)P}, \tag{3.83a}$$

with $\mathbf{L}_{\text{eff}} \equiv (1/\hbar)\mathscr{L}_{\text{eff}}$. Partitioning the Liouville operator as $\mathscr{L} = \mathscr{L}_0 + \mathscr{V}$, and assuming that $\mathscr{L}_0$ commutes with P so that $P\mathscr{L}_0 Q = Q\mathscr{L}_0 P = 0$, the effective Liouville operator is given by

$$P\mathscr{L}_{\text{eff}}(\omega)P = P\mathscr{L}_0 P + P\mathscr{R}(\omega)P, \tag{3.83b}$$

with the self-energy, also known as the relaxation superoperator

$$\mathscr{R}(\omega) = \mathscr{V} + \mathscr{V}\frac{1}{\omega - Q\mathscr{L}Q}Q\mathscr{V}. \tag{3.83c}$$

Other projections of the Green function, analogous to Eqs. (2.87b)–(2.87d), can be written as well.

When the Q space represents a smooth continuum, we can neglect the ω dependence of the effective Liouville operator and simply set $\mathscr{L}_{\text{eff}}(\omega) = \mathscr{L}_{\text{eff}}$. In analogy with Eq. (2.96), Eq. (3.82) then becomes

$$P\dot{\rho} = -\frac{i}{\hbar}P\mathscr{L}_{\text{eff}}P(P\rho). \tag{3.84}$$

If we adopt the effective Hamiltonian, we can write Eq. (2.96) as

$$P\dot{\rho} = -\frac{i}{\hbar}[PH_{\text{eff}}P(P\rho) - (P\rho)PH_{\text{eff}}^{\dagger}P]. \tag{3.85}$$

(Since H_{eff} is non-Hermitian, we have to act with $H_{\text{eff}}^{\dagger}$ from the right.) It is then tempting to assume by comparing these two equations that the effective Liouville operator can be represented by a commutator with an effective Hamiltonian. This is however not generally the case, and Eq. (3.84) is much more general than Eq. (3.85). In Eq. (3.52) we have expressed the Liouville space propagator matrix element as a product of two elements of the Hilbert space propagators. We can write this only when we work in the complete space with all degrees of freedom explicitly included. When working with the reduced density operator using an effective Liouville operator, we cannot generally separate the ket and the bra evolutions in this manner.

An effective Liouville operator can thus represent a much richer dynamics compared with an effective Hamiltonian, since it can describe a correlated dynamics of the bra and the ket.

THE CLASSICAL LIOUVILLE EQUATION AND THE WIGNER REPRESENTATION

The density operator and the Liouville equation are well rooted in *classical* statistical mechanics. One of the major advantages of working in Liouville space is the existence of a well-defined classical limit for all quantities. This is to be contrasted with the wavefunction, which does not have a classical counterpart. In this section we first review briefly the classical Liouville equation and then consider the *Wigner representation*, which provides an elegant semiclassical picture that interpolates between classical and quantum mechanics [12–14].

In classical statistical mechanics we describe the evolution of a system in *phase space* [15] whose coordinates are the positions $\mathbf{q}$ and momenta $\mathbf{p}$ of all particles. The state of the system with complete information may be represented by a point in phase space. For N particles, the phase space is $6N$ dimensional. More generally, for systems with incomplete information, we can define a classical distribution function $\rho_c(\mathbf{q}, \mathbf{p})$, which represents the probability distribution in phase space, i.e., $\rho_c(\mathbf{q}, \mathbf{p})\, d\mathbf{p}\, d\mathbf{q}$ is the probability of the system to be found between $\mathbf{p}$ and $\mathbf{p} + d\mathbf{p}$ and $\mathbf{q}$ and $\mathbf{q} + d\mathbf{q}$.

A classical dynamic variable is most generally a function in phase space $A_c(\mathbf{p}, \mathbf{q})$. Its expectation value is

$$\langle A(t)\rangle_c = \iint d\mathbf{p}\, d\mathbf{q}\, A_c(\mathbf{p}, \mathbf{q})\rho_c(\mathbf{pq}; t). \tag{3.86}$$

Since ρ_c represents the distribution of a conserved quantity, it must satisfy the continuity equation*

$$\frac{\partial \rho_c}{\partial t} = -\nabla\cdot(\rho_c \mathbf{v}), \tag{3.87}$$

where $\mathbf{v}$ is the velocity field in phase space, which is a vector field with components $(\dot{p}_j, \dot{q}_j)$. The ∇ operator represents the divergence in phase space with components $(\partial/\partial q_j, \partial/\partial p_j)$. Using an elementary relation of vector calculus [Eq. (4.A.5)] we have

$$\frac{\partial \rho_c}{\partial t} = -\mathbf{v}\cdot\nabla\rho_c - \rho_c\nabla\cdot\mathbf{v}.$$

Using Hamilton's equations, with H_c being the classical Hamiltonian

$$\left.\begin{aligned} \dot{p}_j &= -\frac{\partial H_c}{\partial q_j}, \\ \dot{q}_j &= \frac{\partial H_c}{\partial p_j}, \end{aligned}\right\} \tag{3.88}$$

we get

$$\nabla\cdot\mathbf{v} = \frac{\partial \dot{q}}{\partial q} + \frac{\partial \dot{p}}{\partial p} = \frac{\partial^2 H_c}{\partial p\, \partial q} - \frac{\partial^2 H_c}{\partial p\, \partial q} = 0.$$

We thus have

$$\frac{\partial \rho_c}{\partial t} = -\sum_j\left[\frac{\partial \rho_c}{\partial q_j}\dot{q}_j + \frac{\partial \rho_c}{\partial p_j}\dot{p}_j\right]. \tag{3.89a}$$

This equation implies that the total time derivative of ρ_c vanishes,

$$\frac{d\rho_c}{dt} = 0, \tag{3.89b}$$

* A hydrodynamic formulation of quantum mechanics in terms of probability densities and currents provides an additional physical perspective on semiclassical approximations [S. K. Ghosh and B. M. Deb, *Phys. Rep.* **92**, 1 (1982).]

which is known as *Liouville's theorem.* Substituting Hamilton's equations in the right-hand side of Eq. (3.89a) we finally get

$$\frac{\partial \rho_c}{\partial t} = \sum_j \left[\frac{\partial H_c}{\partial q_j} \frac{\partial \rho_c}{\partial p_j} - \frac{\partial H_c}{\partial p_j} \frac{\partial \rho_c}{\partial q_j} \right]. \tag{3.90}$$

Equation (3.90) is the *classical Liouville equation.* It can be written in the form

$$\frac{\partial \rho_c}{\partial t} = \mathscr{L}_c \rho_c, \tag{3.91}$$

where $\mathscr{L}_c$ is the classical Liouville operator which is given by the *Poisson bracket* $\{\ ,\ \}$

$$\mathscr{L}_c \rho_c = \sum_j \left[\frac{\partial H_c}{\partial q_j} \frac{\partial \rho_c}{\partial p_j} - \frac{\partial H_c}{\partial p_j} \frac{\partial \rho_c}{\partial q_j} \right] \equiv \{\mathscr{H}_c, \rho_c\}. \tag{3.92}$$

One important feature of Liouville space is that it can naturally recover the classical limit when $\hbar \to 0$. This feature can easily be seen if we properly introduce a phase space (**pq**) representation in Liouville space. There is a variety of phase space representations, which all yield the same classical limit when $\hbar \to 0$ [16–19]. In this book we adopt the *Wigner* phase space representation. Let us consider a quantum system characterized by N coordinates x_j, $j = 1, \ldots, N$, their conjugate momenta $\hat{p}_j = -i\hbar\, \partial/\partial x_j$, and masses m_j. Its Hamiltonian is

$$H = -\sum_j \frac{\hbar^2}{2m_j} \frac{\partial^2}{\partial x_j^2} + W(x_1 \cdots x_N),$$

where W is the potential energy. Hereafter we introduce a vector notation and define the N-dimensional vector $\mathbf{x}$ with components x_j. We next express the density operator in the *coordinate representation,* which is a continuous representation

$$\rho(\mathbf{x}, \mathbf{x}', t) \equiv \langle \mathbf{x} | \rho(t) | \mathbf{x}' \rangle \equiv \langle\langle \mathbf{x}\mathbf{x}' \mid \rho(t) \rangle\rangle. \tag{3.93}$$

The second identity follows from the tetradic notation.

Equation (3.93) can be evaluated by taking an arbitrary basis set $\{|m\rangle\}$ and inserting the completeness condition twice, i.e.,

$$\langle \mathbf{x} | \rho | \mathbf{x}' \rangle \equiv \sum_{nm} \langle \mathbf{x} \mid n \rangle \langle n | \rho | m \rangle \langle m \mid \mathbf{x}' \rangle,$$

or equivalently by inserting once the resolution of the unit operator in Liouville space:

$$\langle\langle \mathbf{x}\mathbf{x}' \mid \rho \rangle\rangle = \sum_{nm} \langle\langle \mathbf{x}\mathbf{x}' \mid nm \rangle\rangle \langle\langle nm \mid \rho \rangle\rangle, \tag{3.94a}$$

which gives

$$\langle \mathbf{x} | \rho | \mathbf{x}' \rangle \equiv \sum_{nm} \rho_{nm} \psi_n(\mathbf{x}) \psi_m^*(\mathbf{x}'). \tag{3.94b}$$

To discuss the classical limit and develop semiclassical approximations for the quantum mechanical density operator, we shall introduce the following definitions of phase space bra and ket

$$|\mathbf{pq}\rangle\rangle \equiv \int_{-\infty}^{\infty} d\mathbf{s} \exp(i\mathbf{p}\cdot\mathbf{s}/\hbar)|\mathbf{q} + \mathbf{s}/2, \mathbf{q} - \mathbf{s}/2\rangle\rangle, \tag{3.95a}$$

$$\langle\langle\mathbf{pq}| \equiv \frac{1}{(2\pi\hbar)^N} \int_{-\infty}^{\infty} d\mathbf{s} \exp(-i\mathbf{p}\cdot\mathbf{s}/\hbar)\langle\langle\mathbf{q} + \mathbf{s}/2, \mathbf{q} - \mathbf{s}/2|. \tag{3.95b}$$

Note that $|\mathbf{pq}\rangle\rangle$ is not a primary (or ordinary) Liouville space vector; it cannot be expressed as an outer product $|\mathbf{p}\rangle\langle\mathbf{q}|$. It is instead, defined by the above transformation, from the coordinate representation $|\mathbf{x}, \mathbf{x}'\rangle\rangle \equiv |\mathbf{q} + \mathbf{s}/2, \mathbf{q} - \mathbf{s}/2\rangle\rangle$, with $\mathbf{q} = (\mathbf{x} + \mathbf{x}')/2$ being the mean coordinate, whereas $\mathbf{p}$ is the momentum conjugate to $\mathbf{s} \equiv \mathbf{x} - \mathbf{x}'$. Using these definitions we immediately have

$$\iint d\mathbf{p}\, d\mathbf{q}|\mathbf{pq}\rangle\rangle\langle\langle\mathbf{pq}| = I, \tag{3.96a}$$

$$\langle\langle\mathbf{pq}\,|\,\mathbf{p}'\mathbf{q}'\rangle\rangle = \delta(\mathbf{p} - \mathbf{p}')\delta(\mathbf{q} - \mathbf{q}'). \tag{3.96b}$$

The expectation value of an arbitrary dynamic variable A is given by

$$\langle A(t)\rangle \equiv \mathrm{Tr}[A\rho(t)] = \langle\langle A^{\dagger}|\rho(t)\rangle\rangle. \tag{3.97}$$

This can be written in the coordinate representation

$$\begin{aligned}\langle A(t)\rangle &= \int_{-\infty}^{\infty}\int_{-\infty}^{\infty} d\mathbf{x}\, d\mathbf{x}'\,\langle\langle A^{\dagger}|\mathbf{x}, \mathbf{x}'\rangle\rangle\langle\langle\mathbf{x}, \mathbf{x}'\,|\,\rho(t)\rangle\rangle \\ &= \int_{-\infty}^{\infty}\int_{-\infty}^{\infty} d\mathbf{x}\, d\mathbf{x}'\; A(\mathbf{x}', \mathbf{x})\rho(\mathbf{x}, \mathbf{x}'; t),\end{aligned} \tag{3.98a}$$

or using the phase space variables Eq. (3.96a)

$$\begin{aligned}\langle A(t)\rangle &= \int_{-\infty}^{\infty}\int_{-\infty}^{\infty} d\mathbf{p}\, d\mathbf{q}\langle\langle A^{\dagger}|\mathbf{pq}\rangle\rangle\langle\langle\mathbf{pq}\,|\,\rho(t)\rangle\rangle \\ &= \int_{-\infty}^{\infty}\int_{-\infty}^{\infty} d\mathbf{p}\, d\mathbf{q}\; A_{\mathrm{w}}(\mathbf{p}, \mathbf{q})\rho_{\mathrm{w}}(\mathbf{pq}; t).\end{aligned} \tag{3.98b}$$

The analogy with Eq. (3.86) is evident. We shall view this as the expectation value in the *Wigner representation.* Since Eqs. (3.98a) and (3.98b) simply correspond to different representations, we shall introduce hereafter a unified notation and recast them in the form

$$\langle A(t)\rangle = \int_{-\infty}^{\infty} d\Gamma A(\Gamma)\rho(\Gamma; t). \tag{3.99}$$

Equation (3.99) should be understood as follows: We either express A and ρ in the coordinate representations, setting $d\Gamma = d\mathbf{x}\, d\mathbf{x}'$, or use the Wigner representation, and then $d\Gamma = d\mathbf{p}\, d\mathbf{q}$. The evaluation of $\langle A(t) \rangle$ then reduces to solving the Liouville equation for $\rho(\Gamma; t)$ and then performing the integration over Γ. This shows that the Wigner transform indeed provides a new representation that can be used to calculate expectation values.

In Eqs. (3.95) we have chosen a different normalization for the bra and the ket and took $\langle\langle \mathbf{pq}|$ to differ from the Hermitian conjugate of $|\mathbf{pq}\rangle\rangle$ by a factor of $(2\pi\hbar)^N$. We could have used a common normalization factor of $(2\pi\hbar)^{-N/2}$ in both definitions. With this normalization they would be Hermitian conjugates, and Eqs. (3.96) will still hold. The present definitions have, however, some important advantages, which will be outlined below, and which justify the little inconvenience of giving up the Hermitian conjugation property of the bra and the ket. To see this let us first consider the density operator in the *Wigner representation* ρ_w

$$\rho_w(\mathbf{q}, \mathbf{p}; t) \equiv \langle\langle \mathbf{pq} \mid \rho(t) \rangle\rangle = \frac{1}{(2\pi\hbar)^N} \int_{-\infty}^{\infty} \langle \mathbf{q} + \mathbf{s}/2 | \rho(t) | \mathbf{q} - \mathbf{s}/2 \rangle \exp\left(-\frac{i}{\hbar} \mathbf{p} \cdot \mathbf{s} \right) d\mathbf{s}, \tag{3.100a}$$

and the inverse transform is

$$\rho(\mathbf{q} + \mathbf{s}/2, \mathbf{q} - \mathbf{s}/2; t) = \int_{-\infty}^{\infty} \rho_w(\mathbf{q}, \mathbf{p}; t) \exp\left(\frac{i}{\hbar} \mathbf{p} \cdot \mathbf{s} \right) d\mathbf{p}. \tag{3.100b}$$

The density operator is normalized as

$$\int_{-\infty}^{\infty} \int_{-\infty}^{\infty} \delta(\mathbf{x} - \mathbf{x}')\, d\mathbf{x}\, d\mathbf{x}'\, \rho(\mathbf{x}, \mathbf{x}'; t) = \int_{-\infty}^{\infty} \int_{-\infty}^{\infty} d\mathbf{q}\, d\mathbf{p}\, \rho_w(\mathbf{q}, \mathbf{p}; t) = 1. \tag{3.101}$$

We first note that ρ_w now has the same units as the classical distribution function ρ_c and is normalized in the same way. One appealing consequence of this normalization is that as $\hbar \to 0$, $\rho_w(\mathbf{q}, \mathbf{p}; t)$ becomes equal to the classical distribution function. Let us turn now to the dynamic variable A. The Wigner representation $A_w(\mathbf{q}, \mathbf{p}; t)$ of an arbitrary operator A is defined by simply replacing ρ with A in Eq. (3.100a), and deleting the $(2\pi\hbar)^{-N}$ factor

$$A_w(\mathbf{q}, \mathbf{p}; t) \equiv \langle\langle A^\dagger(t) \mid \mathbf{pq} \rangle\rangle = \int_{-\infty}^{\infty} \langle \mathbf{q} + \mathbf{s}/2 | A(t) | \mathbf{q} - \mathbf{s}/2 \rangle \exp\left(\frac{i}{\hbar} \mathbf{p} \cdot \mathbf{s} \right) d\mathbf{s}. \tag{3.102}$$

With this normalization, the unit operator in Hilbert space is equal to the number 1 in the Wigner representation, and we have:

$$\begin{aligned} \langle\langle 1 \mid \mathbf{pq} \rangle\rangle &= \int_{-\infty}^{\infty} d\mathbf{s} \exp(i\mathbf{p} \cdot \mathbf{s}/\hbar) \langle\langle 1 \mid \mathbf{q} + \mathbf{s}/2, \mathbf{q} - \mathbf{s}/2 \rangle\rangle \\ &= \int_{-\infty}^{\infty} d\mathbf{s} \exp(i\mathbf{p} \cdot \mathbf{s}/\hbar) \delta(\mathbf{s}) = 1. \end{aligned} \tag{3.103}$$

Furthermore, adopting the present normalization, a quantum operator that depends on either the momenta $\hat{\mathbf{p}}$ or the coordinates $\hat{\mathbf{x}}$ alone will thus retain its form in the Wigner representation, i.e.,

$$[A(\hat{\mathbf{x}})]_w \equiv \langle\langle A^\dagger(\hat{\mathbf{x}}) | \mathbf{pq}\rangle\rangle = A(\mathbf{q}), \tag{3.104a}$$

$$[A(\hat{\mathbf{p}})]_w \equiv \langle\langle A^\dagger(\hat{\mathbf{p}}) | \mathbf{pq}\rangle\rangle = A(\mathbf{p}). \tag{3.104b}$$

As a special case, let us take ρ to be the unit operator. Equation (3.98b) then implies

$$\mathrm{Tr}\, A = \iint \langle \mathbf{x}'|A|\mathbf{x}\rangle \delta(\mathbf{x} - \mathbf{x}')\, d\mathbf{x}\, d\mathbf{x}' = \iint d\mathbf{p}\, d\mathbf{q}\, A(\mathbf{pq}). \tag{3.105}$$

One should be extremely careful when carrying out the Wigner transformation. The Wigner transform of a product of two operators is in general different from the product of their transforms. From Eq. (3.102) it follows that [12]

$$\begin{aligned}(AB)_w &= A_w \exp[(i\hbar/2)T]B_w \\ &= B_w \exp[-(i\hbar/2)T]A_w,\end{aligned} \tag{3.106}$$

where

$$T \equiv \frac{\overleftarrow{\partial}}{\partial \mathbf{q}}\frac{\overrightarrow{\partial}}{\partial \mathbf{p}} - \frac{\overleftarrow{\partial}}{\partial \mathbf{p}}\frac{\overrightarrow{\partial}}{\partial \mathbf{q}}.$$

The arrows indicate the direction of operation of the derivative. To zeroth-order in $\hbar$ we indeed have $(AB)_w = B_w A_w$, but Eq. (3.106) contains corrections to higher powers in $\hbar$.

As an example for the Wigner representation, consider a quantum harmonic oscillator whose Hamiltonian is

$$H = -\frac{\hbar^2}{2m}\frac{\partial^2}{2x^2} + \tfrac{1}{2}m\omega^2 x^2. \tag{3.107}$$

Its canonical quantum density operator [Eq. (3.26)] is

$$\rho_Q(x, x') = [1 - \exp(-\beta\hbar\omega)] \sum_{n=0}^{\infty} \exp(-n\beta\hbar\omega)\psi_n(x)\psi_n^*(x'), \tag{3.108}$$

$\psi_n(x)$ being its nth eigenfunction, with energy $n\hbar\omega$.

The Wigner transform Eq. (3.100a) can be calculated using standard properties of Hermite polynomials. The result is

$$\rho_w(p, q) = \frac{\omega}{2\pi\sigma} \exp\left[-\frac{p^2/m + m\omega^2 q^2}{4\sigma}\right], \tag{3.109a}$$

with

$$\sigma = \tfrac{1}{2}\hbar\omega \coth\left(\frac{\beta\hbar\omega}{2}\right). \tag{3.109b}$$

At low temperatures we have

$$\sigma = \tfrac{1}{2}\hbar\omega, \qquad \beta\hbar\omega \gg 1.$$

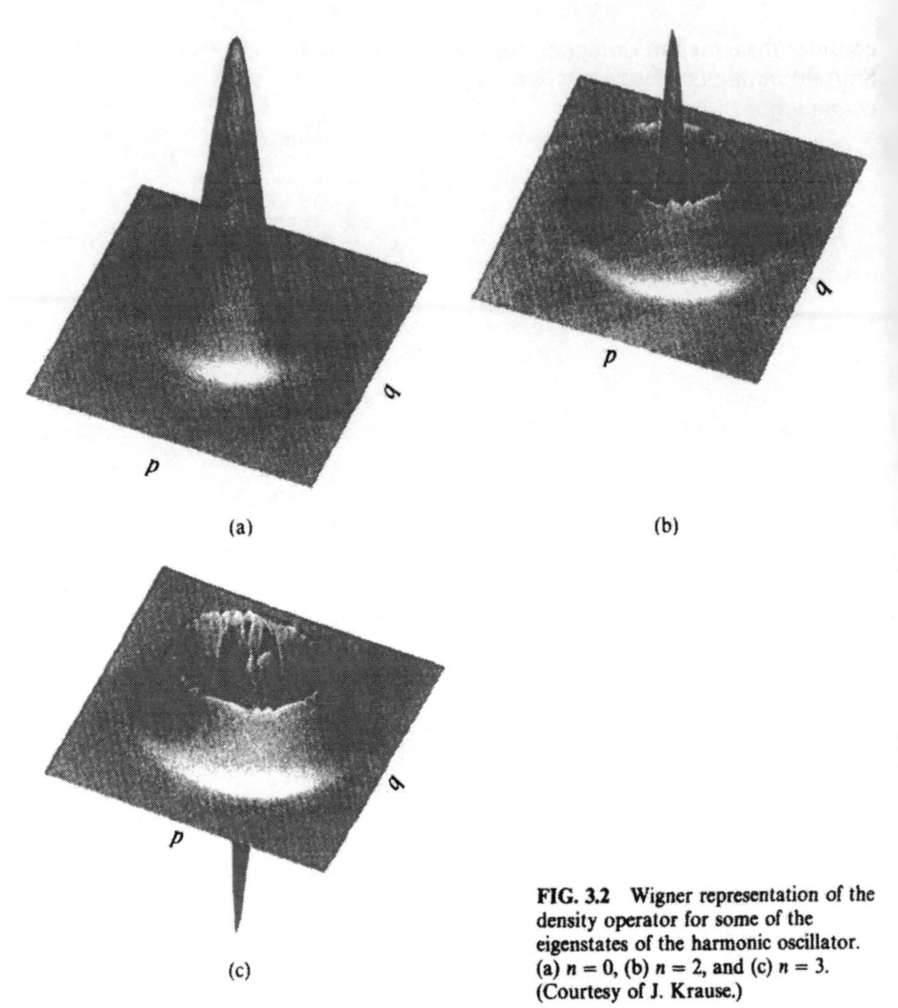

FIG. 3.2 Wigner representation of the density operator for some of the eigenstates of the harmonic oscillator. (a) $n = 0$, (b) $n = 2$, and (c) $n = 3$. (Courtesy of J. Krause.)

ρ_w is then the Wigner transform of the ground state of the harmonic oscillator. In the opposite, high temperature limit we have

$$\sigma = k_B T, \qquad \beta\hbar\omega \ll 1$$

ρ_w is then simply equal to the classical canonical distribution function of the harmonic oscillator, which does not depend on $\hbar$. The density operators in the Wigner representation corresponding to three eigenstates of the harmonic oscillator $|\psi_n\rangle\langle\psi_n|$ with $n = 0$, 2, and 3, are displayed in Figure 3.2.

It is a unique feature of the harmonic oscillator that its equilibrium canonical Wigner function retains its simple Gaussian form at all temperatures. At high temperatures ρ_w is classical, as required by the correspondence principle. At low temperatures, however, the classical distribution function has a zero width $\rho_c(p, q) = \delta(p)\,\delta(q)$ whereas the Wigner function has a width corresponding to the zero point motion.

The significance of the Wigner representation can be better appreciated if we consider the equation of motion for the density operator in this representation. Starting with Eq. (3.106) we immediately get for the Wigner transform of a commutator of two operators

$$[A, B]_w = 2iA(\mathbf{q}, \mathbf{p}) \sin\left(\frac{\hbar}{2} T\right) B(\mathbf{q}, \mathbf{p}). \tag{3.110}$$

Substituting this in the Liouville equation (3.27), we obtain

$$\dot{\rho}_w = \frac{2}{\hbar} H_c \sin\left[\frac{\hbar}{2}\left(\frac{\overleftarrow{\partial}}{\partial \mathbf{q}} \frac{\overrightarrow{\partial}}{\partial \mathbf{p}} - \frac{\overleftarrow{\partial}}{\partial \mathbf{p}} \frac{\overrightarrow{\partial}}{\partial \mathbf{q}}\right)\right] \rho_w. \tag{3.111}$$

This can be recast as

$$\dot{\rho}_w = \mathscr{L}_w \rho_w, \tag{3.112a}$$

with

$$\mathscr{L}_w = \frac{2}{\hbar} H_c \sin(\tfrac{1}{2}\hbar T), \tag{3.112b}$$

and the T operator was introduced in Eqs. (3.106). These equations can be easily expanded in powers of $\hbar$,

$$\sin(\tfrac{1}{2}\hbar T) = \tfrac{1}{2}\hbar T - \tfrac{1}{6}(\tfrac{1}{2}\hbar T)^3 \cdots.$$

Keeping only the leading term, the $\hbar$ factor cancels in Eq. (3.111) and we recover the classical Liouville equation (3.90). In the classical limit, the density operator ρ_w thus becomes equal to the classical phase space distribution function ρ_c. Note, however, that despite the formal analogy with the classical Liouville equation, the Wigner transform provides an exact quantum representation; the density operator as well as all other operators retain their full quantum character and are represented by complex functions in phase space.

Finally, we note that the transformation from the quantum density operator to a function in phase space is not unique, and other transforms that are different from (though related to) the Wigner representation have been suggested [16–20]. They all reduce to the classical density operator when appropriate, but introduce quantum corrections in a different manner.

CONCLUDING COMMENTS

The density operator is not particularly useful for simple quantum systems in a pure state. Using a basis set with N functions, the state vector is defined by N complex numbers c_j, which require $2N$ real parameters. Since the state is normalized $\sum_j |c_j|^2 = 1$ and the wavefunction has an arbitrary phase, we characterize the wavefunction by $2N - 2$ independent parameters. The density operator on the other hand has N real diagonal elements and $N(N-1)$ complex off diagonal elements. Since it is Hermitian $\rho_{jk} = \rho_{kj}^*$, and using the normalization, this amounts to $N^2 - 1$ independent parameters. This is the price we have to pay for working with the density operator. The Liouville space description seems more complicated for the obvious reason that its $\sim N^2$ dimensionality is larger than that of wavefunction space $\sim N$. This is the reason why simple quantum systems are usually described in Hilbert space using wavefunctions. In many applications, however, the rewards far exceed this price: we can describe directly ensemble averages through a mixed state, and construct a reduced description. Moreover, the density operator has a classical analogue, and the Wigner representation provides an elegant connection between classical and quantum mechanics, and lends itself naturally to semiclassical approximations. It is interesting to note that the wavefunction can have an arbitrary overall phase that cancels once physical observables are calculated. The density operator does not depend on that phase, which is another reflection of its more physical nature.

In Chapter 5 we shall show that the density operator provides the most natural and overall simplest description of linear and nonlinear optical spectroscopy since it properly maintains the *time orderings* of various interactions. We shall then introduce the nonlinear response functions and discuss the tremendous advantages of the Liouville space formulation of nonlinear spectroscopy.

It is important to keep in mind that by working with the density operator we are not necessarily in Liouville space. We have at our disposal three options:

1. Use wavefunctions throughout the calculation and never define the density operator.
2. Introduce the density operator but still remain in Hilbert space. In this case we think of the density operator as an operator (like any other operator H, p, etc.) represented by an $N \times N$ matrix. Its time evolution is given then by acting from right and left [Eq. (3.25)].
3. Introduce the density operator and work in Liouville space. It then becomes a vector in a space of higher dimensionality in which all ordinary Hilbert space operators are vectors. Only here we exploit its full power; we no longer think of bras and kets separately but in pairs. Each component of the vector represents the dynamics of a pair, and we maintain a simultaneous book-keeping of the bras and kets. The time evolution is then given by acting from the left [Eq. (3.43)].

Liouville space has different vectors, addition, and scalar product operations than Hilbert space. The Liouville space notation allows us to carry out expansions and formal manipulations very effectively. In some cases it is desirable to express the final physical observables such as the optical response functions in terms of ordinary Hilbert space operators and correlation functions [21]. In these cases the tetradic notation reduces the algebra involved in the intermediate steps considerably. However, in many cases we stay in Liouville space all the way to the end and express the physical observables in terms of tetradic operators. Such an example is the Bloch equations introduced in Appendix 6B or the phase space wavepackets discussed in Chapter 13.

NOTES

1. J. Von Neumann, *Mathematische Grundlagen der Quantenmechanik* (Springer-Verlag, Berlin, 1932); English trans. by R. T. Beyer, *Mathematical Foundations of Quantum Mechanics* (Princeton University Press, Princeton, New Jersey, 1955).

2. U. Fano, *Rev. Mod. Phys.* **29**, 74 (1957).
3. R. Zwanzig, *Lect. Theoret. Phys.* **3**, 106 (1961); *Physica* **30**, 1109 (1964); *J. Chem. Phys.* **33**, 1338 (1960).
4. A. G. Redfield, *Adv. Mag. Reson.* **1**, 1 (1965); J. Jeener, *Advances in Magnetic Resonance*, Vol. 10, J. S. Waugh, Ed. (Academic Press, New York, 1982), p. 1.
5. W. H. Louisell, *Quantum Statistical Properties of Radiation* (Wiley, New York, 1973).
6. U. Fano, *Phys. Rev.* **131**, 259 (1963); in *Lectures on Many-Body Problems*, E. R. Caianiello, Ed. (Academic Press, New York, 1964), p. 217.
7. A. Abragam, *The Principles of Nuclear Magnetism* (Oxford, London, 1961); M. Goldman, *Quantum Description of High-Resolution NMR in Liquids* (Clarendon Press, Oxford, London, 1988).
8. R. R. Ernst, G. Bodenhausen, and A. Wokaun, *Principles of Nuclear Magnetic Resonance in One and Two Dimensions* (Clarendon Press, Oxford, London, 1987).
9. A. Ben-Reuven, *Adv. Chem. Phys.* **33**, 235 (1975).
10. S. Mukamel, *Phys. Rep.* **93**, 1 (1982).
11. H. Mori, *Prog. Theoret. Phys.* **33**, 423; **34**, 399 (1965); M. Tokuyama and H. Mori, *Prog. Theoret. Phys.* **55**, 411 (1976).
12. E. Wigner, *Phys. Rev.* **40**, 749 (1932); M. Hillery, R. F. O'Connell, M. O. Scully, and E. P. Wigner, *Phys. Rep.* **106**, 121 (1984).
13. H. Mori, I. Oppenheim, and J. Ross, in *Studies in Statistical Mechanics*, Vol. 1, J. DeBoer and G. E. Unlenbeck, Eds. (North-Holland, Amsterdam, 1962); J. T. Hynes, J. M. Deutch, C. H. Wang, and I. Oppenheim, *J. Chem. Phys.* **48**, 3085 (1968).
14. K. Imre, E. Ozizmir, M. Rosenbaum, and P. E. Zweifel, *J. Math. Phys.* **8**, 1097 (1967).
15. H. Goldstein, *Classical Mechanics*, 2nd ed. (Addison-Wesley, London, 1981).
16. K. E. Cahill and R. J. Glauber, *Phys. Rev.* **177**, 1857 (1969), ibid 1882 (1969).
17. P. D. Drummond and C. W. Gardiner, *J. Phys. A* **13**, 2353 (1980).
18. W. H. Wells, *Ann. Phys.* **12**, 1 (1961).
19. S. S. Mizrahi, *Physica* **127A**, 241 (1984); **135A**, 237 (1986).
20. C. W. Gardiner, *Handbook of Stochastic Methods for Physics, Chemistry and the Natural Sciences*, Vol. 13, Springer Series in Synergetics (Springer-Verlag, Berlin, 1985).
21. D. Forster, *Hydrodynamic Fluctuations, Broken Symmetry, and Correlation Functions* (Benjamin, New York, 1975).

BIBLIOGRAPHY

R. P. Feynman, *Statistical Mechanics: A Set of Lectures* (Benjamin Cummings, Reading, Pennsylvania, 1982).

L. D. Landau and E. M. Lifshitz, *Statistical Physics* (Addison-Wesley, Reading, Massachusetts, 1969).

For a review of projection operator techniques:

B. J. Berne and G. D. Harp, *Adv. Chem. Phys.* **17**, 63 (1970).

For applications of the density operator in electronic structure:

E. R. Davidson, *Reduced Density Matrices in Quantum Chemistry* (Academic Press, New York, 1976).

R. McWeeney and B. T. Sutcliffe, *Methods of Molecular Quantum Mechanics* (Academic Press, New York, 1976).

R. G. Parr and W. Yang, *Density—Functional Theory of Atoms and Molecules* (Oxford University Press, New York, 1989).

CHAPTER 4

Quantum Electrodynamics, Optical Polarization, and Nonlinear Spectroscopy

This chapter presents a microscopic treatment of the interaction of nonrelativistic quantum systems with the electromagnetic field. We provide a rigorous definition of the polarization operator [Eq. (4.18c) together with (4.19c)] which applies to systems of arbitrary size, without invoking the dipole approximation. The polarization plays a key role in spectroscopy; it controls the radiation-matter interaction as well as the intermolecular forces [Eq. (4.36)]. Most of the derivations of this chapter are not essential for using the rest of the book. Readers can thus skip the first part of this chapter and proceed directly to the semiclassical Maxwell Liouville equations [Eqs. (4.47) and (4.48)] and the subsequent survey of optical signals.

THE MINIMAL COUPLING HAMILTONIAN AND THE RADIATION-MATTER INTERACTION

The electromagnetic field consists of an electric field $\hat{E}(\mathbf{r})$ and a magnetic field $\hat{B}(\mathbf{r})$. In the theory of electromagnetism it is convenient to represent them through the vector potential $\hat{A}(\mathbf{r})$ and the scalar potential $\hat{U}(\mathbf{r})$. The electric and magnetic fields can be obtained from these potentials using the relations:

$$\hat{B}(\mathbf{r}) = \nabla \times \hat{A}(\mathbf{r}) \tag{4.1a}$$

$$\hat{E}(\mathbf{r}) = -\frac{1}{c}\frac{\partial \hat{A}(\mathbf{r})}{\partial t} + \nabla \hat{U}(\mathbf{r}). \tag{4.1b}$$

For basic properties of vector fields see Appendix 4A. (In this chapter we shall label operators by a ^ in order to distinguish them from expectation values. For brevity we shall often omit the ^ in other chapters, whenever there is no ambiguity.) The fields $\hat{E}$ and $\hat{B}$ are physical observables. The potentials $\hat{A}$ and $\hat{U}$ are mathematical constructs that are used to represent the fields [1–4]. Equations (4.1) can be interpreted in the following way: Eq. (4.1a) is the definition of the magnetic field operator $\hat{B}(\mathbf{r})$ through the vector potential operator $\hat{A}(\mathbf{r})$, and Eq. (4.1b) is the equation of motion for the operator $\hat{A}(\mathbf{r})$, since it can be immediately obtained from the Heisenberg equation

$$\frac{\partial \hat{A}(\mathbf{r})}{\partial t} = \frac{i}{\hbar}[\hat{H}, \hat{A}(\mathbf{r})], \tag{4.1c}$$

by evaluating the commutator in the right-hand side of Eq. (4.1c). $\hat{H}$ is the total Hamiltonian, which will be introduced later.

The choice of $\hat{A}$ and $\hat{U}$ is not unique; in fact there are an infinite number of possible $(\hat{A}, \hat{U})$ pairs that generate the same $\hat{E}$ and $\hat{B}$ fields. A transformation involving $\hat{A}$ and $\hat{U}$ that leaves $\hat{E}$ and $\hat{B}$ intact is called a *gauge transformation.* A gauge transformation can be carried out using an arbitrary scalar operator function $\hat{F}(\mathbf{r})$ (i.e., a function of position $\mathbf{r}$ as well as of all coordinates of the material system). The new potentials $\hat{A}'$ and $\hat{U}'$ are related to the old ones by

$$\hat{A}'(\mathbf{r}) = \hat{A}(\mathbf{r}) + \nabla \hat{F}(\mathbf{r}), \tag{4.2a}$$

$$\hat{U}'(\mathbf{r}) = \hat{U}(\mathbf{r}) + \frac{i}{\hbar c}[\hat{H}, \hat{F}(\mathbf{r})]. \tag{4.2b}$$

Using the Heisenberg equation of motion for the operator $\hat{F}(\mathbf{r})$ we can recast Eq. (4.2b) in the form

$$\hat{U}'(\mathbf{r}) = \hat{U}(\mathbf{r}) + \frac{1}{c}\frac{\partial \hat{F}(\mathbf{r})}{\partial t}. \tag{4.2c}$$

Equation (4.2a) together with Eq. (4.2c) represent the standard gauge transformation. It can be easily verified by substitution that $\hat{B}$ and $\hat{E}$ are indeed invariant to this transformation. Equation (4.2a) together with Eq. (4.2b) provide an alternative form for the gauge transformation, which can be interpreted as follows: The gauge fixes the longitudinal part of the vector potential, which is connected to the scalar potential $\hat{U}(\mathbf{r})$ through Eq. (4.1b). Therefore, a change of the longitudinal part of the vector potential given by Eq. (4.2a) should be accompanied by a change in the scalar potential, which can be obtained from Eqs. (4.1b) and (4.1c), and is given by Eq. (4.2b).

A gauge may be chosen for convenience, and the Hamiltonian for a system of charges interacting with the electromagnetic field can be written in several equivalent forms, which depend on the choice of gauge. We start by adopting the *Coulomb gauge* where we choose the vector potential to be purely transverse, i.e.,

$$\nabla \cdot \hat{A}(\mathbf{r}) = 0. \tag{4.3}$$

The electrodynamics of a nonrelativistic material system interacting with the electromagnetic field may be described using the minimal-coupling Hamiltonian, which will be our basic starting point [5–8],

$$\hat{H}_{\text{min}} = \sum_{\alpha} \frac{1}{2m_\alpha}\left[\hat{\mathbf{p}}_\alpha - \frac{q_\alpha}{c}\hat{A}(\hat{\mathbf{r}}_\alpha)\right]^2 + \frac{1}{2}\sum_{\alpha \neq \beta}\frac{q_\alpha q_\beta}{|\hat{\mathbf{r}}_\alpha - \hat{\mathbf{r}}_\beta|} + \hat{H}_{\text{rad}}. \tag{4.4}$$

Here $\hat{\mathbf{r}}_\alpha$, $\hat{\mathbf{p}}_\alpha$, m_α, and q_α denote the position, momentum, mass, and charge of the particle labeled α. The material variables satisfy the usual commutation relations

$$\begin{cases} [\hat{\mathbf{r}}_{\alpha i}, \hat{\mathbf{r}}_{\beta j}] = [\hat{\mathbf{p}}_{\alpha i}, \hat{\mathbf{p}}_{\beta j}] = 0 \\ [\hat{\mathbf{r}}_{\alpha i}, \hat{\mathbf{p}}_{\beta j}] = i\hbar\delta_{\alpha\beta}\delta_{ij} \end{cases} \quad i, j = x, y, z. \tag{4.5}$$

$\hat{A}(\mathbf{r}_\alpha)$ is the vector potential of the electromagnetic field evaluated at the position of the α particle.

The free radiation-field Hamiltonian is given by

$$\hat{H}_{\rm rad} = \frac{1}{8\pi}\int d^3\mathbf{r}[\hat{E}^{\perp 2}(\mathbf{r}) + \hat{B}^2(\mathbf{r})]. \tag{4.6}$$

The canonical field variables are the vector potential $\hat{A}(\mathbf{r})$, and its conjugate momentum, $-(1/4\pi c)\hat{E}^\perp(\mathbf{r})$, where $\hat{E}^\perp(\mathbf{r})$ is the transverse field. They satisfy the commutation relations:

$$[\hat{A}_i(\mathbf{r}), \hat{E}_j^\perp(\mathbf{r}')] = -4\pi i\hbar c\, \delta_{ij}^\perp(\mathbf{r} - \mathbf{r}'), \tag{4.7}$$

where c is the speed of light. We further have

$$\left.\begin{aligned} [\hat{E}_x^\perp(\mathbf{r}), \hat{B}_y(\mathbf{r}')] &= 4\pi i\hbar c \frac{\partial}{\partial z}\delta(\mathbf{r} - \mathbf{r}'), \\ [\hat{A}_i(\mathbf{r}), \hat{A}_j(\mathbf{r}')] &= [\hat{A}_i(\mathbf{r}), \hat{B}_j(\mathbf{r}')] = 0. \end{aligned}\right\} \tag{4.8}$$

Here $i, j = x, y, z$, and $\delta_{ij}^\perp$ is the "transverse delta function" (see Appendix 4A).

We next introduce the dynamic variables representing the quantized transverse electromagnetic field. We denote by $\hat{a}_{\mathbf{k}\lambda}$ and $\hat{a}_{\mathbf{k}\lambda}^\dagger$ the annihilation and creation operator for the $\mathbf{k}\lambda$th mode with wavevector $\mathbf{k}$ and polarization λ [5]. They satisfy the Bose commutation relations

$$\begin{cases} [\hat{a}_{\mathbf{k}\lambda}, \hat{a}_{\mathbf{k}'\lambda'}] = [\hat{a}_{\mathbf{k}\lambda}^\dagger, \hat{a}_{\mathbf{k}'\lambda'}^\dagger] = 0, \\ [\hat{a}_{\mathbf{k}\lambda}, \hat{a}_{\mathbf{k}'\lambda'}^\dagger] = \delta_{\mathbf{k}\mathbf{k}'}\delta_{\lambda\lambda'}. \end{cases} \tag{4.9}$$

Using these operators, the vector potential assumes the form

$$\hat{A}(\mathbf{r}) = \sum_{\mathbf{k}\lambda} c\frac{\varepsilon_k}{\omega_k}[\hat{a}_{\mathbf{k}\lambda}e_{\mathbf{k}\lambda}\exp(i\mathbf{k}\cdot\mathbf{r}) + \hat{a}_{\mathbf{k}\lambda}^\dagger e_{\mathbf{k}\lambda}\exp(-i\mathbf{k}\cdot\mathbf{r})], \tag{4.10a}$$

with $\varepsilon_k = [2\pi\hbar\omega_k/\Omega]^{1/2}$, Ω being the quantization volume, $e_{\mathbf{k}\lambda}$ is a unit vector perpendicular to $\mathbf{k}$, and

$$\omega_k \equiv kc. \tag{4.10b}$$

We recall that the time evolution of any (material and field) operator is given by the Heisenberg equation (2.61d)

$$\dot{\hat{F}} = \frac{i}{\hbar}[\hat{H}, \hat{F}].$$

Using Eqs. (4.1), (4.4), and (4.8) we obtain

$$\hat{E}^\perp(\mathbf{r}) = \sum_{\mathbf{k}\lambda} i\varepsilon_k[\hat{a}_{\mathbf{k}\lambda}e_{\mathbf{k}\lambda}\exp(i\mathbf{k}\cdot\mathbf{r}) - \hat{a}_{\mathbf{k}\lambda}^\dagger e_{\mathbf{k}\lambda}\exp(-i\mathbf{k}\cdot\mathbf{r})], \tag{4.11}$$

and

$$\hat{B}(\mathbf{r}) = \sum_{\mathbf{k}\lambda} i\varepsilon_k[\hat{a}_{\mathbf{k}\lambda}(\tilde{\mathbf{k}} \times e_{\mathbf{k}\lambda}) \exp(i\mathbf{k}\cdot\mathbf{r}) - \hat{a}^\dagger_{\mathbf{k}\lambda}(\tilde{\mathbf{k}} \times e_{\mathbf{k}\lambda}) \exp(-i\mathbf{k}\cdot\mathbf{r})]. \quad (4.12)$$

$\tilde{\mathbf{k}}$ is a unit vector along the propagation direction. Note that the electric field is purely transverse, since the scalar potential that represents longitudinal interactions is included in the Hamiltonian in the form of the electrostatic interactions. Using these variables, the radiation field Hamiltonian in second quantized form becomes

$$\hat{H}_{\text{rad}} = \hbar \sum_{\mathbf{k}\lambda} \omega_{\mathbf{k}}(\hat{a}^\dagger_{\mathbf{k}\lambda}\hat{a}_{\mathbf{k}\lambda} + \tfrac{1}{2}). \quad (4.13)$$

We shall now calculate the electrodynamics of the matter described by this Hamiltonian. Introducing the charge-density operator

$$\hat{\rho}(\mathbf{r}) = \sum_\alpha q_\alpha \delta(\mathbf{r} - \hat{\mathbf{r}}_\alpha), \quad (4.14)$$

and using the Heisenberg equation, we get the continuity equation

$$\dot{\hat{\rho}}(\mathbf{r}, t) = -\nabla\cdot\hat{J}(\mathbf{r}, t) \quad (4.15)$$

with $\hat{J}(\mathbf{r})$ being the *electric current* operator

$$\hat{J}(\mathbf{r}) = \sum_\alpha q_\alpha[\dot{\hat{\mathbf{r}}}_\alpha \delta(\mathbf{r} - \hat{\mathbf{r}}_\alpha) + \delta(\mathbf{r} - \hat{\mathbf{r}}_\alpha)\dot{\hat{\mathbf{r}}}_\alpha]. \quad (4.16)$$

$\dot{\hat{\mathbf{r}}}_\alpha \equiv (i/\hbar)[\hat{H}, \hat{\mathbf{r}}_\alpha]$ is the velocity of the α particle, which in the Coulomb gauge is related to its canonical momentum by

$$m_\alpha \dot{\hat{\mathbf{r}}}_\alpha = p_\alpha - \frac{q_\alpha}{c}\hat{A}(\hat{\mathbf{r}}_\alpha). \quad (4.17)$$

In systems with bound charges, it proves more convenient to work with the polarization operator $\hat{P}(\mathbf{r})$ rather than with the current $\hat{J}(\mathbf{r})$. The polarization is introduced as follows: We assume that the system is made out of molecules with nonoverlapping charge distributions. The coordinates of the particles (electrons and nuclei) may be defined relative to a molecular center of mass (or charge) R_m, which for simplicity is assumed to be fixed.* Then the charge density operator of molecule m is given by

$$\hat{\rho}_m(\mathbf{r}) = \sum_\alpha q_{m\alpha}\delta(\mathbf{r} - \hat{\mathbf{r}}_{m\alpha}).$$

Here $\hat{\mathbf{r}}_{m\alpha}$ is the position operator and $q_{m\alpha}$ is the electric charge of particle α (electron or nucleus) belonging to molecule m. The charge density operator may be partitioned as follows [4]:

$$\hat{\rho}_m(\mathbf{r}) = \rho_0(m; \mathbf{r}) - \nabla\cdot\hat{\mathscr{P}}_m(\mathbf{r}), \quad (4.18a)$$

where

$$\rho_0(m; \mathbf{r}) = \sum_\alpha q_{m\alpha}\delta(\mathbf{r} - \mathbf{R}_m) \quad (4.18b)$$

* For a treatment of molecular motion see reference [9].

is the net charge at the center of mass, and

$$\hat{\mathscr{P}}_m(\mathbf{r}) = \sum_\alpha \int_0^1 du\, q_{m\alpha}(\hat{\mathbf{r}}_{m\alpha} - \mathbf{R}_m)\delta[\mathbf{r} - \mathbf{R}_m - u(\hat{\mathbf{r}}_{m\alpha} - \mathbf{R}_m)] \tag{4.18c}$$

is the polarization field of molecule m. The u-integration is a number integration that ensures the correct coefficients of the multipolar expansion of the polarization operator (see Appendix 4B). Note that Eq. (4.18a) does not define the polarization $\hat{\mathscr{P}}_m$ uniquely; for one thing the transverse part of $\hat{\mathscr{P}}_m$ is completely arbitrary since its divergence vanishes. We have used the freedom in its definition to require that $\hat{\mathscr{P}}_m(\mathbf{r})$ be a radial vector with respect to the molecular center $\mathbf{R}_m$.

The total charge density, net charge, and polarization operators of the material system are then given respectively by summing Eqs. (4.18) over molecules

$$\hat{\rho}(\mathbf{r}) = \sum_m \hat{\rho}_m(\mathbf{r}), \tag{4.19a}$$

$$\rho_0(\mathbf{r}) = \sum_m \rho_0(m; \mathbf{r}), \tag{4.19b}$$

$$\hat{P}(\mathbf{r}) = \sum_m \hat{\mathscr{P}}_m(\mathbf{r}). \tag{4.19c}$$

In all applications considered in this book, the system consists of a collection of *neutral* particles (atoms or molecules); then $\rho_0(m; \mathbf{r}) = 0$. Taking the time derivative of Eq. (4.18a), we have

$$\dot{\hat{\rho}}(\mathbf{r}) = -\nabla \cdot \dot{\hat{P}}(\mathbf{r}). \tag{4.20}$$

Comparing with the continuity equation (4.15) we have

$$\nabla \cdot [\hat{J}(\mathbf{r}) - \dot{\hat{P}}(\mathbf{r})] = 0,$$

which implies that $\hat{J} - \dot{\hat{P}}$ is a transverse field (with zero divergence). We can then write

$$\hat{J}(\mathbf{r}) = \dot{\hat{P}}(\mathbf{r}) + \nabla \times \hat{\mathscr{M}}(\mathbf{r}). \tag{4.21}$$

When $\dot{\hat{P}}$ is evaluated by taking the commutation of $\hat{P}$ with $\hat{H}$, and using the definition of $\hat{J}$, we finally get

$$\hat{\mathscr{M}}(\mathbf{r}) = \sum_m \hat{\mathscr{M}}_m(\mathbf{r}) \tag{4.22a}$$

$$\hat{\mathscr{M}}_m(\mathbf{r}) = \frac{1}{2}\sum_\alpha \int_0^1 du\, q_{m\alpha}\{(\hat{\mathbf{r}}_{m\alpha} - \mathbf{R}_m) \times \dot{\hat{\mathbf{r}}}_{m\alpha}\} u\delta[\mathbf{r} - \mathbf{R}_m - u(\hat{\mathbf{r}}_{m\alpha} - \mathbf{R}_m)]. \tag{4.22b}$$

Equation (4.22b) is the *magnetization density* of molecule m. The two contributions in Eq. (4.21) to the current are thus denoted the polarization current and the magnetization current, respectively [10, 11]. Equations (4.18), (4.19), and (4.22) provide an exact formal definition of the polarization and the magnetization operators, expressed as integrals over the dummy variable u. Upon expanding the $\delta[\mathbf{r} - \mathbf{R}_m - u(\hat{\mathbf{r}}_{m\alpha} - \mathbf{R}_m)]$ factors in powers of u, we can carry out the u-integration term by term and obtain an infinite expansion in the derivatives of the δ function [2, 6]. This *multipolar expansion* is given in Appendix 4B.

THE POWER–ZIENAU TRANSFORMATION AND THE MULTIPOLAR HAMILTONIAN

If we partition the minimal coupling Hamiltonian [Eq. (4.4)] into a pure material part, a field part, and a radiation–matter interaction, we find that the interaction term is

$$\sum_\alpha \left[\frac{-q_\alpha}{cm_\alpha} \hat{\mathbf{p}}_\alpha \cdot \hat{A}(\hat{\mathbf{r}}_\alpha) + \frac{q_\alpha^2}{2m_\alpha c^2} \hat{A}^2(\hat{\mathbf{r}}_\alpha) \right].$$

Because of this form of the interaction, the minimal coupling Hamiltonian is also denoted the p.A Hamiltonian. Two features of this form of the interaction make it somewhat inconvenient. First, the field variables enter through the vector potential, rather than the electric field, which is independent of the choice of a gauge and is more directly related to optical measurements. Second, the interaction is nonlinear in the radiation degrees of freedom and contains an $\hat{A}^2(\hat{\mathbf{r}}_\alpha)$ term. That term is often neglected even though it may affect the optical response in a significant way. It leads to a more complex form of the optical response functions, which can later be simplified using sum rules. For these reasons it is desirable to adopt a different form of the Hamiltonian by performing a canonical transformation on Eq. (4.4). Such a transformation was first proposed by Göppert-Mayer [12] within the dipole approximation and then extended to its most general form by Power and Zienau [13–15]. Unlike a gauge transformation, which affects only field variables, a canonical transformation affects the material as well as the field variables. With this transformation, any operator (material, field or both) is changed as

$$\hat{F}' = \exp(i\hat{S})\hat{F}\exp(-i\hat{S}). \tag{4.23a}$$

For the Power–Zienau transformation $\hat{S}$ is taken to be

$$\hat{S} \equiv (\hbar c)^{-1} \int \hat{P}(\mathbf{r}) \cdot \hat{A}(\mathbf{r})\, d\mathbf{r}. \tag{4.23b}$$

$\hat{P}(\mathbf{r})$ and $\hat{A}(\mathbf{r})$ are given by (4.19c) and (4.10a), respectively.

By performing this transformation, the field and the material variables mix. The new field variables contain an "old matter" contribution and the new material variables contain an "old field" contribution. Life could be very confusing if we do not distinguish clearly between the old and the new variables, and should we attempt to go back to the old variables during a calculation. The way this transformation can be used effectively is to completely forget about the old variables and to treat the new field variables as pure field degrees of freedom, and do the same for the matter. Consequently, all operators written hereafter in this section and in the rest of the book are to be understood as the transformed operators $\hat{F}'$. All commutation rules such as Eqs. (4.5) or (4.7) are preserved by this transformation provided, of course, that all operators are understood as the "new" operators. We also note that the vector potential $\hat{A}(\mathbf{r})$, and the particle coordinates $\hat{\mathbf{r}}_\alpha$ commute with $\hat{S}$, and are therefore unaffected by the transformation. Since the polarization operator $\hat{P}(\mathbf{r})$ depends only on the particle coordinates (and not on their momenta) it is also unaffected by the transformation. The canonical momentum is changed, however, as are the photon annihilation and creation operators. Thus

$$\left.\begin{aligned}
\hat{\mathbf{r}}'_{m\alpha} &= \hat{\mathbf{r}}_{m\alpha}, \\
\hat{\mathbf{p}}'_{m\alpha} &= \hat{\mathbf{p}}_{m\alpha} - \frac{q_{m\alpha}}{c}\hat{A}(\hat{\mathbf{r}}_{m\alpha}) - \int \hat{\mathbf{n}}_\alpha(m;\mathbf{r}) \times \hat{B}(\mathbf{r})\,d\mathbf{r}, \\
\hat{\mathbf{a}}'_{\mathbf{k}\lambda} &= \hat{\mathbf{a}}_{\mathbf{k}\lambda} - \left(\frac{i}{\hbar}\right)\frac{\varepsilon_k}{\omega_k} e_{\mathbf{k}\lambda}\cdot\hat{P}(\mathbf{k}), \\
\hat{\mathbf{a}}'^{\dagger}_{-\mathbf{k}\lambda} &= \hat{\mathbf{a}}^{\dagger}_{-\mathbf{k}\lambda} + \left(\frac{i}{\hbar}\right)\frac{\varepsilon_k}{\omega_k} e_{\mathbf{k}\lambda}\cdot\hat{P}(\mathbf{k}),
\end{aligned}\right\} \tag{4.24}$$

where we have introduced the polarization vector field

$$\hat{\mathbf{n}}_\alpha(m;\mathbf{r}) \equiv q_{m\alpha}(\hat{\mathbf{r}}_{m\alpha} - \mathbf{R}_m)\int_0^1 du\, u\delta[\mathbf{r} - \mathbf{R}_m - u(\hat{\mathbf{r}}_{m\alpha} - \mathbf{R}_m)]. \tag{4.25}$$

Equations (4.24) yield

$$\left.\begin{aligned}
\hat{A}'(\mathbf{r}) &= \hat{A}(\mathbf{r}), \\
\hat{B}'(\mathbf{r}) &= \hat{B}(\mathbf{r}), \\
\hat{E}'^{\perp}(\mathbf{r}) &= \hat{E}^{\perp}(\mathbf{r}) + 4\pi\hat{P}^{\perp}(\mathbf{r}) \equiv \hat{D}^{\perp}(\mathbf{r}), \\
\hat{P}'(\mathbf{r}) &= \hat{P}(\mathbf{r}).
\end{aligned}\right\} \tag{4.26}$$

Note that $\hat{a}_{\mathbf{k}\lambda}$ and $\hat{\mathbf{a}}'_{\mathbf{k}\lambda}$ refer to destruction of different types of photons, since the old material and field degrees of freedom are mixed in the new representation. To avoid confusion with the transverse part of the Maxwell field, namely $\hat{E}^{\perp}(\mathbf{r})$, $\hat{E}'^{\perp}(\mathbf{r})$ is given the symbol $\hat{D}^{\perp}(\mathbf{r})$ and called the *electric displacement operator*. By transformation of the mode expansion Eq. (4.11) we directly obtain, using Eqs. (4.24) and (4.26), the field $\hat{D}^{\perp}(\mathbf{r})$ expanded in field modes.

$$\hat{D}^{\perp}(\mathbf{r}) = i\sum_{\mathbf{k}\lambda}\varepsilon_k[\hat{a}_{\mathbf{k}\lambda}e_{\mathbf{k}\lambda}\exp(i\mathbf{k}\cdot\mathbf{r}) - \hat{a}^{\dagger}_{\mathbf{k}\lambda}e_{\mathbf{k}\lambda}\exp(-i\mathbf{k}\cdot\mathbf{r})]. \tag{4.27}$$

The transformed Hamiltonian,

$$\hat{H}_{\mathrm{mult}} \equiv \exp(i\hat{S})\hat{H}_{\mathrm{min}}\exp(-i\hat{S}), \tag{4.28}$$

is known as the *multipolar Hamiltonian* [14], and is given by

$$\hat{H}_{\mathrm{mult}} = \hat{H}_{\mathrm{mol}} + \hat{V}_{\mathrm{inter}} + \hat{H}_{\mathrm{self}} + \hat{H}_{\mathrm{rad}} + \sum_m \hat{H}_{\mathrm{int}}(m). \tag{4.29}$$

Here the first three terms represent the purely material Hamiltonian. $\hat{H}_{\mathrm{rad}}$ is the pure radiation field Hamiltonian, and the last term represents the radiation–matter interaction. $\hat{H}_{\mathrm{mol}}$ is the molecular Hamiltonian

$$\hat{H}_{\mathrm{mol}} = \sum_m \frac{\hat{\mathbf{p}}^2_{m\alpha}}{2m_\alpha} + \sum_m \hat{V}_m, \tag{4.30a}$$

where the intramolecular potential term $\hat{V}_m$ is defined as

$$\hat{V}_m = \sum_{\alpha<\beta}\frac{q_{m\alpha}q_{m\beta}}{|\hat{\mathbf{r}}_{m\alpha} - \hat{\mathbf{r}}_{m\beta}|}. \tag{4.30b}$$

The second term $\hat{V}_{\text{inter}}$ represents the intermolecular potential energy, which is simply the electrostatic interaction of the charge distributions associated with each molecule, namely

$$\hat{V}_{\text{inter}} = \sum_{m<m'} \int d\mathbf{r} \int d\mathbf{r}' \frac{\hat{\rho}_m(\mathbf{r})\hat{\rho}_{m'}(\mathbf{r}')}{|\mathbf{r}-\mathbf{r}'|}, \tag{4.31a}$$

where we have defined the molecular charge density operator,

$$\hat{\rho}_m(\mathbf{r}) = \sum_\alpha q_{m\alpha}\delta(\mathbf{r} - \hat{\mathbf{r}}_{m\alpha}). \tag{4.31b}$$

The molecules are assumed not to overlap, so that intermolecular electron exchange is neglected.

$\hat{H}_{\text{self}}$ is the self-interaction of the molecular polarization and is related to the modulus square of the transverse part of the electric polarization $\hat{P}(\mathbf{r})$

$$\hat{H}_{\text{self}} = 2\pi \int |\hat{P}^{\perp}(\mathbf{r})|^2\, d\mathbf{r}. \tag{4.32}$$

The transverse polarization $\hat{P}^{\perp}(\mathbf{r})$ can be obtained from the total polarization $\hat{P}(\mathbf{r})$ via Eq. (4.A.9a)

$$\hat{P}_i^{\perp}(\mathbf{r}) = \tfrac{2}{3}\hat{P}_i(\mathbf{r}) - \frac{1}{4\pi}\sum_j \int d\mathbf{r}'\, T_{ij}(\mathbf{r}-\mathbf{r}')\hat{P}_j(\mathbf{r}') \tag{4.33a}$$

where the ijth component of the second rank tensor $\mathbf{T}(\mathbf{r})$ is

$$T_{ij}(\mathbf{r}) = \frac{\delta_{ij} - 3\hat{r}_i\hat{r}_j}{r^3}. \tag{4.33b}$$

Due to the nonlocal nature of $\hat{P}^{\perp}(\mathbf{r})$, the self-interaction term contains both intramolecular and intermolecular interactions.

This concludes the material part of the Hamiltonian. The radiation field Hamiltonian $\hat{H}_{\text{rad}}$ is

$$\hat{H}_{\text{rad}} = \frac{1}{8\pi}\int [\hat{D}^{\perp 2}(\mathbf{r}) + \hat{B}^2(\mathbf{r})]\, d\mathbf{r}, \tag{4.34a}$$

or in second quantized form

$$\hat{H}_{\text{rad}} = \sum_{\mathbf{k}\lambda} \hbar\omega_k(\hat{a}^\dagger_{\mathbf{k}\lambda}\hat{a}_{\mathbf{k}\lambda} + \tfrac{1}{2}). \tag{4.34b}$$

Here $\hat{A}(\mathbf{r})$ is a general coordinate and $-(1/4\pi c)\hat{D}^{\perp}(\mathbf{r})$ is its conjugate momentum. They satisfy the commutation relation

$$[\hat{A}(\mathbf{r}), \hat{D}^{\perp}(\mathbf{r}')] = -4\pi\hbar c i\delta^{\perp}(\mathbf{r}-\mathbf{r}').$$

Finally, the molecular interaction with the electromagnetic field is given by $\hat{H}_{\text{int}}$

$$\hat{H}_{\text{int}}(m) = -\int \hat{P}_m(\mathbf{r})\cdot\hat{D}^{\perp}(\mathbf{r})\, d\mathbf{r} - \int \hat{M}_m(\mathbf{r})\cdot\hat{B}(\mathbf{r})\, d\mathbf{r}$$
$$+ \sum_\alpha \frac{1}{2m_\alpha c^2}\left[\int [\hat{n}_\alpha(m;\mathbf{r}) \times \hat{B}(\mathbf{r})]\, d\mathbf{r}\right]^2. \tag{4.35}$$

This interaction is particularly simple when the magnetic (i.e., second and third) terms are neglected. In this case, which is often justified in optical measurements, the interaction is simply of the form of a scalar product of the polarization and the field, $\hat{P}\cdot\hat{D}^{\perp}$. This form cures the difficulties associated with the minimal coupling Hamiltonian outlined in the beginning of this section, and is the main reason for performing the Power–Zienau transformation. The interaction now depends on the field rather than the vector potential and is linear in the field. It should be noted that the multipolar Hamiltonian can be alternatively derived using a gauge transformation. Instead of the Coulomb gauge, we require that $(\mathbf{r} - R_m)\cdot\hat{A}(\mathbf{r}) = 0$ in the vicinity of a charge distribution centered at the coordinate R_m.

The present form of the electrostatic intermolecular interaction [Eq. (4.31a)] was used by Longuet-Higgins [16] to calculate the intermolecular forces between molecules with nonoverlapping charge distributions, without invoking the dipole approximation. It represents long-range forces in the sense that electron exchange is neglected. For large intermolecular separations and neutral molecules, this will result in the van der Waals (London) $\sim R^{-6}$ interaction. A crossover to a Casimir–Polder dispersion $\sim R^{-7}$ interaction takes place for very large separation (larger than the optical wavelength) [14].

Since the polarization is the key quantity in the optical response, it may be convenient to recast the intermolecular interactions in terms of the polarization density (rather than the charge density). To this end we integrate Eq. (4.31a) by parts and obtain (see Figure 4.1):

$$\hat{V}_{\text{inter}} = \sum_{m<m'} \int d\mathbf{r} \int d\mathbf{r}'\, \hat{\mathscr{P}}_m(\mathbf{r})\cdot T(\mathbf{r}-\mathbf{r}')\cdot\hat{\mathscr{P}}_{m'}(\mathbf{r}'). \tag{4.36}$$

Alternatively, we may write the electrostatic interaction as

$$\hat{V}_{\text{inter}} = \sum_{m<m'} \int d\mathbf{r}\, \hat{\mathscr{P}}^{\perp}_m(\mathbf{r})\cdot\hat{\mathscr{P}}^{\perp}_{m'}(\mathbf{r}) = \sum_{m<m'} \int d\mathbf{r}\, \hat{\mathscr{P}}^{\parallel}_m(\mathbf{r})\cdot\hat{\mathscr{P}}^{\parallel}_{m'}(\mathbf{r}). \tag{4.37}$$

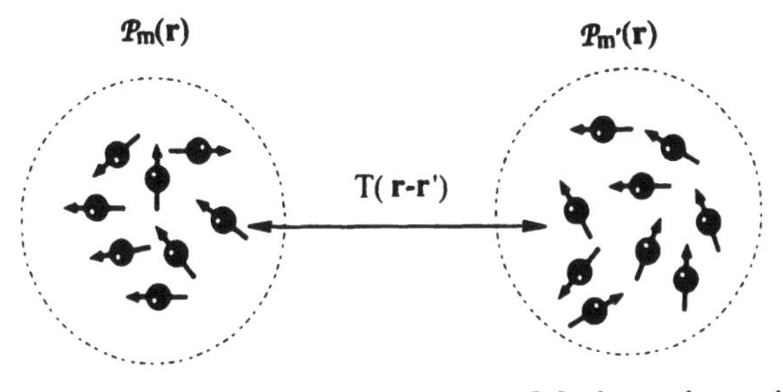

FIG. 4.1 Any charge distribution can be represented as a vector field of the polarization. It can thus be visualized as a continuous distribution of dipoles. The complete electrostatic interaction between two neutral, nonoverlapping particles can then be recast in a dipole–dipole form [Eq. (4.36)]. The dipole approximation is obtained by replacing each particle by a single dipole.

Even though the total polarization field of a single molecule is localized to essentially the region of its charges, satisfying Eq. (4.18c), its transverse and longitudinal parts are delocalized everywhere as can be seen from Eqs. (4.A.10) (this is true for any vector field). The form of Eq. (4.36) is most useful, as it allows us to work with the total polarization field, rather than the transverse field.

The multipolar Hamiltonian is properly expressed in terms of the fields $\hat{D}$ and $\hat{B}$, and not $\hat{E}$ and $\hat{B}$, due to the canonical transformation. We have

$$\hat{D}^{\perp}(\mathbf{r}) = \hat{D}(\mathbf{r}) = \hat{E}(\mathbf{r}) + 4\pi\hat{P}(\mathbf{r}), \tag{4.38}$$

$\hat{D}(\mathbf{r})$ is entirely transverse in neutral systems and $\hat{E}^{\parallel}(\mathbf{r}) = -4\pi\hat{P}^{\parallel}(\mathbf{r})$. The velocity of the $m\alpha$ particle is now related to its canonical momentum by

$$m_{m\alpha}\dot{\hat{\mathbf{r}}}_{m\alpha} = \hat{\mathbf{p}}_{m\alpha} + \frac{1}{c}\int [\hat{n}_{\alpha}(m;\mathbf{r}) \times \hat{B}(\mathbf{r})]\, d\mathbf{r}. \tag{4.39a}$$

In $\hat{H}_{\text{int}}(m)$, $\hat{M}_m(\mathbf{r})$ is the magnetization of the mth molecule:

$$\hat{M}_m(\mathbf{r}) = \sum_{\alpha} \frac{1}{2m_{\alpha}c}[\hat{n}_{\alpha}(m;\mathbf{r}) \times \hat{\mathbf{p}}_{m\alpha} - \hat{\mathbf{p}}_{m\alpha} \times \hat{n}_{\alpha}(m;\mathbf{r})]. \tag{4.39b}$$

This is the correct form of $\hat{M}_m(r)$, since it is expressed in terms of the canonical variables ($\hat{\mathbf{r}}_{m\alpha}$ and the new momentum $\hat{\mathbf{p}}_{m\alpha}$). Note that $\hat{\mathscr{M}}_m(\mathbf{r}) \neq \hat{M}_m(\mathbf{r})$; the former, given by Eq. (4.22), is defined through the particle's velocity, as opposed to the canonical momentum.

The multipolar form of the Hamiltonian provides an elegant and convenient starting point for calculating the electrodynamics of systems of charges interacting with the electromagnetic field. It should be noted that the intermolecular part of the self energy contribution to the Hamiltonian exactly cancels V_{int}, leaving just the intramolecular contribution, so that it is possible to recast the multipolar Hamiltonian in a form in which all direct intermolecular forces have been eliminated; they are all fully retarded and mediated by the radiation field,

$$\hat{V}_{\text{inter}} + \hat{H}_{\text{self}} = 2\pi\sum_{m}\int |\hat{\mathscr{P}}_m^{\perp}(\mathbf{r})|^2\, d\mathbf{r}.$$

This form is useful for calculating, e.g., spontaneous emission processes and long-range intermolecular interactions. We chose to keep the unretarded interaction explicitly, since this form is most suitable for deriving the equations of motion commonly used in nonlinear optics, as will be shown next.

THE COUPLED FIELD AND MATTER EQUATIONS OF MOTION AND THE SEMICLASSICAL HAMILTONIAN

The dynamics of an arbitrary operator $\hat{Q}$ (that can depend on both material and field variables) may be calculated using the Heisenberg equation of motion [Eq. (2.63d)] together with the multipolar Hamiltonian. A word of caution: the self interaction term H_{self} is often assumed to only give rise to a trivial energy shift of the material system, and is neglected. We do not make this approximation, since in the following derivation this term is eliminated identically by other contributions. We have thus [17]

$$
\begin{aligned}
\frac{\hbar}{i}\frac{d\hat{Q}}{dt} = {} & [H_{\text{mol}} + \hat{H}_{\text{rad}} + \hat{V}_{\text{inter}}, \hat{Q}] \\
& - \frac{1}{2}\int \{[\hat{P}(\mathbf{r}), \hat{Q}]\cdot\hat{E}^{\perp}(\mathbf{r}) + \hat{E}^{\perp}(\mathbf{r})\cdot[\hat{P}(\mathbf{r}), \hat{Q}]\}\, d\mathbf{r} \\
& - \frac{1}{2}\int \{[\hat{D}^{\perp}(\mathbf{r}), \hat{Q}]\cdot\hat{P}(\mathbf{r}) + \hat{P}(\mathbf{r})\cdot[\hat{D}^{\perp}(\mathbf{r}), \hat{Q}]\}\, d\mathbf{r} \\
& - \frac{1}{2}\int \{[\hat{M}(\mathbf{r}), \hat{Q}]\cdot\hat{B}(\mathbf{r}) + \hat{B}(\mathbf{r})\cdot[\hat{M}(\mathbf{r}), \hat{Q}]\}\, d\mathbf{r} \\
& - \frac{1}{2}\int \{[\hat{B}(\mathbf{r}), \hat{Q}]\cdot\hat{M}(\mathbf{r}) + \hat{M}(\mathbf{r})\cdot[\hat{B}(\mathbf{r}), \hat{Q}]\}\, d\mathbf{r} \\
& + \sum_m \sum_\alpha \frac{1}{2m_\alpha c^2}\left[\left\{\int \hat{n}_\alpha(m; \mathbf{r}) \times \hat{B}(\mathbf{r})\, d\mathbf{r}\right\}^2, \hat{Q}\right].
\end{aligned}
\tag{4.40}
$$

Here all operators are taken at time t. We shall now consider separately the time evolution of pure field and pure material variables, where this equation simplifies considerably.

Equations of motion for the field; The Maxwell equations

We first use the Heisenberg equation to derive the equation of motion for the displacement. Defining the magnetic displacement field

$$\hat{H}(\mathbf{r}) \equiv \hat{B}(\mathbf{r}) - 4\pi\hat{\mathscr{M}}(\mathbf{r}), \tag{4.41}$$

we have

$$\frac{1}{c}\frac{\partial\hat{D}(\mathbf{r})}{\partial t} = \nabla \times \hat{\mathbf{H}}(\mathbf{r}). \tag{4.41a}$$

Also, since $\hat{D}(\mathbf{r})$ is entirely transverse

$$\nabla\cdot\hat{\mathbf{D}}(\mathbf{r}) = 0. \tag{4.41b}$$

The relations

$$-\frac{1}{c}\frac{\partial\hat{B}(\mathbf{r})}{\partial t} = \nabla \times \hat{E}(\mathbf{r}), \tag{4.41c}$$

$$\nabla\cdot\hat{B}(\mathbf{r}) = 0, \tag{4.41d}$$

follow directly from the definition of the potentials Eq. (4.1). Equations (4.41) are known as the atomic field equations [18, 19]. They are purely microscopic, and it is the statistical averaging of these equations that leads to the macroscopic form of Maxwell's equations. We further neglected the center of mass motion of the molecules, thus neglecting the coupling of the molecular translational motion to the polarization field (the Röntgen current). Upon taking the time derivative of both sides of Eq. (4.41a) [and the curl of Eq. (4.41c)] and substituting Eqs (4.38) and (4.21), we obtain the wave equation

$$\nabla \times \nabla \times \hat{E}(\mathbf{r}) + \frac{1}{c^2}\frac{\partial^2 \hat{E}(\mathbf{r})}{\partial t^2} = -\frac{4\pi}{c^2}\frac{\partial \hat{J}(\mathbf{r})}{\partial t}. \tag{4.42}$$

In the present book we shall consider only nonmagnetic materials for which the magnetization moment may be neglected. Setting $\hat{\mathcal{M}} = 0$ (and thus $\hat{H} = \hat{B}$), we obtain

$$\boxed{\nabla \times \nabla \times \hat{E}(\mathbf{r}) + \frac{1}{c^2}\frac{\partial^2}{\partial t^2}\hat{E}(\mathbf{r}) = -\frac{4\pi}{c^2}\frac{\partial^2 \hat{P}(\mathbf{r})}{\partial t^2}.} \tag{4.43}$$

The general solution of Eq. (4.43) is given in Appendix 4C. For a transverse field, $\nabla \cdot \hat{E} = 0$ and making use of Eq. (4.A.4), the wave equation assumes the simpler form

$$\nabla^2 \hat{E}(\mathbf{r}) - \frac{1}{c^2}\frac{\partial^2 \hat{E}(\mathbf{r})}{\partial t^2} = \frac{4\pi}{c^2}\frac{\partial^2 \hat{P}(\mathbf{r})}{\partial t^2}. \tag{4.44a}$$

The energy flux (intensity) of the field is given by the Poynting vector

$$\hat{S} = \frac{c}{8\pi}(\hat{E} \times \hat{B} - \hat{B} \times \hat{E}). \tag{4.44b}$$

The expectation value of this vector gives the field intensity. Since in a vacuum the electric and magnetic field amplitudes are identical $E = B$, we have [3]

$$I(\mathbf{r}) = \frac{c}{4\pi}|E(\mathbf{r}, t)|^2 = \frac{c}{4\pi}|B(\mathbf{r}, t)|^2. \tag{4.44c}$$

This can be recast in the form

$$I(\mathbf{r}) = cT_0,$$

where T_0 is the field energy density (i.e., energy per unit volume).

$$T_0 = \frac{1}{8\pi}[|E(\mathbf{r}, t)|^2 + |(B(\mathbf{r}, t)|^2].$$

The semiclassical material equations

Having obtained the equation of motion for the field, we shall now consider the Heisenberg equation of motion for a material operator $\hat{Q}(t)$. In this case the third and fifth terms in the right-hand side of Eq. (4.40) vanish since material and field operators commute. If we further neglect the magnetic (fourth and sixth terms) we obtain

$$\frac{\hbar}{i}\frac{d\hat{Q}(t)}{dt} = [\hat{H}_{\text{mol}} + \hat{V}_{\text{inter}}, \hat{Q}(t)] - \frac{1}{2}\int d\mathbf{r}\{[\hat{P}(\mathbf{r}, t), \hat{Q}(t)]\hat{E}^{\perp}(\mathbf{r}, t) + \hat{E}^{\perp}(\mathbf{r}, t)[\hat{P}(\mathbf{r}, t), \hat{Q}(t)]\}. \quad (4.45)$$

The question now arises as to whether we can derive Eq. (4.45) directly using an effective semiclassical Hamiltonian that depends only on the material degrees of freedom, i.e.,

$$\frac{\hbar}{i}\frac{d\hat{Q}(t)}{dt} = [\hat{H}_{\text{eff}}, \hat{Q}(t)]. \quad (4.46a)$$

If the field operators $\hat{E}^{\perp}$ were to commute with the material degrees of freedom, that would be straightforward, and the effective Hamiltonian would simply be

$$\hat{H}_{\text{eff}} \stackrel{?}{=} \hat{H}_{\text{mol}} + \hat{H}_{\text{int}} - \int dr\, \hat{P}(r)\cdot\hat{E}^{\perp}(r). \quad (4.46b)$$

However, in the multipolar Hamiltonian, the displacement $\hat{D}$ rather than the Maxwell field $\hat{E}^{\perp}$ is the "pure field" variable and

$$\hat{E}^{\perp} \equiv \hat{D} - 4\pi\hat{P}^{\perp}.$$

Therefore, $\hat{E}^{\perp}$ is a joint material and field operator, which does not commute with the material degrees of freedom. Consequently, the proposed effective Hamiltonian Eq. (4.46b) is incorrect since it will not yield Eq. (4.45)!

A widely used procedure is to invoke the semiclassical approximation whereby the field is treated classically whereas the quantum nature of the matter degrees of freedom is preserved. We then replace $\hat{E}^{\perp}$ in Eq. (4.45) by its expectation value

$$\hat{E}^{\perp}(\mathbf{r}, t) \rightarrow \langle\hat{E}^{\perp}(\mathbf{r}, t)\rangle \equiv E^{\perp}(\mathbf{r}, t).$$

This way $\hat{E}^{\perp}$ is converted into a c-number, which obviously commutes with any other operator. When this is done, the equation of motion follows from an effective semiclassical Hamiltonian and reads

$$\frac{\hbar}{i}\frac{d\hat{Q}(t)}{dt} = [\hat{H}_{\text{sc}}, \hat{Q}(t)], \quad (4.47a)$$

where

$$\hat{H}_{\text{sc}} \equiv \hat{H}_{\text{mol}} + \hat{V}_{\text{inter}} - \int d\mathbf{r}\, \hat{P}(\mathbf{r})\cdot E^{\perp}(\mathbf{r}, t). \quad (4.47b)$$

It should be noted that H_{sc} would not change had we replaced the total polarization operator $\hat{P}(r)$ by the transverse polarization $\hat{P}^{\perp}(r)$. [The scalar product of $\hat{P}^{\parallel}(\mathbf{r})$ with $E^{\perp}(\mathbf{r})$ vanishes identically.] However, it is usually preferable to work with the total polarization $\hat{P}(\mathbf{r})$, since this operator has a simpler formal expression and it is local, i.e., it essentially vanishes in regions with no charge density. The transverse polarization, in contrast, is nonlocal, as can be seen from Eq. (4.33a). The ability to write a closed "mean field" equation for the material alone is not a trivial result, and should not be taken for granted. In general, only when a separation of timescales exists between two types of degrees of freedom, we can eliminate some of the degrees of freedom and derive an effective Hamiltonian in a reduced space [see Eq. (2.96)]. The Born–Oppenheimer (adiabatic) approximation for separating electronic and nuclear motions is an example.

The semiclassical Hamiltonian provides the basis for most theoretical treatments of the nonlinear response, and will be repeatedly used throughout this book.

THE COUPLED MAXWELL–LIOUVILLE EQUATIONS

Within the semiclassical approximation outlined above, the electric field is replaced by its expectation value and becomes a classical c-number. The dynamics of the system is calculated by solving coupled equations for $E(\mathbf{r})$ [Eq. (4.43)] and for the polarization $P(\mathbf{r})$ which is calculated quantum mechanically using Eq. (4.47).

In the Schrödinger picture, the semiclassical approximation results in the *Maxwell–Liouville equations*

$$\nabla \times \nabla \times E(\mathbf{r}, t) + \frac{1}{c^2}\frac{\partial^2}{\partial t^2} E(\mathbf{r}, t) = -\frac{4\pi}{c^2}\frac{\partial^2 P(\mathbf{r}, t)}{\partial t^2}, \tag{4.48a}$$

$$P(\mathbf{r}, t) = \mathrm{Tr}[\hat{P}(\mathbf{r})\rho(t)], \tag{4.48b}$$

$$\frac{\partial \rho(t)}{\partial t} = -\frac{i}{\hbar}[\hat{H}_{sc}(t), \rho(t)]. \tag{4.48c}$$

The polarization $P(\mathbf{r}, t)$ depends on position $\mathbf{r}$ through the dependence of the operator $\hat{P}(\mathbf{r})$ [Eqs. (4.18c) and (4.19c)] and through the nonequilibrium density operator. In this procedure we solve the coupled equations for P and E self-consistently.

Alternatively, the material part can be calculated by adopting the Heisenberg picture. We then replace Eqs. (4.48b) and (4.48c) by the Heisenberg equation

$$\dot{\hat{P}}(\mathbf{r}, t) = \frac{i}{\hbar}[\hat{H}_{sc}(t), \hat{P}(\mathbf{r}, t)]. \tag{4.49}$$

Equations (4.48a) and (4.49) are not closed for P and E, and, in general, we need consider more dynamic variables in addition to the polarization (otherwise this will be the end of this book!). The way we treat P varies according to the system. In general, the Schrödinger picture is useful for simple systems with a few levels. Many-body effects are more easily treated by truncating the hierarchy generated by the Heisenberg equation. The Heisenberg picture also suggests a physical coupled oscillator representation, which provides a quasiparticle interpretation of the radiative dynamics. The Heisenberg picture will be used in Chapters 16 and 17.

The semiclassical procedure naturally separates the calculation of optical measurements into two steps. The calculation of the polarization P can be made by using Eqs. (4.48b) and (4.48c) [or Eq. (4.49)] alone, and we do not need to consider the Maxwell equation [Eq. (4.48a)] at this stage. The calculation of the optical signal requires, however, the solution of the Maxwell equation for $\hat{E}(\mathbf{r}, t)$. Retardation effects, e.g., spontaneous emission will show up only macroscopically, when the Maxwell equations are solved. This procedure misses some effects related to the retarded nature of the radiation–matter interaction. It also does not account for photon statistics, which requires a quantum description of the field [20]. Despite these limitations, this level of theoretical treatment is often adequate for a broad range of applications, and its simplicity results in its widespread use. A more rigorous procedure will be discussed in Chapter 17.

The Maxwell–Liouville equations constitute the basic workhorse of nonlinear response calculations. A common special case assumes a material two-level system with the addition of relaxation terms to the matter equation (see Chapter 6). The coupled equations are then known as the Bloch–Maxwell equations.

Equation (4.18c) implies that each molecule can be rigorously regarded as a continuous distribution of dipoles introduced through the polarization density. The use of dipoles, rather than charges, greatly simplifies the treatment of the electrodynamics of the system since we can then represent both intermolecular interactions and radiation matter interactions in the same fashion. In the dipole approximation [Eq. (4.B.7)] we add up all these dipoles, and represent each molecule by a single dipole. This amounts to keeping only the zeroth moment of the distribution and yields

$$\hat{\mathscr{P}}_m(\mathbf{r}) = \mu_m \delta(\mathbf{r} - \mathbf{R}_m),$$

with

$$\mu_m = \sum_\alpha q_{m\alpha}(\hat{\mathbf{r}}_{m\alpha} - \mathbf{R}_m).$$

The polarization operator is obtained by substituting these expressions into Eq. (4.19c). It is possible to improve upon the dipole approximation in two ways: we either consider higher moments, which will result in adding higher multipoles (quadrupole, etc.) as shown in Appendix 4B. This is the conventional *multipolar expansion*, which is frequently used in electrostatics and electrodynamics [1, 2]. Alternatively, it may be possible to describe a molecule as a collection of a few dipoles (rather than a single dipole). This is obvious in, e.g., a large molecule with two chromophores. How to define an optimal set of dipoles is an open and interesting question that requires further formal developments. This representation seems particularly suitable for developing computational algorithms for molecular dynamics simulations of spectra in condensed phases.

OPTICAL MEASUREMENTS AND THE POLARIZATION

To classify the various nonlinear techniques, it is a common practice to expand the polarization in powers of the electric field. The full expansion and its microscopic significance will be established in Chapter 5. At this point, we merely wish to illustrate how the polarization is related to the observables in typical optical measurements. To that end we separate P into its linear and nonlinear parts, i.e.,

$$P(\mathbf{r}, t) \equiv P^{(1)}(\mathbf{r}, t) + P_{\mathrm{NL}}(\mathbf{r}, t), \tag{4.50}$$

where $P^{(1)}$ is linear in E and P_{NL} is the nonlinear polarization. It will be shown in Chapter 5 that $P^{(1)}$ can be expanded using the *linear response function* $S^{(1)}(\mathbf{r}, t)$ as follows:

$$P^{(1)}(\mathbf{r}, t) = \int d\mathbf{r}_1 \int_0^t dt_1\, S^{(1)}(\mathbf{r} - \mathbf{r}_1, t - t_1) E(\mathbf{r}_1, t_1). \tag{4.51}$$

Combining Eqs. (4.50) and (4.51) with Eq. (4.48a) we get

$$\nabla \times \nabla \times E(\mathbf{r}, t) + \frac{1}{c^2}\frac{\partial^2}{\partial t^2} \int d\mathbf{r}_1 \int_0^t dt_1\, \varepsilon(\mathbf{r} - \mathbf{r}_1, t - t_1) E(\mathbf{r}_1, t_1) = -\frac{4\pi}{c^2}\frac{\partial^2 P_{\mathrm{NL}}(\mathbf{r}, t)}{\partial t^2}, \tag{4.52a}$$

where the *dielectric function* is given by

$$\varepsilon(\mathbf{r} - \mathbf{r}_1, t - t_1) \equiv \delta(t - t_1)\delta(\mathbf{r} - \mathbf{r}_1) + 4\pi S^{(1)}(\mathbf{r} - \mathbf{r}_1, t - t_1). \tag{4.52b}$$

We next make use of the identity (4.A.4) and the Helmholtz theorem (4.A.6)–(4.A.8), to split Eq. (4.52a) into its transverse and longitudinal components. For the transverse component we get

$$\nabla^2 E^{\perp}(\mathbf{r}, t) - \frac{1}{c^2}\frac{\partial^2}{\partial t^2} \int d\mathbf{r}' \int_0^t dt'\, [\varepsilon(\mathbf{r} - \mathbf{r}', t - t') E(\mathbf{r}', t')]^{\perp} = \frac{4\pi}{c^2}\frac{\partial^2 P_{\mathrm{NL}}^{\perp}(\mathbf{r}, t)}{\partial t^2}. \tag{4.53}$$

The nonlinear polarization P_{NL} carries the complete microscopic information required for the description of any nonlinear-optical process. It may be obtained by solving either the semiclassical Liouville or the Heisenberg equations for the material system. We treat the field $E^{\perp}(\mathbf{r}, t)$ as given, and expand the macroscopic polarization P^{NL} in powers of this field. When this expansion is substituted into Eq. (4.53), the wave equation now becomes a closed nonlinear equation for the Maxwell field E.

In general, calculating the optical signal requires the solution of the coupled set of equations (4.48) that depends on the material, the incoming pulses, as well as the geometry. In some simple cases, it is possible to solve the equations analytically and obtain a closed form expression for the signal. We shall now present the relations between some of the most common optical observables and the polarization.

Linear absorption

One of the simplest optical measurements is the absorption of a weak stationary plane wave propagating along the z direction in a linear and isotropic medium. It can be calculated by setting $P_{\mathrm{NL}} = 0$ and assuming a solution of Maxwell equations of the form

$$E(z, t) = E_0 \exp(ikz - i\omega t). \tag{4.54}$$

Upon the substitution in Maxwell equations (4.53), we get

$$\frac{k^2c^2}{\omega^2} = \varepsilon(\mathbf{k}, \omega), \tag{4.55}$$

with the frequency and the wavevector-dependent dielectric function

$$\varepsilon(\mathbf{k}, \omega) = \int d\mathbf{r} \int_0^\infty dt_1\, \varepsilon(\mathbf{r} - \mathbf{r}', t_1) \exp[-i\mathbf{k}(\mathbf{r} - \mathbf{r}') + i\omega t_1], \tag{4.56}$$

$$\varepsilon(\mathbf{k}, \omega) = 1 + 4\pi\chi^{(1)}(\mathbf{k}, \omega), \tag{4.57}$$

and $\chi^{(1)}(\mathbf{k}, \omega)$ is the linear susceptibility

$$\chi^{(1)}(\mathbf{k}, \omega) \equiv \int d\mathbf{r} \int_0^\infty dt\, \hat{S}^{(1)}(\mathbf{r}, t) \exp(-i\mathbf{k}\cdot\mathbf{r} + i\omega t). \tag{4.58}$$

Due to the isotropy of the medium, $\varepsilon(\mathbf{k}, \omega)$ is a scalar, and we need not consider its tensorial components.

Equation (4.55) can be solved for the dispersion (ω vs. k)

$$\frac{kc}{\omega} = \sqrt{\varepsilon(\mathbf{k}, \omega)} \equiv n(\omega) + i\kappa(\omega).$$

Here n is the *index of refraction* and κ is an *extinction coefficient*. We thus have

$$E(z, t) = E_0 \exp[ik'z - i\omega t - \kappa_a'(\omega)z/2], \tag{4.59}$$

with the renormalized wavevector* k' and the *absorption coefficient* κ_a'

$$k' \equiv \frac{\omega n(\omega)}{c}, \tag{4.60a}$$

$$\kappa_a'(\omega) \equiv \frac{2\omega\kappa(\omega)}{c}. \tag{4.60b}$$

The field intensity $I(z) \equiv |\hat{E}(z, t)|^2$ thus becomes

$$I(z) = I_0 \exp[-\kappa_a'(\omega)z]. \tag{4.61}$$

We now relate $\kappa(\omega)$ to $\chi^{(1)}$. Separating $\chi^{(1)}$ into its real and imaginary parts

$$\chi^{(1)} = \chi' + i\chi''. \tag{4.62}$$

We get

$$\sqrt{1 + 4\pi\chi' + 4\pi i\chi''} \equiv n + i\kappa. \tag{4.63}$$

---

* The fact that the wavevector k' in the medium is different from the vacuum wavevector (ω/c) is known as the extinction theorem or the Ewald Oseen theorem [M. Born and E. Wolf, *Principle of Optics*, 6th ed. (Pergamon, Oxford, 1980)].

Taking the square of the equation and solving for the real and the imaginary parts, we get

$$2n^2 = 1 + 4\pi\chi' + [(1 + 4\pi\chi')^2 + (4\pi\chi'')^2]^{1/2},$$

$$\kappa = 2\pi\chi''/n.$$

For $4\pi\chi'' \ll 1 + 4\pi\chi'$ we get

$$n(\omega) = [1 + 4\pi\chi'(\omega)]^{1/2},$$

$$\kappa(\omega) = 2\pi \frac{\chi''(\omega)}{n(\omega)}.$$

The absorption coefficient is thus related to the imaginary part of $\chi^{(1)}$,

$$\kappa_a'(\omega) = \frac{4\pi\omega}{n(\omega)c} \chi''(\omega). \tag{4.64}$$

For a system of noninteracting absorbers, the absorption is usually expressed in terms of the *absorption cross section* $\kappa_a(\omega)$ defined by

$$I(z) = I_0 \exp[-\kappa_a(\omega)\rho_0 z], \tag{4.65}$$

where ρ_0 is the number density of absorbers. Obviously $\kappa_a(\omega) \equiv \rho_0^{-1}\kappa_a'(\omega)$ has a units of area. For typical dye molecules κ_a is $\sim 0.1\ (\text{nm})^2$. A word of caution: It will be shown in Chapter 17 that optical susceptibilities defined with respect to the transverse field do not include spontaneous emission. Equation (4.65) holds for a thick slab with a high density of molecules where spontaneous emission is absent. For other geometries the solution of the Maxwell equation is more complicated and may not be represented by a single plane wave [Eq. (4.54)].

n-wave mixing

Wave mixing techniques were mentioned in the introduction, and specific examples will be analyzed throughout the book. An $n + 1$-wave mixing process starts with n incoming fields (whether pulsed or stationary) that interact with matter and generate a new field. We shall represent the average electric field corresponding to the incoming modes in the form

$$E(\mathbf{r}, t) = \sum_{j=1}^{n} [E_j(\mathbf{r}, t) \exp(i\mathbf{k}_j\mathbf{r} - i\omega_j t) + E_j^*(\mathbf{r}, t) \exp(-i\mathbf{k}_j\mathbf{r} + i\omega_j t)]. \tag{4.66}$$

To avoid absorptive losses, nonlinear wave mixing experiments are usually performed under conditions where the signal field is outside the absorption spectrum. We can then neglect the imaginary part of $\chi^{(1)}(\omega_s)$, and the dielectric function is real. We further assume that the linear response is local in space for all relevant frequencies so that

$$\varepsilon(\omega_j) \equiv \varepsilon(\mathbf{k} = 0, \omega_j) \equiv n_j^2. \tag{4.67}$$

Here n_j is the refractive index of the medium for the generated light at frequency

$\omega = \omega_j$, and we have

$$k_j \equiv \frac{\omega_j}{c} n_j.$$

The linear polarization was already taken care of by n_j so that the Maxwell equation for the field becomes

$$\nabla \times \nabla \times E(\mathbf{r}, t) + \frac{n^2}{c^2} \frac{\partial^2 E(\mathbf{r}, t)}{\partial t^2} = -\frac{4\pi}{c^2} \frac{\partial^2 P_{\mathrm{NL}}(\mathbf{r}, t)}{\partial t^2}. \tag{4.68}$$

The interaction of the incoming modes with the medium generates the nonlinear polarization $P_{\mathrm{NL}}(\mathbf{r}, t)$, which serves as a source for a new generated field. The microscopic origin of this interaction is the main subject of this book and will be discussed in Chapter 5. For now we simply take it as given, and calculate the optical signal generated by $P_{\mathrm{NL}}(\mathbf{r}, t)$ as a source. The nonlinear polarization will be expanded in the form

$$P_{\mathrm{NL}}(\mathbf{r}, t) = \sum_{n=2,3\ldots} \sum_{\mathrm{s}} P_{\mathrm{s}}^{(n)}(t) \exp(i\mathbf{k}_{\mathrm{s}} \cdot \mathbf{r} - i\omega_{\mathrm{s}} t), \tag{4.69}$$

where $\mathbf{k}_{\mathrm{s}}$ and ω_{s} are any combination of the incoming wavevectors and frequencies

$$\mathbf{k}_{\mathrm{s}} = \pm\mathbf{k}_1 \pm \mathbf{k}_2 \pm \mathbf{k}_3 \cdots \pm \mathbf{k}_n; \quad \omega_{\mathrm{s}} = \pm\omega_1 \pm \omega_2 \pm \omega_3 \cdots \pm \omega_n.$$

We shall consider a geometry where the material occupies a slab $0 < z < l$ with $k_{\mathrm{s}} l \gg 1$. $\mathbf{k}_{\mathrm{s}}$ is taken to be along the z axis. We focus on a single Fourier component of the induced polarization:

$$P_{\mathrm{NL}}(\mathbf{r}, t) = P_{\mathrm{s}}(t) \exp(i\mathbf{k}_{\mathrm{s}} z - i\omega_{\mathrm{s}} t), \tag{4.70}$$

and look for a solution of the form:

$$E(\mathbf{r}, t) = E_{\mathrm{s}}(z, t) \exp(-i\omega_{\mathrm{s}} t + i\mathbf{k}'_{\mathrm{s}} \cdot \mathbf{r}) + c.c., \tag{4.71}$$

with

$$k'_{\mathrm{s}} \equiv \frac{\omega_{\mathrm{s}}}{c} n_{\mathrm{s}}.$$

The difference between k_{s} and k'_{s} is a result of the frequency dispersion of $n(\omega)$ as well as the geometry of the experiment ($\mathbf{k}_{\mathrm{s}}$ is obtained by a vector addition of $\mathbf{k}_n$ whereas ω_{s} is given by a regular sum of ω_n, and for noncollinear geometries $\mathbf{k}_{\mathrm{s}}$ will be different from $\mathbf{k}'_{\mathrm{s}}$). Typically, $\mathbf{k}'_{\mathrm{s}}$ is close to, but not identical to $\mathbf{k}_{\mathrm{s}}$. We further assume that $P_{\mathrm{s}}(t)$ is slowly varying in time compared with the optical period:

$$\left| \frac{\partial}{\partial t} P_{\mathrm{s}}(t) \right| \ll |\omega_{\mathrm{s}} P_{\mathrm{s}}(t)|. \tag{4.72}$$

The spatial dependent part of $E_{\mathrm{s}}(\mathbf{r}, t)$ with the slowly varying temporal envelope satisfies

$$\nabla \times \nabla \times E_{\mathrm{s}}(\mathbf{r}, t) - \frac{n_{\mathrm{s}}^2 \omega_{\mathrm{s}}^2}{c^2} E_{\mathrm{s}}(\mathbf{r}, t) = \frac{4\pi\omega_{\mathrm{s}}^2}{c^2} P_{\mathrm{s}}(t) \exp(i\mathbf{k}_{\mathrm{s}} \cdot \mathbf{r}). \tag{4.73}$$

Using the slowly varying amplitude for $E_s(z, t)$ we get

$$ik'_s \frac{\partial E_s(z)}{\partial z} = -2\pi \frac{\omega_s^2}{c^2} P_s(t) \exp(i\Delta\mathbf{k}\cdot z) \tag{4.74}$$

with $\Delta\mathbf{k} \equiv \mathbf{k}_s - \mathbf{k}'_s$ is the difference between the combination of incoming wavevectors and the wavevector of the generated wave.

The generated light in this direction grows from zero intensity at the beginning of the illuminated region, $z = 0$, and its growth stops at the slab interface $z = l$. The solution to the wave equation (4.74) can be obtained by integrating both sides over z from 0 to l, and has the following form:

$$E_s(l, t) = \frac{2\pi i}{n(\omega_s)} \frac{\omega_s}{c} l P_s(t) \operatorname{sinc}(\Delta k l/2) \exp(i\Delta k l/2), \tag{4.75}$$

where

$$\operatorname{sinc} x \equiv \frac{\sin x}{x}. \tag{4.76}$$

The sinc function (displayed in Figure 4.2) is well localized. For $l \to \infty$ we have $\operatorname{sinc}(\Delta k l/2) \to (2\pi)^3 \delta(\Delta k l)$. The $\operatorname{sinc}(\Delta k l/2)$ function is large only when $\Delta k \cdot l \ll \pi$. This is known as *the phase matching condition.* In some techniques such as degenerate four wave mixing, this condition is automatically satisfied. In other cases it limits the effective interaction length l and thus the magnitude of the signal. The resulting signal is finally given by

$$I_s(t) = \frac{n_s c}{8\pi} |E_s|^2 = \frac{\pi}{2n_s} \frac{\omega_s^2}{c} l^2 |P_s(t)|^2 \operatorname{sinc}^2[(\Delta\mathbf{k}\cdot l/2)]. \tag{4.77}$$

The present expression applies to a simple slab geometry and assumes that the generated signal field is very weak compared with the incoming fields $E_s(l) \ll E_j$, so that the incoming fields are unaffected by the wave mixing process, and are assumed to be given. The propagation problem thus becomes linear and

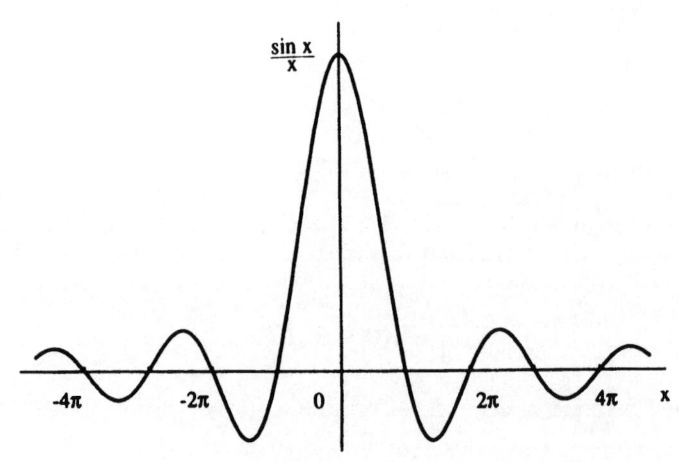

FIG. 4.2 The sinc function, which determines the phase matching [Eq. (4.76)].

may be solved exactly. For optically thick samples, when absorption is incorporated, and for other geometries, one may need to solve the Maxwell–Liouville equations simultaneously for the nonlinear polarization and for the generated fields. In these cases propagation and internal dynamics responsible for the polarization cannot be separated so conveniently. This may complicate the interpretation of optical measurements and gives rise to other fascinating effects such as optical bistability. The role of spontaneous emission and radiative widths in these measurements will be treated in Chapter 17.

Heterodyne detection of wave mixing

The nonlinear wave mixing signal calculated above is proportional to $|P_s(t)|^2$. An important advantage of wave mixing is that the signal can be generated in a new direction, different from the incoming fields. Therefore, unlike other techniques such as absorption spectroscopy, wave mixing have essentially no background, which makes the detection much more sensitive. Despite this advantage, typically the nonlinear polarization and the signal are very weak. An improved sensitivity and signal to noise ratio may be achieved by using a *heterodyne detection*. In this configuration we add a new field denoted the *local oscillator* $E_{\mathrm{LO}}(t)$, which has the same wavevector as the signal field. The term "local oscillator" is derived from radiowaves and is related to an oscillator located in the receiver of the signal. This terminology is somewhat confusing in laser spectroscopy since that local-oscillator is just another light beam, which is no more local than the others.

After the nonlinear process is over, we observe the superposition of the local oscillator field and the signal field. The total detected intensity is then [21]

$$I_{\mathrm{T}}(t) = \frac{n(\omega_s)c}{4\pi}|E_{\mathrm{LO}}(t) + E_{\mathrm{S}}(t)|^2 = I_{\mathrm{LO}}(t) + I_{\mathrm{S}}(t) + 2\frac{n(\omega_s)c}{4\pi}\,\mathrm{Re}[E^*_{\mathrm{LO}}(t)\cdot E_{\mathrm{S}}(t)]. \tag{4.78}$$

Here I_{LO} is the intensity of the local oscillator field and I_{S} is the signal derived earlier. Usually $E_{\mathrm{LO}} \gg E_{\mathrm{S}}$ (the signal is much weaker than the applied local oscillator field), and the second term is negligible compared with the third. The local oscillator intensity $I_{\mathrm{LO}}(t)$ can be easily subtracted, resulting in the heterodyne signal

$$I_{\mathrm{HET}}(t) = [n(\omega_s)c/4\pi]\,\mathrm{Re}[E^*_{\mathrm{LO}}(t)\cdot E_{\mathrm{S}}(t)]. \tag{4.79}$$

A few important points to note are

1. The heterodyne signal is linear rather than quadratic in the (weak) nonlinear polarization $P_s(t)$. It is therefore much stronger and easier to detect.
2. Although the polarization $P(\mathbf{r}, t)$ is real, this is no longer the case in $\mathbf{k}$ space, and $P_s(t)$ is in general complex. By controlling the relative phase of E_{LO} and $P_s(t)$ it may become possible to probe separately the real and the imaginary parts of $P_s(t)$, resulting in a *phase-sensitive detection*. This important aspect will be illustrated in Chapters 10 and 11.
3. For some techniques, the heterodyne signal offers a much simpler interpretation. For example, in photon echo spectroscopy, the selective elimination of inhomogeneous broadening is more general and complete (see Chapter 10).

4. When the signal field is generated along the direction of one of the incoming beams, we have an intrinsic heterodyne detection. Examples are pump-probe absorption (see below), stimulated Raman, and the accumulated photon echo (Chapter 10). The signal further has the same frequency of the carrier wave. Equation (4.79) thus holds with $\omega_s = \omega_{\mathrm{LO}}$ and E_{LO} is one of the fields. Making use of Eq. (4.75), we obtain

$$I(t) = \frac{n(\omega_s)c}{4\pi} \mathrm{Re}[E_2^*(t)\cdot E_S(t)] \propto -(\omega_2 l)\, \mathrm{Im}[E_2^*(t)P_s(t)]. \tag{4.80}$$

Absorption of a weak probe

We shall now calculate the absorption of a weak probe pulse in an optically thin medium following a nonlinear excitation by one (or more) laser pulses. The simplest example is a pump-probe experiment where the system is subjected to two light pulses: the pump pulse and the probe pulse. The following derivation is not limited however to this case, and we can consider an arbitrary excitation even with strong pulses, as long as the probe is weak.

The calculation will be carried out in two ways, which give identical results, although from a different perspective, by focusing either on the material system or on the field.

The matter perspective. The energy of the matter is given by the expectation value of the Hamiltonian

$$\langle W(t)\rangle \equiv \mathrm{Tr}[\hat{H}_{sc}(t)\rho(t)],$$

where the semiclassical Hamiltonian is given by Eq. (4.47b). The rate of absorption of energy is then

$$\frac{d\langle W(t)\rangle}{dt} = \mathrm{Tr}\left[\frac{d\hat{H}_{sc}(t)}{dt}\rho(t)\right] + \mathrm{Tr}\left[\hat{H}_{sc}(t)\frac{d\rho(t)}{dt}\right].$$

It can be easily verified that the second term vanishes identically

$$\mathrm{Tr}\left[\hat{H}_{sc}(t)\frac{d}{dt}\rho(t)\right] = -\frac{i}{\hbar}\mathrm{Tr}\,\hat{H}_{sc}(t)[\hat{H}_{sc}(t), \rho(t)]$$
$$= -\frac{i}{\hbar}\mathrm{Tr}[\hat{H}_{sc}^2(t)\rho(t) - \hat{H}_{sc}(t)\rho(t)\hat{H}_{sc}(t)] = 0.$$

The last equality follows from the cyclic invariance of the trace, which causes the two terms to cancel. Combining this with Eq. (4.47b) we have

$$\frac{d\langle W(t)\rangle}{dt} = -\int d\mathbf{r}\,\frac{\partial E(\mathbf{r},t)}{\partial t}P(\mathbf{r},t). \tag{4.81}$$

We assume that the nonlinear excitation produces a nonlinear polarization $P_s(t)$ with wavevector $\mathbf{k}_s$ and frequency ω_s [Eq. (4.70)]. We shall represent the probe field by a single mode in the expansion [Eq. (4.66)] and denote its wavevector, frequency, and temporal profile, $\mathbf{k}_j$, ω_j, and $E(\mathbf{k}_j, t)$, respectively. Invoking the slowly varying amplitude approximation, we assume that the probe field envelope varies slowly in time compared with its optical period, i.e.,

$$\left| \frac{\partial}{\partial t} E_j(t) \right| \ll |\omega_j E_j(t)|. \tag{4.82}$$

In this case we have

$$\frac{\partial}{\partial t} E_j(\mathbf{r}, t) \approx -i\omega_j[E_j(t)\exp(i\mathbf{k}_j\mathbf{r} - i\omega_j t) - E_j^*(t)\exp(-i\mathbf{k}_j\mathbf{r} + i\omega_j t)].$$

Substituting this into (4.81), we obtain

$$\frac{d\langle W(t)\rangle}{dt} = 2\omega_j \operatorname{Im} \sum_s \{E_j^*(t)P_j(t)\exp[-i(\mathbf{k}_j - \mathbf{k}_s)\mathbf{r} + i(\omega_j - \omega_s)t]$$
$$- E_j(t)P_j(t)\exp[i(\mathbf{k}_j + \mathbf{k}_s)\mathbf{r} - i(\omega_j + \omega_s)t]\}. \tag{4.83}$$

The total energy loss of the transmitted probe field is obtained by taking the average of Eq. (4.83) over an optical cycle, followed by an integration over t. The only surviving contribution is from the first term with $\omega_s = \omega_j$ and $\mathbf{k}_s = \mathbf{k}_j$.

$$S_A(\omega_j, t) = 2\omega_j \operatorname{Im}\lfloor E_j^*(t)P_j(t)\rfloor,$$

which can be recast in the form

$$S_A(\omega_j, t) = 2|E_j(t)|^2\omega_j \operatorname{Im}[P_j(t)/E_j(t)]. \tag{4.84}$$

Integrating over time and dividing by the total incoming energy*

$$\frac{cn(\omega_j)}{2\pi}\int dt\, |E_j(t)|^2,$$

we finally obtain the probe absorption signal

$$S_A(\omega_j) = \frac{4\pi\omega_j}{cn(\omega_j)} \operatorname{Im} \int_{-\infty}^{\infty} dt\, E_j^*(t)P_j(t) \Big/ \int_{-\infty}^{\infty} dt\, |E_j(t)|^2. \tag{4.85}$$

The field perspective. We can derive these results in an alternative way, by looking at the field. Consider a field propagating along the z direction. We shall expand the field and the nonlinear polarization in the form

$$E(z, t) = \varepsilon_j(z, t)\exp(-i\omega_j t + ik_j z) + c.c.,$$

$$P(z, t) = P_j(z, t)\exp(-i\omega_j t + ik_j z) + c.c.$$

* The index of refraction takes into account the effects of reflection, i.e., the energy entering the medium is different from the incident energy by this factor.

In the slowly varying amplitude approximation [Eq. (4.74)] we have

$$\frac{\partial}{\partial z}\varepsilon_j(z,t) = i\frac{2\pi\omega_j}{cn(\omega_j)}P_j(z,t). \tag{4.86}$$

We next represent the field in terms of its amplitude and phase

$$\varepsilon_j(z,t) = \varepsilon_0(z,t)\exp[i\phi(z,t)],$$

and define the field intensity

$$I(z,t) \equiv \frac{c}{8\pi}|\varepsilon_0(z,t)|^2.$$

Substituting these in Eq. (4.86) and multiplying by $\varepsilon_j^*(z,t)$ we get

$$\varepsilon_0\frac{\partial}{\partial z}\varepsilon_0 + i\varepsilon_0^2\frac{\partial}{\partial z}\phi = i\frac{2\pi\omega}{cn(\omega)}\varepsilon_j^*(z,t)P_j(z,t).$$

The real and the imaginary parts of this equation result in the following coupled equations for I and ϕ

$$\frac{\partial I}{\partial z} = -I\frac{4\pi\omega_j}{cn(\omega_j)}\mathrm{Im}[P_j(z,t)/\varepsilon_j(z,t)], \tag{4.87a}$$

$$\frac{\partial\phi}{\partial z} = \frac{2\pi\omega_j}{cn(\omega_j)}\mathrm{Re}[P_j(z,t)/\varepsilon_j(z,t)]. \tag{4.87b}$$

Equation (4.87a) generalizes Eq. (4.64) to allow for a time-dependent field. For slowly varying fields we then get the absorption coefficient.

$$\kappa_a'(\omega_j) = \frac{4\pi\omega_j}{cn(\omega_j)}Im[P_j(t)/\varepsilon_j(t)],$$

which is identical to Eq. (4.84). It also agrees with Eq. (4.64) for linear absorption.

To gain some additional physical insight regarding the significance of Eq. (4.85), let us represent the probe field and the polarization in the form

$$E_j(t) = 2E_0(t)\cos\omega t = E_0(t)\exp(i\omega t) + c.c., \tag{4.88a}$$

and

$$P_j(t) = 2P_0(t)\cos(\omega t + \varphi) = P_0(t)\exp(i\omega t + i\varphi) + c.c., \tag{4.88b}$$

where $P_0(t)$ and $E_0(t)$ are real.

$P_j(t)$, which can have an arbitrary phase with respect to $E(t)$, can be written as a sum of two contributions that are in-phase and out of phase with respect to the field, respectively.

$$P_j(t) = [P_0(t)\cos\varphi]\cos\omega t - [P_0(t)\sin\varphi]\sin\omega t.$$

Upon the substitution of Eqs. (4.88) in Eq. (4.85) we obtain

$$S_A(\omega_j) \propto -\left[\int_{-\infty}^{\infty} E_0(t)P_0(t)\,dt\right]\sin\phi.$$

The probe absorption is thus related to the out-of-phase component of the polarization. It can also be interpreted in terms of a heterodyne detection of the field by itself (self-action).

Finally, we note that the signal may be alternatively obtained by an improved detection scheme, in which the transmitted probe field is dispersed through a monochrometer and the signal is measured as a function of the dispersed frequency ω. The formal expression for the resulting signal S_{disp} may be obtained by solving the Maxwell equations using the polarization field $P^{(3)}$ as a source. It can also be derived by performing a Fourier transform of Eq. (4.85) resulting in

$$S_{\text{A}}(\omega_j) = \int_{-\infty}^{\infty} d\omega \, S_{\text{disp}}(\omega), \tag{4.89}$$

with

$$S_{\text{disp}}(\omega) = \frac{4\pi\omega_j}{cn(\omega)} \frac{\text{Im}\, \tilde{E}_j^*(\omega)\tilde{P}_j(\omega)}{\int d\omega \, |\tilde{E}_j^*(\omega)|^2}. \tag{4.90}$$

Here $\tilde{E}_j(\omega)$ is the probe field amplitude,

$$\tilde{E}_j(\omega) = \int_{-\infty}^{\infty} dt \exp(i\omega t) E_j(t). \tag{4.91}$$

and $\tilde{P}_j^{(3)}(\omega)$ is the ω Fourier component of the third-order polarization, defined in a similar fashion. This equation can also be written as

$$S_{\text{disp}}(\omega) = \frac{4\pi\omega_j}{cn(\omega)} \frac{|\tilde{E}_j(\omega)|^2 \, \text{Im}[\tilde{P}_j(\omega)/\tilde{E}_j(\omega)]}{\int d\omega \, |\tilde{E}_j(\omega)|^2}. \tag{4.92}$$

APPENDIX 4A

Transverse and Longitudinal Vector Fields

Below we present a few definitions and useful relations satisfied by vector fields [4]. The gradient of an ordinary (scalar) function ϕ is

$$\nabla\phi = \frac{\partial\phi}{\partial x}\mathbf{e}_x + \frac{\partial\phi}{\partial y}\mathbf{e}_y + \frac{\partial\phi}{\partial z}\mathbf{e}_z, \tag{4.A.1}$$

$\mathbf{e}_j$ is a unit vector along the j axis $j = x, y, z$.

A vector field $\hat{F}(\mathbf{r})$ is a function that assigns a vector to every point in space. Examples are the electric and the magnetic fields, the electrical current, etc. The divergence of $\hat{F}$, $\nabla\cdot\hat{F}$ is a scalar, defined as

$$\nabla\cdot\hat{F} \equiv \frac{\partial\hat{F}_x}{\partial x} + \frac{\partial\hat{F}_y}{\partial y} + \frac{\partial\hat{F}_z}{\partial z}. \tag{4.A.2}$$

The rotor (curl) of $\hat{F}$, $\nabla\times\hat{F}$ is a vector field defined by the determinant

$$\nabla\times\hat{F} = \begin{vmatrix} \mathbf{e}_x & \mathbf{e}_y & \mathbf{e}_z \\ \dfrac{\partial}{\partial x} & \dfrac{\partial}{\partial y} & \dfrac{\partial}{\partial z} \\ \hat{F}_x & \hat{F}_y & \hat{F}_z \end{vmatrix}$$

$$= \mathbf{e}_x\left(\frac{\partial\hat{F}_z}{\partial y} - \frac{\partial\hat{F}_y}{\partial z}\right) - \mathbf{e}_y\left(\frac{\partial\hat{F}_z}{\partial x} - \frac{\partial\hat{F}_x}{\partial z}\right) + \mathbf{e}_z\left(\frac{\partial\hat{F}_y}{\partial x} - \frac{\partial\hat{F}_x}{\partial y}\right). \tag{4.A.3}$$

Some useful relations of vector calculus are

$$\nabla\times(\nabla\times\hat{F}) = \nabla(\nabla\cdot\hat{F}) - \nabla^2\hat{F} \tag{4.A.4}$$

$$\nabla\cdot(\phi\hat{F}) = \hat{F}\nabla\phi + \phi\nabla\cdot\hat{F} \tag{4.A.5}$$

$$\nabla\cdot(\nabla\times\hat{F}) = 0 \tag{4.A.6}$$

$$\nabla\times(\nabla\phi) = 0.$$

The Helmholtz theorem states that any vector field can be written as a sum of two components, one with zero divergence and the other with zero curl.

$$\hat{F} = \hat{F}^{\perp} + \hat{F}^{\parallel}, \tag{4.A.7}$$

with

$$\nabla \cdot \hat{F}^{\perp} = 0, \tag{4.A.8a}$$

and

$$\nabla \times \hat{F}^{\parallel} = 0. \tag{4.A.8b}$$

$\hat{F}^{\perp}$ is the *transverse* part of $\hat{F}$, whereas $\hat{F}^{\parallel}$ is denoted the *longitudinal* part of the vector field. In real space these components are given by

$$\hat{F}_i^{\perp}(\mathbf{r}) = \int \hat{F}_j(\mathbf{r}')\delta_{ij}^{\perp}(\mathbf{r} - \mathbf{r}')\, d\mathbf{r}' \tag{4.A.9a}$$

$$\hat{F}_i^{\parallel}(\mathbf{r}) = \int \hat{F}_j(\mathbf{r}')\delta_{ij}^{\parallel}(\mathbf{r} - \mathbf{r}')\, d\mathbf{r}', \tag{4.A.9b}$$

with the transverse and longitudinal δ-dyadics given by

$$\delta_{ij}^{\perp}(\mathbf{r} - \mathbf{r}') = \tfrac{2}{3}\delta_{ij}\delta(\mathbf{r} - \mathbf{r}') - \frac{T_{ij}(\mathbf{r} - \mathbf{r}')}{4\pi} \tag{4.A.10a}$$

$$\delta_{ij}^{\parallel}(\mathbf{r} - \mathbf{r}') = \tfrac{1}{3}\delta_{ij}\delta(\mathbf{r} - \mathbf{r}') + \frac{T_{ij}(\mathbf{r} - \mathbf{r}')}{4\pi}. \tag{4.A.10b}$$

$\delta(\mathbf{r} - \mathbf{r}')$ is the Dirac δ-function, and the tensor T is

$$T_{ij}(\mathbf{r}) = \frac{\delta_{ij} - 3\hat{r}_i\hat{r}_j}{r^3}. \tag{4.A.11}$$

Here $\hat{\mathbf{r}}$ is a unit vector in the direction of $\mathbf{r}$ and $r \equiv |\mathbf{r}|$. Note that the transverse and the longitudinal components are related to the original vector field by a *nonlocal* transformation. This implies that the transverse and the longitudinal components of a vector field at a given point in space depend on the actual field everywhere.

Equations (4.A.9) can also be written in the form

$$\hat{F}^{\perp}(\mathbf{r}) = \frac{1}{4\pi}\nabla \times \int d\mathbf{r}' \frac{\nabla' \times \hat{F}(\mathbf{r}')}{|\mathbf{r} - \mathbf{r}'|}, \tag{4.A.12a}$$

$$\hat{F}^{\parallel}(\mathbf{r}) = -\frac{1}{4\pi}\nabla \int d\mathbf{r}' \frac{\nabla' \cdot \hat{F}(\mathbf{r}')}{|\mathbf{r} - \mathbf{r}'|}. \tag{4.A.12b}$$

In momentum space we may resolve a vector field along the direction of the wavevector [the longitudinal component $\hat{F}^{\parallel}(\mathbf{k})$], and transverse to $\mathbf{k}$, $\hat{F}^{\perp}(\mathbf{k})$:

$$\hat{F}_i^{\parallel}(\mathbf{k}) = \hat{k}_i\hat{k}_j\hat{F}_j(\mathbf{k}) \tag{4.A.13a}$$

$$\hat{F}_i^{\perp}(\mathbf{k}) = (\delta_{ij} - \hat{k}_i\hat{k}_j)\hat{F}_j(\mathbf{k}) \tag{4.A.13b}$$

Equations (4.A.9), (4.A.10), and (4.A.13) are related by spatial Fourier transforms.

APPENDIX 4B

The Multipolar Expansion of the Polarization and Magnetization

We have shown that an atom or a molecule may be modeled as a polarization field $\hat{P}(\mathbf{r})$ distributed over space. In the multipolar expansion [1–4] we expand this field in terms of its moments. By retaining only the lowest order terms we may obtain useful and greatly simplified expressions for the optical response. This is done by first performing the formal expansion

$$\delta[\mathbf{r} - \mathbf{R}_m - u(\hat{\mathbf{r}}_{m\alpha} - \mathbf{R}_m)]$$
$$= \{1 - u(\hat{\mathbf{r}}_{m\alpha} - \mathbf{R}_m)\cdot\nabla_r + \tfrac{1}{2}[u(\hat{\mathbf{r}}_{m\alpha} - \mathbf{R}_m)\cdot\nabla_r]^2 + \cdots\}\delta(\mathbf{r} - \mathbf{R}_m), \quad (4.B.1)$$

(which is the Taylor expansion of the Dirac δ function). This expansion can be understood as follows: Let us consider an arbitrary function $\hat{F}(r)$. Multiplying both sides and integrating over r, the left-hand side gives

$$\int d\mathbf{r}\ \hat{F}(\mathbf{r})\delta[\mathbf{r} - \mathbf{R}_m - u(\hat{\mathbf{r}}_{m\alpha} - \mathbf{R}_m)] = \hat{F}[\mathbf{R}_m + u(\hat{\mathbf{r}}_{m\alpha} - \mathbf{R}_m)]. \quad (4.B.2)$$

The right-hand side gives

$$F(\mathbf{R}_m) + F'(\mathbf{R}_m)[\hat{u}(\hat{\mathbf{r}}_{m\alpha} - \mathbf{R}_m)] + \tfrac{1}{2}F''(\mathbf{R}_m)[u(\mathbf{r}_{m\alpha} - \mathbf{R}_m)]^2 + \cdots, \quad (4.B.3)$$

which is simply the Taylor series of the left-hand side. This proves Eq. (4.B.1).

Upon the substitution of the expansion (4.B.1) in Eqs. (4.18c) and (4.22b), the integration over the variable u can be carried out, resulting in the multipolar series for the polarization and magnetization

$$\mathscr{P}_m(\mathbf{r}) = (\hat{\mathscr{P}}_m^{\mathrm{D}}(\mathbf{r}) - \hat{\mathscr{P}}_m^{\mathrm{Q}}(\mathbf{r})\cdot\nabla + \cdots)\delta(\mathbf{r} - \mathbf{R}_m), \quad (4.B.4)$$

$$\hat{\mathscr{M}}_m(\mathbf{r}) = (\hat{\mathscr{M}}_m^{\mathrm{D}}(\mathbf{r}) - \hat{\mathscr{M}}_m^{\mathrm{Q}}(\mathbf{r})\cdot\nabla + \cdots)\delta(\mathbf{r} - \mathbf{R}_m), \quad (4.B.5)$$

with

$$\hat{\mathscr{P}}_m^{\mathrm{D}}(\mathbf{r}) = \sum_\alpha q_{m\alpha}(\hat{\mathbf{r}}_{m\alpha} - \mathbf{R}_m)\delta(\mathbf{r} - \mathbf{R}_m), \qquad \text{electric dipole}$$

$$\hat{\mathscr{P}}_m^{\mathrm{Q}}(\mathbf{r}) = \frac{1}{2}\sum_\alpha q_{m_\alpha}(\hat{\mathbf{r}}_{m_\alpha} - \mathbf{R}_m) \otimes (\hat{\mathbf{r}}_{m_\alpha} - \mathbf{R}_m)\delta(\mathbf{r} - \mathbf{R}_m), \quad \text{electric quadrupole}$$

$$\hat{\mathscr{M}}_m^{\mathrm{D}}(\mathbf{r}) = \sum_\alpha \frac{1}{2m_\alpha c} q_{m\alpha}(\hat{\mathbf{r}}_{m\alpha} - \mathbf{R}_m) \times \hat{\mathbf{p}}_{m\alpha}\delta(\mathbf{r} - \mathbf{R}_m), \qquad \text{magnetic dipole}$$

$$\hat{\mathscr{M}}_m^{\mathrm{Q}}(\mathbf{r}) =$$

$$\sum_\alpha \frac{1}{6m_\alpha c} q_{m\alpha}\{[\hat{\mathbf{r}}_{m\alpha} - \mathbf{R}_m) \times \hat{\mathbf{p}}_{m\alpha}] \otimes (\hat{\mathbf{r}}_{m\alpha} - \mathbf{R}_m) + (\hat{\mathbf{r}}_{m\alpha} - \mathbf{R}_m) \otimes [(\hat{\mathbf{r}}_{m\alpha} - \mathbf{R}_m) \times \hat{\mathbf{p}}_{m\alpha}]\}.$$

magnetic quadrupole (4.B.6)

Here $\otimes$ denotes the tensor product.

Most commonly we shall invoke the *dipole approximation* where we keep only the first electric term in the multipolar expansions. We then write

$$\hat{P}(\mathbf{r}) = \sum_m \mu_m \delta(\mathbf{r} - \mathbf{R}_m),$$

where we have introduced the dipole moment operator for the mth molecule

$$\mu_m \equiv \sum_\alpha q_{m\alpha}(\hat{\mathbf{r}}_{m\alpha} - \mathbf{R}_m). \tag{4.B.7}$$

What is the expansion parameter in the multipolar expansion? If the system interacts with a field with wavevector $\mathbf{k}$, then successive terms will be multiplied by a factor $\mathbf{k}\cdot\mathbf{r}_\alpha$. This shows that the expansion parameter is a/λ, a being the molecular size and λ the optical wavelength. As for the magnetic terms, we note that the magnetic dipole is the same order in a/λ as the electric quadrupole. However, it follows directly from the above definitions that $\mathscr{M}^{\mathrm{D}}/\mathscr{P}^{\mathrm{Q}} \equiv v/c$, v being the electron velocity and c the speed of light. This is the smallness parameter that allows us to neglect the magnetic terms in many materials. The other magnetic term [the third term in Eq. (4.35)] does not contain the velocity explicitly, but the appearance of c^{-2} suggests that it is of the order of v^2/c^2. Indeed, it is possible to estimate its contribution to optical response functions and show that it is also proportional to this smallness parameter, which again justifies neglecting it in many cases. We have thus shown that there are two expansion parameters used in the dipole approximations, a/λ and v/c. The fine structure constant $\alpha \equiv e^2/\hbar c$ is the natural expansion parameter in the electrodynamics of free electrons. It is interesting to note that for the hydrogen atom we have $\alpha \sim a/\lambda \sim v/c$ so that all of these parameters coincide. This is not, however, the case in general, and for systems other than atomic hydrogen, these are three distinct parameters, not necessarily related in a simple way.

APPENDIX 4C

Green Function Solution of the Maxwell Equation

When the polarization $P(\mathbf{r}, t)$ is given, the Maxwell equation is linear and inhomogeneous and can be solved exactly in a closed form [22]. The general solution of Eq. (4.42) is

$$\hat{E}(\mathbf{r}, t) = \hat{E}_{ext}(\mathbf{r}, t) + \int_{-\infty}^{t} dt' \int_{0}^{\Omega} d\mathbf{r}' \, \mathscr{G}(\mathbf{r} - \mathbf{r}'; t - t') \cdot \hat{P}(\mathbf{r}'; t'), \qquad (4.C.1)$$

with the Green function

$$\mathscr{G}(\mathbf{r} - \mathbf{r}'; t - t') = \int_{-\infty}^{\infty} d\omega \, \mathscr{G}(\mathbf{r} - \mathbf{r}'; \omega) \exp(-i\omega|t - t'|), \qquad (4.C.2)$$

and

$$\mathscr{G}_{ij}(\mathbf{r} - \mathbf{r}'; \omega) = \left(\nabla_i \nabla_j + \frac{\omega^2}{c^2} \delta_{ij} \right) \frac{\exp(i\omega|\mathbf{r} - \mathbf{r}'|/c)}{|\mathbf{r} - \mathbf{r}'|}. \qquad (4.C.3)$$

Here the spatial integration runs over the volume Ω, which includes all the polarization (outside Ω it vanishes), and $\hat{E}_{ext}(\mathbf{r}, t)$ is an external field that, by definition, comes from sources not explicitly included in the polarization $\hat{P}$. The electric field given by Eq. (4.C.1) is defined at all points $\mathbf{r}$. The integration in Eq. (4.C.1) has a singularity at the origin $\mathbf{r} = \mathbf{r}'$. In practice it is then necessary to exclude a small sphere around the origin with a radius ε. The contribution of that sphere can then be evaluated exactly (we take the radius to be sufficiently small that the polarization does not vary within the sphere). We can then recast Eq. (4.C.1) in the form

$$\hat{E}(\mathbf{r}, t) = \hat{E}_{ext}(\mathbf{r}, t) + \int_{-\infty}^{t} dt' \int_{\varepsilon(\mathbf{r})}^{\Omega} d\mathbf{r}' \, \mathscr{G}(\mathbf{r} - \mathbf{r}'; t - t') \cdot \hat{P}(\mathbf{r}'; t') - \frac{4\pi}{3} \hat{P}(\mathbf{r}; t),$$

Within the extent of each charge distribution, the contribution to the field is given by the polarization (the final term). The region of integration in Eq. (4.C.1) thus excludes the particular polarization to which $\mathbf{r}$ belongs and hence the second term on the right-hand side gives the field due to all the other molecules. When $\mathbf{r}$ lies outside the region of the polarization, the final term vanishes.

We next partition the total electric field into its transverse and longitudinal components, each defined at all points $\mathbf{r}$, such that

$$\hat{E}^{\perp}(\mathbf{r}, t) = \hat{E}(\mathbf{r}, t) - \hat{E}^{\parallel}(\mathbf{r}, t), \qquad (4.C.4)$$

where the longitudinal part is given by

$$\hat{E}^{\parallel}(\mathbf{r}, t) = -\frac{4\pi}{3}\hat{P}(\mathbf{r}; t) - \int_{\varepsilon(\mathbf{r})}^{\Omega} d\mathbf{r}'\, T(\mathbf{r} - \mathbf{r}')\hat{P}(\mathbf{r}'; t), \tag{4.C.5}$$

which follows using Eq. (4.A.10b). In Eq. (4.C.5) the longitudinal field at point $\mathbf{r}$ has contributions from the local electric polarization and the electrostatic field due to the remaining dipoles outside the region of polarization containing $\mathbf{r}$.

In $\mathbf{k}$, ω space the Green function solution of the Maxwell equations assumes the form

$$\hat{E}(\mathbf{k}, \omega) = \hat{E}_{\text{ext}}(\mathbf{k}, \omega) + \mathscr{G}(\mathbf{k}, \omega)\hat{P}(\mathbf{k}, \omega), \tag{4.C.6}$$

where

$$\mathscr{G}_{ij}(\mathbf{k}, \omega) = 4\pi\,\frac{\delta_{ij}\omega^2 - c^2 k_i k_j}{\omega^2 - c^2|\mathbf{k}|^2}. \tag{4.C.7}$$

NOTES

1. P. Mazur, *Adv. Chem. Phys.* **1**, 309 (1958).
2. S. R. deGroot, *Foundations of Electrodynamics* (North-Holland, Amsterdam, 1972).
3. L. D. Landau and E. M. Lifshitz, *Electrodynamics of Continuous Media*, 2nd ed. (Pergamon, Oxford, 1984).
4. J. D. Jackson, *Classical Electrodynamics*, 2nd ed. (Wiley, New York, 1975).
5. R. Loudon, *The Quantum Theory of Light*, 2nd ed. (Clarendon Press, Oxford, 1983).
6. C. Cohen-Tannoudji, I. DuPont-Roc, and G. Grynberg, *Photons and Atoms: Introduction to Quantum Electrodynamics* (Wiley, New York, 1989).
7. C. Cohen-Tannoudji, J. Dupont-Roc, and G. Grynberg, *Atom-Photon Interactions* (Wiley, New York, 1992).
8. P. W. Milonni, *Phys. Rep.* **25**, 1 (1976).
9. M. Babiker, E. A. Power, and T. Thirunamachandran, *Proc. R. Soc. London* **A338**, 235 (1974); W. P. Healy, *J. Phys.* **A10**, 279 (1977).
10. R. G. Woolley, *Proc. R. Soc. London* **A321**, 557 (1971).
11. E. A. Power and T. Thirunamachandran, *Mathematika* **18**, 240 (1971).
12. M. Göppert-Mayer, *Ann. Phys. (Leipzig)* **9**, 273 (1931).
13. E. A. Power and S. Zienau, *Phil. Trans. R. Soc. London* **A251**, 427 (1959).
14. E. A. Power, *Introductory Quantum Electrodynamics* (Longmans, London, 1964); D. P. Craig, and T. Thirunamachandran, *Molecular Quantum Electrodynamics* (Academic Press, London, 1984); E. A. Power, *Adv. Chem. Phys.* **12**, 167 (1967).
15. W. P. Healy, *Phys. Rev. A* **22**, 2891 (1980); *Nonrelativistic Quantum Electrodynamics* (Academic Press, New York, 1982).
16. H. C. Longuett-Higgins, *Proc. R. Soc. (London)* **A259**, 433 (1960); *Disc. Faraday Soc.* **40**, 7 (1965).
17. J. Knoester and S. Mukamel, *Phys. Rev. A* **41**, 3812 (1990); J. Jenkins and S. Mukamel, *J. Chem. Phys.* **98**, 7046 (1993).
18. S. R. deGroot, *The Maxwell Equations* (North-Holland, Amsterdam, 1969).
19. M. Babiker, E. A. Power, and T. Thirunamachandran, *Proc. R. Soc. (London)* **A332**, 187 (1973).
20. P. Meystre and M Sargent III, *Elements of Quantum Optics* (Springer-Verlag, Berlin, 1990); G. S. Agarwal and R. W. Boyd, *Phys. Rev. A* **38**, 4019 (1988); M. Sargent III, D. A. Holm, and M. S. Zubairy, *Phys. Rev. A* **31**, 3112 (1985).
21. M. D. Levenson and S. S. Kano, *Introduction to Nonlinear Laser Spectroscopy* (Academic Press, New York, 1988).
22. J. Van Kranendonk and J. E. Sipe, in *Progress in Optics XV*, E. Wolf. Ed. (North-Holland, Amsterdam, 1977), p. 245.

BIBLIOGRAPHY

M. Gross and S. Haroche, *Phys. Rep.* **93**, 301 (1982).
R. Glauber, in *Proceedings of the International School of Physics "Enrico Fermi,"* Course **XL 11** (Quantum Optics) (Academic Press, New York, 1969).
W. Heitler, *The Quantum Theory of Radiation*, 3rd ed. (Oxford University Press, London, 1954).
H. A. Lorentz, *The Theory of Electrons* (Leipzig, Teubner, 1916) (reprinted, Dover, New York, 1951).
L. Mandel and E. Wolf, *Optical Coherence and Quantum Optics* (Cambridge University Press, London, 1995).
H. Margenau and N. R. Kestner, *Theory of Intermolecular Forces*, 2nd ed. (Pergamon, New York, 1971).
J. J. Sakurai, *Advanced Quantum Mechanics* (Addison-Wesley, Reading, Massachusetts, 1978).
J. Schwinger, Ed., *Selected Papers on Quantum Electrodynamics* (Dover, New York, 1958).
A. E. Siegman, *Lasers* (University Science Books, Mill Valley, California, 1986).
A. Sommerfeld, *Lectures on Theoretical Physics IV: Optics* (Academic Press, New York, 1964).

Additional reading on the multipolar Hamiltonian

D. L. Andrews, D. P. Craig, and T. Thirunamachandran, *Int. Rev. Phys. Chem.* **8**, 37 (1989).
W. A. Peticolas, *Ann. Rev. Phys. Chem.* **18**, 233 (1967).
E. A. Power and T. Thirunamachandran, *Am. J. Phys.* **46**, 370 (1978).
S. Stenholm *Phys. Rep.* **C6**, 1 (1973).
R. G. Woolley, *Adv. Chem. Phys.* **33**, 153 (1975).

CHAPTER 5

Nonlinear Response Functions and Optical Susceptibilities

CORRELATION FUNCTION EXPRESSIONS FOR THE RESPONSE FUNCTIONS OF A SMALL PARTICLE

The optical polarization $P(\mathbf{r}, t)$ is the only material quantity that appears in the Maxwell equations. It serves as a source for the radiation field. Consequently, a complete knowledge of the optical polarization is sufficient for the interpretation of any time-domain or frequency-domain spectroscopic measurement. Electronic and nuclear motions and relaxation processes will show up in optical measurements only through their effect on the polarization. The calculation of the optical polarization is therefore the primary goal of any theory of optical spectroscopy, and is a key for interpreting optical measurements.* In Chapter 4 we presented a microscopic definition of the polarization operator [Eqs. (4.18c) and (4.19c)] and showed how some typical measurements such as linear absorption and multiwave mixing may be expressed in terms of the polarization. In this chapter we survey the general formalism (nonlinear response theory) that is commonly used in the calculation of the optical polarization and susceptibilities. We shall adopt the semiclassical Liouville–Maxwell level of theoretical description developed in Chapter 4 [Eqs. (4.47) and (4.48)] whereby the transverse radiation field is considered classical whereas the material system is treated quantum mechanically. The polarization can then be computed by assuming that the transverse Maxwell field $E(\mathbf{r}, t)$ is given, and calculating the material response

* Strictly speaking, a complete characterization of the electromagnetic field requires the knowledge of the polarization operator and not merely its expectation value. We are considering here coherent measurements that can be expressed in terms of field amplitudes, and are uniquely determined by the expectation value of the polarization. However, it is not always possible to work with the expectation value of the field itself. Incoherent measurements may require the expectation value of E^2 [P. D. Maker, *Phys. Rev. A* **1**, 923 (1970)]. Also correlation measurements of statistical properties of the field (photon statistics) go beyond the present treatment since they too are related to higher moments of the field [J. Mostowski and M. G. Raymer, in *Contemporary Nonlinear Optics*, G. P. Agarwal and R. W. Boyd, Eds. (Academic Press, Boston, 1992), p. 187; H. J. Carmichael, *An Open Systems Approach to Quantum Optics* (Springer-Verlag, Berlin, 1993)].

to that field. As indicated previously, in doing so we neglectsome effects of retardation of the radiation field such as certain aspects of polariton formation. While these are important effects, they are relevant only under special conditions and it is possible to treat many important properties of the nonlinear response without incorporating them. Most of this book will therefore be based on this level of formulation. More rigorous (and complex) treatments including a quantum electrodynamical description of the field will be discussed in Chapter 17.

Another important assumption made here is that the radiation fields are sufficiently weak to allow a perturbative, order by order, expansion of the response in the fields. Such expansions are appropriate for experiments involving radiation fields that are not too strong, so that the expansion can be truncated at some low power of the fields E^2, E^3, etc.* This "weak field" approach is justified even for powerful lasers since radiation fields used in typical nonlinear measurements are usually much weaker than the internal electric fields. The electric field in the hydrogen atom $E = e/a_0^2$ corresponds to a field intensity of $\sim 10^{17}$ W/cm^2. Spectroscopic measurements using fields of comparable or larger strength might show some important new effects that cannot be treated perturbatively [1, 2]. An excellent description of these phenomena exists using the "dressed state" picture (combined state of the material and the radiation field) [3–5]. Direct numerical integration of the Maxwell–Liouville equations of motion (with the field present) or use of the Floquet expansion may be very useful in some cases [6]. We shall not consider these effects here, and focus on cases in which the perturbative expansion is valid.

Using the semiclassical Maxwell–Liouville equations developed in Chapter 4, the material system evolution is governed by the semiclassical Hamiltonian [Eq. (4.47b)], and the radiation–matter interaction is

$$H_{\text{int}} = -\int E(\mathbf{r}, t) \cdot P(\mathbf{r})\, d\mathbf{r},$$

$P(\mathbf{r})$ being the polarization density, and $E(\mathbf{r}, t)$ is the classical transverse electric field.† This form of the coupling is very general and, so far, we have not even made the dipole approximation.

In general, the interaction with the radiation field creates a complex temporal and spatial excitation profile in the matter. For the sake of clarity we shall first consider the response of a single particle whose size is much smaller than the optical wavelength. This particle could be an atom, a molecule, a semiconductor or a metal nanocrystal. In this case we can adopt the dipole approximation (also known as the long wavelength approximation), where the particle may be represented as a point dipole as far as the field is concerned. We can then focus on the temporal response alone. Using this model, which is very common in the theoretical treatment of the optical response, the introduction of important concepts such as time ordering, Liouville space paths, and response functions is greatly simplified. The spatial aspects, which will be ignored for the moment, will be addressed at the end of this chapter where we shall consider the more general nonlocal response and introduce spatial dispersion and the wavevector dependence of the response functions.

---

* Some formal arguments against this expansion, which imply that it can only be used for short times, were made by N. Van Kampen, *Physica Norvegica* 5, 279 (1971).

† For brevity we eliminate the $\perp$ superscript and denote $E^{\perp}$ as E.

When the dipole approximation is made we can consider a single particle located at $\mathbf{r}$ and write

$$H_{\text{int}}(t) = -E(\mathbf{r}, t) \cdot V, \tag{5.1a}$$

with the dipole operator*

$$V = \sum_{\alpha} q_{\alpha}(\mathbf{r} - \mathbf{r}_{\alpha}), \tag{5.1b}$$

and the sum runs over all the electrons and nuclei α with charges q_{α} and positions $\mathbf{r}_{\alpha}$. In a typical nonlinear optical measurement, the electric field $E(\mathbf{r}, t)$ may be represented as a superposition of several incoming modes (whether pulsed or continuous). The various nonlinear optical spectroscopies differ by the timing, the detuning, and the direction of these modes. At this point, we can keep the radiation field completely general. Explicit forms for the electric field $E(\mathbf{r}, t)$ for specific optical measurements will be introduced and discussed in detail in later chapters.

In general, both the electric field $E(\mathbf{r}, t)$ and the dipole operator V are vectors and Eq. (5.1) contains their scalar product. The optical response functions to be introduced below are, therefore, high-rank tensors. The vector nature is particularly important for the optical response of anisotropic media (e.g., crystals, multilayers) or when studying orientational effects that can be detected by varying the polarization of the light fields (polarization spectroscopy). For the sake of clarity we shall not normally use a vector or tensor notation, and will not specify the components of vector quantities such as E or V, unless this is essential, e.g., in experiments utilizing explicitly the polarization of light.

To calculate the optical polarization, we start at $t = t_0$ and assume that the system is in thermal equilibrium with respect to its Hamiltonian $H \equiv H_{\text{mol}} + V_{\text{inter}}$ [Eq. (4.47b) with $E^{\perp}(\mathbf{r}, t) = 0$], and its canonical density operator is given by

$$\rho(t_0) = \rho(-\infty) = \exp(-\beta H)/\text{Tr} \exp(-\beta H), \tag{5.2}$$

where $\beta = (k_B T)^{-1}$, k_B is the Boltzmann constant, and T is the temperature. Since hereafter we work with expectation values, we no longer label operators by a $\hat{}$. The total Hamiltonian of the system is $H_T = H + H_{\text{int}}(t)$, and its dynamics is described by the Liouville equation

$$\frac{d\rho}{dt} = -\frac{i}{\hbar}[H, \rho] - \frac{i}{\hbar}[H_{\text{int}}(t), \rho].$$

Using Liouville space notation we have

$$\frac{d\rho}{dt} = -\frac{i}{\hbar} \mathscr{L}\rho - \frac{i}{\hbar} \mathscr{L}_{\text{int}}(t)\rho. \tag{5.3}$$

We have introduced the Liouville space operators $\mathscr{L}$, $\mathscr{L}_{\text{int}}$, and $\mathscr{V}$ defined by their

* Since in this section we consider a single particle, we delete the position variable $\mathbf{r}$ and denote $V(\mathbf{r})$ simply as V. Note also that in Chapter 4, the dipole operator V was denoted μ and the position of the particle $\mathbf{r}$ was denoted $\mathbf{R}_m$.

actions on an ordinary operator A, that is,

$$\mathscr{L}A \equiv [H, A], \tag{5.4a}$$

$$\mathscr{L}_{\text{int}}(t)A \equiv [H_{\text{int}}(t), A], \tag{5.4b}$$

$$\mathscr{V}A \equiv [V, A]. \tag{5.4c}$$

The time-dependent density operator will now be expanded in powers of the electric field [7–11], i.e.,

$$\rho(t) \equiv \rho^{(0)}(t) + \rho^{(1)}(t) + \rho^{(2)}(t) + \cdots,$$

where $\rho^{(n)}$ denotes the nth order contribution in the electric field and $\rho^{(0)}(t) = \rho(-\infty)$. Using Eqs. (3.43) and (3.58), we then have

$$\rho^{(n)}(t) = \left(-\frac{i}{\hbar}\right)^n \int_{t_0}^{t} d\tau_n \int_{t_0}^{\tau_n} d\tau_{n-1} \cdots \int_{t_0}^{\tau_2} d\tau_1$$

$$\mathscr{G}(t - \tau_n)\mathscr{L}_{\text{int}}(\tau_n)\mathscr{G}(\tau_n - \tau_{n-1})\mathscr{L}_{\text{int}}(\tau_{n-1}) \cdots \mathscr{G}(\tau_2 - \tau_1)\mathscr{L}_{\text{int}}(\tau_1)\mathscr{G}(\tau_1 - t_0)\rho(t_0), \tag{5.5}$$

where τ_j with $t \geqslant \tau_n \geqslant \cdots \geqslant \tau_1 \geqslant t_0$, represent the actual times of the interactions, and

$$\mathscr{G}(\tau) \equiv \theta(\tau) \exp\left(-\frac{i}{\hbar}\mathscr{L}\tau\right), \tag{5.6}$$

is the Liouville space Green function for the material system in the absence of the radiation field. Here $\theta(\tau)$ is the Heavyside step function [$\theta(\tau) = 1$ for $\tau > 0$ and $\theta(\tau) = 0$ for $\tau < 0$]. Since $\rho(t_0)$ represents the equilibrium density operator, it does not evolve with time when subject to the material Hamiltonian with no field [see discussion before Eq. (3.26a)]. This implies

$$\mathscr{G}(\tau_1 - t_0)\rho(t_0) = \rho(t_0).$$

We further have

$$\mathscr{L}_{\text{int}}(t) = -E(t)\mathscr{V},$$

where $\mathscr{V}$ is the time-independent dipole coupling in Liouville space introduced in Eq. (5.4c), and thus we get

$$\rho^{(n)}(t) = \left(\frac{i}{\hbar}\right)^n \int_{t_0}^{t} d\tau_n \int_{t_0}^{\tau_n} d\tau_{n-1} \cdots \int_{t_0}^{\tau_2} d\tau_1 E(\mathbf{r}, \tau_n)E(\mathbf{r}, \tau_{n-1}) \cdots E(\mathbf{r}, \tau_1)$$

$$\mathscr{G}(t - \tau_n)\mathscr{V}\mathscr{G}(\tau_n - \tau_{n-1})\mathscr{V} \cdots \mathscr{G}(\tau_2 - \tau_1)\mathscr{V}\rho(t_0). \tag{5.7}$$

Finally, we change the time variables (see Figure 5.1):

$$t_1 \equiv \tau_2 - \tau_1, \qquad t_2 \equiv \tau_3 - \tau_2 \cdots \qquad \text{and } t_n \equiv t - \tau_n$$

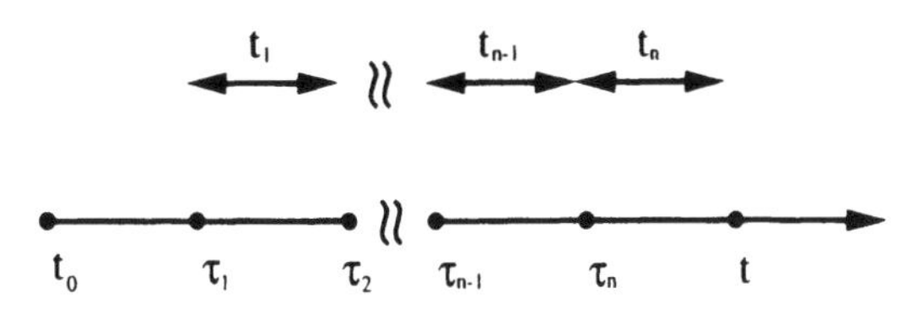

FIG. 5.1 Time variables used in Eqs. (5.7) and (5.8). The density operator is calculated at time t. $\tau_1 \leqslant \tau_2 \cdots \leqslant \tau_n$ are the time-ordered points when the interactions between the radiation field and matter takes place. $t_1 \cdots t_n$ are the time intervals between these interactions. t_1 is the earliest time interval, t_n is the last.

and then send $t_0 \to -\infty$, resulting in

$$\rho^{(n)}(t) = \left(\frac{i}{\hbar}\right)^n \int_0^\infty dt_n \int_0^\infty dt_{n-1} \cdots \int_0^\infty dt_1\, \mathscr{G}(t_n)\mathscr{V}\mathscr{G}(t_{n-1})\mathscr{V} \cdots \mathscr{G}(t_1)\mathscr{V}\rho(-\infty)$$
$$\times E(\mathbf{r}, t - t_n)E(\mathbf{r}, t - t_n - t_{n-1}) \cdots E(\mathbf{r}, t - t_n - t_{n-1} \cdots - t_1). \qquad (5.8)$$

Here, $t_n > 0$ represent the time intervals between successive interactions.

Equation (5.8) can alternatively be rearranged by introducing the operator $V(\tau)$ [and its Liouville space analogue $\mathscr{V}(\tau)$] in the interaction picture

$$V(\tau) \equiv \exp\left(\frac{i}{\hbar} H\tau\right) V \exp\left(-\frac{i}{\hbar} H\tau\right), \qquad (5.9a)$$

$$\mathscr{V}(\tau) \equiv \exp\left(\frac{i}{\hbar} \mathscr{L}\tau\right) \mathscr{V} \exp\left(-\frac{i}{\hbar} \mathscr{L}\tau\right). \qquad (5.9b)$$

It should be noted that for an arbitrary operator A

$$\mathscr{V}(\tau)A = [V(\tau), A], \qquad (5.9c)$$

as can be verified by a direct substitution.

Using these definitions, Eq. (5.8) can be recast into the form

$$\rho^{(n)}(t) = \left(\frac{i}{\hbar}\right)^n \int_0^\infty dt_n \int_0^\infty dt_{n-1} \cdots \int_0^\infty dt_1$$
$$\mathscr{G}(t_n + t_{n-1} + \cdots + t_1)\mathscr{V}(t_{n-1} + \cdots + t_1) \cdots \mathscr{V}(t_1)\mathscr{V}(0)\rho(-\infty)$$
$$\times E(\mathbf{r}, t - t_n)E(\mathbf{r}, t - t_n - t_{n-1}) \cdots E(\mathbf{r}, t - t_n - t_{n-1} \cdots - t_1). \qquad (5.10)$$

The position and time-dependent polarization $P(\mathbf{r}, t)$, is given by the expectation value of the dipole operator V:

$$P(\mathbf{r}, t) = \mathrm{Tr}[V\rho(t)] = \langle\langle V | \rho(t)\rangle\rangle. \qquad (5.11)$$

Upon the substitution of Eq. (5.5) into Eqs. (5.11), we obtain the Taylor expansion of the polarization in powers of the radiation field $E(\mathbf{r}, t)$

$$P(\mathbf{r}, t) = P^{(1)}(\mathbf{r}, t) + P^{(2)}(\mathbf{r}, t) + P^{(3)}(\mathbf{r}, t) + \cdots \tag{5.12a}$$

with

$$P^{(n)}(\mathbf{r}, t) \equiv \langle\langle V | \rho^{(n)}(t) \rangle\rangle. \tag{5.12b}$$

$P^{(n)}$ denotes the polarization to nth order in the radiation field. We assume that the polarization vanishes at thermal equilibrium so that $P^{(0)} = 0$. Note that $P^{(n)}$ contains products of $n + 1$ dipole operator factors V.

As indicated in Chapter 4, each order in this expansion represents a different class of optical measurements. The linear polarization $P^{(1)}$ is responsible for linear optics, $P^{(2)}$ represents second-order nonlinear processes such as frequency sum generation, and $P^{(3)}$ is the third-order polarization that enters in a broad variety of techniques including four wave mixing and pump-probe spectroscopy. Inserting the perturbative expansion of $\rho(t)$ [Eqs. (5.8)] into (5.12) we obtain

$$P^{(n)}(\mathbf{r}, t) = \int_0^\infty dt_n \int_0^\infty dt_{n-1} \cdots \int_0^\infty dt_1\, S^{(n)}(t_n, t_{n-1}, \ldots, t_1)$$

$$E(\mathbf{r}, t - t_n)E(\mathbf{r}, t - t_n - t_{n-1}) \cdots E(\mathbf{r}, t - t_n - t_{n-1} \cdots - t_1), \tag{5.13}$$

with

$$S^{(n)}(t_n, t_{n-1}, \ldots, t_1) \equiv$$

$$\left(\frac{i}{\hbar}\right)^n \langle\langle V | \mathcal{G}(t_n)\mathcal{V}\mathcal{G}(t_{n-1})\mathcal{V} \cdots \mathcal{G}(t_1)\mathcal{V} | \rho(-\infty) \rangle\rangle. \tag{5.14}$$

The nth-order *nonlinear response function*, $S^{(n)}$, carries the complete microscopic information necessary for the calculation of optical measurements. Using Eq. (5.10), it can be recast into the alternative form

$$S^{(n)}(t_n, t_{n-1}, \ldots, t_1) = \left(\frac{i}{\hbar}\right)^n \theta(t_1)\theta(t_2) \cdots \theta(t_n)$$

$$\times \langle\langle V(t_n + \cdots + t_1) | \mathcal{V}(t_{n-1} + \cdots + t_1) \cdots \mathcal{V}(t_1)\mathcal{V}(0) | \rho(-\infty) \rangle\rangle. \tag{5.15}$$

Switching back to Hilbert space, making use of Eq. (5.9c), and acting with all commutators to the right, we have

$$S^{(n)}(t_n, t_{n-1}, \ldots, t_1) = \left(\frac{i}{\hbar}\right)^n \theta(t_1)\theta(t_2) \cdots \theta(t_n)$$

$$\times \langle V(t_n + \cdots + t_1)[V(t_{n-1} + \cdots + t_1), [\cdots [V(t_1), [V(0), \rho(-\infty)]] \cdots]] \rangle \tag{5.16a}$$

or, by acting to the left

$$S^{(n)}(t_n, t_{n-1}, \ldots, t_1) = \left(\frac{i}{\hbar}\right)^n \theta(t_1)\theta(t_2)\cdots\theta(t_n)$$

$$\times \langle[[[\cdots[V(t_n + \cdots + t_1), V(t_{n-1} + \cdots + t_1)]\cdots], V(t_1)], V(0)]\rho(-\infty)\rangle. \tag{5.16b}$$

Equations (5.14) through (5.16) present four equivalent forms for the nonlinear response function. These expressions are written in different spaces. Equations (5.14) and (5.15) are written in Liouville space and treat ρ as a vector. Equations (5.16), on the other hand, treat ρ as a matrix and are expressed in terms of a combination of ordinary correlation functions in Hilbert space. Although we have used the Liouville space expression for the response functions in the derivation of Eqs. (5.16), the actual evaluation of these expressions is carried out in Hilbert space. In fact, these expressions could be derived by a direct integration of Eq. (3.24) without introducing Liouville space notation at all! In the coming chapters we shall develop Hilbert space as well as Liouville space Green function techniques for the evaluation of the response function.

Since $P^{(n)}(\mathbf{r}, t)$ is a physical observable, and is given by the expectation value of a Hermitian operator, it is real. The same argument also holds for the electric field $E(\mathbf{r}, t)$. Consequently, the response functions $S^{(n)}$ must be real as well. This can be easily verified using Eqs. (5.14) through (5.16) in which the various terms appear in complex conjugate pairs, as will be demonstrated shortly. Note that the polarization at a given time can depend only on the electric field at earlier times. This reflects the principle of *causality* [12], which implies that the cause must precede the effect and, therefore, $S^{(n)}$ vanishes should any of its time arguments t_j become negative

$$S^{(n)}(t_n, t_{n-1} \cdots t_1) = 0 \qquad \text{if } t_j < 0. \tag{5.17}$$

This is guaranteed by the $\theta(t_j)$ factors in $S^{(n)}$. Bearing this in mind, we can change all the lower limits of integration in Eq. (5.13) from 0 to $-\infty$. The consequences of causality in the frequency domain will be discussed in Chapter 6.

This concludes our presentation of the basic formal definitions of the optical response functions. In the coming section we shall first examine more closely the three lowest order optical response functions and discuss the various contributions to these quantities in terms of *Liouville space pathways* [13]. For clarity, we first consider the response of a particle that is smaller than an optical wavelength. In this case the response is local in space and we can focus merely on its temporal (or spectral) characteristics. Following that, we consider a system of arbitrary size and derive the most general expressions for the nonlinear response functions that are nonlocal in space and time. We then expand the field in modes, recast the nonlinear response in momentum (**k**) space, and introduce the wavevector and frequency-dependent optical susceptibilities. Subsequently, we comment on the relationship among frequency-domain and time-domain measurements, and conclude by summarizing the merits of working in Liouville space. Two additional alternative methods for evaluating the response function, which are based on either calculating the wavefunction (rather than the density operator), or using equations of motion in the Heisenberg picture, will be surveyed in

Appendixes 5A and 5B, respectively. Although for the most part we are going to use the Liouville space approach, it is instructive to compare with these other commonly used methods. Such comparison will be made in Chapter 6, where we apply these results to the nonlinear response of a small particle.

LIOUVILLE SPACE PATHWAYS IN THE TIME DOMAIN

Since the nonlinear response function $S^{(n)}$ contains the Liouville space dipole operator $\mathscr{V}$ to nth order, and since each $\mathscr{V}$ represents a commutator, which can act either from the left or from the right, the nonlinear response function $S^{(n)}$ will contain 2^n terms once these commutators are evaluated. It is easy to show that in practice only half of the 2^n terms are independent, and need to be considered; the contribution of the other half is simply the complex conjugate of the former. In general, by reversing all "right" and "left" choices of $\mathscr{V}$ in a given term, we obtain its complex conjugate. The overall $\pm$ sign is determined by simply counting the number of times $\mathscr{V}$ acts from the right, since there is a $(-)$ factor associated with each "right" choice. We shall now survey the resulting terms, denoted the Liouville space pathways, for the three lowest order response functions.

Linear response

The linear response function that controls all linear optical measurements is given by Eq. (5.14) or (5.15)

$$S^{(1)}(t_1) = \frac{i}{\hbar}\langle\langle V|\mathscr{G}(t_1)\mathscr{V}|\rho(-\infty)\rangle\rangle, \tag{5.18a}$$

$$S^{(1)}(t_1) = \frac{i}{\hbar}\,\theta(t_1)\langle\langle V(t_1)|\mathscr{V}(0)|\rho(-\infty)\rangle\rangle, \tag{5.18b}$$

or, alternatively when using Eq. (5.16),

$$S^{(1)}(t_1) = \frac{i}{\hbar}\,\theta(t_1)\langle[V(t_1), V(0)]\rho(-\infty)\rangle. \tag{5.18c}$$

We thus get

$$S^{(1)}(t_1) = \frac{i}{\hbar}\,\theta(t_1)[J(t_1) - J^*(t_1)], \tag{5.19}$$

where

$$J(t_1) \equiv \langle V(t_1)V(0)\rho(-\infty)\rangle, \tag{5.20a}$$

$$J^*(t_1) = \langle V(0)V(t_1)\rho(-\infty)\rangle, \tag{5.20b}$$

and the asterisk denotes the complex conjugate.

The two terms in (5.19) represent the two Liouville space pathways that contribute to the linear response.

Second-order response

The calculation of the second-order response starts with

$$S^{(2)}(t_2, t_1) = \left(\frac{i}{\hbar}\right)^2 \langle\langle V|\mathcal{G}(t_2)\mathcal{V}\mathcal{G}(t_1)\mathcal{V}|\rho(-\infty)\rangle\rangle, \tag{5.21}$$

or, the Hilbert space form,

$$S^{(2)}(t_2, t_1) = \left(\frac{i}{\hbar}\right)^2 \theta(t_1)\theta(t_2)\langle[[V(t_2 + t_1), V(t_1)], V(0)]\rho(-\infty)\rangle. \tag{5.22}$$

We now repeat the same steps made for the linear response. Since (5.21) has two $\mathcal{V}$ factors, and each can act either from the left or from the right, $S^{(2)}$ will altogether have four Liouville space paths,

$$S^{(2)}(t_2, t_1) = \left(\frac{i}{\hbar}\right)^2 \theta(t_1)\theta(t_2) \sum_{\alpha=1}^{2} [Q_\alpha(t_2, t_1) + Q_\alpha^*(t_2, t_1)], \tag{5.23}$$

where

$$\left.\begin{aligned} Q_1(t_2, t_1) &= \langle V(t_1 + t_2)V(t_1)V(0)\rho(-\infty)\rangle, \\ Q_2(t_2, t_1) &= -\langle V(t_1)V(t_1 + t_2)V(0)\rho(-\infty)\rangle. \end{aligned}\right\} \tag{5.24}$$

In an isotropic medium, the even order response functions vanish by symmetry. This can be shown as follows: If we reverse the direction of all fields, we expect the polarization to reverse its direction as well so that $P^{(2n)} \to -P^{(2n)}$. On the other hand, Eq. (5.13) depends on E^{2n}, which does not change sign and $P^{(2n)} \to P^{(2n)}$. The only way these two relations hold simultaneously is when $P^{(2n)}$ vanishes identically. Measurements of the second-order response are particularly interesting in organized anisotropic media such as interfaces, quantum wells, and monolayers. The lowest nonvanishing nonlinear response function in an isotropic medium is $S^{(3)}$, which will be considered next.

Third-order response

The calculation starts with Eq. (5.14):

$$S^{(3)}(t_3, t_2, t_1) = \left(\frac{i}{\hbar}\right)^3 \langle\langle V|\mathcal{G}(t_3)\mathcal{V}\mathcal{G}(t_2)\mathcal{V}\mathcal{G}(t_1)\mathcal{V}|\rho(-\infty)\rangle\rangle. \tag{5.25}$$

The three $\mathcal{V}$ factors can act either from the left or from the right, resulting in $2^3 = 8$ terms. Proceeding along the same lines used earlier we have

$$S^{(3)}(t_3, t_2, t_1) = \left(\frac{i}{\hbar}\right)^3 \theta(t_1)\theta(t_2)\theta(t_3) \times \langle[[[V(t_3 + t_2 + t_1), V(t_2 + t_1)], V(t_1)], V(0)]\rho(-\infty)\rangle, \tag{5.26}$$

which gives

$$S^{(3)}(t_3, t_2, t_1) = \left(\frac{i}{\hbar}\right)^3 \theta(t_1)\theta(t_2)\theta(t_3) \sum_{\alpha=1}^{4} [R_\alpha(t_3, t_2, t_1) - R_\alpha^*(t_3, t_2, t_1)], \tag{5.27}$$

where

$$R_1(t_3, t_2, t_1) = \langle V(t_1)V(t_1 + t_2)V(t_1 + t_2 + t_3)V(0)\rho(-\infty)\rangle, \quad (5.28a)$$

$$R_2(t_3, t_2, t_1) = \langle V(0)V(t_1 + t_2)V(t_1 + t_2 + t_3)V(t_1)\rho(-\infty)\rangle, \quad (5.28b)$$

$$R_3(t_3, t_2, t_1) = \langle V(0)V(t_1)V(t_1 + t_2 + t_3)V(t_1 + t_2)\rho(-\infty)\rangle, \quad (5.28c)$$

$$R_4(t_3, t_2, t_1) = \langle V(t_1 + t_2 + t_3)V(t_1 + t_2)V(t_1)V(0)\rho(-\infty)\rangle. \quad (5.28d)$$

LIOUVILLE SPACE PATHWAYS IN THE FREQUENCY DOMAIN

In continuous wave (cw) experiments, it is natural to recast the response functions in the frequency domain (rather than in the time domain). We shall first introduce a convention for Fourier transform to be used throughout the book. We shall denote an arbitrary function $F(t)$ and its Fourier transform $\tilde{F}(\omega)$.

$$F(t) = \frac{1}{2\pi}\int_{-\infty}^{\infty} d\omega\, \tilde{F}(\omega)\exp(-i\omega t), \quad (5.29a)$$

$$\tilde{F}(\omega) = \int_{-\infty}^{\infty} dt\, F(t)\exp(i\omega t). \quad (5.29b)$$

In addition, we recall the one-sided Fourier transform used for the Green function [Eq. (3.66)].

$$\theta(t)\mathcal{G}(t) = -\frac{1}{2\pi i}\int_{-\infty}^{\infty} d\omega\, \mathcal{G}(\omega)\exp(-i\omega t), \quad (5.30a)$$

$$\mathcal{G}(\omega) = -i\int_{0}^{\infty} dt\, \mathcal{G}(t)\exp(i\omega t). \quad (5.30b)$$

For simplifying the notation, we use the same symbol for the time-domain and the frequency-domain functions in Eqs. (5.30). This should not create any confusion since the argument (t or ω) will always be specified. It is important, however, to distinguish between the ordinary transform $\tilde{F}(\omega)$ and the one-sided one $F(\omega)$. We have already used the one-sided transform in the definitions of the Green functions, i.e., the relations between $G(t)$ and $G(E)$ [Eqs. (2.71)] and $\mathcal{G}(t)$ and $\mathcal{G}(\omega)$ [Eqs. (3.66)]. The extra i factor in Eqs. (5.30) is introduced to keep the conventional definitions of Green functions.

The nth order response function in the frequency domain is defined as

$$S^{(n)}(\Omega_n, \Omega_{n-1}, \ldots, \Omega_1) \equiv \int_0^{\infty} dt_n \int_0^{\infty} dt_{n-1} \cdots \int_0^{\infty} dt_1\, S^{(n)}(t_n, t_{n-1}, \ldots, t_1)$$
$$\times \exp(i\Omega_n t_n + i\Omega_{n-1}t_{n-1} + \cdots + i\Omega_1 t_1). \quad (5.31a)$$

Using (5.14) we then have

$$S^{(n)}(\Omega_n, \Omega_{n-1}, \ldots, \Omega_1) \equiv \left(-\frac{1}{\hbar}\right)^n \langle\langle V|\mathscr{G}(\Omega_n)\mathscr{V}\mathscr{G}(\Omega_{n-1})\mathscr{V}\cdots\mathscr{G}(\Omega_1)\mathscr{V}|\rho(-\infty)\rangle\rangle. \tag{5.31b}$$

Here $\mathscr{G}(\Omega)$ is the Liouville space Green function in the frequency domain [Eq. (3.65)]. By introducing a Fourier transform of the field $\tilde{E}(\mathbf{r}, \omega)$ [see Eqs. (5.29)], the nonlinear polarization in real space can be written in the form

$$P^{(n)}(\mathbf{r}, t) = \frac{1}{(2\pi)^n}\int d\omega_1 \int d\omega_2 \cdots \int d\omega_n\, S^{(n)}(\omega_1 + \omega_2 + \cdots + \omega_n, \ldots, \omega_1 + \omega_2, \omega_1)$$
$$\times\, \tilde{E}(\mathbf{r}, \omega_1)\tilde{E}(\mathbf{r}, \omega_2)\cdots\tilde{E}(\mathbf{r}, \omega_n)\exp(-i\omega_s t), \tag{5.32}$$

with $\omega_s \equiv \omega_1 + \omega_2 + \cdots + \omega_n$.

We shall now review the Liouville space paths corresponding to the response up to third order.

Linear response

The Liouville space expression for the linear response function in the frequency domain is

$$S^{(1)}(\omega_1) = \left(-\frac{1}{\hbar}\right)\langle\langle V|\mathscr{G}(\omega_1)\mathscr{V}|\rho(-\infty)\rangle\rangle. \tag{5.33}$$

By performing the one-sided Fourier transform, the response functions in the frequency domain can be expressed as

$$S^{(1)}(\omega_1) = \left(-\frac{1}{\hbar}\right)[J(\omega_1) + J^*(-\omega_1)], \tag{5.34a}$$

where

$$J(\omega_1) \equiv -i\int_0^\infty dt_1 \exp(i\omega_1 t_1)J(t_1). \tag{5.34b}$$

Second-order response

The Liouville space expression is

$$S^{(2)}(\omega_1 + \omega_2, \omega_1) = \left(-\frac{1}{\hbar}\right)^2 \langle\langle V|\mathscr{G}(\omega_1 + \omega_2)\mathscr{V}\mathscr{G}(\omega_1)\mathscr{V}|\rho(-\infty)\rangle\rangle. \tag{5.35}$$

Alternatively, using Eqs. (5.23) we have

$$S^{(2)}(\omega_1 + \omega_2, \omega_1) = \left(-\frac{1}{\hbar}\right)^2 \sum_{\alpha=1}^{2} [Q_\alpha(\omega_1 + \omega_2, \omega_1) + Q_\alpha^*(-\omega_1 - \omega_2, -\omega_1)], \tag{5.36a}$$

where

$$Q_\alpha(\omega_1 + \omega_2, \omega_1) \equiv (-i)^2 \int_0^\infty dt_2 \int_0^\infty dt_1 \exp[i(\omega_1 + \omega_2)t_2 + i\omega_1 t_1]Q_\alpha(t_2, t_1). \tag{5.36b}$$

Third-order response

The Liouville space expression is

$$S^{(3)}(\omega_1 + \omega_2 + \omega_3, \omega_1 + \omega_2, \omega_1)$$
$$= \left(-\frac{1}{\hbar}\right)^3 \langle\langle V|\mathscr{G}(\omega_1 + \omega_2 + \omega_3)\mathscr{V}\mathscr{G}(\omega_1 + \omega_2)\mathscr{V}\mathscr{G}(\omega_1)\mathscr{V}|\rho(-\infty)\rangle\rangle. \quad (5.37)$$

In terms of correlation functions [Eqs. (5.27)] we have

$$S^{(3)}(\omega_1 + \omega_2 + \omega_3, \omega_1 + \omega_2, \omega_1)$$
$$= \left(-\frac{1}{\hbar}\right)^3 \sum_{\alpha=1}^{4} [R_\alpha(\omega_1 + \omega_2 + \omega_3, \omega_1 + \omega_2, \omega_1)$$
$$+ R_\alpha^*(-\omega_1 - \omega_2 - \omega_3, -\omega_1 - \omega_2, -\omega_1)], \quad (5.38a)$$

where

$$R_\alpha(\omega_1 + \omega_2 + \omega_3, \omega_1 + \omega_2, \omega_1)$$
$$\equiv (-i)^3 \int_0^\infty dt_3 \int_0^\infty dt_2 \int_0^\infty dt_1 R_\alpha(t_3, t_2, t_1)$$
$$\times \exp[i(\omega_1 + \omega_2 + \omega_3)t_3 + i(\omega_1 + \omega_2)t_2 + i\omega_1 t_1]. \quad (5.38b)$$

The frequency domain response functions $S^{(n)}$ are not symmetric with respect to the permutation of the frequency arguments. The reason is that although we switched to the frequency domain, we still maintain a bookkeeping of time ordering (ω_1 interacts first, etc.). This is not necessary in the frequency domain and it is possible to define symmetrized response functions that are called *polarizabilities* and denoted α, β, γ, etc., i.e.,

$$\alpha(\omega) \equiv S^{(1)}(\omega),$$

$$\beta(\omega_1, \omega_2) \equiv \frac{1}{2!} \sum_p S^{(2)}(\omega_1 + \omega_2, \omega_1),$$

$$\gamma(\omega_1, \omega_2, \omega_3) \equiv \frac{1}{3!} \sum_p S^{(3)}(\omega_1 + \omega_2 + \omega_3, \omega_1 + \omega_2, \omega_1),$$

where the p summation is over all the $n!$ permutations of the frequencies $\omega_1, \omega_2, \ldots, \omega_n$. α is the linear polarizability, β is the first hyperpolarizability, γ is the second hyperpolarizability, etc. Because of the invariance of the integration in (5.32) to a permutation of frequencies, we can substitute the polarizabilities in this equation (α for $S^{(1)}$, β for $S^{(2)}$, and γ for $S^{(3)}$) without affecting the result. The hyperpolarizabilities are completely symmetric with respect to the permutations of their frequency arguments and are therefore easier to calculate in some cases.

NONLOCAL EXPRESSIONS FOR THE OPTICAL RESPONSE OF EXTENDED SYSTEMS

So far we calculated the response functions for a simple model of an isolated small particle. The response is then formulated in the time (or frequency) domain and since the particle is essentially a point particle (as far as the field is concerned), the spatial variation of the field is not important. We shall now generalize these results and present closed formal expressions for the response functions for a system with arbitrary size, including explicit vector notation for the field and the polarization. To that end we go back to (5.1a) and rewrite it as

$$H_{\text{int}} = -\int d\mathbf{r}\, E(\mathbf{r}, t)\cdot V(\mathbf{r}), \tag{5.39}$$

$V(\mathbf{r}) \equiv \hat{P}(\mathbf{r})$ being a polarization operator. To maintain a consistent notation with the previous sections, we denote the polarization operator in this section by $V(\mathbf{r})$. A straightforward generalization of Eq. (5.13) gives

$$P^{(n)}_{\nu_s}(\mathbf{r}, t) = \int d\mathbf{r}_n \int d\mathbf{r}_{n-1}\cdots\int d\mathbf{r}_1 \int_0^\infty dt_n \int_0^\infty dt_{n-1}\cdots\int_0^\infty dt_1 \sum_{\nu_1\cdots\nu_n} S^{(n)}{}_{\nu_1\cdots\nu_n\nu_s}(\mathbf{r}; \mathbf{r}_n\cdots\mathbf{r}_1,$$
$$t_n\cdots t_1)E_{\nu_n}(\mathbf{r}_n, t-t_n)E_{\nu_{n-1}}(\mathbf{r}_{n-1}, t-t_n-t_{n-1})\cdots E_{\nu_1}(\mathbf{r}_1, t-t_n\cdots-t_1), \tag{5.40}$$

with the nonlinear response function [Eq. (5.14)]

$$S^{(n)}_{\nu_1\ldots\nu_n\nu_s}(\mathbf{r}; \mathbf{r}_n\cdots\mathbf{r}_1, t_n\cdots t_1) =$$
$$\left(\frac{i}{\hbar}\right)^n \langle\langle V_{\nu_s}(\mathbf{r})|\mathcal{G}(t_n)\mathcal{V}_{\nu_n}(\mathbf{r}_n)\mathcal{G}(t_{n-1})\mathcal{V}_{\nu_{n-1}}(\mathbf{r}_{n-1})\cdots\mathcal{G}(t_1)\mathcal{V}_{\nu_1}(\mathbf{r}_1)|\rho(-\infty)\rangle\rangle, \tag{5.41}$$

and

$$\mathcal{V}_\nu(\mathbf{r})A \equiv [V_\nu(\mathbf{r}), A] \equiv [\hat{P}_\nu(\mathbf{r}), A].$$

Here $E_{\nu_i}(\mathbf{r}_j, t)$ is the field polarized in the ν_j direction with $\nu_j \equiv x, y, z$. $S^{(n)}$ describes the ν_s component of the polarization induced at point $\mathbf{r}$ and time t, by n interactions with the field, taking place at times

$$t-t_n, t-t_n-t_{n-1}, \ldots, t-t_n-t_{n-1}\cdots-t_1$$

and points $\mathbf{r}_1, \mathbf{r}_2\cdots\mathbf{r}_n$.

The explicit dependence of the nonlinear response on $\mathbf{r}_n$ reflects the *nonlocal* nature of the response in space, in addition to the nonlocal temporal response, which is contained in the dependence on t_n. The spatial dependence and the explicit inclusion of Cartesian components are the only differences between Eq. (5.40) and Eq. (5.13). Due to spatial correlations in the material system, resulting, e.g., from intermolecular forces or delocalized electronic states, an interaction with the field at one point can affect the polarization at other points. Similarly, Eq. (5.16b) assumes the form

$$S^{(n)}_{\nu_1\ldots\nu_n\nu_s}(\mathbf{r}; \mathbf{r}_n, \ldots \mathbf{r}_1, t_n, \ldots t_1) = \theta(t_1)\theta(t_2)\cdots\theta(t_n)\left(\frac{i}{\hbar}\right)^n$$
$$\langle[[\cdots[V_{\nu_1}(\mathbf{r}_1, t_n+\cdots+t_1), V_{\nu_2}(\mathbf{r}_2, t_{n-1}+\cdots+t_1)]\cdots, V_{\nu_n}(\mathbf{r}_n, t_1)], V_{\nu_s}(\mathbf{r}, 0)]\rho(-\infty)\rangle, \tag{5.42}$$

with

$$V_\nu(\mathbf{r}, t) \equiv \exp\left(\frac{i}{\hbar}Ht\right)V_\nu(\mathbf{r}) \exp\left(-\frac{i}{\hbar}Ht\right).$$

Transforming Eq. (5.40) to the frequency domain we obtain

$$P_{\nu_s}^{(n)}(\mathbf{r}, t) = \frac{1}{(2\pi)^n}\int d\mathbf{r}_n \int d\mathbf{r}_{n-1} \cdots \int d\mathbf{r}_1 \int d\omega_n \int d\omega_{n-1} \cdots \int d\omega_1$$

$$\sum_{\nu_1 \ldots \nu_n} \chi_{\nu_s\nu_1\ldots\nu_n}^{(n)}(\mathbf{r}; \mathbf{r}_1 \cdots \mathbf{r}_n, -\omega_s; \omega_1, \ldots, \omega_n)$$

$$\times\, \tilde{E}_{\nu_n}(\mathbf{r}_n, \omega_n) \cdots \tilde{E}_{\nu_1}(\mathbf{r}_1, \omega_1) \exp(-i\omega_s t), \tag{5.43a}$$

$$\chi_{\nu_s\nu_1\ldots\nu_n}^{(n)}(\mathbf{r}; \mathbf{r}_1 \cdots \mathbf{r}_n; -\omega_s; \omega_1 \cdots \omega_n)$$

$$= \frac{1}{n!}\sum_p S_{\nu_1\ldots\nu_n\nu_s}^{(n)}(\mathbf{r}; \mathbf{r}_n \cdots \mathbf{r}_1, \omega_1 + \cdots + \omega_n, \ldots, \omega_1 + \omega_2, \omega_1), \tag{5.43b}$$

$$S_{\nu_1\ldots\nu_n\nu_s}^{(n)}(\mathbf{r}; \mathbf{r}_n, \ldots, \mathbf{r}_1, \Omega_n, \Omega_{n-1} \cdots \Omega_1)$$

$$= \left(-\frac{1}{\hbar}\right)^n \langle\langle V_{\nu_s}(\mathbf{r})|\mathcal{G}(\Omega_n)\mathcal{V}_{\nu_n}(\mathbf{r}_n)\mathcal{G}(\Omega_{n-1})\mathcal{V}_{\nu_{n-1}}(\mathbf{r}_{n-1}) \cdots \mathcal{G}(\Omega_1)\mathcal{V}_{\nu_1}(\mathbf{r}_1)|\rho(-\infty)\rangle\rangle, \tag{5.43c}$$

and

$$\omega_s \equiv \omega_1 + \omega_2 + \cdots + \omega_n.$$

The nth order optical susceptibility $\chi^{(n)}$ is an $n + 1$ rank tensor. $\sum_p$ is a sum over all permutations of $(\mathbf{r}_1\omega_1\nu_1), \ldots, (\mathbf{r}_n\omega_n\nu_n)$. This sum makes $\chi^{(n)}$ invariant to interchange of any of its $r_j\omega_j\nu_j$ indices.

NONLINEAR RESPONSE IN MOMENTUM (k) SPACE

So far, all our expressions were given in real space as a function of the position **r**. This offers the best microscopic understanding of the physical processes underlying the optical response, since it is usually localized in real space. However, the transverse electric field is macroscopic and the optical response of macroscopic and homogeneous systems is most commonly represented in momentum (**k**) space rather than in real space.

In analogy with the definition of the polarizabilities, we can now introduce optical susceptibilities in which the response functions are symmetrized with respect to all frequencies and wavevectors. Using the Fourier transform

$$E(\mathbf{r}, t) = \frac{1}{(2\pi)^4}\int_{-\infty}^{\infty} d\omega \int d\mathbf{k} \exp(-i\omega t + i\mathbf{k}\cdot r)\tilde{E}(\mathbf{k}, \omega),$$

we can recast the polarization into the form

$$P^{(n)}_{\nu_s}(\mathbf{r}, t) = \frac{1}{(2\pi)^{4n}} \sum_{\nu_1\nu_2\ldots\nu_s} \int d\mathbf{k}_1 \cdots \int d\mathbf{k}_n \int d\omega_1 \cdots \int d\omega_n$$

$$\chi^{(n)}_{\nu_1\ldots\nu_n\nu_s}(-\mathbf{k}_s - \omega_s; \mathbf{k}_1\omega_1, \ldots, \mathbf{k}_n\omega_n) \exp(-i\omega_s t + i\mathbf{k}_s\cdot\mathbf{r})$$

$$\times \tilde{E}_{\nu_1}(\mathbf{k}_1, \omega_1)\tilde{E}_{\nu_2}(\mathbf{k}_2, \omega_2) \cdots \tilde{E}_{\nu_n}(\mathbf{k}_n, \omega_n), \qquad (5.44a)$$

with

$$\chi^{(n)}_{\nu_1\ldots\nu_n\nu_s}(-\mathbf{k}_s - \omega_s; \mathbf{k}_1\omega_1, \mathbf{k}_2\omega_2, \ldots, \mathbf{k}_n\omega_n)$$

$$\equiv \frac{1}{n!}\sum_p S^{(n)}_{\nu_1\ldots\nu_n\nu_s}(\mathbf{k}_1, \mathbf{k}_2, \ldots, \mathbf{k}_n, \omega_1 + \cdots + \omega_n, \omega_1 + \cdots + \omega_{n-1}, \ldots, \omega_1), \qquad (5.44b)$$

and

$$S^{(n)}_{\nu_1\ldots\nu_n\nu_s}(\mathbf{k}_1\Omega_1, \mathbf{k}_2\Omega_2, \ldots, \mathbf{k}_n\Omega_n)$$

$$\equiv \left(-\frac{1}{\hbar}\right)^n \langle\langle V_{\nu_s}(\mathbf{k}_s)|\mathscr{G}(\Omega_n)\mathscr{V}_{\nu_n}(\mathbf{k}_n)\mathscr{G}(\Omega_{n-1}) \cdots \mathscr{V}_{\nu_2}(\mathbf{k}_2)\mathscr{G}(\Omega_1)\mathscr{V}_{\nu_1}(\mathbf{k}_1)|\rho(-\infty)\rangle\rangle. \qquad (5.44c)$$

Here

$$\omega_s \equiv \omega_1 + \cdots + \omega_n; \; \mathbf{k}_s \equiv \mathbf{k}_1 + \cdots + \mathbf{k}_n.$$

$V_{\nu_j}(\mathbf{k}_j)$ is defined by the Fourier decomposition similar to Eq. (5.44a)

$$V_{\nu_j}(\mathbf{k}_j) = \int d\mathbf{r}\, V_{\nu_j}(\mathbf{r}) \exp(-i\mathbf{k}_j\cdot\mathbf{r}),$$

and $\mathscr{V}_{\nu_j}(\mathbf{k}_j)$ is its Liouville space counterpart. It is possible to expand Eq. (5.44c) in Liouville space paths, in complete analogy with what we did for the single particle; the only difference is that V now depends on the wavevector $\mathbf{k}$. Like the polarizabilities, the susceptibilities are completely invariant to any permutation of the field arguments $(\omega_j\mathbf{k}_j\nu_j)$. Note that we can substitute $S^{(n)}$ for $\chi^{(n)}$ in Eq. (5.44a) without affecting the result. $\chi^{(n)}$ contains in general $n!\,2^n$ terms corresponding to the 2^n Liouville space paths and the $n!$ permutations of the fields. The real space and k space susceptibilities are connected by the spatial Fourier transform

$$\chi^{(n)}_{\nu_1\ldots\nu_n\nu_s}(-\mathbf{k}_s - \omega_s; \mathbf{k}_1\omega_1, \ldots, \mathbf{k}_n\omega_n)$$

$$= \int d\mathbf{r}_1 \cdots \int d\mathbf{r}_n\, \chi^{(n)}_{\nu_1\ldots\nu_n\nu_s}(\mathbf{r}; \mathbf{r}_1 \cdots \mathbf{r}_n; -\omega_s; \omega_1 \cdots \omega_n)$$

$$\times \exp[i\mathbf{k}_1\cdot(\mathbf{r}_1 - \mathbf{r}) + \cdots + i\mathbf{k}_n\cdot(\mathbf{r}_n - \mathbf{r})]. \qquad (5.45)$$

In an nth order optical process we need consider only n components (modes) of the field in **k** space,

$$E_\nu(\mathbf{r}, t) = \sum_{j=1}^{n} [E_{j,\nu}(t)\exp(i\mathbf{k}_j\cdot\mathbf{r}) + E^*_{j,\nu}(t)\exp(-i\mathbf{k}_j\cdot\mathbf{r})], \tag{5.46a}$$

so that

$$\tilde{E}_\nu(\mathbf{k}, \omega) = (2\pi)^3 \sum_{j=1}^{n} E_{j,\nu}(\omega)\delta(\mathbf{k} - \mathbf{k}_j). \tag{5.46b}$$

Here $E_{j,\nu}(t)$ represents the temporal envelope of the jth mode, with polarization direction ν. Since we retain here the time dependence of the field envelopes $E_{j,\nu}(t)$, this is not a Fourier transform. The expansion [Eq. (5.46a)] rather constitutes a mixed time and frequency representation of the field. Mathematically, this is redundant and one may justifiably object to this wasteful abuse of Fourier transforms (in a genuine Fourier transform we would add more modes to the expansion and the coefficients become time independent). Nevertheless, it makes physical sense. Usually the envelopes $E_{j,\nu}(t)$ vary on a much slower timescale than their optical periods ω_j^{-1}, and $\tilde{E}_\nu(\mathbf{k}, \omega)$ consists of well resolved features corresponding to the various modes. This suggests various approximations, such as the slowly varying envelope described in Chapter 4, which can be naturally introduced using this expansion in modes.

We shall be interested in calculating a particular Fourier component of the nonlinear polarization, i.e.,

$$\tilde{P}_\nu^{(n)}(\mathbf{r}, \omega_s) = \sum_s \tilde{P}_{s\nu}^{(n)}(\omega_s)\exp(i\mathbf{k}_s\cdot\mathbf{r}), \tag{5.47}$$

where $\mathbf{k}_s$ and ω_s are given by any combination of the incoming fields $\mathbf{k}_s = \pm\mathbf{k}_1 + \cdots \pm \mathbf{k}_n$ and $\omega_s = \pm\omega_1 + \cdots \pm \omega_n$. Hereafter we shall adopt the + sign $\mathbf{k}_s = \mathbf{k}_1 + \cdots + \mathbf{k}_n$ and $\omega_s = \omega_1 + \cdots + \omega_n$. Any other component can be obtained by simply changing the signs of one or more components $(\omega_j, \mathbf{k}_j) \to (-\omega_j, -\mathbf{k}_j)$ and $E_j \to E_j^*$. Making use of Eqs. (5.40) and (5.44), we obtain

$$\tilde{P}_{s\nu}^{(n)}(\omega_s) = \frac{n!}{(2\pi)^{n-1}} \int d\omega_1' \cdots \int d\omega_n' \delta(\omega_s - \omega_s') \chi^{(n)}_{\nu\nu_1\ldots\nu_n}(-\mathbf{k}_s - \omega_s'; \mathbf{k}_1\omega_1' \cdots \mathbf{k}_n\omega_n')$$
$$\times \tilde{E}_{1\nu_1}(\omega_1') \cdots \tilde{E}_{n\nu_n}(\omega_n'). \tag{5.48}$$

The number of terms appearing in the calculation of $P^{(n)}$ is very large, considering the various Liouville space pathways combined with the permutations over fields. Fortunately, once a specific choice of $\mathbf{k}_s$ is made for a particular resonant nonlinear technique, only a few of these terms will be dominant. The reason is that each change in sign of $\mathbf{k}_j \to -\mathbf{k}_j$ is also accompanied by reversing the sign of the corresponding frequency $\omega_j \to -\omega_j$ in Eq. (5.45). The various pathways differ by the action of $\mathscr{V}$ from the left or the right. This naturally changes the sign of relevant material frequencies. Consequently, some of the contributions to $P^{(n)}(\mathbf{k}_s, t)$ contain highly oscillatory terms where the molecular and field frequencies add, whereas in other contributions these frequencies subtract. The former terms make a very small contribution and may be neglected. This approximation, known as the rotating wave approximation, simplifies the calculation considerably as will be demonstrated repeatedly in the book, when we carry out a detailed analysis of various techniques. We further note that although $P(\mathbf{r}, t)$ is real, $P(\mathbf{k}_s, t)$ is complex and has both real and imaginary parts. The measurement of the modulus and the phase of the complex polarization in $\mathbf{k}$ space using heterodyne detection was discussed in Chapter 4 and will be illustrated later for specific applications.

In an ideal frequency-domain optical process, the envelopes of all applied fields are time independent.

$$E_{j\nu}(t) = E_{j\nu} \exp(-i\omega_j t), \tag{5.49a}$$

or

$$\tilde{E}_{j,\nu}(\omega_j') = 2\pi E_{j,\nu}\delta(\omega_j' - \omega_j), \qquad j = 1, \ldots, n. \tag{5.49b}$$

In this case, Eq. (5.47) reduces to

$$P_{\nu_s}^{(n)}(\mathbf{k}_s, t) = n! \sum_{\nu_1 \ldots \nu_n} \chi^{(n)}_{\nu_1 \ldots \nu_n \nu_s}(-\mathbf{k}_s - \omega_s; \mathbf{k}_1\omega_1, \mathbf{k}_2\omega_2, \ldots, \mathbf{k}_n\omega_n) \cdot E_{1,\nu_1} E_{2,\nu_2} \cdots E_{n,\nu_n}. \tag{5.50}$$

For the sake of clarity we hereafter drop the tensor notation. We shall use it in Chapter 15 where we discuss polarization spectroscopy.

OPTICAL RESPONSE OF A HOMOGENEOUS MEDIUM OF NONINTERACTING PARTICLES

The relationship between microscopic single-particle response functions and macroscopic susceptibilities is a complex many body problem that does not have a simple general solution. Various procedures for addressing this important issue will be introduced in Chapters 16 and 17. For now let us confine ourselves to the simplest case and assume that the material system consists of a collection of particles that interact with the radiation field and may also be coupled to some external bath degrees of freedom but that do not directly interact with each other. This is always the case when the number density of particles ρ_0 is sufficiently low. The following results can thus be thought of as the low-density limit of the optical susceptibilities. The polarization density may be expressed as the sum over the dipole operators for all particles and we have

$$H_{\text{int}}(t) = -\sum_m E(\mathbf{R}_m, t) \cdot V_m,$$

where V_m is the dipole operator of the particle labeled m and located at $\mathbf{R}_m$. We can then treat the polarizations at different points $P(\mathbf{r})$ as independent degrees of freedom, and the optical response is local in space (i.e., the polarization at a given point $\mathbf{r}$ depends only on the field acting at that point).

$$S^{(n)}(\mathbf{r}; \mathbf{r}_n \cdots \mathbf{r}_1, t_n \cdots t_1) = S^{(n)}(t_n \cdots t_1)\delta(\mathbf{r} - \mathbf{r}_n)\delta(\mathbf{r} - \mathbf{r}_{n-1}) \cdots \delta(\mathbf{r} - \mathbf{r}_1),$$

where $S^{(n)}(t_n \cdots t_1)$ are the single particle response functions introduced earlier in this chapter. We can therefore focus on a single particle, located at $\mathbf{r}$, and the total macroscopic polarization can be simply calculated by multiplying its polarization by the number density of particles ρ_0 (not to be confused with the density operator). Similarly, the susceptibilities $\chi^{(n)}$ do not depend explicitly on the wavevectors in this case. The linear susceptibility is

$$\chi^{(1)}(-\omega_1;\omega_1) = -\frac{1}{\hbar}\rho_0[J(\omega_1) + J^*(-\omega_1)]. \tag{5.51}$$

Hereafter we adopt abbreviated notation for the linear susceptibility and denote it as $\chi^{(1)}(\omega_1)$.

It is common to use the *dielectric function* $\varepsilon(\omega)$ instead of the linear response function. $\varepsilon(\omega)$ is defined by [13]

$$\varepsilon(\omega) \equiv 1 + 4\pi\chi^{(1)}(\omega). \tag{5.52a}$$

The displacement field is then given by $D^{(1)} = \varepsilon(\omega)E$. Combining this with Eq. (5.51) we have

$$\varepsilon(\omega) = 1 - 4\pi\frac{\rho_0}{\hbar}[J(\omega_1) + J^*(-\omega)]. \tag{5.52b}$$

The second order susceptibility is given by

$$\chi^{(2)}(-\omega_S;\omega_1,\omega_2) = \frac{1}{2!}\rho_0\sum_p S^{(2)}(\omega_1+\omega_2,\omega_1), \tag{5.53}$$

where $S^{(2)}$ is given by Eq. (5.36a). The third-order susceptibility is

$$\chi^{(3)}(-\omega_S;\omega_1,\omega_2,\omega_3) = \frac{-1}{3!}\rho_0\sum_p S^{(3)}(\omega_1+\omega_2+\omega_3,\omega_1+\omega_2,\omega_1), \tag{5.54}$$

where $S^{(3)}$ is given by Eq. (5.38a).

The optical polarization in a macroscopic system $P(\mathbf{k}_s, t)$ always depends on the wavevector. When the density of particles is low, the response is local; the response functions $S^{(n)}$ and the susceptibilities $\chi^{(n)}$ do not depend on the wavevector, and the only wavevector dependence of $P(\mathbf{k}_s, t)$ is through the radiation field. This is no longer the case when particles are allowed to interact, or when the electronic states of the system are delocalized, since then there is an additional nonlocal dependence on wavevectors of the response functions themselves! The calculation then becomes a much more complex many-body problem, and new and fascinating effects related to spatial coherences and exciton dynamics and transport may show up. One difficulty is that the field acting on a given particle is not equal to the average field in the medium. This is known as the local field problem. Consequently, even though the particles may not interact directly, their common interaction with the field creates some dynamic correlations that make this a complex many-body problem. Nonlocal response will be addressed in Chapters 16 and 17.

TIME VERSUS FREQUENCY DOMAIN TECHNIQUES

The formal expressions for the optical polarization in terms of nonlinear response functions provide a natural general framework for the calculation of optical measurements. The nonlinear response function $S^{(n)}$ [Eq. (5.14)] or its Fourier transform [Eq. (5.31b)] constitutes an intrinsic material property that contains the complete microscopic information necessary for the calculation of any nth order nonlinear optical process. The details of a particular experiment are contained in the external fields, $E_j(t)$, and in the particular choice of the observation mode $\mathbf{k}_s$. The optical polarization is calculated by convoluting the response function with the external fields, and choosing $\mathbf{k}_s$ [Eq. (5.43)]. It is only at this stage that the distinction is made among the various spectroscopic techniques.

When the experiment is performed using pulses with finite durations (and bandwidths), we need to integrate the polarization over all the time intervals [Eq. (5.13)], or frequencies [Eq. (5.32)] in order to calculate the optical response. There are, however, two limiting cases representing *ideal time-domain or ideal frequency-domain* spectroscopies whereby these integrations can be eliminated, and the final expression simplifies considerably.

In an ideal time-domain optical process, all the applied fields are pulsed, with durations short compared to any material dynamics timescale; their temporal envelopes $E_j(\tau)$ can then be represented by

$$E_j(\tau) = E_j\delta(\tau - \tau_j)\exp(-i\omega_j\tau_j), \qquad j = 1, \ldots, n \tag{5.55}$$

with $\tau_1 < \cdots < \tau_{n-1} < \tau_n < t$. Substituting Eq. (5.55) in Eq. (5.40), we obtain

$$P^{(n)}(\mathbf{k}_S, t) = S^{(n)}(t_n, t_{n-1}, \ldots, t_1)E_n E_{n-1} \cdots E_1 \exp[-i(\omega_1 + \cdots + \omega_n)t]$$
$$\times \exp[i(\omega_1 + \omega_2 + \cdots + \omega_n)t_n + \cdots + i(\omega_1 + \omega_2)t_2 + i\omega_1 t_1], \tag{5.56}$$

where $t_j \equiv \tau_{j+1} - \tau_j$, $j = 1, \ldots, n-1$, and $t_n \equiv t - \tau_n$.

The nonlinear polarization is in this case directly proportional to the nonlinear response function, with the time arguments being the delays between successive pulses. The n-wave mixing signal Eq. (4.77) is proportional to

$$I_s \sim |S^{(n)}(t_n, t_{n-1}, \ldots, t_1)|^2. \tag{5.57}$$

In the other extreme of an ideal frequency-domain experiment, the field envelopes do not depend on time [Eq. (5.50)] and the polarization is proportional to the susceptibilities $\chi^{(n)}$ [Eq. (5.48)]. The n-wave signal is then given by

$$I_s \sim |\chi^{(n)}(-\omega_s; \omega_1, \omega_2, \ldots, \omega_n)|^2. \tag{5.58}$$

A word of caution: The two limits are not symmetric, and there is a fundamental difference between them. The reason is that it is, in principle, impossible to generate a light pulse shorter than an optical period, and even very short pulses ($\sim$20 fs) are still long compared to optical periods. Equation (5.57) as it stands correspond to unrealistic pulses that are a "mathematical" δ function. A "physical" δ function pulse is short compared with all relevant material timescales, but is still long compared with optical periods. For such pulses we should invoke the *rotating wave approximation* (RWA) [14] and select only resonant terms in $S^{(n)}$ where the optical frequency is essentially cancelled by a material frequency of opposite sign, and neglect terms that oscillate at optical frequencies. Depending on the choice of $\mathbf{k}_s$ and the particular spectroscopic technique, we should therefore use in Eq. (5.57) only the few terms in $S^{(n)}$ that survive under the RWA. This, of course, simplifies the calculation and the interpretation of time-domain measurements since they require fewer contributions. On the other hand, the calculation is not universal and the surviving terms should be selected on a case by case basis. *It is incorrect to eliminate the time integrations and use the complete expression for* $S^{(n)}$ in Eq. (5.57). This problem does not exist in the frequency domain expression, since there is no fundamental limitation on the spectral resolution and in principle it is possible to use ideal monochromatic beams. A mathematical δ function in frequency is an achievable goal by a competent experimentalist. A bandwidth of $\sim$kHz that corresponds to a relative bandwidth over frequency of $\sim 10^{-12}$ is not uncommon, and laser bandwidths of 1 Hz are also achievable. The non-RWA terms will be very small and they can be neglected if we so wish. For monochromatic beams we thus have a choice between a universal and lengthy expression for $\chi^{(n)}$ or a specific and more compact expression with fewer terms. We do not enjoy that luxury in ideal impulsive time-domain measurements, where we need to be specific and select the right terms.

We further note that in time-domain experiments we may control the temporal order of the various radiative interactions. In a frequency-domain experiment, however, all time orderings contribute. For example, the response functions $S^{(2)}(t_2, t_1)$ and $S^{(3)}(t_3, t_2, t_1)$ contain 4 and 8 terms, respectively, whereas the corresponding frequency-domain susceptibilities contain $4 \times 2 = 8$ and $8 \times 6 = 48$ terms. Time-domain experiments are therefore usually easier to interpret. On the other hand, some interesting interference effects may show up in frequency-domain experiments. This will be demonstrated, for example, in our discussion of ground-state and excited-state CARS in Chapter 9.

As indicated earlier, the form of Eqs. (5.16) guarantees that $S^{(1)}(t_1)$, $S^{(2)}(t_2, t_1)$ and $S^{(3)}(t_3, t_2, t_1)$ are real functions. In fact, it can be easily shown that all $S^{(n)}$ are real. This follows from their definitions [Eq. (5.13)] since they relate two real quantities, the electric field $E(\mathbf{r}, t)$ and the polarization $P^{(n)}(\mathbf{r}, t)$. The individual contributions (Liouville space paths) to the response function are, however, complex. In frequency-domain measurements, or once a particular wavevector $\mathbf{k}_s$ is chosen for the signal, the relevant polarization $P^{(3)}(\mathbf{k}_s, t)$ is complex. Consequently, the optical susceptibilities are complex as well. Let us comment now on the physical significance of the real and the imaginary parts of $\chi^{(n)}$. To that end we define a zero-order field

$$E_s^0 \equiv \varepsilon_s \cos(\omega_s t - k_s \mathbf{r}),$$

with amplitude

$$\varepsilon_s \equiv E_1 E_2 \cdots E_n.$$

E_s^0 can be thought of as the source for the nonlinear polarization $P^{(n)}$, and we can recast Eq. (5.50) into the form

$$P^{(n)}(\mathbf{r}, t) = \chi^{(n)}(-\omega_s; \omega_1, \ldots, \omega_n)\varepsilon_s \exp(i\omega_s t - i\mathbf{k}_s \cdot \mathbf{r}) + c.c.$$

we next separate $\chi^{(n)}$ into its real and imaginary parts

$$\chi^{(n)} = \mathrm{Re}\,\chi^{(n)} + i\,\mathrm{Im}\,\chi^{(n)}.$$

Combining the above two equations we get

$$P^{(n)}(\mathbf{r}, t) = (\mathrm{Re}\chi^{(n)})\varepsilon_s \cos(\omega_s t - \mathbf{k}\cdot\mathbf{r}) + (\mathrm{Im}\chi^{(n)})\varepsilon_s \sin(\omega_s t - \mathbf{k}\cdot\mathbf{r}). \quad (5.59)$$

The real part of $\chi^{(n)}$ thus gives the component of $P^{(n)}$ that is in phase with the inducing field E_s^0 whereas the imaginary part gives the out of phase component. Methods for probing separately the real and the imaginary parts of the response functions are available and will be discussed later. Heterodyne detection with phase locked pulses is one example (see Chapter 10).

WHY LIOUVILLE SPACE?

Liouville space descriptions are widely used in the theory of ordinary linear optical lineshapes. They are common in magnetic resonance and in the description of nonlinear optics of few-level systems. For example, the Bloch equations describe the evolution of a two-level system interacting with a strong radiation field and with a thermal bath. The effects of the bath are incorporated via relaxation parameters representing level (T_1) relaxation and coherence (T_2) dephasing (see Appendix 6B). These equations also operate in Liouville space. However, they cannot be applied to the spectroscopy of complex systems such as solvated dyes or semiconductors that require a more microscopic treatment. The nonlinear optics literature is loaded with attempts to force any physical system, no matter how complex, into a two-level system that can be treated by the Bloch equations. Such attempts have naturally a limited success and do not provide much of a physical insight. In the present formulation, on the other hand, we have used the Liouville space formulation to express the nonlinear response functions in terms of multitime correlation functions. $J(t_1)$, $Q_\alpha(t_2, t_1)$, and $R_\alpha(t_3, t_2, t_1)$ are the auxiliary linear, second-order, and third-order nonlinear response functions, which can be expressed in terms of two-point, three-point, and four-point dipole correlation functions, respectively.

Correlation functions provide a natural link between theory and experiment. They can be formally defined without alluding to a particular technique, and, consequently, they unify a large body of information and clarify the fundamental relationships among various techniques. The present formalism that connects the various linear and nonlinear optical measurements to the proper response functions requires the development of efficient systematic approaches to calculate the response functions for a variety of realistic physical systems including molecules, semiconductors, and metals. The correlation functions can be evaluated by numerous methods, e.g., the density expansion for pressure broadening in the gas phase and in liquids, the cumulant expansion for phonon broadening, semiclassical and molecular dynamics simulations, dielectric continuum models, etc. These applications will be presented in the coming chapters.

We shall now discuss the physical picture offered by the density operator and elaborate on the main advantages of Liouville space.

Intuitive picture and semiclassical approximations

Physical observables are directly and linearly related to the density operator. Consequently, every step and intermediate quantity appearing in the density operator description has a simple physical meaning and a clear classical analogue. This is to be contrasted with wavefunction-based theories that calculate a transition amplitude, which by itself is not an observable quantity. The signal is related to sums of products of such amplitudes. Moreover, the density operator, which can be represented as a wavepacket in Liouville space, has a well-defined classical limit, namely the classical distribution function in phase space $\rho_c(\mathbf{pq}; t)$. The Wigner representation of the density operator offers the opportunity to develop a beautiful and simple semiclassical picture that interpolates between the fully quantum and classical pictures. Wavefunctions, on the other hand, do not have a clear classical counterpart (although there are, of course, very powerful semiclassical approximations for the wavefunction such as the WKB approximation) [15].

Reduced descriptions and thermal averaging: mixed states

The advantages of the Liouville space description are apparent for large systems at finite temperatures. In the Liouville space form we calculate directly the thermally averaged signal by propagating a density operator representing a mixed state (rather than a pure state). No further averaging over thermal distributions is necessary. Furthermore, in the condensed phase we must develop a *reduced description* in which we follow explicitly only the dynamics of a few selected and relevant degrees of freedom; the remaining degrees of freedom are treated as a thermal bath using methods of quantum statistical mechanics. The bath may consist of some intermolecular vibrational degrees of freedom that are weakly coupled to the electronic transition, solvent degrees of freedom, optical and acoustic phonons in semiconductors, local modes for impurity spectra, etc. The Brownian oscillator model, to be introduced in Chapter 8, provides a nice example for a system with a simple reduced description. The density operator offers a practical way for developing such a reduced description, which allows us to perform all the necessary averagings directly. Using the wavefunction we have to consider each member of the ensemble, calculate its dynamics, and average the result at the end. When this is done, we find that the effects of the bath on the various Liouville space terms are profoundly different, which further demonstrates why we need to reformulate the problem in Liouville space, where these terms are naturally separated.

There is a tremendous experimental and theoretical interest in studies of geometrically confined atomic, molecular semiconductor, and metal systems such as nanostructures and clusters. Some of the key questions in these studies are: when is a system large enough to acquire bulk properties and what are the relevant material coherence sizes that dominate the optical response. By formulating the problem in Liouville space we can treat small and large systems in the same fashion, and address these issues in a clear way. We thus obtain a unified formulation of systems ranging from isolated atoms and molecules to condensed phases.

Time-ordering and Liouville space pathways

When the density operator is expanded in powers of an external perturbation, such as an incoming electric field, the resulting expansion separates naturally into several contributions, each representing a different time ordering of the various interactions. The time variables appearing in the expansion of the density operator are chronologically ordered and represent the time intervals between successive interactions. This is not so when the calculation is performed by expanding the wavefunction. The time variables appearing in the wavefunction description are not fully ordered and consequently have a much less transparent physical interpretation (see Appendix 5A). The complete and direct bookkeeping of time ordering makes the description using the density operator most intuitive and directly relevant to experiment. As an example, in Chapter 13 we shall develop the doorway-window picture for pump-probe spectroscopy. The resulting intuitive separation of the calculation into preparation, evolution, and detection stages follows naturally from the density matrix formulation.

The nonlinear response function is calculated by summing over the various possible pathways in Liouville space, which contribute to the polarization. This amounts to a path integral in Liouville space [16, 17]. Path integrals have been extensively used as a useful tool for numerical computations of mixed quantum-classical calculations. The density operator formulation strongly suggests a similar development for phase space-based numerical procedures. Graphic visualization of these paths is provided by the coupling scheme or the double-sided Feynman diagrams, which will be introduced in Figure 6.3.

Finally, since the various Liouville space pathways are complex quantities, they may interfere when added. This interference sometimes results in dramatic effects such as the generation of fluorescence, and extra resonances in four wave mixing. These will be discussed in Chapter 9.

Relationships among different nonlinear spectroscopies

The nonlinear response function can be applied to a broad range of nonlinear optical measurements, which differ by the temporal sequences of pulses as well as their frequencies and wavevectors. The response function enables us to calculate all these spectroscopies and compare their information content. Since the nonlinear response functions are successively probing higher order correlation functions, they necessarily contain more information as the order n is increased. This can be utilized, e.g., to eliminate homogeneous broadening selectively, as is done in photon echoes or hole burning (Chapter 10), which is in principle impossible using the linear response. When calculating these measurements using wavefunctions we need to repeat the calculation for every new technique. The Liouville space response functions offer, therefore, a compact and a unified treatment of a large variety of spectroscopies. For example, the relationship among coherent and spontaneous Raman techniques, fluorescence, hole burning, and photon echo, etc. can be rigorously established without alluding to any specific model for the matter; the formal correlation function expressions are sufficient, and specific models can be introduced only at the very end of the calculation. An important example is the popular classification of line broadening mechanisms as either "homogeneous" or "inhomogeneous." This classification is based on the assumption of a separation of timescales between the relevant inverse spectral width and the underlying timescales of the system. In the case of the Brownian oscillator model, which may represent an arbitrary timescale, a unified treatment of both cases can be formulated and solved using the Liouville space formulation (see Chapter 8). A continuous transition from homogeneous to inhomogeneous behavior as the material timescales are varied can be described, and the above classification becomes obsolete. Finally, there is a fundamental interest in the establishment of exact Fourier transform relationships among frequency-domain and time-domain techniques. This issue is addressed very generally and clearly by the Liouville space treatment.

The role of dephasing processes

Diagonal and off diagonal elements of the density operator are called populations and coherences, respectively. When an off diagonal element evolves in time, it acquires a phase since its evolution from the left (ket) and the right (bra) is governed by different nuclear Hamiltonians. This phase depends on the state of the bath. When we perform an ensemble average of these elements over the distribution of the bath degrees of freedom, this variable phase results in a damping of these elements. The damping of off diagonal elements of the density operator resulting from phase (as opposed to amplitude) fluctuations is called dephasing or phase relaxation. Dephasing processes can be described only in Liouville space, where we distinguish between diagonal and off diagonal elements of the density operator. We then follow simultaneously the evolution of the bra and of the ket and maintain the bookkeeping of their joint state. Dephasing processes directly affect all spectroscopic observables. Since different Liouville space pathways represent distinct sequences of populations and coherences, they will be affected differently by nuclear motions and dephasing processes. This further demonstrates why we need to formulate the problem in Liouville space, where these terms are naturally separated. It should be emphasized that in a complete description of a system, where we follow explicitly the dynamics of all degrees of freedom, we do not need to introduce the concept of dephasing, and a wavefunction description is of course correct. However, this is impractical for realistic condensed-phase systems where a reduced description is essential. A beautiful analogy is the concept of entropy. The entropy of any system described by the density operator ρ can be defined as [18–20]

$$S = -k_B \operatorname{Tr} \rho \ln \rho. \tag{5.60}$$

Since any function of ρ also satisfies the Liouville equation, and since the trace of a commutator vanishes, we immediately have that the entropy is independent on time.

$$\dot{S} \equiv \frac{i}{\hbar} k_B \operatorname{Tr}[H, \rho \ln \rho] = 0. \tag{5.61a}$$

This is, however, not the case for the reduced density operator σ [see Eq. (3.22)]. The system's entropy is defined as

$$S = -k_B \operatorname{Tr}_S \sigma \ln \sigma. \tag{5.61b}$$

This entropy does change with time due to dephasing and relaxation processes induced by the bath, reflecting the loss of information in a reduced description. In a complete dynamic description of a system we do not need to introduce entropy; however, on the greatly reduced thermodynamic level, entropy is essential for a proper description of the system. This is illustrated in Eqs. (5.61) where the entropy of the entire system, "the universe," is time independent whereas the entropy of the subsystem is a significant and useful measure of the amount of missing information about the system. In principle it is possible to calculate any physical process without ever considering the entropy. However, it is hard to imagine our understanding of macroscopic phenomena without the second law of thermodynamics. The same arguments apply to dephasing processes.

APPENDIX 5A

Wavefunction versus Density-Operator Formulation of the Nonlinear Response

As indicated in Chapter 3, quantum dynamics can be adequately formulated without introducing the density operator, but using the wavefunction instead. The optical polarization, for example, can be calculated by using a perturbative expansion of the wavefunction. Using the density operator offers, however, many advantages, and in complex systems in condensed phases we are forced in practice to use it.

At this point we address more fully the relationship between the Liouville space and the Hilbert space formulations of the optical response. We shall make a specific comparison for the calculation of $P^{(3)}$. We recall that

$$P^{(3)}(\mathbf{r}, t) \equiv \mathrm{Tr}[V\rho^{(3)}(\mathbf{r}, t)].$$

When the system is in a pure state we have

$$\rho(\mathbf{r}, t) = |\psi(\mathbf{r}, t)\rangle\langle\psi(\mathbf{r}, t)|$$

where ψ is the wavefunction. The $\mathbf{r}$ index reminds us that the molecular density operator depends on the phase of the field, which varies with position $\mathbf{r}$. Using these definitions it is clear that $\rho^{(3)}$ can result from expanding the bra and the ket separately to various orders in the field, keeping the sum of their orders to be 3, i.e.,

$$\rho^{(3)}(\mathbf{r}, t) = |\psi^{(3)}(\mathbf{r}, t)\rangle\langle\psi^{(0)}(\mathbf{r}, t)| + |\psi^{(2)}(\mathbf{r}, t)\rangle\langle\psi^{(1)}(\mathbf{r}, t)| + h.c.$$

where $|\psi^{(n)}\rangle$ is the expansion of the wavefunction to nth order in the field. Substituting this in the expression for $P^{(3)}$ we get

$$P^{(3)}(\mathbf{r}, t) = \langle\psi^{(3)}(t)|V|\psi^{(0)}(t)\rangle + \langle\psi^{(2)}(t)|V|\psi^{(1)}(t)\rangle + c.c. \tag{5.A.1}$$

Using time-dependent perturbation theory we have

$$|\psi^{(0)}(\mathbf{r}, t)\rangle = |\psi(-\infty)\rangle,$$

$$|\psi^{(1)}(\mathbf{r}, t)\rangle = \frac{i}{\hbar}\int_{-\infty}^{t} d\tau_1\, E(\mathbf{r}, \tau_1)G(t - \tau_1)V|\psi(-\infty)\rangle,$$

$$|\psi^{(2)}(\mathbf{r}, t)\rangle = \left(\frac{i}{\hbar}\right)^2 \int_{-\infty}^{t} d\tau_2 \int_{-\infty}^{\tau_2} d\tau_3\, E(\mathbf{r}, \tau_2)E(\mathbf{r}, \tau_3)G(t - \tau_2)VG(\tau_2 - \tau_3)V|\psi(-\infty)\rangle.$$

$$|\psi^{(3)}(\mathbf{r}, t)\rangle = \left(\frac{i}{\hbar}\right)^3 \int_{-\infty}^{t} d\tau_1 \int_{-\infty}^{\tau_1} d\tau_2 \int_{-\infty}^{\tau_2} d\tau_3\, E(\mathbf{r}, \tau_1)E(\mathbf{r}, \tau_2)E(\mathbf{r}, \tau_3)$$
$$\times G(t - \tau_3)VG(\tau_3 - \tau_2)VG(\tau_2 - \tau_1)V|\psi(-\infty)\rangle. \qquad (5.A.2)$$

Here

$$G(t) = \theta(t)\exp\left(-\frac{i}{\hbar}Ht\right)$$

is the Green function in Hilbert space. $\rho^{(3)}$ and $P^{(3)}$ have naturally four terms. This is to be contrasted with the direct Liouville space expansion of the density operator, which gives

$$\rho^{(3)}(\mathbf{r}, t) \equiv \left(\frac{i}{\hbar}\right)^3 \int_0^{\infty} dt_3 \int_0^{\infty} dt_2 \int_0^{\infty} dt_1\, E(\mathbf{r}, t - t_3)E(\mathbf{r}, t - t_2 - t_3)$$
$$\times E(\mathbf{r}, t - t_1 - t_2 - t_3)\mathscr{G}(t_3)\mathscr{V}\mathscr{G}(t_2)\mathscr{V}\mathscr{G}(t_1)\mathscr{V}|\psi(-\infty)\rangle\langle\psi(-\infty)|.$$

Since $\rho^{(3)}$ contains the Liouville space dipole operator $\mathscr{V}$ to third order, and since each $\mathscr{V}$ represents a commutator, which can act either from the left or from the right, the nonlinear response function $S^{(3)}$ will contain $2^3 = 8$ terms once these commutators are evaluated. How are the four terms in the Hilbert space expansion (5.A.1) related to the eight terms in the Liouville space expansion? The first term in (5.A.1) is identical with R_1 of Eqs. (5.28). Let us consider the second term more closely. When Eqs. (5.A.2) are substituted in the matrix element $\langle\psi^{(2)}(t)|V|\psi^{(1)}(t)\rangle$ we see that its evaluation requires a triple integration over τ_1, τ_2, and τ_3. τ_1 represents the time in which the radiation field interacted with the ket $|\psi^{(1)}(t)\rangle$, whereas the two interactions with the bra $\langle\psi^{(2)}(t)|$ take place at times τ_2 and τ_3. τ_2 and τ_3 are time ordered (namely $\tau_2 \geq \tau_3$). However, τ_1 varies independently and does not have a specific time ordering with respect to τ_2 and τ_3. In order to keep track of the relative time ordering of the ket (τ_1) and the bra (τ_2, τ_3) interactions, we shall separate the τ_1 integration into three contributions:

$$\int_{-\infty}^{t} d\tau_1 = \int_{-\infty}^{\tau_3} d\tau_1 + \int_{\tau_3}^{\tau_2} d\tau_1 + \int_{\tau_2}^{\infty} d\tau_1. \qquad (5.A.3)$$

When this is done $\langle\psi^{(2)}(t)|V|\psi^{(1)}(t)\rangle$ will be split into three terms which correspond to R_2, R_3, and R_4 of Eqs. (5.28). Similarly, its complex conjugate $\langle\psi^{(1)}(t)|V|\psi^{(2)}(t)\rangle$ will also be split into three terms. Altogether $P^{(3)}$ will thus contain eight terms, in agreement with the Liouville space expression. We note that in the Liouville space form we integrate over t_1, t_2, and t_3, which are the time intervals between successive interactions. In the wavefunction form τ_1, τ_2, and τ_3 are the times of the actual interactions with the fields.

In summary, the Hilbert space expression yields the same eight terms obtained from the Liouville space expansion after the substitution of Eqs. (5.A.2) and (5.A.3). *In the Liouville space formulation we maintain a simultaneous book-keeping of the interactions with the ket and with the bra and each of the resulting terms has a complete well-defined time ordering.* Each time ordering is calculated separately as a distinct Liouville space pathway (with the corresponding Feynman diagram, as will be shown in the next chapter). This is not the case for the wavefunction calculation where we focus on amplitudes, and the various time orderings of the ket and the bra interactions are lumped together. The different terms in this case simply reflect the order of the interactions within the bra and the ket (but not the relative time ordering of bra and ket interactions!). When the system interacts with a thermal bath, the eight terms represent very different physical processes, and their separate treatment is absolutely crucial. The density operator separates these terms directly and naturally without the need for any change of time variables.

In conclusion we note that these arguments are not limited to $P^{(3)}$ and can be easily extended to any order. In general, we have

$$P(\mathbf{r}, t) = \langle\psi(\mathbf{r}, t)|V|\psi(\mathbf{r}, t)\rangle, \tag{5.A.4}$$

and

$$P^{(n)}(\mathbf{r}, t) = \sum_{m=0}^{n} \langle\psi^{(n-m)}(\mathbf{r}, t)|V|\psi^{(m)}(\mathbf{r}, t)\rangle, \tag{5.A.5}$$

with

$$\psi^{(n)}(\mathbf{r}, t) = \left(\frac{i}{\hbar}\right)^n \int_{-\infty}^{t} d\tau_n \int_{-\infty}^{\tau_n} d\tau_{n-1} \cdots \int_{-\infty}^{\tau_2} d\tau_1\, E(\mathbf{r}, \tau_1)E(\mathbf{r}, \tau_2)\cdots E(\mathbf{r}, \tau_n)$$
$$\times\, G(t-\tau_n)VG(\tau_n-\tau_{n-1})V\cdots G(\tau_2-\tau_1)V|\psi(-\infty)\rangle. \tag{5.A.6}$$

The connection between the wavefunction and the density operator expressions can be established also for a general (not necessarily pure) state, by simply averaging over the distribution of initial states at the end of the calculation

$$P(\mathbf{r}, t) = \sum_a P(a)\langle\psi_a(\mathbf{r}, t)|V|\psi_a(\mathbf{r}, t)\rangle.$$

APPENDIX 5B

Calculating the Response Functions Using the Heisenberg Representation

The Heisenberg picture provides an alternative method for calculating the response function. It has numerous advantages, particularly for complex many-body systems (as will be demonstrated in Chapter 17). In order to derive the response function using this picture, we first write

$$P^{(n)}(t) = \langle\langle\rho(-\infty)|V^{(n)}(t)\rangle\rangle. \tag{5.B.1}$$

$V(t)$ is the dipole coupling in the Heisenberg picture, which satisfies the Heisenberg equation of motion,

$$\dot{V} = \frac{i}{\hbar}\mathscr{L}V + \frac{i}{\hbar}\mathscr{L}_{\text{int}}(t)V. \tag{5.B.2}$$

The general solution of this equation can be written by introducing the Green function

$$\mathscr{G}^{\dagger}(t) = \theta(t)\exp_{+}\left[\frac{i}{\hbar}\int_0^t d\tau\,\mathscr{L}(\tau)\right].$$

We then have

$$V(t) = \mathscr{G}^{\dagger}(t)\exp_{+}\left[\frac{i}{\hbar}\int_0^t d\tau\,\mathscr{G}(\tau)\mathscr{L}_{\text{int}}(\tau)\mathscr{G}^{\dagger}(\tau)\right]V(0),$$

so that

$$V^{(n)}(t) = \left(\frac{i}{\hbar}\right)^n \int_{t_0}^t d\tau_1 \int_{t_0}^{\tau_1} d\tau_2 \cdots \int_{t_0}^{\tau_{n-1}} d\tau_n$$

$$\mathscr{G}^{\dagger}(t-\tau_1)\mathscr{L}_{\text{int}}(\tau_1)\mathscr{G}^{\dagger}(\tau_1-\tau_2)\mathscr{L}_{\text{int}}(\tau_2)\cdots\mathscr{G}^{\dagger}(\tau_{n-1}-\tau_n)\mathscr{L}_{\text{int}}(\tau_n)\mathscr{G}^{\dagger}(\tau_n)|V(0)\rangle\rangle. \tag{5.B.3}$$

We next multiply by $\rho(-\infty)$ from the left and take a trace. The last $\mathscr{G}^{\dagger}(t-\tau_1)$ factor then disappears since $\rho(-\infty)$ does not evolve with the unperturbed Hamiltonian. We further substitute Eqs. (5.1b) and (5.4b) and get

$$\begin{aligned} P^{(n)}(t) &\equiv \langle\langle\rho(-\infty)|V^{(n)}(t)\rangle\rangle \\ &= \left(-\frac{i}{\hbar}\right)^n \int_{t_0}^t d\tau_1 \int_{t_0}^{\tau_1} d\tau_2 \cdots \int_{t_0}^{\tau_{n-1}} d\tau_n\, E(\tau_1)E(\tau_2)\cdots E(\tau_n) \\ &\quad \times \langle\langle\rho(-\infty)|\mathscr{V}(-\tau_1)\mathscr{V}(-\tau_2)\cdots\mathscr{V}(-\tau_n)|V(0)\rangle\rangle, \end{aligned} \tag{5.B.4}$$

where $\mathscr{V}(\tau)$ was defined in Eq. (5.9b). The correlation function is invariant to a time translation. Adding τ_1 to all time arguments, we have

$$\langle\langle\rho(-\infty)|\mathscr{V}(0)\mathscr{V}(\tau_1-\tau_2)\cdots\mathscr{V}(\tau_1-\tau_n)|V(\tau_1)\rangle\rangle.$$

We next change the time variables (see Figure 5.1) and send $t_0 \to -\infty$, and get

$$P^{(n)}(t) = \left(-\frac{i}{\hbar}\right)^n \int_0^\infty dt_n \int_0^\infty dt_{n-1} \cdots \int_0^\infty dt_1$$
$$E(t-t_n)E(t-t_n-t_{n-1})\cdots E(t-t_1-\cdots-t_n)$$
$$\times\langle\langle\rho(-\infty)|\mathscr{V}(0)\mathscr{V}(t_1)\mathscr{V}(t_2+t_1)\cdots\mathscr{V}(t_{n-1}+\cdots+t_1)|V(t_n+t_{n-1}+\cdots+t_1)\rangle\rangle. \tag{5.B.5}$$

Taking the Hermitian conjugate of Eq. (5.B.5) will result in Eq. (5.15). This demonstrates the equivalence of the Schrödinger picture and the Heisenberg picture.

NOTES

1. M. Gavrila, Ed. *Advances in Atomic Molecular and Optical Physics*, Suppl. 1 (1993).
2. K. C. Kulander, *Phys. Rev. A* **35**, 445 (1987).
3. S. Haroche, *Ann. Phys.* **6**, 189; 327 (1971); C. Cohen-Tannoudji, in *Cargese Lectures in Physics*, Vol. 2 (Gordon and Breach, New York, 1967); C. Cohen-Tannoudji, in *Frontiers in Laser Spectroscopy*, Vol. 1, R. Balian, S. Haroche, and S. Libermann, Eds. (North-Holland, Amsterdam, 1977), p. 1.
4. B. R. Mollow, *Phys. Rev.* 188 (1969); *Phys. Rev. A* **12**, 1919 (1975); **15**, 1023 (1977).
5. B. J. Herman, J. H. Eberly, and M. G. Raymer, *Phys. Rev. A* **39**, 3447 (1989).
6. B. W. Shore, *The Theory of Coherent Atomic Excitation*, Vol. 1, 2 (Wiley, New York, 1990).
7. L. Onsager and S. Machlup, *Phys. Rev.* **91**, 1505 (1953).
8. R. Kubo and K. Tomita, *J. Phys. Soc. Jpn.* **9**, 888 (1954); *R. Kubo, J. Phys. Soc. Jpn.* **12**, 570 (1957).
9. J. A. Armstrong, N. Bloembergen, J. Ducuing, and P. S. Pershan, *Phys. Rev.* **127**, 1918 (1962).
10. P. N. Butcher, *Nonlinear Optical Phenomena* (Ohio State University Engineering Publications, Columbus, 1965).
11. I. Oppenheim, *Progress in Theoretical Physics Suppl.*, K. Kawasaki, Y. Kuramoto, and H. Okamoto, Eds. (1990), p. 369.
12. L. V. Keldysh, D. A. Kirzhnitz, and A. A. Maradudin, *The Dielectric Function of Condensed Systems* (North-Holland, Amsterdam, 1989).
13. S. Mukamel, *Phys. Reports* **93**, 1 (1982); S. Mukamel and R. F. Loring, *J. Opt. Soc. Am. B* **3**, 595 (1986); S. Mukamel, *Annu. Rev. Phys. Chem.* **41**, 647 (1990).
14. W. H. Louisell, *Quantum Statistical Properties of Radiation* (Wiley, New York, 1973).
15. L. D. Landau and E. M. Lifshitz, *Quantum Mechanics Non-Relativistic Theory*, 2nd ed. (Pergamon Press, London, 1965).
16. N. Hashitsume, M. Mori, and T. Takahashi, *J. Phys. Soc. Jpn.* **55**, 1887 (1986); N. Hashitsume, K. Naito, and A. Washiwo, *J. Phys. Soc. Jpn.* **59**, 464 (1990).
17. V. N. Popov, *Functional Integrals in Quantum Field Theory and Statistical Physics* (D. Reidel, Dordrecht, 1983).

18. E. T. Jaynes, *Phys. Rev.* **106**, 620 (1957); **108**, 171 (1957).
19. A. Katz, *Principles of Statistical Mechanics. The Information Theory Approach* (Freeman, San Francisco, 1967) [Eq. (5.60)].
20. W. H. Zurek, in *Frontiers of Nonequilibrium Statistical Physics*, P. Meystre and M. O. Scully, Eds. (Plenum, New York, 1986); W. G. Unruh and W. H. Zurek, *Phys. Rev. D* **40**, 1071 (1989).

CHAPTER 6

The Optical Response Functions of a Multilevel System with Relaxation

In Chapter 5 we developed formal expressions for optical response functions and susceptibilities. These were recast either in terms of Liouville space Green functions, or using ordinary multitime correlation functions in Hilbert space. The derivation was made without alluding to any specific model for the matter. In this chapter we consider a model system consisting of a collection of noninteracting small particles, such as a dilute solution of a dye molecule. The formal expressions of Chapter 5 will be applied to calculate the first-, the second-, and the third-order response functions $S^{(1)}$, $S^{(2)}$, and $S^{(3)}$, as well as the corresponding susceptibilities, using the exact eigenstates of the particle.

All spectroscopic measurements can eventually be related to level positions and transition dipole moments. Highly resolved spectra of simple quantum systems (atoms, and small molecules) are commonly analyzed in terms of their eigenstates [1–3]. Eigenstate-based expressions are always very helpful in gaining a physical insight, and in deriving some exact formal properties of response functions. Typically, in condensed phases, we never attempt to find the global eigenstates of the entire system. Only the eigenstates of a few degrees of freedom are calculated explicitly, and the rest of the system (denoted a thermal bath) is incorporated using correlation function time-domain methods. We thus divide the spectroscopic information into level positions (frequencies), and linewidths (relaxation rates) [4–6], which represent the contributions of many unresolved states. In this chapter we simply add a phenomenological relaxation rate to represent the bath. More microscopic semiclassical treatments of bath degrees of freedom using the coordinate representation will be presented in the next chapter.

We shall first consider the linear response of an isolated particle (with no relaxation) and calculate it in two ways; in Hilbert space, and using Liouville space Green functions. The calculation offers a clear insight on the significance of the Liouville space pathways. We shall discuss the Kramers–Kronig relations and the fluctuation–dissipation theorem. Following that, we evaluate the linear and the nonlinear response functions including relaxation. We shall do so in three ways, which are based on either solving the Liouville equation for the density operator, or the Schrödinger equation for the wavefunction, or the Heisenberg equation for the polarization. In the absence of relaxation these methods yield identical though different looking expressions, in which the Liouville space paths combine in various ways and their interferences, as reflected in one, two, and three photon resonances, show up differently. The type of relaxation and dephasing allowed in each of these schemes is not the same and therefore these results may be different once relaxation is incorporated. We shall compare and analyze these various forms and their range of applicability. Graphic representations of the underlying Liouville space paths including double sided Feynman diagrams are introduced. We then establish the connection with the anharmonic oscillator model for the electronic polarization, which is commonly used in phenomenological descriptions of the nonlinear response. Following that, we introduce a more realistic population relaxation matrix, which conserves the total population, and which is used in the Bloch equations, and show how it affects the nonlinear response. Finally, we provide a microscopic definition of homogeneous, inhomogeneous, and intermediate dephasing using Liouville space Green functions and a weak coupling model to a thermal bath. The relaxation superoperator is calculated in Appendix 6A. A derivation and discussion of the optical Bloch equations are given in Appendix 6B.

LINEAR RESPONSE AND THE KRAMERS–KRONIG RELATIONS

Linear spectroscopy is the simplest and the most elementary tool for exploring the structure and dynamic processes in physical systems through their interaction with the radiation field. When the applied field is sufficiently weak, we retain only the leading (first) term in the expansion [Eq. (5.12)], resulting in the linear response function $S^{(1)}(t)$ and the dielectric function $\varepsilon(\omega)$ (which is related to the linear susceptibility $\chi^{(1)}(\omega)$ through Eq. (5.52). Phenomena such as propagation, reflection, refraction, transmission, and absorption of light are all related to the linear response [7].

Consider the eigenstates of a single small particle

$$H|\nu\rangle = \varepsilon_\nu|\nu\rangle \qquad \nu = a, b, \ldots \tag{6.1}$$

The dipole operator represented in this basis set is

$$V = \sum_{a,b} \mu_{ab}|a\rangle\langle b|. \tag{6.2a}$$

We further introduce the transition frequencies

$$\hbar\omega_{\nu\nu'} \equiv \varepsilon_\nu - \varepsilon_{\nu'}. \tag{6.2b}$$

We assume that the system has no permanent dipoles so that the diagonal elements $\mu_{aa} = 0$. Using these definitions, we have for the time-domain response function [Eq. (5.19)]

$$J(t) = \sum_{a,b} P(a)|\mu_{ab}|^2 \exp(-i\omega_{ba}t), \tag{6.3a}$$

$$S^{(1)}(t) = \frac{2}{\hbar}\theta(t) \sum_{a,b} P(a)|\mu_{ab}|^2 \sin \omega_{ba}t. \tag{6.3b}$$

In the frequency domain we have

$$J(\omega) = \sum_{a,b} P(a) \frac{|\mu_{ab}|^2}{\omega - \omega_{ba} + i\varepsilon}, \tag{6.4}$$

and the linear susceptibility Eq. (5.51) becomes

$$\chi^{(1)}(\omega) = -\frac{1}{\hbar} \rho_0 \sum_{a,b} P(a)|\mu_{ab}|^2 \left[\frac{1}{\omega - \omega_{ba} + i\varepsilon} + \frac{1}{-\omega - \omega_{ba} - i\varepsilon} \right]. \tag{6.5a}$$

Here ρ_0 is the number of particles per unit volume, and $P(a)$ is the thermal population of the $|a\rangle$ state

$$P(a) \equiv \exp(-\beta\varepsilon_a) \Big/ \sum_a \exp(-\beta\varepsilon_a),$$

where $\beta \equiv 1/k_B T$.

Equation (6.5a) can be rearranged in various alternative forms that highlight different formal aspects. Combining the two terms we get

$$\chi^{(1)}(\omega) = \frac{1}{\hbar} \rho_0 \sum_{a,b} P(a) \frac{2|\mu_{ab}|^2 \omega_{ba}}{\omega_{ba}^2 - (\omega + i\varepsilon)^2}. \tag{6.5b}$$

Alternatively, by interchanging the dummy indices a and b in the second term we can recast it in the form

$$\chi^{(1)}(\omega) = \frac{1}{\hbar} \rho_0 \sum_{a,b} [P(b) - P(a)] \frac{|\mu_{ab}|^2}{\omega - \omega_{ba} + i\varepsilon}. \tag{6.5c}$$

The response function $S^{(1)}(t)$ is finite only for $t > 0$ and it vanishes for $t < 0$. This reflects the principle of *causality*; the polarization at a given time may depend only on the electric field at earlier times (the "cause" must precede the "effect"). The signature of causality in the frequency domain is that the poles of $\chi^{(1)}(\omega)$ are all in the lower half complex ω plane, i.e., $\omega = \omega_{ba} - i\varepsilon$. By performing contour integration (see Chapter 2) it immediately follows that

$$\frac{1}{2\pi} \int d\omega\, \chi^{(1)}(\omega) \exp(-i\omega t) = \rho_0 S^{(1)}(t), \tag{6.6}$$

which vanishes for $t < 0$.

An important consequence of causality is that the real and the imaginary parts of $\chi^{(1)}$ are not independent. If we define

$$\chi^{(1)}(\omega) = \chi'(\omega) + i\chi''(\omega), \tag{6.7a}$$

we have

$$\chi'(\omega) = \frac{1}{\pi} \mathscr{P}\!\!\mathscr{P} \int \frac{\chi''(\omega')}{\omega' - \omega} d\omega', \tag{6.7b}$$

and

$$\chi''(\omega) = -\frac{1}{\pi} \mathscr{P}\!\!\mathscr{P} \int \frac{\chi'(\omega')}{\omega' - \omega} d\omega', \tag{6.7c}$$

where $\mathscr{PP}$ stands for the principal part of the integral [see Eq. (2.93a)]. These are known as the *Kramers–Kronig relations* [7, 8]. They allow us to calculate the complete complex susceptibility $\chi^{(1)}(\omega)$ using either $\chi'(\omega)$ or $\chi''(\omega)$ alone. They are frequently used to check the consistency of calculations or measurements. Some extensions of these relations to nonlinear susceptibilities have been suggested [9].

It was shown in Eqs. (4.64) and (4.65) that the absorption cross section is proportional to the imaginary part of the susceptibility. Using Eqs. (6.4) we thus get for the imaginary part of $J(\omega)$ $[J(\omega) \equiv J'(\omega) + iJ''(\omega)]$

$$J''(\omega) = \pi \sum_{a,b} P(a)|\mu_{ab}|^2 \delta(\omega - \omega_{ba}), \tag{6.8a}$$

whereas from Eq. (6.5c) we get

$$\chi''(\omega) = \frac{\pi}{\hbar} \rho_0 \sum_{a,b} [P(a) - P(b)]|\mu_{ab}|^2 \delta(\omega - \omega_{ba}). \tag{6.8b}$$

Equation (6.8a) is known as the *Fermi Golden Rule*. Equation (6.8b) has a simple physical interpretation. The interaction with the radiation field results in absorption as well as stimulated emission. $\chi''(\omega)$ represents the net absorption, which is the difference of the rates of both processes. This equation further shows that two equally populated levels with $P(a) = P(b)$ (e.g., at infinite temperature) do not contribute to the response function or to the absorption lineshapes, since absorption and stimulated emission exactly cancel in this case.

Equations (6.8) have some interesting symmetries. We first note that χ'' is an odd function of frequency, i.e.,

$$\chi''(-\omega) = -\chi''(\omega). \tag{6.9a}$$

Using the canonical distribution function we have

$$P(a) - P(b) = P(a)[1 - P(b)/P(a)] = [1 - \exp(-\beta\hbar\omega_{ba})]P(a).$$

When this is substituted in Eq. (6.8b) and using Eq. (6.8a) we get

$$\chi''(\omega) = \frac{1}{\hbar} \rho_0[1 - \exp(-\beta\hbar\omega)]J''(\omega). \tag{6.9b}$$

Combining Eq. (6.9a) and (6.9b) we obtain

$$J''(-\omega) = \exp(-\beta\hbar\omega)]J''(\omega). \tag{6.9c}$$

Equation (6.9c) can be also obtained directly from Eq. (6.8a).

Equation (6.9c) is the *detailed balance condition* [10] and Eq. (6.9b) is the *fluctuation–dissipation theorem* [11, 12], which provides an exact relationship between absorption (i.e., energy dissipation) rate and thermal fluctuations represented by the correlation function $J(t)$. This theorem is remarkable in its universality; the only assumption made is that the response is linear, and no further assumption is necessary regarding the physical nature of the system. J'' and χ'' are depicted schematically in Figure 6.1.

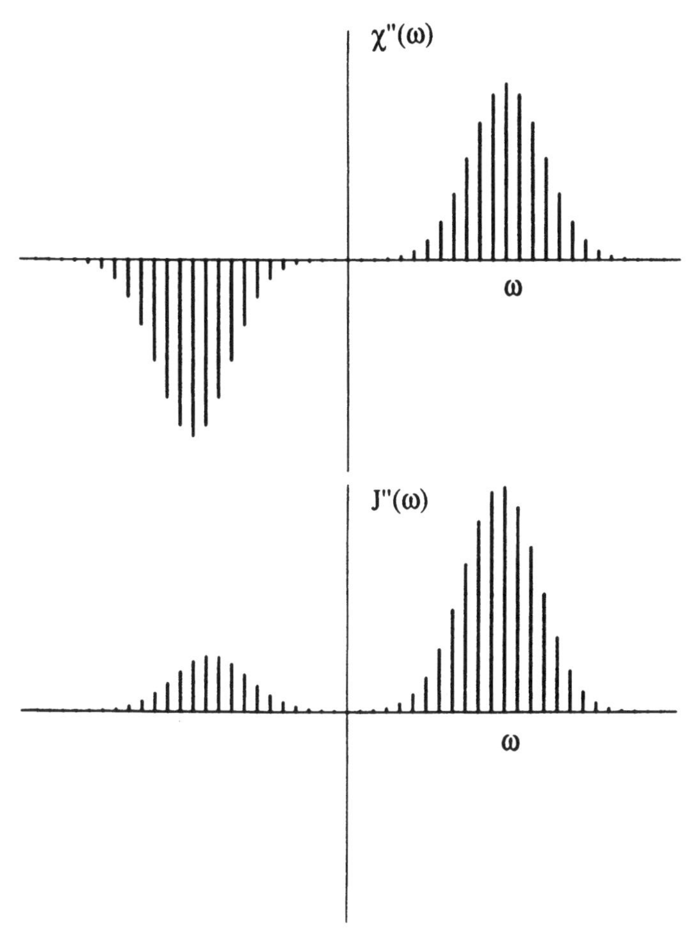

FIG. 6.1 Schematic form of the spectral density of the correlation function $J''(\omega)$ and of the response function $\chi''(\omega)$. The asymmetric form of J'' [Eq. (6.9c)] and the antisymmetric form of χ'' [Eq. (6.9a)] reflect the detailed balance condition and the fluctuation dissipation theorem.

EIGENSTATES REVISITED: LINEAR RESPONSE IN LIOUVILLE SPACE

So far we have calculated the linear response function by evaluating the ordinary correlation function $J(t)$. This calculation was not actually performed in Liouville space. The Liouville space and the density operator were merely used to derive the formal expression, but the actual expansion in eigenstates was made in Hilbert space. We shall now use the eigenstate representation of the Liouville space Green function to rederive these results. This will demonstrate the equivalence of both approaches and will pave the way for the use of the same Green functions in the evaluation of the nonlinear response function, where the Liouville space form provides a natural and compact time-ordered expansion.

We start with the Liouville space expression for $\chi^{(1)}$ [see Eq. (5.33)]:

$$\chi^{(1)}(\omega) = -\frac{1}{\hbar}\rho_0\langle\langle V|\mathscr{G}(\omega)\mathscr{V}|\rho(-\infty)\rangle\rangle. \tag{6.10}$$

The evaluation of Eq. (6.10) requires calculating the Liouville space matrix elements of the equilibrium density operator $\rho(-\infty)$, the Green function $\mathscr{G}(\omega)$, and the dipole coupling $\mathscr{V}$ and V. We shall now evaluate these quantities. In Liouville space $\rho(-\infty)$ is represented by the following vector

$$|\rho(-\infty)\rangle\rangle = \sum_a P(a)|aa\rangle\rangle.$$

We next turn to the Green function. The eigenstates of the Hamiltonian can be used to construct the eigenstates of the Liouville operator, and we have

$$\mathscr{L}|ab\rangle\rangle = \hbar\omega_{ab}|ab\rangle\rangle.$$

Similarly, the Liouville space Green function is diagonal, with matrix elements

$$\langle\langle cd|\mathscr{G}(\omega)|ab\rangle\rangle = \mathscr{G}_{cd,cd}(\omega)\delta_{ac}\delta_{bd},$$

and

$$\mathscr{G}_{cd,cd}(\omega) = \frac{1}{\omega - \omega_{cd} + i\varepsilon}. \tag{6.11}$$

We have thus

$$\mathscr{G}(\omega) = \sum_{c,d}|cd\rangle\rangle\mathscr{G}_{cd,cd}(\omega)\langle\langle cd|.$$

When these relations are substituted in Eq. (6.10) we get

$$\chi^{(1)}(\omega) = -\frac{1}{\hbar}\rho_0\sum_{a,c,d}P(a)\langle\langle V|cd\rangle\rangle\mathscr{G}_{cd,cd}(\omega)\langle\langle cd|\mathscr{V}|aa\rangle\rangle. \tag{6.12}$$

We shall now evaluate the dipole matrix elements. The Liouville space dipole operator $\mathscr{V}$ has two contributions

$$\langle\langle cd|\mathscr{V}|aa\rangle\rangle = V_{ca}\delta_{da} - V^*_{da}\delta_{ca}.$$

These contributions correspond to its action from the left (ket) and from the right (bra) side, respectively. We thus get

$$\mathscr{V}|aa\rangle\rangle = \sum_b(\mu_{ba}|ba\rangle\rangle - \mu_{ab}|ab\rangle\rangle).$$

Combining these with Eqs. (6.10) and (6.12) we obtain

$$\chi^{(1)}(\omega) = -\frac{1}{\hbar}\rho_0\sum_{a,b}P(a)[\langle\langle V|ba\rangle\rangle\mathscr{G}_{ba,ba}(\omega)\langle\langle ba|\mathscr{V}|aa\rangle\rangle$$
$$+ \langle\langle V|ab\rangle\rangle\mathscr{G}_{ab,ab}(\omega)\langle\langle ab|\mathscr{V}|aa\rangle\rangle].$$

Using the definition of the Liouville space matrix elements [Eq. (3.31)] we have

$$\langle\langle ba|\mathscr{V}|aa\rangle\rangle = \mu_{ba},$$

$$\langle ab|\mathscr{V}|aa\rangle\rangle = -\mu_{ba}^{*},$$

$$\langle\langle V|ba\rangle\rangle \equiv \mathrm{Tr}(V|b\rangle\langle a|) = \mu_{ab},$$

$$\langle\langle V|ab\rangle\rangle \equiv \mathrm{Tr}(V|a\rangle\langle b|) = \mu_{ba}.$$

When these equations are substituted in the above expression for $\chi^{(1)}(\omega)$ we get

$$\chi^{(1)}(\omega) = -\frac{1}{\hbar}\rho_0 \sum_{a,b} P(a)|\mu_{ab}|^2[\mathscr{G}_{ba,ba}(\omega) - \mathscr{G}_{ab,ab}(\omega)]. \tag{6.13}$$

Upon the substitution of the Liouville space Green function matrix elements [Eqs. (6.11)] in Eq. (6.13) we recover Eq. (6.5a).

We have thus shown that the two terms in the expression for the response function actually correspond to two distinct "pathways" in Liouville space, which differ by whether the first interaction is with the bra or the ket. The Liouville space pathways for the nonlinear response function will be calculated next.

RESPONSE FUNCTIONS OF A MULTILEVEL MANIFOLD WITH RELAXATION

Consider the multilevel Hamiltonian (Eq. 6.1). We assume that the system is further coupled to a thermal bath whose effects will be incorporated via a *reduced equation of motion* for the density operator. A general discussion of such equations is given in Appendix 6A. To keep the model as general as possible without introducing unnecessary details, we shall adopt the following simple reduced equations of motion.

$$\dot{\rho}_{\nu\nu'} = (-i\omega_{\nu\nu'} - \Gamma_{\nu\nu'})\rho_{\nu\nu'}, \tag{6.14}$$

with

$$\Gamma_{\nu\nu'} = \tfrac{1}{2}(\gamma_\nu + \gamma_{\nu'}) + \hat{\Gamma}_{\nu\nu'}. \tag{6.15}$$

Here $\rho_{\nu\nu'}$ is a matrix element of the reduced density operator (traced over the bath). γ_ν and $\gamma_{\nu'}$ are the inverse lifetimes of the ν and ν' levels. This relaxation matrix does not conserve the total probability (i.e., the trace of the density operator) and is adopted here mainly for its simplicity. $\hat{\Gamma}_{\nu\nu'}$ is the *pure dephasing rate* for the $\nu\nu'$ transition. For $\nu = \nu'$ we have $\hat{\Gamma}_{\nu\nu} = 0$. More general relaxation matrices are considered in Appendix 6A.

Equation (6.14) corresponds to a system described by an effective Liouville operator and its general solution is

$$\rho_{\nu\nu'}(t) = \mathscr{G}_{\nu\nu',\nu\nu'}(t)\rho_{\nu\nu'}(0),$$

where the Liouville space Green function $\mathscr{G}(t)$ assumes the form

$$\mathscr{G}(t) = \theta(t)\sum_{\nu,\nu'} |\nu\nu'\rangle\rangle I_{\nu\nu'}(t)\langle\langle \nu\nu'|. \tag{6.16a}$$

Here ν, $\nu' = a, b, c, d, \ldots$ and we have introduced the auxiliary function

$$I_{\nu\nu'}(t) \equiv \theta(t)\exp(-i\omega_{\nu\nu'}t - \Gamma_{\nu\nu'}t). \tag{6.16b}$$

Equations (6.16) provide a straightforward generalization of Eq. (3.47a) to include relaxation. The linear response function is calculated by the substitution of Eqs. (6.16) in Eq. (5.18a). We then obtain Eq. (5.19) with

$$J(t) = \sum_{a,b} P(a)|\mu_{ab}|^2 I_{ab}(t). \tag{6.17}$$

This generalizes Eq. (6.3a) to include relaxation.

The second-order response is calculated by substituting our Green function in Eq. (5.21). We then recover Eq. (5.23) with

$$\left.\begin{aligned} Q_1(t_2, t_1) &= \sum_{a,b,c} P(a)\mu_{ac}\mu_{cb}\mu_{ba}I_{ca}(t_2)I_{ba}(t_1) \\ Q_2(t_2, t_1) &= -\sum_{a,b,c} P(a)\mu_{ac}\mu_{cb}\mu_{ba}I_{bc}(t_2)I_{ba}(t_1). \end{aligned}\right\} \tag{6.18}$$

Turning now to the third-order response, we substitute our Green function in Eq. (5.25) and obtain Eq. (5.27) with

$$\left.\begin{aligned} R_1(t_3, t_2, t_1) &= \sum_{a,b,c,d} P(a)\mu_{ab}\mu_{bc}\mu_{cd}\mu_{da}I_{dc}(t_3)I_{db}(t_2)I_{da}(t_1), \\ R_2(t_3, t_2, t_1) &= \sum_{a,b,c,d} P(a)\mu_{ab}\mu_{bc}\mu_{cd}\mu_{da}I_{dc}(t_3)I_{db}(t_2)I_{ab}(t_1), \\ R_3(t_3, t_2, t_1) &= \sum_{a,b,c,d} P(a)\mu_{ab}\mu_{bc}\mu_{cd}\mu_{da}I_{dc}(t_3)I_{ac}(t_2)I_{ab}(t_1), \\ R_4(t_3, t_2, t_1) &= \sum_{a,b,c,d} P(a)\mu_{ab}\mu_{bc}\mu_{cd}\mu_{da}I_{ba}(t_3)I_{ca}(t_2)I_{da}(t_1). \end{aligned}\right\} \tag{6.19}$$

In order to calculate the optical susceptibilities, we introduce the Green function in the frequency domain [Eq. (3.65)]

$$\mathcal{G}(\omega) = \sum_{\nu,\nu'} |\nu\nu'\rangle\rangle I_{\nu\nu'}(\omega)\langle\langle\nu\nu'|, \tag{6.20a}$$

with the frequency domain (complex) lineshape function:

$$I_{\nu\nu'}(\omega) \equiv -i\int_0^\infty d\tau\, I_{\nu\nu'}(\tau)\exp(i\omega\tau).$$

Using Eqs. (6.16b), we then get

$$I_{\nu\nu'}(\omega) \equiv \frac{1}{\omega - \omega_{\nu\nu'} + i\Gamma_{\nu\nu'}}. \tag{6.20b}$$

Equations (6.20) generalize Eq. (3.70a) to include relaxation. The linear susceptibility Eq. (6.5c) now assumes the form

$$\chi^{(1)}(\omega) = -\frac{1}{\hbar}\rho_0 \sum_{a,b} [P(a) - P(b)]|\mu_{ab}|^2 I_{ab}(\omega), \tag{6.21}$$

which can be recast as

$$\chi^{(1)}(\omega) = \frac{1}{\hbar}\rho_0 \sum_{a,b} P(a)\frac{2|\mu_{ab}|^2\omega_{ba}}{\omega_{ba}^2 - (\omega + i\Gamma_{ab})^2}.$$

The real and the imaginary parts of $\chi^{(1)}$ in the vicinity of a particular resonance are displayed in Figure 6.2. χ'' has a typical Lorentzian profile with a maximum at the resonance frequency. χ' has a dispersive form and it vanishes on resonance. Substituting Eqs. (6.18) and (6.19) in Eqs. (5.53) and (5.54) we obtain the nonlinear susceptibilities

$$\begin{aligned}\chi^{(2)}(-\omega_s; \omega_1, \omega_2) = \frac{1}{2}\left(\frac{1}{\hbar}\right)^2 \rho_0 \sum_p \sum_{a,b,c} P(a)\mu_{ab}\mu_{bc}\mu_{ca} \\ \times [I_{ca}(\omega_1 + \omega_2)I_{ba}(\omega_1) - I_{bc}(\omega_1 + \omega_2)I_{ba}(\omega_1) \\ + I_{ab}(\omega_1 + \omega_2)I_{ac}(\omega_1) - I_{bc}(\omega_1 + \omega_2)I_{ac}(\omega_1)],\end{aligned} \tag{6.22}$$

$$\begin{aligned}\chi^{(3)}(-\omega_s; \omega_1, \omega_2, \omega_3) = \frac{1}{6}\left(-\frac{1}{\hbar}\right)^3 \rho_0 \sum_p \sum_{a,b,c,d} P(a)\mu_{ab}\mu_{bc}\mu_{cd}\mu_{da} \\ \times [I_{dc}(\omega_1 + \omega_2 + \omega_3)I_{db}(\omega_1 + \omega_2)I_{da}(\omega_1) \\ + I_{dc}(\omega_1 + \omega_2 + \omega_3)I_{db}(\omega_1 + \omega_2)I_{ab}(\omega_1) \\ + I_{dc}(\omega_1 + \omega_2 + \omega_3)I_{ac}(\omega_1 + \omega_2)I_{ab}(\omega_1) \\ + I_{ba}(\omega_1 + \omega_2 + \omega_3)I_{ca}(\omega_1 + \omega_2)I_{da}(\omega_1) \\ - I_{cb}(\omega_1 + \omega_2 + \omega_3)I_{db}(\omega_1 + \omega_2)I_{ab}(\omega_1) \\ - I_{cb}(\omega_1 + \omega_2 + \omega_3)I_{db}(\omega_1 + \omega_2)I_{da}(\omega_1) \\ - I_{cb}(\omega_1 + \omega_2 + \omega_3)I_{ca}(\omega_1 + \omega_2)I_{da}(\omega_1) \\ - I_{ad}(\omega_1 + \omega_2 + \omega_3)I_{ac}(\omega_1 + \omega_2)I_{ab}(\omega_1)].\end{aligned} \tag{6.23}$$

The p summation was introduced in Chapter 5 and denotes a sum over all permutations of the frequencies, ω_1, ω_2, and ω_3.

Since the nth order response functions and $\chi^{(n)}$ contain n dipole interaction factors $\mathscr{V}$, they will have 2^n terms corresponding to the various possible choices of the V's to act from the left or the right. The physical significance of Eqs. (6.21)–(6.23) can be better understood by adopting a graphic representation. We shall use three types of graphs that are complementary and provide a useful bookkeeping device for the various terms and interactions in the calculation of the optical polarization. These are given in Figures 6.3, 6.4, and 6.5 for $\chi^{(3)}$, $\chi^{(2)}$, and $\chi^{(1)}$, respectively. We shall explain these pictorial representations by focusing on $\chi^{(3)}$. First we consider the density operator coupling scheme (Figure 6.3(A)) [13]. We start at $|\rho(-\infty)\rangle\rangle = |aa\rangle\rangle$, which is shown in the upper left-hand corner. A horizontal (vertical) bond represents an interaction V acting from the

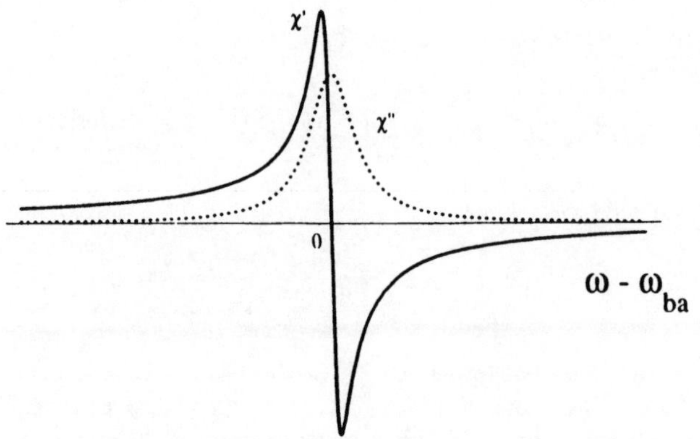

FIG. 6.2 The real and the imaginary parts of the linear susceptibility $\chi^{(1)}(\omega)$ in the vicinity of an optical resonance ω_{ba} [Eq. (6.21)].

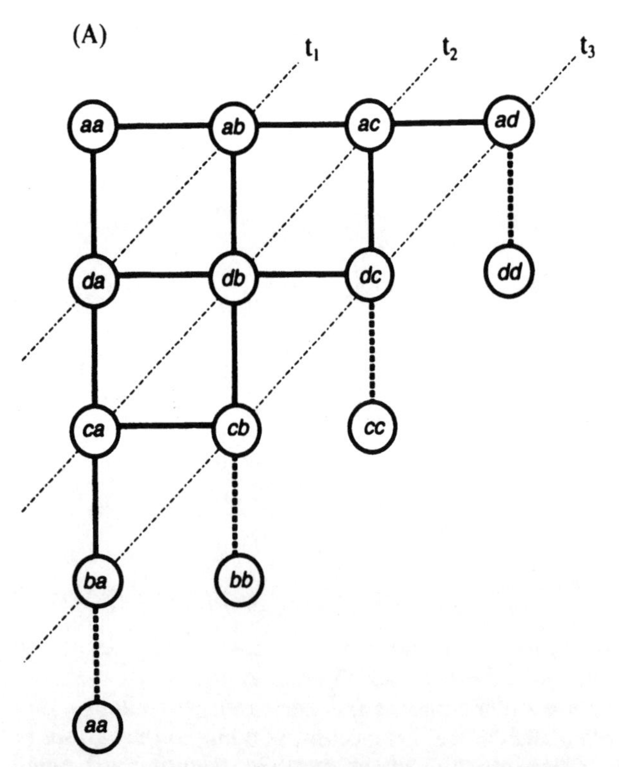

FIG. 6.3 (A) Liouville space coupling scheme for the third-order response function [Eqs. (6.23)]. Solid lines denote radiative coupling V, horizontal (vertical) lines represent action of V from the right (left). Starting at *aa*, after three perturbations, the system finds itself along the dashed line. The dotted lines represent the last V, which acts from the left. At the end of four perturbations, the system is in a diagonal state (*aa*, *bb*, *cc*, or *dd*). The number of three-bond pathways leading to *ad*, *ba*, *dc*, and *cb* is 1, 1, 3, and 3, respectively. Altogether, there are, therefore, eight pathways.

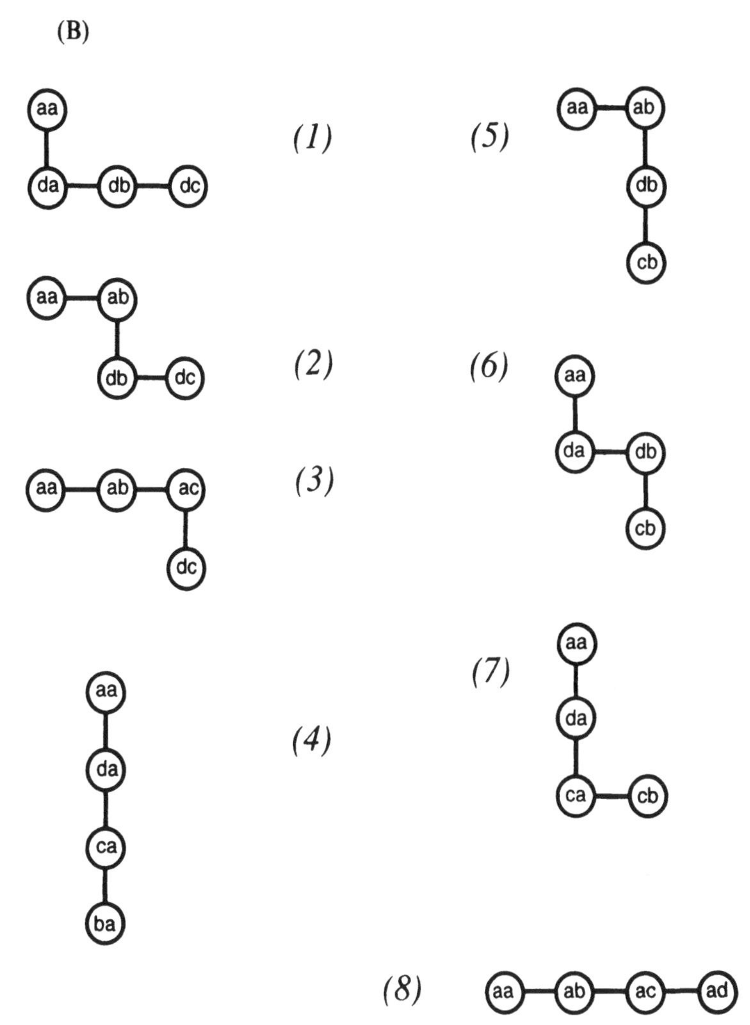

FIG. 6.3 *(continued)* **(B)** The eight Liouville space pathways that contribute to the third-order response function (6.3a). The eight terms in Eqs. (6.23) correspond, respectively, to pathways (i)–(viii) *(continued)*.

right (left). After the first interaction (which takes place at time $t - t_1 - t_2 - t_3$), the system finds itself in either of the states $|ab\rangle\rangle$ or $|da\rangle\rangle$ (note that b and d are dummy indices that run over the entire manifold of states). The system then evolves for a period t_1, interacts again (at time $t - t_2 - t_3$), evolves for a period t_2, interacts again at time $t - t_3$, and evolves for a period t_3. Then, at time t, the polarization is calculated by operating with V from the left and performing a trace. The eight Liouville space pathways that contribute to $\chi^{(3)}$ are displayed in Figure 6.3(B). A third type of representation of all the 48 terms in Eq. (6.23) can be obtained by using the *double-sided Feynman diagrams* [14, 15]. The representation of Figures 6.3(A) and 6.3(B) just shows the time ordering associated with each term. The Feynman diagrams displayed in Figure 6.3(C)

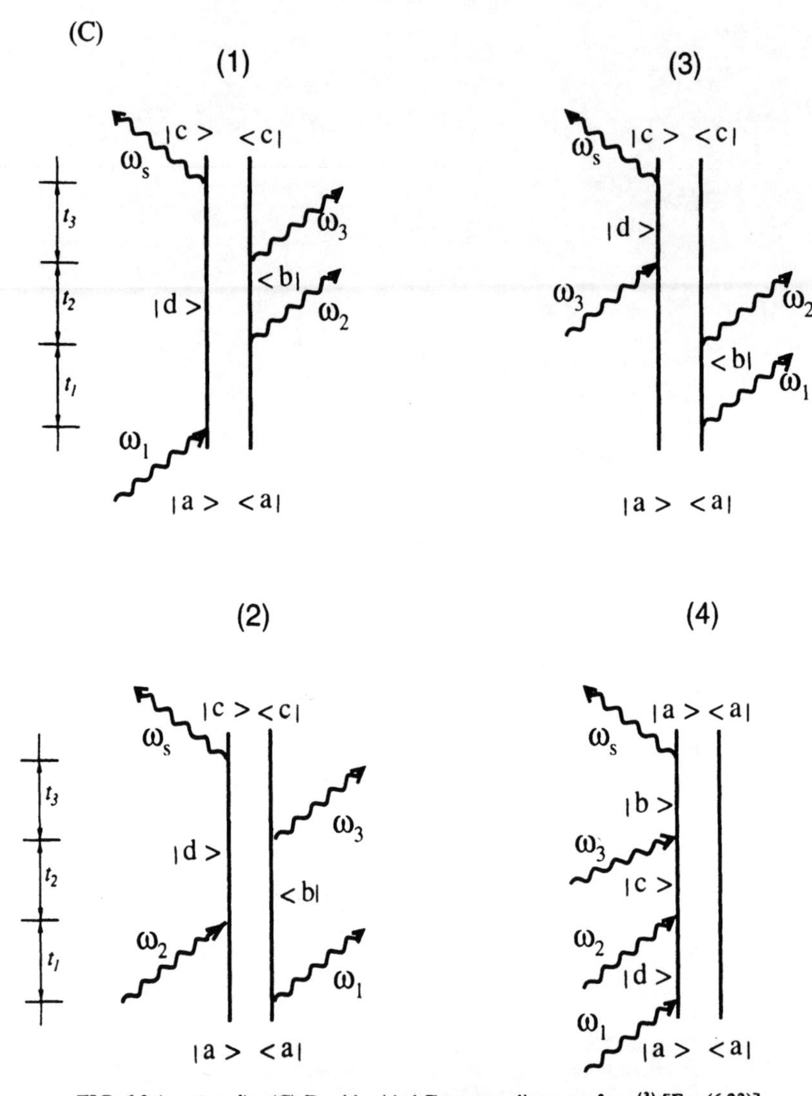

FIG. 6.3 (*continued*) (C) Double-sided Feynman diagrams for $\chi^{(3)}$ [Eq. (6.23)].

show the complete terms (time ordering, choice of frequencies, and their sign). The double-sided diagrams are introduced by the following rules:

1. The density operator is represented by two vertical lines. The line on the left represents the ket and the line on the right represents the bra.
2. Time runs vertically from bottom to top.
3. An interaction with the radiation field is represented by a wavy line.
4. Each diagram has an overall sign of $(-1)^n$ where n is the number of interactions from the right (bra). This is because each time an interaction V acts from the right in a commutator, it carries a minus sign.

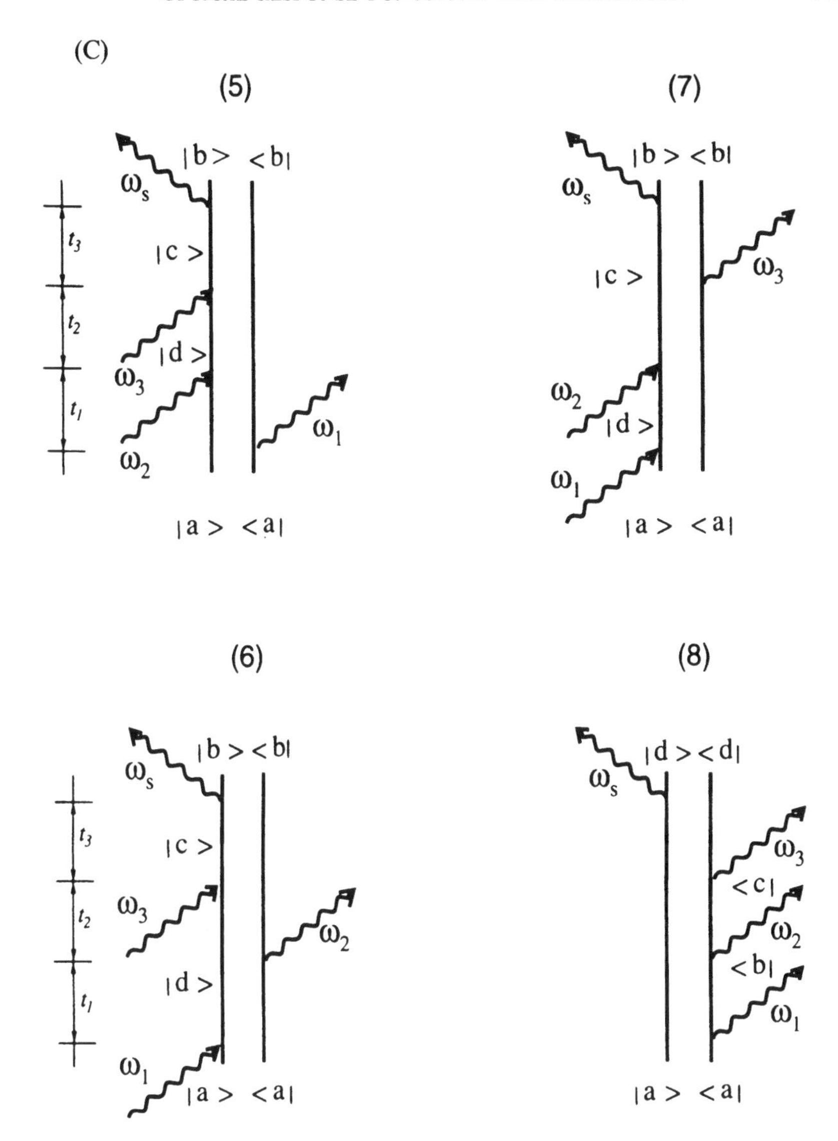

FIG. 6.3 (C) *(continued)*.

5. Each interaction is assigned an arrow and labelled by the corresponding field $\omega_j(>0)$. An arrow pointing to the right and labeled ω_j represents a contribution of $E_j \exp(-i\omega_j t + i\mathbf{k}_j \cdot \mathbf{r})$ to the polarization. An arrow pointing to the left represents a $E_j^* \exp(i\omega_j t - i\mathbf{k}_j \cdot \mathbf{r})$ contribution.
6. The overall frequency of the term is the sum of the three frequencies ω_j, ω_k, and ω_l with their appropriate signs.

In a quantum description of the radiation–matter interaction (see Chapter 4), a negative frequency $-\omega_j$ is associated with the photon annihilation operator a

whereas a positive frequency is connected with the photon creation operator $a^\dagger$. The sign convention of ω_j is such that incoming arrows represent photon absorption while outgoing arrows represent photon emission. An incoming arrow on the left or an outgoing arrow on the right represents the action of the photon annihilation operator. Conversely, an outgoing arrow on the left or an incoming arrow on the right represents the action of the photon creation operator.

In Figure 6.3(c) we display the eight Feynman diagrams, which correspond, respectively, to the eight terms in Eq. (6.23). The complete set of 48 diagrams can be obtained by performing the $3! = 6$ permutations over the fields ω_1, ω_2, and ω_3.

Figure 6.4 contains the corresponding graphic illustrations for $\chi^{(2)}$. Here we have only two time intervals t_1 and t_2, four Liouville space paths, and eight Feynman diagrams. For $\chi^{(1)}$ (Figure 6.5) we have a single time interval and two Liouville space paths and Feynman diagrams.

It is a common practice in the literature to use one-sided diagrams with field arrows going up and down. This type of diagram is of very limited value since it is essential to keep track of the bra and the ket. Absorption of a photon ω_j on the left (ket) or emission on the right (bra) result in a positive frequency $\exp(i\omega_j t)$. Emission from the left or absorption from the right will result in a $\exp(-i\omega_j t)$ factor. An arrow going up signifying "absorption" does not represent a well-defined contribution to the optical response since we need also to know

(A)

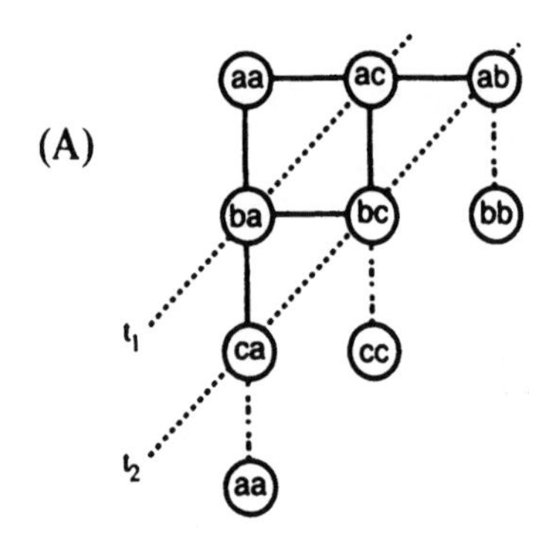

(B)

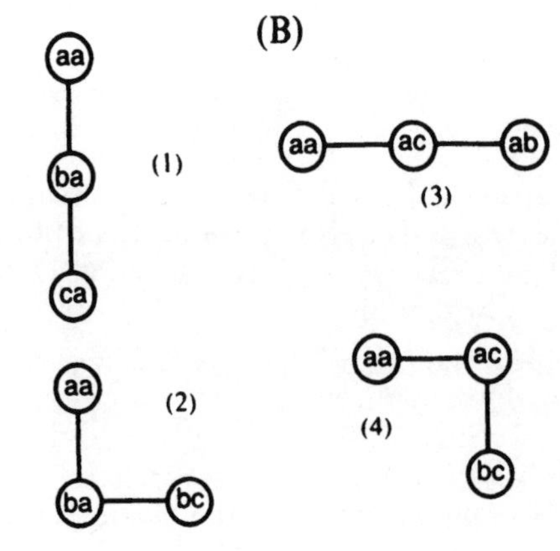

FIG. 6.4 (A) Liouville space coupling scheme for the second-order response function [Eq. (6.22)]. (B) The four Liouville space pathways that contribute to the second-order response function. The four terms in Eq. (6.22) correspond respectively to pathways (i)–(iv).

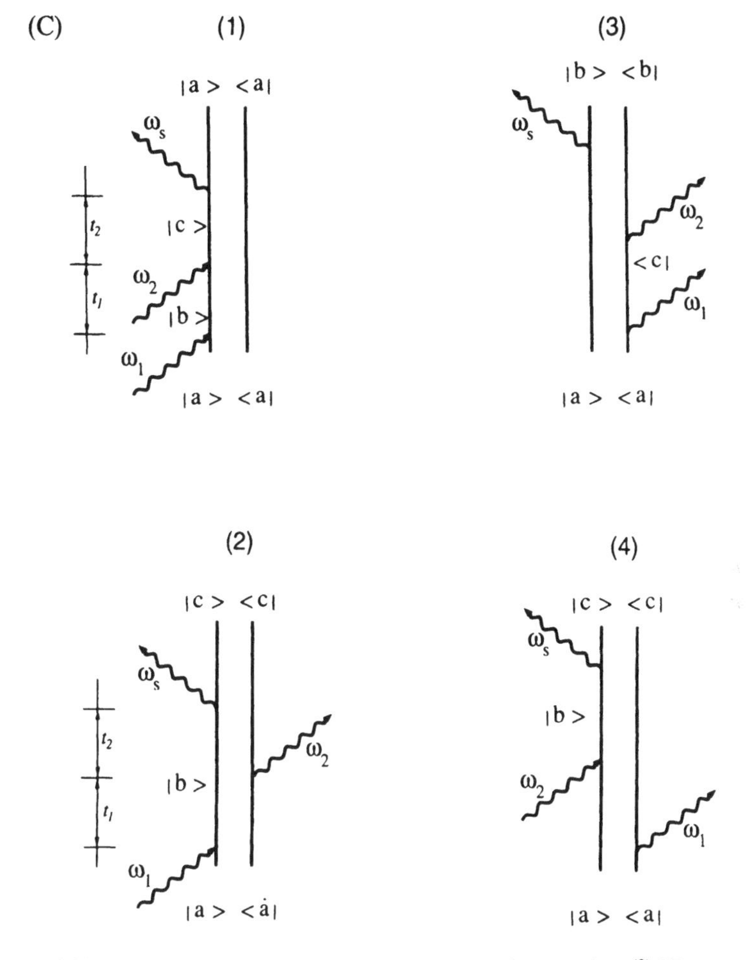

FIG. 6.4 (*continued*) (C) Double-sided Feynman diagrams for $\chi^{(2)}$ [Eq. (6.22)].

whether it acts on the ket or the bra. This information is not provided by one-sided diagrams. Another point to be noted is that although the expression for $\chi^{(n)}$ is to $n + 1$th order in the dipole operator, it is related to $\rho^{(n)}$ and not to $\rho^{(n+1)}$. The reason is that the last interaction appears only when we take the expectation value, and is not part of the expansion of $\rho^{(n)}$. We have arbitrarily chosen to act with the last interaction from the left in Eq. (5.11). Due to the cyclic permutation symmetry of the trace, this result will not change had we performed the last action from the right. ($\mathrm{Tr}[V\rho(t)] = \mathrm{Tr}[\rho(t)V]$). It is therefore incorrect to interpret $V\rho^{(n)}$ as the state of the system after $n + 1$ interactions. Let us consider for example pathway (1) in Figure 6.3(B). In Figure 6.3(A) we acted with V at the end from the left and obtained $|cc\rangle\rangle$. Had we acted from the right we would obtain $|dd\rangle\rangle$. It is then incorrect to infer from Figure 6.3(A) that the state of the system at the end of the process is $|cc\rangle\rangle$.

We note that the nonlinear susceptibility $\chi^{(3)}$ contains a combination of

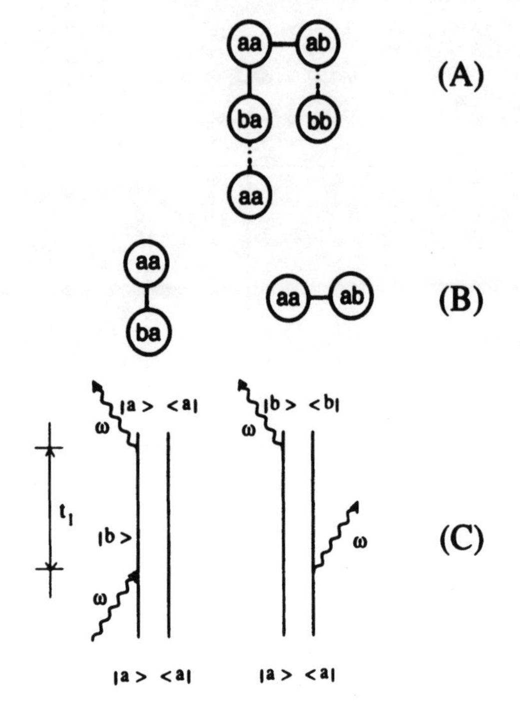

FIG. 6.5 (A) Liouville space coupling scheme for the linear response function [Eq. (6.21)]. (B) The two Liouville space pathways that contribute to the response function. The two terms in Eq. (6.21) correspond, respectively, to pathways (i) and (ii). (C) Double-sided Feynman diagrams for $\chi^{(1)}$ [Eq. (6.21)].

various factors in which a single field frequency ω_j or a combination of two $\omega_j \pm \omega_k$ or three $\omega_j \pm \omega_k \pm \omega_m$ field frequencies is resonant with a material transition frequency $\omega_{\nu\nu'}$. These factors represent single-photon, two-photon, and three-photon resonances, which originate from the t_1, the t_2, and the t_3 integrations, respectively. We should bear in mind that these terms may interfere when added so that Eqs. (6.22) or (6.23) may sometimes be misleading. For example, it will be shown in Chapter 9 that certain two photon resonances which show up in individual Liouville space paths for $\chi^{(3)}$, cancel due to interference when all the paths are added. We should therefore be extremely cautious not to rely on "typical" terms in the nonlinear response and we should always sum them carefully before identifying a particular resonance.

Finally, we note that in many cases it is possible to select only a few terms that dominate the nonlinear susceptibilities and neglect the rest. Consider, for example, a two electronic level system (see Figure 7.1), and assume that all laser beams are tuned near resonance with the electronic transition frequency ω_{eg}^0. Some of the 48 terms in Eq. (6.23) contain processes where a photon is absorbed (an incoming arrow) and the system is excited (from g to e), or conversely a photon is emitted (an outgoing arrow) and the system is deexcited (from e to g). These processes are resonant and make a significant contribution to the signal with a denominator $\sim(\omega_j - \omega_{eg}^0)^{-1}$. If at least one interaction violates this condition (e.g., a photon is absorbed and the system gets deexcited), the result will be a highly oscillatory off resonant term, which will make a negligible contribution with a denominator $\sim(\omega_j + \omega_{eg}^0)^{-1}$. We can then invoke the *rotating wave approximation* (RWA) [16] and retain only the former terms. The

RWA will be repeatedly used in the discussion of resonant techniques. The double-sided Feynman diagrams allow us to select the surviving terms by a simple inspection.

THE NONLINEAR RESPONSE FUNCTIONS CALCULATED USING THE WAVEFUNCTION IN HILBERT SPACE

In Appendix 5A we outlined the procedure for computing the optical response by following the time evolution of the wavefunction. We showed that the final result contains fewer terms compared with the density operator calculation, since we do not maintain a complete bookkeeping of time ordering in this case. Two limitations of this procedure should be noted. We assume that the system is in a pure state at all times and hence can be described by a wavefunction. This requires that (1) initially the system is in a pure state, $|\psi_a\rangle$ and (2) the relaxation matrix contains no pure dephasing contributions $\hat{\Gamma}_{\nu\nu'} = 0$; otherwise a pure state will evolve into a statistical mixture. Condition (1) can be easily relaxed by adding an average over initial states at the end of the calculation. Condition (2) is, however, a fundamental limitation of a wavefunction description.

We shall now employ this procedure for calculating $\chi^{(2)}$ and $\chi^{(3)}$ of our multilevel system, assuming $\hat{\Gamma}_{\nu\nu'} = 0$. Using the notation of Appendix 5A, we obtain for the second-order polarization

$$P^{(2)}(\mathbf{r}, t) = \langle\psi_a^{(2)}(t)|V|\psi_a^{(0)}\rangle + \langle\psi_a^{(0)}(t)|V|\psi_a^{(2)}\rangle + \langle\psi_a^{(1)}(t)|V|\psi_a^{(1)}(t)\rangle, \tag{6.24}$$

and for third order

$$P^{(3)}(\mathbf{r}, t) = \langle\psi_a^{(3)}(t)|V|\psi_a^{(0)}\rangle + \langle\psi_a^{(2)}(t)|V|\psi_a^{(1)}(t)\rangle + h.c. \tag{6.25}$$

Making use of Eq. (2.59b), and the expansion of the field into modes [Eq. (5.46), omitting the field polarization indices], we can calculate the wavefunction order by order. To first order we have

$$|\psi_a^{(1)}(t)\rangle = -\frac{1}{\hbar}\sum_j\sum_b \frac{E_j\,\mu_{ba}\exp(-i\omega_j t)}{\omega_j - \omega_{ba} + i\Gamma_{ba}}|b\rangle - \frac{1}{\hbar}\sum_j\sum_b \frac{E_j^*\mu_{ba}\exp(i\omega_j t)}{-\omega_j - \omega_{ba} + i\Gamma_{ba}}|b\rangle, \tag{6.26}$$

where we have set the initial state energy $\varepsilon_a = 0$ without affecting the generality of this expression. In a similar way we can calculate the higher order terms $|\psi_a^{(2)}\rangle$ and $|\psi_a^{(3)}\rangle$. Substituting these in Eqs. (6.24), and adding a thermal average over initial states $|\psi_a\rangle$ and a sum over the permutations of the fields, we finally get [17]

$$\chi^{(2)}(-\omega_s; \omega_1, \omega_2) = \frac{1}{2}\left(\frac{1}{\hbar}\right)^2 \rho_0 \sum_p \sum_{a,b,c} P(a)\mu_{ab}\mu_{bc}\mu_{ca}$$

$$\times [I_{ab}(\omega_1)I_{ac}(\omega_1+\omega_2) + I_{ca}(\omega_1)I_{ba}(\omega_1+\omega_2) - I_{ab}(\omega_1)I_{ca}(\omega_2)], \tag{6.27}$$

and

$$\chi^{(3)}(-\omega_s; \omega_3, \omega_2, \omega_1) = \frac{1}{6}\left(-\frac{1}{\hbar}\right)^3 \rho_0 \sum_p \sum_{a,b,c,d} P(a)\mu_{ab}\mu_{bc}\mu_{cd}\mu_{da}$$
$$\times [-I_{ab}(\omega_1)I_{ac}(\omega_1+\omega_2)I_{ad}(\omega_1+\omega_2+\omega_3)$$
$$+ I_{ab}(\omega_1)I_{ac}(\omega_1+\omega_2)I_{da}(\omega_3)$$
$$+ I_{ba}(\omega_1+\omega_2+\omega_3)I_{ca}(\omega_1+\omega_2)I_{da}(\omega_1)$$
$$- I_{ab}(\omega_3)I_{ca}(\omega_1+\omega_2)I_{da}(\omega_1)]. \quad (6.28)$$

We reiterate that it is impossible to incorporate pure dephasing processes within a wavefunction formulation. Therefore, in Eqs. (6.27) and (6.28) $I_{\nu\nu'}(\omega)$ is given by Eq. (6.20b) with $\hat{\Gamma}_{\nu\nu'} = 0$.

THE NONLINEAR RESPONSE FUNCTIONS CALCULATED USING THE HEISENBERG EQUATIONS OF MOTION

A third method for calculating the optical response was outlined in Appendix 5B. In this method we derive a hierarchy of equations of motion for dynamic variables (rather than for the density operator). For the present multilevel system this can be done as follows: Let us assume that the system is initially in the state $|a\rangle$. We first construct a complete set of operators:

$$B_n^\dagger \equiv |n\rangle\langle a| \qquad n \neq a, \quad (6.29)$$

n represents all states other than the initial state $|a\rangle$. These operators satisfy the commutation relations

$$\left.\begin{aligned} &[B_n, B_m] = [B_n^\dagger, B_m^\dagger] = 0, \\ &[B_n^\dagger, B_m] = B_n^\dagger B_m - \left(1 - \sum_l{}' B_l^\dagger B_l\right)\delta_{nm}, \end{aligned}\right\} \quad (6.30)$$

where $\sum'$ stands for a sum over all states excluding a. Note that

$$B_n B_n^\dagger = |a\rangle\langle a| = 1 - \sum_l{}' B_l^\dagger B_l. \quad (6.31)$$

The polarization operator is given by

$$\hat{P} = \sum_n{}' [\mu_{an} B_n + \mu_{na} B_n^\dagger] + \sum_{n,m}{}' \mu_{nm} B_n^\dagger B_m + \mu_{aa} |a\rangle\langle a|, \quad (6.32)$$

with the dipole matrix elements

$$\mu'_{nm} \equiv \begin{cases} \mu_{nm} & n \neq m \\ \mu_{nn} - \mu_{aa} & n = m. \end{cases} \quad (6.33)$$

We next write the Heisenberg equations of motion and take the expectation value, resulting in

$$\frac{1}{i}\langle\dot{B}_n\rangle = (\omega_{an} + i\Gamma_{an})\langle B_n\rangle + \frac{1}{\hbar}\mu_{na}E(t)$$

$$-\frac{1}{\hbar}E(t)\sum_m{}' [\mu_{ma}\langle B_m^\dagger\rangle\langle B_n\rangle - \mu'_{nm}\langle B_m\rangle + \mu_{na}\langle B_m^\dagger\rangle\langle B_m\rangle]. \quad (6.34)$$

In the derivation of Eq. (6.34) we have assumed that pure dephasing processes are negligible so that $\Gamma_{an} = \frac{1}{2}(\gamma_a + \gamma_n)$. This assumption ensures that the system remains in a pure state at all times. To third order in the field we can then factorize the expectation values $\langle B_m^+ B_n\rangle = \langle B_m^+\rangle\langle B_n\rangle$. A more detailed discussion of this type of factorization is given in Chapter 17.

By transforming Eq. (6.34) to the frequency domain and using the expansion of the field into modes [Eq. (5.46), omitting the field polarization indices] we can then solve the equation order by order in the field. To second order we have

$$\tilde{B}_n^{(1)}(\omega_1) = -\frac{1}{\hbar}\frac{\mu_{na}E_1}{\omega_1 - \omega_{na} + i\Gamma_{an}}, \quad (6.35a)$$

$$\tilde{B}_n^{(2)}(\omega_1 + \omega_2) = \frac{1}{2\hbar^2}\sum_p\sum_m{}' \frac{\mu_{nm}\mu_{ma}E_1E_2}{(\omega_1 + \omega_2 - \omega_{na} + i\Gamma_{an})(\omega_1 - \omega_{ma} + i\Gamma_{am})}. \quad (6.35b)$$

Substituting in Eq. (6.32), collecting terms, and adding a thermal average over initial states a, we get

$$\chi^{(2)}(-\omega_s; \omega_1, \omega_2) = \frac{1}{2}\left(\frac{1}{\hbar}\right)^2 \rho_0 \sum_p\sum_a P(a) \sum_{b,c}{}' \mu_{ab}\mu_{bc}\mu_{ca}$$
$$\times [I_{ab}(\omega_1)I_{ac}(\omega_1 + \omega_2) + I_{ca}(\omega_1)I_{ba}(\omega_1 + \omega_2) - I_{ab}(\omega_1)I_{ca}(\omega_2)]. \quad (6.36)$$

Proceeding along the same lines, we can solve for $B^{(3)}$, and when substituting in Eq. (6.32) we get

$$\chi^{(3)}(-\omega_s; \omega_1, \omega_2, \omega_3)$$
$$= \frac{1}{6}\left(\frac{1}{\hbar}\right)^3 \rho_0 \sum_p\sum_a P(a) \sum_{b,c,d}{}' \mu_{ab}\mu_{bc}\mu_{cd}\mu_{da}[I_{ab}(\omega_1)I_{ac}(\omega_1 + \omega_2)I_{ad}(\omega_1 + \omega_2 + \omega_3)$$
$$- I_{ab}(\omega_1)I_{ac}(\omega_1 + \omega_2)I_{da}(\omega_3) - I_{ba}(\omega_1 + \omega_2 + \omega_3)I_{ca}(\omega_1 + \omega_2)I_{da}(\omega_1)$$
$$+ I_{ab}(\omega_3)I_{ca}(\omega_1 + \omega_2)I_{da}(\omega_1)] + \frac{1}{6}\left(\frac{1}{\hbar}\right)^3 \rho_0 \sum_p\sum_a P(a) \sum_{b,c}{}' \mu_{ab}\mu_{ba}\mu_{ac}\mu_{ca}$$
$$\times [I_{ba}(\omega_1 + \omega_2 + \omega_3)I_{ba}(\omega_3)I_{ac}(\omega_1) - I_{ba}(\omega_1 + \omega_2 + \omega_3)I_{ac}(\omega_2)I_{ca}(\omega_1)$$
$$- I_{ab}(\omega_1 + \omega_2 + \omega_3)I_{ab}(\omega_3)I_{ca}(\omega_1) + I_{ab}(\omega_1 + \omega_2 + \omega_3)I_{ca}(\omega_2)I_{ac}(\omega_1)]. \quad (6.37)$$

For an isolated system with $\Gamma_{\nu\nu'} = 0$, Eqs. (6.36) and (6.37) can be considerably simplified, and recast in the form

$$\chi^{(2)}(-\omega_s; \omega_1, \omega_2) = \frac{1}{2}\left(\frac{1}{\hbar}\right)^2 \rho_0 \sum_{p_3} \sum_a P(a) \sum_{b,c}{}' \mu_{ab}\mu_{bc}\mu_{ca} I_{ba}(\omega_1 + \omega_2) I_{ca}(\omega_1), \tag{6.38}$$

where $\sum_{p_3}$ denotes the sum over all permutations of the three frequencies $-\omega_s$, ω_1, and ω_2.

Similarly, we get the third-order susceptibility

$$\chi^{(3)}(-\omega_s; \omega_1, \omega_2, \omega_3) =$$

$$\frac{1}{6}\left(-\frac{1}{\hbar}\right)^3 \rho_0 \sum_{p_4} \sum_a P(a) \sum_{b,c,d}{}' \mu_{ad}\mu_{dc}\mu_{cb}\mu_{ba} I_{da}(\omega_1 + \omega_2 + \omega_3) I_{ca}(\omega_1 + \omega_2) I_{ba}(\omega_1)$$

$$+ \frac{1}{6}\left(-\frac{1}{\hbar}\right)^3 \sum_{p_4} \sum_a P(a) \sum_{b,c}{}' \mu_{ab}\mu_{ba}\mu_{ac}\mu_{ca} I_{ba}(\omega_1 + \omega_2 + \omega_3) I_{ca}(\omega_1) I_{ac}(\omega_2), \tag{6.39}$$

where $\sum_{p_4}$ denotes the sum over permutations of the four frequencies $-\omega_s$, ω_1, ω_2, and ω_3.

These results were first derived by Orr and Ward [18] by a tedious renormalized perturbation theory, starting with the wavefunction expression [Eqs. (6.27) and (6.28)]. They are particularly useful for calculating off resonant susceptibilities, e.g., when we set all frequencies $\omega_j = 0$. In this case, the density operator and the wavefunction expressions contain some terms with vanishing denominators. $\chi^{(n)}(0)$, however, does not diverge and the divergencies cancel when all of these terms are carefully added. The exclusion of a in the $\sum'$ summation in Eqs. (6.38) and (6.39) results from these cancellations, and these equations contain no diverging terms. The built-in cancellation is an important advantage of this form. This advantage becomes even more apparent when treating many-body susceptibilities of interacting particles. In this case it is very difficult to calculate the global density operator of the entire system, yet the Heisenberg equation provides a simple method for carrying out these calculations, as will be shown in Chapter 17.

In concluding this section, we note that we have considered here a hierarchy of possible relaxation matrices. Equations (6.38) and (6.39) are valid only in the absence of any relaxation $\Gamma_{\nu\nu'} = 0$. The wavefunctions [Eqs. (6.27) and (6.28)] and the Heisenberg [Eqs. (6.36) and (6.37)] forms hold also in the presence of finite lifetimes but no pure dephasing $\Gamma_{\nu\nu'} = 1/2(\gamma_\nu + \gamma_{\nu'})$. The density operator expressions [Eqs. (6.22) and (6.23)] allow also the addition of pure dephasing $\Gamma_{\nu\nu'} = 1/2(\gamma_\nu + \gamma_{\nu'}) + \Gamma_{\nu\nu'}$. It is possible to include pure dephasing processes in the Heisenberg picture. This requires its extension by adding more dynamic variables including products of B_n and $B_m^\dagger$ operators. This extension will be carried out in Chapter 17.

THE ANHARMONIC OSCILLATOR PICTURE FOR THE OPTICAL POLARIZATION

The linear optical response can be interpreted by modeling the material system as a collection of uncoupled *harmonic oscillators.* This picture was first proposed by Lorentz [19, 20], and further developed by Fano [21]. Using Eq. (6.2), the optical polarization for our multilevel system is given by

$$P(t) \equiv \mathrm{Tr}[V\rho(t)] = \sum_{a,b} P_{ab}(t),$$

where

$$P_{ab} \equiv \mu_{ba}\rho_{ab} + \mu_{ab}\rho_{ba}.$$

We next consider the equations of motion of the density operator

$$\dot{\rho}_{ab} = -i\omega_{ab}\rho_{ab} + \frac{i}{\hbar}E(t)\mu_{ab}(\rho_{bb} - \rho_{aa}) - \Gamma_{ab}\rho_{ab}, \tag{6.40a}$$

$$\dot{\rho}_{ba} = i\omega_{ab}\rho_{ba} - \frac{i}{\hbar}E(t)\mu_{ba}(\rho_{bb} - \rho_{aa}) - \Gamma_{ab}\rho_{ba}, \tag{6.40b}$$

$$\dot{\rho}_{aa} = -\frac{i}{\hbar}E(t)(\mu_{ab}\rho_{ba} - \mu_{ba}\rho_{ab}) - \gamma_a\rho_{aa} + \gamma_b\rho_{bb}, \tag{6.40c}$$

$$\dot{\rho}_{bb} = \frac{i}{\hbar}E(t)(\mu_{ab}\rho_{ba} - \mu_{ba}\rho_{ab}) + \gamma_a\rho_{aa} - \gamma_b\rho_{bb}, \tag{6.40d}$$

These equations were obtained by combining Eqs. (3.28) with the relaxation superoperator derived in Appendix 6A. γ_a and γ_b represent population relaxation (T_1) rates [see Eq. (6.A.4)] whereas Γ_{ab} is the dephasing rate [denoted $\Gamma_{ab,ab}$ in Eq. (6.A.11)]. We next introduce the following new variables

$$Q_{ab}(t) \equiv -i(\mu_{ba}\rho_{ab} - \mu_{ab}\rho_{ba}),$$

and the population difference

$$W_{ab} \equiv \rho_{aa} - \rho_{bb}.$$

Transforming the equations of motion of the density operator to the new variables we get

$$\dot{P}_{ab} = \omega_{ab}Q_{ab} - \Gamma_{ab}P_{ab}, \tag{6.41a}$$

$$\dot{Q}_{ab} = -\omega_{ab}P_{ab} + \frac{2}{\hbar}E(t)|\mu_{ba}|^2 W_{ba} - \Gamma_{ab}Q_{ab}, \tag{6.41b}$$

$$\dot{W}_{ab} = \frac{2}{\hbar}E(t)Q_{ab} - \Gamma_1(W_{ab} - \bar{W}_{ab}), \tag{6.41c}$$

where the population relaxation rate $\Gamma_1 \equiv \gamma_a + \gamma_b$, and $\bar{W}_{ab} \equiv (\gamma_b - \gamma_a)/(\gamma_b + \gamma_a)$ is the equilibrium value of W_{ba}. For optical transitions with $\omega_{ba} \gg k_B T$ we have $\bar{W}_{ab} = 1$.

Equation (6.41) together with the initial conditions $P_{ab}(-\infty) = Q_{ab}(-\infty) = 0$ and $W_{ab}(-\infty) = \bar{W}_{ab}$ describe the dynamics of the system. By expanding the variables P_{ab}, Q_{ab}, and $W_{ab} - \bar{W}_{ab}$ in a power series in the field, these equations can be solved order by order. By inspection it is clear that the leading term in the expansion of P_{ab} and Q_{ab} is first order in the field whereas $W_{ab} - \bar{W}_{ab}$ is at least second order in the field. To first order we thus set $W_{ab} = \bar{W}_{ab}$. By further taking a time derivative of both sides of Eq. (6.41a) and substituting $\dot{Q}_{ab}$ from Eq. (6.41b), we get the linearized equations

$$\ddot{P}_{ab} + \omega_{ab}^2 P_{ab} + 2\Gamma_{ab}\dot{P}_{ab} = \frac{2}{\hbar} E(t)|\mu_{ab}|^2 \omega_{ba} \bar{W}_{ab}, \tag{6.42}$$

where we have used the fact that for optical transitions $\Gamma_{ab} \ll \omega_{ab}$ and neglected a Γ_{ab}^2 term. Equation (6.42) describes a driven harmonic oscillator where P_{ab} is its coordinate and Q_{ab} is the conjugate momentum. (Warning: This notation may be somewhat confusing since P usually denotes the momentum and Q is the coordinate.) In the hydrogen atom the polarization is proportional to the coordinate of the electron; the electron thus behaves as a harmonic oscillator. More generally it is the *polarization* and not the electron that is modeled as an oscillator. Switching to the frequency domain and averaging over the distribution of initial states we get

$$\chi^{(1)}(\omega) = \frac{1}{\hbar} \rho_0 \sum_{a,b} P(a) \frac{2|\mu_{ab}|^2 \omega_{ba}}{\omega_{ba}^2 - \omega^2 - 2i\omega\Gamma_{ab}}. \tag{6.43}$$

Equation (6.43) (with $\Gamma_{ab} = 0$) is identical to Eq. (6.5b). Equation (6.42) establishes a harmonic (Drude) oscillator picture that provides an exact representation for the linear response; the contribution of each pair of levels coupled by the dipole operator to the linear susceptibility is identical to that of a harmonic oscillator. This equivalence is highlighted by the following notation. Consider a small particle with Z electrons (e.g., an atom). Its dipole operator is given by

$$V = -e \sum_j x_j,$$

x_j being the x coordinate of electron j and e is the electron charge. Since the Hamiltonian H depends on the momentum only through the kinetic energy $p_j^2/2m$, we immediately have the double commutator

$$[[x_j, H], x_j] = \frac{\hbar^2}{2m}.$$

Since the coordinates and momenta of different electrons commute, we further have

$$[[\sum_j x_j, H], \sum_j x_j] = \frac{\hbar^2}{2m} Z.$$

Taking the diagonal $\nu\nu$ matrix element of both sides, where ν is an eigenstate of H with energy ε_ν, we get the following sum rule, which is valid for an arbitrary state ν

$$\sum_{\nu'} f_{\nu\nu'} = Z,$$

with

$$f_{vv'} \equiv \frac{2m}{e^2\hbar^2}(\varepsilon_{v'} - \varepsilon_v)|\mu_{vv'}|^2$$
$$= \frac{2m}{e^2\hbar}\omega_{v'v}|\mu_{vv'}|^2,$$

being the *oscillator strength* of the vv' transition. This is known as the Thomas–Reiche–Kuhn sum rule (or the oscillator strength sum rule) [22]. $f_{vv'}$ is dimensionless and real. When v is the ground state, then $\omega_{v'v}$ and $f_{vv'}$ are positive. With this notation the right-hand side of Eq. (6.42) becomes $(e^2 f_{ab}/m)\bar{W}_{ab}E(t)$, and $\hbar$ drops out, as expected for a classical picture. Equation (6.43) can now be written as

$$\chi^{(1)}(\omega) = \frac{e^2\rho_0}{m}\sum_{a,b} P(a)\frac{f_{ab}}{\omega_{ba}^2 - \omega^2 - 2i\omega\Gamma_{ab}}.$$

This form is identical to the response of a collection of harmonic oscillators.

It is tempting to push this analogy further to the nonlinear regime, and to calculate the nonlinear susceptibilities using an *anharmonic oscillator* model. Such models have been proposed since the early days of nonlinear optics [23]. We thus generalize Eq. (6.42) and write

$$\ddot{P}_{ab} + \omega_{ab}^2 P_{ab} + 2\Gamma_{ab}\dot{P}_{ab} = \frac{2}{\hbar}E(t)|\mu_{ab}|^2\omega_{ba}\bar{W}_{ab} + A_{\mathrm{NL}}, \tag{6.44}$$

where A_{NL} is a nonlinear (anharmonic) part. It is essential to add a nonlinear contribution A_{NL} since the harmonic oscillator is linear, and for $A_{\mathrm{NL}} = 0$ all of its nonlinear susceptibilities vanish identically $\chi^{(2)} = \chi^{(3)} = \cdots = 0$.

The intuitive form for A_{NL}, proposed using the analogy with a classical anharmonic oscillator is [23]

$$A_{\mathrm{NL}} = \lambda_2 P_{ab}^2 + \lambda_3 P_{ab}^3 + \cdots, \tag{6.45}$$

with λ_j are the anharmonities. The mapping of the problem into anharmonic oscillators has important practical implications on optical properties in condensed phases, since it provides an invaluable physical insight and may allow a new interpretation in terms of quasiparticles. We shall address this point further in Chapters 16 and 17. Moreover, by treating the electronic degrees of freedom as oscillators we can couple them more naturally to nuclear degrees of freedom, which constitute another set of oscillators. The incorporation of nuclear notions thus becomes much more straightforward compared with the eigenstate representation, and lends itself more easily to semiclassical approximations.

We shall now attempt to derive the anharmonic oscillator equation and assess its applicability. Starting with Eqs. (6.41), taking a derivative of Eq. (6.41a) and substituting Eq. (6.41b) for $\dot{Q}_{ab}$ we get:

$$\ddot{P}_{ab} + \omega_{ab}^2 P_{ab} + 2\Gamma_{ab}\dot{P}_{ab} = \frac{2}{\hbar}E(t)\omega_{ba}|\mu_{ab}|^2 W_{ab}, \tag{6.46a}$$

$$\dot{W}_{ab} = \frac{2E(t)}{\hbar\omega_{ab}}(\dot{P}_{ab} + \Gamma_{ab}P_{ab}) - \Gamma_1(W_{ab} - \bar{W}_{ab}). \tag{6.46b}$$

From this result we can directly conclude that in order to calculate optical nonlinearities, we can no longer consider the polarization as the only dynamic variable. We need to add more variables (W_{ab} in this case), which become relevant, and the simple anharmonic oscillator model cannot be generally justified. This problem becomes even more serious when interactions among particles are considered, since nonlocal interparticle coherences become relevant as well, and the number of necessary variables becomes prohibitively large (see Chapter 17). Nevertheless, we shall try to bring these equations as close as possible to a simple oscillator form. Let us neglect pure dephasing processes and retain only T_1 relaxation with $\Gamma_1 = \Gamma_{ab}$. The system can then be considered to be in a pure state so that (see Chapter 3)

$$|\rho_{ab}|^2 = \rho_{aa}\rho_{bb}. \tag{6.47}$$

Combining this with $\rho_{aa} + \rho_{bb} = 1$, we then have

$$W_{ab} = \sqrt{1 - 4|\rho_{ab}|^2}. \tag{6.48}$$

From the definitions of P_{ab} and Q_{ab} we get

$$\rho_{ab} = \frac{P_{ab} + iQ_{ab}}{2\mu_{ba}}.$$

Using Eq. (6.41) we have

$$Q_{ab} = \frac{\dot{P}_{ab} + \Gamma_{ab}P_{ab}}{\omega_{ab}}.$$

Combining the above two equations results in

$$\rho_{ab} = \frac{P_{ab} + i\dot{P}_{ab}/\omega_{ab} + i\Gamma_{ab}P_{ab}/\omega_{ab}}{2\mu_{ab}}. \tag{6.49}$$

Since electronic population relaxation rates ($<10^{12}\,\mathrm{s}^{-1}$) are typically much smaller than optical frequencies ($\sim 10^{15}\,\mathrm{s}^{-1}$) then $\Gamma_{ab} \ll \omega_{ab}$, and we can safely neglect the last term in the numerator. Equations (6.48) and (6.49) yield

$$W_{ab} = \sqrt{1 - |P_{ab} + i\dot{P}_{ab}/\omega_{ab}|^2/|\mu_{ab}|^2}.$$

The last (nonlinear) term in Eqs. (6.46a) contains an E factor. To order E^3 we thus need to expand W_{ba} to order E^2, and obtain

$$W_{ab} \cong 1 - \frac{|P_{ab} + i\dot{P}_{ab}/\omega_{ab}|^2}{2|\mu_{ab}|^2}.$$

Equation (6.46a) then assumes the form of Eq. (6.44) with the nonlinear part

$$A_{\mathrm{NL}} = \frac{E(t)}{\hbar\omega_{ab}}|\omega_{ab}P_{ab} + i\dot{P}_{ab}|^2. \tag{6.50}$$

We have thus accomplished our goal and derived a closed equation for P_{ab} (with no additional dynamic variables). The following points should be noted in comparing this result with the phenomenological anharmonic oscillator expression [Eq. (6.45)].

1. Equations (6.44) together with (6.50) hold only for a system is in a pure state (pure dephasing processes must be neglected).
2. The nonlinearity [Eq. (6.50)] does not look at all like the phenomenological model [Eq. (6.45)]. We find no $\sim P_{ab}^2$ type of nonlinearity. Instead we have $E(t)P_{ab}^2$ and $E(t)\dot{P}_{ab}^2$ terms. The anharmonicity is thus a function of both the "position" (P_{ab}) and the "velocity" ($\dot{P}_{ab}$) of the oscillator, as well as of the electric field $E(t)$.
3. The main reason for the failure of the simple oscillator picture is that optical nonlinearities depend on more relevant dynamic variables (in addition to the polarization). The number of oscillators thus increases as nonlinearities of higher order are considered. For linear optics suffice it to look at the polarization. For $\chi^{(3)}$ we need to include the population variables W_{ab} as well. It is generally impossible to eliminate these additional variables and construct a closed description in terms of the polarization alone. Moreover, even when this can be done, we obtain different, field-dependent, nonlinearities. A generalized oscillator picture is extremely valuable in the treatment of many body effects in optical nonlinearities, as will be shown in Chapters 16 and 17.

$\chi^{(3)}$ WITH A REALISTIC POPULATION-RELAXATION MATRIX

The formal expressions derived in Chapter 5 allow us to incorporate a general relaxation mechanism. In Eq. (6.14) we used the simplest model. The T_1 relaxation matrix used there does not conserve populations (i.e., the trace of the density operator) since it simply assumes that every level has a finite lifetime and decays outside the system. The incorporation of a more general relaxation matrix requires additional details of the system and the bath. Let us consider a two-level system $|a\rangle$ and $|b\rangle$, and assume a simple master equation for population relaxation [10]

$$\left.\begin{aligned} \dot{\rho}_{aa} &= -\gamma_a \rho_{aa} + \gamma_b \rho_{bb}, \\ \dot{\rho}_{bb} &= -\gamma_b \rho_{bb} + \gamma_a \rho_{aa}, \end{aligned}\right\} \tag{6.51}$$

where γ_a and γ_b satisfy the detailed balance condition [see Eq. (6.A.4)].

The Green function solution of these equations is

$$\begin{pmatrix} \rho_{aa}(t) \\ \rho_{bb}(t) \end{pmatrix} = \begin{pmatrix} \mathcal{G}_{aa,aa}(t) & \mathcal{G}_{aa,bb}(t) \\ \mathcal{G}_{bb,aa}(t) & G_{bb,bb}(t) \end{pmatrix} \begin{pmatrix} \rho_{aa}(0) \\ \rho_{bb}(0) \end{pmatrix}. \tag{6.52}$$

This relaxation scheme as well as the time evolutions of ρ_{aa} and ρ_{bb} are displayed in Figure 6.6. In general, the Green function should have four indices. In the previous applications it was diagonal; we thus used an abbreviated notation where $\mathcal{G}_{ab}$ stands for $\mathcal{G}_{ab,ab}$, etc. Once we include a T_1 relaxation matrix we can no longer avoid using tertradic notation. Substituting these Green functions in Eq. (5.25) we obtain, for the nonlinear response function,

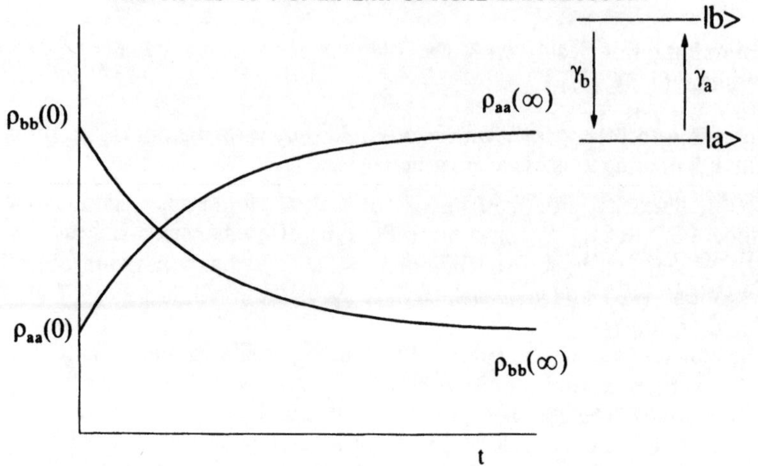

FIG. 6.6 Relaxation of a two-level system to thermal equilibrium as described by the master equation [Eq. (6.51)].

$$R(t_3, t_2, t_1) = |\mu_{ab}|^4[P(a) - P(b)] \times [I_{ba}(t_3) - I_{ab}(t_3)]\hat{I}(t_2)[I_{ba}(t_1) + I_{ab}(t_1)], \quad (6.53)$$

where

$$\hat{I}(t_2) \equiv \mathcal{G}_{aa,aa}(t_2) + \mathcal{G}_{bb,bb}(t_2) - \mathcal{G}_{aa,bb}(t_2) - \mathcal{G}_{bb,aa}(t_2).$$

The solution of the master equation (6.52) gives

$$\hat{I}(t_2) = 2\exp(-\gamma t_2), \quad (6.54)$$

with

$$\gamma \equiv \gamma_a + \gamma_b.$$

In the frequency domain we then get [24]

$$\chi^{(3)}(-\omega_s; \omega_1, -\omega_2, \omega_3) = \frac{1}{6}\left(\frac{1}{\hbar}\right)^3 \rho_0 \sum_p |\mu_{ab}|^4[P(a) - P(b)] \times [I_{ba}(\omega_s) - I_{ab}(\omega_s)] \frac{2}{\omega_1 - \omega_2 + i\gamma} [I_{ba}(\omega_1) + I_{ab}(\omega_1)]. \quad (6.55)$$

Here we have considered the signal frequency $\omega_s = \omega_1 - \omega_2 + \omega_3$, which corresponds to, e.g., degenerate four wave mixing.

For comparison, the relaxation model used in Eqs. (6.23) implies

$$\mathcal{G}_{bb,aa}(t) = \mathcal{G}_{aa,bb}(t) = 0,$$

we then had

$$\hat{I}(t_2) = \exp(-\gamma_a t_2) + \exp(-\gamma_b t_2). \quad (6.56)$$

Equation (6.23) can thus be obtained from Eq. (6.55) by changing ω_2 to $-\omega_2$ and making the simple substitution:

$$\frac{2}{\omega_1 - \omega_2 + i\gamma} \to \frac{1}{\omega_1 - \omega_2 + i\gamma_a} + \frac{1}{\omega_1 - \omega_2 + i\gamma_b}. \quad (6.57)$$

The right-hand side of Eq. (6.57) diverges when $\omega_1 = \omega_2$ even when population relaxation is included, since the ground state has an infinitely long lifetime

($\gamma_a = 0$). The more realistic master equation [the left-hand side of Eq. (6.57)] cures this problem since once relaxation is included, γ is finite and $\chi^{(3)}$ does not diverge. We note that the radiative lifetime should not be included in $\chi^{(3)}$ (see Chapter 17), and the relaxation considered here is nonradiative.

Finally, we note that the present relaxation matrix is identical to the one used in the celebrated Bloch equations (see Appendices 6A and 6B). In the Bloch equations we further make the RWA (which we did not make here). $\chi^{(3)}$ for the Bloch equations is thus given by Eq. (6.55) with the neglect of the non-RWA terms [i.e., setting $I_{ab}(\omega_1) = I_{ab}(\omega_s) = 0$].

HOMOGENEOUS, INHOMOGENEOUS, AND INTERMEDIATE DEPHASING

Relaxation and dephasing were introduced earlier in this chapter through phenomological reduced equations of motion. It is common to denote this type of relaxation "homogeneous broadening" as opposed to "inhomogeneous broadening" resulting from a static distribution of transition frequencies. Theories of spectral line broadening traditionally make a distinction between homogeneous and inhomogeneous contributions to spectral line shapes. When different molecules absorb at different frequencies because of different local environments or different initial states, the line is said to be inhomogeneously broadened. This broadening is static in nature, reflecting the spread in transition frequencies and carries no dynamic information. In contrast, homogeneous broadening arises from an interaction with a bath with a very fast time scale, reflecting rapid fluctuations in the local environment. As far as the radiation field is concerned, all molecules on the average appear identical.

To clarify the microscopic origin of these two types of broadening mechanisms and consider more general cases, which are neither homogeneous nor inhomogenous, we shall focus on a specific model system. Consider a two electronic level molecule coupled to a thermal bath. The combined Hamiltonian of the system and the bath is

$$H = |g\rangle[H_g(\mathbf{q}_s) + h_g(\mathbf{q}_B)]\langle g| + |e\rangle[\omega_{eg}^0 + H_e(\mathbf{q}_s) + h_e(\mathbf{q}_B)]\langle e|. \tag{6.58}$$

Here the ground-state Hamiltonian is partitioned into a system part H_g, which depends on the system coordinates $\mathbf{q}_s$, and a bath part h_g, which depends on the bath coordinates $\mathbf{q}_B$. Similar partitioning of the excited-state Hamiltonians into H_e and h_e is made. The interaction between the system and the bath is modeled through the difference between h_g and h_e, which implies that the bath eigenstates are different, depending on the electronic state of the system. This Hamiltonian does not couple the bath directly to the molecular vibrational degrees of freedom. The key physical assumption underlying this Hamiltonian is that the bath is primarily sensitive to the molecular charge distribution, which may be very different for the ground $|g\rangle$ and the electronically excited $|e\rangle$ states. The interaction of the bath with the molecule is only weakly dependent on the nuclear coordinates, and this dependence, which results in bath-induced vibrational relaxation and dephasing, is ignored. Typically, the pure-dephasing rate between different electronic levels (i.e., electronic dephasing) is much larger than vibrational dephasing between levels belonging to the same electronic manifold. This is the rationale for the present approximation. Vibrational dephasing may play an important role in some techniques (e.g., Raman spectroscopy). However, for the sake of clarity in the following discussion we chose to neglect it. The manifold of vibronic levels belonging to the ground electronic state will be denoted $|a\rangle, |c\rangle, \ldots$. The electronically excited state manifold will be denoted $|b\rangle, |d\rangle, \ldots$. (The system is depicted in Figure 7.1.) We then have

$$(H_g + h_g)|\nu\alpha\rangle = (\varepsilon_\nu + \varepsilon_\alpha)|\nu\alpha\rangle \qquad \nu = a, c, \ldots$$

and

$$(\omega_{eg} + H_e + h_e)|\nu\beta\rangle = (\varepsilon_\nu + \varepsilon_\beta)|\nu\beta\rangle \qquad \nu = b, d, \ldots.$$

Here $a, b, c, d, \ldots$ stand for the collection of all system quantum numbers, whereas α and β represent the bath quantum numbers. The molecule is assumed to be initially at thermal equilibrium in the ground-state manifold, and its density operator is the direct product of the system (σ_g) and the bath (ρ_g) components, that is

$$\rho(-\infty) = \sigma_g(\mathbf{q}_s)\rho_g(\mathbf{q}_B),$$

with

$$\sigma_g = \exp(H_g/k_B T)/\mathrm{Tr}\exp(-H_g/k_B T) = \sum_a |a\rangle P(a)\langle a|,$$

$$\rho_g = \exp(-h_g/k_B T)/\mathrm{Tr}\exp(-h_g/k_B T) = \sum_a |\alpha\rangle \rho_g(\alpha)\langle\alpha|.$$

Here

$$P(a) = \exp(-\varepsilon_a/k_B T)\Big/\sum_a \exp(-\varepsilon_a/k_B T),$$

$$\rho_g(\alpha) = \exp(-\varepsilon_a/k_B T)\Big/\sum_\alpha \exp(-\varepsilon_a/k_B T).$$

Adopting Liouville space notation, we write the equilibrium density operator in the form

$$|\rho(-\infty)\rangle\rangle = \sum_{a,\alpha} P(a)\rho_g(\alpha)|{}^{\alpha\alpha}_{aa}\rangle\rangle.$$

Here $|\nu\nu'\rangle\rangle$ is the Liouville space vector corresponding to $|\nu\rangle\langle\nu'|$. $|{}^{\alpha\alpha}_{aa}\rangle\rangle$ stands for the direct product of the system $|aa\rangle\rangle$ state and the bath $|\alpha\alpha\rangle\rangle$ state.

The molecular electronic dipole operator couples vibronic states belonging to different electronic states. We then have

$$V = \sum_{\substack{\nu=a,c\\ \nu'=b,d}} [\mu_{\nu\nu'}|\nu\rangle\langle\nu'| + \mu_{\nu'\nu}|\nu'\rangle\langle\nu|],$$

where the summation runs over the entire manifolds of ground and electronically excited states. For the present model, which has no permanent dipole moments, the second-order response vanishes. We shall then focus on the first- and third-order functions. Making use of Eq. (5.14), we obtain for the optical response functions

$$S^{(1)}(t) = \frac{i}{\hbar}\sum_{a,b} P(a)|\mu_{ab}|^2[\langle\mathcal{G}_{ab}(t)\rho_g\rangle - \langle\mathcal{G}_{ba}(t)\rho_g\rangle], \tag{6.59}$$

and

$$
\begin{aligned}
S^{(3)}(t_3, t_2, t_1) = \left(\frac{i}{\hbar}\right)^3 \sum_{a,b,c,d} & P(a)\mu_{ab}\mu_{bc}\mu_{cd}\mu_{da} \\
& \times [\langle \mathcal{G}_{dc}(t_3)\mathcal{G}_{db}(t_2)\mathcal{G}_{da}(t_1)\rho_g\rangle + \langle \mathcal{G}_{dc}(t_3)\mathcal{G}_{db}(t_2)\mathcal{G}_{ab}(t_1)\rho_g\rangle \\
& + \langle \mathcal{G}_{dc}(t_3)\mathcal{G}_{ac}(t_2)\mathcal{G}_{ab}(t_1)\rho_g\rangle + \langle \mathcal{G}_{ba}(t_3)\mathcal{G}_{ca}(t_2)\mathcal{G}_{da}(t_1)\rho_g\rangle \\
& - \langle \mathcal{G}_{ad}(t_3)\mathcal{G}_{ac}(t_2)\mathcal{G}_{ab}(t_1)\rho_g\rangle - \langle \mathcal{G}_{cb}(t_3)\mathcal{G}_{db}(t_2)\mathcal{G}_{ab}(t_1)\rho_g\rangle \\
& - \langle \mathcal{G}_{cb}(t_3)\mathcal{G}_{db}(t_2)\mathcal{G}_{da}(t_1)\rho_g\rangle - \langle \mathcal{G}_{cb}(t_3)\mathcal{G}_{ca}(t_2)\mathcal{G}_{da}(t_1)\rho_g\rangle], \qquad (6.60)
\end{aligned}
$$

where the eight terms correspond to pathways (1)–(8) of Figure 6.3, respectively. The following notation was introduced in these equations,

$$\mathcal{G}_{\nu\nu'}(t) \equiv \langle\langle \nu\nu'|\mathcal{G}(t)|\nu\nu'\rangle\rangle_S = \mathrm{Tr}_S[|\nu'\rangle\langle\nu|\mathcal{G}(t)|\nu\rangle\langle\nu'|].$$

The subscript S signifies that this is a matrix element in the system space only, that is, we perform a partial trace over the system degrees of freedom, and $\mathcal{G}_{\nu\nu'}(t)$ is still a Liouville space operator in the bath phase space. The angular brackets $\langle\cdots\rangle$ denote averaging over the bath, that is,

$$\langle\mathcal{G}\rangle = \sum_{\alpha,\alpha'} \langle\langle\alpha'\alpha'|\mathcal{G}|\alpha\alpha\rangle\rangle\rho_g(\alpha). \qquad (6.61)$$

This is the usual thermodynamic averaging, which implies averaging over initial states $|\alpha\alpha\rangle\rangle$ and summing over final states $|\alpha'\alpha'\rangle\rangle$. To clarify the notation, we shall write explicitly one of the terms in Eq. (6.60):

$$
\begin{aligned}
\langle \mathcal{G}_{ad}(t_3)\mathcal{G}_{ac}(t_2)\mathcal{G}_{ab}(t_1)\rho_g\rangle = \sum_{\substack{\alpha,\alpha_1,\alpha_2,\\ \alpha_3,\alpha_4,\alpha'}} & \langle\langle {}^{\alpha'\alpha'}_{a\;d}|\mathcal{G}(t_3)|{}^{\alpha_3\alpha_4}_{a\;d}\rangle\rangle\langle\langle {}^{\alpha_3\alpha_4}_{a\;c}|\mathcal{G}(t_2)|{}^{\alpha_1\alpha_2}_{a\;c}\rangle\rangle \\
& \times \langle\langle {}^{\alpha_1\alpha_2}_{a\;b}|\mathcal{G}(t_1)|{}^{\alpha\alpha}_{ab}\rangle\rangle\rho_g(\alpha) \qquad (6.62)
\end{aligned}
$$

In the frequency domain, the optical susceptibilities are given by

$$\chi^{(1)}(\omega) = \frac{1}{\hbar}\rho_0 \sum_{a,b} P(a)|\mu_{ab}|^2[\langle\mathcal{G}_{ba}(\omega)\rho_g\rangle - \langle\mathcal{G}_{ab}(\omega)\rho_g\rangle], \qquad (6.63a)$$

$$
\begin{aligned}
& \chi^{(3)}(-\omega_1 + \omega_2 - \omega_3; \omega_3, -\omega_2, \omega_1) \\
& = \left(-\frac{1}{\hbar}\right)^3 \rho_0 \sum_p \sum_{a,b,c,d} P(a)\mu_{ab}\mu_{bc}\mu_{cd}\mu_{da} \\
& \quad \times [\langle \mathcal{G}_{dc}(\omega_1 - \omega_2 + \omega_3)\mathcal{G}_{db}(\omega_1 - \omega_2)\mathcal{G}_{da}(\omega_1)\rho_g\rangle \\
& \quad + \langle \mathcal{G}_{dc}(\omega_1 - \omega_2 + \omega_3)\mathcal{G}_{db}(\omega_1 - \omega_2)\mathcal{G}_{ab}(\omega_1)\rho_g\rangle \\
& \quad + \langle \mathcal{G}_{dc}(\omega_1 - \omega_2 + \omega_3)\mathcal{G}_{ac}(\omega_1 - \omega_2)\mathcal{G}_{ab}(\omega_1)\rho_g\rangle \\
& \quad + \langle \mathcal{G}_{ba}(\omega_1 - \omega_2 + \omega_3)\mathcal{G}_{ca}(\omega_1 - \omega_2)\mathcal{G}_{da}(\omega_1)\rho_g\rangle \\
& \quad - \langle \mathcal{G}_{ad}(\omega_1 - \omega_2 + \omega_3)\mathcal{G}_{ac}(\omega_1 - \omega_2)\mathcal{G}_{ab}(\omega_1)\rho_g\rangle \\
& \quad - \langle \mathcal{G}_{cb}(\omega_1 - \omega_2 + \omega_3)\mathcal{G}_{db}(\omega_1 - \omega_2)\mathcal{G}_{ab}(\omega_1)\rho_g\rangle \\
& \quad - \langle \mathcal{G}_{cb}(\omega_1 - \omega_2 + \omega_3)\mathcal{G}_{db}(\omega_1 - \omega_2)\mathcal{G}_{da}(\omega_1)\rho_g\rangle \\
& \quad - \langle \mathcal{G}_{cb}(\omega_1 - \omega_2 + \omega_3)\mathcal{G}_{ca}(\omega_1 - \omega_2)\mathcal{G}_{da}(\omega_1)\rho_g\rangle]. \qquad (6.63b)
\end{aligned}
$$

Here again, the eight terms correspond, respectively, to pathways (1)–(8) of Figure 6.3, and the angular brackets $\langle \cdots \rangle$ were defined in Eq. (6.62).

Equations (6.60) and (6.63) can be used to provide a microscopic definition of homogeneous and inhomogeneous line broadening. When the bath correlation time is very short compared with the inverse linewidth, the average of the product of Green functions may be factorized into the product of the averaged Green functions, i.e.,

$$\langle \mathscr{G}_{ad}(t_3)\mathscr{G}_{ac}(t_2)\mathscr{G}_{ab}(t_1)\rho_g\rangle \cong \langle \mathscr{G}_{ad}(t_3)\rho_g\rangle\langle \mathscr{G}_{ac}(t_2)\rho_g\rangle\langle \mathscr{G}_{ab}(t_1)\rho_g\rangle \tag{6.64}$$

and similarly in the frequency domain.

We are now in a position to give a microscopic definition of homogeneous and inhomogeneous dephasing mechanisms. We shall denote any broadening mechanism for which this factorization applies as "homogeneous." In the previous discussion we had simply set $\langle \mathscr{G}_{\nu\nu'}(t)\rangle = I_{\nu\nu'}(t)$ [Eq. (6.16b)]. However, more general forms of the averaged Green functions (not necessarily exponential) may be used. Inhomogeneous broadening is the other extreme, where all three Green functions depend on some parameters (Γ), and the entire product should be averaged over the distribution $W(\Gamma)$ of these parameters

$$\chi^{(1)}(\omega) = \int \chi_{\mathrm{H}}^{(1)}(\omega; \Gamma) W(\Gamma)\, d\Gamma, \tag{6.65a}$$

and

$$\chi^{(3)}(-\omega_s; \omega_1 + \omega_2 + \omega_3, \omega_1 + \omega_2, \omega_1)$$
$$= \int \chi_{\mathrm{H}}^{(3)}(-\omega_s; \omega_1 + \omega_2 + \omega_3, \omega_1 + \omega_2, \omega_1; \Gamma) W(\Gamma)\, d\Gamma. \tag{6.65b}$$

Static inhomogeneous broadening may thus be incorporated by convoluting the homogeneous response functions calculated in the absence of inhomogeneous broadening, with the distribution of these parameters. What makes this mechanism "inhomogeneous" is the fact that the Γ parameters maintain their values throughout the optical process, which makes the Green functions fully correlated.

More generally, neither of these limits holds, and the dynamics of the bath cannot be considered either very fast or very slow compared with the relevant timescales t_1, t_2, and t_3. The line broadening is then intermediate and one should evaluate the complete four point correlation function. This makes the calculation more difficult, but the calculation is worthwhile, since it carries more detailed and meaningful information in the intermediate case. This generalization will be carried out in the next two chapters using a semiclassical representation of the response functions.

The absorption lineshape is given by combining Eqs. (4.64) with (5.34b) and (5.51). We define a lineshape function normalized to a unit area

$$\sigma_a(\omega) \equiv \chi''(\omega) \Big/ \int d\omega'\, \chi''(\omega'), \tag{6.66a}$$

with

$$\int \sigma_a(\omega)\, d\omega = 1.$$

In the above definition we have neglected the weak variation of ω and the index of refraction $n(\omega)$ in Eq. (4.65) across the absorption band. We can then write

$$\sigma_a(\omega) = J''(\omega)\Big/\int d\omega'\, J''(\omega'). \tag{6.66b}$$

Using Eq. (6.21) and focusing on a particular pair of levels g and e separated by an optical frequency $\omega_{eg} \gg k_B T$, we get the *Lorentzian* profile

$$\sigma_a(\omega) = \frac{1}{\pi}\frac{\Gamma}{(\omega - \omega_{eg})^2 + \Gamma^2}. \tag{6.67}$$

Another common profile that in many cases represents inhomogeneous broadening is the Gaussian

$$\sigma_a(\omega) = \frac{1}{\sqrt{2\pi}\Delta}\exp[-(\omega - \omega_{eg})^2/2\Delta]^2. \tag{6.68}$$

Finally, we note that when using simple phenomenological equations such as the Bloch equations, all optical nonlinearities depend on the same few parameters (frequencies, dipole matrix elements, and relaxation rates). It is then impossible to classify the various nonlinear techniques by their information content. The correlation function approach clearly shows how high nonlinearities involving the averages of products of several Green functions contain more detailed microscopic information about the system.

EXAMPLES OF SINGLE-FREQUENCY TECHNIQUES

Many important applications of the nonlinear response functions and susceptibilities to time domain and frequency domain techniques will be covered in the coming chapters. At this point we shall discuss a few simple examples in which all fields have the same frequency.

1. The second harmonic generation signal is given by

$$S_{\text{SHG}} = |\chi^{(2)}(-2\omega - 2\mathbf{k}; \mathbf{k}\omega, \mathbf{k}\omega)|^2.$$

   This is the simplest nonlinear technique, and was the first experimental demonstration of wave mixing [25].

2. The third harmonic generation is described by [26]

$$S_{\text{THG}} = |\chi^{(3)}(-3\omega - 3\mathbf{k}; \mathbf{k}\omega, \mathbf{k}\omega, \mathbf{k}\omega)|^2.$$

   The absolute magnitude of $\chi^{(3)}$ for THG calculated for sodium atoms is displayed in Figure 6.7.

3. Two photon absorption [27] is a special case of pump-probe spectroscopy when the pump is equal to the probe

$$S_{\text{TPA}} = \text{Im}\, \chi^{(3)}(-\omega - \mathbf{k}; \mathbf{k}\omega, -\mathbf{k} - \omega, \mathbf{k}\omega).$$

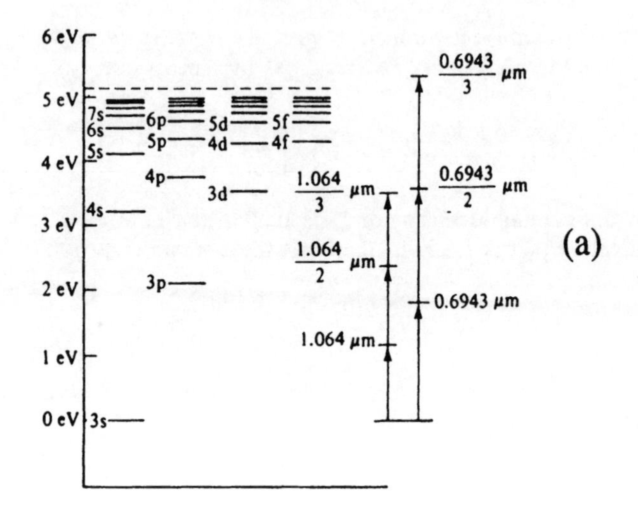

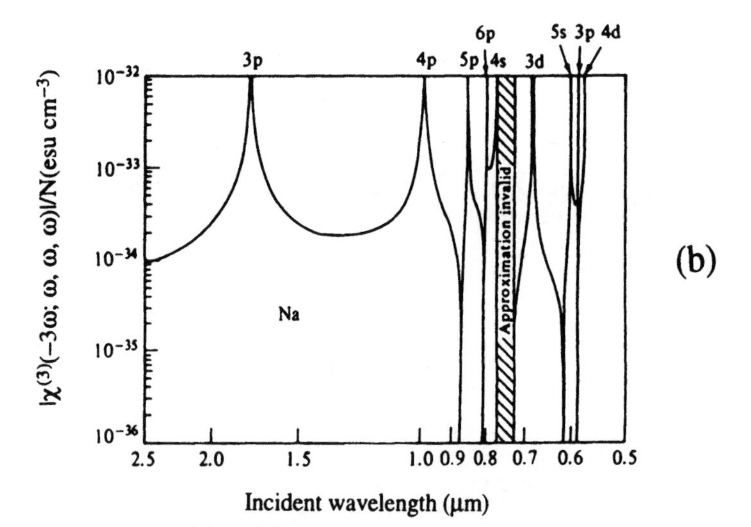

FIG. 6.7 (a) Energy level scheme of sodium. (b) Calculated third-harmonic hyperpolarizability per atom $\chi^{(3)}/N$. [From R. B. Miles and S. E. Harris, "Optical third-harmonic generation in alkali metal vapors," *IEEE J. Quant. Elec.* **QE-9**, 470 (1973).]

4. Nonlinear index of refraction [28]. The propagation of a strong off resonant field in a medium can be described in terms of an intensity-dependent index of refraction. This is responsible, e.g., for phenomena such as self-focusing. We thus have

$$n = n_0 + n_2 I,$$

with the field intensity $I = (n_0 c/2\pi)|E|^2$, and

$$n_2 = \frac{12\pi^2}{n_0^2 c} \operatorname{Re} \chi^{(3)}(-\omega - \mathbf{k}; \mathbf{k}\omega, -\mathbf{k} - \omega, \mathbf{k}\omega).$$

5. Degenerate four wave mixing [29]

$$S_{\mathrm{D4WM}} = |\chi^{(3)}(-\omega - \mathbf{k}_2; \mathbf{k}_2\omega, -\mathbf{k}_1 - \omega, \mathbf{k}_1\omega)|^2.$$

In this technique (also known as phase conjugation) the system interacts with two counter propagating beams with wavevectors $\mathbf{k}_1$ and $-\mathbf{k}_1$. A third beam $\mathbf{k}_2$ is then backscattered in the direction $-\mathbf{k}_2$. Since the scattered field amplitude is the complex conjugate of the incoming field, its spatial profile is reversed and in a nonuniform static medium the backscattered field does not suffer from abberations; their effects in the forward and backward propagation processes exactly cancel.

APPENDIX 6A

Reduced Equations of Motion for the Density Operator

The model system of a particle (e.g., a molecule) interacting with a thermal bath can be treated using various degrees of sophistication [30, 31]. A common approach is to consider the reduced density operator of the system. This way we focus on a few relevant degrees of freedom and trace over the bath. The key to a successful application of this method is the possibility of calculating the reduced density operator directly. (If we first calculate the full density operator and then trace over the bath, we have not really simplified the problem!) In this Appendix we derive simple reduced equations of motion by calculating the relaxation superoperator. We can then introduce the notions of population relaxation and dephasing. These equations may be used to calculate the nonlinear response function.

We start by partitioning the Hamiltonian in the form

$$H = H_S(\mathbf{q}_S) + H_B(\mathbf{q}_B) + H_{SB}(\mathbf{q}_S, \mathbf{q}_B).$$

H_S is the system Hamiltonian with eigenstates

$$H_S|\nu\rangle = \varepsilon_\nu|\nu\rangle \qquad \nu = a, b, c, d, \ldots,$$

H_B is the bath Hamiltonian with eigenstates

$$H_B|\alpha\rangle = \varepsilon_\alpha|\alpha\rangle.$$

Using the basis set of system eigenstates we get

$$H_S = \sum_{\nu=a,b,c,d} |\nu\rangle\varepsilon_\nu\langle\nu|,$$

$$H_{SB} = \sum_\nu |\nu\rangle G_\nu(\mathbf{q}_B)\langle\nu| + \sum_{\nu\neq\nu'} |\nu\rangle F_{\nu\nu'}(\mathbf{q}_B)\langle\nu'|.$$

The dipole moment depends only on the system degrees of freedom and can be expanded in the form

$$V = \sum_{\nu,\nu'} \mu_{\nu\nu'}|\nu\rangle\langle\nu'|.$$

$F_{\nu\nu'}$ is an off-diagonal element that is responsible for population relaxation whereas G_ν is diagonal and induces pure dephasing. We shall consider their effects assuming a weak coupling of the system and the bath, and utilizing the relaxation equations derived formally in Chapter 3 using projection operator techniques.

The system density operator, defined by

$$\sigma \equiv \mathrm{Tr}_{\mathrm{B}}\, \rho,$$

satisfies the reduced equation of motion

$$\dot{\sigma}_{ab} = -i\omega_{ab}\sigma_{ab} - \sum_{c,d}\Gamma_{ab,cd}\sigma_{cd} + \frac{i}{\hbar}E(t)\sum_{c}(\mu_{ac}\sigma_{cb} - \mu_{cb}\sigma_{ac}).$$

The effect of the bath is contained in the tetradic relaxation superoperator $\Gamma_{ab,cd}$, which will be calculated next.

Relaxation of population: The master equation

Taking the matrix elements of the relaxation superoperator Eq. (3.82) and (3.83) in the population subspace related to diagonal elements of the density operator, and expanding them to second order in the system–bath interaction, we get [32]

$$\Gamma_{bb,aa}(\tau) = \sum_{\alpha\beta}\langle\langle{}^{\beta\beta}_{bb}|L_{\mathrm{SB}}\exp(-iL_0\tau)L_{\mathrm{SB}}|{}^{\alpha\alpha}_{aa}\rangle\rangle P(\alpha).$$

Here L_0 and L_{SB} are the Liouville operators corresponding to the zero order Hamiltonian $H_0 = H_{\mathrm{S}} + H_{\mathrm{B}}$ and H_{SB}, respectively. Evaluating the Liouville space matrix element then yields

$$\Gamma_{bb,aa}(\tau) = \exp(i\omega_{ab}\tau)\sum_{\alpha\beta}|\langle b\beta|F|a\alpha\rangle|^2\exp(i\omega_{\alpha\beta}\tau)P(\alpha) + c.c.,$$

$$\Gamma_{bb,aa}(\tau) = \exp(i\omega_{ab}\tau)\langle F(\tau)F(0)\rangle + c.c.,$$

$$\Gamma_{bb,aa} = \pi\sum_{\alpha\beta}P(\alpha)|\langle b\beta|F|a\alpha\rangle|^2\delta(\varepsilon_a + \varepsilon_\alpha - \varepsilon_b - \varepsilon_\beta).$$

Similarly, we obtain for the other matrix elements

$$\Gamma_{aa,bb} = \pi\sum_{\alpha\beta}P(\beta)|\langle a\alpha|F|b\beta\rangle|^2\delta(\varepsilon_a + \varepsilon_\alpha - \varepsilon_b - \varepsilon_\beta), \tag{6.A.1}$$

$$\Gamma_{aa,aa} = -\Gamma_{bb,aa}, \tag{6.A.2a}$$

$$\Gamma_{bb,bb} = -\Gamma_{aa,bb}. \tag{6.A.2b}$$

In the right-hand side of Eq. (6.A.1) we can substitute

$$P(\beta) \equiv P(\alpha)\frac{P(\beta)}{P(\alpha)} = P(\alpha)\exp[-(\varepsilon_\beta - \varepsilon_\alpha)/k_{\mathrm{B}}T]$$

$$= P(\alpha)\exp\lfloor\hbar\omega_{ba}/k_{\mathrm{B}}T).$$

The last equality follows from the conservation of energy $\varepsilon_a + \varepsilon_\alpha = \varepsilon_b + \varepsilon_\beta$ implied by the δ function. Substituting of this in Eq. (6.A.1) we get

$$\Gamma_{bb,aa} = \exp(-\hbar\omega_{ba}/k_{\mathrm{B}}T)\Gamma_{aa,bb}. \tag{6.A.3}$$

This important relation is the *detailed balance condition.*

In summary, denoting

$$\gamma_a \equiv \Gamma_{bb,aa},$$

and

$$\gamma_b \equiv \Gamma_{aa,bb},$$

the reduced equation of motion for level populations becomes

$$\begin{pmatrix} \dot{\rho}_{aa} \\ \dot{\rho}_{bb} \end{pmatrix} = -\begin{pmatrix} \gamma_a & -\gamma_b \\ -\gamma_a & \gamma_b \end{pmatrix} \begin{pmatrix} \rho_{aa} \\ \rho_{bb} \end{pmatrix}, \tag{6.A.4}$$

with

$$\gamma_b/\gamma_a = \exp(-\hbar\omega_{ba}/k_B T).$$

By adding both equations we get

$$\dot{\rho}_{aa} + \dot{\rho}_{bb} = 0,$$

which implies the conservation of probability, $\rho_{aa} + \rho_{bb} = 1$.

By subtracting these equations, and defining

$$W_{ab} \equiv \rho_{aa} - \rho_{bb},$$

we get

$$\dot{W}_{ab} = -\Gamma_1(W_{ab} - \bar{W}_{ab}), \tag{6.A.5a}$$

with

$$\Gamma_1 \equiv \gamma_a[1 + \exp(-\hbar\omega_{ba}/k_B T)], \tag{6.A.5b}$$

$$\bar{W}_{ab} \equiv \frac{\gamma_b - \gamma_a}{\gamma_b + \gamma_a} = \tanh(\hbar\omega_{ba}/k_B T). \tag{6.A.5c}$$

The solution of Eq. (6.A.5) is

$$W_{ab}(t) = \bar{W}_{ab} + \exp(-\Gamma_1 t)[W_{ab}(0) - \bar{W}_{ab}]. \tag{6.A.6}$$

In the absence of radiation field the system will relax to thermal equilibrium and attain the same temperature of the bath. Note that at $t = 0$, $W_{ab}(t) = W_{ab}(0)$, and as $t \to \infty$, $W_{ab}(t) = \bar{W}$.

A reduced equation for the diagonal elements of the density operators is called a master equation [10]. In general, for an N level system Eq. (6.A.4) can be generalized as

$$\dot{\rho}_{\nu\nu} = -\sum_{\nu'} \Gamma_{\nu\nu'} \rho_{\nu'\nu'}. \tag{6.A.7}$$

The relaxation matrix always satisfies the following relations

1. Detailed balance
$$\Gamma_{\nu\nu'}/\Gamma_{\nu'\nu} = \exp(-\hbar\omega_{\nu\nu'}/k_B T), \tag{6.A.8}$$

2. The sum of the elements of the relaxation matrix in each column is zero. This guarantees the conservation of probability
$$\Gamma_{\nu'\nu'} = -\sum_{\nu} \Gamma_{\nu\nu'}. \tag{6.A.9}$$

The sum of the elements of the relaxation matrix in each column is zero.

Pure dephasing processes

We now turn to calculate the relaxation superoperator for the off-diagonal elements of the density operator. We again expand the relaxation superoperator Eq. (3.82) and (3.83) to second order in the system–bath interaction:

$$\Gamma_{ab,ab}(\tau) = -\sum_{\alpha\beta} \langle\langle {}^{\beta\beta}_{ab}| L_{SB} \exp(-iL_0\tau) L_{SB} |{}^{\alpha\alpha}_{ab}\rangle\rangle P(\alpha).$$

Let us first consider the contribution of the off diagonal coupling F:

$$\Gamma_{ab,ab} = \tfrac{1}{2}(\Gamma_{aa,bb} + \Gamma_{bb,aa}) = \tfrac{1}{2}(\gamma_a + \gamma_b).$$

However, this is not the entire story since there is an additional contribution coming from the diagonal system-bath coupling (G). Let us define

$$G_{\nu\nu'} \equiv \tfrac{1}{2}(G_\nu - G_{\nu'}),$$

$$\tilde{G}_{\nu\nu'} \equiv \tfrac{1}{2}(G_\nu + G_{\nu'}),$$

so that

$$H_{SB} = G_{\nu\nu'}(|\nu\rangle\langle\nu| - |\nu'\rangle\langle\nu'|) + \tilde{G}_{\nu\nu'}(|\nu\rangle\langle\nu| + |\nu'\rangle\langle\nu'|).$$

Consider the following element of the relaxation superoperator

$$\Gamma_{ab,ab}(\tau) = 2\exp(-i\omega_{ab}\tau)[\langle G_{\nu\nu'}(\tau)G_{\nu\nu'}(0)\rangle + c.c.].$$

Since σ_{ab} has a high frequency motion we shall transform to the rotating frame

$$\tilde{\sigma}_{ab}(t) \equiv \exp(i\omega_{ab}t)\sigma_{ab}(t).$$

Upon transforming the equation of motion

$$\dot{\sigma}_{ab} = -i\omega_{ab}\sigma_{ab} - \int_0^t d\tau\, \Gamma_{ab,ab}(t-\tau)\sigma_{ab}(\tau),$$

we get

$$\dot{\tilde{\sigma}}_{ab} = -\int_0^t d\tau\, \tilde{\Gamma}_{ab,ab}(t-\tau)\tilde{\sigma}_{ab}(\tau).$$

Making a Markovian approximation (separation of time scales) we get

$$\left.\begin{aligned}
\dot{\tilde{\sigma}}_{ab} &= -\hat{\Gamma}_{ab}\tilde{\sigma}_{ab},\\
\hat{\Gamma}_{ab} &= 2\int_0^\infty [\langle G_{ab}(\tau)G_{ab}(0)\rangle + c.c.],\\
\dot{\sigma}_{ab} &= -i\omega_{ab}\sigma_{ab} - \hat{\Gamma}_{ab}\sigma_{ab},\\
\hat{\Gamma}_{ab} &= 2\sum_{\alpha\beta} |\langle a\beta|G|a\alpha\rangle - \langle b\beta|G|b\alpha\rangle|^2\delta(\varepsilon_\alpha - \varepsilon_\beta).
\end{aligned}\right\} \quad (6.A.10)$$

$\Gamma_{ab,ab} \equiv \hat{\Gamma}_{ab}$ is denoted the *pure dephasing rate.*

In summary,

$$\Gamma_{ab,ab} = \tfrac{1}{2}(\gamma_a + \gamma_b) + \hat{\Gamma}_{ab}. \tag{6.A.11}$$

The off-diagonal elements decay faster than expected from the decay of the diagonal elements. This can be rationalized as follows. For a pure state we have

$$|\rho_{ab}| = \sqrt{\rho_{aa}\rho_{bb}},$$

so that

$$\begin{aligned}\dot{\rho}_{ab} &= \tfrac{1}{2}\dot{\rho}_{aa}\sqrt{\rho_{bb}/\rho_{aa}} + \tfrac{1}{2}\dot{\rho}_{bb}\sqrt{\rho_{aa}/\rho_{bb}} \\ &= -\tfrac{1}{2}(\gamma_a + \gamma_b)\rho_{ab}.\end{aligned}$$

If we let ρ_{aa} and ρ_{bb} decay, without allowing ρ_{ab} to decay we shall violate the inequality

$$\sqrt{\rho_{aa}\rho_{bb}} \geq |\rho_{ab}|,$$

which proves our point. Adopting NMR terminology, γ_a and γ_b represent the decay of the amplitude (population) and $\Gamma_1 \equiv 1/T_1$, where T_1 the longitudinal relaxation time. The pure dephasing rate $\hat{\Gamma}$ is due to phase relaxation (loss of phase). The total dephasing rate is denoted $\Gamma_{ab,ab} \equiv 1/T_2$ where T_2 is the transverse relaxation time.

APPENDIX 6B

The Optical Bloch Equations

The Bloch equations [33] are the basic equations of motion traditionally used for the interpretation of nuclear magnetic resonance and nonlinear optical spectroscopies. They are based on the following assumptions:

1. We specialize to the optical response of a two-level system.
2. We consider a near resonance excitation and invoke the rotating wave approximation (RWA), neglecting highly oscillatory terms in the equations of motion. We can then transform the Liouville equation to a form with a time-independent effective Liouville operator.
3. Relaxation is incorporated using the reduced equations of motion derived in Appendix 6A, including a population relaxation (T_1) and pure dephasing (T_2) relaxation times.

The derivation goes as follows [34]. We first consider a two-level system interacting with a strong monochromatic field (neglecting the bath for now)

$$H = |a\rangle\varepsilon_a\langle a| + |b\rangle\varepsilon_b\langle b| - \mu E(t)[|a\rangle\langle b| + |b\rangle\langle a|].$$

The electric field is given by $E(t) = E_0 \cos \omega t$ and it is assumed to be near resonant, i.e., $\omega \sim \omega_{ba}$.

In the absence of a field, we have

$$\rho_{ab}(t) = \exp(i\omega_{ba}t)\rho_{ab}(0)$$

and

$$\rho_{ba}(t) = \exp(-i\omega_{ba}t)\rho_{ba}(0).$$

We next define the new variables

$$\tilde{\rho}_{ab}(t) \equiv \exp(-i\omega t)\rho_{ab}(t),$$

$$\tilde{\rho}_{ba}(t) \equiv \exp(i\omega t)\rho_{ba}(t).$$

$\tilde{\rho}_{ab}$ and $\tilde{\rho}_{ba}$ are expected to be slowly varying since for a near resonant field the additional frequency factors almost cancel the natural evolution frequency. The Liouville equation then reads

$$\begin{pmatrix} \dot{\tilde{\rho}}_{ab} \\ \dot{\tilde{\rho}}_{ba} \\ \dot{\rho}_{aa} \\ \dot{\rho}_{bb} \end{pmatrix} = -i \begin{bmatrix} \Delta & 0 & -\tilde{\Omega}(t) & \tilde{\Omega}(t) \\ 0 & -\Delta & \tilde{\Omega}^*(t) & -\tilde{\Omega}(t) \\ -\tilde{\Omega}^*(t) & \tilde{\Omega}(t) & 0 & 0 \\ \tilde{\Omega}^*(t) & -\tilde{\Omega}(t) & 0 & 0 \end{bmatrix} \begin{pmatrix} \tilde{\rho}_{ab} \\ \tilde{\rho}_{ba} \\ \rho_{aa} \\ \rho_{bb} \end{pmatrix},$$

where the frequency detuning Δ is given by

$$\hbar\Delta \equiv \hbar\omega + \varepsilon_a - \varepsilon_b$$

and

$$\tilde{\Omega}(t) \equiv \Omega[1 + \exp(2i\omega t)],$$

with $\hbar\Omega \equiv \mu E_0$. We now proceed in the following steps:

1. $\tilde{\Omega}(t)$ contains a time-independent (zero frequency) term and a term oscillating at twice the field frequency 2ω. The effect of the latter term is expected to be small since it oscillates much faster than any other frequency in the problem. This can be verified by using some typical numbers. For a field of 10 MW/cm^2 with $\mu = 0.1$ Debye,* we have $\Omega = 1.45 \times 10^{10}$ s^{-1}. ω_{ab} and ω are $\sim 10^{15}$–10^{16} s^{-1}. We thus neglect the time-dependent part and write $\tilde{\Omega}(t) \cong \Omega$. This assumption may break for very intense fields (note that laser fields with intensities of up to 10^{19} W/cm^2 are available!).

   The resulting Liouville equation can now be recast in the form

$$\dot{\tilde{\rho}} = -\frac{i}{\hbar}[H_{\text{eff}}, \tilde{\rho}],$$

   where we have introduced the effective Hamiltonian

$$H_{\text{eff}} \equiv \begin{pmatrix} \hbar\Delta & \hbar\Omega \\ \hbar\Omega & 0 \end{pmatrix}.$$

   The transformed density operator thus satisfies an identical equation to that of a two-level system with a static coupling (Eq. 2.25). The radiation field adds the coupling $\hbar\Omega = \mu E_0$ and shifts the two level frequency (dressed states).

2. Since $\rho_{aa} + \rho_{bb} = 1$ (conservation of probability), we can eliminate one population variable and use three instead of the four elements of the density operator. We further separate $\tilde{\rho}_{ab}$ into its real and imaginary parts and define

$$u \equiv \tilde{\rho}_{ab} + \tilde{\rho}_{ba}$$

$$v \equiv i(\tilde{\rho}_{ab} - \tilde{\rho}_{ba})$$

$$w \equiv \rho_{aa} - \rho_{bb}.$$

   Here u, v, and w are real. w represents the population difference and u and v are the real and the imaginary parts of the the optical coherence.

3. We add T_1 and T_2 processes using the relaxation superoperator developed in Appendix 6A.

   The optical Bloch equations finally read

$$\begin{aligned} \dot{u} &= -\Delta v - \Gamma_2 u, \\ \dot{v} &= \Delta u - 2\Omega w - \Gamma_2 v, \\ \dot{w} &= 2\Omega v - \Gamma_1(w - \bar{w}), \end{aligned} \tag{6.B.1}$$

* 1 debye (D) $\equiv 10^{-18}$ esu cm. The dipole moment of an electron and a proton 1 Å apart is 4.8 D.

where $\Gamma_2 = 1/2\Gamma_1 + \hat{\Gamma}$. Since all the coefficients in these equations are time independent, we can obtain a general solution for an arbitrary field strength. Note, however, that since we have invoked the RWA, the equations are limited to near resonance situations. In the absence of relaxation, the solution results in the Rabi oscillations [see Eq. (2.29)]. Because of the relaxation, the nature of the time evolution changes dramatically and the system reaches a steady state.

The steady-state solution is obtained by setting $\dot{u} = \dot{v} = \dot{w} = 0$. We then get

$$u_{ss} = \frac{\bar{w}\cdot\Delta\cdot 2\Omega}{\Delta^2 + \Gamma_2^2 + (\Gamma_2/\Gamma_1)(2\Omega)^2}, \tag{6.B.2}$$

$$v_{ss} = \frac{-2\bar{w}\Gamma_2\Omega}{\Delta^2 + \Gamma_2^2 + (\Gamma_2/\Gamma_1)(2\Omega)^2}, \tag{6.B.3}$$

$$w_{ss} = \frac{\bar{w}[\Gamma_2^2 + \Delta^2]}{\Delta^2 + \Gamma_2^2 + (\Gamma_2/\Gamma_1)(2\Omega)^2}, \tag{6.B.4}$$

The stationary absorption lineshape is given by [35]

$$I(\omega) = 2\Omega v_{ss} = \frac{(2\Omega)^2\Gamma_2\bar{w}}{\Delta^2 + \Gamma_2^2 + (\Gamma_2/\Gamma_1)(2\Omega)^2}, \tag{6.B.5}$$

The following points can now be made:

1. For $\bar{w} = 0$ there is *no absorption*. At infinite temperature the absorption and stimulated emission exactly cancel, resulting in no net absorption.

2. The lineshape is a Lorentzian with fwhm 2κ where

$$\kappa^2 = \Gamma_2^2 + \frac{\Gamma_2}{\Gamma_1}(2\Omega)^2.$$

For weak fields ($\Omega \ll \Gamma_2$) we have $I(\omega) \propto (2\Omega)^2$, and $\kappa = \Gamma_2$. For strong fields the line broadens (power broadening) and saturates, i.e., the absorption rate does not grow linearly with the field. For very strong fields it becomes independent on field intensity and is limited by the T_1 relaxation rate $I(\omega) = \Gamma_1\bar{W}$.

3. Coherent and incoherent driving: Suppose Γ_2 is sufficiently large so that u and v relax very rapidly compared to the driving time. We can then assume that they reach a steady state, and set $\dot{u} = \dot{v} = 0$. We then have

$$-\Delta v - \Gamma_2 u = 0,$$

$$\Delta u - 2\Omega w - \Gamma_2 v = 0,$$

whose solution is

$$u = \frac{2\Omega\Delta w}{\Delta^2 + \Gamma_2^2}; \qquad v = -\frac{2\Omega w\Gamma_2}{\Delta^2 + \Gamma_2^2}.$$

Upon the substitution of these expressions in the equation of motion for w we obtain the ordinary rate equation (also known as a master equation)

$$\frac{dw}{dt} = -(-2\Omega)^2 \frac{\Gamma_2}{\Delta^2 + \Gamma_2^2} w - \Gamma_1(w - \bar{w}). \tag{6.B.6}$$

This is the incoherent driving limit, which is described by a simple rate equation. As a consistency check for the assumptions leading to the rate equation we require

$$\Gamma_2 \gg \frac{(2\Omega^2)\Gamma_2}{\Delta^2 + \Gamma_2^2},$$

which implies

$$\Delta^2 + \Gamma_2^2 \gg (2\Omega)^2.$$

The rate equation is valid only on a coarse-grained time scale and it never strictly holds at $t = 0$. The Bloch equations thus interpolate continuously between the limits of coherent driving ($\Gamma_1, \Gamma_2 \to 0$) and incoherent driving, where the simple rate equation applies ($\Gamma_2 \to \infty$).

4. A closed form expression can be derived for the complete time evolution. To that end we introduce a vector σ with components $\sigma_1 = u$, $\sigma_2 = v$, and $\sigma_3 = w$, i.e.,

$$\sigma \equiv \begin{pmatrix} u \\ v \\ w \end{pmatrix}.$$

We further define another vector $\Delta\sigma \equiv \sigma - \sigma_{ss}$, i.e.,

$$\Delta\sigma \equiv \begin{pmatrix} u - u_{ss} \\ v - v_{ss} \\ w - w_{ss} \end{pmatrix}.$$

The optical Bloch equations now assume the form:

$$\begin{pmatrix} \Delta\dot{\sigma}_1 \\ \Delta\dot{\sigma}_2 \\ \Delta\dot{\sigma}_3 \end{pmatrix} = \begin{pmatrix} -\Gamma_2 & -\Delta & 0 \\ \Delta & -\Gamma_2 & -2\Omega \\ 0 & 2\Omega & -\Gamma_1 \end{pmatrix} \begin{pmatrix} \Delta\sigma_1 \\ \Delta\sigma_2 \\ \Delta\sigma_3 \end{pmatrix}.$$

This equation is isomorphous to the Liouville equation, with a time-independent Liouville operator, and no inhomogeneous terms. In a matrix form we can write this equation as

$$\Delta\dot{\sigma} = \mathscr{F}\Delta\sigma.$$

The solution of this equation is

$$\Delta\sigma(t) = \mathscr{U}(t)\Delta\sigma(0),$$

with

$$\mathscr{U}(t) = \exp(\mathscr{F}t).$$

Writing down the matrix elements we have

$$\sigma_j(t) = \sigma_j^{ss} + \sum_{j'} \mathscr{U}_{jj'}(t)[\sigma_{j'}(0) - \sigma_{j'}^{ss}],$$

with

$$\mathscr{U}_{jj'}(0) = \delta_{jj'}.$$

Obviously $\sigma_j(t) = \sigma_j(0)$ at $t = 0$ and $\sigma_j(t) = \sigma_j^{ss}$ as $t \to \infty$. The evolution operator can also be written in the form

$$\mathscr{U}_{jj'}(t) = A_{jj'} \exp(-\delta t) + (B_{jj'} \cos \alpha t + C_{jj'} \sin \alpha t) \exp(-\gamma t)$$

where all parameters A, B, δ, α, and γ can be calculated from the Bloch equations. $\gamma \pm i\alpha$ and δ are the eigenvalues of the $\mathscr{F}$ matrix.

NOTES

1. T. W. Hänsch, A. L. Schawlow, and G. W. Series, *Sci. Am.* **240**, 94 (1979).
2. G. Herzberg, *Electronic Spectra of Polyatomic Molecules* (Van Nostrand, Toronto, 1966).
3. C. E. Porter, *Statistical Theory of Spectra: Fluctuations* (Academic Press, New York, 1965).
4. J. H. Van Vleck and V. F. Weisskopf, *Rev. Mod. Phys.* **17**, 227 (1945).
5. R. G. Breene, *Theories of Spectral Lineshapes* (Wiley, New York, 1981).
6. N. Mataga and T. Kubata, *Molecular Interactions and Electronic Spectra* (Dekker, New York, 1970); N. Schwentner, E. G. Koch, and J. Jortner, *Electronic Excitations in Condensed Rare Gases* (Springer, New York, 1985).
7. R. Loudon, *The Quantum Theory of Light*, 2nd ed. (Clarendon Press, Oxford, 1983).
8. H. M. Nusenzveig, *Causality and Dispersion Relations* (Academic, New York, 1972).
9. M. Sheik-Bahai, D. C. Hutchings, D. J. Hagan, and E. W. Van Styland, *J. Quant. Electron.* **27**, 1296 (1991).
10. N. G. Van Kampen, *Stochastic Processes in Physics and Chemistry* (North-Holland, Amsterdam, 1981).
11. R. Kubo, M. Toda, and N. Hashitsume, *Statistical Physics II: Nonequilibrium Statistical Mechanics* (Springer-Verlag, New York, 1985).
12. H. B. Callen, *Thermodynamics* (Wiley, New York, 1960).
13. S. Mukamel, *Phys. Rev. A* **28**, 3480 (1983); *Adv. Chem. Phys.* **70**, Part I, 165 (1988).
14. T. K. Yee and T. K. Gustafson, *Phys. Rev. A* **18**, 1597 (1978).
15. S. A. J. Druet and J. P. E. Taran, *Progr. Quant. Electron.* **7**, 1 (1981).
16. W. H. Louisell, *Quantum Statistical Properties of Radiation* (Wiley, New York, 1973).
17. N. Bloembergen, H. Lotem, and R. T. Lynch, *Indian J. Pure Appl. Phys.* **16**, 151 (1978).
18. J. F. Ward, *Rev. Mod. Phys.* **37**, 1 (1965); B. J. Orr and J. F. Ward, *Mol. Phys.* **20**, 513 (1971).
19. H. A. Lorentz, *The Theory of Electrons* (Dover, New York, 1952).
20. L. Rosenfeld, *Theory of Electrons* (North-Holland, Amsterdam, 1951).
21. U. Fano, *Rev. Mod. Phys.* **45**, 553 (1974).

22. H. A. Bethe and E. E. Salpeter, *Quantum Mechanics of One- and Two-Electron Atoms* (Plenum, New York, 1977).
23. N. Bloembergen, *Nonlinear Optics* (Benjamin, New York, 1965).
24. R. W. Boyd and S. Mukamel, *Phys. Rev. A* **29**, 1973 (1984).
25. P. A. Franken, A. E. Hill, C. W. Peters, and G. Weinreich, *Phys. Rev. Lett.* **7**, 118 (1961).
26. R. B. Miles and S. E. Harris, *IEEE J. Quant. Electron.* **9**, 470 (1973); D. C. Hanna, M. A. Yuratich, and D. Cotter, *Nonlinear Optics of Free Atoms and Molecules* (Springer-Verlag, Berlin, 1979).
27. W. Kaiser and C. G. B. Garrett, *Phys. Rev. Lett.* **8**, 404 (1961).
28. B. Ya. Zel'dovich, N. F. Pilipetsky, and V. V. Shkunov, *Principles of Phase Conjugation* (Springer-Verlag, Berlin, 1985); G. P. Agrawal, *Nonlinear Fiber Optics* (Academic Press, Boston, 1989).
29. J. Nilsen and A. Yariv, *J. Opt. Soc. Am.* **71**, 180 (1981); J. Nilsen, N. S. Gluck, and A. Yariv, *Opt. Lett.* **6**, 380 (1981).
30. A. G. Redfield, *Adv. Mag. Reson.* **1**, 1 (1965).
31. R. Zwanzig, *Lect. Theoret. Phys.* **3**, 106 (1961); *Physica* **30**, 1109 (1964); *J. Chem. Phys.* **33**, 1338 (1960).
32. S. Mukamel, *Chem. Phys.* **37**, 33 (1979).
33. F. Bloch, *Phys. Rev.* **70**, 460 (1946).
34. L. Allen and J. H. Eberly, *Optical Resonance and Two-Level Atoms* (Wiley, New York, 1975).
35. R. Karplus and J. Schwinger, *Phys. Rev.* **73**, 1020 (1948).

CHAPTER 7

Semiclassical Simulation of the Optical Response Functions

High resolution spectroscopists often refer to broadened lineshapes as "bad spectroscopy"; their ultimate goal is to resolve all the material eigenstates and classify them in terms of symmetries and types of motions. In some cases it is indeed possible to resolve broadened lines by improving the spectral resolution. However, in many cases this goal is unattainable since intrinsic (e.g., radiative) line broadening may be larger than the spacing between levels. A time domain picture, which is no less intuitive than the eigenstate picture, may then be more appropriate. Broadened spectral lines should not be viewed as inferior and signatures of failure and incompetence, and their interpretation constitutes a fundamental challenge in condensed matter spectroscopy.

In the previous chapter we calculated the nonlinear response functions for a multilevel particle. The influence of the environment on the particle (e.g., interaction with nonabsorbing solvent molecules in a solution or with lattice vibrations in semiconductors) is in general represented by multitime correlation functions [Eq. (5.14)]. When the typical correlation time of the bath degrees of freedom is much shorter than the timescales associated with the inverse linewidth, the various propagation periods are totally uncorrelated and each Liouville space path may be factorized into a product of functions of the various evolution periods t_j [see Eq. (6.64)]. In the opposite (static) limit of infinitely long bath correlation time, the various propagation periods are fully correlated, and each path needs to be averaged over the static distribution of bath configurations [see Eq. (6.65)]. These two extreme situations are denoted homogeneous and inhomogeneous broadening, respectively. Homogeneous broadening is usually incorporated by adding a pure dephasing rate $\hat{\Gamma}$ to the Liouville equation (see, e.g., the optical Bloch equations), which results in a Lorentzian lineshape. Inhomogeneous broadening is incorporated by convoluting the response functions or susceptibilities with a static (e.g., Gaussian) distribution of transition frequencies. Although this classification has proved extremely useful in analyzing a wide range of nonlinear optical phenomena, there are numerous situations in which the assumption of separation of timescales on which it is based does not hold. The failure of this classification is often evident in nonlinear spectroscopies, which can eliminate inhomogeneous broadening and are thus more sensitive to dynamic details. Much insight can be gained into the effects of a bath with a finite timescale by considering microscopic models that interpolate between the limits of homogeneous line broadening (fast modulation of the transition energy) and inhomogeneous line broadening (a static distribution of transition energies).

In this chapter we examine the correlation function expressions of the response functions. Semiclassical approximations [1–6] make it possible to evaluate the response functions using classical trajectories, without calculating the global eigenstates of the entire system. Classical molecular dynamics simulations provide an effective tool in the interpretation of nonlinear spectroscopy in the condensed phase, where it is most appropriate to concentrate on lineshapes that span many eigenstates and contain coarse grained information. On the other hand, an eigenstate-based interpretation is called for in simple systems such as atoms in the gas phase. Intermediate size (mesoscopic) systems (e.g., clusters and nanostructures) can be approached from both ends and are usually the most challenging, since neither the small size nor the large size methods may be easily applied.

Comparison of simulations with spectroscopic measurements can provide invaluable physical insight. Moreover, simulations can be used to test in a controlled way the various assumptions made in simplified solvable models such as the multimode Brownian oscillator model to be developed in Chapter 8. A primary motivation for using semiclassical methods is that an exact quantum mechanical treatment of a system with many degrees of freedom (N) involves matrix diagonalizations, where the computational work scales exponentially with N. The computational effort required by the classical methods is dominated by the force evaluations necessary to run a trajectory, which scale as N^2. These techniques represent a good compromise between accuracy and computational cost, and are capable of addressing some key questions raised by resonant electronic spectroscopy in the condensed phase. These include

1. *Nuclear dynamics:* Nuclear motions can be either coherent or overdamped in character, and can occur when the system is in an electronic coherence or when it is in an electronic population. Each type of nuclear motion causes distinct spectroscopic effects.

2. *Distribution of system timescales:* condensed phase systems can have a broad, continuous distribution of timescales, ranging from the femtosecond regime to seconds. The classification of line broadening as either inhomogeneous or homogeneous is therefore not always possible.

3. *Distribution of transition frequencies:* The chromophore can experience a distribution of static solvent environments. The resulting distribution of transition frequencies is often Gaussian. However, non-Gaussian distributions are both important and common.

4. *Zero point notion:* Semiclassical simulations can take into account some limited but important vibrational quantum effects. Zero point motion and vibrational quantization can be accounted for by adopting a harmonic reference system, and then building in quantum effects based on that system.

LINEAR RESPONSE OF A TWO ELECTRONIC LEVEL SYSTEM: SEMICLASSICAL SIMULATIONS AND PHASE AVERAGING

In this section we present a systematic derivation of semiclassical approximations for the linear response. Consider a particle with two electronic states, a ground state, $|g\rangle$ and an electronically excited state $|e\rangle$, which are coupled to nuclear degrees of freedom denoted by $\mathbf{q}$ (Figure 7.1). The adiabatic Hamiltonian is then given by

$$H = |g\rangle H_g(\mathbf{q})\langle g| + |e\rangle H_e(\mathbf{q})\langle e|, \tag{7.1a}$$

where

$$H_g(\mathbf{q}) = T(\mathbf{q}) + W_g(\mathbf{q}), \tag{7.1b}$$

$$H_e(\mathbf{q}) = \hbar\omega_{eg}^0 + T(\mathbf{q}) + W_e(\mathbf{q}). \tag{7.1c}$$

Here $T(\mathbf{q})$ is the nuclear kinetic energy and W_g and W_e are the adiabatic potential functions for both states [7]. ω_{eg}^0 is the electronic gap between the minima of both potentials. The nuclear coordinates ($\mathbf{q}$) can represent, e.g., intramolecular vibrations as well as motions of a bath (solvent, host matrix, crystal, etc.). The electronic transition dipole operator is given by

$$V = V_{ge}(\mathbf{q})|g\rangle\langle e| + V_{eg}(\mathbf{q})|e\rangle\langle g|.$$

This two-level model will be used repeatedly throughout this book to develop explicit expressions for the nonlinear response functions using various levels of sophistication. A word about notation: $V_{ge}(\mathbf{q})$ is a matrix element over the electronic degrees of freedom but is still an operator in the nuclear space. When we further take the matrix elements of V_{ge} in the nuclear space, we get the complete matrix elements of the dipole operator, which we shall denote $\mu_{vv'}$, with v and v' representing vibronic levels (see Chapter 6). To highlight this distinction between a nuclear operator and a number, we use different symbols (V_{ge} or $\mu_{vv'}$, respectively) for these quantities.

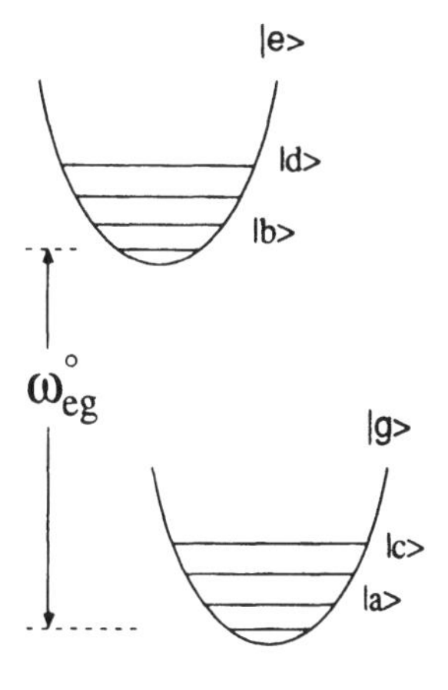

FIG. 7.1 The molecular vibronic level scheme for a two electron level system. $|a\rangle$ and $|c\rangle$ are vibronic levels corresponding to the ground electronic manifold $|g\rangle$. $|b\rangle$ and $|d\rangle$ are vibronic levels corresponding to the excited electronic manifold $|e\rangle$. ω_{eg}^0 is the 0–0 electronic transition frequency.

An electronic transition is usually in the visible or the ultraviolet, and the electronic energy gap is much larger $k_B T$. We can therefore assume that initially ($t \to -\infty$) the system is in thermal equilibrium in its ground electronic state, i.e.,

$$\rho(-\infty) = |g\rangle \rho_g \langle g|,$$

where ρ_g is the equilibrium ground state nuclear density operator

$$\rho_g \equiv \exp(-\beta H_g)/\mathrm{Tr}[\exp(-\beta H_g)].$$

Our first goal is to develop semiclassical computational schemes for calculating the linear response function which avoid the eigenstate representation. We shall therefore expand the response function in the electronic basis set, keeping the nuclear Hamiltonian in an operator form. Expanding Eq. (5.18) we get

$$S^{(1)}(t_1) = \frac{i}{\hbar} \langle\langle V|eg\rangle\rangle_0 \langle\langle eg|\mathcal{G}(t_1)|eg\rangle\rangle_0 \langle\langle eg|\mathcal{V}|\rho(-\infty)\rangle\rangle_0$$

$$+ \frac{i}{\hbar} \langle\langle V|ge\rangle\rangle_0 \langle\langle ge|\mathcal{G}(t_1)|ge\rangle\rangle_0 \langle\langle ge|\mathcal{V}|\rho(-\infty)\rangle\rangle_0.$$

The 0 subscript reminds us that all quantities defined here are still operators or superoperators in the nuclear space, since we have only expanded in the electronic basis set. This equation can be also written as

$$S^{(1)}(t_1) = \frac{i}{\hbar} [\langle\langle V_{eg}|\mathcal{G}_{eg}(t_1)|V_{eg}\rho_g\rangle\rangle - \langle\langle V_{ge}|\mathcal{G}_{ge}(t_1)|\rho_g V_{ge}\rangle\rangle], \tag{7.2}$$

which yields

$$S^{(1)}(t_1) = \frac{i}{\hbar} \theta(t_1)[J(t_1) - J^*(t_1)], \tag{7.3}$$

with

$$J(t_1) = \left\langle \exp\left(\frac{i}{\hbar} H_g t_1\right) V_{ge} \exp\left(-\frac{i}{\hbar} H_e t_1\right) V_{eg}\rho_g \right\rangle. \tag{7.4a}$$

For $t_1 > 0$ we can write

$$J(t_1) = \langle\langle V_{eg}|\mathcal{G}_{eg}(t_1)|V_{eg}\rho_g\rangle\rangle. \tag{7.4b}$$

In Equations (7.4a) and hereafter in this chapter, the angular brackets $\langle \cdots \rangle$ denote a quantum mechanical trace over all the nuclear degrees of freedom. We have further introduced the Green function $\mathcal{G}_{nm}(t)$ defined by its action on an ordinary nuclear operator A

$$\mathcal{G}_{nm}(t)A \equiv \theta(t) \exp\left(-\frac{i}{\hbar} H_n t\right) A \exp\left(\frac{i}{\hbar} H_m t\right). \tag{7.5}$$

This is an abbreviated notation, since $\mathcal{G}_{nm}(t)$ should actually be labeled as a tetradic operator $\mathcal{G}_{nm,nm}(t)$.

Equation (7.4b) can be interpreted as follows. The system is initially at thermal equilibrium in the ground electronic state and its nuclear density operator is ρ_g. Then the dipole operator acts on it, converting it to $|e\rangle V_{eg}\rho_g\langle g|$. Electronically, the system is now in an *optical coherence* $|e\rangle\langle g|$, and $V_{eg}\rho_g$ is still a wavepacket in the nuclear space. The time evolution of the coherence is given by $\mathscr{G}_{eg}(t)$. Finally, we calculate the expectation value of the dipole by multiplying by V_{ge} and taking a trace. The Fourier transform of the resulting correlation function gives the linear response function in the frequency domain.

We can recast Eq. (7.2) in a different form that is more suitable for a semiclassical expansion. To that end we shall first introduce a collective nuclear coordinate U, representing the electronic energy gap,

$$U \equiv H_e - H_g - \hbar\omega_{eg}. \tag{7.6}$$

Here ω_{eg} is a parameter whose choice is completely arbitrary and does not affect our results; it will allow a more convenient representation of the response function by separating out the high optical frequency. We can choose for example $\omega_{eg} = \omega_{eg}^0$; however, a better choice in terms of the thermally averaged electronic energy gap will be made after we introduce the concept of moments of a lineshape [see discussion following Eqs. (7.10e)].

Using these definitions and Eq. (7.A.2) we have

$$\mathscr{G}_{eg}(t_1)A = \exp(-i\omega_{eg}t_1)\mathscr{G}_{gg}(t_1)\exp_+\left[-\frac{i}{\hbar}\int_0^{t_1} d\tau\, U(\tau)\right]A, \tag{7.7a}$$

where A is an arbitrary nuclear operator, and $U(\tau)$ is the electronic energy gap in the Heisenberg picture with respect to ground state dynamics:

$$U(\tau) \equiv \exp\left(\frac{i}{\hbar}H_g\tau\right)U\exp\left(-\frac{i}{\hbar}H_g\tau\right). \tag{7.7b}$$

Here $\exp_+$ is the *time-ordered exponential* defined in Chapter 2

$$\exp_+\left[-\frac{i}{\hbar}\int_0^t d\tau\, U(\tau)\right] = 1 - \frac{i}{\hbar}\int_0^t d\tau\, U(\tau) + \left(-\frac{i}{\hbar}\right)^2\int_0^t d\tau_1\int_0^{\tau_1} d\tau_2\, U(\tau_1)U(\tau_2) + \cdots. \tag{7.7c}$$

Combining Eqs. (7.4b) and (7.7a), we get

$$J(t_1) = \exp(-i\omega_{eg}t_1)\left\langle\left\langle V_{eg}\middle|\mathscr{G}_{gg}(t_1)\exp_+\left[-\frac{i}{\hbar}\int_0^{t_1} d\tau\, U(\tau)\right]\middle|V_{eg}\rho_g\right\rangle\right\rangle,$$

which can be recast as

$$J(t_1) = \exp(-i\omega_{eg}t_1)\left\langle V_{ge}(t_1)\exp_+\left[-\frac{i}{\hbar}\int_0^{t_1} d\tau\, U(\tau)\right]V_{eg}\rho_g\right\rangle, \tag{7.8}$$

where the time evolution of $V_{ge}(t_1)$ is with respect to the ground state Hamiltonian [see Eq. (7.7b)]

$$V_{ge}(t_1) \equiv \exp\left(\frac{i}{\hbar} H_g t_1\right) V_{ge} \exp\left(-\frac{i}{\hbar} H_e t_1\right).$$

Equation (7.8) is our final expression for $J(t_1)$. In this equation, the time evolution of the coherence is factorized into the propagation with respect to a *reference Hamiltonian* (which we chose in this case to be the ground state Hamiltonian) and a time-dependent phase $U(\tau)$. The choice of H_g as the reference Hamiltonian makes sense since linear response is intimately related to ground state fluctuations.

We note that for the present model, where the system is initially in the ground electronic state and we allow only transitions between electronic states, all frequencies ω_{ba} in the eigenstate expansion [Eq. (6.3a)] are positive. Equation (6.3a) is useful for relatively small systems with a few degrees of freedom where the absorption spectrum reveals directly the system frequencies ω_{ba} and the dipole matrix elements. For complex systems and in condensed phases, it is not realistic to expect a complete knowledge of the eigenstates, and an alternative semiclassical time domain "eigenstate-free" picture is more practical. Equation (7.8) lends itself naturally to developing semiclassical approximations for linear spectroscopy. In the classical phase averaging approximation, quantum nuclear dynamics of $V_{ge}(t)$ and $U(t)$ is replaced by a classical trajectory on the ground state potential. Since we treat $U(t)$ as a classical quantity, we may replace the time-ordered exponential by an ordinary exponential. The time-dependent transition dipole and phase can then be evaluated using classical trajectories

$$\dot{q}_j = p_j/m_j,$$

$$\dot{p}_j = -\frac{\partial W_g(\mathbf{q})}{\partial q_j},$$

and with the initial condition as $t \to -\infty$, $q_j = q_{0,j}$ and $p_j = p_{0,j}$. We then have

$$\left.\begin{aligned} U(t) &= U[\mathbf{q}(t)], \\ V_{ge}(t) &= V_{ge}[\mathbf{q}(t)], \\ J(t_1) &= \exp(-i\omega_{eg}t_1)\left\langle V_{ge}(t_1)\exp\left[-\frac{i}{\hbar}\int_0^{t_1} U(\tau)\,d\tau\right]V_{eg}(0)\rho_g\right\rangle. \end{aligned}\right\} \tag{7.9a}$$

This expression may be calculated using an ensemble of trajectories on the ground state,

$$J(t_1) = \exp(-i\omega_{eg}t_1)\iint d\mathbf{p}_0\, d\mathbf{q}_0\, V_{ge}[\mathbf{q}(t_1)]$$

$$\exp\left\{-\frac{i}{\hbar}\int_0^{t_1} d\tau\, U[\mathbf{q}(\tau)]\right\} V_{eg}(\mathbf{q}_0)\rho_g(\mathbf{p}_0, \mathbf{q}_0). \tag{7.9b}$$

Additional semiclassical approximations and simulation techniques for the linear as well as the nonlinear response functions, including a Liouville space wave-packet procedure, will be discussed in Chapters 8 and 12.

MOMENTS OF THE LINEAR ABSORPTION

In some cases, the simulation of the entire response function is quite tedious. It is then common to set a more modest goal; the lineshape is characterized using a few parameters, and an attempt is made to compute these parameters. Convenient parameterization is in terms of the *moments* of the lineshape, which will be introduced below [4, 8–11].

For our two-level model the (normalized) absorption lineshape [Eq. (6.66b)] gives

$$\sigma_a(\omega) = [2\pi\langle|V_{eg}|^2\rho_g\rangle]^{-1}\int_{-\infty}^{\infty} dt\, \exp(i\omega t)\langle V_{ge}\mathcal{G}_{eg}(t)V_{eg}\rho_g\rangle. \quad (7.9c)$$

Note that

$$\langle V_{ge}\mathcal{G}_{eg}(t)V_{eg}\rho_g\rangle = \langle V_{ge}(t)V_{eg}(0)\rho_g\rangle.$$

We shall recast the absorption lineshape in the form

$$\sigma_a(\omega) = \frac{1}{2\pi\tilde{J}(0)}\int_{-\infty}^{\infty} dt\, \tilde{J}(t)\exp[i(\omega - \omega_{eg})t],$$

with

$$\tilde{J}(t) \equiv J(t)\exp(i\omega_{eg}t).$$

This differs from Eq. (5.34b) in the following:

1. We use $\tilde{J}(t)$ instead of $J(t)$ and correct for that by changing ω to $\omega - \omega_{eg}$; ω_{eg} is an arbitrary parameter at this point.
2. Using the symmetry of $J(-t) = J^*(t)$ we changed the lower limit of integration from 0 to $-\infty$.
3. We divided $\sigma_a(\omega)$ by $\tilde{J}(0)$. Using the Fourier representation of the Dirac δ function

   $$\int d\omega\, \exp(i\omega t) = 2\pi\delta(t),$$

   it then follows that

   $$\int \sigma_a(\omega)\, d\omega = \tilde{J}(0)/\tilde{J}(0) = 1,$$

   so that the lineshape is now normalized to a unit area.

$\sigma_a(\omega)$ is a positive function, which is normalized to unity and therefore has all the characteristics of a probability distribution. We shall now define its nth moment with respect to the reference frequency ω_{eg}

$$M_n \equiv \int d\omega\, (\omega - \omega_{eg})^n\sigma_a(\omega), \qquad n = 0, 1, 2. \quad (7.10a)$$

The zeroth moment gives the normalization of the lineshape, the first moment measures the average frequency, whereas the second moment provides an estimate of the linewidth.

As an example, for the Gaussian lineshape [Eq. (6.68)] all odd moments M_{2n+1} vanish identically. As for the even moments we have $M_0 = 1$, $M_2 = \Delta^2$, $M_4 = 3\Delta^2$, and, in general,

$$M_{2n} = 1.3 \cdots (2n-1)\Delta^{2n}.$$

For the Lorentzian profile [Eq. (6.67)] the moments are not very useful since we have $M_0 = 1$, but all other moments diverge due to the long ω^{-2} tail. In practice, the Lorentzian profile usually applies only in the vicinity of the line center, and in the far wings the lineshape falls off much faster than ω^{-2}, so that the moments do exist. It is possible to introduce a frequency cutoff, i.e., define a lineshape that has a Lorentzian profile in a limited frequency range and vanishes outside that range. In this case all moments are well defined. Alternatively, we can measure the linewidth via the full width at half maximum (fwhm), which can always be defined. For the Gaussian it is $2\Delta\sqrt{2\ln 2} \cong 2.35\Delta$, and for the Lorentzian it is 2Γ. In some cases, e.g., doped semiconductors, or dyes in solution, the absorption lineshape has an exponential dependence on frequency in the linewings $\sim\exp(-\omega/\omega_0)$. This is known as the Urbach tail and several models, based on quadratic coupling to phonons or on disorder, have been proposed to account for this lineshape. An illustrative example of a Urbach tail is shown in Figure 7.2.*

Spectral moments can be computed as follows: using a well-known theorem from the theory of distributions, which can be easily verified, we have

$$M_n = 2\pi i^n \left.\frac{d^n \tilde{J}(t)}{dt^n}\right|_{t=0} \tag{7.10b}$$

This can be alternatively written by defining

$$\tilde{H} \equiv |g\rangle H_g(\mathbf{q})\langle g| + |e\rangle(H_e(\mathbf{q}) - \hbar\omega_{eg})\langle e|$$

($\tilde{H}$ differs from H by the ω_{eg} factor). We further introduce the corresponding Liouville operator by $\tilde{\mathscr{L}}A \equiv [\tilde{H}, A]$, and obtain

$$\tilde{J}(t) = \left\langle V_{ge} \exp\left(-\frac{i}{\hbar}\tilde{\mathscr{L}}t\right)V_{eg}\rho_g\right\rangle.$$

We thus have

$$M_n = 2\pi\left(\frac{1}{\hbar}\right)^n \langle V_{ge}\tilde{\mathscr{L}}^n V_{eg}\rho_g\rangle,$$

or

$$M_n = 2\pi\left(\frac{1}{\hbar}\right)^n \langle \underbrace{[[[V_{ge}, \tilde{H}], \tilde{H}]\cdots \tilde{H}]}_{n \text{ times}} V_{eg}\rho_g\rangle. \tag{7.10c}$$

* Shown is a fluorescence excitation spectrum, which in the absence of nonradiative decay channels is identical to the linear absorption.

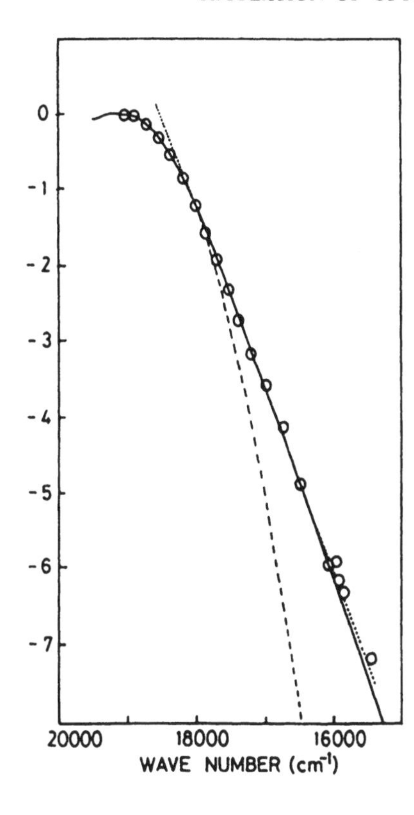

FIG. 7.2 Logarithmic plot of the fluorescence excitation spectrum of rhodamine 6G in water at 290 K (open circles). To see the profile of the spectral shape function, the spectrum has been divided by the photon energy $h\nu$. The dashed and the dotted lines show the fittings by Gaussian and exponential curves, respectively. The solid line has been obtained from the fluorescence spectrum $F(\nu)$ measured by a photon counting method through the Einstein relation $A(\nu) \propto \nu^{-2}F(\nu)\exp[(h\nu - z)/k_B T]$, where z is the energy of the zero-phonon line, k_B is Boltzmann constant, and T is the absolute temperature. [Adapted from S. Kinoshita and T. Kushida, "Stochastic behavior of the dynamic Stokes shift in Rhodamine 6G," *J. Phys. Soc. Jpn.* **56**, 4162 (1987).]

The nth moment can thus be calculated using the short time dynamics of $\tilde{J}$. Specifically, for the lowest two moments we obtain

$$M_0 = 1, \tag{7.10d}$$

$$M_1 = \omega_{eg}^0 - \omega_{eg} + \frac{1}{\hbar}\frac{\langle (W_e - W_g)|V_{eg}|^2\rho_g\rangle}{\langle |V_{eg}|^2\rho_g\rangle} + \frac{1}{\hbar}\frac{\langle [V_{ge}, T]V_{eg}\rho_g\rangle}{\langle |V_{eg}|^2\rho_g\rangle}. \tag{7.10e}$$

The third term in the right-hand side is the thermally averaged energy gap U, whereas the fourth term is a dynamic contribution, T being the nuclear kinetic energy operator.

In conclusion, the following points need to be made:

1. The calculation of the moments requires only very short time dynamics [Eq. (7.10b)] or can be even made using equilibrium (static) simulations [Eq. (7.10c)]. It is therefore considerably less demanding compared with the direct computation of the entire lineshape.
2. Using a few moments it is possible to represent the lineshape in terms of a continued fraction [9].
3. A natural choice of ω_{eg} can be made such that the first moment vanishes $M_1 = 0$. This choice will be adopted hereafter.
4. The higher moments give a large weight to the linewings, which have a less favorable signal-to-noise ratio and are usually less accurately known than the line center. This limits the number of moments that can be extracted accurately from experiment.

5. Mathematically, even a complete knowledge of all the moments M_n does not necessarily determine the lineshape.* Nevertheless, by combining the moments with some physical knowledge about the system, it is in practice possible to construct useful approximations for the lineshape.

HILBERT VERSUS LIOUVILLE SPACE REPRESENTATION OF THE THIRD-ORDER RESPONSE FUNCTION

Since the second-order response function $S^{(2)}$ vanishes for the present two electronic-level model, which contains no permanent dipoles, we shall focus on the third-order response.

The third-order nonlinear response function [Eq. (5.27)] is given by a sum of four terms, which when taking electronic matrix elements of Eq. (5.25) become

$$\left.\begin{aligned} R_1(t_3, t_2, t_1) &\equiv \langle\langle V_{eg}|\mathcal{G}_{eg}(t_3)\mathcal{V}_{eg,ee}\mathcal{G}_{ee}(t_2)\mathcal{V}_{ee,eg}\mathcal{G}_{eg}(t_1)\mathcal{V}_{eg,gg}|\rho_g\rangle\rangle, \\ R_2(t_3, t_2, t_1) &\equiv \langle\langle V_{eg}|\mathcal{G}_{eg}(t_3)\mathcal{V}_{eg,ee}\mathcal{G}_{ee}(t_2)\mathcal{V}_{ee,ge}\mathcal{G}_{ge}(t_1)\mathcal{V}_{ge,gg}|\rho_g\rangle\rangle, \\ R_3(t_3, t_2, t_1) &\equiv \langle\langle V_{eg}|\mathcal{G}_{eg}(t_3)\mathcal{V}_{eg,gg}\mathcal{G}_{gg}(t_2)\mathcal{V}_{gg,ge}\mathcal{G}_{ge}(t_1)\mathcal{V}_{ge,gg}|\rho_g\rangle\rangle, \\ R_4(t_3, t_2, t_1) &\equiv \langle\langle V_{eg}|\mathcal{G}_{eg}(t_3)\mathcal{V}_{eg,gg}\mathcal{G}_{gg}(t_2)\mathcal{V}_{gg,eg}\mathcal{G}_{eg}(t_1)\mathcal{V}_{eg,gg}|\rho_g\rangle\rangle, \end{aligned}\right\} \quad (7.11)$$

where the Liouville space Green function $\mathcal{G}_{nm}(t)$ was defined in Eq. (7.5). The corresponding Liouville space pathways are displayed in Figure 7.3.

The nonlinear response function has the following physical interpretation in Liouville space: the first interaction with the radiation field (which takes place at time $t - t_1 - t_2 - t_3$) sets up an optical coherence in the sample; $\mathcal{G}_{eg}(t_1)$ [or $\mathcal{G}_{ge}(t_1)$] represents the evolution and dephasing of this coherence. The second interaction (at time $t - t_2 - t_3$) converts the optical coherence into either a population on the ground electronic state ($|g\rangle$) or a population on the excited electronic state ($|e\rangle$). During the time t_2 between the second and third interactions, the system's evolution is given by either $\mathcal{G}_{ee}(t_2)$ or $\mathcal{G}_{gg}(t_2)$. Subsequently, the third interaction (at time $t - t_3$) creates again an electronic coherence that evolves for the t_3 period [$\mathcal{G}_{eg}(t_3)$]. Finally, at time t the polarization is calculated by acting with V from the left and performing a trace. The four contributions R_α represent distinct Liouville space pathways that maintain a simultaneous bookkeeping of the evolution of the bra and the ket of the density operator. A pictorial representation of Eq. (7.11) is shown in Figure 7.3.

The time evolution during the t_2 period is given by the Green function $\mathcal{G}_{ee}(t_2)$ (for R_1 and R_2) or $\mathcal{G}_{gg}(t_2)$ (for R_3 and R_4). This represents an ordinary quantum mechanical propagation on the excited or the ground state potential surface, respectively. During the time intervals t_1 and t_3, the system is in an *optical coherence* and its time evolution is represented by $\mathcal{G}_{eg}(t)$ or $\mathcal{G}_{ge}(t)$. In this case the evolution from the right and from the left is with a different Hamiltonian,

* Some examples of probability distributions that are different, even though they have the same moments to all orders, are well known [2].

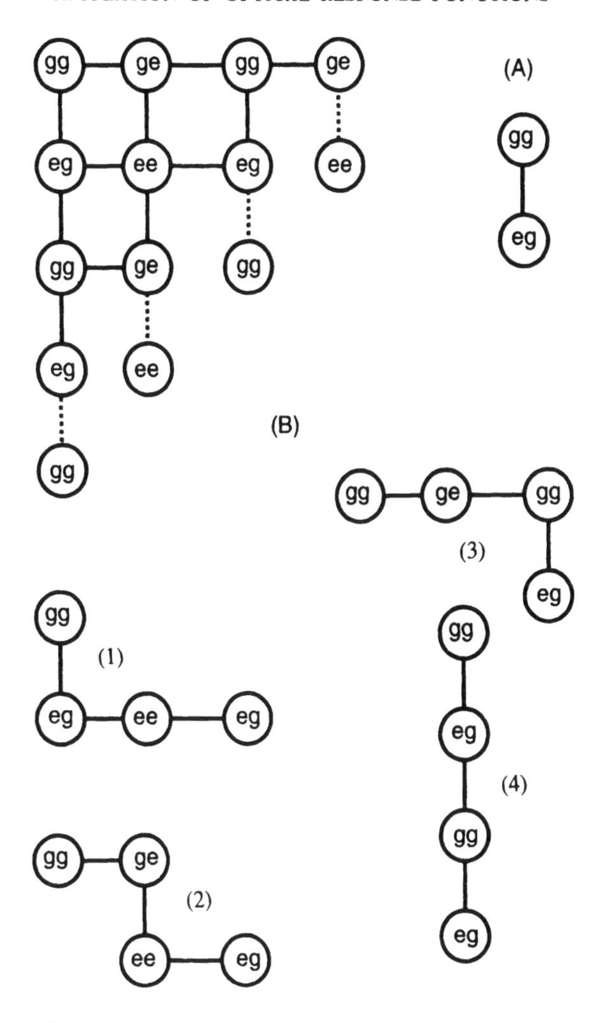

FIG. 7.3 The Liouville space coupling scheme and the pathways contributing to the linear and the nonlinear response function of a two electronic level system. These type of diagrams are explained in Figure 6.3. The pathway labeled (A) contributes to the linear response function $J(t_1)$ [Eq. (7.4b)]. The four pathways labeled (B) [(1), (2), (3), and (4)] correspond, respectively, to R_1, R_2, R_3, R_4 of Eqs. (7.11) and contribute to the nonlinear response function.

and it does not have an obvious classical analogue. It is the later type of propagation that makes the semiclassical evaluation of molecular response functions less straightforward.

Equations (7.11) can alternatively be written using ordinary Hilbert space (as opposed to Liouville space) propagators, i.e.,

$$
\left.
\begin{aligned}
R_1(t_3,t_2,t_1) &= \left\langle \exp\left(\frac{i}{\hbar}H_g t_1\right) V_{ge} \exp\left(\frac{i}{\hbar}H_e t_2\right) V_{eg} \exp\left(\frac{i}{\hbar}H_g t_3\right) V_{ge} \right. \\
&\quad \left. \times \exp\left[-\frac{i}{\hbar}H_e(t_1+t_2+t_3)\right] V_{eg}\rho_g \right\rangle, \\
R_2(t_3,t_2,t_1) &= \left\langle V_{ge} \exp\left[\frac{i}{\hbar}H_e(t_1+t_2)\right] V_{eg} \exp\left(\frac{i}{\hbar}H_g t_3\right) V_{ge} \right. \\
&\quad \left. \times \exp\left[-\frac{i}{\hbar}H_e(t_2+t_3)\right] V_{ge} \exp\left(-\frac{i}{\hbar}H_g t_1\right) \rho_g \right\rangle, \\
R_3(t_3,t_2,t_1) &= \left\langle V_{ge} \exp\left(\frac{i}{\hbar}H_e t_1\right) V_{eg} \exp\left[\frac{i}{\hbar}H_g(t_2+t_3)\right] V_{ge} \right. \\
&\quad \left. \times \exp\left(-\frac{i}{\hbar}H_e t_3\right) V_{eg} \exp\left[-\frac{i}{\hbar}H_g(t_1+t_2)\right] \rho_g \right\rangle, \\
R_4(t_3,t_2,t_1) &= \left\langle \exp\left[\frac{i}{\hbar}H_g(t_1+t_2+t_3)\right] V_{ge} \exp\left(-\frac{i}{\hbar}H_e t_3\right) V_{eg} \right. \\
&\quad \left. \times \exp\left(-\frac{i}{\hbar}H_g t_2\right) V_{ge} \exp\left(-\frac{i}{\hbar}H_e t_1\right) V_{eg}\rho_g \right\rangle.
\end{aligned}
\right\}
\qquad (7.12)
$$

We now introduce the four point (Hilbert space) correlation function

$$
F(\tau_1,\tau_2,\tau_3,\tau_4) \equiv \langle V_{ge}(\tau_1)V_{eg}(\tau_2)V_{ge}(\tau_3)V_{eg}(\tau_4)\rangle, \qquad (7.13)
$$

with

$$
V_{ge}(\tau) \equiv \exp\left(\frac{i}{\hbar}H_g\tau\right) V_{ge} \exp\left(-\frac{i}{\hbar}H_e\tau\right),
$$

$$
V_{eg}(\tau) = V_{ge}^{\dagger}(\tau) = \exp\left(\frac{i}{\hbar}H_e\tau\right) V_{eg} \exp\left(-\frac{i}{\hbar}H_g\tau\right).
$$

Since the correlation function is invariant to time translation (i.e., changing all time arguments $\tau_j \to \tau_j + \tau$), we can always set one of the time arguments to zero. Using this symmetry, we obtain

$$
\left.
\begin{aligned}
R_1(t_3,t_2,t_1) &= F(t_1, t_1+t_2, t_1+t_2+t_3, 0), \\
R_2(t_3,t_2,t_1) &= F(0, t_1+t_2, t_1+t_2+t_3, t_1), \\
R_3(t_3,t_2,t_1) &= F(0, t_1, t_1+t_2+t_3, t_1+t_2), \\
R_4(t_3,t_2,t_1) &= F(t_1+t_2+t_3, t_1+t_2, t_1, 0),
\end{aligned}
\right\}
\qquad (7.14)
$$

Taking the Hermitian conjugate of Eq. (7.13) we can show that

$$
F(\tau_1,\tau_2,\tau_3,\tau_4) = F^*(\tau_4,\tau_3,\tau_2,\tau_1). \qquad (7.15)
$$

Comparing the two forms [Eqs. (7.11) and (7.14)] provides an instructive example of the utility of Liouville space. In the Liouville space form we follow step by step the dynamics underlying each path. The concepts of coherence and dephasing are essential in understanding condensed phase spectroscopy; the fact that the Liouville space formulation brings these features to the forefront is a compelling reason for its use. In the Hilbert space form, on the other hand, the physical picture of coherence vs. population dynamics is hidden in the scrambling of the time arguments.

SEMICLASSICAL SIMULATIONS OF THE NONLINEAR RESPONSE IN LIOUVILLE SPACE: PHASE AVERAGING

Semiclassical methods for simulating condensed phase spectra are well developed, particularly for spectroscopies that involve purely nuclear motions, e.g., infrared spectra and off resonant light scattering [3]. The unifying feature of these techniques is that the experimental signal is expressed in terms of the Fourier transform of a time correlation function on the ground state potential surface. Infrared spectroscopy probes the dipole autocorrelation function, while light scattering probes the polarizability autocorrelation function. These correlation functions have a well-defined classical limit where the relevant (dipole or polarizability) operator is replaced by its classical analogue, and the time evolution is calculated classically.

Equations (7.4) and (7.11) express the electronic response functions in terms of combinations of dipole correlation functions. These correlation functions, however, do not have a classical limit, and this creates a fundamental difficulty in their classical simulation. The problem may be appreciated by noting that these correlation functions depend on nuclear dynamics when the system is in an electronic coherence whereby the right (bra) side of the density operator is in a different electronic state than the left (ket) side. The time evolution of electronic coherences involves a different Hamiltonian from the left and the right and does not have a simple classical analogue. The singularity of electronic correlation functions as $\hbar \to 0$ is apparent using Eqs. (7.5) [by noting that $\exp[-(i/\hbar)H_j t]$; $j = g, e$ is formally ill-defined as $\hbar \to 0$]. For $\mathcal{G}_{nm}(t)$ with $n = m$, the singularity of the bra and the ket evolution cancels as $\hbar \to 0$, resulting in the classical propagation as given by the classical Liouville operator. However, for $n \neq m$, the dependence on $\hbar$ does not have a classical limit in a strict mathematical sense. The following semiclassical method will therefore be based on a subjective choice of a reference model, rather than on a direct expansion in $\hbar$. It is left to physical intuition to develop reasonable semiclassical approximations for electronic spectroscopy [13].*

The Liouville space representation provides a natural starting point for constructing semiclassical approximations for the response function. We shall recast Eqs. (7.11) in the form of a time evolution with respect to a reference Hamiltonian, accompanied by an accumulating time-dependent phase. Possible choices of a reference Hamiltonian for the semiclassical expansion of the coherence Green function $\mathcal{G}_{eg}(t)$ are discussed in Appendix 7A. This form immediately suggests a simple semiclassical approximation and provides a clear insight on the physical significance of the nonlinear response function and the relevance of the reference system.

* An alternative procedure is to use semiclassical approximations for the wave-function in Hilbert space [J. H. Van Vleck, *Proc. Nat. Acad. Sci.* **14**, 178 (1928); M. C. Gutzwiller, *J. Math. Phys.* **10**, 1004 (1969); W. H. Miller, *Adv. Chem. Phys.* **25**, 69 (1974); E. J. Heller, *Lectures in the 1989 NATO Les Houches Summer School on Chaos and Quantum Physics*, edited by M-J. Giannoni, A. Voros, and J. Zinn-Justin (Elsevier, Amsterdam, 1991), p. 54]. The density operator may then be calculated using pairs of trajectories, which follow the evolution of the bra and the ket separately. In this case we need not introduce a reference Hamiltonian.

As explained in Appendix 7A, the expansion of the coherence Green function $\mathscr{G}_{eg}(t)$ is based on a reference system that needs to be chosen carefully. The natural reference for the linear response is the ground state. Equation (7.9b), which expresses the linear response function in terms of a time-ordered exponential energy gap, is based on this reference. This form provides a semiclassical interpretation in terms of the evolution with respect to the ground state and leads to several very useful approximations, such as the cumulant expansion, and the classical Condon approximation as will be shown in Chapter 8.

The choice of a reference Hamiltonian for simulating the nonlinear response functions is less straightforward. For the R_3 and R_4 pathways, the system is in the ground state during t_2 period; we therefore chose H_g as a reference during the entire calculation. For R_1 and R_2, since the system starts in ρ_g, we choose H_g as a reference during t_1. However, during t_2 the system evolves with the excited state Hamiltonian H_e. For long times t_2 it should therefore relax to a new density operator ρ_e representing thermal equilibrium with respect to H_e. We thus choose H_e as a reference during the t_2 and t_3 periods. In summary, we introduce the following two reference Hamiltonians:

$$\left.\begin{aligned} H_a &\equiv H_g, \\ H_b(\tau) &\equiv \begin{cases} H_g & 0 < \tau < t_1 \\ H_e & t_1 < \tau < t_1 + t_2 + t_3. \end{cases} \end{aligned}\right\} \tag{7.16}$$

H_a will be used for pathways R_3 and R_4 and H_b for R_1 and R_2. The corresponding time evolution operators are

$$\left.\begin{aligned} G_a(t) &\equiv \theta(t)\exp\left(-\frac{i}{\hbar}H_g t\right), \\ G_b(t) &\equiv \theta(t)\exp_+\left[-\frac{i}{\hbar}\int_0^t d\tau\, H_b(\tau)\right]. \end{aligned}\right\} \tag{7.17}$$

We further introduce the interaction picture operators:

$$\left.\begin{aligned} U_j(t) &\equiv G_j^\dagger(t) U G_j(t) \\ V_j(t) &\equiv G_j^\dagger(t) V_{ge} G_j(t) \\ V_j^\dagger(t) &\equiv G_j^\dagger(t) V_{eg} G_j(t) \qquad j = a, b. \end{aligned}\right\} \tag{7.18}$$

In Appendix 7B we calculate the response function using these reference Hamiltonians, resulting in Eqs. (7.19).

A fully classical approximation can be derived from this result, in complete analogy with the linear response [Eq. (7.9b)]. We simply replace all time-ordered exponentials $\exp_\pm$ by an ordinary exponential, and all operators by classical functions of phase space. We can then compute Eqs. (7.19) using classical trajectories. It should be noted, however, that this classical approximation does depend on $\hbar$. This is therefore not a classical limit in a strict formal sense but rather a reasonable classical approximation.

$$R_1(t_3, t_2, t_1) = \exp(-i\omega_{eg}t_1 - i\omega_{eg}t_3) \times \left\langle V_b(t_1)V_b^\dagger(t_1+t_2)\exp_-\left[-\frac{i}{\hbar}\int_{t_1+t_2}^{t_1+t_2+t_3} d\tau\, U_b(\tau)\right] \times V_b(t_1+t_2+t_3)\exp_+\left[-\frac{i}{\hbar}\int_0^{t_1} d\tau\, U_b(\tau)\right]V_b^\dagger(0)\rho_g\right\rangle, \quad (7.19a)$$

$$R_2(t_3, t_2, t_1) = \exp(i\omega_{eg}t_1 - i\omega_{eg}t_3) \times \left\langle V_b(0)\exp_-\left[\frac{i}{\hbar}\int_0^{t_1} d\tau\, U_b(\tau)\right]V_b^\dagger(t_1+t_2) \times \exp_-\left[-\frac{i}{\hbar}\int_{t_1+t_2}^{t_1+t_2+t_3} d\tau\, U_b(\tau)\right]V_b(t_1+t_2+t_3)V_b^\dagger(t_1)\rho_g\right\rangle, \quad (7.19b)$$

$$R_3(t_3, t_2, t_1) = \exp(i\omega_{eg}t_1 - i\omega_{eg}t_3) \times \left\langle V_a(0)\exp_-\left[\frac{i}{\hbar}\int_0^{t_1} d\tau\, U_a(\tau)\right]V_a^\dagger(t_1)V_a(t_1+t_2+t_3) \times \exp_+\left[-\frac{i}{\hbar}\int_{t_1+t_2}^{t_1+t_2+t_3} d\tau\, U_a(\tau)\right]V_a^\dagger(t_1+t_2)\rho_g\right\rangle, \quad (7.19c)$$

$$R_4(t_3, t_2, t_1) = \exp(-i\omega_{eg}t_1 - i\omega_{eg}t_3) \times \left\langle V_a(t_1+t_2+t_3)\exp_+\left[-\frac{i}{\hbar}\int_{t_1+t_2}^{t_1+t_2+t_3} d\tau\, U_a(\tau)\right] \times V_a^\dagger(t_1+t_2)V_a(t_1)\exp_+\left[-\frac{i}{\hbar}\int_0^{t_1} d\tau\, U_a(\tau)\right]V_a^\dagger(0)\rho_g\right\rangle. \quad (7.19d)$$

THE STATIC LIMIT: INHOMOGENEOUS BROADENING AND SPECTRAL DIFFUSION

We shall consider now the simplest classical approximation and assume that all nuclear degrees of freedom are essentially static, and their motions can be neglected for the relevant times of interest. This can be realized, e.g., if the nuclei are sufficiently heavy. Starting with Eq. (7.7a), the time evolution of $U(\tau)$ reflects the phase fluctuations coming from different members of the ρ_g ensemble. In general, when nuclear motions are taken into account, they will make an additional contribution to the evolution of $U(\tau)$. In this section we completely neglect these motions and consider only the former source of evolution. By doing so, $\mathcal{G}_{gg}(t) \simeq 1$, $U(\tau)$ becomes time independent, and Eq. (7.7a) gives

$$\mathcal{G}_{eg}(t)A \cong \exp(-i\omega_{eg}t)\exp\left(-\frac{i}{\hbar}Ut\right)A. \tag{7.20}$$

We shall denote the relevant nuclear phase space by Γ and set

$$\omega_{eg}(\Gamma) \equiv \omega_{eg} + \frac{1}{\hbar}U = \omega_{eg}^{0} + \frac{1}{\hbar}[W_e(\Gamma) - W_g(\Gamma)].$$

We then have

$$\mathcal{G}_{eg}(t)A \cong \exp[-i\omega_{eg}(\Gamma)t]A. \tag{7.21}$$

This is the static (classical) limit. Equation (7.21) could be derived directly from Eq. (7.5) by simply neglecting the nuclear kinetic energy in H_g and H_e and assuming that all operators commute.

The quantum trace in this case reduces simply to a classical static averaging over phase space Γ

$$\langle A \rangle \equiv \int \rho_g(\Gamma)A(\Gamma)\,d\Gamma,$$

where A is an arbitrary operator. [Note that by definition $\langle U \rangle = 0$ so that $\langle \omega_{eg}(\Gamma) \rangle = \omega_{eg}$.]

Let us consider now the linear response. Substituting Eq. (7.21) in (7.4b) we get

$$\langle J(t_1) \rangle = \int d\Gamma\, \rho_g(\Gamma)|V_{ge}(\Gamma)|^2 \exp[-i\omega_{eg}(\Gamma)t_1].$$

In the frequency domain we similarly have

$$J(\omega) = \int d\Gamma \frac{1}{\omega - \omega_{eg}(\Gamma) + i\varepsilon}|V_{ge}(\Gamma)|^2\rho_g(\Gamma),$$

and the normalized absorption lineshape becomes

$$\sigma_a(\omega) = \int d\Gamma\, \delta[\omega - \omega_{eg}(\Gamma)]|V_{eg}(\Gamma)|^2\rho_g(\Gamma) \Big/ \int d\Gamma |V_{eg}(\Gamma)|^2\rho_g(\Gamma). \tag{7.22}$$

Equation (7.22) corresponds to the inhomogeneous line broadening limit. It can be interpreted as follows: when the bath is classical (heavy and slow degrees of freedom), no kinetic energy is transferred to it upon optical excitation. Conservation of (potential) energy then implies that the absorption of an ω photon can take place only in nuclear configurations where the potential energy difference between the two states is equal to ω, i.e., $\omega = \omega_{eg}(\Gamma)$. The absorption lineshape $\sigma_a(\omega)$ is then simply obtained by summing (integrating) over the contributions of various configurations. In molecular spectroscopy this is known as the classical Franck-Condon principle [14]. Equation (7.22) can be alternatively recast in the form

$$\sigma_a(\omega) = \sum_{\Gamma^*} |V_{eg}(\Gamma^*)|^2\rho_g(\Gamma^*) \frac{\delta[\omega - \omega_{eg}(\Gamma^*)]}{|\partial\omega_{eg}(\Gamma)/\partial\Gamma|_{\Gamma=\Gamma^*}} \Big/ \int d\Gamma |V_{eg}(\Gamma)|^2\rho_g(\Gamma), \tag{7.23}$$

where Γ^* are the nuclear configurations where $\omega_{eg}(\Gamma^*) = \omega$.

The classical Frank-Condon principle applies only when the linewidth is very broad compared with $\hbar/\tau_m$, τ_m being the typical nuclear dynamics timescale (fast dephasing). In other words, the absorption is inhomogeneous and does not depend on the nuclear masses. The mapping between configuration space and the spectrum is very simple in this case, and Eq. (7.23) forms the basis for the celebrated inversion procedure where the absorption spectrum with a knowledge of the ground state potential is used to obtain the excited state potential. This classical procedure is commonly used for the interpretation of broad lineshapes, e.g., in molecular photodissociation spectroscopy [6, 15, 16]. The observation that this limit is obtained by simply using Eq. (7.21) for the coherence Green function will be used shortly to extend this case to nonlinear spectroscopy.

The fluorescence lineshape within the classical approximation can be obtained from the above expressions by simply replacing $\rho_g(\Gamma)$ with $\rho_e(\Gamma)$,

$$\sigma_f(\omega) = \int d\Gamma |V_{eg}(\Gamma)|^2 \delta[\omega - \omega_{eg}(\Gamma)]\rho_e(\Gamma) \Big/ \int d\Gamma |V_{eg}(\Gamma)|^2 \rho_e(\Gamma).$$

A simple limiting case is obtained when the potentials W_g and W_e are identical $W_g = W_e$ so that $U = 0$, and we get $\sigma_a(\omega) = \sigma_f(\omega) = \delta(\omega - \omega_{eg})$. In this case the nuclear degrees of freedom are completely decoupled from the electronic transition, and do not show up in the spectrum.

We next turn to the third-order response. The static limit holds when the timescale of molecular motions is slow compared with the inverse absorption linewidth. In this limit the nuclei do not move when the system is in an electronic coherence (i.e., the t_1 and the t_3 periods). Substituting Eq. (7.21) in Eq. (7.11), we obtain the following approximation for the nonlinear response function:

$$R_1(t_3, t_2, t_1) = \exp(-i\omega_{eg}t_1 - i\omega_{eg}t_3) \times \left\langle |V_b(0)V_b(t_2)|^2 \exp\left[-\frac{i}{\hbar} t_3 U_b(t_2)\right] \exp\left[-\frac{i}{\hbar} t_1 U_b(0)\right] \rho_g \right\rangle, \quad (7.24a)$$

$$R_2(t_3, t_2, t_1) = \exp(i\omega_{eg}t_1 - i\omega_{eg}t_3) \times \left\langle |V_b(0)V_b(t_2)|^2 \exp\left[-\frac{i}{\hbar} t_3 U_b(t_2)\right] \exp\left[\frac{i}{\hbar} t_1 U_b(0)\right] \rho_g \right\rangle, \quad (7.24b)$$

$$R_3(t_3, t_2, t_1) = \exp(i\omega_{eg}t_1 - i\omega_{eg}t_3) \times \left\langle |V_a(0)V_a(t_2)|^2 \exp\left[-\frac{i}{\hbar} t_3 U_a(t_2)\right] \exp\left[\frac{i}{\hbar} t_1 U_a(0)\right] \rho_g \right\rangle, \quad (7.24c)$$

$$R_4(t_3, t_2, t_1) = \exp(-i\omega_{eg}t_1 - i\omega_{eg}t_3) \times \left\langle |V_a(0)V_a(t_2)|^2 \exp\left[-\frac{i}{\hbar} t_3 U_a(t_2)\right] \exp\left[-\frac{i}{\hbar} t_1 U_a(0)\right] \rho_g \right\rangle, \quad (7.24d)$$

Were the nuclei truly static, their motions could be totally ignored, and we could simply set $t_2 = 0$ in these equations. However, the temporal variation of $U_a(t_2)$ and $U_b(t_2)$ results in time-dependent spectral shifts known as spectral diffusion processes. These will be analyzed in the next chapter using the Brownian oscillator model. Spectroscopic signatures of spectral diffusion in photon echo and hole burning spectroscopy will be discussed in Chapters 10 and 13.

APPENDIX 7A

The Coherence Green Function

The coherence Green function does not have a classical analogue. It can be, however, factorized into an ordinary Green function of a reference system and a time-dependent phase. This representation, which will be developed below, is most valuable for the derivation of semiclassical approximations.

Consider the coherence Green function acting on an arbitrary ordinary nuclear operator A

$$\mathcal{G}_{nm}(t)A \equiv \exp\left(-\frac{i}{\hbar} H_n t\right) A \exp\left(\frac{i}{\hbar} H_m t\right).$$

Adopting a reference Hamiltonian H_j (whose choice is arbitrary at this point) we write

$$H_n \equiv H_j + \bar{W}_j,$$

where $\bar{W}_j \equiv W_n - W_j$ is the difference of the potential functions of the actual and the reference Hamiltonians. Making use of the time-ordered exponential we next write

$$\exp\left(-\frac{i}{\hbar} H_n t\right) \equiv \exp\left(-\frac{i}{\hbar} H_j t\right) \exp_+\left[-\frac{i}{\hbar} \int_0^t d\tau\, \bar{W}_j(\tau)\right],$$

where $\bar{W}_j(\tau)$ is the $\bar{W}_j$ operator in the interaction picture with respect to H_j, i.e.,

$$\bar{W}_j(\tau) \equiv \exp\left(\frac{i}{\hbar} H_j \tau\right) \bar{W}_j \exp\left(-\frac{i}{\hbar} H_j \tau\right).$$

We can now repeat the same steps for the H_m Green function. We first write

$$H_m \equiv H_j + W'_j,$$

with $W'_j \equiv W_m - W_j$ and

$$\exp(iH_m t) = \exp_-\left[\frac{i}{\hbar} \int_0^t d\tau\, W'_j(\tau)\right] \exp(iH_j t),$$

with

$$W'_j(t) \equiv \exp\left(\frac{i}{\hbar} H_j t\right) W'_j \exp\left(-\frac{i}{\hbar} H_j t\right).$$

Combining these equations we get

$$\mathcal{G}_{nm}(t)A \equiv \mathcal{G}_{jj}(t) \exp_+\left[-\frac{i}{\hbar} \int_0^t d\tau\, \bar{W}_j(\tau)\right] A \exp_-\left[\frac{i}{\hbar} \int_0^t d\tau\, W'_j(\tau)\right]. \quad (7.A.1)$$

The coherence propagation is thus split into a propagation with respect to the reference Hamiltonian $H_j[\mathscr{G}_{jj}(t)]$ and time-dependent phases whose evolution is again determined by H_j. The proper choice of H_j should be based on physical grounds and is not unique. Let us consider now two convenient choices. Taking $H_j = H_m$ we have $\bar{W}_m = H_n - H_m$ and $W'_m = 0$ and we have

$$\mathscr{G}_{nm}(t)A = \mathscr{G}_{mm}(t)\exp_+\left[-\frac{i}{\hbar}\int_0^t d\tau\, \bar{W}_m(\tau)\right]A. \tag{7.A.2}$$

Alternatively, we can choose $H_j = H_n$ so that $W'_n = H_m - H_n$ and $\bar{W}_n = 0$,

$$\mathscr{G}_{nm}(t)A = \mathscr{G}_{nn}(t)A\exp_-\left[\frac{i}{\hbar}\int_0^t d\tau\, W'_n(\tau)\right]. \tag{7.A.3}$$

These choices will be used in the derivation of semiclassical approximations for the response functions.

APPENDIX 7B

Derivation of Eqs. (7.19)

We start with the first (t_1) evolution period in Eqs. (7.11), and express it in terms of a time-ordered exponential of the electronic energy gap, choosing the ground state Hamiltonian H_g as a reference. Using Eq. (7.A.2) we get

$$\mathcal{G}_{eg}(t_1)\mathcal{V}_{eg,gg}\rho_g = \mathcal{G}_{gg}(t_1)\exp_+\left[-\frac{i}{\hbar}\int_0^{t_1} d\tau\, U(\tau)\right]\mathcal{V}_{eg,gg}\rho_g. \tag{7.B.1}$$

Similarly, using Eq. (7.A.3) we have

$$\mathcal{G}_{ge}(t_1)\mathcal{V}_{ge,gg}\rho_g = \mathcal{G}_{gg}(t_1)\mathcal{V}_{ge,gg}\rho_g \exp_-\left[\frac{i}{\hbar}\int_0^{t_1} d\tau\, U(\tau)\right]. \tag{7.B.2}$$

All four contributions R_1–R_4 end with a common $G_{eg}(t_3)$ operator. Formally it can be written using a ground state reference in the form:

$$\mathcal{G}_{eg}(t_3)A = \mathcal{G}_{gg}(t_3)\exp_+\left[-\frac{i}{\hbar}\int_0^{t_3} d\tau\, U(\tau)\right]A, \tag{7.B.3}$$

or using the excited state reference

$$\mathcal{G}_{eg}(t_3)A = \mathcal{G}_{ee}(t_3)A\exp_-\left[\frac{i}{\hbar}\int_0^{t_3} d\tau\, U_e(\tau)\right], \tag{7.B.4}$$

with

$$U_e(\tau) \equiv \exp\left(\frac{i}{\hbar}H_e\tau\right)U\exp\left(-\frac{i}{\hbar}H_e\tau\right). \tag{7.B.5}$$

By choosing the reference Hamiltonian [Eqs. (7.16)] for the coherence periods we obtain

$$\begin{aligned} R_1(t_3,t_2,t_1) &= \exp(-i\omega_{eg}t_1 - i\omega_{eg}t_3) \\ &\times \left\langle\!\!\left\langle V\middle|\mathcal{G}_{ee}(t_3)\exp_-\left[-\frac{i}{\hbar}\int_0^{t_3} d\tau\, U_e(\tau)\right]\mathcal{V}_{eg,ee}\mathcal{G}_{ee}(t_2)\mathcal{V}_{ee,eg}\mathcal{G}_{gg}(t_1)\right.\right. \\ &\times \left.\left.\exp_+\left[-\frac{i}{\hbar}\int_0^{t_1} d\tau\, U(\tau)\right]\mathcal{V}_{eg,gg}\middle|\rho_g\right\rangle\!\!\right\rangle, \end{aligned} \tag{7.B.6a}$$

$$R_2(t_3, t_2, t_1) = \exp(i\omega_{eg}t_1 - i\omega_{eg}t_3)$$
$$\times \left\langle\left\langle V|\mathcal{G}_{ee}(t_3)\exp_-\left[-\frac{i}{\hbar}\int_0^{t_3} d\tau\, U_e(\tau)\right]\mathcal{V}_{eg,ee}\mathcal{G}_{ee}(t_2)\mathcal{V}_{ee,eg}\mathcal{G}_{gg}(t_1)\right.\right.$$
$$\left.\left.\times [\mathcal{V}_{ge,gg}\rho_g]\exp_-\left[\frac{i}{\hbar}\int_0^{t_1} d\tau\, U(\tau)\right]\right\rangle\right\rangle, \tag{7.B.6b}$$

$$R_3(t_3, t_2, t_1) = \exp(i\omega_{eg}t_1 - i\omega_{eg}t_3)$$
$$\times \left\langle\left\langle V|\mathcal{G}_{ee}(t_3)\exp_+\left[-\frac{i}{\hbar}\int_0^{t_3} d\tau\, U(\tau)\right]\mathcal{V}_{eg,gg}\mathcal{G}_{gg}(t_2)\mathcal{V}_{gg,ge}\mathcal{G}_{gg}(t_1)\right.\right.$$
$$\left.\left.\times [\mathcal{V}_{ge,gg}\rho_g]\exp_-\left[\frac{i}{\hbar}\int_0^{t_1} d\tau\, U(\tau)\right]\right\rangle\right\rangle, \tag{7.B.6c}$$

$$R_4(t_3, t_2, t_1) = \exp(-i\omega_{eg}t_1 - i\omega_{eg}t_3)$$
$$\times \left\langle\left\langle V|\mathcal{G}_{gg}(t_3)\exp_+\left[-\frac{i}{\hbar}\int_0^{t_3} d\tau\, U(\tau)\right]\mathcal{V}_{eg,gg}\mathcal{G}_{gg}(t_2)\mathcal{V}_{gg,eg}\mathcal{G}_{gg}(t_1)\right.\right.$$
$$\left.\left.\times \exp_+\left[-\frac{i}{\hbar}\int_0^{t_1} d\tau\, U(\tau)\right]\mathcal{V}_{eg,gg}\rho_g\right\rangle\right\rangle, \tag{7.B.6d}$$

Each R_α factor in Eqs. (7.B.6) contains three population Green functions $\mathcal{G}_{gg}(\tau)$ or $\mathcal{G}_{ee}(\tau)$. By acting with them to the left, we shift all time arguments by τ using the H_a Hamiltonian [Eq. (7.16)] for R_3 and R_4 or H_b for R_1 and R_2. Equations (7.B.6) then result in Eqs. (7.19).

NOTES

1. P. W. Anderson, *Phys. Rev.* **86**, 809 (1952).
2. J. Szudy and W. E. Baylis, *J. Quant. Spectrosc. Radiat. Transfer* **15**, 641 (1975).
3. B. J. Berne and R. Pecora, *Dynamic Light Scattering* (Wiley, Toronto, 1976).
4. R. G. Gordon, *Adv. Magn. Reson.* **3**, 1 (1968); R. G. Gordon, *J. Math. Phys.* **9**, 655 (1968); R. G. Gordon, *Adv. Chem. Phys.* **15**, 79 (1969).
5. E. J. Heller, *J. Chem. Phys.* **62**, 1544 (1975); *Acct. Chem. Res.* **14**, 368 (1981).
6. S. Mukamel, *J. Chem. Phys.* **77**, 173 (1982); *Phys. Rep.* **93**, 1 (1982).
7. M. Born and K. Huang, *Dynamical Theory of Crystal Lattices* (Oxford, London, 1954).
8. J. H. Van Vleck, *Phys. Rev.* **72**, 1168 (1948).
9. G. Brosso and G. Pastori Parravicini, *Adv. Chem. Phys.* **62**, 81 (1985); *Adv. Chem. Phys.* **62**, 133 (1985).
10. C. Lanczos, *Applied Analysis* (Prentice-Hall, Englewood Cliffs, New Jersey, 1956).
11. C. T. Corcokan and P. W. Langhoff, *J. Math. Phys.* **18**, 651 (1977).
12. I. Oppenheim, K. E. Shuler, and G. H. Weiss, *Stochastic Processes in Chemical Physics: The Master Equation* (MIT Press, Cambridge, Massachusetts, 1977), p. 25.
13. L. E. Fried and S. Mukamel, *Adv. Chem. Phys.* **84**, 435 (1993); *J. Chem. Phys.* **93**, 3063 (1990).
14. G. Herzberg, *Spectra of Diatomic Molecules* (Van Nostrand, Toronto, 1950), p. 390.
15. A. A. Gordus and R. B. Bernstein, *J. Chem. Phys.* **22**, 790 (1954).
16. J. P. Bergsma, P. H. Berens, K. R. Wilson, D. R. Fredkin, and E. J. Heller, *J. Phys. Chem.* **88**, 612 (1984).

CHAPTER 8

The Cumulant Expansion and the Multimode Brownian Oscillator Model

So far we presented two methods for calculating the response functions. The first (Chapter 6) is based on an expansion in eigenstates, and the second (Chapter 7) is semiclassical and requires the calculation of classical trajectories. In this chapter we develop a third method that is based on the cumulant expansion. This method is particularly suitable for semiclassical simulations. It also provides an exact solution for an important class of models with Gaussian statistics, e.g., when the electronic system is coupled to a harmonic bath. In this case the expansion can be truncated at second order and higher cumulants vanish due to the central limit theorem. This includes the Brownian oscillator model of nuclear dynamics, which is widely used to represent local vibrations of the system as well as collective bath coordinates. Moreover, a large number of phenomenological models commonly used in nonlinear optics, e.g., stochastic models, homogeneous Lorentzian, and inhomogeneous Gaussian line broadening can be obtained as limiting cases of this model. It therefore provides an important unifying framework for the analysis of optical response functions. The cumulant expansion in conjunction with other techniques (density expansion, semiclassical approximations, and molecular dynamic simulations) provides a powerful tool for the analysis of optical response.

We shall first write the response functions of the two-level model of Chapter 7 in the Condon approximation, where we can combine all the Liouville space paths and derive a compact formal expression. We then solve for the response functions using the cumulant expansion truncated at second order, and discuss various useful symmetries of the nuclear spectral density that controls the optical response. These results are applied to several models for nuclear dynamics: a discrete and a continuous set of harmonic modes, and the multimode Brownian oscillator model, which has numerous important limiting cases. Other applications include dielectric fluctuations in a polar solvent, and pressure broadening in the gas phase. We next develop the inhomogeneous cumulant expansion, which allows us to incorporate static inhomogeneous broadening in addition to the homogeneous part. Finally, we present closed form expressions of $\chi^{(1)}$ and $\chi^{(3)}$ for a single overdamped Brownian oscillator mode, which provides an important example of spectral diffusion.

THE CONDON APPROXIMATION FOR THE OPTICAL RESPONSE OF A TWO-LEVEL SYSTEM

Consider the two electronic level system discussed in Chapter 7, characterized by the Hamiltonian Eq. (7.1). The electronic transition moment depends on the nuclear degrees of freedom through the parametric dependence of the electronic wavefunctions on these coordinates. This dependence is weak in many cases (e.g., molecular systems) since the adiabatic separation of electronic and nuclear motions is usually quite good (exceptions are systems with electronic degeneracy, such as Jahn Teller systems) [1]. We can then expand the transition moment in a Taylor series

$$V_{eg}(\mathbf{q}) = V_{eg}(\mathbf{q}_0) + V'_{eg}(\mathbf{q}_0)(\mathbf{q} - \mathbf{q}_0) + \cdots,$$

where $\mathbf{q}_0$ is some reference nuclear configuration. Hereafter in this chapter we shall invoke the Condon approximation [2] and assume that the $\mathbf{q}$ dependence is negligible, so that we can retain only the first term in the expansion. $V_{eg}(\mathbf{q})$, which is an operator in the nuclear subspace, thus becomes a simple number, and, for brevity, we set $V_{eg}(\mathbf{q}_0) = 1$. Equation (7.4b) then reduces to

$$J(t_1) = \langle \mathcal{G}_{eg}(t_1)\rho_g \rangle, \tag{8.1}$$

which yields

$$S^{(1)}(t_1) = \frac{i}{\hbar}[\langle \mathcal{G}_{eg}(t_1)\rho_g \rangle - \langle \mathcal{G}_{ge}(t_1)\rho_g \rangle]. \tag{8.2}$$

The third-order response functions [Eqs. (7.11)] similarly become

$$R_1(t_3, t_2, t_1) = \langle \mathcal{G}_{eg}(t_3)\mathcal{G}_{ee}(t_2)\mathcal{G}_{eg}(t_1)\rho_g \rangle, \tag{8.3a}$$

$$R_2(t_3, t_2, t_1) = \langle \mathcal{G}_{eg}(t_3)\mathcal{G}_{ee}(t_2)\mathcal{G}_{ge}(t_1)\rho_g \rangle, \tag{8.3b}$$

$$R_3(t_3, t_2, t_1) = \langle \mathcal{G}_{eg}(t_3)\mathcal{G}_{gg}(t_2)\mathcal{G}_{ge}(t_1)\rho_g \rangle, \tag{8.3c}$$

$$R_4(t_3, t_2, t_1) = \langle \mathcal{G}_{eg}(t_3)\mathcal{G}_{gg}(t_2)\mathcal{G}_{eg}(t_1)\rho_g \rangle. \tag{8.3d}$$

Substituting Eqs. (8.3) in Eq. (5.27) we obtain

$$\begin{aligned} S^{(3)}(t_3, t_2, t_1) = {} & \left(\frac{i}{\hbar}\right)^3 \langle [\mathcal{G}_{eg}(t_3) - \mathcal{G}_{ge}(t_3)]\mathcal{G}_{ee}(t_2)[\mathcal{G}_{eg}(t_1) + \mathcal{G}_{ge}(t_1)]\rho_g \rangle \\ & + \left(\frac{i}{\hbar}\right)^3 \langle [\mathcal{G}_{eg}(t_3) - \mathcal{G}_{ge}(t_3)]\mathcal{G}_{gg}(t_2)[\mathcal{G}_{eg}(t_1) + \mathcal{G}_{ge}(t_1)]\rho_g \rangle. \end{aligned} \tag{8.4}$$

The third-order response function thus consists of a sum of two terms. The first contains four pathways in which the system is in the excited state during the middle time interval t_2, and the second contains four pathways in which the system is in the ground state during that period.

In Chapter 7 we introduced a semiclassical expansion of the response function. In the classical limit we can then combine the various Liouville space paths to obtain a compact expression for the entire response function. In Eq. (7.6) we defined a collective energy gap coordinate U that depends on the parameter ω_{eg}. We then indicated that ω_{eg} can be chosen so that the first moment of the lineshape [Eq. (7.10e)] should vanish. In the Condon approximation this implies that ω_{eg} is equal to the thermally averaged electronic energy gap.

$$\omega_{eg} \equiv \langle (H_e - H_g)\rho_g \rangle = \omega_{eg}^0 + \frac{1}{\hbar}\langle (W_e - W_g)\rho_g \rangle. \tag{8.5a}$$

With this definition, we have for the collective electronic gap coordinate

$$U \equiv H_e - H_g - \hbar\omega_{eg} = W_e(\mathbf{q}) - W_g(\mathbf{q}) - \langle (W_e - W_g)\rho_g \rangle. \tag{8.5b}$$

Using Eqs. (7.9a) together with Eqs. (8.1) and (8.2) we get

$$S^{(1)}(t) = \frac{2}{\hbar}\,\theta(t)\left\langle \sin\left[\int_0^t d\tau\; \varepsilon_a(\tau)\right]\rho_g \right\rangle. \tag{8.6}$$

Similarly, Eqs. (8.4) together with Eqs. (7.19) yield

$$\begin{aligned} S^{(3)}(t_3, t_2, t_1) = {} & \frac{4}{\hbar^3}\,\theta(t_1)\theta(t_2)\theta(t_3)\left\langle \sin\left[\int_{t_1+t_2}^{t_1+t_2+t_3} \varepsilon_b(\tau)\, d\tau\right]\cos\left[\int_0^{t_1} d\tau\; \varepsilon_b(\tau)\right]\rho_g \right\rangle \\ & + \frac{4}{\hbar^3}\,\theta(t_1)\theta(t_2)\theta(t_3)\left\langle \sin\left[\int_{t_1+t_2}^{t_1+t_2+t_3} \varepsilon_a(\tau)\, d\tau\right]\cos\left[\int_0^{t_1} d\tau\; \varepsilon_a(\tau)\right]\rho_g \right\rangle. \end{aligned} \tag{8.7}$$

Here

$$\varepsilon_a(\tau) \equiv \omega_{eg} + \frac{1}{\hbar}\, U_a(\tau)$$

and

$$\varepsilon_b(\tau) \equiv \omega_{eg} + \frac{1}{\hbar}\, U_b(\tau)$$

represent a classical time evolution of the electronic gap using the two reference Hamiltonians introduced in Eqs. (7.16)–(7.18). The moments of the absorption lineshape in the Condon approximation become

$$\left.\begin{aligned} M_0 &= 1, \\ M_1 &= 0, \\ M_n &= \langle U^n \rho_g \rangle. \end{aligned}\right\} \tag{8.8}$$

The elimination of the first moment results from the choice of ω_{eg} [Eq. (8.5a)].

THE CUMULANT EXPANSION

We shall now evaluate the response functions perturbatively using the cumulant expansion to second order. The following calculation follows the same steps used in the derivation of the Magnus expansion in Chapter 2, except that the expansion is now performed on the correlation function rather than on the time evolution operator. We start by invoking the Condon approximation on Eq. (7.9a), setting $V_{eg} = V_{ge} = 1$. We then obtain

$$J(t_1) = \exp(-i\omega_{eg}t_1)\left\langle \exp_+\left[-\frac{i}{\hbar}\int_0^{t_1} d\tau\, U(\tau)\right]\rho_g\right\rangle, \tag{8.9}$$

where the time evolution is with respect to the ground state Hamiltonian

$$U(\tau) \equiv \exp\left(\frac{i}{\hbar}H_g\tau\right)U\exp\left(-\frac{i}{\hbar}H_g\tau\right).$$

Expanding Eq. (8.9) perturbatively in U, we have

$$J(t_1) = \exp(-i\omega_{eg}t_1)\left[1 - \frac{i}{\hbar}\int_0^{t_1} d\tau\langle U(\tau)\rho_g\rangle \right.$$
$$\left. + \left(-\frac{i}{\hbar}\right)^2 \int_0^{t_1} d\tau_2 \int_0^{t_2} d\tau_1 \langle U(\tau_2)U(\tau_1)\,\rho_g\rangle + \cdots\right] \tag{8.10a}$$

Note that the term linear in U does not depend on time. This can be seen by switching to the Schrödinger picture

$$\langle U(\tau)\rho_g\rangle = \langle U\rho_g(\tau)\rangle = \langle U\rho_g(0)\rangle.$$

The last equality simply implies that the equilibrium ground state distribution does not evolve with time when subjected to the ground state Hamiltonian. From the definition of U [Eq. (8.5b)] we see that this term actually vanishes identically.

The cumulant expansion is obtained by postulating

$$J(t_1) = \exp[-\mathscr{F}(t_1)], \tag{8.11}$$

with an expansion of $\mathscr{F}(t_1)$ in powers of U,

$$\mathscr{F}(t_1) \equiv \mathscr{F}_1(t_1) + \mathscr{F}_2(t_1) + \cdots.$$

When Eq. (8.11) is expanded in a Taylor series in $\mathscr{F}$, $\mathscr{F}$ is expanded in powers of U, and the terms are collected in orders of U, we can compare it with Eq. (8.10a) and obtain expressions for $\mathscr{F}_1$, $\mathscr{F}_2$, etc. Truncating at $\mathscr{F}_2$, we obtain the second-order cumulant approximation

$$S^{(1)}(t_1) = \frac{i}{\hbar}\theta(t_1)[J(t_1) - J^*(t_1)], \tag{8.12a}$$

$$J(t_1) = \exp[-i\omega_{eg}t_1 - g(t_1)], \tag{8.12b}$$

with

$$g(t) \equiv \int_0^t d\tau_2 \int_0^{\tau_2} d\tau_1\, C(\tau_1), \tag{8.13a}$$

and

$$C(\tau_1) \equiv \frac{1}{\hbar^2}\langle U(\tau_1)U(0)\rho_g\rangle. \tag{8.13b}$$

The calculation of the third-order response function using the second-order cumulant expansion is carried out in Appendix 8A [3]. This results in the following expression for the four point dipole correlation function [Eq. (7.13)]:

$$\begin{aligned} F(\tau_1, \tau_2, \tau_3, \tau_4) = {} & \exp[-i\omega_{eg}(\tau_1 - \tau_2 + \tau_3 - \tau_4)] \\ & \times \exp[-g(\tau_1 - \tau_2) + g(\tau_1 - \tau_3) - g(\tau_2 - \tau_3) \\ & - g(\tau_1 - \tau_4) + g(\tau_2 - \tau_4) - g(\tau_3 - \tau_4)]. \end{aligned} \tag{8.14}$$

Upon the substitution of the proper time arguments [Eq. (7.14)] in Eq. (8.14), we obtain the final expression for the third-order response function:

$$\begin{aligned} S^{(3)}(t_3, t_2, t_1) = {} & \left(\frac{i}{\hbar}\right)^3 \theta(t_1)\theta(t_2)\theta(t_3) \\ & \times \sum_{\alpha=1}^{4} [R_\alpha(t_3, t_2, t_1) - R_\alpha^*(t_3, t_2, t_1)], \end{aligned} \tag{8.15a}$$

$$\begin{aligned} R_1(t_3, t_2, t_1) = {} & \exp(-i\omega_{eg}t_1 - i\omega_{eg}t_3) \\ & \times \exp[-g^*(t_3) - g(t_1) - f_+(t_3, t_2, t_1)], \end{aligned} \tag{8.15b}$$

$$\begin{aligned} R_2(t_3, t_2, t_1) = {} & \exp(i\omega_{eg}t_1 - i\omega_{eg}t_3) \\ & \times \exp[-g^*(t_3) - g^*(t_1) + f_+^*(t_3, t_2, t_1)], \end{aligned} \tag{8.15c}$$

$$\begin{aligned} R_3(t_3, t_2, t_1) = {} & \exp(i\omega_{eg}t_1 - i\omega_{eg}t_3) \\ & \times \exp[-g(t_3) - g^*(t_1) + f_-^*(t_3, t_2, t_1)], \end{aligned} \tag{8.15d}$$

$$\begin{aligned} R_4(t_3, t_2, t_1) = {} & \exp(-i\omega_{eg}t_1 - i\omega_{eg}t_3) \\ & \times \exp[-g(t_3) - g(t_1) - f_-(t_3, t_2, t_1)], \end{aligned} \tag{8.15e}$$

where

$$f_-(t_3, t_2, t_1) \equiv g(t_2) - g(t_2 + t_3) - g(t_1 + t_2) + g(t_1 + t_2 + t_3), \tag{8.16a}$$

$$f_+(t_3, t_2, t_1) \equiv g^*(t_2) - g^*(t_2 + t_3) - g(t_1 + t_2) + g(t_1 + t_2 + t_3). \tag{8.16b}$$

THE SPECTRAL DENSITY AND ITS SYMMETRIES

The two-time correlation function of the electronic energy gap U, $C(t)$, is the key quantity that carries all the microscopic information necessary for calculating the optical response functions, within the second-order cumulant approximation. We shall now explore various ways of rewriting this correlation function and examine its symmetries. We first note that $C(t)$ is complex and satisfies

$$C(-t) = C^*(t). \tag{8.17}$$

We next separate it into its real and imaginary parts:

$$C(t) = C'(t) + iC''(t), \tag{8.18a}$$

with

$$C'(t) \equiv \frac{1}{2\hbar^2}[\langle U(t)U(0)\rho_g\rangle + \langle U(0)U(t)\rho_g\rangle], \tag{8.18b}$$

$$C''(t) \equiv -\frac{i}{2\hbar^2}[\langle U(t)U(0)\rho_g\rangle - \langle U(0)U(t)\rho_g\rangle]. \tag{8.18c}$$

Both $C'(t)$ and $C''(t)$ are real. Also $C'(t) = C'(-t)$ is an even function of time whereas $C''(t) = -C''(-t)$ is odd.

The symmetries of these correlation functions become more apparent if we expand them in the vibronic eigenstates of the ground state Hamiltonian H_g (Figure 7.1). The main point of using the cumulant expansion is to provide a semiclassical procedure for evaluating the necessary correlation functions without calculating the nuclear eigenstates, as will be shown below. Nevertheless, even if we do not intend to use eigenstate expansions in actual calculations, they provide a useful insight on the general formal properties of these correlation functions. We thus have

$$C(t) = \frac{1}{\hbar^2}\sum_{a,c} P(a)|U_{ac}|^2 \exp(-i\omega_{ca}t), \tag{8.19a}$$

$$C'(t) = \sum_{a,c} \frac{P(c) + P(a)}{2\hbar^2}|U_{ac}|^2 \cos(\omega_{ca}t), \tag{8.19b}$$

$$C''(t) = \sum_{a,c} \frac{P(c) - P(a)}{2\hbar^2}|U_{ac}|^2 \sin(\omega_{ca}t). \tag{8.19c}$$

In the frequency domain, we introduce the spectral density

$$\tilde{C}(\omega) \equiv \int_{-\infty}^{\infty} dt \exp(i\omega t)C(t) = 2\mathrm{Re}\int_0^{\infty} dt \exp(i\omega t)C(t). \tag{8.20}$$

Upon the substitution of Eq. (8.18a) in (8.20) we get

$$\tilde{C}(\omega) = \tilde{C}'(\omega) + \tilde{C}''(\omega), \tag{8.21a}$$

with

$$\tilde{C}'(\omega) \equiv \int_{-\infty}^{\infty} dt\, \exp(i\omega t) C'(t) = 2 \int_0^{\infty} dt\, \cos(\omega t) C'(t), \tag{8.21b}$$

with

$$\tilde{C}''(\omega) \equiv -i \int_{-\infty}^{\infty} dt\, \exp(i\omega t) C''(t) = 2 \int_0^{\infty} dt\, \sin(\omega t) C''(t), \tag{8.21c}$$

Both $\tilde{C}'(\omega)$ and $\tilde{C}''(\omega)$ are real. $\tilde{C}'(\omega) = \tilde{C}'(-\omega)$ is even, and $\tilde{C}''(\omega) = -\tilde{C}''(-\omega)$ is odd.

Using the eigenstate representation we have

$$\tilde{C}(\omega) = \frac{2\pi}{\hbar^2} \sum_{a,c} P(a) |U_{ac}|^2 \delta(\omega - \omega_{ca}), \tag{8.22a}$$

$$\tilde{C}'(\omega) = \frac{\pi}{\hbar^2} \sum_{a,c} [P(a) + P(c)] |U_{ac}|^2 \delta(\omega - \omega_{ca}), \tag{8.22b}$$

$$\tilde{C}''(\omega) = \frac{\pi}{\hbar^2} \sum_{a,c} [P(a) - P(c)] |U_{ac}|^2 \delta(\omega - \omega_{ca}). \tag{8.22c}$$

$\tilde{C}(\omega)$ satisfies the detailed-balance condition [see Eq. (6.9c)]

$$\tilde{C}(-\omega) = \exp(-\beta\hbar\omega) \tilde{C}(\omega). \tag{8.23}$$

We further have

$$\tilde{C}'(\omega) = \frac{1 + \exp(-\beta\hbar\omega)}{2} \tilde{C}(\omega),$$

$$\tilde{C}''(\omega) = \frac{1 - \exp(-\beta\hbar\omega)}{2} \tilde{C}(\omega).$$

We thus get

$$\tilde{C}'(\omega) = \coth(\beta\hbar\omega/2) \tilde{C}''(\omega) \tag{8.24a}$$

and

$$\tilde{C}(\omega) = [1 + \coth(\beta\hbar\omega/2)] \tilde{C}''(\omega). \tag{8.24b}$$

The relation between $\tilde{C}'(\omega)$ and $\tilde{C}''(\omega)$ [Eq. (8.24a)] reflects the fluctuation–dissipation theorem, which connects the response functions with the correlation functions [see Eqs. (6.9)]. The eigenstate expansions further show that both $\tilde{C}(\omega)$ and $\tilde{C}'(\omega)$ are nonnegative.

The fluctuation–dissipation theorem provides a general connection between $C'(t)$ and $C''(t)$, and shows that they are not independent. In general, $C(t)$ is complex and does not have a simple classical analogue. However, $\tilde{C}'(\omega)$ and $\tilde{C}''(\omega)$ can be interpreted classically. Since for a harmonic bath $\tilde{C}''(\omega)$ is independent of temperature [see Eq. (8.38b)], it is somewhat simpler than $\tilde{C}'(\omega)$ and we can use it as the basic quantity that characterizes the bath. We shall therefore express the entire $C(t)$ function in terms of its imaginary part

$$C(t) = \int_{-\infty}^{\infty} d\omega \cos(\omega t) \coth(\beta\hbar\omega/2)\tilde{C}''(\omega) - i\int_{-\infty}^{\infty} d\omega \sin(\omega t)\tilde{C}''(\omega).$$

Upon the substitution of Eq. (8.24a) in this result, we can express $C(t)$ in terms of its real part, $\tilde{C}'(t)$.

We shall now combine all of these relations and express $g(t)$ in terms of $\tilde{C}''(\omega)$ in a way that is convenient for numerical calculations. It follows directly from Eqs (8.13b) that

$$g(t) = -\frac{1}{2\pi}\int_{-\infty}^{\infty} d\omega \frac{\tilde{C}(\omega)}{\omega^2}[\exp(-i\omega t) + i\omega t - 1],$$

or

$$g(t) = \frac{1}{2\pi}\int_{-\infty}^{\infty} d\omega \frac{1-\cos(\omega t)}{\omega^2}\coth(\beta\hbar\omega/2)\tilde{C}''(\omega) + \frac{i}{2\pi}\int_{-\infty}^{\infty} d\omega \frac{\sin(\omega t) - \omega t}{\omega^2}\tilde{C}''(\omega). \quad (8.25)$$

This result can be also recast in the form

$$g(t) = \Delta^2 \int_0^t d\tau_2 \int_0^{\tau_2} d\tau_1\, M'(\tau_1) - i\lambda \int_0^t d\tau[1 - M''(\tau)], \quad (8.26)$$

$$M'(t) = \frac{1}{\pi\Delta^2}\int_0^{\infty} d\omega\, \tilde{C}''(\omega)\coth(\beta\hbar\omega/2)\cos(\omega t), \quad (8.27a)$$

$$M''(t) = \frac{1}{\pi\lambda}\int_0^{\infty} d\omega \frac{\tilde{C}''(\omega)}{\omega}\cos(\omega t), \quad (8.27b)$$

$$\Delta^2 = \frac{1}{\pi}\int_0^{\infty} d\omega\, \tilde{C}''(\omega)\coth(\beta\hbar\omega/2), \quad (8.28a)$$

$$\lambda = \frac{1}{\pi}\int_0^{\infty} d\omega \frac{\tilde{C}''(\omega)}{\omega}. \quad (8.28b)$$

Here Δ and λ are two static coupling strength parameters whereas $M'(t)$ and $M''(t)$ are two auxiliary real functions with $M'(0) = M''(0) = 1$ and $\dot{M}'(0) = \dot{M}''(0) = 0$. These quantities can also be related to $C(t)$ in the time domain, by combining Eqs. (8.13a), (8.18a), and (8.26). For the real part we have

$$M'(t) = C'(t)/C'(t=0),$$

$$\Delta^2 = C'(t=0),$$

whereas for the imaginary part we obtain

$$\int_0^t d\tau\, C''(\tau) \equiv \lambda[1 - M''(t)],$$

so that

$$\lambda = \int_0^\infty C''(\tau)\, d\tau,$$

$$M''(t) = 1 - \int_0^t d\tau\, C''(\tau) \Big/ \int_0^\infty d\tau\, C''(\tau).$$

$M'(t)$ can be calculated from $M''(t)$ by first calculating $\tilde{C}''(\omega)$ using Eq. (8.27b) and then substituting in Eq. (8.27a),

$$M'(t) = (\lambda/\Delta^2) \int_0^\infty d\omega\, \omega \cos(\omega t) \coth(\hbar\omega\beta/2) \int_0^\infty d\tau\, M''(\tau) \cos(\omega\tau). \quad (8.29)$$

In the high temperature limit $\beta\hbar\omega \ll 1$, we have $\coth(\beta\hbar\omega/2) \cong 2k_B T/\hbar\omega$, so that

$$\Delta^2 = \frac{2\lambda k_B T}{\hbar}, \quad (8.30a)$$

$$M'(t) = M''(t) \equiv M(t). \quad (8.30b)$$

We then get

$$g(t) = \frac{2\lambda k_B T}{\hbar} \int_0^t d\tau_2 \int_0^{\tau_2} d\tau_1\, M(\tau_1) - i\lambda \int_0^t d\tau[1 - M(\tau)],$$

$$M(t) = \frac{1}{\pi\lambda} \int_0^\infty d\omega \frac{\tilde{C}''(\omega)}{\omega} \cos(\omega t), \qquad \beta\hbar\omega \ll 1$$

$$\lambda = \frac{1}{\pi} \int_0^\infty d\omega \frac{\tilde{C}''(\omega)}{\omega}. \quad (8.31)$$

In this limit, the $g(t)$ function depends on a single coupling strength (λ) and a single relaxation function $M(t)$. Semiclassical methods for the evaluation of these correlation functions, based on the results of Chapter 7, will be presented later in this chapter.

COUPLING TO HARMONIC VIBRATIONS

Discrete set of oscillators: coherent nuclear motions

Nuclear motions may often be modeled as small amplitude vibrations. The electronic system may be coupled to a number of vibrational modes representing, e.g., intramolecular vibrations, local intermolecular modes, and collective solvent modes. Impurity centers in solids and excitons in semiconductors are coupled to acoustic and optical phonon modes. Nuclear motions that couple to the electronic transition of our particle can often be modeled as independent harmonic modes whose equilibrium position is displaced between the two electronic states. In this model, the Hamiltonian assumes the form [4–8]

$$H = |g\rangle H_g \langle g| + |e\rangle H_e \langle e|, \tag{8.32a}$$

with

$$H_g = \sum_j \left[\frac{p_j^2}{2m_j} + \tfrac{1}{2} m_j \omega_j^2 q_j^2 \right], \tag{8.32b}$$

$$H_e = \hbar\omega_{eg}^0 + \sum_j \left[\frac{p_j^2}{2m_j} + \tfrac{1}{2} m_j \omega_j^2 (q_j + d_j)^2 \right], \tag{8.32c}$$

and p_j, q_j, and m_j represent the momentum, the coordinate, and the mass, respectively, of the jth nuclear mode. In spectroscopic applications it is a common practice to switch to dimensionless coordinates. We thus define the dimensionless momentum $\not{p}_j \equiv (\hbar\omega_j m_j)^{-1/2} p_j$ and coordinate $\not{q}_j \equiv (m_j\omega_j/\hbar)^{1/2} q_j$ of the jth nuclear mode, and $\not{d}_j \equiv (m_j\omega_j/\hbar)^{1/2} d_j$ is the dimensionless displacement of the equilibrium configuration of this mode between the two electronic states. The Hamiltonians then assume the form

$$H_g = \tfrac{1}{2} \sum_j \hbar\omega_j (\not{p}_j^2 + \not{q}_j^2), \tag{8.33a}$$

$$H_e = \hbar\omega_{eg}^0 + \tfrac{1}{2} \sum_j \hbar\omega_j [\not{p}_j^2 + (\not{q}_j + \not{d}_j)^2]. \tag{8.33b}$$

The electronic energy gap is in this case

$$\omega_{eg} = \omega_{eg}^0 + \tfrac{1}{2} \sum_j \not{d}_j^2 \omega_j \tag{8.33c}$$

and

$$U = \hbar \sum_j \omega_j \not{d}_j \not{q}_j. \tag{8.33d}$$

For this model, all cumulants higher than the second vanish identically, as can be shown using Wick's theorem [9], and the second-order cumulant expression for the response function is exact. We then have

$$C(t) = \sum_j \omega_j^2 S_j [(\bar{n}_j + 1) \exp(-i\omega_j t) + \bar{n}_j \exp(i\omega_j t)], \tag{8.34}$$

where we have introduced the Huang–Rhys factor $S_j \equiv \tfrac{1}{2}\not{d}_j^2$, which is a dimensionless parameter representing the coupling strength of the nuclear degrees of freedom to the electronic transition. This factor can be directly obtained from absorption and fluorescence spectra [see Eqs. (8.43) and (8.44)], since it controls the vibrational progression that accompanies an electronic transition.

$$\bar{n}_j = [\exp(\beta\hbar\omega_j) - 1]^{-1}, \tag{8.35}$$

is the thermally averaged occupation number of the jth mode. By partitioning the correlation function into its real and imaginary parts we have

$$C'(t) = \sum_j S_j\omega_j^2 \coth(\beta\hbar\omega_j/2)\cos(\omega_j t) \tag{8.36a}$$

and

$$C''(t) = -\sum_j S_j\omega_j^2 \sin(\omega_j t). \tag{8.36b}$$

In the frequency domain we get the spectral density

$$\tilde{C}(\omega) = \sum_j S_j\omega_j^2[(\bar{n}_j + 1)\delta(\omega - \omega_j) + \bar{n}_j\delta(\omega + \omega_j)]. \tag{8.37}$$

The even and the odd components of the spectral density are given by

$$\tilde{C}'(\omega) = \sum_j S_j\omega_j^2 \coth(\beta\hbar\omega_j/2)[\delta(\omega - \omega_j) + \delta(\omega + \omega_j)], \tag{8.38a}$$

$$\tilde{C}''(\omega) = \sum_j S_j\omega_j^2[\delta(\omega - \omega_j) - \delta(\omega + \omega_j)]. \tag{8.38b}$$

Note that $C''(\omega)$ is independent of temperature for this model. Equation (8.25b) together with Eq. (8.38b) result in the following expression for the line broadening function:

$$g(t) = \sum_j S_j[(\coth(\beta\hbar\omega_j/2)(1 - \cos(\omega_j t)) + i(\sin(\omega_j t) - \omega_j t)]. \tag{8.39}$$

Alternatively, we can recast $g(t)$ in the form of Eq. (8.26) with

$$\Delta^2 = \sum_j S_j\omega_j^2 \coth(\beta\hbar\omega_j/2) \equiv \sum_j \Delta_j^2 \tag{8.40a}$$

$$\lambda = \sum_j S_j\omega_j \equiv \sum_j \lambda_j, \tag{8.40b}$$

$$M'(t) = \frac{\sum_j S_j\omega_j^2 \coth(\beta\hbar\omega_j/2)\cos(\omega_j t)}{\sum_j S_j\omega_j^2 \coth(\beta\hbar\omega_j/2)}, \tag{8.41a}$$

and

$$M''(t) = \frac{\sum_j S_j\omega_j \cos(\omega_j t)}{\sum_j S_j\omega_j}. \tag{8.41b}$$

As shown in Appendix 8B, the absorption lineshape is given by

$$\sigma_a(\omega) = \frac{1}{\pi}\,\mathrm{Re}\int_0^\infty dt\, \exp[i(\omega - \omega_{eg})t - g(t)], \tag{8.42a}$$

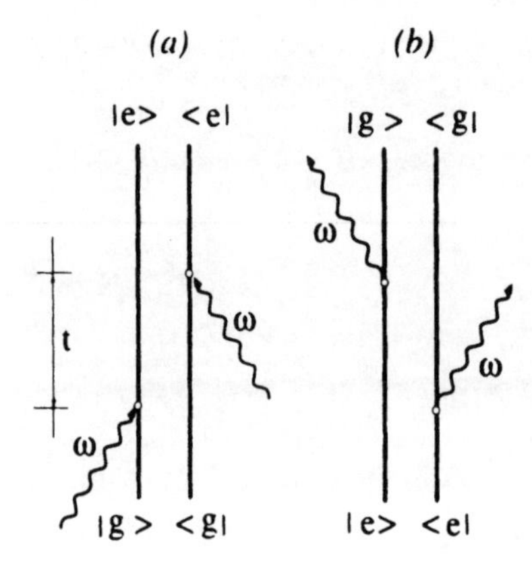

FIG. 8.1 Double-sided Feynman diagrams representing (a) absorption and (b) relaxed fluorescence.

and the relaxed-fluorescence (emission) spectrum is

$$\sigma_f(\omega) = \frac{1}{\pi} \operatorname{Re} \int_0^\infty dt \exp[i(\omega - \omega_{eg} + 2\lambda)t - g^*(t)]. \tag{8.42b}$$

Double-sided Feynman diagrams representing absorption and emission are displayed in Figure 8.1. The first moment of σ_a vanishes [Eq. (8.8)] whereas the first moment of σ_f is -2λ. The fluorescence is thus red shifted with respect to the linear absorption and the shift, 2λ, is known as the Stokes shift. Moreover, for the present model σ_f is the mirror image of σ_a around $\omega_{eg}^0 = \omega_{eg} - \lambda$, i.e.,

$$\sigma_a(\omega_{eg} + \Delta\omega) = \sigma_f(\omega_{eg} - 2\lambda - \Delta\omega).$$

For a single mode, the absorption lineshape assumes the form

$$\sigma_a(\omega) = \exp[-S_j \coth(\beta\hbar\omega_j/2)] \times \sum_{n=-\infty}^{\infty} \exp(n\beta\hbar\omega_j/2) I_n[S_j\sqrt{\coth^2(\beta\hbar\omega_j/2) - 1}]\delta(\omega - \omega_{eg} - n\omega_j), \tag{8.43}$$

where $I_n(z)$ is the modified Bessel function. The relaxed-fluorescence spectrum is also given by Eq. (8.43) with simply replacing the $\delta(\omega - \omega_{eg} - n\omega_j)$ factor by $\delta(\omega - \omega_{eg} + 2\lambda + n\omega_j)$. The spectral density, together with the absorption and the relaxed fluorescence lineshapes for a single oscillator system, will be displayed later in this chapter, after we add relaxation and obtain the Brownian oscillator model.

At zero temperature we get

$$\sigma_a(\omega) = \exp(-S_j) \sum_{n=0}^{\infty} \frac{S_j^n}{n!} \delta(\omega - \omega_{eg} - n\omega_j), \tag{8.44a}$$

and

$$\sigma_f(\omega) = \exp(-S_j) \sum_{n=0}^{\infty} \frac{S_j^n}{n!} \delta(\omega - \omega_{eg} + 2\lambda + n\omega_j), \tag{8.44b}$$

which is a simple Poisson distribution. The transition dipole matrix elements μ_{nm} for this model, as well as for more general harmonic potentials with a different frequency ω_j in the ground and in the excited state, are well known [10, 11].

These lineshapes can therefore be easily calculated also using the sum over states expressions of Chapter 6.

Continuous distribution of oscillators: dephasing and relaxation

In many cases it is possible to represent a condensed phase system in terms of a continuous distribution of oscillators. By changing the summation in Eq. (8.33) to integration we get

$$H_g = \frac{\hbar}{2}\int_0^\infty d\omega\, W(\omega)\omega(\mathbf{p}[\omega]^2 + \mathbf{q}[\omega]^2), \tag{8.45a}$$

$$H_e = \hbar\omega_{eg}^0 + \frac{\hbar}{2}\int_0^\infty d\omega\, W(\omega)\omega\{[\mathbf{p}[\omega]]^2 + [\mathbf{q}[\omega] + \mathbf{d}(\omega)]^2\}. \tag{8.45b}$$

Here $W(\omega)\,d\omega$ denotes the number of oscillators between ω and $\omega + d\omega$. $\mathbf{p}[\omega]$ and $\mathbf{q}[\omega]$ are infinite-dimensional vectors parameterized by the continuous index ω. $\mathbf{d}(\omega)$ represents the displacement of an oscillator with a frequency ω. We further define the Huang–Rhys factor $S(\omega) \equiv \frac{1}{2}d^2(\omega)$. $g(t)$ is then given by Eq. (8.39), where the summation is replaced by an integration. The optical response functions can be recast in the form of Eq. (8.26) with

$$\tilde{C}''(\omega) = \tfrac{1}{2}W(\omega)S(\omega)\omega^2 = \tfrac{1}{4}W(\omega)d^2(\omega)\omega^2. \tag{8.46}$$

To illustrate these results, let us consider the following model

$$\tilde{C}''(\omega) = 2\lambda\frac{\omega\Lambda}{\omega^2 + \Lambda^2}. \tag{8.47}$$

This simple example represents the overdamped Brownian oscillator, as will be shown below. For this model $M''(t)$ is a single exponential $M''(t) = \exp(-\Lambda t)$. $M'(t)$ can be calculated by performing a contour integration on Eq. (8.27a). We then get [12–14]

$$\begin{aligned} g(t) &= g'(t) + ig''(t), \\ g''(t) &= -(\lambda/\Lambda)[\exp(-\Lambda t) + \Lambda t - 1], \\ g'(t) &= (\lambda/\Lambda)\cot(\hbar\beta\Lambda/2)[\exp(-\Lambda t) + \Lambda t - 1] \\ &\quad + \frac{4\lambda\Lambda}{\hbar\beta}\sum_{n=1}^{\infty}\frac{\exp(-\nu_n t) + \nu_n t - 1}{\nu_n(\nu_n^2 - \Lambda^2)}, \\ \nu_n &\equiv \frac{2\pi}{\hbar\beta}n. \end{aligned} \tag{8.48}$$

Here ν_n are known as the Matsubara frequencies.

In the high temperature limit, Eq. (8.48) simplifies and assumes the form

$$g(t) = \frac{2\lambda k_B T}{\hbar\Lambda^2}[\exp(-\Lambda t) + \Lambda t - 1] - i(\lambda/\Lambda)[\exp(-\Lambda t) + \Lambda t - 1], \qquad k_B T \gg \hbar\Lambda. \quad (8.49)$$

This can also be obtained directly from Eq. (8.31).

Let us consider a few limiting cases. The nature of the optical response depends crucially on the following dimensionless parameter:

$$\kappa \equiv \Lambda/\Delta = \left(\frac{\hbar\Lambda^2}{2\lambda k_B T}\right)^{1/2}. \quad (8.50)$$

When $\kappa \ll 1$, nuclear dynamics (with timescale Λ^{-1}) are slow compared with the magnitude of fluctuations (coupling strength). In this case, we can expand Eq. (8.49) in a Taylor series in powers of Λt, resulting in the short-time approximation:

$$g(t) = \frac{1}{\hbar}\lambda k_B T t^2. \quad (8.51)$$

The absorption and fluorescence lineshapes then assume the Gaussian profiles:

$$\sigma_a(\omega) = (2\pi\Delta^2)^{-1/2}\exp[-(\omega - \omega_{eg})^2/2\Delta^2], \quad (8.52a)$$

$$\sigma_f(\omega) = (2\pi\Delta^2)^{-1/2}\exp[-(\omega - \omega_{eg} + 2\lambda)^2/2\Delta^2], \quad (8.52b)$$

From Eq. (8.5a) and Appendix 8B we have $\omega_{eg} = \omega_{eg}^0 + \lambda$, and the linewidth parameter is

$$\Delta^2 = \frac{2\lambda k_B T}{\hbar}. \quad (8.52c)$$

This is known as the slow modulation (static, inhomogeneous) limit. Equations (8.52) show that the fluorescence maximum is shifted by 2λ to the red compared with the absorption lineshape. This Stokes shift is readily observed in molecular photophysics [15] and in impurity spectra in crystals (i.e., the small polaron problem) [16].

In the other extreme $\kappa \gg 1$, nuclear dynamics are fast compared with the coupling strength. In this case, the $\exp(-\Lambda t)$ in the right-hand side of Eq. (8.49) vanishes very rapidly and may be ignored. At long times we further have $\Lambda t - 1 \cong \Lambda t$ and we get

$$g(t) = \hat{\Gamma}t - i\lambda t, \quad (8.53a)$$

with

$$\hat{\Gamma} \equiv \frac{\lambda k_B T}{\hbar\Lambda}, \quad (8.53b)$$

When Eq. (8.53a) is substituted into (8.42), we obtain,

$$\sigma_a(\omega) = \sigma_f(\omega) = \frac{1}{\pi}\frac{\hat{\Gamma}}{(\omega - \omega_{eg}^0)^2 + \hat{\Gamma}^2}. \quad (8.54)$$

This is known as the fast modulation (homogeneous) limit [17]. In this case, the Stokes shift vanishes since the bath motions are so fast that the radiation field simply observes an averaged two-level system. Both absorption and fluorescence lineshapes assume a simple Lorentzian form with fwhm of $2\hat{\Gamma}$. Eq. (8.47) is one of the limiting cases of the Brownian oscillator model. The spectral density together with the absorption and relaxed fluorescence lineshapes for this model will be displayed later in this chapter.

In summary, as κ is varied, the optical absorption or fluorescence lineshapes change continuously from a Gaussian with a Stokes shift of 2λ in the static limit to a Lorentzian with no Stokes shift in the fast modulation limit. For intermediate values of κ, the line cannot be simply classified as either homogeneous or inhomogeneous.

Magnetic resonance lineshapes in solids and glasses are very broad due to inhomogeneous contributions. As the temperature is raised, nuclear motions make the line homogeneous, and the linewidth narrows from $\sim\Delta$ to $\sim\Delta^2/\Lambda$. This effect, also known as motional narrowing, was discovered by Bloembergen, Purcell, and Pound [18], and is the reason why NMR lines in liquids are usually narrow, which makes high resolution studies possible.

Equations (8.52) and (8.54) represent universal behaviors that go beyond the specific form of the spectral density [Eq. (8.47)], and can be obtained directly from Eq. (8.31). Assuming slow dynamics we set $M(t) = 1$ and obtain Eqs. (8.52). For fast dynamics we assume $M(t) = \Lambda^{-1}\delta(t)$ with

$$\Lambda^{-1} = \int_0^\infty d\tau\, M(\tau), \tag{8.55}$$

which yields Eq. (8.54). More generally we can state that in the inhomogeneous limit the characteristic nuclear time scale $(1/\Lambda)$ is much longer than the observed inverse linewidth, and the converse is true for the homogeneous limit.

When the electronic system is coupled to several independent degrees of freedom then the spectral density and $g(t)$ are additive and we have

$$g(t) = \sum_j g_j(t), \tag{8.56}$$

It then follows from Eqs. (8.12) and (8.15) that $J(t)$ and the contribution of each Liouville space path R_α can be factorized into a product of independent terms. This factorization is rigorous and is independent on the cumulant expansion. Assuming that the spectral density is a sum of one slow (inhomogeneous) and one fast (homogeneous) contribution we obtain by substituting Eqs. (8.51) and (8.53) in Eq. (8.56)

$$J(t) = \exp(-i\omega_{eg}t - \hat{\Gamma}t)\exp(-\tfrac{1}{2}\Delta^2 t^2). \tag{8.57}$$

The absorption lineshape [Eq. (8.42a)] assumes a Voigt profile (a convolution of a Gaussian and a Lorentzian), and the nonlinear response function is simply given by,

$$R_1(t_3, t_2, t_1) = R_4(t_3, t_2, t_1) = \exp[-i\omega_{eg}(t_1 + t_3) - \hat{\Gamma}(t_1 + t_3)] \times \exp[-\tfrac{1}{2}\Delta^2(t_1 + t_3)^2], \tag{8.58a}$$

$$R_2(t_3, t_2, t_1) = R_3(t_3, t_2, t_1) = \exp[i\omega_{eg}(t_1 - t_3) - \hat{\Gamma}(t_1 + t_3)] \times \exp[-\tfrac{1}{2}\Delta^2(t_3 - t_1)^2]. \tag{8.58b}$$

It should be noted that when the nuclei are static, the response function can be calculated exactly without performing the cumulant expansion. The line broadening then need not necessarily be Gaussian. Setting $U(t) = U$ independent of t in Eqs. (7.24) we get

$$R_1(t_3, t_2, t_1) = R_4(t_3, t_2, t_1)$$
$$= \exp[-i\omega_{eg}(t_1 + t_3)]\left\langle \exp\left[-\frac{i}{\hbar} U(t_1 + t_3)\right]\rho_g\right\rangle, \quad (8.59a)$$

$$R_2(t_3, t_2, t_1) = R_3(t_3, t_2, t_1)$$
$$= \exp[i\omega_{eg}(t_1 - t_3)]\left\langle \exp\left[\frac{i}{\hbar} U(t_1 - t_3)\right]\rho_g\right\rangle. \quad (8.59b)$$

Equations (8.59) are not limited to the Gaussian form. It further follows from these expressions that the response function does not depend on the time interval t_2. This is to be understood since in the static limit, the only evolution is of the phase. During the t_2 period the system is in an electronic population and does not acquire a phase. In general, when the bath correlation time is finite, the system can undergo spectral diffusion during the t_2 interval. Spectral diffusion processes are responsible for phenomena such as the time-dependent Stokes shift, and will be discussed further in Chapter 13.

The central limit theorem states that the sum of a large number of independent random variables has a Gaussian distribution, provided the second moment of the individual random variables is finite [19]. A random walk is the typical example for this remarkable fundamental theorem of statistics. When the second moment of the individual random variables does not exist, it is possible to generalize the theorem and obtain other types of interesting unusual distributions. These are related to Levy walks [20]. Since the interaction with the environment is the sum of many small contributions, it is reasonable to expect, based on the central limit theorem, that the line profile should assume a Gaussian form.* However, in general, the inhomogeneously broadened line can assume any form and is not necessarily a Gaussian, and a homogeneous line need not always be a Lorentzian. It is therefore impossible (and useless) to attempt to determine the homogeneous or inhomogeneous nature of the spectral line broadening based on the lineshape alone! The form (Lorentzian vs. Gaussian) is suggestive but by no means conclusive. Nonlinear optical techniques can, however, resolve this issue unambiguously. Equations (8.59) show that R_1 and R_4 depend only on the combination $t_1 + t_3$, whereas R_2 and R_3 depend only on $t_1 - t_3$. These dependencies form the basis for the elimination of inhomogeneous broadening in the hole-burning and the photon echo techniques, which will be discussed in Chapters 10 and 11. Applications to off resonant techniques will require extending the present expression to include the variation of the transition dipole with nuclear coordinates (non-Condon effects). This will be carried out in Chapter 14.

* Molecular dynamics calculations for a solute in liquid water provide a nice illustration [J. S. Bader and D. Chandler, *Chem. Phys. Lett.* **157**, 501 (1989)].

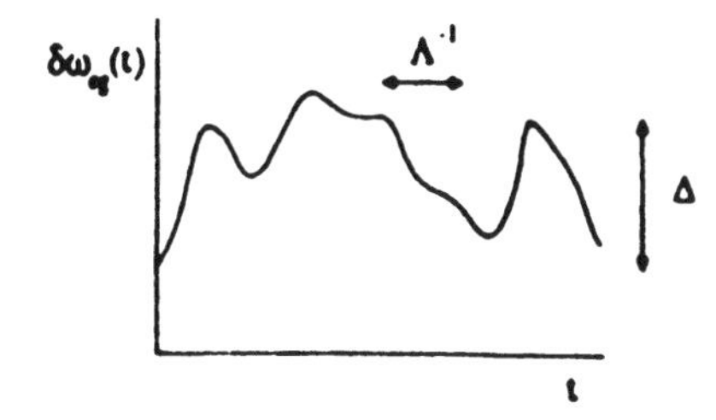

FIG. 8.2 The stochastic process $\delta\omega_{eg}(t)$ represents the time-dependent molecular electronic energy gap. Δ represents the magnitude of the fluctuations, and Λ^{-1} represents their time scale.

Stochastic models are commonly used in the modeling of line broadening resulting from coupling to a bath with a finite timescale [17, 18, 21]. The effects of the bath are accounted for by assuming that due to the random force exerted on the system by the bath, the electronic energy gap ω_{eg} becomes a stochastic function of time. In the Hamiltonian [Eq. (7.1)], we then replace ω_{eg}^0 by

$$\omega_{eg}^0 \to \omega_{eg}^0 + \delta\omega_{eg}(t)$$

where $\delta\omega_{eg}(t)$ is a random function of time (see Figure 8.2). Assuming that it represents a stochastic Gaussian Markov modulation of the electronic energy gap we have

$$\langle \delta\omega_{eg}(t) \rangle = 0 \tag{8.60a}$$

and

$$\langle \delta\omega_{eg}(t)\delta\omega_{eg}(0) \rangle = \Delta^2 \exp(-\Lambda t). \tag{8.60b}$$

Here, the angular brackets $\langle \cdots \rangle$ denote averaging over the stochastic process, Δ is the root-mean-squared amplitude of the frequency fluctuations, and Λ^{-1} is their correlation time. Equations (8.60) can be rationalized starting with our microscopic model [Eq. (7.1)] by taking the bath degrees of freedom to be classical and assuming that the bath is sufficiently large that its motions are independent of the state of the system. By switching to the interaction picture (with respect to the bath Hamiltonian), ω_{eg}^0 becomes time dependent. This choice of a stochastic Hamiltonian is based on the assumption that the bath couples mainly to the electronic degrees of freedom, so that the ground-state and the excited-state vibronic manifolds are being stochastically modulated with respect to each other, but no modulation occurs for frequencies of transitions belonging to the same electronic manifold, that is, vibrational dephasing is neglected. This is often a realistic assumption. The linear response function for this model is given by Eq. (8.12) with the neglect of the imaginary part of $g(t)$ [setting $\lambda = 0$ in Eq. (8.26)]. Obviously the model does not satisfy the fluctuation–dissipation theorem and it completely misses the Stokes shift. The reason is that it allows the bath to affect the system but does not allow the system to affect the bath. This is in contrast to the microscopic oscillator model, which consistently incorporates the system–bath interaction. The full width at half maximum Γ_0 of the absorption lineshape is displayed in Figure 8.3 versus κ [Eq. (8.50)]. It may be adequately fitted by the Padé approximant [22]

$$\Gamma_0/\Delta = \frac{2.355 + 1.76\kappa}{1 + 0.85\kappa + 0.88\kappa^2}. \tag{8.60c}$$

The stochastic model applies when the bath temperature is sufficiently high, so that the Stokes shift is much smaller than the linewidth, and may be neglected.

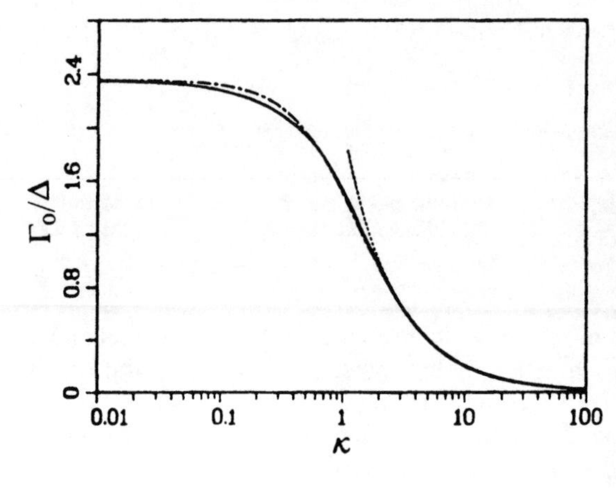

FIG. 8.3 The full-width at half-maximum (Γ_0/Δ) of the stochastic lineshape function $\sigma_a(\omega)$ vs. κ. Solid line, exact curve (calculated numerically): dot-dashed line, the Padé approximant [Eq. (8.60c)]; dotted line, the limiting form $\Gamma_0/\Delta = 2\kappa$, which holds in the fast modulation $\kappa \gg 1$ limit. [From J. Sue, Y. J. Yan, and S. Mukamel, "Raman Excitation Profiles of Polyatomic Molecules in Condensed Phases. A Stochastic Theory," *J. Chem. Phys.*, **85**, 462 (1986).]

Stochastic models are not limited to the present Gaussian model [Eq. (8.60)] and other stochastic models such as the two-state jump model [7] are also commonly used.

THE MULTIMODE BROWNIAN OSCILLATOR MODEL

So far we have used a harmonic model for nuclear motions in two ways. We first considered high frequency isolated vibrations that undergo coherent (oscillatory) motion. We then studied a continuous distribution of oscillators that results in irreversible relaxation. The linearly displaced Brownian harmonic oscillator model unifies these two pictures and interpolates between the coherent and the damped nuclear motions. It therefore provides a general and convenient way for incorporating the coupling of nuclear motions to optical transitions. The system is taken to be a two electronic-level system with a ground state $|g\rangle$ and an excited state $|e\rangle$ with some primary nuclear coordinates q_j coupled linearly to the electronic system. The Brownian oscillator Hamiltonian (also known as the Spin-Boson Hamiltonian) is given by

$$H = |g\rangle H_g \langle g| + |e\rangle H_e \langle e| + H', \tag{8.61a}$$

where H_g and H_e are given by Eqs. (8.32). The nuclear degrees of freedom are in turn coupled to a bath consisting of a set of harmonic oscillators with coordinates x_n and momenta p_n. The interaction between the primary oscillators and the nth bath oscillator is assumed to be linear with a coupling strength c_n. H' describes the bath oscillators and their coupling to the primary oscillators and is given by [13, 22, 23]*

$$H' = \sum_n \left[\frac{p_n^2}{2m_n} + \frac{m_n \omega_n^2}{2} \left(x_n - \sum_j \frac{c_{nj} q_j}{m_n \omega_n^2} \right)^2 \right]. \tag{8.61b}$$

* The choice of harmonic potentials for the primary coordinates allows the incorporation of a bath with an arbitrary spectral density and temperature. By restricting the model for dissipation it may be possible to treat anharmonic potentials [see Eq. (12.49)].

The total system is assumed to be initially at equilibrium in the ground electronic state:

$$\rho_g = |g\rangle\langle g| \exp[-\beta(H_g + H')]/\mathrm{Tr}\{\exp[-\beta(H_g + H')]\},$$

where $\beta \equiv 1/k_B T$ is the inverse temperature.

The second-order cumulant provides an exact solution for this harmonic model. The electronic energy gap is $U = \sum_j m_j \omega_j^2 d_j q_j$. We further define the correlation function of the primary coordinates

$$C_j(t) \equiv \langle q_j(t) q_j(0) \rho_g \rangle, \tag{8.62a}$$

and we have

$$C(t) = \sum_j \xi_j^2 C_j(t); \qquad \xi_j \equiv \frac{m_j \omega_j^2 d_j}{\hbar}. \tag{8.62b}$$

In analogy with Eqs. (8.18), we separate $C_j(t)$ into its real and imaginary parts

$$C_j'(t) \equiv \tfrac{1}{2}[\langle q_j(t) q_j(0) \rho_g \rangle + \langle q_j(0) q_j(t) \rho_g \rangle], \tag{8.63a}$$

$$C_j''(t) = -\tfrac{i}{2}[\langle q_j(t) q_j(0) \rho_g \rangle - \langle q_j(0) q_j(t) \rho_g \rangle]. \tag{8.63b}$$

We similarly define these quantities in the frequency domain [see Eqs. (8.21)] $\tilde{C}_j(\omega)$, $\tilde{C}_j'(\omega)$, and $\tilde{C}_j''(\omega)$. We then have

$$\tilde{C}''(\omega) = \sum_j \tilde{C}_j''(\omega); \qquad \tilde{C}_j''(\omega) = \xi_j^2 \tilde{C}_j''(\omega). \tag{8.64a}$$

The correlation function for this model can be calculated using various methods, e.g., path integral techniques, resulting in

$$\tilde{C}_j''(\omega) = \frac{\hbar}{2m_j} \frac{\omega \gamma_j(\omega)}{(\omega_j^2 + \omega \Sigma_j(\omega) - \omega^2)^2 + \omega^2 \gamma_j^2(\omega)}. \tag{8.64b}$$

Here $\gamma_j(\omega)$ and $\Sigma_j(\omega)$ represent the relaxation and level-shift respectively induced by coupling to the bath. They are related by the Kramers-Kronig relation.

$$\gamma_j(\omega) = \frac{\pi}{m_j} \sum_n \frac{c_{nj}^2}{2m_n \omega_n^2} [\delta(\omega - \omega_n) + \delta(\omega + \omega_n)], \tag{8.64c}$$

$$\Sigma_j(\omega) = -\frac{1}{\pi} \mathrm{Re} \int_{-\infty}^{\infty} d\omega' \frac{\gamma_j(\omega')}{\omega' - \omega}. \tag{8.64d}$$

This result can be alternatively obtained by the following simplified derivation, which also provides an additional physical insight. We start again with the Hamiltonian H_s (Eq. (8.61a)) but instead of modeling the bath explicitly through H' we simply assume that each mode undergoes a Brownian motion satisfying the generalized Langevin equation [24–26]:

$$m_j\ddot{q}_j(t) + m_j\omega_j^2 q_j(t) + m_j \int_{-\infty}^{t} d\tau\, \gamma_j(t-\tau)\dot{q}_j(\tau) = f_j(t) + F_j(t). \tag{8.65a}$$

Here γ_j represents a time-dependent friction (with "memory"), and $f_j(t)$ is a Gaussian stochastic random force representing the effect of the bath degrees of freedom on the jth mode. The friction and the random force satisfy the relations

$$\langle f_j(t)\rangle = 0, \tag{8.65b}$$

$$\langle f_j(t)f_k(\tau)\rangle = \delta_{jk} 2m_j k_B T\gamma_j(t-\tau). \tag{8.65c}$$

Here $\langle\cdots\rangle$ denotes an average over the stochastic variables and δ_{jk} is the Kronecker delta. Equation (8.65b) implies that the average of the random force is zero. Equation (8.65c) can be derived using the fluctuation–dissipation theorem. $F_j(t)$ is an external driving force, which is a given function of time and does not depend on the random force. It represents a coupling term $F_j(t)q_j$, which is added to the Hamiltonian. In the absence of the driving force, the Langevin equation describes how a classical oscillator with an arbitrary initial distribution in phase space relaxes to thermal equilibrium. Standard methods for calculating for the time-dependent distribution function are available [23–25].

In applying the classical Langevin equation to real quantum systems [27], we face the following problem: The quantum correlation function $C_j(t)$ is complex and satisfies $C_j(-t) = C_j^*(t)$. The classical correlation function is real and even $C_j(-t) = C_j(t)$. We clearly cannot directly equate the two. To overcome this difficulty, we shall adopt the following strategy: We use Eq. (8.65a) to calculate $\tilde{C}_j''(\omega)$ and then use Eqs. (8.27) and (8.28) to construct a consistent expression for $g(t)$, which satisfies the fluctuation–dissipation theorem, and all the symmetries of the quantum correlation function. We start our calculation by looking at the ensemble average $\langle q_j(t)\rangle$. Since the Langevin equation is linear, $\langle q_j(t)\rangle$ satisfies the same equation without the f_j factor (note that $\langle f_j(t)\rangle = 0$). Equation (8.65a) can then be solved by making a Fourier transform of both sides. We define a Fourier transform

$$\langle\tilde{q}_j(\omega)\rangle \equiv \int_{-\infty}^{\infty} dt\, \exp(i\omega t)\langle q_j(t)\rangle,$$

and similarly we define $\tilde{\gamma}_j(\omega)$ and $\tilde{F}_j(\omega)$. Equation (8.65a) then yields

$$\langle\tilde{q}_j(\omega)\rangle = \alpha_j(\omega)\tilde{F}_j(\omega), \tag{8.66a}$$

with the susceptibility

$$\alpha_j(\omega) = \frac{1}{m_j}\frac{1}{-\omega^2 + \omega_j^2 - i\omega\gamma_j(\omega)}. \tag{8.66b}$$

Using our response function we have

$$\alpha_j(\omega) = \int_0^\infty dt \left\langle -\frac{i}{\hbar}[q_j(t), q_j(0)]\rho_g \right\rangle \exp(i\omega t). \tag{8.66c}$$

Comparing Eq. (8.66c) with Eq. (8.63b) immediately yields

$$\tilde{C}_j''(\omega) = i\frac{\hbar}{2}[\alpha_j(\omega) - \alpha_j^*(\omega)].$$

This together with Eq. (8.66b) results in Eq. (8.64b).

The variation of $\tilde{\gamma}_j(\omega)$ with frequency reflects the timescales of the thermal motions of the bath. Assuming that these motions are very fast compared with the oscillator motion, we can set $\tilde{\gamma}_j(\omega) = \gamma_j$ independent of ω and $\Sigma_j = 0$. For this case (also known as ohmic dissipation) [14, 23], we can perform the Fourier transform analytically using contour integration, resulting in a closed expression for the correlation function in the time domain. We then have

$$\left.\begin{aligned} g(t) &= \sum_j g_j(t) \\ g_j(t) &= \xi_j^2 \int_0^t d\tau_1 \int_0^{\tau_1} d\tau_2\, C_j'(\tau_2) + i\xi_j^2 \int_0^t d\tau_1 \int_0^{\tau_1} d\tau_2\, C_j''(\tau_2). \end{aligned}\right\} \tag{8.67a}$$

$\tilde{C}''(\omega)$ has two poles in the lower ω plane at $\omega = -i\phi_j, -i\phi_j'$ with

$$\phi_j = \frac{\gamma_j}{2} + i\zeta_j,$$

$$\phi_j' = \frac{\gamma_j}{2} - i\zeta_j,$$

$$\zeta_j = \sqrt{\omega_j^2 - \gamma_j^2/4}.$$

We then have

$$C_j''(t) = -\frac{\hbar}{2m_j}\cdot\frac{1}{\zeta_j}\exp(-\gamma_j|t|/2)\sin(\zeta_j t), \tag{8.67b}$$

$$\begin{aligned} C_j'(t) = {} & \frac{\hbar}{4m_j\zeta_j}[\coth(i\phi_j'\hbar\beta/2)\exp(-\phi_j' t) - \coth(i\phi_j\hbar\beta/2)\exp(-\phi_j t)] \\ & - \frac{2\gamma_j}{m_j\beta}\sum_{n=1}^\infty \frac{\nu_n \exp(-\nu_n t)}{(\omega_j^2 + \nu_n^2)^2 - \gamma_j^2\nu_n^2}, \end{aligned} \tag{8.67c}$$

with the Matsubara frequencies [14]

$$\nu_n \equiv \frac{2\pi}{\hbar\beta}n.$$

When Eqs. (8.67b) and (8.67c) are substituted in Eq. (8.67a), the double-time integration can be trivially carried out since the time dependence is exponential, and we obtain a closed form expression for $g_j(t)$.

Depending on the relative magnitude of ω_j and γ_j, the Brownian oscillator model can represent various types of nuclear motions. Typical forms of $\tilde{C}_j''(\omega)$ and $C_j''(t)$ are displayed in Figure 8.5. The limiting cases for this model are summarized below.

High frequency (underdamped) modes

For high frequency modes, the relaxation rate (friction) γ_j is typically small compared with the frequency ω_j. In the absence of friction $\gamma_j = 0$, the oscillator experiences a coherent motion, and we have

$$\left.\begin{aligned} C_j''(t) &= -\frac{\hbar}{2m_j\omega_j}\sin(\omega_j t), \\ C_j'(t) &= \frac{\hbar}{2m_j\omega_j}\coth(\beta\hbar\omega_j/2)\cos(\omega_j t), \end{aligned}\right\} \tag{8.68}$$

we then recover Eqs. (8.39).

When γ_j is finite but small $\gamma_j < 2\omega_j$, then ζ_j is real and the correlation function shows damped oscillations. Equation (8.67c) simplifies considerably at high temperatures, where we have

$$\begin{aligned} C'(t) &= \frac{1}{2im_j\beta\zeta_j\omega_j^2}[\phi_j'\exp(-\phi_j t) - \phi_j\exp(-\phi_j' t)] \\ &= \frac{1}{m_j\beta\omega_j^2}\left[\cos(\zeta_j t) + \left(\frac{\gamma_j}{2\xi_j}\right)\sin(\zeta_j t)\right]\exp(-\gamma_j t/2); \qquad k_B T \gg \hbar\gamma_j. \end{aligned} \tag{8.69}$$

Strongly overdamped modes

Macroscopic solvation coordinates have usually very low characteristic frequencies compared with the friction, and they are strongly overdamped. For $\gamma_j \gg 2\omega_j$ we have

$$\tilde{C}_j''(\omega) = 2\lambda_j'\frac{\omega\Lambda_j}{\omega^2 + \Lambda_j^2}, \tag{8.70a}$$

where

$$\Lambda_j \equiv \omega_j^2/\gamma_j,$$

and

$$\lambda_j' \equiv \frac{\hbar}{2m_j\omega_j^2}.$$

We then have

$$C_j''(t) = -i\lambda_j'\Lambda_j\exp(-\Lambda_j t), \tag{8.70b}$$

$$C_j'(t) = \lambda_j'\Lambda_j\cot(\beta\hbar\Lambda_j/2)\exp(-\Lambda_j t) + \frac{4\Lambda_j\lambda_j'}{\hbar\beta}\sum_{n=1}^{\infty}\frac{\nu_n\exp(-\nu_n t)}{\nu_n^2 - \Lambda_j^2}, \tag{8.70c}$$

where ν_n were introduced in Eq. (8.67c). This case was analyzed previously starting with the spectral density Eq. (8.47), and an explicit expression for $g_j(t)$ is given in Eq. (8.48), with $\Lambda = \Lambda_j$ and

$$\lambda_j = \xi_j^2 \frac{\hbar}{2m_j\omega_j^2} = \xi_j^2 \lambda_j'.$$

If in addition the temperature is high so that $k_B T \gg \hbar\Lambda_j$, we further have

$$C_j''(t) = \lambda_j' \Lambda_j \exp(-\Lambda_j t), \tag{8.71a}$$

and

$$C_j'(t) = \frac{2\lambda_j' k_B T}{\hbar} \exp(-\Lambda_j t). \tag{8.71b}$$

In this case $g_j(t)$ is given by Eq. (8.49).

The various limiting cases of the Brownian oscillator model are summarized in Figure 8.4. The spectral densities and the corresponding correlation functions for an underdamped, intermediate, and overdamped Brownian oscillator are displayed in Figure 8.5. The absorption and relaxed fluorescence spectra for

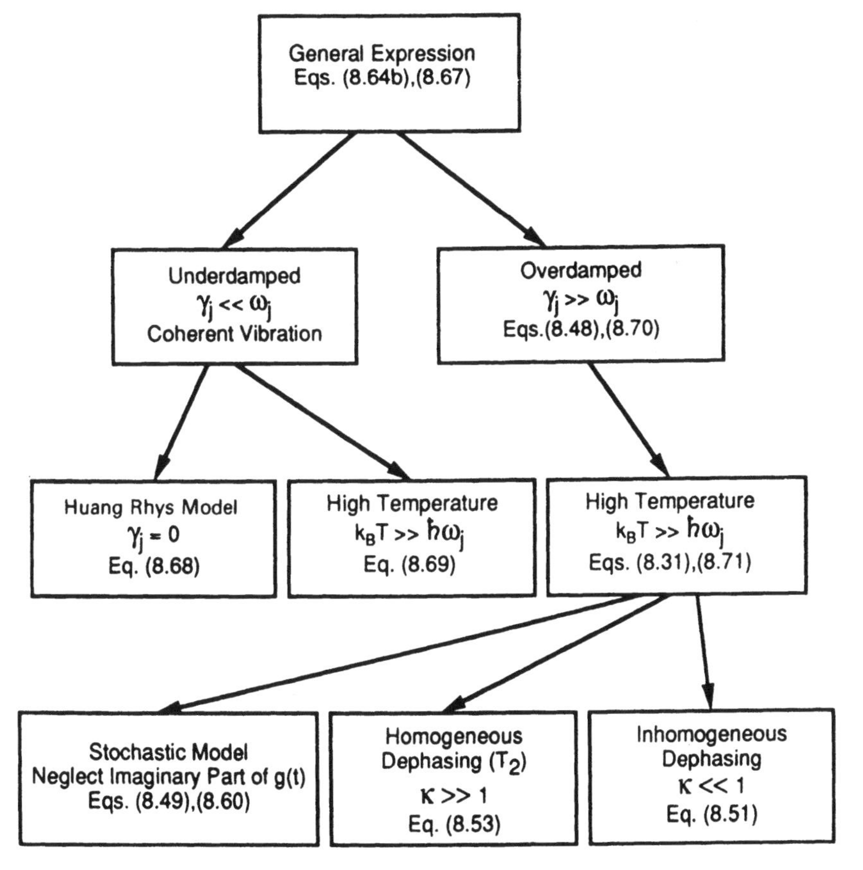

FIG. 8.4 Limiting cases of the Brownian oscillator model.

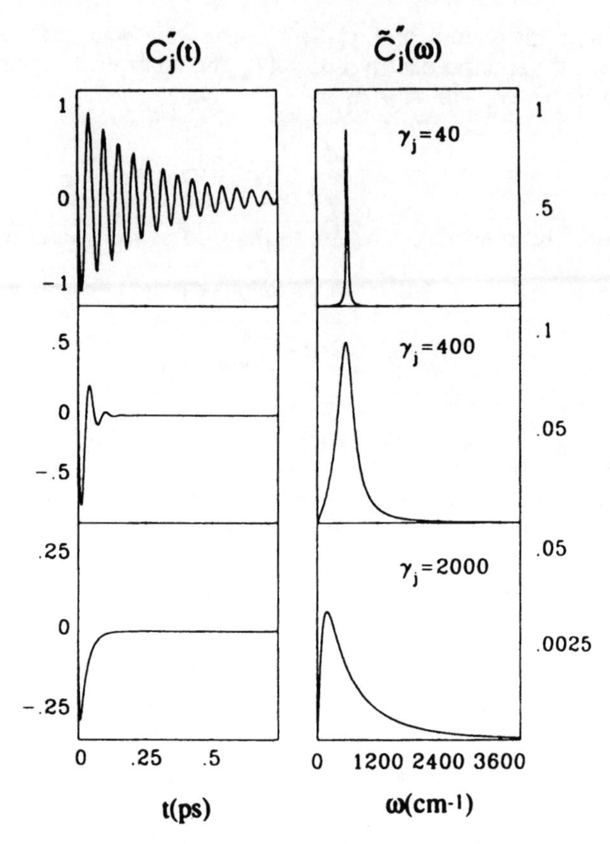

FIG. 8.5 Spectral densities (right column) and correlation functions (left column) for the Brownian oscillator model $\omega_j = 600\ \text{cm}^{-1}$, $d_j = 4$. From top to bottom, γ_j is increased as indicated and the oscillator is underdamped, intermediate, and overdamped respectively. The vertical scale is in arbitrary units and shows the relative magnitudes of the various curves.

these limiting cases are shown in Figures 8.6, 8.7, and 8.8 respectively. The interpolation of the absorption lineshape for the overdamped oscillator between the Lorentzian and Gaussian limits is illustrated in Figure 8.9. The Stokes shift is negligible in this case, and the fluorescence coincides with the absorption.

The Brownian oscillator model is used in many disciplines to represent collective modes subject to friction. These range from the dynamics of earthquakes, noise in electric circuits, vibrations of mechanical structures, and the dynamics of chemical and nuclear reactions. It is also identical to the Drude oscillator model introduced earlier for optical polarizarion; in that case ω_j is the optical frequency and typically $\gamma_j \ll \omega_j$ (underdamped motion). Here we use it to represent nuclear motions, and both overdamped and underdamped cases are of physical interest. The model can interpolate continuously between high frequency and overdamped modes and provides a convenient means for parameterizing experimental data as well as classical simulations. This parameterization may assist in identifying the relevant collective coordinates that dominate the optical response.

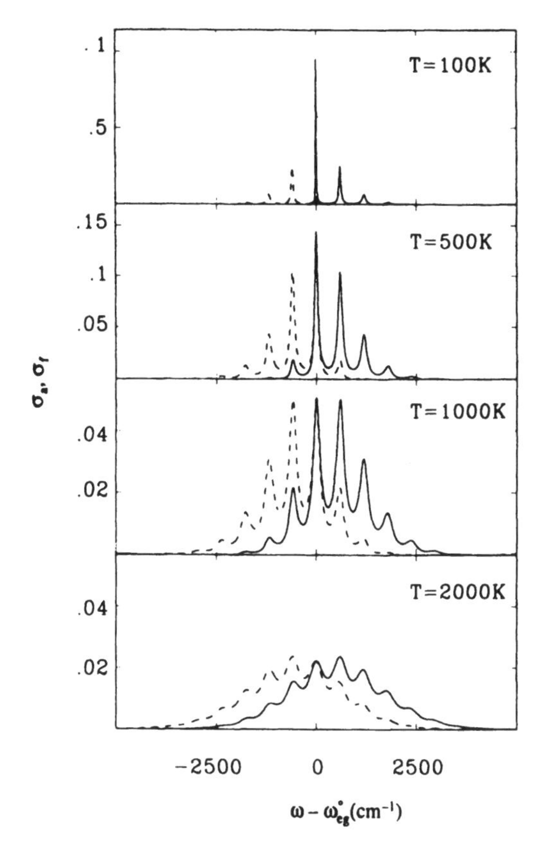

FIG. 8.6 The linear absorption (solid curve) and relaxed fluorescence (dashed curve) lineshapes of a two electronic level system coupled to a single underdamped Brownian oscillator mode at different temperatures, as indicated. The frequency, the damping, and the dimensionless displacement are taken to be $\omega_j = 600$ [cm^{-1}], $\gamma_j = 40$ [cm^{-1}], and $d_j = 2.0$.

The Brownian oscillator can always be represented as a continuous distribution of coherent oscillators with no damping. Depending on the physical picture we wish to adopt, we may prefer to represent the system either in terms of a few Brownian oscillators or many ordinary oscillators (with no damping). Mathematically, the two models are identical. By using the cumulant expansion we are in practice mapping our system onto a collection of harmonic oscillators. In the overdamped limit, the dynamics does not depend separately on ω_j and γ_j but rather on the combination $\Lambda_j \equiv \omega_j^2/\gamma_j$. In the studies of dyes in solution Λ_j is the inverse timescale of solvent relaxation and λ_j is the solvent reorganization frequency. In Chapter 12 we shall consider the dynamics of the Brownian oscillator in phase space using the Wigner representation. We shall then show that in the overdamped limit the momentum relaxes rapidly, and is maintained at thermal equilibrium at all times. We can then drop it out of the problem, and describe the dynamics using Gaussian wavepackets in coordinate space. This behavior is described by the Smoluchowski equation.

Equations (8.67) can be generalized in various ways. It is possible to keep the frequency dependence of the friction $\gamma_j(\omega)$, which implies that the Langevin equation has a memory. Later in this chapter we shall develop the inhomogeneous cumulant expansion, which incorporates inhomogeneous broadening of arbitrary shape into the cumulant expansion.

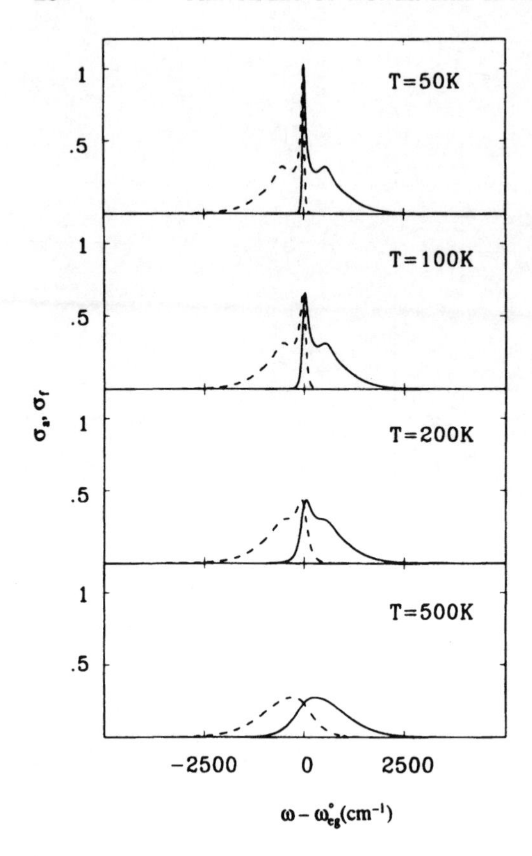

FIG. 8.7 Same as Figure 8.6 except that the damping is intermediate $\gamma_j = 400\ \text{cm}^{-1}$.

ADDITIONAL APPLICATIONS OF THE CUMULANT EXPANSION

Dielectric fluctuations in a polar medium [28–31]

Many optical measurements are conducted on molecules in a polar environment (e.g., solvents, glasses, polymers, crystals, and proteins). In this case it is possible to use an electrostatic model for the coupling of the molecule to the medium. By treating that coupling perturbatively, we can then relate the optical properties to the spectrum of equilibrium dielectric fluctuations of the medium, which in turn can be calculated using its frequency and wavevector-dependent dielectric function. Since the latter quantity can be measured or computed independently, we can predict the outcome of optical measurements using this basic information. This type of modeling has been successfully applied for predicting the time-dependent Stokes shift of solvated dyes, which will be discussed in Chapter 13. In addition, it is commonly used in the calculation of transport properties of a charged particle in the medium (the polaron model), and in the calculation of electron transfer rates [15, 16].

The model is constructed as follows: We consider a two-level molecule that has a different electronic charge distribution in the ground and in the excited state, embedded in a polar medium. The charge distribution creates a fixed

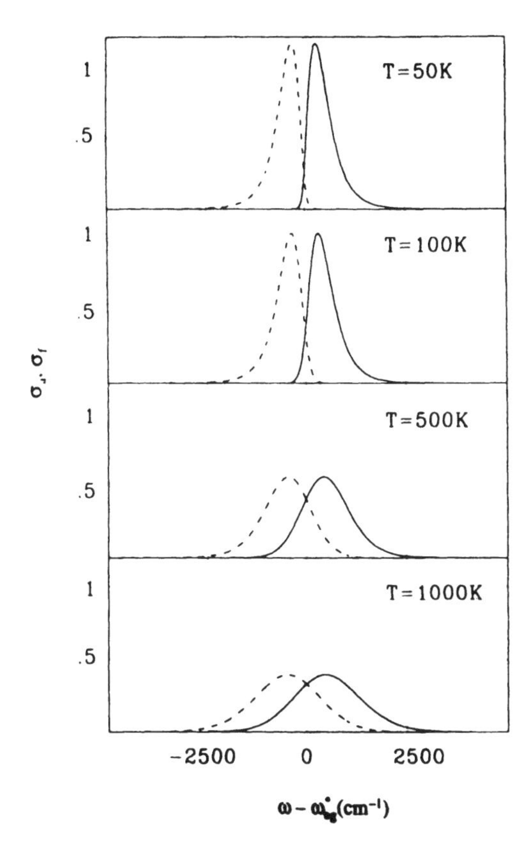

FIG. 8.8 Same as Figure 8.6 except that the oscillator is overdamped with $\gamma_j = 2000\ \text{cm}^{-1}$.

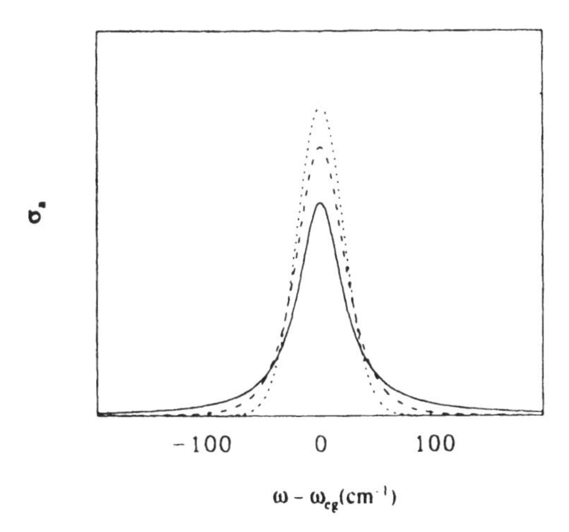

FIG. 8.9 Linear absorption lineshapes for the overdamped Brownian oscillator model. The various curves are for different values of the ratio $\kappa = \Delta/\Lambda$. $\kappa = 0.1$, 1, 10 from top to bottom, changing from the static (Gaussian) limit to the fast modulation (Lorentzian) limit. The full width at half maximum is $\Gamma = 50\ \text{cm}^{-1}$, and the temperature $T = 1000°\text{K}$.

electrostatic field denoted $D_j(\mathbf{r})$ with $j = g, e$. Let us denote the dipole moment of the solvent per unit volume at point $\mathbf{r}$ by $\hat{P}(\mathbf{r})$. The electrostatic interaction with the solvent is then given by

$$W_g = -\int d\mathbf{r}\, D_g(\mathbf{r})\cdot\hat{P}(\mathbf{r}),$$

$$W_e = -\int d\mathbf{r}\, D_e(\mathbf{r})\cdot[\hat{P}(\mathbf{r}) - \langle\hat{P}(\mathbf{r})\rangle],$$

$D_j(\mathbf{r})$ is a given function, whereas $\hat{P}(\mathbf{r})$ is an operator that depends on all the solvent degrees of freedom (translations and rotations). At equilibrium, and in the absence of the solute the expectation value of $\hat{P}(\mathbf{r})$ vanishes since the medium is isotropic.

The electronic energy gap (solvation) coordinate [Eq. (8.5b)] is now given by

$$U = -\int d\mathbf{r}\, D_{eg}(\mathbf{r})\cdot[\hat{P}(\mathbf{r}) - \langle\hat{P}(\mathbf{r})\rangle],$$

where $D_{eg}(\mathbf{r})$ is the difference in the electric fields at point $\mathbf{r}$ created by the system's charge distribution when it is in the excited electronic state $|e\rangle$ and in the ground electronic state $|g\rangle$

$$D_{eg}(\mathbf{r}) \equiv D_e(\mathbf{r}) - D_g(\mathbf{r}).$$

In the present model, U represents a collective solvation coordinate that depends on all solvent degrees of freedom. The solvation coordinate is a sum of a large number of small contributions from individual solvent molecules. Using the central limit theorem we then expect its statistical properties to be Gaussian so that the second-order cumulant expansion should be directly applicable.

The optical response for this model can be calculated using Eqs. (8.31), where the parameters λ and $M(t)$ may be expanded perturbatively in the solvent–solute interaction. To lowest order (linear response) they may be related to the longitudinal dielectric fluctuations of the solvent, which are described by its polarization correlation function. The calculation is carried out in Appendix 8C, and we finally get [32]

$$\hbar\lambda = \frac{1}{64\pi^4}\int d\mathbf{k}\, |D_{eg}(\mathbf{k})|^2\left[\frac{1}{\varepsilon(\mathbf{k},\infty)} - \frac{1}{\varepsilon(\mathbf{k},0)}\right], \tag{8.72a}$$

$$M(t) = \frac{1}{2\pi i}\,\frac{\displaystyle\int d\mathbf{k}\, |D_{eg}(\mathbf{k})|^2 \int_{-\infty}^{\infty}\frac{d\omega}{\omega}\exp(i\omega t)\left[\frac{1}{\varepsilon(\mathbf{k},\omega)} - \frac{1}{\varepsilon(\mathbf{k},\infty)}\right]}{\displaystyle\int d\mathbf{k}\, |D_{eg}(\mathbf{k})|^2\left[\frac{1}{\varepsilon(\mathbf{k},\infty)} - \frac{1}{\varepsilon(\mathbf{k},0)}\right]}. \tag{8.72b}$$

Equations (8.72) apply to a solvent with an arbitrary dielectric function $\varepsilon(\mathbf{k},\omega)$, in the high temperature limit.

Usually the complete wavevector and frequency-dependent dielectric function is not known, although some theoretical effort was focused on this problem. In most cases only its long wavelength limit $\varepsilon(\mathbf{k} = 0, \omega) \equiv \varepsilon(\omega)$ is readily available. Assuming that $\varepsilon(\mathbf{k}, \omega)$ is independent of $\mathbf{k}$, which amounts to neglecting the local solvent structure, Eqs. (8.72) reduce to the results of the *dielectric continuum approximation* [33, 34]:

$$\hbar\lambda = \frac{1}{64\pi^4}\int d\mathbf{k}\,|D_{eg}(\mathbf{k})|^2[1/\varepsilon_\infty - 1/\varepsilon_0] = \frac{1}{64\pi^4}\int d\mathbf{r}\,|D_{eg}(\mathbf{r})|^2[1/\varepsilon_\infty - 1/\varepsilon_0], \tag{8.73a}$$

$$M(t) = \frac{1}{2\pi i}\int_{-\infty}^{\infty}\frac{d\omega}{\omega}\exp(i\omega t)\frac{1/\varepsilon(\omega) - 1/\varepsilon_\infty}{1/\varepsilon_\infty - 1/\varepsilon_0}. \tag{8.73b}$$

Here ε_0 is the static ($\omega = 0$), and ε_∞ is the high-frequency (optical) value of $\varepsilon(\omega)$. Equations (8.72) or (8.73) together with Eq. (8.31) provide closed expressions for the optical response functions and relate them to the solvent dielectric function. They constitute a theoretical basis for the study of solvent dielectric fluctuation effects in linear and nonlinear optical spectroscopy.

We shall now derive explicit expressions for λ and $M(t)$ using a simple model. For calculating the Stokes shift parameter λ we assume a particle with a permanent dipole moment that changes upon electronic excitation from μ_g to μ_e. Since the point dipole model does not hold at short distances and leads to an unphysical divergence of λ, we shall assume a continuous charge distribution over a sphere with radius R, which has the dipole moment μ_g or μ_e. Equation (8.73a) then yields

$$\lambda = R^{-3}|\mu_e - \mu_g|^2(1/\varepsilon_\infty - 1/\varepsilon_0). \tag{8.74}$$

The solvent dynamics are contained in $M(t)$, which is the normalized correlation function of the solvation coordinate [30]. The dielectric function $\varepsilon(\omega)$ is well documented for a large number of solvents. The simplest form is the Debye model, which is characterized by a single relaxation time, the Debye timescale τ_D:

$$\varepsilon(\omega) = \varepsilon_\infty + \frac{\varepsilon_0 - \varepsilon_\infty}{1 + i\omega\tau_D}. \tag{8.75a}$$

Substituting Eq. (8.75a) into Eq. (8.73b), we obtain

$$M(t) = \exp(-t/\tau_L), \tag{8.75b}$$

where $\tau_L \equiv \tau_D(\varepsilon_\infty/\varepsilon_0)$ is the longitudinal solvent relaxation timescale.

Typical values of $(\varepsilon_\infty/\varepsilon_0)$ for polar solvents are ~ 0.1, so that the longitudinal timescales is much faster than the Debye relaxation time τ_D. Other forms for $\varepsilon(\omega)$ (Cole–Cole, Cole–Davidson, the Williams Watts, etc.) contain a multitude of timescales.

The Anderson–Talman model and the density expansion

When an atom in the gas phase interacts with foreign atoms at low pressure, its coupling to the surrounding medium can be described in terms of uncorrelated binary collisions. The theory of collisional broadening (also known as pressure broadening) of spectral lines is widely used for the analysis of high resolution spectra in the gas phase [35–39].

Consider a two-level atom interacting with the radiation field and with a gas of structureless bath particles (perturbers). A typical example is Na atoms in Argon gas. We denote the atomic levels $|g\rangle$ and $|e\rangle$ with frequency ω_{eg}. In the Anderson–Talman model we neglect the interperturber interactions. The exact linear response function of the system can then be calculated as follows: we first consider a system consisting of the atom plus a single perturber in a volume Ω. Neglecting the center-of-mass motion, we have

$$H_g = -\frac{\hbar^2}{2\mu}\frac{\partial^2}{\partial q^2} + W_g(q)$$

and

$$H_e = -\frac{\hbar^2}{2\mu}\frac{\partial^2}{\partial q^2} + W_e(q),$$

where q is the absorber–perturber relative coordinate, and μ is their reduced mass. We assume that the transition dipole moment does not depend on internuclear separation, and use the Condon approximation form for the correlation functions $J(t)$ [Eq. (8.1)]. We thus write

$$J_1(t) = \exp(-i\omega_{eg}^0 t)\left[1 - \frac{1}{\Omega} g_1(t)\right],$$

with

$$g_1(\tau) = \Omega\,\mathrm{Tr}\left\{\left[1 - \exp\left(\frac{i}{\hbar}H_e\tau\right)\exp\left(-\frac{i}{\hbar}H_g\tau\right)\right]\rho_g\right\}, \tag{8.76}$$

and ρ_g is the canonical distribution of initial states. Here the subscript 1 signifies that these are single perturber quantities. Equation (8.76) can be interpreted as follows: the perturbation can occur only when the absorber and the perturber are interacting. Since the interaction range is microscopic (of the order of a few Å), the fraction of time in which they interact is $O(1/\Omega)$. The Ω factor in the definition of $g(\tau)$ simply cancels that $1/\Omega$ dependence, so that $g_1(\tau)$ is independent on Ω. The single perturber linear absorption in this case has the form

$$\sigma_1(\omega) = \frac{1}{\pi}\,\mathrm{Re}\int_0^\infty d\tau\left[1 - \frac{1}{\Omega}g_1(\tau)\right]\exp[-i(\omega - \omega_{eg}^0)\tau].$$

We next define the *phase shift* in a binary collision

$$\phi(t) \equiv \int_0^t d\tau\, U(\tau), \tag{8.77a}$$

where $U(\tau)$ is a classical function of time. We then have [see Eq. (7.9a)]

$$g_1(t) = \langle[1 - \cos\phi(t)]\rho_g\rangle - i\langle[\sin\phi(t)\rho_g\rangle. \tag{8.77b}$$

In a macroscopic system consisting of N noninteracting perturbers in the volume Ω, the contributions of the various perturbers are independent so that

$$J(t) = \exp(-i\omega_{eg}^{0}t)\left[1 - \frac{1}{\Omega}g_1(t)\right]^N.$$

In the thermodynamic limit $N \to \infty$, $\Omega \to \infty$ and a fixed perturber density $N/\Omega \equiv \rho_0$ we then have

$$J(t) = \exp[-i\omega_{eg}^{0}t - \rho_0 g_1(t)], \tag{8.78}$$

which finally yields for the absorption lineshape*

$$\sigma_a(\omega) = \frac{1}{\pi}\,\mathrm{Re}\int_0^\infty d\tau\,\exp[-i(\omega - \omega_{eg}^{0})\tau - \rho_0 g_1(\tau)]. \tag{8.79}$$

This exactly solvable model thus provides physical motivation for carrying out the cumulant expansion. In the gas phase lineshapes are usually very narrow (typical pressure broadening is ~20 MHz/Torr), compared with the duration of a collision of ~0.1 ps. The homogeneous limit thus usually holds. In this case it is known as the impact limit of the binary collision approximation.

If we include the interactions among perturbers, the response function can no longer be calculated exactly. However, it is possible to use the cluster expansion where the expansion parameter is the perturber density ρ_0

$$g(t) = \rho_0 g_1(t) + \rho_0^2 g_2(t) + \cdots$$

where the contributions $g_2, g_3, \ldots$ correspond to two-perturber, three-perturber spectra, etc., which scale with higher powers of the density ρ_0^2, ρ_0^3, etc. Performing a cluster expansion to lowest order in perturber density, the absorption lineshape $\sigma_a(\omega)$ is expressed in terms of the single perturber quantity $g_1(t)$. The single perturber dynamics is the only microscopic information relevant for the absorption lineshape at low densities. This expansion can be also modified to apply for spectra of a solute in the liquid phase, by incorporating the liquid pair distribution function [40, 41].

THE INHOMOGENEOUS CUMULANT EXPANSION AND SEMICLASSICAL SIMULATIONS

We shall now outline a numerical procedure for implementing the cumulant expansion in a microscopic semiclassical simulation. Since the cumulant expansion in the inhomogeneous limit can predict only a Gaussian lineshape, we shall extend it by introducing the inhomogeneous cumulant expansion [42, 43], which provides for an arbitrary inhomogeneous distribution of transition frequencies $[W(\omega_{eg})]$, while simultaneously treating the response of the bath to the system.

* The exponential relatioship between $g_1(t)$ and $J(t)$ is reminiscent of the relationship between the free energy and the partition function [11]. The reason is the same: for noninteracting particles, $g(t)$ (or the free energy) is additive whereas $J(t)$ (or the partition function) assumes the form of a product of independent contributions.

In the inhomogeneous cumulant expansion, we divide the nuclear degrees of freedom into a fast and a slow group. Due to the slow modes the phase space of the system is divided into slowly interconverting regions. For each configuration of the slow modes, the fast degrees of freedom are treated via the cumulant expansion. The resulting expression is subsequently averaged over the static (inhomogeneous) distribution of configurations. This provides an exact treatment of the slow modes. We shall denote the static degrees of freedom collectively by Γ, and their distribution by $W(\Gamma)$, with $\int W(\Gamma)\,d\Gamma = 1$.*

We shall first perform the cumulant expansion for a given Γ. All quantities can then be expressed in terms of the ensemble averaged electronic energy gap

$$\omega_{eg}(\Gamma) = \omega_{eg}^{0} + \frac{1}{\hbar}\,\mathrm{Tr}_0[(W_e - W_g)\rho_g], \tag{8.80}$$

and the correlation function

$$C(t;\Gamma) = \frac{1}{\hbar^2}\,\mathrm{Tr}_0[\exp(iH_g t)U\exp(-iH_g t)U\rho_g]. \tag{8.81}$$

As an example, for the multimode Brownian oscillator model, Γ can include the parameters ω_j, γ_j, and λ_j. We further introduce the line broadening function $g(t;\Gamma)$:

$$g(t;\Gamma) \equiv \int_0^t d\tau_2 \int_0^{\tau_2} d\tau_1\, C(\tau_1;\Gamma). \tag{8.82}$$

After performing the cumulant expansion we carry out the Γ integration. For the linear response we have

$$J(t) = \int d\Gamma\, W(\Gamma)\exp[-i\omega_{eg}(\Gamma)t - g(t;\Gamma)], \tag{8.83}$$

whereas for the nonlinear response we obtain

$$R_1(t_3,t_2,t_1) = \int d\Gamma\, W(\Gamma)\exp[-i\omega_{eg}(\Gamma)(t_1+t_3)]$$
$$\times \exp[-g^*(t_3;\Gamma) - g(t_1;\Gamma) - f_+(t_3,t_2,t_1;\Gamma)], \tag{8.84a}$$

$$R_2(t_3,t_2,t_1) = \int d\Gamma\, W(\Gamma)\exp[-i\omega_{eg}(\Gamma)(t_3-t_1)]$$
$$\times \exp[-g^*(t_3;\Gamma) - g^*(t_1;\Gamma) - f_+^*(t_3,t_2,t_1;\Gamma)], \tag{8.84b}$$

$$R_3(t_3,t_2,t_1) = \int d\Gamma\, W(\Gamma)\exp[-i\omega_{eg}(\Gamma)(t_3-t_1)]$$
$$\times \exp[-g(t_3;\Gamma) - g^*(t_1;\Gamma) + f_-^*(t_3,t_2,t_1;\Gamma)], \tag{8.84c}$$

* The distribution of Γ can also be discrete. An important example is a two-well oscillator model for the bath where Γ can assume two values. This model is frequently used in the analysis of spectral lineshapes in glasses [P. W. Anderson, B. I. Halperin, and C. M. Varma, *Phil. Mag.* **25**, 1 (1972)].

$$R_4(t_3, t_2, t_1) = \int d\Gamma\, W(\Gamma) \exp[-i\omega_{eg}(\Gamma)(t_1 + t_3)]$$
$$\times \exp[-g(t_3; \Gamma) - g(t_1; \Gamma) - f_-(t_3, t_2, t_1; \Gamma)]. \quad (8.84d)$$

where

$$f_-(t_3, t_2, t_1; \Gamma) \equiv g(t_2; \Gamma) - g(t_2 + t_3; \Gamma) - g(t_1 + t_2; \Gamma) + g(t_1 + t_2 + t_3; \Gamma), \quad (8.85a)$$

$$f_+(t_3, t_2, t_1; \Gamma) \equiv g^*(t_2; \Gamma) - g^*(t_2 + t_3; \Gamma) - g(t_1 + t_2; \Gamma) + g(t_1 + t_2 + t_3; \Gamma). \quad (8.85b)$$

In these expressions, homogeneous broadening enters through $\omega_{eg}(\Gamma)$ and $g(t; \Gamma)$ and the Γ integration takes inhomogeneous broadening into account. The important point is that the Γ averaging is carried out at the end; we first exponentiate, then average. This is, of course, the rigorous way to perform any static averaging.

We now present a semiclassical procedure for calculating optical lineshapes via the inhomogeneous cumulant expansion. We need to calculate the classical correlation function:

$$C_c(t) = \frac{1}{\hbar^2} \int d\mathbf{q}\, d\mathbf{p}\, U(\mathbf{q}[t]) U(\mathbf{q}) \rho_g(\mathbf{q}, \mathbf{p}), \quad (8.86a)$$

where ρ_g is the classical ground state density operator,

$$\rho_g(\mathbf{q}, \mathbf{p}) = \exp[-\beta H_g(\mathbf{q}, \mathbf{p})] \Big/ \int d\mathbf{q}\, d\mathbf{p} \exp[-\beta H_g(\mathbf{q}, \mathbf{p})], \quad (8.86b)$$

and H_g is the classical ground state Hamiltonian. $\mathbf{q}[t]$ is the coordinate of the phase point $(\mathbf{q}, \mathbf{p})$ propagated forward classically for a time t. The phase space integration is over the fast degrees of freedom.

A convenient and numerically effective method for computing classical correlation functions, first proposed by Einstein [44], is known as the Wiener–Khinchin theorem (Appendix 8D). The first step is to generate initial conditions $(\mathbf{q}, \mathbf{p})$ from the classical Boltzmann distribution. This can be done through a variety of methods (e.g., molecular dynamics equilibration or Monte Carlo equilibration). We then run classical trajectories using these initial conditions and calculate $U(t)$ along each trajectory for the time interval $[-\tau, \tau]$. The next step is to calculate the Fourier transform of $U(t)$, $\tilde{U}(\omega)$ [see Eqs. (5.29)]. The classical correlation function is finally given in the frequency domain

$$\tilde{C}_c(\omega) = \frac{1}{2\tau\hbar^2} |\tilde{U}(\omega)|^2. \quad (8.87)$$

Making use of Eqs. (8.24a) in the high temperature limit $\beta\hbar\omega/2 \ll 1$ (which is the condition for the classical approximation to hold), we have

$$\tilde{C}'_c(\omega) = \frac{1}{2\tau\hbar^2} |\tilde{U}(\omega)|^2, \quad (8.88a)$$

$$\tilde{C}''_c(\omega) = \frac{1}{2\tau\hbar^2} \frac{\hbar\omega}{2k_B T} |\tilde{U}(\omega)|^2. \quad (8.88b)$$

In Appendix 8E we show that in the time domain, the classical correlation and response functions are related by

$$C_c''(t) = \frac{1}{k_B T} \frac{d}{dt} C_c(t). \tag{8.89}$$

Equation (8.88b) can be alternately derived by a Fourier transformation of Eq. (8.89).

In using the classical correlation function to approximate a quantum correlation function, we encounter precisely the same problem we faced in our discussion of the Brownian oscillator. The only difference is that for the Brownian oscillator the classical dynamics is described by the classical Langevin equation whereas here the information is obtained using classical trajectories. The problem is that the quantum correlation function $C(t)$ is complex and satisfies $C(-t) = C^*(t)$ whereas classical correlation functions are real and even $C_c(-t) = C_c(t)$. A procedure that overcomes this difficulty is to use the classical calculation to compute either the real part $C'(t)$ or the imaginary part $C''(t)$ of the correlation function, and then use the fluctuation–dissipation theorem to calculate the other part. For the harmonic oscillator model C_c'' is identical to its quantum counterpart C'' at all temperatures. In fact, in this case C_c'' is independent of temperature, and thus the high temperature limit always applies. It then makes sense to assume

$$\tilde{C}''(\omega) \cong \tilde{C}_c''(\omega). \tag{8.90}$$

By combining Eqs. (8.88) and (8.90) with Eqs. (8.24) we finally obtain

$$\tilde{C}'(\omega) = \coth\left(\frac{\hbar\omega}{2k_B T}\right)\tilde{C}_c''(\omega), \tag{8.91a}$$

$$\tilde{C}''(\omega) = \tilde{C}_c''(\omega). \tag{8.91b}$$

Other approaches for constructing consistent semiclassical approximations for quantum correlation functions have been proposed [45–47]. Standard molecular dynamics procedures can be used in these calculations [48].

Our semiclassical procedure, which is exact for a harmonic system, thus goes as follows: The classical simulation yields $\tilde{C}_c(\omega)$ [Eq. (8.87)]. Equations (8.91) are then used to calculate $\tilde{C}'(\omega)$ and $\tilde{C}''(\omega)$, which can be substituted in Eqs. (8.26)–(8.28) to obtain the quantities that enter into the cumulant expression. We then use Eqs. (8.83) and (8.84) to find the trajectory's contribution to the desired optical response function and repeat for many independent trajectories until satisfactory convergence is reached.

In applying the inhomogeneous cumulant expansion we must partition phase space into noninterconverting (or very slowly interconverting) regions, and calculate the quantum correlation function $C(t)$ within each region. This can be done by running a trajectory for a time that is short compared to the interconversion times, and long compared to the fast motions (e.g., typical vibrational period). Trajectories with different initial conditions then represent different realizations of Γ. If the interconversion process is too rapid to meet these

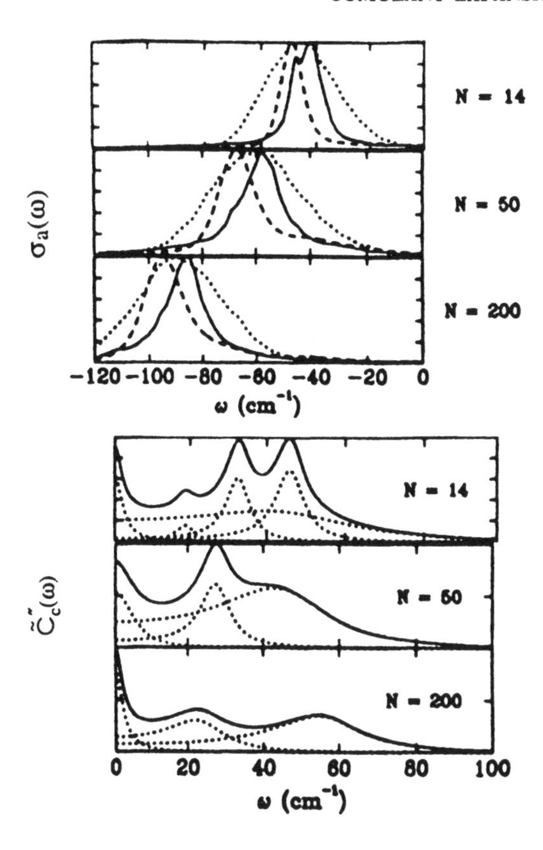

FIG. 8.10 Calculated absorption spectra of benzene–Ar_N clusters. Solid line, phase averaging formula. Dashed line, second-order cumulant approximation. Dotted line, static approximation. ω is plotted relative to the gas phase 0–0 absorption frequency ω_{eg}^0. [Adopted from L. E. Fried and S. Mukamel, "Simulation of Nonlinear Electronic Spectroscopy in the Condensed Phase," *Adv. Chem. Phys.* **84**, 435 (1993).]

FIG. 8.11 Calculated spectral densities for benzene–Ar_N clusters. Solid line, the best fit of $C_c(\omega)$ to a three-mode Brownian oscillator model. Dotted lines, the individual contributions to $C_c(\omega)$ from each Brownian oscillator. [Adopted from L. E. Fried and S. Mukamel, "Simulation of Nonlinear Electronic Spectroscopy in the Condensed Phase," *Adv. Chem. Phys.* **84**, 435 (1993).]

requirements, the configurations should not be treated as producing an inhomogeneous distribution.

In Figure 8.10 we compare the cumulant, the static and the phase averaging results obtained using classical simulations of a benzene molecule embedded in an argon cluster. In Figure 8.11 we show how the resulting spectral densities can be decomposed into a few Brownian oscillators.

One of the main advantages of the Brownian oscillator model is that it gives a simple picture of a few damped collective coordinates. Also, the analytic expressions for the response functions can easily be used to predict the outcome of a variety of nonlinear spectroscopies. Comparison of the parameter values obtained from different experiments should provide a valuable check on the applicability of the Brownian oscillator model. Brownian oscillator parameters can be extracted from classical simulations that are justified at sufficiently high temperatures. For harmonic systems, the classical procedure is exact at all temperatures. To incorporate inhomogeneous broadening we simply assume that the oscillator parameters carry a dependence on initial condition: $\omega_j = \omega_j(\Gamma)$, $\gamma_j = \gamma_j(\Gamma)$, $\lambda_j = \lambda_j(\Gamma)$, $\omega_{eg} = \omega_{eg}(\Gamma)$. $W(\Gamma)$ in Eqs. (8.83) and (8.84) then becomes $W(\omega_j, \gamma_j, \lambda_j, \omega_{eg})$, i.e., the density of oscillators having these parameters. Ordinary inhomogeneous broadening is commonly introduced by assuming a distribution of the electronic transition frequencies $W(\omega_{eg})$. This inhomogeneous distribution of oscillators provides a more general line broadening mechanism. In the present formulation, the transition frequency ω_{eg} can be correlated with the underlying nuclear dynamics. The conventional use of $W(\omega_{eg})$ assumes that the transition frequency is uncorrelated with nuclear dynamics.

OPTICAL SUSCEPTIBILITIES OF A MULTILEVEL SYSTEM INTERACTING WITH A MEDIUM WITH AN ARBITRARY TIMESCALE

Earlier in this chapter we analyzed the optical response of an electronic two level system coupled to a Brownian oscillator representing nuclear dynamics. In the strongly overdamped limit [Eq. (8.49)] this provides a simple model for electronic dephasing that properly takes into account the coupling to a bath with an arbitrary timescale, and therefore interpolates continuously between the homogeneous and the inhomogeneous limits of line broadening. We shall now calculate the frequency-domain optical susceptibilities $\chi^{(1)}$ and $\chi^{(3)}$ for this model.

Let us consider first the linear response function. We start by defining the auxiliary functions

$$J_n(t) \equiv \exp[-g(t)][1 - \exp(-\Lambda t)]^n, \tag{8.92}$$

and in the frequency domain

$$J_n(\omega) \equiv -i \int_0^\infty dt \exp(i\omega t - \gamma t/2) J_n(t). \tag{8.93}$$

Note that our usual spectral density is $J(\omega) = J_0(\omega)$.

It is possible to derive a continued fraction expression for $J(\omega)$ [49]. Using Eq. (8.93), we obtain the following recurrence relations:

$$i(\Delta^2/\Lambda - i\lambda)J_1(\omega) = 1 - (\omega - \lambda + i\gamma/2)J_0(\omega),$$

$$(\Delta^2/\Lambda - i\lambda)J_{n+1}(\omega) = n\Lambda J_{n-1}(\omega) + i[\omega - \lambda + i(n\Lambda + \gamma/2)]J_n(\omega), \qquad \text{for } n > 0. \tag{8.94}$$

Solving these equations, results in the continued fraction form for $J(\omega) \equiv J_0(\omega)$:

$$J(\omega) = \cfrac{1}{\omega - \lambda + i\gamma/2 - \cfrac{\Delta^2 - i\lambda\Lambda}{\omega - \lambda + i(\Lambda + \gamma/2) - \cfrac{2(\Delta^2 - i\lambda\Lambda)}{\omega - \lambda + i(2\Lambda + \gamma/2) - \cdots}}} \tag{8.95}$$

This expression is particularly suitable for numerical computations.

An alternative useful expression for $J_n(\omega)$ may be derived directly from Eqs. (8.92) by the substitution $y = 1 - \exp(-\Lambda t)$. We then get

$$J_n(\omega) = \frac{n!}{i\Lambda(\frac{1}{2}(\hat{\Gamma} + \gamma)/\Lambda - i\lambda/\Lambda - i\omega/\Lambda)_{n+1}} \times M(n + 1, \tfrac{1}{2}(\hat{\Gamma} + \gamma)/\Lambda - i\lambda/\Lambda - i\omega/\Lambda + n + 1, z). \tag{8.96}$$

$M(a, b, x)$ is the confluent hypergeometric function [50],

$$M(a, b, z) \equiv \frac{\Gamma(b)}{\Gamma(b-a)\Gamma(a)} \int_0^1 dt\, e^{zt}\, t^{a-1}(1-t)^{b-a-1} = \sum_{m=0}^{\infty} \frac{(a)_m z^m}{m!(b)_m}, \tag{8.97a}$$

with $\hat{\Gamma} \equiv 2\Delta^2/\Lambda$, z is defined by

$$z \equiv (\Delta/\Lambda)^2 - i\lambda/\Lambda, \tag{8.97b}$$

and

$$(b)_m \equiv b(b+1)\cdots(b+m-1), \qquad (b)_0 = 1. \tag{8.97c}$$

$\Gamma(x)$ represents the gamma function with $\Gamma(m+x) = (x)_m\Gamma(x)$. In particular we have for $n = 0$

$$J_0(\omega) = \frac{1}{\omega + \lambda + i(\hat{\Gamma} + \gamma)/2} \sum_{m=0}^{\infty} \frac{[(\Delta/\Lambda)^2 - i\lambda/\Lambda]^m}{(\frac{1}{2}(\hat{\Gamma} + \gamma)/\Lambda - i\lambda/\Lambda - i\omega/\Lambda + 1)_m}, \tag{8.98}$$

which is equivalent to Eq. (5.34b) together with Eq. (8.12b) and Eq. (8.49).

Let us assume that in addition to the overdamped mode, the system has a vibronic manifold (see Figure 7.1) where a, c denote the ground state vibronic manifold and b, d denote the excited state manifold. We then have

$$J(t) = J^{\text{I}}(t)J^{\text{II}}(t) \tag{8.99}$$

where $J^{\text{I}}(t)$ [Eq. (6.17)] represents the vibronic manifold and $J^{\text{II}}(t)$ [Eq. (8.12b)] together with Eq. (8.49) represents the overdamped mode. Combining these we get

$$J(\omega) = \sum_{a,b} P(a)\mu_{ab}\mu_{ba}J_0(\omega - \omega_{ba}). \tag{8.100}$$

$\chi^{(1)}$ is then obtained by substituting this in Eq. (5.51).

We next turn to the calculation of $\chi^{(3)}$. One important and extremely useful property of the Liouville space representation is that when the Hamiltonian consists of a sum of contributions of different degrees of freedom, then each of the four terms contributing to $S^{(3)}$ can be factorized into a product of terms corresponding to each contribution. For the present model, we therefore have

$$R_\alpha(t_3, t_2, t_1) = R_\alpha^{\text{I}}(t_3, t_2, t_1)R_\alpha^{\text{II}}(t_3, t_2, t_1), \qquad \alpha = 1, \ldots, 4. \tag{8.101}$$

R_α^{I} represent the contribution of the multilevel system and were given by Eq. (6.19) with $\Gamma_{vv'} = 0$. R_α^{II} represent the contribution of the bath mode and are given by Eqs. (8.15) together with (8.49).

The evaluation of the frequency-domain susceptibility $\chi^{(3)}$ requires a triple Fourier transform. The necessary integrations can be readily carried out in the factorization approximation [Eq. (6.64)] whereby $R_\alpha(t_3, t_2, t_1)$ factorize into a product of three functions of t_1, t_2, and t_3, respectively. In that case we simply need three single integrations, and $\chi^{(3)}$ assumes the form [Eq. (6.23)]. This is not the case, however, for the present model. Upon the substitution of Eq. (8.49) in

(8.16) we get

$$f_-(t_3, t_2, t_1) = z \exp(-\Lambda t_2)[1 - \exp(-\Lambda t_1)][1 - \exp(-\Lambda t_3)], \quad (8.102a)$$

$$f_+(t_3, t_2, t_1) = z \exp(-\Lambda t_2)[z^*/z - \exp(-\Lambda t_1)][1 - \exp(-\Lambda t_3)]. \quad (8.102b)$$

We next expand the nonlinear response functions in powers of $\exp(-\Lambda t_2)$. This will be done by a Taylor expansion of Eqs. (8.15) in $f_\pm$, i.e.,

$$\exp(f_\pm) = 1 + f_\pm + 1/2(f_\pm)^2 + \cdots.$$

Substituting this in Eq. (8.101), the nonlinear response function becomes

$$R_1(t_3, t_2, t_1) = \sum_{n=0}^{\infty} \frac{(-z)^n}{n!} \exp(-n\Lambda t_2) J'_n(t_1) J^*_n(t_3)$$
$$\times \sum_{a,b,c,d} P(a)\mu_{ab}\mu_{bc}\mu_{cd}\mu_{da} \exp[-i\omega_{dc}t_3 - i\omega_{db}t_2 - i\omega_{da}t_1), \quad (8.103a)$$

$$R_2(t_3, t_2, t_1) = \sum_{n=0}^{\infty} \frac{z^{*n}}{n!} \exp(-n\Lambda t_2) J'^*_n(t_1) J^*_n(t_3)$$
$$\times \sum_{a,b,c,d} P(a)\mu_{ab}\mu_{bc}\mu_{cd}\mu_{da} \exp[-i\omega_{dc}t_3 - i\omega_{db}t_2 - i\omega_{ab}t_1), \quad (8.103b)$$

$$R_3(t_3, t_2, t_1) = \sum_{n=0}^{\infty} \frac{z^{*n}}{n!} \exp(-n\Lambda t_2) J^*_n(t_1) J_n(t_3)$$
$$\times \sum_{a,b,c,d} P(a)\mu_{ab}\mu_{bc}\mu_{cd}\mu_{da} \exp[-i\omega_{dc}t_3 - i\omega_{ac}t_2 - i\omega_{ab}t_1), \quad (8.103c)$$

$$R_4(t_3, t_2, t_1) = \sum_{n=0}^{\infty} \frac{(-z)^n}{n!} \exp(-n\Lambda t_2) J_n(t_1) J_n(t_3)$$
$$\times \sum_{a,b,c,d} P(a)\mu_{ab}\mu_{bc}\mu_{cd}\mu_{da} \exp[-i\omega_{ba}t_3 - i\omega_{ca}t_2 - i\omega_{da}t_1). \quad (8.103d)$$

Here

$$J'_n(t) \equiv \exp[-g(t)][z^*/z - \exp(-\Lambda t)]^n = J_n(t) + \sum_{k=1}^{n} \binom{n}{k} (z^*/z - 1)^k J_{n-k}(t),$$
$$n = 0, 1, 2, \ldots. \quad (8.104)$$

Equations (8.103) are particularly useful for switching to the frequency domain since each term in the sum over n is factorized as a product of a function of t_1, t_2, t_3. The triple Fourier transformation [Eq. (5.38b)] can then be carried out, resulting in [49]

$$R_1(\omega_j + \omega_k + \omega_q, \omega_j + \omega_k, \omega_j) = -\sum_{n=0}^{\infty} \sum_{a,b,c,d} P(a)\mu_{ab}\mu_{bc}\mu_{cd}\mu_{da}$$
$$\times \frac{(-z)^n}{n!} \frac{J^*_n(\omega_{dc} - \omega_j - \omega_k - \omega_q) J'_n(\omega_j - \omega_{da})}{\omega_j + \omega_k - \omega_{db} + i(n\Lambda + \gamma)},$$
$$(8.105a)$$

$$R_2(\omega_j + \omega_k + \omega_q, \omega_j + \omega_k, \omega_j) = \sum_{n=0}^{\infty} \sum_{a,b,c,d} P(a)\mu_{ab}\mu_{bc}\mu_{cd}\mu_{da} \times \frac{z^{*n}}{n!} \frac{J_n^*(\omega_{dc} - \omega_j - \omega_k - \omega_q)J_n'^*(\omega_{ab} - \omega_j)}{\omega_j + \omega_k - \omega_{db} + i(n\Lambda + \gamma)}, \tag{8.105b}$$

$$R_3(\omega_j + \omega_k + \omega_q, \omega_j + \omega_k, \omega_j) = -\sum_{n=0}^{\infty} \sum_{a,b,c,d} P(a)\mu_{ab}\mu_{bc}\mu_{cd}\mu_{da} \times \frac{z^{*n}}{n!} \frac{J_n(\omega_j + \omega_k + \omega_q - \omega_{dc})J_n^*(\omega_{ab} - \omega_j)}{\omega_j + \omega_k - \omega_{ac} + i(n\Lambda + \gamma')}, \tag{8.105c}$$

$$R_4(\omega_j + \omega_k + \omega_q, \omega_j + \omega_k, \omega_j) = \sum_{n=0}^{\infty} \sum_{a,b,c,d} P(a)\mu_{ab}\mu_{bc}\mu_{cd}\mu_{da} \times \frac{(-1)^n z^n}{n!} \frac{J_n(\omega_j + \omega_k + \omega_q - \omega_{ba})J_n(\omega_j - \omega_{da})}{\omega_j + \omega_k - \omega_{ca} + i(n\Lambda + \gamma')}, \tag{8.105d}$$

Here $J_n(\omega)$ were defined in Eq. (8.93) and $J_n'(\omega)$ are given by

$$J_n'(\omega) \equiv -i \int_0^{\infty} dt \exp(i\omega t - \gamma t/2) J_n'(t)$$
$$= J_n(\omega) + \sum_{k=1}^{n} \binom{n}{k} (z^*/z - 1)^k J_{n-k}(\omega). \tag{8.106}$$

$\chi^{(3)}$ is finally given by substituting these in Eqs. (5.38a) and (5.54).

The small parameter that controls the n summations in Eqs. (8.103) or (8.105) is $1/\Lambda$, and the nth term contains a $1/n\Lambda$ factor. When the bath correlation time is sufficiently short, we can retain only the $n = 0$ terms (the $n\Lambda$ denominators with $n = 0$ are much smaller than those with finite n). In this case we recover the factorization approximation introduced in Eq. (6.64) (homogeneous broadening). In the other extreme (slow damping $\Lambda \to 0$) we have inhomogeneous broadening. The n summations then converge very slowly. Rather than performing this summation we can alternatively simply average the result over the static distribution of ω_{eg} [see Eqs. (6.65)].

APPENDIX 8A

The Cumulant Expansion for $F(\tau_1, \tau_2, \tau_3, \tau_4)$

Using the time ordered exponentials (see Chapter 2) and the Condon approximation (page 210) we have

$$V_{ge}(\tau) = \exp(iH_g\tau)\exp(-iH_e\tau)$$
$$= \exp(-\omega_{eg}\tau)\exp_+\left[-i\int_0^\tau d\tau'\, U(\tau')\right] \quad (8.A.1)$$

and

$$V_{eg}(\tau) = \exp(iH_e\tau)\exp(-iH_g\tau) = \exp(i\omega_{eg}\tau)\exp_-\left[i\int_0^\tau d\tau'\, U(\tau')\right]. \quad (8.A.2)$$

Upon the substitution in Eq. (7.13) we get

$$F(\tau_1, \tau_2, \tau_3, \tau_4) = \exp[-i\omega_{eg}(\tau_1 - \tau_2 + \tau_3 - \tau_4)]$$
$$\times \left\langle \exp_+\left[-i\int_0^{\tau_1} d\tau_1'\, U(\tau_1')\right]\exp_-\left[i\int_0^{\tau_2} d\tau_2'\, U(\tau_2')\right]\right.$$
$$\left.\times \exp_+\left[-i\int_0^{\tau_3} d\tau_3'\, U(\tau_3')\right]\exp_-\left[i\int_0^{\tau_4} d\tau_4'\, U(\tau_4')\right]\right\rangle. \quad (8.A.3)$$

Note that ω_{eg} is the renormalized frequency [Eq. (8.5a)].

Expanding to second order in U we have

$$F(\tau_1, \tau_2, \tau_3, \tau_4) = \exp[-i\omega_{eg}(\tau_1 - \tau_2 + \tau_3 - \tau_4)]$$
$$\times [1 - g(\tau_1) - g^*(\tau_2) - g(\tau_3) - g^*(\tau_4) + h(\tau_1, \tau_2) - h(\tau_1, \tau_3)$$
$$+ h(\tau_2, \tau_3) + h(\tau_1, \tau_4) - h(\tau_2, \tau_4) + h(\tau_3, \tau_4) + \cdots]. \quad (8.A.4)$$

Here

$$g(\tau_1) \equiv \int_0^{\tau_1} d\tau \int_0^{\tau} d\tau' \langle U(\tau)U(\tau')\rangle \quad (8.A.5a)$$

and

$$h(\tau_1, \tau_2) \equiv \int_0^{\tau_1} d\tau \int_0^{\tau_2} d\tau' \langle U(\tau)U(\tau')\rangle. \quad (8.A.5b)$$

The time arguments in the expression for $g(\tau)$ are ordered, i.e., $\tau \geq \tau'$. In contrast, for $h(\tau_1, \tau_2)$, the τ and the τ' arguments are not ordered. However, by simple algebraic manipulations, h can be written as a combination of three g functions with different time arguments. This is illustrated in Figure 8.12.

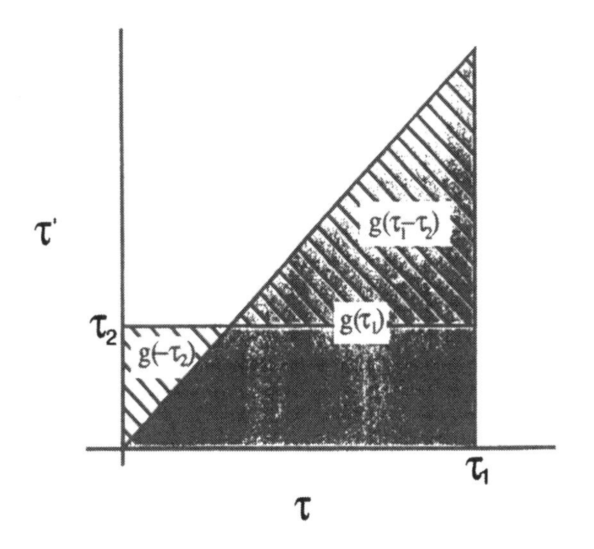

FIG. 8.12 Separation of the non-time-ordered integral Eq. (8.A.5b) into three time-ordered contributions allows us to express $h(\tau_1, \tau_2)$ in terms of the function $g(\tau)$.

We then get

$$h(\tau_1, \tau_2) = g(\tau_1) - g(\tau_1 - \tau_2) + g(-\tau_2). \tag{8.A.6}$$

Combining Eqs. (8.A.4) and (8.A.6), we get

$$\begin{aligned} F(\tau_1, \tau_2, \tau_3, \tau_4) = {} & \exp[-i\omega_{eg}(\tau_1 - \tau_2 + \tau_3 - \tau_4)] \\ & \times [1 - g(\tau_1 - \tau_2) + g(\tau_1 - \tau_3) - g(\tau_2 - \tau_3) - g(\tau_1 - \tau_4) \\ & + g(\tau_2 - \tau_4) - g(\tau_3 - \tau_4) + \cdots]. \end{aligned} \tag{8.A.7}$$

Equation (8.14) is finally obtained by making a cumulant expansion to second order, which amounts to exponentiating this expression.

APPENDIX 8B

Absorption and Emission Lineshapes: The Stokes Shift

Fluorescence emission is an important spectroscopic technique that provides complementary information to linear absorption. The relaxed fluorescence and the absorption lineshapes are given by very similar expressions. We shall now discuss the relationship between the two, for a two-electronic level system.

In a fluorescence measurement the system is prepared in the excited state. The internal and nuclear degrees of freedom usually relax very rapidly compared with the radiative lifetime. The emission lineshape can then be calculated using the Fermi Golden Rule with the initial density operator $\rho(-\infty) = |e\rangle\rho_e\langle e|$. ρ_e represents the nuclear degrees of freedom at equilibrium in the electronically excited state, and is given by the expression for ρ_g with interchanging H_g by H_e (see Figure 9.3). We thus have

$$\sigma_f(\omega) = \left[\sum_b P(b)\langle b|V^2|b\rangle\right]^{-1} \sum_{a,b} P(b)|V_{ab}|^2 \delta(\omega - \omega_{eg} - \omega_{ba}), \tag{8.B.1}$$

with

$$P(b) \equiv \langle b|\rho_e|b\rangle. \tag{8.B.2}$$

A more rigorous derivation of Eq. (8.B.1), together with a discussion of its limitations, and an extension to situations where the nuclear relaxation is not necessarily fast, as well as to time and frequency resolved fluorescence will be made in Appendix 9A.

Within the classical Condon approximation, the fluorescence lineshape can be obtained using Eqs. (8.5) and (8.9) by replacing ρ_g and ρ_e, we define

$$\omega'_{eg} = \langle (H_e - H_g)\rho_e\rangle = \omega^0_{eg} + \frac{1}{\hbar}\langle (W_e - W_g)\rho_e\rangle, \tag{8.B.3}$$

$$U' \equiv H_e - H_g - \hbar\omega'_{eg} = W_e - W_g - \frac{1}{\hbar}\langle (W_e - W_g)\rho_e\rangle, \tag{8.B.4}$$

and

$$J'(t_1) = \exp(-i\omega'_{eg}t_1)\left\langle \exp_+\left[-\frac{i}{\hbar}\varepsilon\int_0^{t_1} d\tau\, U'(\tau)\right]\rho_e\right\rangle,$$

where

$$U'(\tau) = \exp\left(\frac{i}{\hbar}H_e\tau\right)U'\exp\left(-\frac{i}{\hbar}H_e\tau\right).$$

We then get

$$\sigma_f(\omega) = \frac{1}{\pi} \mathrm{Re} \int_0^\infty dt \exp[i(\omega - \omega_{eg})t - g_f(-t)], \tag{8.B.5}$$

with

$$g_f(t) = \int_0^t d\tau_2 \int_0^{\tau_2} d\tau_1\, C_f(\tau_1), \tag{8.B.6}$$

$$C_f(\tau_1) = \frac{1}{\hbar^2} \langle U'(\tau_1) U'(0) \rho_e \rangle. \tag{8.B.7}$$

When the spectrum is evaluated to first order in the cumulant expansion, setting $C(\tau) = C'(\tau) = 0$, we get $\sigma_a(\omega) = \delta(\omega - \omega_{eg})$ and $\sigma_f(\omega) = \delta(\omega - \omega'_{eg})$; $\omega_{eg} - \omega'_{eg}$ is the shift between the absorption, centered around ω_{eg} and the emission centered around ω'_{eg}. This is known as the *Stokes shift*.

When the interaction with the solvent is weak we can evaluate the Stokes shift perturbatively. To that end we use the identity

$$\exp(-\beta H_e) = \exp(-\beta H_g) - \int_0^\beta d\alpha \exp[-(\beta - \alpha)H_g] U \exp(-\alpha H_e).$$

This identity, when iterated, generates an expansion similar to Eq. (2.58) with $H = H_e$, $H_0 = H_g$, $H' = U$, and $it = \beta$. Truncating to first order in U we obtain

$$\begin{aligned} \omega'_{eg} - \omega_{eg} &= \langle U \rho_g \rangle - \langle U \rho_e \rangle \\ &= \int_0^\beta d\alpha \sum_{a,c} \exp[-(\beta - \alpha)E_a] |U_{ac}|^2 \exp(-\alpha E_c) \\ &= \sum_{a,c} \frac{P(a) - P(c)}{\hbar \omega_{ac}} |U_{ac}|^2 = \frac{\hbar}{\pi} \int_{-\infty}^\infty \frac{\tilde{C}''(\omega)}{\omega} d\omega = 2\hbar\lambda, \end{aligned}$$

where the last identity follows from Eq. (8.28b). We have thus shown that

$$2\lambda \equiv \frac{1}{\hbar} [\langle U \rho_g \rangle - \langle U \rho_e \rangle].$$

$\langle U\rho_g \rangle (\langle U \rho_g \rangle)$ represent the spectral shift of the absorption (fluorescence) between the isolated molecule and that in a medium. Their difference, the Stokes shift, is usually more readily available experimentally (for molecules that do fluorescence!). λ is therefore a more useful parameter than $\langle U \rho_g \rangle$. In the theory of charge transfer in polar solvents, λ is called the solvent reorganization energy [15].

The Stokes shift is related to the first moment of the lineshapes. We now consider the actual lineshapes. In general, there is no simple relation between $g(t)$ and $g_f(t)$; the fluorescence and absorption lineshapes can therefore be very different. However, to second order in U we have

$$g_f(-t) = g^*(t).$$

Using the last two relations, the linear absorption and the fluorescence lineshapes are finally given by

$$\sigma_a(\omega) = \frac{1}{\pi} \text{Re} \int_0^\infty dt \exp[i(\omega - \omega_{eg})t] \exp[-g(t)], \tag{8.B.8}$$

$$\sigma_f(\omega) = \frac{1}{\pi} \text{Re} \int_0^\infty dt \exp[i(\omega - \omega_{eg} + 2\lambda)t] \exp[-g^*(t)]. \tag{8.B.9}$$

Note that $\sigma_f(\omega_{eg} - 2\lambda - \Delta\omega) = \sigma_a(\omega_{eg} + \Delta\omega)$ so that in this case the absorption and the emission lineshapes are simply related by a mirror image symmetry around $\omega_{eg}^0 \equiv \omega_{eg} - \lambda$.

APPENDIX 8C

Brownian Oscillator Parameters for a Polar Solvent

We start with the correlation function of the longitudinal polarization [51]

$$C_{pp}(\mathbf{k}, t) = \int \langle \hat{P}(\mathbf{r}, t)\cdot\hat{\mathbf{k}}\hat{P}(0, 0)\cdot\hat{\mathbf{k}}\rangle \exp(i\mathbf{k}\cdot\mathbf{r})\, d\mathbf{r}.$$

We further introduce the static polarization structure factor $S(\mathbf{k})$

$$S(\mathbf{k}) \equiv C_{pp}(\mathbf{k}, 0) = \int \langle \hat{P}(\mathbf{r}, 0)\cdot\hat{\mathbf{k}}\hat{P}(0, 0)\cdot\hat{\mathbf{k}}\rangle \exp(i\mathbf{k}\cdot\mathbf{r})d\mathbf{r},$$

We are using here the longitudinal polarization since the static displacement fields D_j are longitudinal. Otherwise, we need to consider the transverse component as well. The spatial Fourier transforms of the displacement fields are

$$D_j(\mathbf{k}) \equiv \int d\mathbf{r}\, \exp(i\mathbf{k}\mathbf{r})D_j(\mathbf{r}), \qquad j = \text{g, e and ge}.$$

Using these quantities we obtain

$$\hbar^2\Delta^2 = \frac{1}{(2\pi)^3}\int d\mathbf{k}\, |D_{eg}(\mathbf{k})|^2 S(\mathbf{k}).$$

Eq. (8.30a) gives

$$\lambda = \frac{\hbar\Delta^2}{2k_B T},$$

and the correlation funtion is given by

$$M(t) = \frac{1}{(2\pi)^3}\frac{1}{\Delta^2}\int d\mathbf{k}\, |D_{eg}(\mathbf{k})|^2 C_{pp}(\mathbf{k}, t).$$

A general relationship exists between the polarization correlation function $C_{pp}(\mathbf{k}, t)$ and the wavevector and frequency-dependent dielectric dielectric function of the solvent, $\varepsilon(\mathbf{k}, \omega)$. This relationship is based on the fluctuation–dissipation theorem. In the high temperature limit $k_B T \gg \hbar\omega$ we obtain

$$C_{pp}(\mathbf{k}, t) \cong -\frac{k_B T}{i8\pi^2}\int_{-\infty}^{\infty}\frac{d\omega}{\omega}\exp(i\omega t)\left[\frac{1}{\varepsilon(\mathbf{k},\omega)} - \frac{1}{\varepsilon(\mathbf{k},\infty)}\right].$$

The static structure factor is then given by

$$S(\mathbf{k}) \cong \frac{k_B T}{i8\pi^2}\int_{-\infty}^{\infty}\frac{d\omega}{\omega}\left[\frac{1}{\varepsilon(\mathbf{k},\infty)} - \frac{1}{\varepsilon(\mathbf{k},\omega)}\right]$$
$$= \frac{k_B T}{4\pi}\left[\frac{1}{\varepsilon(\mathbf{k},0)} - \frac{1}{\varepsilon(\mathbf{k},\infty)}\right].$$

Combining these expressions finally results in Eqs. (8.72).

APPENDIX 8D

Classical Correlation Functions and the Wiener–Khinchin Theorem

Correlation functions are important dynamic quantities whose calculation is of considerable interest. Experimental spectroscopic observables can always be expressed in terms of appropriate correlation functions, and their calculation is a key step in the interpretation of experiments.

Given two dynamic operators A and B and the equilibrium density operator of the system ρ_{eq}, we define their two time quantum mechanical correlation function

$$\langle A(t)B(0)\rangle \equiv \mathrm{Tr}\left[A \exp\left(-\frac{i}{\hbar}Ht\right)B\rho_{eq}\exp\left(\frac{i}{\hbar}Ht\right)\right], \qquad (8.D.1)$$

or making use of the Liouville operator, we get

$$\langle A(t)B(0)\rangle \equiv \mathrm{Tr}\left[A \exp\left(-\frac{i}{\hbar}\mathscr{L}t\right)B\rho_{eq}\right], \qquad (8.D.2)$$

with

$$\rho_{eq} \equiv \frac{\exp(-\beta H)}{\mathrm{Tr}\exp(-\beta H)}. \qquad (8.D.3)$$

Classically, correlation functions are calculated using a time average. The classical correlation function of two dynamic variables A and B is defined by

$$C_{AB}(t) \equiv \lim_{\tau\to\infty}\frac{1}{2\tau}\int_{-\tau}^{\tau} d\tau'\, A(t)B(t+\tau'). \qquad (8.D.4)$$

In practice, we usually measure (or calculate) $A(t)$ and $B(t)$ only during a finite (and long) time interval $[-\tau, \tau]$. We therefore define

$$A_\tau(t) = \begin{cases} A(t) & -\tau \le t \le \tau \\ 0 & \text{else.} \end{cases}$$

We next construct the following quantity

$$C_{AB}^{\tau}(t) \equiv \frac{1}{2\tau}\int_{-\tau}^{\tau} A_\tau(t+\tau')B_\tau(\tau')\, d\tau'. \qquad (8.D.5)$$

From these definitions it follows that

$$\left.\begin{aligned} C_{AB}(t) &= \lim_{\tau\to\infty} C^{\tau}_{AB}(t). \\ \tilde{A}_{\tau}(\omega) &= \int_{-\infty}^{\infty} A_{\tau}(t)\exp(-i\omega t)\,dt, \\ \tilde{B}_{\tau}(\omega) &= \int_{-\infty}^{\infty} B_{\tau}(t)\exp(-i\omega t)\,dt. \end{aligned}\right\} \tag{8.D.6}$$

Since the variables of interest $A(t)$, $B(t)$ are real we have

$$\tilde{A}_{\tau}(-\omega) = \tilde{A}^{*}_{\tau}(\omega); \qquad \tilde{B}_{\tau}(-\omega) = \tilde{B}^{*}_{\tau}(\omega).$$

We further define

$$\tilde{C}^{\tau}_{AB}(\omega) = \int_{-\infty}^{\infty} C^{\tau}_{AB}(t)\exp(-i\omega t)\,dt,$$

$$\tilde{C}_{AB}(\omega) = \int_{-\infty}^{\infty} C_{AB}(t)\exp(-i\omega t)\,dt.$$

Hence

$$\tilde{C}_{AB}(\omega) = \lim_{\tau\to\infty} \tilde{C}^{\tau}_{AB}(\omega) \tag{8.D.7}$$

Using these definitions we have

$$\begin{aligned} \tilde{C}^{\tau}_{AB}(\omega) &= \frac{1}{2\tau}\int_{-\infty}^{\infty}\int_{-\infty}^{\infty} \exp(-i\omega\tau')A_{\tau}(t)B_{\tau}(t+\tau')\,dt\,d\tau' \\ &= \frac{1}{2\tau}\int_{-\infty}^{\infty}\int_{-\infty}^{\infty} A_{\tau}(t)B_{\tau}(t+\tau')\exp(i\omega t)\exp[-i\omega(t+\tau')]\,dt\,d\tau' \\ &= \frac{1}{2\tau}\int_{-\infty}^{\infty} A_{\tau}(t)\exp(i\omega t)\left[\int_{-\infty}^{\infty} B_{\tau}(t+\tau')\exp[-i\omega(t+\tau')]\,d\tau'\right]dt \\ &= \frac{1}{2\tau}\tilde{A}_{\tau}(-\omega)\tilde{B}_{\tau}(-\omega) \end{aligned}$$

We finally get for the Fourier transform of the correlation function

$$\tilde{C}_{AB}(\omega) = \lim_{\tau\to\infty}\frac{1}{2\tau}\tilde{A}^{*}_{\tau}(\omega)\tilde{B}_{\tau}(\omega), \tag{8.D.8}$$

and when $A = B$

$$\tilde{C}_{A}(\omega) = \lim_{\tau\to\infty}\frac{1}{2\tau}|\tilde{A}_{\tau}(\omega)|^{2}. \tag{8.D.9}$$

This relation is known as the Wiener–Khinchin theorem. Using fast Fourier transform techniques [52], it provides a very effective way for calculating classical correlation functions.

APPENDIX 8E

Relation between Classical Correlation Functions and Response Functions

We start with a general response function for two operators A and B that may depend on both the coordinates and the momenta of our system

$$S_{AB}(t) \equiv -\frac{i}{\hbar}\langle [A(t), B(0)]\rho(-\infty)\rangle .$$

For optical response we have $A = B = V$ [see Eq. (5.18c)], however, the following derivation applies to a general linear response function. Using the cyclic permutation invariance of the trace, we can recast this in the form

$$S_{AB}(t) = \frac{i}{\hbar}\langle [A(t), \rho(-\infty)]B(0)\rangle . \tag{8.E.1}$$

We next switch to the Wigner representation. Making use of Eq. (3.110) we have

$$\frac{i}{\hbar}[A(t), \rho(-\infty)]_w = -\frac{2}{\hbar}A_w \sin\left(\frac{\hbar}{2}T\right)\rho_w(-\infty).$$

The T operator represents the Poisson brackets and was defined in Eq. (3.106). In the classical limit, we retain only the zeroth-order term in $\hbar$ and we have

$$\frac{i}{\hbar}[A(t), \rho(-\infty)]_w = -AT\rho_c .$$

A in the right-hand side is now the classical operator, and ρ_c is the classical equilibrium density operator

$$\rho(-\infty) = \rho_c(\mathbf{p}, \mathbf{q}) = \frac{\exp[-\beta H(\mathbf{p}, \mathbf{q})]}{\int\int d\mathbf{p}\, d\mathbf{q} \exp[-\beta H(\mathbf{p}, \mathbf{q})]},$$

where $H(\mathbf{p}, \mathbf{q})$ is the classical Hamiltonian. Since the T operator includes only first derivatives, and since ρ_c is a function of the Hamiltonian we can write

$$\mathrm{AT}\rho_c = \frac{\partial \rho_c}{\partial H}(ATH),$$

which yields

$$AT\rho_c = -\beta\rho_c(ATH) = -\beta\rho_c \frac{dA}{dt},$$

where we have used the classical Liouville equation. Combining these equations we get

$$\frac{i}{\hbar}[A(t), \rho(-\infty)]_w = \beta\rho_c(\mathbf{p}, \mathbf{q}) \frac{dA}{dt}.$$

Substituting this in Eq. (8.E.1) we have

$$S_{AB}(t) = \beta \iint d\mathbf{p}\, d\mathbf{q}\, \rho_c(\mathbf{p}, \mathbf{q}) B(\mathbf{pq}; 0) \frac{d}{dt} A(\mathbf{pq}; t),$$

which finally yields

$$-\frac{i}{\hbar}\mathrm{Tr}\{[A(t), B(0)]\rho(-\infty)\} = \frac{1}{k_B T}\frac{d}{dt}\mathrm{Tr}[A(t)B(0)\rho(-\infty)]. \tag{8.E.2}$$

Equation (8.89) follows directly from Eq. (8.E.2) by substituting $A = B = U/\hbar$.

NOTES

1. H. C. Longuet-Higgins, *Adv. Spectrosc.* **2**, 429 (1961).
2. G. Herzberg, *Spectra of Diatomic Molecules*, 2nd ed. (Van Nostrand, New York, 1950), p. 199.
3. S. Mukamel, *Phys. Rev. A* **26**, 617 (1982).
4. K. Huang and S. Rhys, *Proc. Soc. (London)* **A204**, 406 (1950).
5. S. Pekar, *Zh. Eksp. Teor. Fiz.* **20**, 510 (1950); *Usp. Fiz. Nauk* **50**, 193 (1953).
6. M. Lax, *J. Chem. Phys.* **20**, 1752 (1952).
7. R. Kubo and Y. Toyozawa, *Prog. Theoret. Phys.* **13**, 160 (1955).
8. J. J. Markham, *Rev. Mod. Phys.* **31**, 956 (1959).
9. A. A. Abrikosov, L. P. Gorkov, and I. E. Dzyaloshinski, *Methods of Quantum Field Theory in Statistical Physics* (Dover Publications, New York, 1975).
10. J. Katriel, *J. Phys. B* **3**, 1315 (1970).
11. R. Kubo, *Statistical Mechanics* (North-Holland, Amsterdam, 1971).
12. W. B. Bosma, Y. J. Yan, and S. Mukamel, *Phys. Rev. A* **42**, 6920 (1990).
13. Y. Tanimura and S. Mukamel, *Phys. Rev.* E **47**, 118 (1993).
14. U. Weiss, *Quantum Dissipative Systems* (World Scientific, Singapore, 1993).
15. R. A. Marcus and N. Sutin, *Biochim. Biophys. Acta* **811**, 265 (1985).
16. T. Holstein, *Ann. Phys. (New York)* **8**, 325 (1959); **8**, 343 (1959); *Condensed Matter Physics*, The Theodore D. Holstein Symposium, R. L. Orbach, Ed. (Springer-Verlag, Berlin, 1987).
17. R. Kubo, in *Fluctuation Relaxation and Resonance in Magnetic Systems*, D. Ter Haar, Ed. (Oliver and Boyd, Edinburgh, 1962); R. Kubo, *Adv. Chem. Phys.* **15**, 101 (1969).
18. N. Bloembergen, E. M. Purcell, and R. V. Pound, *Phys. Rev.* **73**, 679 (1948).

19. W. Feller, *An Introduction to Probability Theory and Its Applications*, Vol. **1** (Wiley, New York, 1966).
20. E. W. Montroll and M. F. Shlesinger, in *Studies in Statistical Mechanics*, J. Lebowitz and E. W. Montroll, Eds. (North-Holland, Amsterdam, 1984); M. F. Shlesinger, G. M. Zaslavsky, and J. Klafter, *Nature* (*London*) **363**, 31 (1993).
21. P. W. Anderson and P. R. Weiss, *Rev. Mod. Phys.* **25**, 269 (1953).
22. S. Mukamel, *Adv. Chem. Phys.* **70**, Part I, 165 (1988); *Annu. Rev. Phys. Chem.* **41**, 647 (1990); Y. J. Yan and S. Mukamel, *J. Chem. Phys.* **89**, 5160 (1988).
23. A. O. Caldeira and A. J. Leggett, *Physic* **121A**, 587 (1983); H. Grabert, P. Schramm, and G. L. Ingold, *Phys. Rep.* **168**, 115 (1988).
24. S. Chandrasekhar, *Rev. Mod. Phys.* **15**, 1 (1943).
25. M. C. Wang and G. E. Uhlenbeck, *Rev. Mod. Phys.* **17**, 323 (1945).
26. S. A. Adelman and J. D. Doll, *J. Chem. Phys.* **63**, 4908 (1975); S. A. Adelman, *J. Chem. Phys.* **64**, 124 (1976).
27. R. Benguria and M. Kac, *Phys. Rev. Lett.* **46**, 1 (1980).
28. P. Debye, *Polare Moleküle* (Hirzel, Leipzig) (1929); *Polare Molecules* (Chemical Catalog, New York, 1929); *Polar Molecules* (Dover, New York, 1945).
29. H. Frohlich, *Theory of Dielectrics* (Oxford University, Oxford, 1958).
30. C. J. F. Bottcher, *Theory of Electric Polarization*, Vol. I (Elsevier, Amsterdam, 1973); C. J. F. Bottcher and P. Bordewijk, *Theory of Electric Polarization*, Vol. II (Elsevier, Amsterdam, 1978).
31. P. Madden and D. Kivelson, *Adv. Chem. Phys.* **56**, 467 (1984).
32. Y. J. Yan and S. Mukamel, *J. Phys. Chem.* **93**, 6991 (1999); L. E. Fried and S. Mukamel, *J. Chem. Phys.* **93**, 932 (1990).
33. M. Maroncelli and G. R. Fleming, *J. Chem. Phys.* **89**, 5044 (1988); M. Maroncelli, *J. Mol. Liq.* **57**, 1 (1993).
34. P. F. Barbara and W. Jarzeba, *Adv. Photochem.* **15**, 1 (1990).
35. H. Margenau, *Phys. Rev.* **48**, 755 (1935); **82**, 156 (1951).
36. P. W. Anderson, *Phys. Rev.* **86**, 809 (1952); P. E. Anderson and J. D. Talman, *Bell Telephone System Tech. Publ.* **3117** (1956).
37. H. Baranger, *Phys. Rev.* **111**, 481 (1958); **112**, 855 (1958).
38. E. L. Lewis, *Phys. Rep.* **58**, 1 (1980).
39. A. Royer, *Phys. Rev. A* **6**, 1741 (1972); **7**, 1078 (1973); **22**, 1625 (1980).
40. I. Messing, B. Raz, and J. Jortner, *J. Chem. Phys.* **66**, 2239 (1977).
41. S. Mukamel, *Phys. Rev. A* **28**, 3480 (1983).
42. R. Islampour and S. Mukamel, *J. Chem. Phys.* **80**, 54870 (1984).
43. L. E. Fried and S. Mukamel, *Adv. Chem. Phys.* **84**, 435 (1993).
44. A. Einstein, *Arch. Sci. Phys. Natur.* **37**, 254 (1914) [reproduced in *IEEE ASSP Mag.* **4**, 6 (1987)].
45. P. Egelsteff, *Adv. Phys.* **11**, 203 (1962).
46. B. Berne, J. Jortner, and R. G. Gordon, *J. Chem. Phys.* **47**, 1600 (1967).
47. J. Borysow, M. Moraldi, and L. Frommhold, *Mol. Phys.* **56**, 913 (1985).
48. M. P. Allen and D. J. Tildesley, *Computer Simulations of Liquids* (Oxford Science Publications, Oxford, 1987); M. P. Allen and D. J. Tildley, Eds., *Computer Simulations in Chemical Physics*, NATO ASI Series, Vol. 397 (Kluwer, 1993).
49. S. Mukamel and Y. J. Yan, in *Recent Trends in Raman Spectroscopy*, S. B. Banerjee and S. S. Jha, Eds. (World Scientific, Singapore, 1989), p. 160.
50. A. Abramowitz and A. R. Stegum, *Handbook of Mathematical Functions* (Dover, New York, 1970).
51. M. Sparpaglione and S. Mukamel, *J. Chem. Phys.* **88**, 3263 (1988); *J. Chem. Phys.* **88**, 4300 (1988).
52. E. O. Brigham, *The Fast Fourier Transform* (Prentice-Hall, Englewood Cliffs, New Jersey, 1974).

CHAPTER 9

Fluorescence, Spontaneous-Raman, and Coherent-Raman Spectroscopy

Spontaneous light emission (SLE) spectroscopy constitutes an important class of widely used techniques. The experiment is illustrated schematically in Figure 1.1: a single laser beam (either pulsed or stationary) interacts with the system, and the emitted light is detected [1–3]. The spectral, temporal, polarization, and angular characteristics of the scattered light may then be studied. SLE measurements are routinely performed on a broad range of systems including atoms in the gas phase [4], complex molecules in solution [5], ultracold isolated molecules in supersonic beams [6], proteins [7], molecular crystals [8], and semiconductors [9].

The reader may wonder whether SLE belongs in the realm of nonlinear spectroscopy. At first glance, SLE is a *linear* technique since the scattered light intensity is proportional to the incident light intensity (provided the latter is sufficiently weak). However, in this chapter we show that the SLE process may be described using four point correlation functions of the dipole operator [10] and its calculation resembles that of the third-order nonlinear response functions $R_\alpha(t_3, t_2, t_1)$. It is therefore closely related to $\chi^{(3)}$, and can be thought of as a nonlinear spectroscopy, although the signal is linear in the incoming field intensity, and in a strict sense the technique is "linear." The reason for this apparent inconsistency is that SLE involves only one strong classical mode of the radiation field, namely the incoming field. The scattered radiation mode on the other hand has no photons initially (i.e., it is in the vacuum state) and has a single photon at the end of the process. The SLE process can be envisioned as a consequence of vacuum fluctuations of the radiation field [11]. This mode should therefore be treated quantum mechanically, and, unlike a classical field, it does not show up in the signal through its amplitude. This is why the nonlinear nature of the process is not apparent by merely looking at the power dependence of the relevant polarization on the field.

The SLE signal may sometimes be divided into a spontaneous Raman and a fluorescence component. The Raman spectrum is a direct coherent scattering process involving optical polarization, whereas fluorescence is an incoherent

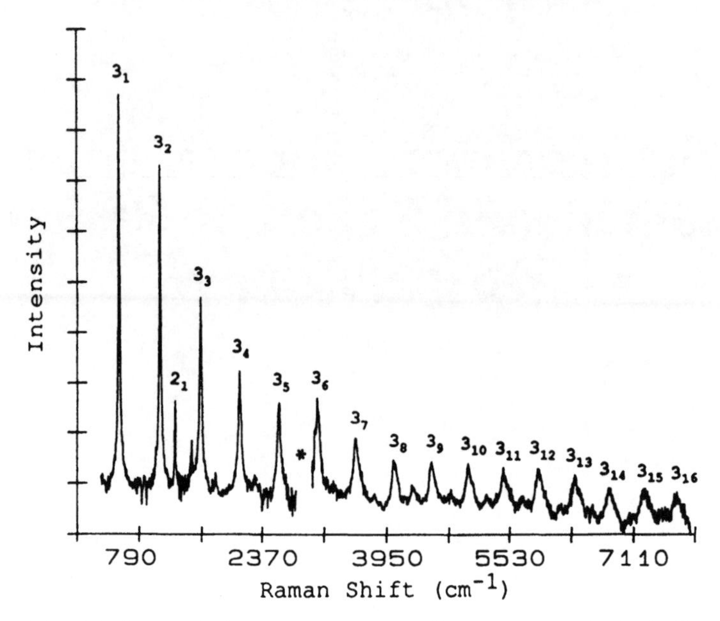

FIG. 9.1 Resonance Raman spectrum of CH_3I in hexane obtained with excitation at 266 nm. The region of interference due to subtraction of the strong hexane CH stretching band is marked with an asterisk. [From A. B. Myers and F. Markel, "Vibrational dephasing and frequency shifts for highly excited vibrations of methyl iodide in solution," *Chem. Phys.* **149**, 21 (1990).]

sequential process that can be viewed as an excitation followed by emission. Typical Raman and fluorescence spectra in solution are displayed in Figures 9.1 and 9.2. The two spectroscopies are usually treated very differently and seem to have little in common. Traditionally, Raman spectra were measured using off resonant excitation, whereas fluorescence is necessarily a resonant technique. As an electronic resonance is approached, both components may coexist. The observation of Raman lines under resonant conditions may obscure the precise distinction between the two. To avoid confusion we use the term SLE to denote the total emission spectrum. The present formulation of these processes in terms of the nonlinear response functions will enable us to treat both techniques in a unified fashion, and compare their information content. Fluorescence and Raman spectra emerge as two components of the same process rather than as disconnected processes. They provide a nice illustration for a sequential versus a coherent optical process, which appear also in other techniques, such as pump-probe spectroscopy, to be discussed in Chapter 13. Coherent Raman spectroscopy (CARS), which is closely related to spontaneous Raman, will also be discussed in this chapter.

ABSORPTION OF A QUANTUM FIELD

The calculation of SLE will be made using a quantum mechanical description of the radiation field. To familiarize ourselves with this description, we shall first show how the linear absorption lineshape [Eq. (7.9c)] can be calculated using a

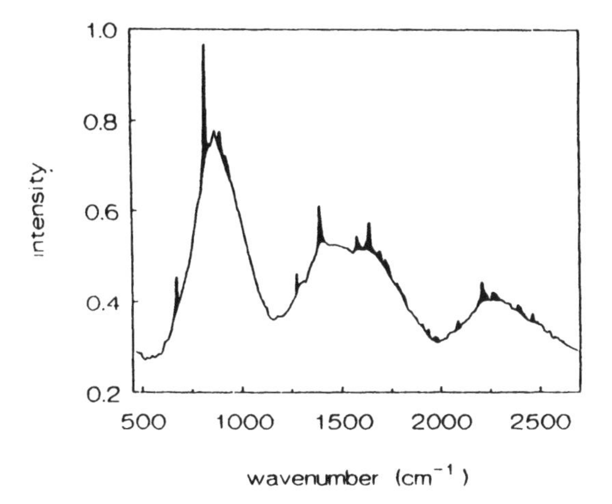

FIG. 9.2 Room temperature resonance light scattering spectra for the $S_1 \leftarrow S_0$ transition of azulene in cyclohexane at zero laser detuning from the absorption maximum. The sharp (dark) Raman features are clearly distinct from the broad underlying fluorescence bands. Solvent Raman signals were subtracted. [Reproduced from E. T. J. Nibbering, K. Duppen, and D. A. Wiersma, "Optical dephasing in solution: A line shape and resonance light scattering study of azulene in isopentane and cyclohexane," *J. Chem. Phys.* **93**, 5477 (1990).]

quantum representation of the field, and recover our previous results obtained for a classical field. The exercise will demonstrate how to work with the quantum field in the calculation of SLE.

In a linear frequency-domain absorption measurement we consider a single-mode, stationary and monochromatic quantum electromagnetic field with frequency ω_S described by the operator (see Chapter 4),

$$E(\mathbf{r}, t) = E_S \exp(i\mathbf{k}_S \cdot \mathbf{r} - i\omega_S t) + E_S^* \exp(-i\mathbf{k}_S \cdot \mathbf{r} + i\omega_S t), \tag{9.1}$$

where

$$\left. \begin{aligned} E_S &= i(2\pi\hbar\omega_S/\Omega)^{1/2} a_S, \\ E_S^* &= -i(2\pi\hbar\omega_S/\Omega)^{1/2} a_S^\dagger. \end{aligned} \right\} \tag{9.2}$$

Here $a_S^\dagger$ (a_S) is the creation (annihilation) operator for the Sth mode (see Chapter 4), and we have adopted a box normalization with Ω being the quantization volume. These operators satisfy the Bose commutation rule $[a_S, a_S^\dagger] = 1$. Strictly speaking, Eq. (9.1) corresponds to the displacement field D. However, in the present discussion we consider a very dilute sample of noninteracting molecules whereby the macroscopic polarization is very small, and to lowest order in molecular density we can set $D = E$.

The total (material + field) initial density operator is given by

$$|\rho_T(-\infty)\rangle\rangle = |\rho(-\infty)\rangle\rangle |\rho_R(-\infty)\rangle\rangle, \tag{9.3}$$

where $|\rho(-\infty)\rangle\rangle$ is the equilibrium material density operator and $|\rho_R(-\infty)\rangle\rangle$ represents the state of the incident field which has n_S photons in the incoming mode, that is,

$$|\rho_R(-\infty)\rangle\rangle = |n_S, n_S\rangle\rangle.$$

In a linear absorption experiment, we measure the steady-state rate of change of the occupation number of this mode. The operator representing this quantity is

$$\mathcal{N}_S \equiv \frac{d}{dt} a_S^\dagger a_S = \frac{i}{\hbar} \mathcal{L}_{int} a_S^\dagger a_S = \frac{i}{\hbar} [H_{int}, a_S^\dagger a_S], \tag{9.4}$$

where the radiation–matter interaction H_{int} was defined in Eq. (5.1a).

The absorption lineshape $\sigma_a(\omega_S)$ is given by the linear response to the stationary external field. We thus need to evaluate the density operator to first order in $\mathcal{L}_{int}$. We then have

$$\sigma_a(\omega_S) \propto \langle\langle \mathcal{N}_S | \rho_T^{(1)} \rangle\rangle.$$

Substituting Eq. (5.5) and eliminating some numerical constants, we obtain

$$\sigma_a(\omega_S) \propto (-i)^2 \int_0^\infty dt_1 \langle\langle a_S^\dagger a_S | \mathcal{L}_{int}(t) \mathcal{G}(t_1) \mathcal{L}_{int}(t - t_1) | \rho_T(-\infty) \rangle\rangle. \tag{9.5}$$

Note that the trace in this expression is over the material as well as the field degrees of freedom.

We next substitute Eq. (9.1) in the dipole interaction term (Eq. 5.1b) and partition it into its positive and negative frequency components:

$$H_{int}(t) \equiv V^+ \exp(-i\mathbf{k}_S \cdot \mathbf{r} + i\omega_S t) + V^- \exp(i\mathbf{k}_S \cdot \mathbf{r} - i\omega_S t), \tag{9.6}$$

with $V^+ \equiv V a_S^\dagger$ and $V^- \equiv V^* a_S$. We further define the corresponding Liouville space operators $\mathcal{V}^+$ and $\mathcal{V}^-$. It can be easily verified that the only nonvanishing terms in Eq. (9.5) will contain one V^+ and one V^- factor. Terms with two V^+ or two V^- factors will vanish once the trace over the field is taken

$$\sigma_a(\omega_S) \propto -i\langle\langle a_S^\dagger a_S | \mathcal{V}^- \mathcal{G}(\omega) \mathcal{V}^+ | \rho_T(-\infty) \rangle\rangle$$
$$+ i\langle\langle a_S^\dagger a_S | \mathcal{V}^+ \mathcal{G}(\omega) \mathcal{V}^- | \rho_T(-\infty) \rangle\rangle.$$

By acting with the last $\mathcal{V}$ factor to the left we get

$$a_S^\dagger a_S \mathcal{V}^- = V^-,$$

$$a_S^\dagger a_S \mathcal{V}^+ = V^+.$$

Performing the trace over the radiation field degrees of freedom we obtain

$$\sigma_a(\omega_S) \propto \text{Im}\langle\langle V | \mathcal{G}(\omega) \mathcal{V} | \rho(-\infty) \rangle\rangle.$$

Now the trace is over the material degrees of freedom alone since the radiation mode was already traced out. This agrees with Eq. (7.9c) together with Eq. (6.10), derived using a classical field.

SPONTANEOUS LIGHT EMISSION SPECTROSCOPY

In an SLE experiment, the electromagnetic field may be decomposed into two modes:

$$E(\mathbf{r}, t) = \sum_{j=\mathrm{L,S}} [E_j(t) \exp(i\mathbf{k}_j \cdot \mathbf{r} - i\omega_j t) + E_j^*(t) \exp(-i\mathbf{k}_j \cdot \mathbf{r} + i\omega_j t)]. \tag{9.7}$$

The incoming field $j = \mathrm{L}$ will be treated classically; hence its temporal envelope $E_{\mathrm{L}}(t)$ is a classical function of time. The scattered field mode amplitude E_{S} (which is time independent) is generated by spontaneous emission and will be treated quantum mechanically [Eq. (9.2)]. Unlike the linear absorption experiment treated above, the scattered field is now initially in the vacuum (zero-photon) state and its density operator is

$$|\rho_{\mathrm{R}}(-\infty)\rangle\rangle = |n_{\mathrm{S}} = 0, n_{\mathrm{S}} = 0\rangle\rangle. \tag{9.8}$$

The operator representing the photon emission rate is given by Eq. (9.4), and the time and frequency resolved photon emission rate is given by $\langle\langle \mathcal{N}_{\mathrm{S}} | \rho_{\mathrm{T}}(t)\rangle\rangle$.* At first glance, there is a problem with the present definition of the time and frequency resolved spectrum, since it apparently violates the Heisenberg uncertainty relation for the emitted photon $\Delta\omega_s \Delta t \geq 1/2$. The problem is resolved by realizing that there is an intrinsic uncertainty in the actual time where the photon is emitted. This point will be discussed in Chapter 13 for pump-probe spectroscopy, and the same arguments hold for the SLE spectrum as well.

When the total system + field density operator $\rho_{\mathrm{T}}(t)$ is expanded perturbatively in H_{int}, we find that the lowest order contributing to S_{SLE} is $\rho_{\mathrm{T}}^{(3)}$. We thus have

$$\langle\langle \mathcal{N}_{\mathrm{S}} | \rho_{\mathrm{T}}(t)\rangle\rangle = \frac{1}{\hbar^4} \int_0^\infty dt_1 \int_0^\infty dt_2 \int_0^\infty dt_3$$

$$\langle\langle a_{\mathrm{S}}^\dagger a_{\mathrm{S}} | \mathcal{L}_{\mathrm{int}}(t) \mathcal{G}(t_3) \mathcal{L}_{\mathrm{int}}(t - t_3) \mathcal{G}(t_2) \mathcal{L}_{\mathrm{int}}(t - t_2 - t_3) \mathcal{G}(t_1) \mathcal{L}_{\mathrm{int}}(t - t_1 - t_2 - t_3) | \rho_{\mathrm{T}}(-\infty)\rangle\rangle. \tag{9.9}$$

For a general n-photon process, the amplitude is nth order in H_{int} (or $\mathcal{L}_{\mathrm{int}}$) and the intensity (amplitude squared) must be $2n$th order in this interaction. This explains why SLE, which is a two-photon process ($n = 2$), contains four material–field interactions and is closely related to the third-order response function; the emitted light amplitude is second order in the radiation–matter interaction, first order with respect to the incident mode, and first order with respect to the emitted mode.

Equation (9.9) is formally similar (but not identical to) Eqs. (5.13) and (5.14). Notable differences are the appearance of the quantum field creation and annihilation operators and the fact that the last interaction $\mathcal{L}_{\mathrm{int}}(t)$ is a Liouville space commutator rather than an ordinary Hilbert space operator. The interpretation of Eq. (9.9) is as follows: the system starts at $t = -\infty$ with a density operator $\rho_{\mathrm{T}}(-\infty)$ [Eq. (9.3) together with Eq. (9.8)]. It then interacts four times

* In order to compare with experiment we need to take into account the finite spectral and temporal resolution of the detection device. This can be done by a proper convolution of the bare spectrum defined here with spectral filter and time gate profiles of the detector [H. Stolz, *Time Resolved Light Scattering from Excitons* (Springer-Verlag, Berlin, 1994)].

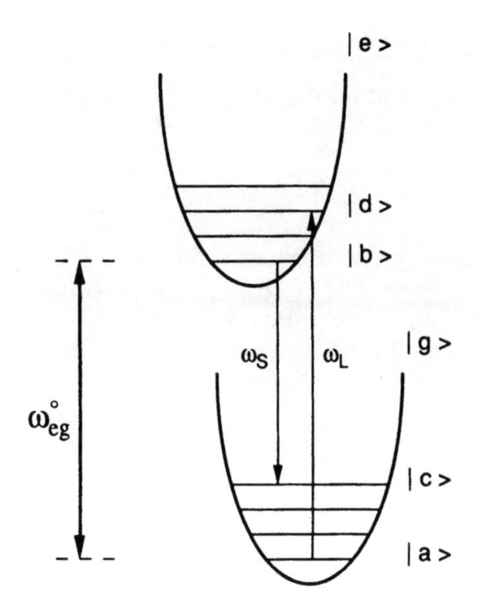

FIG. 9.3 A two-electronic-level scheme for spontaneous light emission (Raman and fluorescence) process.

with the electromagnetic field at times $t - t_1 - t_2 - t_3$, $t - t_2 - t_3$, $t - t_3$, and t. During the intervals between interactions (t_1, t_2, and t_3) its evolution is described by $\mathscr{G}(\tau)$. Finally, the SLE spectrum at time t is given by the expectation value of $\mathscr{N}_S$, which represents the photon emission rate.

We shall now apply this formal expression to calculate the SLE spectrum for the two electronic level system introduced in Chapter 7 (Figure 9.3). Equation (9.9) has many terms when evaluated since each $\mathscr{L}_{int}$ factor is a commutator that can act either from the left or from the right, and can represent coupling with either the incident or the scattered field modes. However, using the following arguments we can greatly reduce the number of terms.

1. Two of the $\mathscr{L}_{int}$ terms must be with the ω_L mode and two are with ω_s, otherwise the signal vanishes.
2. For each mode, one interaction has to act from the left and one from the right.
3. Within the rotating wave approximation (RWA), the first interaction $\mathscr{L}_{int}(t - t_1 - t_2 - t_3)$ must be with the ω_L mode (it has to represent photon absorption, and mode s is initially in the vacuum state). This approximation further eliminates many off resonant terms.
4. The last interaction $\mathscr{L}_{int}(t)$ must be with the ω_s mode. This can be directly seen by acting with $\mathscr{L}_{int}(t)$ to the left and noting that the ω_L part of $\mathscr{L}_{int}$ commutes with $a_S^\dagger a_S$ and will not contribute. We then have

$$\langle\langle a_S^\dagger a_S | \mathscr{L}_{int}(t) = \langle\langle V a_S^\dagger \exp(i\omega_S t) + V a_S \exp(-i\omega_S t)|.$$

The photon emission rate now becomes

$$\langle\langle \mathscr{N}_S \,|\, \rho_T(t)\rangle\rangle = \frac{1}{\hbar^4}\frac{2\pi\hbar\omega_S}{\Omega} S_{SLE}(\omega_L, \omega_S, t), \tag{9.10a}$$

with

$$S_{\mathrm{SLE}}(\omega_{\mathrm{L}}, \omega_{\mathrm{S}}, t) = 2\,\mathrm{Re} \int_0^\infty dt_3 \int_0^\infty dt_2 \int_0^\infty dt_1$$

$$\{E_{\mathrm{L}}(t - t_1 - t_2 - t_3)E_{\mathrm{L}}^*(t - t_2 - t_3) \exp(i\omega_{\mathrm{L}} t_1 + i\omega_{\mathrm{S}} t_3) R_1(t_3, t_2, t_1)$$

$$+ E_{\mathrm{L}}^*(t - t_1 - t_2 - t_3)E_{\mathrm{L}}(t - t_2 - t_3) \exp(-i\omega_{\mathrm{L}} t_1 + i\omega_{\mathrm{S}} t_3) R_2(t_3, t_2, t_1)$$

$$+ E_{\mathrm{L}}^*(t - t_1 - t_2 - t_3)E_{\mathrm{L}}(t - t_3)$$

$$\times \exp[-i\omega_{\mathrm{L}} t_1 - i(\omega_{\mathrm{L}} - \omega_{\mathrm{S}})t_2 + i\omega_{\mathrm{S}} t_3] R_3(t_3, t_2, t_1)\}. \quad (9.10b)$$

Making use of our multilevel expression for the nonlinear response function [Eq. (6.60)] we have

$$S_{\mathrm{SLE}}(\omega_{\mathrm{L}}, \omega_{\mathrm{S}}, t) = 2\,\mathrm{Re} \sum_{a,b,c,d} P(a)\mu_{ab}\mu_{bc}\mu_{cd}\mu_{da} \int_0^\infty dt_3 \int_0^\infty dt_2 \int_0^\infty dt_1$$

$$\{\langle \mathcal{G}_{cb}(t_3)\mathcal{G}_{db}(t_2)\mathcal{G}_{ab}(t_1)\rangle$$

$$\times \exp[-i\omega_{\mathrm{S}} t_3 - i\omega_{\mathrm{L}} t_1] E_{\mathrm{L}}^*(t - t_1 - t_2 - t_3)E_{\mathrm{L}}(t - t_2 - t_3)$$

$$+ \langle \mathcal{G}_{dc}(t_3)\mathcal{G}_{db}(t_2)\mathcal{G}_{ab}(t_1)\rangle$$

$$\times \exp[i\omega_{\mathrm{S}} t_3 - i\omega_{\mathrm{L}} t_1] E_{\mathrm{L}}^*(t - t_1 - t_2 - t_3)E_{\mathrm{L}}(t - t_2 - t_3)$$

$$+ \langle \mathcal{G}_{dc}(t_3)\mathcal{G}_{ac}(t_2)\mathcal{G}_{ab}(t_1)\rangle$$

$$\times \exp[i\omega_{\mathrm{S}} t_3 - i(\omega_{\mathrm{L}} - \omega_{\mathrm{S}})t_2 - i\omega_{\mathrm{L}} t_1] E_{\mathrm{L}}^*(t - t_1 - t_2 - t_3)E_{\mathrm{L}}(t - t_3)\}. \quad (9.11)$$

The six pathways that contribute to the SLE process are illustrated by the density operator coupling scheme shown in Figure 9.4a. In Figure 9.4b we display three of these pathways, labeled as (i), (ii), and (iii), which correspond, respectively, to the three terms in the right-hand side of Eq. (9.11). The corresponding double-sided Feynman diagrams are shown in Figure 9.4c. The other three pathways are the complex conjugates to these pathways. In pathways (i) and (ii), the system first interacts twice with the ω_{L} field at times $t - t_1 - t_2 - t_3$ and $t - t_2 - t_3$ and then twice with the ω_{S} mode at times $t - t_3$ and t. This represents a sequential process (mode L first, mode S second), which is related to fluorescence. Pathway (iii) represents a coherent Raman process where the system interacts successively with ω_{L}, ω_{S}, ω_{L}, and ω_{S}.

FREQUENCY DOMAIN SLE

Time resolved fluorescence will be analyzed in Chapter 13 using a phase space wavepacket representation. In the remainder of this chapter we restrict our discussion to stationary SLE experiments, where the envelope of the incoming field $E_{\mathrm{L}}(t)$ is independent of time. We define the differential photon scattering cross section $\sigma_{\mathrm{SLE}}(\omega_{\mathrm{L}}, \omega_{\mathrm{S}})$ as follows: $\sigma_{\mathrm{SLE}}(\omega_{\mathrm{L}}, \omega_{\mathrm{S}})\,d\omega_{\mathrm{S}}$ is the number of photons

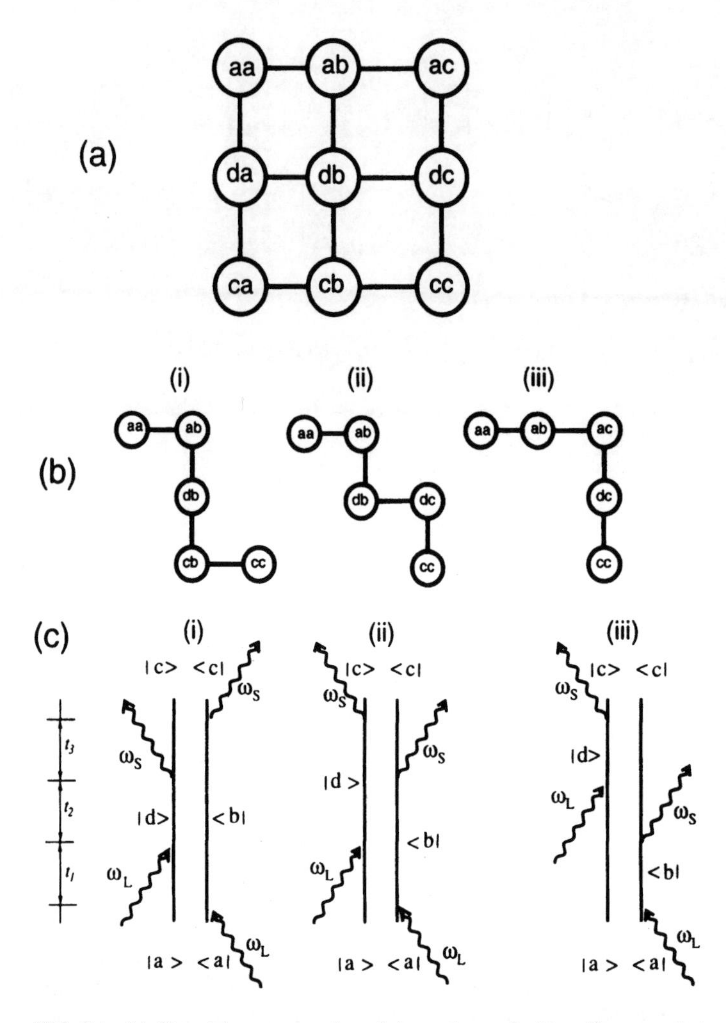

FIG. 9.4 (a) Pictorial representation of the pathways in Liouville space that contribute to SLE spectra. The SLE process is obtained by all pathways that start at *aa* and end at *cc* in fourth order (four bonds). There are six pathways that contribute. However, owing to symmetry, we need consider only the three pathways shown in b. The other three are obtained by a complex conjugation and permutation of b and d. (b) The three Liouville space pathways that contribute to SLE spectra. The three terms in Eqs. (9.11) or (9.12) correspond, respectively, to pathways (i), (ii), and (iii). Pathways (i) and (ii) contribute only to fluorescence, while (iii) contributes to both fluorescence and Raman. (c) Double-sided Feynman diagrams for the three Liouville space pathways contributing to spontaneous light emission.

emitted between ω_S and $\omega_S + d\omega_S$ per unit time, divided by the incident photon flux. The scattered differential power is then given by

$$I_S(\omega_L, \omega_S) = I_0(\omega_L) z \rho_0 \sigma_{SLE}(\omega_L, \omega_S).$$

I_0 is the incident power, z is the interaction pathlength, and ρ_0 is the number density of molecules. We then have

$$\sigma_{SLE}(\omega_L, \omega_S) = \langle \cos^2 \theta_L \rangle \langle \cos^2 \theta_S \rangle \frac{2\Omega\omega_S^2}{\pi^2 c^3} \frac{\hbar\omega_L}{(c/4\pi)|2E_L|^2} \langle\langle \mathcal{N}_S | \rho(t) \rangle\rangle.$$

The $\langle \cos^2 \theta \rangle$ terms represent averaging over molecular orientations, θ being the angle between the molecular dipole moment and the electric field polarization. For an isotropic medium $\langle \cos^2 \theta \rangle = 1/3$. In general, this term should be replaced by a four-point correlation function of molecular orientation, and the SLE spectrum can also detect the rotational dynamics, provided polarized light and detection are used. We postpone the treatment of rotational dynamics to Chapter 15. The third factor comes from the summation over all field modes with frequency ω_S. Using the density of field modes in a cavity with volume Ω we have $\sum_{k_s} \to \Omega \int (\omega_S^2/\pi^2 c^3)\, d\omega_S$. The factor of 2 comes from the summation over the two possible polarizations of the emitted light. The fourth term represents the division by the incoming photon flux [Eq. (4.44c)]. Note that the amplitude of the electric field in real space $E(\mathbf{r}, t)$ [Eq. (9.7)] is $|2E_L|$.

Making use of Eqs. (9.10), we finally obtain the differential photon scattering cross section*

$$\sigma_{SLE}(\omega_L, \omega_S) = \frac{4\omega_L \omega_S^3}{9\hbar^2 c^4} S_{SLE}(\omega_L, \omega_S),$$

where

$$\begin{aligned} S_{SLE}(\omega_L, \omega_S) = 2\,\mathrm{Re} \int_0^\infty dt_3 \int_0^\infty dt_2 \int_0^\infty dt_1\, \{&\exp(i\omega_L t_1 + i\omega_S t_3) R_1(t_3, t_2, t_1) \\ &+ \exp(-i\omega_L t_1 + i\omega_S t_3) R_2(t_3, t_2, t_1) \\ &+ \exp[-i\omega_L t_1 - i(\omega_L - \omega_S)t_2 + i\omega_S t_3] R_3(t_3, t_2, t_1)\}. \end{aligned} \tag{9.12a}$$

This equation can be alternatively written in terms of the nonlinear response function in the frequency domain:

$$S_{SLE}(\omega_L, \omega_S) = 2\,\mathrm{Im}[R_1(\omega_S, 0, \omega_L) + R_2(\omega_S, 0, -\omega_L) + R_3(\omega_S, \omega_S - \omega_L, -\omega_L)]. \tag{9.12b}$$

Expanding in the electronic eigenstates [Eq. (7.11)] we get

$$R_1(\omega_S, 0, \omega_L) = \langle\langle V | \mathcal{G}_{eg}(\omega_S) \mathcal{V}_{eg,ee} \mathcal{G}_{ee}(0) \mathcal{V}_{ee,eg} \mathcal{G}_{eg}(\omega_L) \mathcal{V}_{eg,gg} | \rho_g \rangle\rangle,$$

$$R_2(\omega_S, 0, -\omega_L) = \langle\langle V | \mathcal{G}_{eg}(\omega_S) \mathcal{V}_{eg,ee} \mathcal{G}_{ee}(0) \mathcal{V}_{ee,ge} \mathcal{G}_{ge}(-\omega_L) \mathcal{V}_{ge,gg} | \rho_g \rangle\rangle,$$

$$R_3(\omega_S, \omega_S - \omega_L, -\omega_L) = \langle\langle V | \mathcal{G}_{eg}(\omega_S) \mathcal{V}_{eg,gg} \mathcal{G}_{gg}(\omega_S - \omega_L) \mathcal{V}_{gg,ge} \mathcal{G}_{ge}(-\omega_L) \mathcal{V}_{ge,gg} | \rho_g \rangle\rangle.$$

* We are neglecting here the frequency dispersion of the index of refraction. Otherwise this formula should be multiplied by $n(\omega_S)/n(\omega_L)$.

Further expansion in nuclear eigenstates [Eq. (6.63)] finally yields

$$S_{\rm SLE}(\omega_{\rm L}, \omega_{\rm S}) = 2\,{\rm Im} \sum_{a,b,c,d} P(a)\mu_{ab}\mu_{bc}\mu_{cd}\mu_{da}$$

$$[\langle \mathscr{G}_{cb}(-\omega_s)\mathscr{G}_{db}(0)\mathscr{G}_{ab}(-\omega_{\rm L})\rho_g\rangle + \langle \mathscr{G}_{dc}(\omega_s)\mathscr{G}_{db}(0)\mathscr{G}_{ab}(-\omega_{\rm L})\rho_g\rangle$$
$$+ \langle \mathscr{G}_{dc}(\omega_s)\mathscr{G}_{ac}(\omega_s - \omega_{\rm L})\mathscr{G}_{ab}(-\omega_{\rm L})\rho_g\rangle]. \tag{9.13}$$

Equations (9.12) and (9.13) suggest that the stationary SLE spectrum is closely related to $\chi^{(3)}$,

$$S_{\rm SLE}(\omega_{\rm L}, \omega_{\rm S}) \stackrel{?}{=} {\rm Im}\,\chi^{(3)}(-\omega_{\rm S}; \omega_{\rm S}, -\omega_{\rm L}, \omega_{\rm L}).$$

All the terms appearing in Eq. (9.13) indeed belong to $\chi^{(3)}$. However, $\chi^{(3)}$ contains additional terms, which do not contribute to spontaneous emission. The difference can be attributed to the quantum nature of the scattered mode, which is initially in the vacuum state. In the SLE process we need to act once with $a_{\rm S}^\dagger$ from the left and once with $a_{\rm S}$ from the right in order to bring the system to the one photon state $|1\rangle\langle 1|$. Acting with $a_{\rm S}^\dagger$ from the right or $a_{\rm S}$ from the left will give zero. In calculating $\chi^{(3)}$, on the other hand, we treat the field classically, since its occupation number is large, and we can therefore act with the positive and negative frequency parts of the field (E and E^*) from the left and from the right. Consequently, due to the quantum nature of the emitted field, not all the terms that contribute to $\chi^{(3)}$ are relevant to SLE, and we cannot simply express $S_{\rm SLE}$ in terms of $\chi^{(3)}$ as suggested by the above expression. Another difference is that $\chi^{(3)}$ to a first approximation does not depend on spontaneous emission, whereas radiative decay has to be included in $S_{\rm SLE}$. This will be discussed further following Eqs. (9.14).

Equations (9.12) allow us to apply all the results and models developed for the nonlinear response function in earlier chapters for the calculation of the SLE. Consider the response function of a multilevel system with homogeneous broadening (Figure 9.3). Making use of Eq. (6.23) we obtain

$$S_{\rm SLE}(\omega_{\rm L}, \omega_{\rm S}) = \sum_{a,c} P(a) K_{ca}(\omega_{\rm L}, \omega_{\rm S}), \tag{9.14a}$$

where

$$K_{ca}(\omega_{\rm L}, \omega_{\rm S}) = K_{\rm I} + K_{\rm II} + K_{\rm III}, \tag{9.14b}$$

and where

$$K_{\rm I} = -i \sum_{b,d} \mu_{ab}\mu_{bc}\mu_{cd}\mu_{da} \frac{1}{\omega_{ba} - \omega_{\rm L} + i\Gamma} \frac{1}{\omega_{bd} + i\gamma} \frac{1}{\omega_{bc} - \omega_{\rm S} + i\Gamma} + c.c., \tag{9.14c}$$

$$K_{\rm II} = -i \sum_{b,d} \mu_{ab}\mu_{bc}\mu_{cd}\mu_{da} \frac{1}{\omega_{bd} + i\gamma} \frac{1}{\omega_{ba} - \omega_{\rm L} + i\Gamma} \frac{1}{\omega_{cd} - \omega_{\rm S} + i\Gamma} + c.c., \tag{9.14d}$$

$$K_{\rm III} = -i \sum_{b,d} \mu_{ab}\mu_{bc}\mu_{cd}\mu_{da} \frac{1}{\omega_{ba} - \omega_{\rm L} + i\Gamma} \frac{1}{\omega_{ca} + \omega_{\rm S} - \omega_{\rm L} + i\varepsilon} \frac{1}{\omega_{cd} + \omega_{\rm S} + i\Gamma} + c.c., \tag{9.14e}$$

In Eqs. (9.14), we have assumed

$$\gamma_a = \gamma_c = 0,$$

which implies an infinite lifetime for vibronic states belonging to the ground electronic state,

$$\gamma_b = \gamma_d = \gamma,$$

$$\Gamma = 1/2\gamma + \hat{\Gamma}.$$

Here $P(a)$ is the equilibrium population of $|a\rangle$, and γ_j is the inverse lifetime of state $|j\rangle$. ε is a small positive number, and at the end of the calculation we should set $\varepsilon \to 0$. In an SLE process, a molecule initially in the state $|a\rangle$ with energy ε_a absorbs a photon ω_L, emits a photon ω_S, and ends up in the state $|c\rangle$ with energy ε_c. $|b\rangle$ and $|d\rangle$ denote a manifold of vibronic states belonging to an electronically excited state, whereas $|a\rangle$ and $|c\rangle$ usually belong to the ground electronic state. A complete expression for a multilevel vibronic system including vibrational relaxation and coupling to an overdamped Brownian oscillator with an arbitrary relaxation timescale Λ^{-1} can be derived using Eq. (9.12b), and is presented in Appendix 9A. In the fast modulation limit it reduces to Eqs. (9.14).

The excited state inverse lifetime γ could be radiative or nonradiative in nature. The inclusion of radiative decay is a delicate matter since the response functions and susceptibilities defined with respect to the transverse field are purely material quantities and should not contain the radiative decay. However, in the present model E_L should be understood as the external incoming field rather than the transverse field and consequently we need to incorporate radiative decay in the Hamiltonian. This will be clarified in Chapter 17; at this point we simply add the radiative decay rate phenomenologically.

As a check, we note that in the absence of nonradiative decay channels, the system emits one photon for every absorbed photon (i.e., the emission quantum yield is unity). The integrated SLE cross section should therefore be equal to the linear absorption cross section

$$\int_0^\infty d\omega_S\, \sigma_{SLE}(\omega_L, \omega_S) = \kappa_a(\omega_L),$$

where $\kappa_a(\omega_L)$ is the photon absorption cross section [Eq. (4.64)]

$$\kappa_a(\omega_L) = \frac{4\pi\omega_L}{3n(\omega_L)\hbar c}\,[J(\omega_L) + J^*(-\omega_L)].$$

Absorption cross sections for intense molecular transitions are typically ~ 1 Å^2. The cross section for a single resonance Raman line is $\sim 10^{-8}$ Å^2, whereas the total resonance Raman scattering cross section can reach the order of $\sim 10^{-6}$ Å^2.

FLUORESCENCE AND RAMAN SPECTROSCOPY

It is sometimes possible to distinguish between two types of contributions to spontaneous emission in condensed phases [4, 10, 12–20]. These are denoted Raman and fluorescence, respectively. The fluorescence component is also referred to as "hot luminescence" or "redistribution." The Raman component

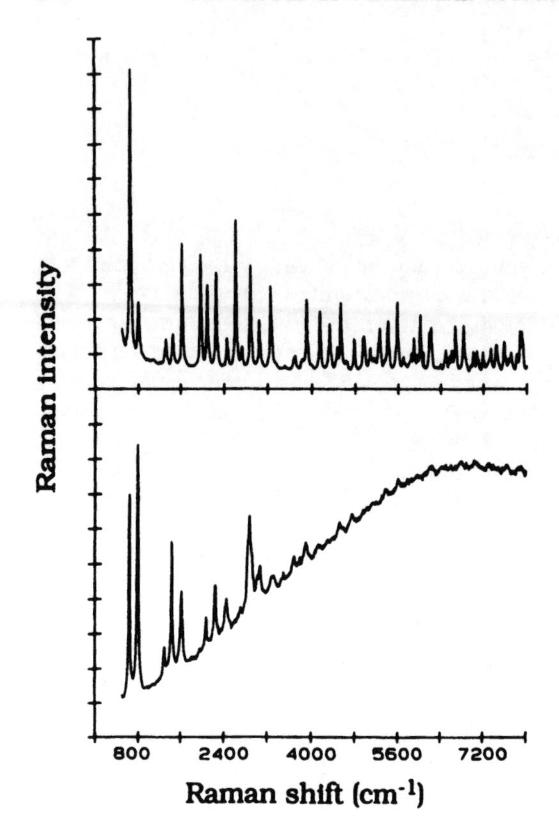

FIG. 9.5 Comparison of resonance Raman spectra of CS_2 in the vapor phase (top) and in hexadecane solvent (bottom), showing fluorescence component caused by pure dephasing. [From A. B. Myers, "Femtosecond molecular dynamics probed through resonance Raman intensities," *J. Opt. Soc. Am.* B 7, 1664 (1990).]

consists of relatively narrow emission lines centered at $\omega_S = \omega_L - \omega_{ca}$, where a⟩ and $|c\rangle$ are any pair of ground-state vibronic levels, and their widths are typically narrower than a few cm^{-1}. Fluorescence lineshapes in solution are much broader (typical width of a few hundred cm^{-1}), and they are peaked at $\omega_S = \omega_{bc}$, where levels $|b\rangle$ and $|c\rangle$ belong to the excited and the ground electronic states, respectively (see Figures 9.2 and 9.5). A clear distinction between these two components may be made by tuning the incident frequency ω_L. The Raman lines will shift linearly with ω_L ($\omega_s = \omega_L - \omega_{ca}$) whereas the fluorescence component will remain roughly in the same position.

The origin of these components can be clarified and a deeper insight regarding the role of dephasing can be obtained by a close examination of Eqs. (9.14). In the $K_{\rm III}$ term we can use the identity

$$\frac{1}{\omega_{ca} + \omega_S - \omega_L + i\varepsilon} = \mathscr{P}\mathscr{P}\left(\frac{1}{\omega_S - \omega_L + \omega_{ca}}\right) - i\pi\delta(\omega_S - \omega_L + \omega_{ca}), \quad (9.15)$$

where $\mathscr{P}\mathscr{P}$ stands for the principal part (see Chapter 2). The $\delta(\omega_s - \omega_L + \omega_{ca})$ term is reponsible for the Raman component, whereas all the remaining terms in Eqs. (9.14) constitute the "fluorescence" spectrum.

Using Eq. (9.15) we can now rearrange Eqs. (9.14) and obtain [10]

$$S_{\rm SLE}(\omega_{\rm L}, \omega_{\rm S}) = S_{\rm RAMAN}(\omega_{\rm L}, \omega_{\rm S}) + S_{\rm FL}(\omega_{\rm L}, \omega_{\rm S}), \tag{9.16}$$

with

$$S_{\rm RAMAN}(\omega_{\rm L}, \omega_{\rm S}) = 2\pi \sum_{a,c} P(a)|\chi_{ca}(\omega_{\rm L})|^2 \delta(\omega_{ca} + \omega_{\rm S} - \omega_{\rm L}), \tag{9.17a}$$

$$\chi_{ca}(\omega_{\rm L}) = \sum_b \frac{\mu_{cb}\mu_{ba}}{\omega_{ab} + \omega_{\rm L} + i\Gamma}, \tag{9.17b}$$

and

$$S_{\rm FL}(\omega_{\rm L}, \omega_{\rm S}) = \sum_{\substack{a,c\\b,d}} P(a)\mu_{ab}\mu_{bc}\mu_{cd}\mu_{da} \times \frac{2\hat{\Gamma}}{\omega_{bd} + i\gamma} \frac{1}{\omega_{ba} - \omega_{\rm L} + i\Gamma} \frac{1}{\omega_{da} - \omega_{\rm L} - i\Gamma} \times \left[\frac{1}{\omega_{dc} - \omega_{\rm S} - i\Gamma} - \frac{1}{\omega_{bc} - \omega_{\rm S} + i\Gamma}\right]. \tag{9.18}$$

Equations (9.16)–(9.18) are our final expression for the homogeneously broadened SLE signal. When the molecule has only one excited $|b\rangle$ state (i.e., $b = d$), it further simplifies to

$$S_{\rm SLE}(\omega_{\rm L}, \omega_{\rm S}) = 2\pi \sum_{a,c} |\mu_{ab}|^2 |\mu_{bc}|^2 P(a) \times \frac{1}{(\omega_{ba} - \omega_{\rm L})^2 + \Gamma^2} \left[\delta(\omega_{ca} + \omega_{\rm S} - \omega_{\rm L}) + \frac{2\hat{\Gamma}}{\gamma} \frac{\Gamma/\pi}{(\omega_{bc} - \omega_{\rm S})^2 + \Gamma^2}\right]. \tag{9.19}$$

We shall now discuss the main features of SLE spectra as given by these expressions.

1. Equation (9.16) naturally partitions the spectrum into a Raman and a fluorescence component. Equations (9.17) are known as the Kramers–Heisenberg expression for Raman scattering. The Raman amplitude χ_{ca} is usually obtained using a second-order perturbative calculation of the wavefunction. In the absence of pure dephasing ($\hat{\Gamma} = 0$), an interesting destructive interference takes place, whereby the $\mathscr{PP}$ contribution of $K_{\rm III}$ [Eq. (9.15)] cancels exactly the contributions $K_{\rm I} + K_{\rm II}$. The fluorescence then vanishes, $S_{\rm FL} = 0$, and the total emission is of the Raman type.

In the absence of dephasing, we can define a Raman amplitude χ_{ca} [Eq. (9.17b)] and express the cross section $S_{SLE}(\omega_L, \omega_S)$ in terms of its modulus square $|\chi_{ca}|^2$. SLE spectra taken, e.g., at low pressure gas phase, in supersonic beams, and on impurities in crystals at low temperatures may, therefore, be adequately interpreted using the Kramers–Heisenberg expression. This is no longer the case once dephasing processes are incorporated. The pure dephasing rate $\hat{\Gamma}$ has a dual role in Eq. (9.16). First it broadens the Kramers–Heisenberg terms by contributing to the width Γ. In addition, it "redistributes" the emitted intensity by reducing the coherent Raman component and generating the sequential fluorescence component instead. This results in dephasing-induced (fluorescence) terms given by the second term in the square brackets, which are proportional to $\hat{\Gamma}$. The fundamental difference is that now the process does not have an "amplitude." This reflects the fact that due to the coupling with the bath, the system is no longer in a pure state and cannot be represented by a wave function; a density operator description is then essential. Stochastic fluctuations in the laser field (e.g., phase fluctuations) can also sometimes be incorporated as dephasing, and may induce a fluorescence emission even in the absence of dephasing in the matter.

In Figure 9.6 we display a two-dimensional fluorescence spectrum that highlights the correlation between the incident and the scattered frequencies.

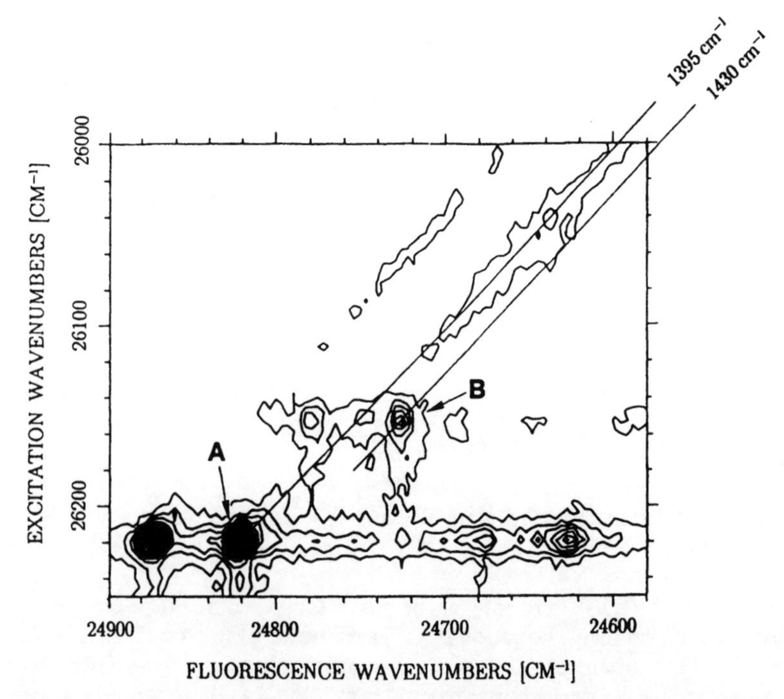

FIG. 9.6 2D total luminescence spectra of tetrabenzonaphthalene in *n*-hexane at 4.2 K covering the $S_{10} \leftarrow S_{00}$ region in excitation. Shown are vibronic transitions at about $\nu_{vib}(S_0) = 1400\ \text{cm}^{-1}$ in fluorescence. [K. Palewska, E. C. Meister, and U. P. Wild, "Spectroscopic evidence for the coexistence of two stereoisomers of tetrabenzonaphthalene in Shpol'skii-type matrices at 4.2 k," *Chem. Phys.* **138**, 115 (1989).]

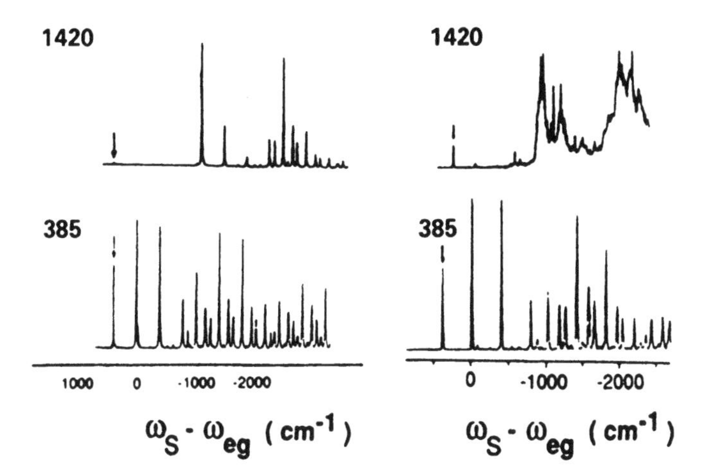

FIG. 9.7 Comparison of a harmonic model calculation of the fluorescence spectra of jet-cooled anthracene with two different excess vibrational energies as indicated (in cm^{-1}) (left panel) with experiment (right panel). The arrow in each spectrum indicates the laser excitation position. Bath modes do not enter in these calculations. [Reproduced from Y. Yan, K. Shan, and S. Mukamel, "Intramolecular dephasing and vibrational redistribution in the dispersed fluorescence of ultracold molecules: Application to the anthracene," *J. Chem. Phys.* **87**, 2021 (1987).]

In Figure 9.7 we show SLE spectra of ultracold anthracene with two excess vibrational energies in the excited state (385 and 1420 cm^{-1}). In the left panels we show calculated spectra obtained using the displaced oscillator model including only the strong optically active modes. The agreement is very good at the lower energy and fails completely at the higher energy, owing to the onset of intramolecular vibrational redistribution induced by coupling to the other molecular modes. A calculation of the fluorescence of the same molecule excited with various temperatures with and without vibrational redistribution is shown in Figure 9.8.

2. It should be recognized that there is a subjective element in the distinction between Raman and fluorescence. In principle it is possible to incorporate the bath degrees of freedom in our system. In this case pure dephasing no longer exists, $\hat{\Gamma} = 0$, and the entire SLE spectrum can be calculated using the Kramers–Heisenberg formula [Eq. (9.17)], where the a, b, c indices now denote the global states of the system and bath. On the other hand, if we adopt a reduced description of the system, keeping only a few degrees of freedom explicitly, dephasing has to be added and the same spectrum can be partitioned into a Raman and a fluorescence component. Theoretically, the distinction is thus intimately connected with the choice of a basis set and level of description. In simple systems it is natural to adopt a complete description (everything is Raman), whereas in complex systems this is virtually impossible, and a reduced description that makes the distinction between Raman and fluorescence is more adequate.

3. The *fluorescence excitation spectrum* (or *action spectrum*) is obtained by measuring the total light emission vs. the incident frequency [6]. If the emission

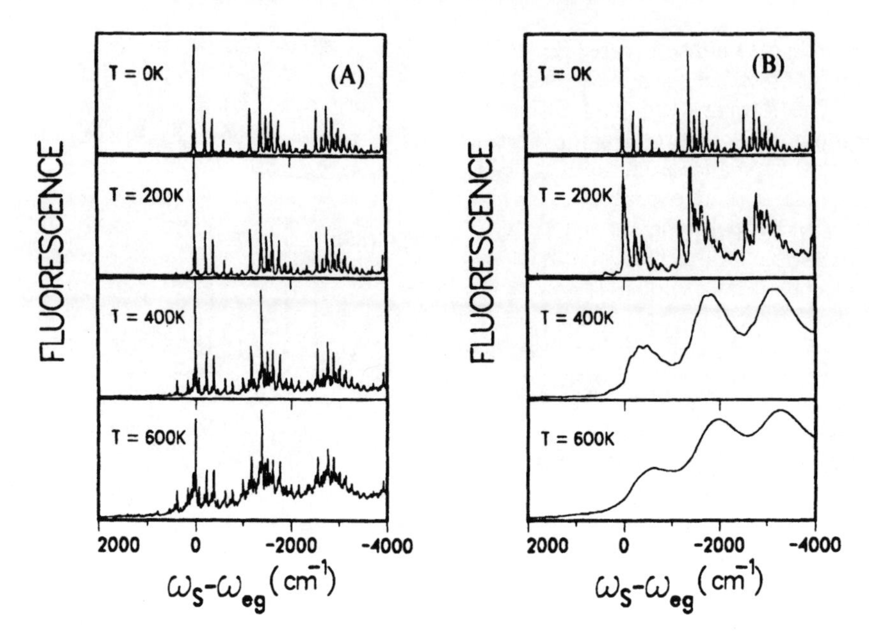

FIG. 9.8 Calculated emission spectra of anthracene when the molecular vibrations are equilibriated with a heat bath at different temperatures from 0 K up to 600 K, as indicated. (A) Using only the 17 strongly optically active modes. (B) Using all the 66 modes. The bath was included using a statistical model. [Reproduced from Y. Yan, K. Shan, and S. Mukamel, "Intramolecular dephasing and vibrational redistribution in the dispersed fluorescence of ultracold molecules: Application to anthracene," *J. Chem. Phys.* **87**, 2021 (1987).]

quantum yield is 1 (i.e., no competing nonradiative channels) then the excitation spectrum is identical to the absorption. In many cases excitation measurements are more sensitive than absorption since they have a lower background. An example of a precise fluorescence excitation measurement of the absorption lineshape in the far wings is shown in Figure 7.2. The excitation spectrum can be calculated by integrating Eq. (9.16) over ω_S, which yields

$$\begin{aligned} S_{\text{exc}}(\omega_L) &\equiv \int S_{\text{SLE}}(\omega_L, \omega_S)\, d\omega_S \\ &= 2\pi \sum_{\substack{a,c \\ b,d}} \mu_{ab}\mu_{bc}\mu_{cd}\mu_{da} P(a) \left[1 + \frac{2i\hat{\Gamma}}{\omega_{bd} + i\gamma}\right] \frac{1}{\omega_{ba} - \omega_L + i\Gamma} \frac{1}{\omega_{da} - \omega_L - i\Gamma}. \end{aligned} \tag{9.20}$$

4. In general we can define the *Raman yield* as the fraction of photons emitted in the Raman component

$$Y_R(\omega_L) = \frac{\int d\omega_S\, S_{\text{RAMAN}}(\omega_L, \omega_S)}{\int d\omega_S\, S_{\text{SLE}}(\omega_L, \omega_S)}. \tag{9.21}$$

Equation (9.20) shows that when the system has only a single relevant excited state ($b = d$, and $\omega_{bd} = 0$), the Raman yield is $2\hat{\Gamma}/\gamma$. This is also typically the ratio in the multistate case [Eq. (9.20)]. Usually in condensed phases $\hat{\Gamma}/\gamma \gg 1$, and the fluorescence term is expected to dominate the SLE spectrum.

The Raman yield may be used to distinguish between homogeneous and inhomogeneous broadening. Physically, we expect the Raman yield to be unity for off resonance excitation, since the excited state should not be populated in this case, and we expect to see no fluorescence. Equation (9.16) predicts, however, a Raman yield that is independent on the ω_L detuning. This points out a limitation of the present model. In general, the dephasing rate is frequency dependent $\hat{\Gamma}(\omega)$ and it vanishes at large detunings. Only over a limited frequency range (given by the inverse bath timescale) can it be assumed frequency independent. The vanishing of $\hat{\Gamma}(\omega)$ at large detunings guarantees the disappearance of the fluorescence and $Y_R \to 1$. Taking $\hat{\Gamma}(\omega)$ to be frequency independent implies an infinitely short timescale of the bath. When the overdamped Brownian oscillator model (with a finite timescale) is used (Appendix 9A) this problem is cured and that expression can be safely used for arbitrary detunings [20]. In Figure 9.9 we show the SLE spectra for a molecule with a single high frequency vibrational mode. The bath is in this case represented by the stochastic model of line broadening [Eqs. (8.60)]. The fluorescence component clearly disappears when the excitation detuning is between successive vibronic levels. The ratio of the fluorescence to Raman intensities for the azulene system of Figure 9.2 is shown in Figure 9.10.

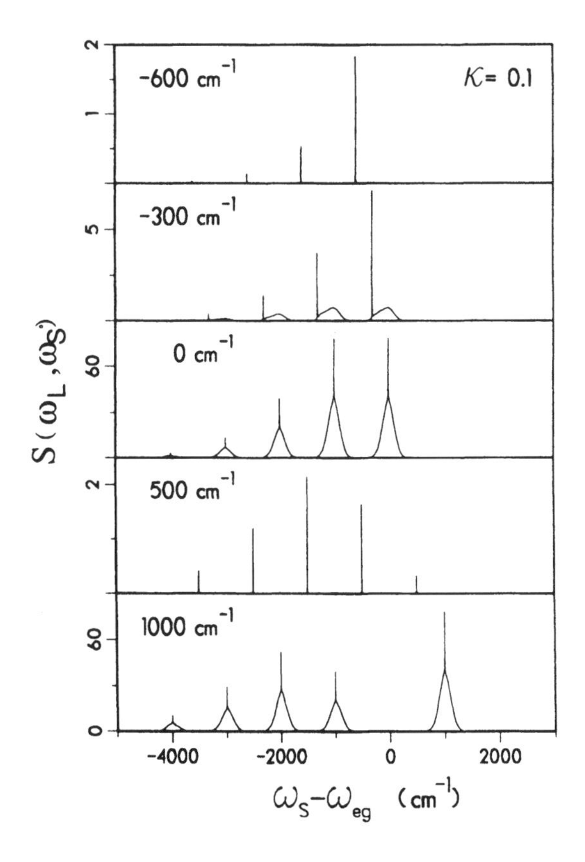

FIG. 9.9 The SLE line shape for a single mode with $\omega = 1000\ \text{cm}^{-1}$, $d = 1.4$, and $\gamma = 2\ \text{cm}^{-1}$ coupled to a stochastic bath with parameters $\Delta = 130\ \text{cm}^{-1}$ and $\Lambda = 13\ \text{cm}^{-1}$, which correspond to $\Gamma_0 = 250\ \text{cm}^{-1}$ and $\kappa = 0.1$. Dispersed emission vs. ω_s is shown for several values of the detunings of $\omega_L - \omega_{eg}$, as indicated in each panel. [Reproduced from Y. J. Yan and S. Mukamel, "Molecular fluorescence and near resonance Raman yield as a probe for solvation dynamics," *J. Chem. Phys.* **86**, 6085 (1987).]

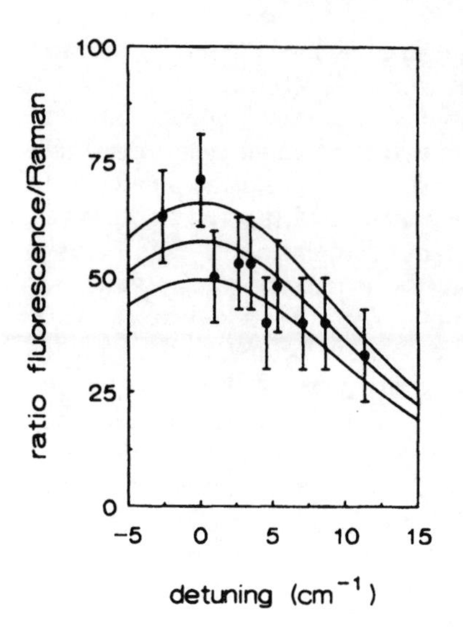

FIG. 9.10 Ratio of the fluorescence to Raman intensities (filled circles) for the azulene system of Figure 9.2 as a function of frequency detuning of the excitation laser. The calculated dependence of this ratio is shown using three values of the lifetime (from top to bottom); 1.6, 1.4, and 1.2 ps. The measured lifetime was found to be 1.6 ps.

5. Raman excitation profiles are obtained by tuning ω_L and detecting the narrow Raman component at $\omega_S - \omega_L = \omega_{ac}$ [19]. Each pair of ground vibronic states $|a\rangle$ and $|c\rangle$ has its distinct excitation profile, which is proportional to $|\chi_{ca}(\omega_L)|^2$. In Figure 9.11 we display Raman excitation profiles of azulene in CS_2, together with a fit to a seven mode harmonic model including a stochastic model for the solvent. These fits were used to obtain the stochastic parameters of the solvent. In Figure 9.12 we show the stochastic solvent parameters and their variations with temperature, obtained by fitting the absorption and the fluorescence lineshapes of the azulene system of Figure 9.2.

6. The pure dephasing term broadens the excitation and the dispersed emission spectra (S_{SLE} vs. ω_s) resulting in a Lorentzian profile without affecting the lifetime. The microscopic rationale for this broadening is that our system can exchange energy with the bath, and conservation of energy for the pure system $\delta(\omega_{ca} + \omega_S - \omega_L)$ no longer holds.

7. Raman and fluorescence components can be also distinguished in the time domain. Using Eq. (9.11) it can be shown that the Raman emission follows the excitation pulse temporal profile, whereas the fluorescence is delayed and its time profile determined by radiative and nonradiative decay rates [4, 16].

8. In the present analysis of SLE, we used the simplest model of line broadening (pure homogeneous electronic dephasing and lifetime broadening). In Appendix 9A we generalize these results to include broadening induced by a single overdamped Brownian oscillator model. The result given in Eq. (9.A.4) contains an extra summation over an index n. By keeping only the $n = 0$ term we recover Eq. (9.14). The finite timescale of the oscillator thus gives additional terms (with $n \neq 0$) that peak at the positions of the Raman lines, but are broader. They are known as "broad Raman" [17]. In practice they usually merge with the fluorescence and simply modify its lineshape. A broad Raman component is also obtained if we include a finite vibrational relaxation rate, even if electronic dephasing is homogeneous with a fast bath timescale. A more general expression

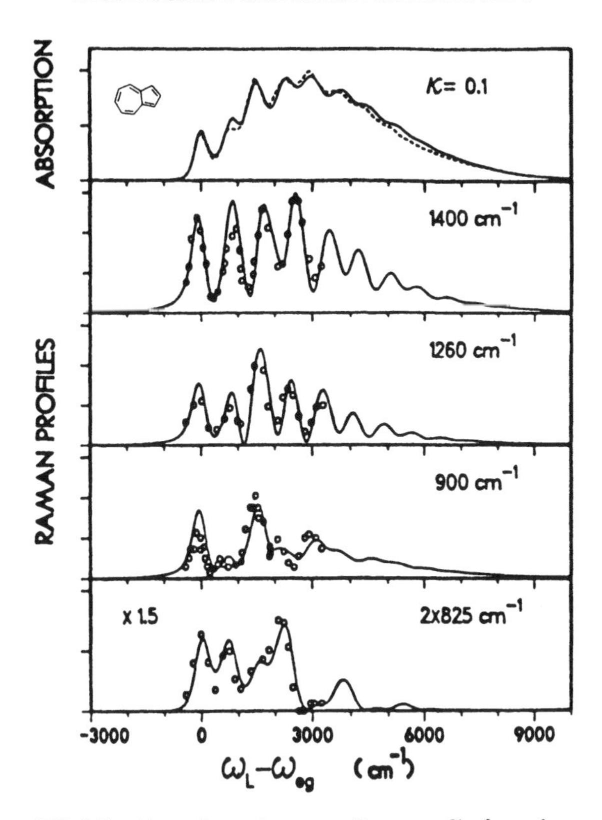

FIG. 9.11 Absorption and resonance Raman profiles for azulene in CS_2 at 300 K. The solid lines are theoretical curves computed using a seven-mode stochastic harmonic model. The absorption curve (upper panel) is $\omega\sigma_a(\omega)$, where $\sigma_a(\omega)$ is given in Eq. (7.9c). The dashed line represents experimental data. The calculated Raman profiles (lower four panels) are compared with the experiment of C. K. Chan et al. [*J. Chem. Phys.* **82**, 4813 (1985)] for four different Raman transitions, as indicated in each panel. The broadening parameters are $\Delta = 180\ \text{cm}^{-1}$, $\Lambda = 18\ \text{cm}^{-1}$, and $\Gamma_0 = 408\ \text{cm}^{-1}$. $\gamma = 0$ and $\omega_{eg} = 14{,}286\ \text{cm}^{-1}$. [From J. Sue, Y. J. Yan, and S. Mukamel, "Raman excitation profiles of polyatomic molecules in condensed phases. A stochastic theory," *J. Chem. Phys.* **85**, 462 (1986).]

that contains vibrational relaxation as well as an overdamped Brownian oscillator mode with an arbitrary timescale is given by Eq. (9.A.1).

RESONANT COHERENT RAMAN SPECTROSCOPY

Frequency domain coherent anti-Stokes Raman spectroscopy (CARS)

Certain four wave mixing measurements may show up narrow two-photon resonances that closely resemble spontaneous Raman spectra [21]. We shall now

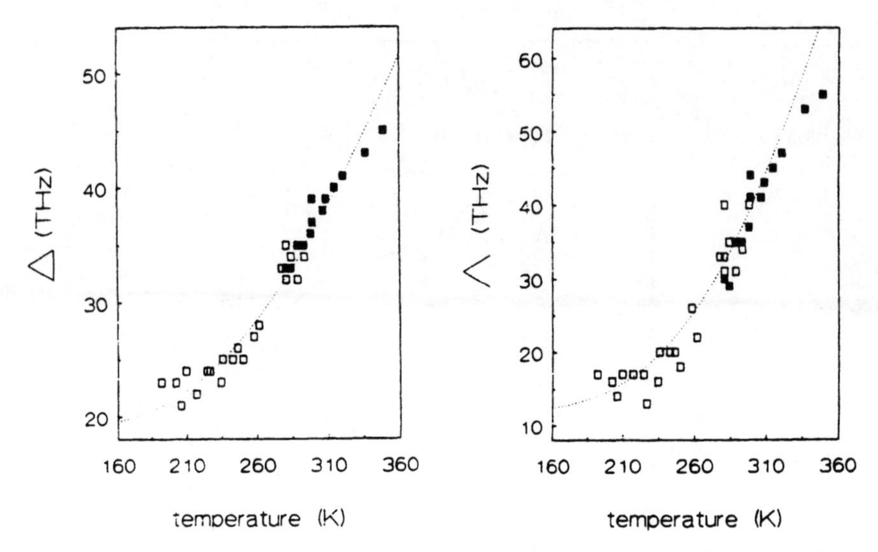

FIG. 9.12 Temperature dependence of the modulation strength Δ (left) and the inverse correlation time Λ (right) for the $S_1 \leftarrow S_0$ transition of azulene in cyclohexane (filled symbols) and in isopentane (open symbols) (shown to extend the temperature range). The system and reference are the same as Figure 9.2..

consider these coherent Raman measurements known as coherent anti-Stokes Raman spectroscopy (CARS), and coherent Stokes Raman spectroscopy (CSRS). Coherent Raman spectroscopy is a four wave mixing technique involving two incoming fields (i.e., $\mathbf{k}_3 = \mathbf{k}_1$), and the signal mode is

$$\mathbf{k}_s = 2\mathbf{k}_1 - \mathbf{k}_2; \qquad \omega_s = 2\omega_1 - \omega_2.$$

Since two fields are identical, there are only three (rather than six) permutations of frequencies. Writing the frequency permutations explicitly in Eq. (5.54), we get

$$\begin{aligned}\chi^{(3)}_{\text{CARS}} &\equiv \chi^{(3)}(-\omega_s; \omega_1, -\omega_2, \omega_1)\\ &= \tfrac{1}{4}\rho_0[S^{(3)}(2\omega_1 - \omega_2, \omega_1 - \omega_2, \omega_1) + S^{(3)}(2\omega_1 - \omega_2, \omega_1 - \omega_2, -\omega_2)\\ &\quad + S^{(3)}(2\omega_1 - \omega_2, 2\omega_1, \omega_1)],\end{aligned} \tag{9.22}$$

where $S^{(3)}$, given in Eq. (5.38a), is a sum of eight terms. In coherent Raman spectroscopy, we look for *two-photon resonances* in the signal, which occur when $\omega_1 - \omega_2$ equals an energy difference between two ground-state or excited-state vibrational states, i.e., $\omega_1 - \omega_2 = \pm\omega_{ca}$ or $\omega_1 - \omega_2 = \pm\omega_{db}$, respectively (see Figure 9.4). The terminology of CARS and CSRS refers to the cases where $\omega_1 > \omega_2$ or $\omega_2 > \omega_1$, respectively. Since there is no difference between the theoretical treatment of the two, we shall focus on the CARS resonances, $\omega_1 - \omega_2 = \omega_{ca}$ (ground-state CARS), and $\omega_1 - \omega_2 = \omega_{db}$ (excited-state CARS).

For the sake of clarity, we shall restrict the present analysis to the two level model with homogeneous broadening used above in the analysis SLE. From Eq. (9.22), it is clear that CARS resonances can come only from the middle Green function (the second frequency argument). The third term

$S^{(3)}(2\omega_1 - \omega_2, 2\omega_1, \omega_1)$ cannot contribute to CARS, since its two-photon resonances are at $2\omega_1$ and not at $\omega_1 - \omega_2$. We shall therefore ignore this term and consider only the terms containing ground state resonances $I_{ca}(\omega_1 - \omega_2)$ and excited state resonances $I_{db}(\omega_1 - \omega_2)$ [see Eq. (6.23)]. We shall denote them χ_{ca} and χ_{db}, respectively. Using Eqs. (5.58), we have for the CARS signal generated at $\mathbf{k}_S = 2\mathbf{k}_1 - \mathbf{k}_2$

$$S_{\text{CARS}}(\omega_1, \omega_2) = |\chi^{(3)}_{\text{CARS}}(-\omega_s; \omega_1, -\omega_2, \omega_1)|^2, \tag{9.23a}$$

with

$$\chi^{(3)}_{\text{CARS}} = \frac{1}{4}\left(\frac{1}{\hbar}\right)^3 \rho_0 \left[\sum_{a,c} \phi_{ca}(\omega_1, \omega_2) + \sum_{b,d} \phi_{db}(\omega_1, \omega_2)\right]. \tag{9.23b}$$

For the ground-state CARS, we find eight terms containing an $\omega_1 - \omega_2 \pm \omega_{ca}$ denominator

$$\phi_{ca}(\omega_1, \omega_2) = \sum_{b,d} \mu_{ab}\mu_{bc}\mu_{cd}\mu_{da}[I_{ba}(2\omega_1 - \omega_2) + I^*_{bc}(\omega_2 - 2\omega_1)]$$
$$\times \frac{P(a)I_{da}(\omega_1) + P(a)I_{da}(-\omega_2) - P(c)I^*_{dc}(-\omega_1) - P(c)I^*_{dc}(\omega_2)}{\omega_1 - \omega_2 - \omega_{ca} + i\Gamma_{ac}}. \tag{9.24a}$$

where the subscript *ca* denotes that these are ground-state $(\omega_1 - \omega_2 = \omega_{ca})$ resonances.

The excited state resonances come from the eight terms in Eq. (9.22) which contain an $\omega_1 - \omega_2 - \omega_{db}$ denominator

$$\phi_{db}(\omega_1, \omega_2) = \sum_{a,c} P(a)\mu_{ab}\mu_{bc}\mu_{cd}\mu_{da}[I_{dc}(2\omega_1 - \omega_2) + I^*_{bc}(\omega_2 - 2\omega_1)]$$
$$\times \frac{I_{da}(\omega_1) + I_{da}(-\omega_2) - I^*_{ba}(-\omega_1) - I^*_{ba}(\omega_2)}{\omega_1 - \omega_2 - \omega_{db} + i\Gamma_{db}}. \tag{9.24b}$$

We shall now invoke the rotating wave approximation (RWA), and retain only resonant terms in which all denominators contain a difference of a field frequency and a molecular optical frequency, and neglect all terms where at least one denominator is antiresonant.

For ground-state CARS, we have

$$\phi_{ca}(\omega_1, \omega_2) = \sum_{b,d} P(a)\mu_{ab}\mu_{bc}\mu_{cd}\mu_{da} \frac{I_{ba}(2\omega_1 - \omega_2)[I_{da}(\omega_1) - \exp(-\omega_{ca}/k_B T)I^*_{dc}(\omega_2)]}{\omega_1 - \omega_2 - \omega_{ca} + i\Gamma_{ac}}, \tag{9.25a}$$

and for excited state CARS

$$\phi_{db}(\omega_1, \omega_2) = \sum_{a,c} P(a)\mu_{ab}\mu_{bc}\mu_{cd}\mu_{da} \frac{I_{dc}(2\omega_1 - \omega_2)[I_{da}(\omega_1) - I^*_{ba}(\omega_2)]}{\omega_1 - \omega_2 - \omega_{db} + i\Gamma_{db}}. \tag{9.25b}$$

Equations (9.25) show that both the excited-state and the ground-state CARS arise from two pathways that may interfere. To make the interference in the excited state resonances more apparent, we rearrange part of Eq. (9.25b) in the form

$$\frac{I_{da}(\omega_1) - I^*_{ba}(\omega_2)}{\omega_1 - \omega_2 - \omega_{db} + i\Gamma_{db}} = \frac{1}{-\omega_2 - \omega_{ab} + i\Gamma_{ab}}\left[1 + \frac{i\hat{\Gamma}}{\omega_1 - \omega_2 - \omega_{db} + i\Gamma_{db}}\right],$$

with

$$\hat{\Gamma} \equiv \Gamma_{ab} + \Gamma_{ad} - \Gamma_{bd} = \gamma_a + \hat{\Gamma}_{ab} + \hat{\Gamma}_{ad} - \hat{\Gamma}_{bd}.$$

As is clearly shown by the second term in the square brackets, the excited-state CARS resonances are induced by dephasing. When $|a\rangle$ is the actual ground vibronic state, $\gamma_a = 0$, and $\hat{\Gamma}$ is then a combination of pure-dephasing widths that vanishes in the absence of pure dephasing. The two pathways thus interfere destructively, and, in the absence of pure dephasing, the CARS excited-state resonances disappear. When pure dephasing is added, the destructive interference is no longer complete, and a dephasing-induced resonance appears. Such resonances have been observed experimentally. They have been denoted PIER 4 (pressure-induced extra resonance in four-wave mixing) [22] or DICE (dephasing-induced coherent emission) [23]. This example is a clear demonstration of how dangerous it is to infer the existence of resonances by simply looking at a "typical" term in $\chi^{(3)}$. Due to inteferences, all the relevant terms have to be carefully combined before any conclusion can be drawn. Equation (9.25b) may suggest that ω_{db} resonances always exist in $\chi^{(3)}$; the following equation shows, however, the exact cancellation of these resonances due to interference in the absence of pure dephasing.

The appearance of new resonances in optical measurements due to dephasing is reminiscent of the generation of the fluorescence, again due to the removal of an exact cancellation between Liouville space paths. Both effects constitute a dramatic illustration of the role of interference in the nonlinear response. In Chapter 17 we shall consider another dephasing-induced resonance: degenerate four wave mixing. It will be calculated for a many body system using equations of motion, but the origin of the resonance is very similar to CARS and it shows up also for the single molecule model considered here.

Substituting the explicit form of $I_{vv'}(\omega)$ in Eqs. (9.23) and (9.24) we have

$$\phi_{ca}(\omega_1, \omega_2) = \phi'_{ca}(\omega_1, \omega_2)\frac{1}{\omega_1 - \omega_2 - \omega_{ca} + i\Gamma_{ca}}, \tag{9.26a}$$

$$\phi_{db}(\omega_1, \omega_2) \cong \phi'_{db}(\omega_1, \omega_2)\frac{1}{\omega_1 - \omega_2 - \omega_{db} + i\Gamma_{db}}, \tag{9.26b}$$

where

$$\phi'_{ca}(\omega_1, \omega_2) = \sum_{b,d} P(a)\mu_{ab}\mu_{bc}\mu_{cd}\mu_{da}\frac{1}{2\omega_1 - \omega_2 - \omega_{ba} + i\Gamma_{ba}}$$
$$\times\left[\frac{1}{\omega_1 - \omega_{da} + i\Gamma_{da}} - \frac{1}{\omega_2 - \omega_{dc} - i\Gamma_{dc}}\exp(-\omega_{ca}/k_B T)\right], \tag{9.27a}$$

and

$$\phi'_{db}(\omega_1, \omega_2) = i\hat{\Gamma} \sum_{a,c} P(a)\mu_{ab}\mu_{bc}\mu_{cd}\mu_{da}$$
$$\times \frac{1}{2\omega_1 - \omega_2 - \omega_{dc} + i\Gamma_{dc}} \frac{1}{\omega_1 - \omega_{da} + i\Gamma_{da}}$$
$$\times \frac{1}{-\omega_2 - \omega_{ab} + i\Gamma_{ab}} \frac{1}{\omega_1 - \omega_2 - \omega_{db} + i\Gamma_{db}}, \qquad (9.27b)$$

where we have retained only the part of ϕ_{bd} that contains the $\omega_1 - \omega_2 - \omega_{db}$ resonance.

For molecules in solution, the vibrational dephasing rates, Γ_{ac} and Γ_{bd} are usually much smaller than a characteristic electronic line broadening, which is typically a few hundred cm^{-1}. A Lorentzian profile with a width of Γ_{ac} or Γ_{bd} may therefore be approximated by a delta function. We then write

$$\left|\frac{1}{\omega_1 - \omega_2 + \omega_{ca} + i\Gamma_{ca}}\right|^2 \cong \frac{\pi\delta(\omega_1 - \omega_2 - \omega_{ca})}{\Gamma_{ca}},$$
$$\left|\frac{1}{\omega_1 - \omega_2 + \omega_{db} + i\Gamma_{db}}\right|^2 \cong \frac{\pi\delta(\omega_1 - \omega_2 - \omega_{db})}{\Gamma_{db}}.$$

Using these approximations, and ignoring background terms, we can recast S_{CARS} in the form

$$S_{\mathrm{CARS}} \propto \frac{\pi}{\Gamma_{ca}} \sum_{a,c} |\phi'_{ca}(\omega_1, \omega_2)|^2 \delta(\omega_1 - \omega_2 - \omega_{ca})$$
$$+ \frac{\pi}{\Gamma_{db}} \sum_{b,d} |\phi'_{db}(\omega_1, \omega_2)|^2 \delta(\omega_1 - \omega_2 - \omega_{db}), \qquad (9.28)$$

$|\phi'_{ca}(\omega_1, \omega_1 - \omega_{ca})|^2$ is the *coherent Raman excitation profile* corresponding to a ground-state resonance, and $|\phi'_{db}(\omega_1, \omega_1 - \omega_{db})|^2$ is the coherent Raman excitation profile for an excited-state resonance. These profiles are easily measured by probing the intensity of a particular Raman line ($\omega_1 - \omega_2$ fixed) as a function of ω_1. For comparison, $|\chi_{ca}(\omega_L)|^2$ is the spontaneous Raman excitation profile. Calculated CARS excitation profiles are shown in Figure 9.13.

Picosecond CARS

Equations (9.27) and (9.28) reveal a fundamental difference between ground-state and excited-state CARS resonances. The interference in Eq. (9.28) occurs between two pathways in which the first interaction occurs, respectively, with ω_1 and with $-\omega_2$. In a frequency-domain experiment we have no control over the relative order in time of both interactions; both pathways then contribute equally, and they interfere. The situation is quite different when the CARS experiment is carried out in the time domain. In this mode, CARS is a particular transient

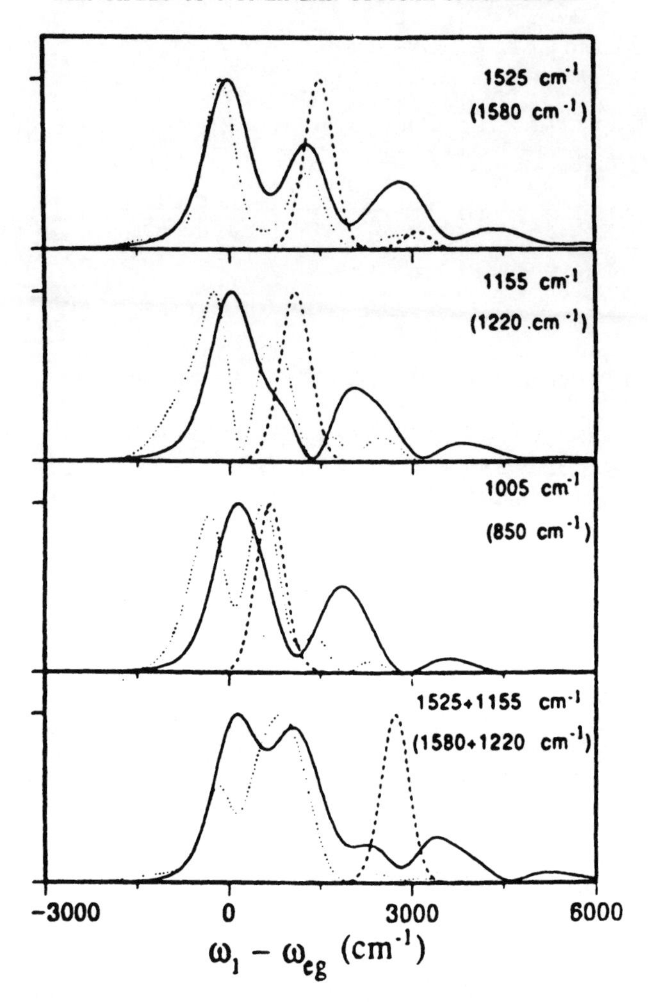

FIG. 9.13 Calculated Raman excitation profiles of β-carotene. A harmonic three-mode model was used with ground state frequencies $\omega_1'' = 1525$, $\omega_2'' = 1155$, and $\omega_3'' = 1005\ \text{cm}^{-1}$ and excited state frequencies $\omega_1' = 1580$, $\omega_2' = 1220$, and $\omega_3' = 850\ \text{cm}^{-1}$. The displacements are $d_1 = 1.12$, $d_2 = 0.95$, and $d_3 = 0.65$. The bath parameters are $\Delta = 362\ \text{cm}^{-1}$ and $\Lambda = 109\ \text{cm}^{-1}$ corresponding to $\kappa = 0.3$ and fwhm $\Gamma_0 = 760\ \text{cm}^{-1}$. In each panel we show the spontaneous Raman (solid line), the Raman excitation profile corresponding to ground-state resonance (dotted line), and the corresponding excited-state resonance (dashed line). The Raman frequency ω_{ca} is indicated in each panel; the corresponding excited-state frequency ω_{db} is indicated in parentheses. All curves have been normalized to the same maximum peak value. [Adapted from S. Mukamel, "Solvation effects on four-wave mixing and spontaneous Raman and fluorescence lineshapes of polyatomic molecules," *Adv. Chem. Phys.* **70**, 165 (1988).]

grating technique. A general discussion of time-domain resonant gratings will be given in Chapter 11. In Chapter 14 we shall consider off resonant Raman spectroscopy where only the ground-state resonances are observed. We shall then present an in-depth analysis from the cw to the impulsive (femtosecond) limits. The following discussion will therefore be brief, and is primarily intended to focus on the issue of interference in excited state CARS.

A time-domain CARS experiment is performed by sending two time-coincident pulses with wavevectors $\mathbf{k}_1$ and $\mathbf{k}_2$. After a variable delay, τ, a second $\mathbf{k}_1$ pulse is applied, and the total coherent emission at $\mathbf{k}_s = 2\mathbf{k}_1 - \mathbf{k}_2$ is detected [24, 25]. We assume that all pulses have the same shape, $E(t)$, and that the pulse duration is short compared with τ. The radiation field is given by

$$E(\mathbf{r}, t) = E(t + \tau)\exp[i\mathbf{k}_1\mathbf{r} - i\omega_1 t] + E(t + \tau)\exp[i\mathbf{k}_2\mathbf{r} - i\omega_2 t] + E(t)\exp[i\mathbf{k}_1\mathbf{r} - i\omega_1 t] + c.c. \tag{9.29}$$

Since thc last interaction has to be with ω_1, we find that there are only two permutations of frequencies that contribute to Eq. (5.45):

$$\begin{aligned} P(\mathbf{k}_s, t) = \int_0^\infty dt_3 \int_0^\infty dt_2 \int_0^\infty dt_1\, & S^{(3)}(t_3, t_2, t_1)E(t - t_3)\exp[i(2\omega_1 - \omega_2)t_3] \\ & \times \{E^*(t + \tau - t_2 - t_3)E(t + \tau - t_1 - t_2 - t_3)\exp[i\omega_1 t_1 + i(\omega_1 - \omega_2)t_2] \\ & + E(t + \tau - t_2 - t_3)E^*(t + \tau - t_1 - t_2 - t_3) \\ & \times \exp[-i\omega_2 t_1 + i(\omega_1 - \omega_2)t_2]\}. \end{aligned} \tag{9.30}$$

Equation (9.30), which is the time-domain analog of Eq. (9.27), contains 16 terms. We now make the following assumptions: The pulse envelope $E(t)$ is sufficiently long that its spectral bandwidth is narrow enough to select a particular resonance ($\omega_1 - \omega_2 = \omega_{db}$ or $\omega_1 - \omega_2 = \omega_{ac}$). We further invoke the RWA, so that the same terms that contribute in the frequency domain to Eqs. (9.23) and (9.24) will contribute here. On the other hand, we take the pulses to be sufficiently short that we can neglect the matter evolution during the pulses. This allows us to set $E(t) = \delta(t)$. We then get for the ground-state CARS

$$P_{ca}(\mathbf{k}_s, t) = \sum_{a,b,c,d} P(a)\mu_{ab}\mu_{bc}\mu_{cd}\mu_{da} I_{ba}(t - \tau)I_{ca}(\tau)\exp[i(2\omega_1 - \omega_2)t - i\omega_1\tau], \tag{9.31a}$$

and for the excited-state CARS

$$P_{db}(\mathbf{k}_s, t) = 2\sum_{a,b,c,d} P(a)\mu_{ab}\mu_{bc}\mu_{cd}\mu_{da} I_{dc}(t - \tau)I_{db}(\tau)\exp[i(2\omega_1 - \omega_2)t - i\omega_1\tau]. \tag{9.31b}$$

The frequency-domain interference of Eq. (9.25b) is no longer present. The system does not have time to evolve between the first two interactions, and the two pathways that contribute to Eq. (9.31b) give an identical contribution (rather than cancel!). When the signal is probed as a function of τ at $t = \tau$, we have

$$S_{ca}(\tau) \propto |I_{ca}(\tau)|^2 = \exp(-2\Gamma_{ca}\tau)$$

and

$$S_{db}(\tau) \propto |I_{db}(\tau)|^2 = \exp(-2\Gamma_{db}\tau).$$

Time-domain CARS can be used to probe excited-state resonances even in the absence of pure dephasing, since the destructive interference of frequency-domain CARS does not apply [26]. The present analysis clarifies the origin of this difference.

APPENDIX 9A

Spontaneous Light Emission with Spectral Diffusion and Vibrational Relaxation

In Chapter 8 we derived an expression for $\chi^{(3)}$ for a system coupled to an overdamped Brownian oscillator that represents spectral diffusion [Eq. (8.105)]. By combining that expression with Eqs. (9.12) we immediately obtain a closed expression for the SLE lineshape for this model [24]. In the following expression we have also included vibrational relaxation in the ground-state and in the excited-state manifolds using a master equation [see Eqs. (6.A.7)–(6.A.9)] with a tetradic relaxation matrix Γ_{VR}. We thus have

$$\begin{aligned}
S_{\text{SLE}}(\omega_{\text{L}}, \omega_{\text{S}}) &= 2\,\text{Im} \sum_{a,c,b\neq d} P(a)\mu_{ab}\mu_{bc}\mu_{cd}\mu_{da} \sum_{n=0}^{\infty} \frac{z^{*n}}{n!} \\
&\times \left[\frac{(-1)^{n+1} J_n(\omega_{bc} - \omega_{\text{S}} + i\Gamma_{bc}) J_n'^{*}(\omega_{\text{L}} - \omega_{ba} + i\Gamma_{ba})}{\omega_{bd} + i(\Gamma_{db} + n\Lambda)} \right. \\
&\left. + \frac{J_n^{*}(\omega_{dc} - \omega_{\text{S}} + i\Gamma_{dc}) J_n'^{*}(\omega_{ab} + \omega_{\text{L}} + i\Gamma_{ab})}{-\omega_{db} + i(\Gamma_{db} + n\Lambda)} \right] \\
&+ 2\,\text{Im} \sum_{a\neq c,b,d} P(a)\mu_{ab}\mu_{bc}\mu_{cd}\mu_{da} \\
&\times \sum_{n=0}^{\infty} \frac{z^{*n}}{n!} \left[-\frac{J_n(\omega_{\text{S}} - \omega_{dc} + i\Gamma_{dc}) J_n^{*}(\omega_{ab} + \omega_{\text{L}} + i\Gamma_{ab})}{\omega_{\text{S}} - \omega_{\text{L}} - \omega_{ac} + i(\Gamma_{ac} + n\Lambda)} \right] \\
&+ 2\,\text{Im} \sum_{a,b,c,d} P(a)\mu_{ab}\mu_{bc}\mu_{cd}\mu_{da} \sum_{n=0}^{\infty} \frac{z^{*n}}{n!} \\
&\times [(-1)^{n+1} J_n(\omega_{bc} - \omega_{\text{S}} + i\Gamma_{bc}) J_n'^{*}(\omega_{\text{L}} - \omega_{ba} + i\Gamma_{ba})(-i\Gamma_{\text{VR}}^{-1})_{bb,dd} \\
&+ J_n^{*}(\omega_{dc} - \omega_{\text{S}} + i\Gamma_{dc}) J_n'^{*}(\omega_{ab} + \omega_{\text{L}} + i\Gamma_{ab})(-i\Gamma_{\text{VR}}^{-1})_{bb,dd} \\
&- J_n(\omega_{\text{S}} - \omega_{dc} + i\Gamma_{dc}) J_n^{*}(\omega_{ab} + \omega_{\text{L}} + i\Gamma_{ab})(\omega_{\text{S}} - \omega_{\text{L}} + i\Gamma_{\text{VR}})_{aa,cc}^{-1}].
\end{aligned} \tag{9.A.1}$$

Using Eq. (8.102), the absorption lineshape for this model is given by

$$\sigma_a(\omega_{\text{L}}) \propto \sum_{a,b} P(a)|\mu_{ab}|^2 J''(\omega_{\text{L}} - \omega_{ba}). \tag{9.A.2}$$

When the system attains thermal equilibrium in the excited state prior to the emission process, the relaxed fluorescence can be calculated using Eqs. (8.B.8) and (8.B.9)

$$\sigma_f(\omega_s) \propto \sum_{a,b} P(b)|\mu_{ab}|^2 J''(\omega_s - \omega_{ba} + 2\lambda). \tag{9.A.3}$$

Eqs. (9.A.1) reduced to (9.A.3) when vibrational relaxation is fast compared with the radiative decay timescale. In a similar way, it is possible to include the present vibrational relaxation model in the third-order nonlinear response and to generalize Eqs. (8.105), replacing the two photon resonance terms with the inverse of the relaxation matrix Γ_{VR}.

In the absence of vibrational relaxation we have

$$S_{SLE}(\omega_L, \omega_S) = 2\,\mathrm{Im} \sum_{a,b,c,d} P(a)\mu_{ab}\mu_{bc}\mu_{cd}\mu_{da} \sum_{n=0}^{\infty} \frac{z^{*n}}{n!}$$
$$\times \left[\frac{(-1)^{n+1} J_n(\omega_{bc} - \omega_S + i\Gamma_{bc}) J_n'^{*}(\omega_L - \omega_{ba} + i\Gamma_{ba})}{\omega_{bd} + i(\Gamma_{bd} + n\Lambda)} \right.$$
$$+ \frac{J_n^{*}(\omega_{dc} - \omega_S + i\Gamma_{dc}) J_n'^{*}(\omega_{ab} + \omega_L + i\Gamma_{ab})}{-\omega_{db} + i(\Gamma_{db} + n\Lambda)}$$
$$\left. - \frac{J_n(\omega_S - \omega_{dc} + i\Gamma_{dc}) J_n^{*}(\omega_{ab} + \omega_L + i\Gamma_{ab})}{\omega_S - \omega_L - \omega_{ac} + i(\Gamma_{ac} + n\Lambda)} \right]. \tag{9.A.4}$$

The summation over n in these expressions represents the finite timescale of the oscillator and the expansion parameter is Λ/Γ. For $\Lambda/\Gamma \gg 1$ we can truncate the expansion, retaining only the $n = 0$ term. We then recover Eqs. (9.14). As the bath timescale $\hbar/\Lambda$ becomes slower, we have to keep more terms in the n expansion. In the static limit $\Lambda/\Gamma \ll 1$, it might be preferential to convolute the homogeneous expression with a Gaussian distribution of ω_{eg} [Eqs. (6.65)] rather than carry out the n summation to infinity.

NOTES

1. Proceedings of the 1993 International Conference on Luminescence, *J. Luminesc.* **60, 61** (1994).
2. J. B. Birks, Ed. *Organic Molecular Photophysics*, Vol. 2 (Wiley, London, 1975).
3. D. Lee and A. C. Albrecht, in *Advances in Infrared and Raman Spectroscopy*, Vol. 12, R. J. H. Clark and R. E. Hester, Eds. (Wiley-Heyden, New York, 1985), pp. 170–213.
4. D. L. Rousseau and P. F. Williams, *J. Chem. Phys.* **64**, 3519 (1976).
5. G. R. Fleming, *Chemical Applications of Ultrafast Spectroscopy* (Oxford, London, 1986).
6. W. R. Lambert, P. M. Felker, and A. H. Zewail, *J. Chem. Phys.* **75**, 5958 (1981); P. M. Felker and A. H. Zewail, *Adv. Chem. Phys.* **70**, 265 (1988).
7. J. R. Lakowicz, *Principles of Fluorescence Spectroscopy* (Plenum, New York, 1983).
8. R. M. Hochstrasser and C. A. Nyi, *J. Chem. Phys.* **70**, 1112 (1979).
9. M. Cardona and G. Guntherodt, Eds. *Light Scattering in Solids VI* (Springer-Verlag, Berlin, 1991).
10. S. Mukamel, *Adv. Chem. Phys.* **70** (Part I), 165 (1988); S. Mukamel, *J. Chem. Phys.* **82**, 5398 (1985).
11. P. W. Milonni, *Phys. Rep.* **25**, 1 (1976).

12. V. Hizhnykov and I. Tehver, *Phys. Status Solid* **21**, 755 (1967); *Opt. Commun.* **32**, 419 (1980).

13. D. L. Huber, *Phys. Rev.* **158**, 843 (1967); **170**, 418 (1968); **178**, 93 (1969).

14. A. Omont, E. W. Smith, and J. Cooper, *Astrophys. J.* **175**, 185 (1972).

15. Y. Toyozawa, *J. Phys. Soc. Jpn.* **41**, 400 (1976); A. Kotani and Y. Toyozawa, *J. Phys. Soc. Jpn.* **41**, 1699 (1976).

16. S. Mukamel, A. Ben-Reuven, and J. Jortner, *Phys. Rev. A* **12**, 947 (1975); *J. Chem. Phys.* **64**, 3971 (1976).

17. T. Takagahara, E. Hanamura, and R. Kubo, *J. Phys. Soc. Jpn.* **43**, 802, 811, 1522 (1977); M. Aihara, *Phys. Rev. B* **25**, 53 (1982).

18. K. Burnett, *Phys. Rep.* **118**, 339 (1985).

19. A. B. Myers and R. A. Mathies, in *Biological Applications of Raman Spectroscopy*, Vol. 2, T. G. Spiro, Ed. (Wiley, New York, 1987), p. 1.

20. Y. J. Yan and S. Mukamel, *J. Chem. Phys.* **85**, 5908 (1986); **86**, 6085 (1987); S. Mukamel and Y. J. Yan, in *Recent Trends in Raman Spectroscopy*, S. B. Banerjee and S. S. Jha, Eds. (World Scientific, Singapore, 1989), p. 160.

21. N. Bloembergen, *Am. J. Phys.* **35**, 989 (1967).

22. A. R. Bogdan, M. W. Downer, and N. Bloembergen, *Phys. Rev. A* **24**, 623 (1981); L. J. Rothberg and N. Bloembergen, *Phys. Rev. A* **30**, 820 (1984); L. Rothberg, in *Progress in Optics*, Vol. 24, E. Wolf, Ed. (North-Holland, Amsterdam, 1987), p. 38.

23. J. R. Andrews and R. M. Hoschstrasser, *Chem. Phys. Lett.* **82**, 381 (1981).

24. A. Laubereau and W. Kaiser, *Rev. Mod. Phys.* **50**, 607 (1978).

25. D. P. Weitekamp, K. Duppen, and D. A. Wiersma, *Phys. Rev. A* **27**, 3089 (1983).

26. S. Mukamel and R. G. Loring, *J. Opt. Soc. Am. B* **3**, 595 (1986).

CHAPTER 10

Selective Elimination of Inhomogeneous Broadening: Photon Echoes

In Chapter 6 we first introduced the concepts of homogeneous and inhomogeneous broadening and presented their formal definitions in terms of the factorization of multitime correlation functions. In Chapter 7 we used a semiclassical approximation for the response functions in the coordinate representation to show how inhomogeneous broadening and spectral diffusion are obtained in the static limit. Later, in Chapter 8 we showed that the overdamped Brownian oscillator model can interpolate between these limits by varying its relaxation timescale Λ^{-1}. We further provided a simple procedure, the inhomogeneous cumulant expansion [see Eqs. (8.83) and (8.84)], which allows the incorporation of inhomogeneous broadening in the simulation of optical response functions. Since inhomogeneous broadening merely implies a static averaging over some distribution of slow degrees of freedom Γ, the procedure may be straightforwardly extended without alluding to the cumulant expansion. We can write

$$J(t) = \int d\Gamma \; W(\Gamma) J_{\mathrm{H}}(t; \Gamma), \tag{10.1}$$

where J_{H} denotes the homogeneous part of J. This generalizes Eq. (8.83), which contains a specific approximation for J_{H}.

In condensed phases, the variation in local environments alters the optical transition frequencies ω_{eg} of individual molecules. In many cases, the distribution of ω_{eg} constitutes the dominant inhomogeneous broadening mechanism. We then assume that the averaged correlation function of a two level particle [Eq. (7.8)] can be factorized as follows:

$$J(t) = J_{\mathrm{H}}(t)\chi(t), \tag{10.2}$$

with

$$J_{\mathrm{H}}(t_1) \equiv \exp(-i\langle\omega_{eg}\rangle_1 t_1)\langle\langle V_{eg}|\mathcal{G}_{gg}(t_1)\exp_+\left[-\frac{i}{\hbar}\int_0^{t_1} d\tau\; U(\tau)\right]|V_{eg}\rho_g\rangle\rangle, \tag{10.3}$$

and

$$\chi(t) \equiv \langle \exp[-i(\omega_{eg}(\Gamma) - \langle\omega_{eg}\rangle_I)t]\rangle_I$$
$$\equiv \int_{-\infty}^{\infty} d\omega_{eg} \exp[-i(\omega_{eg} - \langle\omega_{eg}\rangle_I)t] W(\omega_{eg} - \langle\omega_{eg}\rangle_I). \quad (10.4)$$

Here the subscript I reminds us that this averaging is with respect to W, the inhomogeneous distribution function of the electronic transition frequency, centered at $\langle\omega_{eg}\rangle_I$. From Eq. (10.4) we have

$$W(\omega) = \frac{1}{2\pi}\int_{-\infty}^{\infty} dt \exp(i\omega t)\chi(t). \quad (10.5)$$

$\chi(t)$ is typically a rapidly decaying function that represents *inhomogeneous dephasing*, i.e., the total macroscopic polarization decays since the contributions of various molecules go out of phase. J_H is the *homogeneous* contribution, which contains valuable structural information (level positions and transition dipoles) as well as dynamic information related to relaxation processes. We then get

$$S^{(1)}(t) = \frac{i}{\hbar}\theta(t)[J_H(t)\chi(t) - J_H^*(t)\chi^*(t)]. \quad (10.6)$$

The rapid macroscopic decay of $\chi(t)$ often dominates the slower microscopic homogeneous contribution $J_H(t)$, making $J(t)$ insensitive to $J_H(t)$. As discussed in Chapter 6, the absorption lineshape is related to the Fourier transform of $J(t)$. The insensitivity of $J(t)$ to the homogeneous contribution translates into a similar insensitivity of the absorption lineshape. Substituting Eqs. (10.2) and (10.5) into Eq. (6.66b), we have

$$\sigma_a(\omega) = \int_{-\infty}^{\infty} \sigma_a^H(\omega - \omega')W(\omega')\, d\omega'. \quad (10.7)$$

The static distribution of the electronic transition frequency ω_{eg} thus enters the absorption lineshape as a *convolution* with a broad envelope $W(\omega')$ whose typical width for a dye in solution is hundreds of wavenumbers. The inhomogeneous linewidth is often much broader than all other features in the spectrum. Consequently, most useful structural and dynamic information is hidden underneath this broad envelope. This makes it impossible to extract this information out of the linear absorption spectrum and the dielectric function in condensed phases [1, 2]. This is typical for spectra in solutions, liquids, glasses, proteins, and molecular crystals. In some cases it is possible to overcome this problem by working with carefully prepared systems, e.g., ultracold molecules in supersonic beams [3], as shown in Figure 10.1, or mixed organic crystals at cryogenic temperatures, which reduce the effects of inhomogeneous broadening. When it becomes narrower than the homogeneous contributions, we can set $W(\omega) = \delta(\omega)$ and Eq. (10.7) reduces to $\sigma_a(\omega) = \sigma_a^H(\omega)$. Fluorescence excitation spectra observed using very small irradiation volumes and high ($\sim$100 MHz) spectral resolution have the capacity to observe the spectrum of a single molecule underneath an inhomogeneous line [4–6]. Some remarkable examples are shown in Figures 10.2 and 10.3. Near field microscopy is another technique that can select individual molecules with a spatial resolution of tens of nanometers, smaller than the

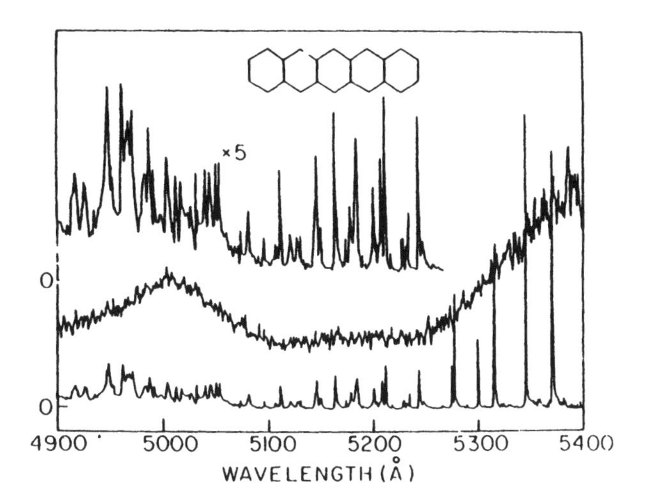

FIG. 10.1 Fluorescence excitation spectrum of the $S_0(^1A_{1g}) \rightarrow S_1(^1B_{2u})$ transition in isolated pentacene molecule cooled in a supersonic expansion. The structureless spectrum is a high temperature sequence-congested spectrum. The structured (lower and upper) spectra corresponds to the ultracold molecule. The exciting dye laser has spectral bandwidth $0.3\,\text{cm}^{-1}$. [Adapted from A. Amirav, U. Even, and J. Jortner, "Intramolecular dynamics of an isolated ultracold large molecule," *Opt. Commun.* **32**, 266 (1980).]

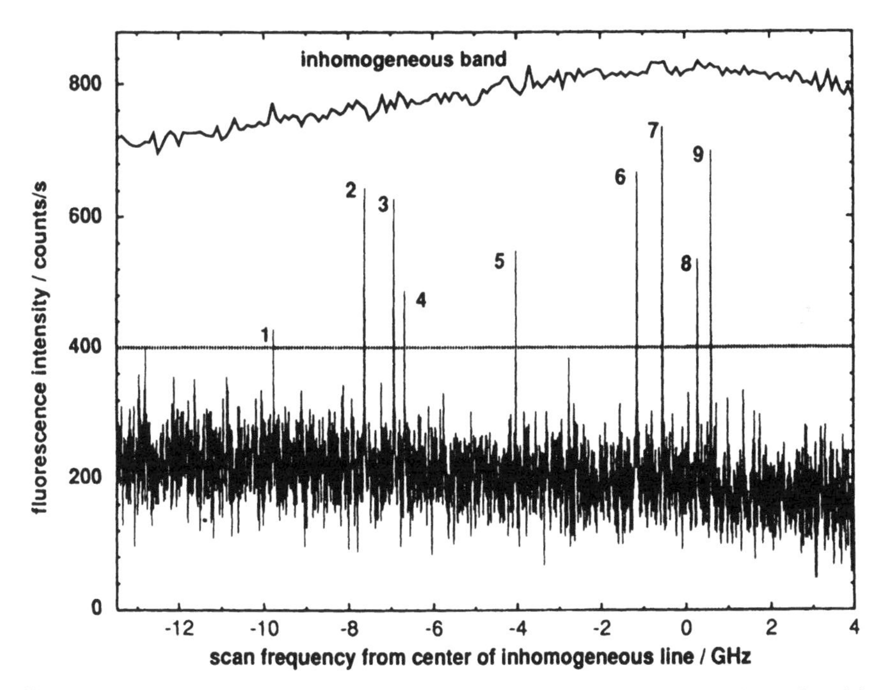

FIG. 10.2 Single molecule fluorescence excitation spectra of pentacene in *p*-terphenyl near the origin of the inhomogeneously broadened O_1 site. The dotted line represents an arbitrary threshold to distinguish signals from single molecules from the background. The excitation spectrum of a more concentrated sample is shown above to indicate the shape of the inhomogeneous absorption band. [U. P. Wild, F. Güttler, M. Pirotta, and A. Renn, "Single molecule spectroscopy: Stark effect of pentacene in p-terphenyl," *Chem. Phys. Lett.* **193**, 451 (1992).]

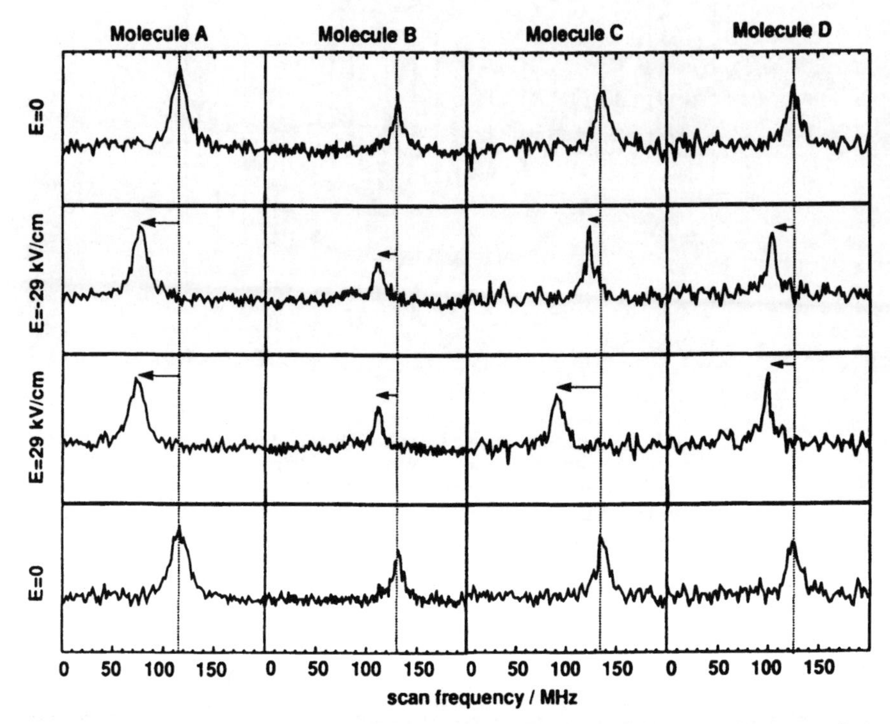

FIG. 10.3 Fluorescence excitation scans of the same system of Figure 10.2. Shown are four molecules measured on the red side of the O_1 site absorption at opposite applied electric fields. The electric-field-induced transition energy shifts are indicated by arrows. Note the different shifts for the individual molecules at the same electric field strength. Also compare the spectral shifts at positive and negative electric field strength, especially for molecule C.

wavelength of light [7]. The technique uses a subwavelength-sized source (e.g., the tip of an optical fiber) located very close to the sample. The diffraction limit no longer applies since the light source is used in the near field regime (closer than an optical wavelength).

For systems that can be described by the second-order cumulant expansion (e.g., the Brownian oscillator model), inhomogeneous broadening has a Gaussian profile whereas homogeneous broadening is Lorentzian, i.e.,

$$\chi(t) = \exp(-\Delta^2 t^2/2), \tag{10.8a}$$

$$J_{\rm H}(t) = \exp(-i\omega_{eg}t - \hat{\Gamma}t), \tag{10.8b}$$

which in the frequency domain yields

$$W(\omega) = \frac{1}{\sqrt{2\pi}\,\Delta} \exp[-\omega^2/2\Delta^2] \tag{10.8c}$$

and

$$\sigma_a^{\rm H}(\omega) = \frac{1}{\pi} \frac{\hat{\Gamma}}{(\omega - \omega_{eg})^2 + \hat{\Gamma}^2}. \tag{10.8d}$$

In Eqs. (10.8b), (10.8d), and hereafter in this book, we simplify the notation and denote $\langle\omega_{eg}\rangle_I$ by ω_{eg}.

The lineshape given by Eqs. (10.7) together with (10.8) is known as the Voigt profile. It appears in, e.g., pressure broadening in the gas phase. The Lorentzian then represents the contribution of fast collisions (the impact limit) whereas the Gaussian results from Doppler broadening (Gaussian distribution of speeds) [8]. In general, the conditions for validity of a simple reduced equation of motion (without memory) (Appendix 6A) justify a Lorentzian profile whereas a Gaussian may be rationalized in many cases using the central limit theorem. A simulation of an inhomogeneously broadened system as a function of the number of molecules is shown in Figure 10.4. It demonstrates how the Gaussian profile is obtained for a large system with many molecules. The overdamped Brownian oscillator model interpolates continuously between these limits, by varying the bath timescale. In general, as is evident from the formal definitions given in Chapter 6, the inhomogeneous profile is not necessarily a Gaussian, whereas the homogeneous profile is not necessarily a Lorentzian. In fact, a Lorentzian inhomogeneous profile resulting from dipole–dipole interactions among randomly distributed absorbers does exist in doped crystals and glasses [2].

Since homogeneous and inhomogeneous contributions enter the linear response in a symmetric fashion (a convolution in the frequency domain or a product in the time domain) then, by definition, linear optical measurements cannot distinguish between them; the linear response is invariant under the interchange of the roles of $J_H(t)$ and $\chi(t)$! This fundamental limitation, however, seldom prevents researchers from speculating about the magnitude of the homogeneous linewidth based solely on the linear absorption lineshape. If one assumes a certain profile for each of the contributions, then it may become possible to extract both contributions by a deconvolution of the lineshape. Such procedures are subjective and rely on the model used for the system. Moreover, their accuracy becomes poor when one of the contributions is dominant.

Certain nonlinear optical techniques have the capacity to selectively eliminate inhomogeneous broadening, thus providing valuable dynamic information even when linear optical measurements are totally useless. To get a feel for that capacity, let us evaluate the inhomogeneous contribution to the third-order

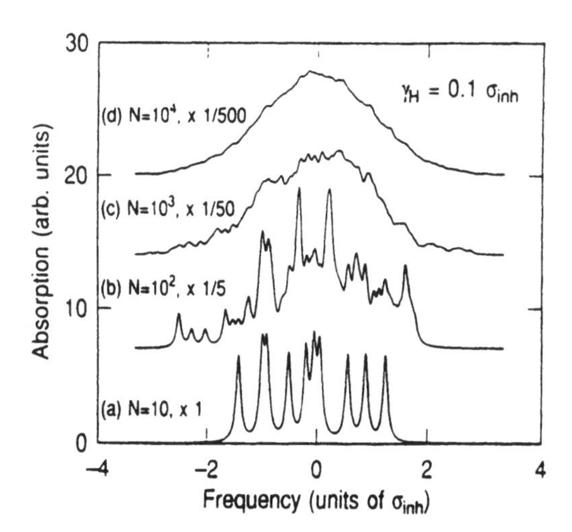

FIG. 10.4 Simulated absorption spectra with different total number of absorbers N, using a Gaussian random variable to select center frequencies. Traces (a) through (d) correspond to N values of 10, 100, 1000, and 10,000, respectively. The homogeneous width is taken to be one-tenth of the inhomogeneous width. [W. E. Moerner and T. Basche, "Optical spectroscopy of single impurity molecules in solids," *Angew. Chem.* **105**, 537 (1993).]

nonlinear response function. By performing an ensemble average of Eq. (7.11) or (7.19) over the distribution of ω_{eg} we get

$$R_\alpha(t_3, t_2, t_1) = R_\alpha^{\mathrm{H}}(t_3, t_2, t_1)\chi(t_3 \pm t_1). \tag{10.9}$$

Here R_α^{H}, the homogeneous part of R_α, is given by Eqs. (7.19) with ω_{eg} taken as $\langle\omega_{eg}\rangle_{\mathrm{I}}$, the electronic transition frequency averaged over the inhomogeneous distribution. The χ function [Eq. (10.4)] comes from the ensemble average of the $\exp[-i\omega_{eg}(t_3 \pm t_1)]$ factor. The sign depends on the path α (a plus sign for R_1 and R_4 and a minus sign for R_2 and R_3). Combining these with Eqs. (5.27), we finally get [9]

$$\begin{aligned} S^{(3)}(t_3, t_2, t_1) = {} & \left(\frac{i}{\hbar}\right)^3 \theta(t_1)\theta(t_2)\theta(t_3)[R_2^{\mathrm{H}}(t_3, t_2, t_1) + R_3^{\mathrm{H}}(t_3, t_2, t_1)]\chi(t_3 - t_1) \\ & + \left(\frac{i}{\hbar}\right)^3 \theta(t_1)\theta(t_2)\theta(t_3)[R_1^{\mathrm{H}}(t_3, t_2, t_1) + R_4^{\mathrm{H}}(t_3, t_2, t_1)]\chi(t_3 + t_1) + c.c. \end{aligned} \tag{10.10}$$

Here, as in Eq. (10.2), we have factorized the total response function into two parts: R_α^{H} ($\alpha = 1, \ldots, 4$) represent the dynamic contributions to the third-order response function $S^{(3)}$ in the absence of inhomogeneous broadening, whereas $\chi(t)$ represents the static (inhomogeneous) contributions. The nonlinear response function depends on three time variables, and the homogeneous and the inhomogeneous contributions depend on these variables in very different ways. This opens up various opportunities for the selective elimination of the latter. We note that the term "inhomogeneous broadening" in this chapter refers strictly to a static distribution of transition frequencies. One can think of other types of inhomogeneities as discussed, e.g., in the inhomogeneous cumulant expansion in Chapter 8, where the Brownian oscillator parameters have some static distribution. In that case, however, the response functions cannot be factorized in the form of Eqs. (10.5) or (10.10), and such inhomogeneities are not eliminated by the techniques described below.

Hole burning [5] and fluorescence line narrowing [10] are common spectroscopic techniques used to probe molecular dynamics and optical dephasing processes by selectively eliminating inhomogeneous broadening. These techniques, discussed in Chapter 13, involve a selective narrow band excitation of a subgroup of molecules, which are then monitored by measuring the spontaneous emission (fluorescence line narrowing) or the absorption of a weak probe (hole burning). Only a small fraction of the molecules within the broad inhomogeneous distribution is selectively investigated in this case, resulting in a partial elimination of inhomogeneous broadening. Another class of measurements that is not sensitive to inhomogeneous broadening is Raman or impulsive pump-probe spectroscopy [11]. Here, the molecular dynamics during the t_2 evolution period, whereby the system is in an electronic population (ρ_{gg} or ρ_{ee}) rather than a coherence (ρ_{eg}), is being probed. During that period there are, of course, no electronic dephasing processes. This is why Raman lines are so narrow (compared with fluorescence) or show up as vibrational coherent quantum beats in femtosecond Raman or pump-probe spectroscopy. Stationary Raman was analyzed in Chapter 9 and impulsive Raman and pump-probe techniques will be considered in Chapters 11 and 14. These examples demonstrate the broad range of options offered by time-domain and frequency-domain nonlinear techniques for eliminating inhomogeneous broadening.

In this chapter we shall analyze photon echo experiments [12], in which the elimination of inhomogeneous broadening is of very different origin. In contrast to the other line narrowing techniques mentioned above, the broad-band pulsed excitation in photon echo spectroscopy is nonselective, and usually the entire inhomogeneous distribution is excited. The ability of the echo technique to eliminate inhomogeneous contributions to the signal is the result of a combined effect of two carefully designed periods of evolution where the optical inhomogeneous dephasing in the first period is exactly cancelled by a rephasing process in the second period. Picosecond [12–15] and femtosecond echo [16–18] experiments have been widely used to obtain the homogeneous lineshape in various systems including crystals, molecular gases at low pressure, liquids, and glasses. Photon echoes are the optical analogues of the spin echoes predicted and observed in nuclear magnetic resonance by Hahn [19]. The standard description of spin echoes uses a strong field that saturates the system (π and $\pi/2$ pulses) [20]. The following treatment, which is based on the response function, shows that photon echo is essentially a weak field effect that can be described as a four-wave mixing process. By using intense (and short) fields the signal becomes stronger but its shape and essential properties do not change. We note that photon echo spectroscopy eliminates the broadening of a single molecule in a bath. In nuclear magnetic resonance there exist more sophisticated multiple pulse techniques that can eliminate intermolecular dipole–dipole interactions [20, 21]. Strong field methods for the selective elimination of interactions (spin decoupling) have been developed as well, and have their analogues in optical spectroscopy [22].

In the following we shall first survey the possible three-pulse four-wave techniques, and identify the configurations leading to photon echoes. We next present a unified theory for the various photon echo techniques, which include the two-pulse photon echo (PE), the three-pulse (stimulated) photon echo (SPE), the accumulated photon echo (APE), and the incoherent photon echo. The latter two are discussed in Appendices 10A and 10B. All of these echo signals can be expressed as a convolution of a same material correlation function [Eq. (10.15)] with the proper sequence of excitation fields. Following that, we consider ideal impulsive echo experiments that use ultrashort pulses, in which the echo signals are directly related to the material correlation function, and all time integrations may be eliminated. We next apply the multimode Brownian oscillator model, which can account for quantum beats resulting from high frequency vibrations as well as low frequency overdamped nuclear modes, for calculating the echo signal. Finally, we establish the connection between photon echo and a different ultrafast technique: spontaneous light emission measured using two pulses with control over their phases (phase locked SLE).

CLASSIFICATION OF TIME-DOMAIN RESONANT FOUR-WAVE MIXING TECHNIQUES

When the distinction between time-domain and frequency-domain spectroscopies was first made in Chapter 5 [Eqs. (5.57) and (5.58)], we indicated that even ideal

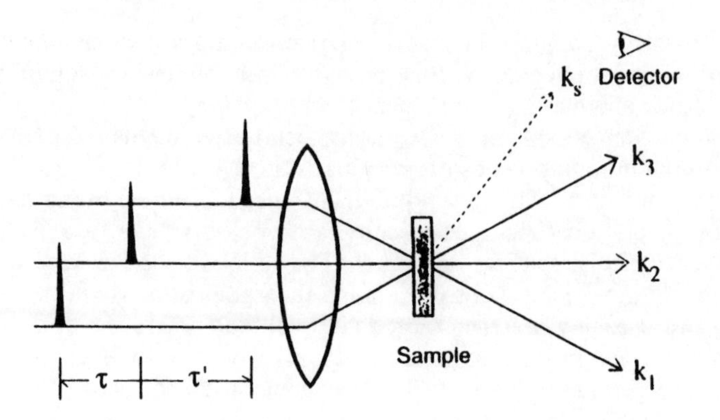

FIG. 10.5 Layout of a three-pulse time-domain four-wave mixing experiment. Time ordering is enforced. The system interacts first with $\mathbf{k}_1$ then $\mathbf{k}_2$ and finally with $\mathbf{k}_3$. The signal in the $\mathbf{k}_s$ direction is observed as a function of the relative delay times τ and τ'.

impulsive time-domain experiments do have some frequency-domain character since light pulses are never short compared with optical frequencies. Consequently, in resonant techniques we invoke the *rotating wave approximation* (RWA) and select the relevant, slowly varying terms. We shall now consider ideal time-domain four wave mixing measurements, performed using three well-separated pulses, and show how by controlling the time ordering of the pulses, and by selecting the wavevector of the signal, we may choose a particular group of Liouville space paths. Consider the external field (Figure 10.5)

$$E(\mathbf{r}, t) = E_1(t + \tau + \tau') \exp(i\mathbf{k}_1\mathbf{r} - i\omega_1 t) + E_2(t + \tau) \times \exp(i\mathbf{k}_2\mathbf{r} - i\omega_2 t) + E_3(t) \exp(i\mathbf{k}_3\mathbf{r} - i\omega_3 t) + c.c. \quad (10.11)$$

Here, $E_j(t)$ denotes the temporal envelope of the jth incident pulse, while ω_j denotes its mean frequency. The three incident pulses are delayed by the time intervals τ' and τ. The possible four wave mixing signals could show up in all of the eight directions $\mathbf{k}_s = \pm\mathbf{k}_1 \pm \mathbf{k}_2 \pm \mathbf{k}_3$, where every combination of signs is, in principle, possible. In practice, however, only some of them are dominant.

We shall consider a two electronic level system and assume that all frequencies are near resonance $\omega_1, \omega_2, \omega_3 \sim \omega_{eg}$. For any combination of pulses, some of the terms in $P^{(3)}$ will contain differences of optical and electronic transition frequencies $(\omega_j - \omega_{eg})$ and others will contain sums of these frequencies $(\omega_j + \omega_{eg})$. The later type of terms will be highly oscillatory and upon performing the time integrations will make a negligible contribution to the signal. When these terms are neglected, we obtain the rotating wave approximation. The RWA was first introduced in Chapter 5 and discussed following the introduction of double-sided Feynman diagrams in Chapter 6. Invoking the RWA, we require that during each of the propagation periods (t_1, t_2, and t_3) the material and field frequencies come only in combinations such as $\exp[\pm i(\omega_j - \omega_{eg})t]$, and neglect terms where at least one propagation is of the form $\exp[\pm i(\omega_j + \omega_{eg})t]$. (In addition, phase matching, resulting from the frequency dispersion of the index of refraction, contributes also to the wavevector selectivity. This was discussed in Chapter 4 and will not be considered here.) It is then clear that the first two interactions must have an opposite sign of frequency (one interaction with E, the other with E^*). Otherwise, during the t_2 period we will get a highly oscillatory term at twice the optical frequency. This also means that in the RWA the first two field interactions create gratings with wavevectors $\pm(\mathbf{k}_2 - \mathbf{k}_1)$. Therefore, the dominant four wave mixing signals come at four possible directions. $\pm\mathbf{k}_3 \pm (\mathbf{k}_2 - \mathbf{k}_1)$.

The surviving components of the nonlinear polarization are then given by

$$\begin{aligned}
&P^{(3)}(\mathbf{k}_a \equiv \mathbf{k}_3 + \mathbf{k}_2 - \mathbf{k}_1, t)\\
&\quad = \left(\frac{i}{\hbar}\right)^3 \int_0^\infty dt_3 \int_0^\infty dt_2 \int_0^\infty dt_1\, [R_2^{\mathrm{H}}(t_3, t_2, t_1) + R_3^{\mathrm{H}}(t_3, t_2, t_1)]\\
&\quad\quad \times \chi(t_3 - t_1) E_3(t - t_3) E_2(t + \tau - t_3 - t_2) E_1^*(t + \tau + \tau' - t_3 - t_2 - t_1)\\
&\quad\quad \times \exp[i(\omega_3 + \omega_2 - \omega_1)t_3 + i(\omega_2 - \omega_1)t_2 - i\omega_1 t_1]
\end{aligned} \tag{10.13}$$

and

$$\begin{aligned}
&P^{(3)}(\mathbf{k}_b \equiv \mathbf{k}_3 - \mathbf{k}_2 + \mathbf{k}_1, t)\\
&\quad = \left(\frac{i}{\hbar}\right)^3 \int_0^\infty dt_3 \int_0^\infty dt_2 \int_0^\infty dt_1\, [R_1^{\mathrm{H}}(t_3, t_2, t_1) + R_4^{\mathrm{H}}(t_3, t_2, t_1)]\\
&\quad\quad \times \chi(t_3 + t_1) E_3(t - t_3) E_2^*(t + \tau - t_3 - t_2) E_1(t + \tau + \tau' - t_3 - t_2 - t_1)\\
&\quad\quad \times \exp[i(\omega_3 - \omega_2 + \omega_1)t_3 - i(\omega_2 - \omega_1)t_2 + i\omega_1 t_1].
\end{aligned} \tag{10.14}$$

The other two components ($-\mathbf{k}_a$ and $-\mathbf{k}_b$) may be obtained by complex conjugation since $P^{(3)}(-\mathbf{k}, t) = [P^{(3)}(\mathbf{k}, t)]^*$.

The double-sided Feynman diagrams corresponding to these components are given in Figure 10.6, and in Table 10.1 we summarize the possible combinations of frequencies and wavevectors.

It should be emphasized that the present discussion is limited to our two level model. In a three-level model we can have, for example, the resonant combination $\mathbf{k}_1 + \mathbf{k}_2 + \mathbf{k}_3$ and $\omega_1 + \omega_2 + \omega_3$, which is responsible for third harmonic generation. In addition, the temporal control over the order of pulses greatly simplifies the results and offers maximum selectivity. If the pulses do overlap (short delays τ, τ' compared with their temporal envelopes) we can get additional contributions. For example, the combination $\mathbf{k}_1 + \mathbf{k}_2 - \mathbf{k}_3$ may also become allowed. This extension can be made by a simple permutation of $\mathbf{k}_1$, $\mathbf{k}_2$, and $\mathbf{k}_3$ and will be considered in the next chapter.

TABLE 10.1 Wavectors, Frequency Factors, and Paths for Resonant Four-Wave Mixing

Signal wavectors $\mathbf{k}_s$	t_1 period	t_2 period	t_3 period	Paths
$\mathbf{k}_a = -\mathbf{k}_1 + \mathbf{k}_2 + \mathbf{k}_3$	$-\omega_1 + \omega_{eg}$	$-\omega_1 + \omega_2$	$-\omega + \omega_2 + \omega_3 - \omega_{eg}$	R_2, R_3
$\mathbf{k}_b = \mathbf{k}_1 - \mathbf{k}_2 + \mathbf{k}_3$	$\omega_1 - \omega_{eg}$	$\omega_1 - \omega_2$	$\omega_1 - \omega_2 + \omega_3 - \omega_{eg}$	R_1, R_4
$-\mathbf{k}_a = \mathbf{k}_1 - \mathbf{k}_2 - \mathbf{k}_3$	$\omega_1 - \omega_{eg}$	$\omega_1 - \omega_2$	$\omega_1 - \omega_2 - \omega_3 + \omega_{eg}$	R_2^*, R_3^*
$-\mathbf{k}_b = -\mathbf{k}_1 + \mathbf{k}_2 - \mathbf{k}_3$	$-\omega_1 + \omega_{eg}$	$-\omega_1 + \omega_2$	$-\omega_1 + \omega_2 - \omega_3 + \omega_{eg}$	R_1^*, R_4^*

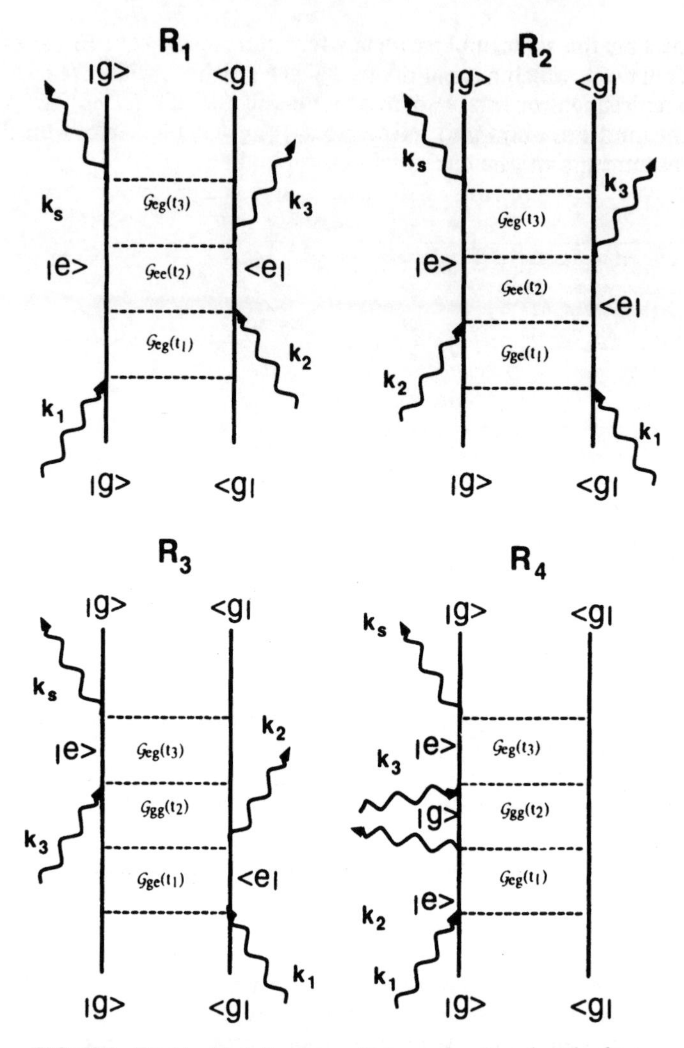

FIG. 10.6 Double-sided Feynman diagrams for time-domain four-wave mixing in a two-level system. Shown are the terms that survive the rotating wave approximation. The propagator $\mathcal{G}$ for each time interval is shown.

In echo experiments we detect the signal with a wavevector

$$\pm \mathbf{k}_a \equiv \pm(\mathbf{k}_3 + \mathbf{k}_2 - \mathbf{k}_1),$$

which selects the R_2 and R_3 paths (see Table 10.1). We note that the static (inhomogeneous) contribution to the four Liouville space paths is $\sim\chi(t_3 - t_1)$ for R_2 and R_3, and $\sim\chi(t_3 + t_1)$ for R_1 and R_4. The echo experiment selects only two of them, R_2 and R_3, which are multiplied by the inhomogeneous dephasing term $\chi(t_3 - t_1)$. In the following sections we shall show precisely under what conditions the inhomogeneous contribution $\chi(t)$ is eliminated completely in an echo experiment. The essence of this selectivity is the rephasing process at $t_1 = t_3$ which generates the echo.

By selecting the signal in the direction $\pm\mathbf{k}_b$, we may probe the other two terms, R_1 and R_4, which are multiplied by the inhomogeneous factor $\chi(t_3 + t_1)$, show no rephasing processes, and do not contribute to the echo. This constitutes a different (and less common) four-wave mixing technique [23]. Examples of such measurements will be discussed in the next chapter.

TWO- AND THREE-PULSE PHOTON ECHO SPECTROSCOPIES

There are several variations of photon echo techniques. These will be considered below and expressed in terms of the nonlinear response function. Since the combination $R_2^{\mathrm{H}} + R_3^{\mathrm{H}}$ will repeatedly appear in the following analysis, we denote it the *echo response function*, and define

$$\mathscr{R}(t_3, t_2, t_1) \equiv R_2^{\mathrm{H}}(t_3, t_2, t_1) + R_3^{\mathrm{H}}(t_3, t_2, t_1). \tag{10.15}$$

Stimulated (three-pulse) echoes with a homodyne or a heterodyne detection

In a stimulated photon echo measurement, three short laser pulses with wavevectors $\mathbf{k}_1$, $\mathbf{k}_2$, and $\mathbf{k}_3$ are sequentially applied to the system, and the signal is generated in the direction $\mathbf{k}_a = \mathbf{k}_3 + \mathbf{k}_2 - \mathbf{k}_1$, with center frequency $\omega_a = \omega_3 + \omega_2 - \omega_1$ (cf. Figure 10.5). (Most commonly one may choose $\mathbf{k}_3 = \mathbf{k}_2$ so that $\mathbf{k}_a = 2\mathbf{k}_2 - \mathbf{k}_1$). The pulse sequence is shown in Figure 10.7 and the corresponding double-sided Feynman diagrams are given in Figure 10.6. The macroscopic polarization with wavevector $\mathbf{k}_a$ induced by the external fields is given by Eq. (10.13). In the homodyne detection scheme, the stimulated echo signal S_{SPE} is given by the total integrated signal in the $\mathbf{k}_a$ direction, i.e.,

$$S_{\mathrm{SPE}}(\tau', \tau) = \int_0^\infty dt\, |P^{(3)}(\mathbf{k}_a, t)|^2. \tag{10.16}$$

In a heterodyne-detected stimulated photon echo (HSPE) measurement, the generated field is mixed with a new local oscillator field

$$E_{\mathrm{LO}}(\mathbf{r}, t) = E_{\mathrm{LO}}(t - \tau') \exp(i\mathbf{k}_{\mathrm{LO}}\cdot\mathbf{r} - i\omega_{\mathrm{LO}}t - i\psi) + c.c.$$

where $E_{\mathrm{LO}}(t - \tau')$ denotes the envelope of the local oscillator pulse, which is centered at $t = \tau'$, and has a carrier frequency ω_{LO}. ψ is the phase shift of the local oscillator field with respect to the product fields $E_3E_2E_1^*$. The resulting signal [see Eq. (4.79)] is [24]

$$\begin{aligned} S_{\mathrm{HSPE}}(\tau', \tau, \psi) &= -2\,\mathrm{Im} \int_{-\infty}^{\infty} dt\, E_{\mathrm{LO}}^*(t - \tau')P^{(3)}(\mathbf{k}_a, t) \exp[i\psi + i(\omega_{\mathrm{LO}} - \omega_a)t] \\ &= 2\,\mathrm{Re} \int_{-\infty}^{\infty} dt \int_0^\infty dt_3 \int_0^\infty dt_2 \int_0^\infty dt_1\, \mathscr{R}(t_3, t_2, t_1)\chi(t_3 - t_1) \\ &\quad \times E_{\mathrm{LO}}^*(t-\tau')E_3(t-t_3)E_2(t+\tau-t_3-t_2)E_1^*(t+\tau+\tau'-t_3-t_2-t_1) \\ &\quad \times \exp[i\psi + i(\omega_{\mathrm{LO}}-\omega_a)t + i\omega_a t_3 + i(\omega_2-\omega_1)t_2 - i\omega_1 t_1]. \end{aligned} \tag{10.17}$$

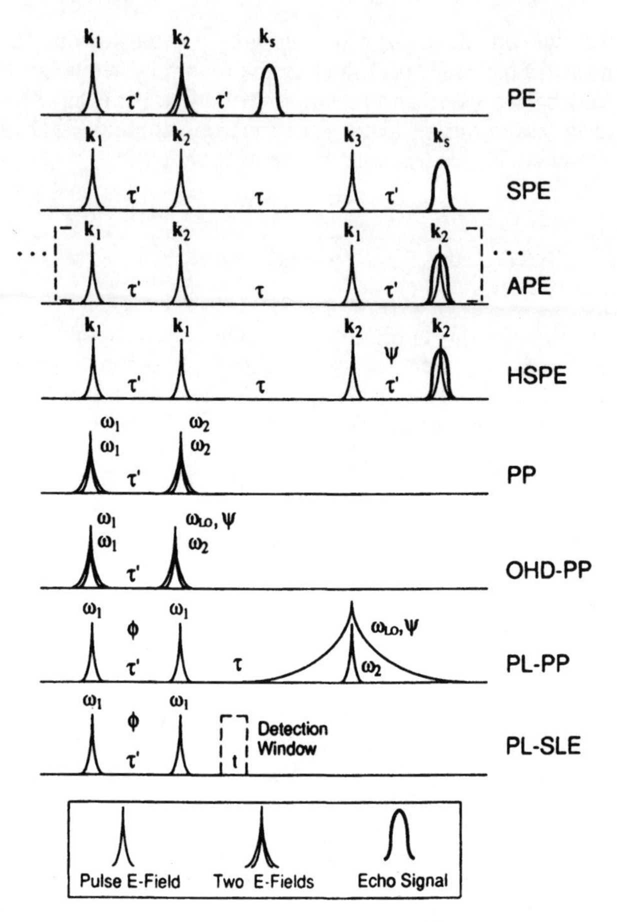

FIG. 10.7 Pulse sequences for the various $P^{(3)}$ techniques. PE, photon echo; SPE, stimulated photon echo; APE, accumulated photon echo; HSPE, heterodyne-detected stimulated photon echo; PP, pump-probe; OHD-PP, optical heterodyne-detected pump-probe; PL-PP, phase-locked pump-probe; PL-SLE, phase-locked spontaneous light emission. The pump-probe techniques will be surveyed in Chapter 11. In APE and HSPE, the fourth pulse coincides spatially and temporally with the echo signal. The phase relations between the two consecutive pulses are adjusted as ϕ and ψ in HSPE, PL-PP, and PL-SLE. On the other hand, the optical phase of the local oscillator (of frequency, ω_{LO}) is described as ψ in OHD-PP and PL-PP. [From M. Cho, N. F. Scherer, G. R. Fleming, and S. Mukamel, "Photon echoes and related four wave mixing spectroscopies using phase-locked pulses," *J. Chem. Phys.* 96, 5618 (1992).]

A limitation of the homodyne detection resides in the detection of the intensity $|E_s|^2$ as opposed to the amplitude E_s of the echo field. Measurements of the echo amplitude would linearize the signal, as is the case for pump-probe spectroscopies that detect the probe beam modified by the pump. By using a strong local oscillator field, the heterodyne interference signal is made much stronger than the homodyne signal. This is especially important when examining the dynamics of dilute and weakly absorbing molecules. Furthermore, it is possible to selectively measure the real and the imaginary parts of the nonlinear response function, which correspond to the absorptive and dispersive components of the optically induced material response, by controlling the phase shift of the local oscillator field with respect to $E_3E_2E_1^*$. This phase-sensitive detection will be discussed below.

Two-pulse photon echoes

The two-pulse photon echo is a special case of the stimulated photon echo in which the third pulse coincides with the second pulse. Its signal, therefore, can be obtained from Eqs. (10.16) and (10.17) by simply setting $E_3(t) = E_2(t)$, $\mathbf{k}_3 = \mathbf{k}_2$, $\mathbf{k}_a = 2\mathbf{k}_2 - \mathbf{k}_1$, $\omega_3 = \omega_2$ and $\tau = 0$: (cf. Figure 10.5)

$$S_{PE}(\tau') = S_{SPE}(\tau', \tau = 0). \tag{10.18}$$

Other variants of echo techniques, namely the accumulated echoes, obtained using either pulsed or incoherent fields, carry similar information, and are discussed in Appendices 10A and 10B.

IMPULSIVE PHOTON ECHOES

A three-pulse echo observed in a dye in solution was displayed in Figure 1.11. An additional example for CdSe nanocrystals is shown in Figure 10.8. Photon echo is an ideal time-domain technique that can be best carried out using very short (impulsive) pulses. No frequency resolution of the light pulses is necessary. This is different from other impulsive techniques such as pump-probe spectroscopy where some spectral resolution is essential. In the previous section we presented general expressions for photon echo spectroscopy that account for arbitrary molecular dynamic timescales and pulse durations. We shall now consider ideal ultrafast echo experiments for which the calculations and their interpretation are greatly simplified.

In the *impulsive limit*, the laser pulses are assumed short compared with any material timescale (vibrational periods, the timescales of homogeneous dephasing, solvent motions, inverse detunings, etc.). The four-fold integrations in Eq. (10.17) can be then trivially carried out, and the impulsive heterodyne stimulated echo is given by

$$S_{HSPE}(\tau', \tau, \psi) = 2\,\mathrm{Re}\{C\mathscr{R}(\tau', \tau, \tau')\exp(i\psi)\exp[i(\omega_{LO} - \omega_1)\tau' + i(\omega_2 - \omega_1)\tau]\}$$

where the parameter

$$C = \int_{-\infty}^{\infty} dt \int_0^{\infty} dt_3 \int_0^{\infty} dt_2 \int_0^{\infty} dt_1\, \chi(t_3 - t_1)$$
$$\times E_{LO}^*(t - \tau')E_3(t - t_3)E_2(t + \tau - t_3 - t_2)E_1^*(t + \tau + \tau' - t_3 - t_2 - t_1), \tag{10.19}$$

generally depends also on τ and τ'. However, in the impulsive limit, the pulses are well separated and τ and τ' are large compared to the pulse widths. In this case, the lower limits in the above t_1, t_2, and t_3 integrations can be replaced

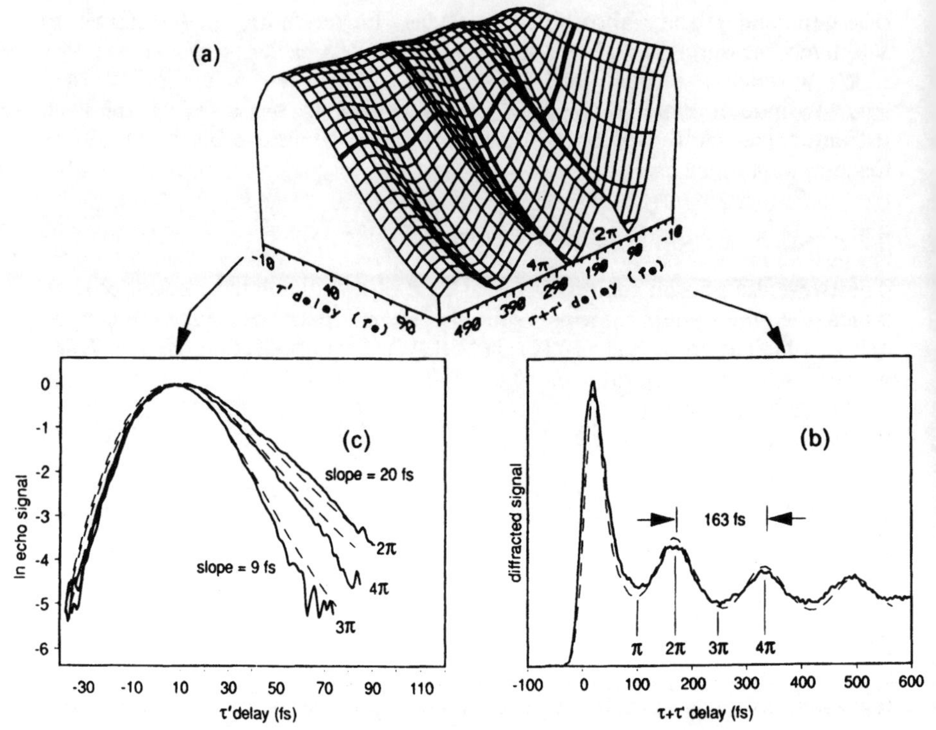

FIG. 10.8 (a) Calculated three-pulse echo signal in the presence of a 200 cm^{-1} phonon mode as a function of the delay between pulse one and two (τ') and between pulse one and three ($\tau + \tau'$). (b) Three pulse echo response of 22 A CdSe nanocrystals as a function of $\tau + \tau'$ with $\tau' = 33$ fs. Dashed line is a theoretical Brownian oscillator fit with one coherent phonon mode ($\omega = 205\ cm^{-1}$, $d = 0.7$, damping time $\Lambda^{-1} = 0.7$ ps and dephasing time $\Gamma^{-1} = 85$ fs). (c) Three-pulse echo response of 22 A CdSe nanocrystals as a function of τ' with the third pulse in phase ($\tau + \tau' = 2\pi/\omega, 4\pi/\omega$) and out of phase ($\tau + \tau' = 3\pi/\omega$) with the coherent phonon mode. Dashed lines are theoretical fits. [From R. W. Schoenlein, D. M. Mittleman, J. J. Shiang, A. P. Alivisatos, and C. V. Shank, "Investigation of femtosecond electronic dephasing in CdSe nanoscrystals using quantum-beat-suppressed photon echoes," *Phys. Rev. Lett.* **70**, 1014 (1993).]

by $-\infty$ and we obtain

$$C = \int d\omega_0\, \chi(\omega_0) E^*_{LO}(\omega_0) E_3(\omega_0) E_2(\omega_0) E^*_1(\omega_0).$$

In the degenerate case when $\omega_1 = \omega_2 = \omega_3 = \omega_{LO}$ and $\hat{E}_1(\omega_0) = \hat{E}_2(\omega_0) = \hat{E}_3(\omega_0) = \hat{E}_{LO}(\omega_0)$, C is a constant real number and we have (setting $C = 1$)

$$\begin{aligned} S_{HSPE}(\tau', \tau, \psi) &= 2\,\mathrm{Re}[\mathscr{R}(\tau', \tau, \tau') \exp(i\psi)] \\ &= \left\{ \begin{array}{ll} 2\,\mathrm{Re}[\mathscr{R}(\tau', \tau, \tau')] & \text{for } \psi = 0 \\ -2\,\mathrm{Im}[\mathscr{R}(\tau', \tau, \tau')] & \text{for } \psi = \pi/2 \end{array} \right\}. \end{aligned} \qquad (10.20)$$

The inhomogeneous dephasing contribution $\chi(t)$ is totally eliminated in Eq. (10.20).

A complete record of the complex echo response function can thus be obtained using phase-controlled pulse pairs in the heterodyne-detected stimulated photon echo experiment. The real (imaginary) part of the nonlinear response functions is obtained from the in-phase (in-quadrature) components.

We reiterate the point made in Chapter 5 regarding the calculation of impulsive measurements. It is crucial to first make the RWA and select the relevant terms for the specific technique out of the total nonlinear response function. Only then can we assume that the pulses are very short and eliminate the time integrations. (Otherwise we implicitly make the unphysical assumption that the pulses are shorter than the optical period.) We further note that in the impulsive limit it is possible to use Eqs. (10.19) and (10.20) even for very strong pulses where the signal can no longer be described by $\chi^{(3)}$. The reason is that no interesting dynamics can occur during the pulse if it is sufficiently short. The only effect of the pulse intensity is to add a prefactor that changes the magnitude of the signal, but not its shape.

We shall now discuss the generation of the echo using the stimulated echo technique. At time $t = -(\tau + \tau')$, the initial ground state equilibrium density operator ρ_{gg} is excited by the first impulsive pulse to prepare an optical coherence ρ_{ge}, whose subsequent evolution is described by the Green function $\mathscr{G}_{ge}(\tau')$. At the time $t = -\tau$ the system interacts with the second impulsive pulse, creating either an electronic ground state population ρ_{gg} or an excited population ρ_{ee}. The time evolution of these nonequilibrium population states is described by the Green functions $\mathscr{G}_{gg}(\tau)$ and $\mathscr{G}_{ee}(\tau)$, respectively. At time $t = 0$ the system finally interacts with the third impulsive pulse, which again prepares the system in an optical coherence, and the stimulated echo arises from the free rephasing processes described by the Green function $\mathscr{G}_{eg}(t)$. At $t = \tau'$, we have, $\mathscr{G}_{eg}(\tau') = [\mathscr{G}_{ge}(\tau')]^{\dagger}$, the two Green functions representing the coherence propagation periods simply become Hermitian conjugates, the effect of inhomogeneous dephasing is eliminated, and the echo reaches its maximum $\chi(t - \tau' = 0)$. For $t > \tau'$, the echo decays due to electronic dephasing. The selective elimination of inhomogeneous dephasing by rephasing processes of each member of the inhomogeneous ensemble is the key characteristic of echo experiments. The essential ingredient is the existence of two propagation periods $\langle \mathscr{G}_{eg}(\tau')\mathscr{G}_{ge}(\tau')\rangle$, in which dephasing and rephrasing take place, respectively.

We next turn to homodyne echo techniques (whether two pulse or three pulse). In the impulsive limit, Eq. (10.16) yields

$$S_{SPE}(\tau', \tau) = \int_0^{\infty} dt\, |\mathscr{R}(t, \tau, \tau')|^2 |\chi(t - \tau')|^2. \tag{10.21}$$

Although inhomogeneous broadening is eliminated at $t = \tau'$ (which gives rise to the echo), the signal has contributions from all times t and therefore inhomogeneous broadening is not completely eliminates in Eq. (10.21). This can be accomplished, however, under somewhat more restrictive conditions. In addition to the impulsive limit we need assume a second simplifying condition, namely large inhomogeneous broadening whereby the inhomogeneous dephasing time, or the inverse inhomogeneous linewidth, is short compared to all the dynamic timescales of the system. This is usually the case in condensed phase spectroscopies. (Fortunately, this is also the situation where the elimination of inhomogeneous broadening is most badly needed.) Under this condition, $\chi(t - \tau') \sim \delta(t - \tau')$, and neglecting some prefactors, the stimulated echo signal is given by

$$S_{SPE}(\tau', \tau) = |\mathscr{R}(\tau', \tau, \tau')|^2. \tag{10.22}$$

[The two-pulse echo signal $S_{PE}(\tau')$ is simply given by Eq. (10.22) by setting $\tau = 0$.]

These results point out another clear advantage of heterodyne detection [Eq. (10.20)], where the complete elimination of inhomogeneous broadening requires only impulsive excitation and holds irrespective of the magnitude of inhomogeneous broadening. Since phase control allows us to measure separately the real and the imaginary parts of the response function, we can express the stimulated photon echo in terms of the HSPE:

$$S_{SPE}(\tau', \tau) = [S_{HSPE}(\tau', \tau, \psi = 0)]^2 + [S_{HSPE}(\tau', \tau, \psi = \pi/2)]^2. \quad (10.23)$$

BROWNIAN OSCILLATOR ANALYSIS OF IMPULSIVE PHOTON ECHOES: FROM QUANTUM BEATS TO SPECTRAL DIFFUSION

The simplest theory of photon echo spectroscopy is based on the Bloch equations and assumes a lineshape with two broadening mechanisms: homogeneous broadening, which originates from very fast motions of the surrounding medium and results in a Lorentzian lineshape, and inhomogeneous broadening. Combining Eqs. 10.9, 10.15 with the Voigt Lineshape (Eq. (8.58b)) we obtain for the integrand in Eq. (10.21)

$$|\mathscr{R}(t, \tau, \tau')|^2 |\chi(t - \tau')|^2 = \exp(-2\hat{\Gamma}t - 2\hat{\Gamma}\tau') \exp[-\Delta^2(t - \tau')^2]. \quad (10.24a)$$

For very large inhomogeneous width ($\Delta \to \infty$) we get a sharply peaked echo at $t = \tau'$

$$|\mathscr{R}(t, \tau, \tau')|^2 |\chi(t - \tau')|^2 \propto \exp[-4\hat{\Gamma}\tau'] \delta(t - \tau'), \quad (10.24b)$$

A comment is now in place regarding a somewhat confusing semantics problem. The phrase "photon echo" is commonly used for the class of techniques discussed here, although, as shown above, actual echo is generated only in the presence of a large inhomogeneous broadening. Thus a photon echo measurement does not necessarily show an echo! Whether to reserve the term "photon echo" only to situations where the technique indeed shows an echo is a matter of taste, which is yet unsettled.

Typical condensed phase systems are characterized by a multitude of relaxation timescales, and the simple classification of line broadening mechanisms as either homogeneous or inhomogeneous is inadequate. In fact, any broadening mechanism can be considered inhomogeneous at short times and may become homogeneous for longer times. The classification thus depends also on the relevant temporal timescale of a particular measurement. The microscopic approach that is based on the response function does not depend on such classification and can be applied to systems with an arbitrary distribution of timescales. The multimode Brownian oscillator model, for example, provides a very useful means for modeling such systems. It can account for quantum beats, as well as homogeneous dephasing, inhomogeneous dephasing, and the spectral diffusion in a unified fashion. We shall now apply this model to calculate the impulsive ordinary photon echo signal. Using Eqs. (10.8a), (10.15), (10.21), and (8.15) we have

$$S_{\mathrm{PE}}(\tau') = \int_0^\infty dt \exp[-\Delta^2(t-\tau')^2] \times \exp[-4g'(t) - 4g'(\tau') + 2g'(t+\tau')]. \quad (10.25)$$

Here we have separated $g(t)$ into its real and imaginary parts i.e., $g(t) \equiv g'(t) + ig''(t)$. The t argument refers to the time of detection. The first factor represents the inhomogeneous contribution of $\chi(t)$. For a multimode system characterized by a very large inhomogeneous broadening, the photon echo signal is sharply peaked at $t = \tau$, and the total signal integrated over t is given by

$$S_{\mathrm{PE}}(\tau') = \prod_j S_{\mathrm{PE}}^{(j)}(\tau'), \quad (10.26a)$$

where the product runs all over the system modes, and

$$S_{\mathrm{PE}}^{(j)}(\tau') = \exp[-8g_j'(\tau') + 2g_j'(2\tau')]. \quad (10.26b)$$

Equations (10.26) constitute our final expression for the ideal impulsive two-pulse echo signal of a multimode Brownian oscillator system. It was shown in Chapter 8 that this model applies to a large variety of situations represented by the limiting forms of $g(t)$ [Eqs. (8.67)–(8.71)]. We shall consider now the two extreme situations. High frequency optically active modes are typically underdamped, i.e., the friction is small compared with the frequency $\gamma_j \ll \omega_j$. Assuming a completely coherent motion with $\gamma_j = 0$, we can substitute Eq. (8.39) in Eq. (10.26) resulting in

$$S_{\mathrm{PE}}^{(j)}(\tau') = \exp[-8(\bar{n}_j + \tfrac{1}{2})S_j(1 - \cos\omega_j\tau')^2]. \quad (10.27)$$

Equation (10.27) illustrates how coherent underdamped motions show up in impulsive echoes as quantum beats.

An additional insight can be provided in this case by using the sum over states expression for the response function [Eq. (6.19)]. We then have

$$S_{\mathrm{PE}}^{(j)}(\tau') = \left| \sum_{a,b,c,d} P(a)\mu_{ab}\mu_{bc}\mu_{cd}\mu_{da} \exp(-i\omega_{dc}\tau' - i\omega_{ab}\tau') \right|^2. \quad (10.28)$$

Equation (10.28) generalizes Eq. (10.27), which is a limited to a harmonic mode with the same frequency in the ground and in the excited state. Equation (10.28) shows that when the vibrational frequency of mode j in the electronically excited state is different from that in the ground electronic state, the echo beats will contain all possible combinations of these two different frequencies, modulated by electronic dephasing. This is a direct consequence of the fact that the echo technique probes the evolution of an electronic coherence, which depends on both electronic states. Quantum beats show up also in other impulsive techniques such as pump-probe spectroscopy (Chapter 11) and stimulated Raman (Chapter 14). These, however, reflect the coherent nuclear motions in electronic populations (be it the excited or the ground electronic state), and are modulated by vibrational dephasing processes, which are typically much slower than electronic dephasing.

The other extreme is that of overdamped motions with $\gamma_j \gg \omega_j$. This is typically the case for low frequency and solvent modes. Upon the substitution of Eq. (8.49) in Eq. (10.26b) we get

$$S_{\rm PE}^{(j)}(\tau') = \exp\left\{-(8\lambda_j k_B T/\Lambda_j) \int_0^{\tau'} d\tau_1 \, [1 - \exp(-\Lambda_j \tau_1)]^2\right\}. \tag{10.29}$$

This case has several limiting behaviors. In the fast modulation (homogeneous) limit where the nuclear correlation time is negligibly small compared with the delay $\Lambda_j^{-1} \ll \tau'$, Eq. (10.29) reduces to

$$S_{\rm PE}^{(j)}(\tau') = \exp(-4\hat{\Gamma}\tau'). \tag{10.30}$$

This is the well-known relation between the echo signal and the homogeneous dephasing rate $\hat{\Gamma} = 2\lambda_j k_B T/\Lambda_j \equiv 1/T_2$, which can be directly derived using the Bloch equations. In the slow modulation limit where $\Lambda_j^{-1} \gg \tau'$, we may expand the exponent of Eq. (10.29) to lowest nonvanishing order, resulting in

$$S_{\rm PE}^{(j)}(\tau') = \exp[-(\tau'/T_j)^3], \tag{10.31a}$$

with the characteristic decay time

$$T_j \equiv \left(\frac{3}{8k_B T \lambda_j \Lambda_j}\right)^{1/3}. \tag{10.31b}$$

This time profile is a signature of spectral diffusion processes [1, 4, 25]. High frequency modes thus show up as quantum beats, whereas the signatures of overdamped modes are irreversible decay.

Model calculations of two pulse echoes for the multimode Brownian oscillator are displayed in Figures 10.9 and 10.10 [9]. In Figure 10.9, we show the photon echo signal as a function of both t and τ' for a four-mode system. The figure shows an initial decay, followed by quantum beats resulting from the modulation of the electronic polarization by vibronic coherences. The echo signal is centered at $t = \tau'$. The first (free induction decay) peak was cut off to better show the other peaks; it is approximately 50 times as high as is shown in the figure.

Figure 10.10 shows the natural logarithm of the total photon echo signal [Eq. (10.25)] for the same four-mode system (solid line). The other two curves are echo traces for different values of the solvent mode correlation time, Λ_3^{-1}: The short-dashed line has $\Lambda_3^{-1} = 10.5$ ps and the long-dashed line has $\Lambda_3^{-1} = 105$ ps. All other parameters are the same as in the solid curve. A larger Λ_3^{-1} means that the solvent relaxation of mode 3 is "more homogeneous"; thus, we expect that systems with smaller values of Λ_3^{-1} will be characterized by slower echo decays. For mode 3 of the three curves in Figure 10.13, we have $T_j = 66.95$, and 210 fs. respectively; this explains the fast initial decay of the curves relative to $\exp[-\Lambda_3\tau]$. It should also be noted that the friction of the two underdamped modes also contributes a (homogeneous) decay to the curves. In addition to these decays, the three curves show oscillations at the frequencies of the 600 and 1500 cm^{-1} modes. The insert gives the linear absorption (solid line) and fluorescence (dashed line) spectra for the system. Unlike the echo signal, these curves do not change appreciably with changing Λ_3, and thus carry information only about the total

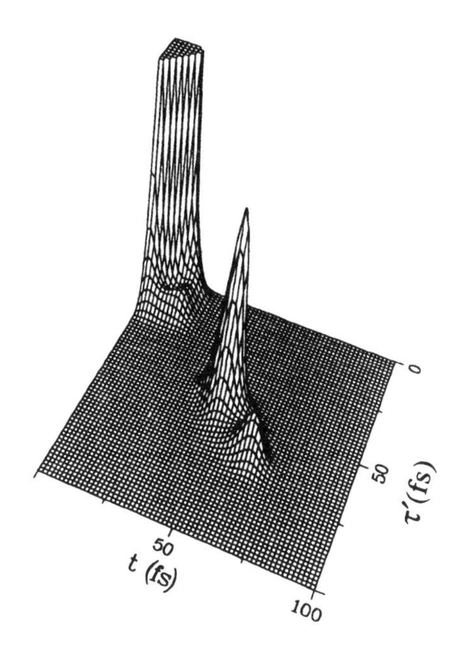

FIG. 10.9 Photon-echo signal plotted as a function of t and τ, for a four-mode system (for parameters see Figure 11.7). In the figure, we see an initial decay, followed by oscillations. These are quantum beats, resulting from the modulation of the electronic polarization by vibronic coherences. Because of the inhomogeneous mode, the echo signal is centered at $t = \tau$. The first peak (the free-induction decay) has been cut off to better show the subsequent peaks; it is approximately 50 times as high as is shown in the figure. [From W. B. Bosma, Y. J. Yan, and S. Mukamel, "Impulsive pump-probe and photon echo spectroscopies of dye molecules in condensed phases," *Phys. Rev. A* **42**, 6920 (1990).]

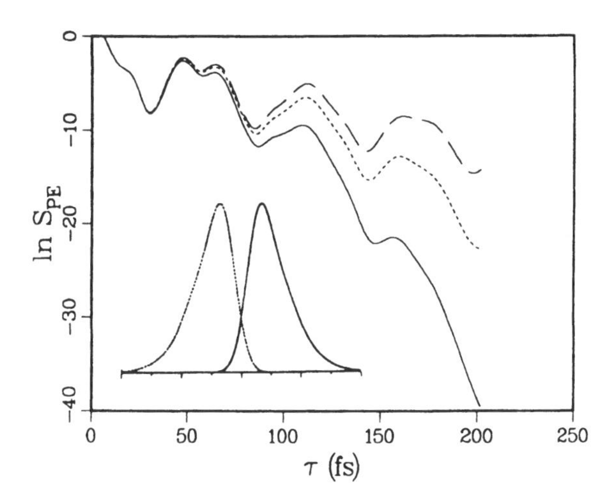

FIG. 10.10 Calculated impulsive two-pulse echo for a three-mode system with two coherent modes $\omega_1 = 600\,\text{cm}^{-1}$, $d_1 = 0.4$, $\omega_2 = 1800\,\text{cm}^{-1}$, $d_2 = 0.5$. In addition, it has an overdamped mode with $\Delta_3 = 510\,\text{cm}^{-1}$. The solid, dashed, and long dash curves correspond to $\Lambda_3/\Delta_3 = 10^{-4}$, 5×10^{-4}, and 10^{-3}, respectively. The inset shows the linear absorption (solid curve) and fluorescence (dotted curve) for this system. The frequency scale is $2000\,\text{cm}^{-1}$ between ticks. These lineshapes are broad and structureless and virtually identical for the three values of Λ_3/Δ_3. The figure demonstrates the greater sensitivity of the photon echo to nuclear motions compared with linear spectroscopy. [From W. B. Bosma, Y. J. Yan, and S. Mukamel, "Impulsive pump-probe and photon echo spectroscopies of dye molecules in condensed phases," *Phys. Rev. A* **42**, 6920 (1990).]

broadening of the system, with no information about the solvent relaxation timescales. Fits of the Brownian oscillator model to three pulse echo in solution and in semiconductor nanocrystals were shown in Figures 1.11 and 10.8, respectively. Several distinct optical measurements were conducted on the dye resorufin in a polar solvent (DMSO). The experimental data, all fitted with the same multimode Brownian oscillator model, are given in Figure 10.11. Figure 10.11a shows the fit to the absorption and relaxed fluorescence lineshapes and Figure 10.11b shows the fit to a two-pulse echo measurement. An additional experiment conducted on the same system is a four-wave mixing using chirped pulses whose frequencies are varied continuously and swept through the resonance. That experiment probes different aspects of the dynamics and may also be reproduced very well with the same model (Figure 10.11c).

We next turn our attention to the three-pulse echo, where the Brownian oscillator model yields the following expressions for the relevant nonlinear response functions:

$$\mathscr{R}(\tau', \tau, \tau') = R_2^{\rm H}(\tau', \tau, \tau') + R_3^{\rm H}(\tau', \tau, \tau'), \tag{10.32a}$$

$$R_2^{\rm H}(\tau', \tau, \tau') = \exp[-2g^*(\tau') + g(\tau) - g(\tau + \tau') - g^*(\tau + \tau') + g^*(\tau + 2\tau')], \tag{10.32b}$$

$$R_3^{\rm H}(\tau', \tau, \tau') = \exp[-g^*(\tau') - g(\tau') + g^*(\tau) - 2g^*(\tau + \tau') + g^*(\tau + 2\tau')]. \tag{10.32c}$$

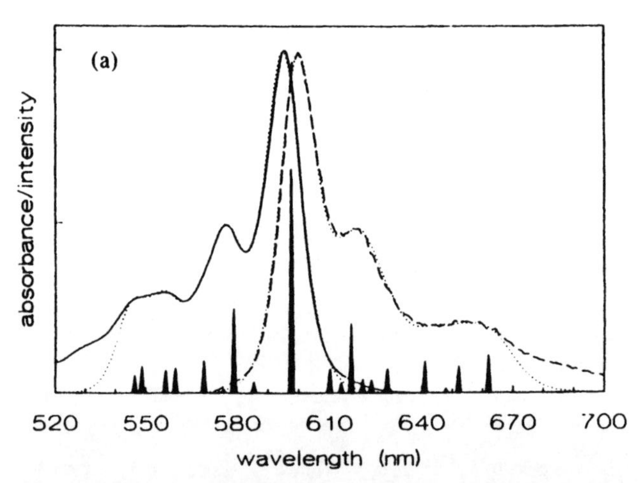

FIG. 10.11 (a) Absorption (solid line) and emission (dashed line) spectra of resorufin dissolved in DMSO at room temperature. The Stokes shift between the origin transition maxima is 114 cm^{-1}. The dotted lines are fits based on the level structure and transition moments shown in black below the curve. The lineshape is based on dynamics determined by a strongly overdamped oscillator with $\Delta = 41$ THz, $\Lambda = 27$ THz, and $\lambda = 23$ THz. In the high temperature limit this latter value is related to Δ as discussed in the text. [E. J. Nibbering, D. A. Wiersma, and K. Duppen, "Ultrafast electronic fluctuation and solvation in liquids," *Chem. Phys.* **183**, 167 (1994); "Femtosecond non-Markovian optical dynamics in solution," *Phys. Rev. Lett.* **66**, 2464 (1991).]

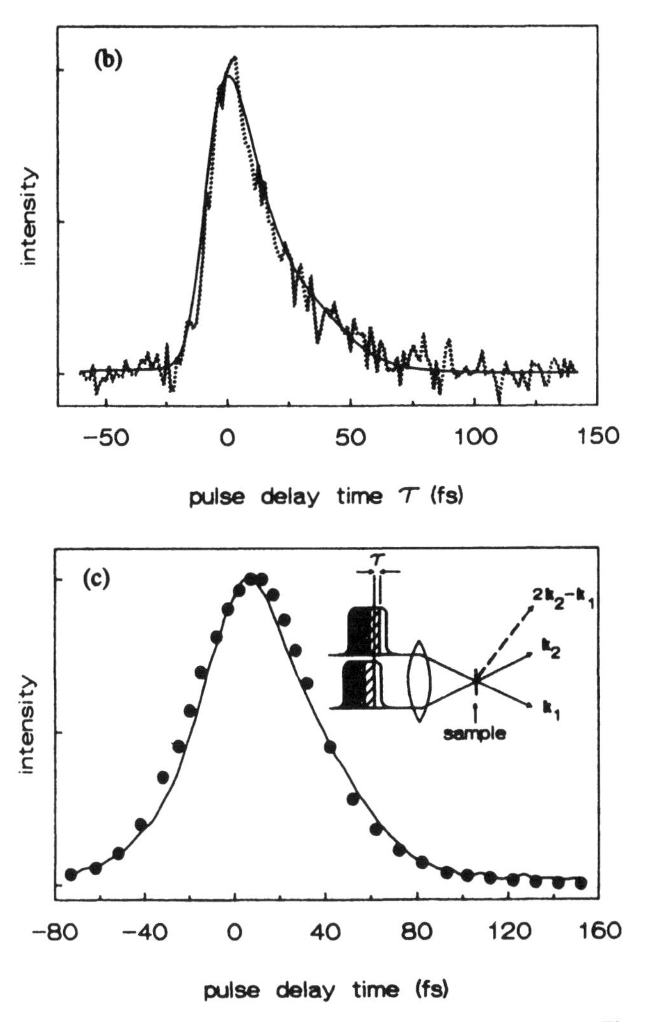

FIG. 10.11 (*continued*) (b) Photon echo signal for the same system. The solid line is a Brownian oscillator fit using the same parameters of (a). (c) Experimental results for a chirped four-wave mixing measurement of the same system (solid circles). The curve is a fit using the same Brownian oscillator parameters of (a). The pulse duration is ca. 1 ps. The chirp rate is 0.5 THz/fs, and the fields sweep through the resonances in about 100 fs. [K. Duppen, F. de Haan, E. T. J. Nibbering and D. A. Wiersma, "Chirped four-wave mixing," *Phys. Rev. A.* **47**, 5120 (1993).]

(In the two-pulse case we simply set $\tau = 0$.) Equations (10.32) show how spectral diffusion processes that take place during the delay τ period may be probed using the three-pulse echo. Some remarkable examples of spectral diffusion of a single molecule observed using fluorescence excitation measurements are displayed in Figures 10.1, 10.2, and 13.6. A direct signature of spectral diffusion is provided by the variation of the apparent dephasing rate observed in a three-pulse echo measurement with the delay τ between pulses 2 and 3. Spectral diffusion processes

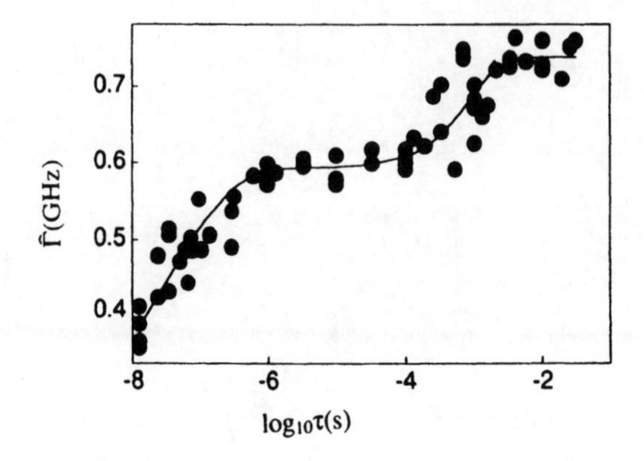

FIG. 10.12 Dependence of the effective pure dephasing rate on the delay between pulses 2 and 3 (i.e., the waiting time τ) in a three pulse echo measurement. The system is Zn-substituted cytochrome *c* in a 3/1 vol. glycerol/water matrix at 1.8 K. [D. Thorn Leeson, O. Berg, D. A. Wiersma, "Low-temperature protein dynamics studied by the long-lived stimulated photon echo," *J. Phys. Chem.* **98**, 3913 (1994).] The solid line is a fit to a distribution of relaxation rates. This distribution is hyperbolic for relaxation rates faster than 1 MHz due to coupling of the chromophore to the glassy matrix. This accounts for the increase of the dephasing rate in the sub-μs region. The increase in the ms region is due to conformational relaxation of the protein represented by a sharp distribution centered at 1 kHz.

that take place during this period change the environment of the chromophore and affect the dephasing rate. An example for the protein cytochrome *c* studied over a large dynamic range of time delays is given in Figure 10.12.

SPONTANEOUS LIGHT EMISSION USING PHASE-LOCKED PULSES

The ability to control the relative phases of sequences of femtosecond optical pulses has opened up some exciting possibilities for optical measurements. We have already demonstrated the utility of phase control in heterodyne-detected spontaneous echoes. We shall now calculate the spontaneous light emission signal created by a pair of phase-locked pulses. This experiment [26], as we shall see, is closely related to the photon echo. The pulse configuration of phase-locked spontaneous light emission (PLSLE) is shown in Figure 10.7. The phase shift of the second pulse with respect to the first pulse is denoted ϕ, the two-pulse delay time is τ', and we have

$$E(\mathbf{r}, t) = E_1(t + \tau') \exp(i\mathbf{k}_1 \cdot \mathbf{r} - i\omega_1 t) + E_2(t) \exp(i\mathbf{k}_1 \cdot \mathbf{r} - i\omega_1 t - i\phi) + c.c.$$

The PLSLE signal is defined as the spontaneous light emission signal that results from one interaction with each field, E_1 and E_2. This can be easily measured by subtracting from the total light emission, the SLE induced by pulse 1 alone and pulse 2 alone

$$S_{\mathrm{SLE}}^{\mathrm{PL}}(\omega_1, \omega_2, \phi, \tau', t) = 2\,\mathrm{Re} \int_0^\infty dt_3 \int_0^\infty dt_2 \int_0^\infty dt_1 \{R_1^{\mathrm{H}}(t_3, t_2, t_1)\chi(t_3 + t_1)$$
$$\times E_2^*(t - t_3 - t_2)E_1(t + \tau' - t_3 - t_2 - t_1)\exp(i\omega_2 t_3 + i\omega_1 t_1 + i\phi) +$$
$$R_2^{\mathrm{H}}(t_3, t_2, t_1)\chi(t_3 - t_1)E_2(t - t_3 - t_2)E_1^*(t + \tau' - t_3 - t_2 - t_1)$$
$$\times \exp(i\omega_2 t_3 - i\omega_1 t_1 - i\phi) +$$
$$R_3^{\mathrm{H}}(t_3, t_2, t_1)\chi(t_3 - t_1)E_2(t - t_3)E_1^*(t + \tau' - t_3 - t_2 - t_1)$$
$$\times \exp[i\omega_2 t_3 - i(\omega_1 - \omega_2)t_2 - i\omega_1 t_1 - i\phi]\}. \quad (10.33)$$

The ordinary SLE spectrum may be obtained from Eq. (10.33) by taking $E_1 = E_2$ and setting $\phi = 0$, $\tau' = 0$. In the simplest mode, one measures the signal integrated over the observation time t. Using fast detection it should be possible to measure its variation with t as well.

The interpretation of Eq. (10.33) closely resembles that of ordinary SLE. The first pulse creates an optical coherence. In the cases of pathways R_1 and R_2, the second phase-locked pulse generates population in the excited state. Finally, the system interacts twice with the outgoing mode, which results in spontaneous fluorescence. On the other hand, the third pathway R_3 involves a ground state population evolution t_2 period and is clearly related to the spontaneous Raman process. Double-sided diagrams representing the Liouville pathways R_1, R_2, and R_3, were shown in Figure 9.4.

When both excitation pulses are impulsive, Eq. (10.33) reduces to

$$S_{\mathrm{SLE}}^{\mathrm{PL}}(\omega_1, \omega_2, \phi, \tau', t) = 2\,\mathrm{Re}\Big\{\exp(-i\omega_1\tau' - i\phi) \int_0^\infty d\tau\, \theta(t - \tau)R_1^{\mathrm{H}}(t - \tau, \tau, \tau')$$
$$\times \chi(t - \tau + \tau')\exp[i\omega_2(t - \tau)]\Big\}$$
$$+ 2\,\mathrm{Re}\Big\{\exp(-i\omega_1\tau' - i\phi) \int_0^\infty d\tau\, \theta(t - \tau)R_2^{\mathrm{H}}(t - \tau, \tau, \tau')$$
$$\times \chi(t - \tau - \tau')\exp[i\omega_2(t - \tau)]\Big\}$$
$$+ 2\,\mathrm{Re}\Big\{\exp(-i\omega_1\tau' - i\phi) \int_0^\infty d\tau\, \theta(\tau' - \tau)R_3^{\mathrm{H}}(t, \tau, \tau' - \tau)$$
$$\times \chi(t - \tau' + \tau)\exp[i\omega_2(t + \tau)]\Big\}. \quad (10.34)$$

In the large inhomogeneous broadening limit $\chi(t) \cong \delta(t)$, we finally obtain

$$S_{\mathrm{SLE}}^{\mathrm{PL}}(\omega_1, \omega_2, \phi, \tau', t) = 2\,\mathrm{Re}\{\exp[i(\omega_2 - \omega_1)\tau' - i\phi]$$
$$\times [\theta(\tau')\theta(t - \tau')R_2^{\mathrm{H}}(\tau', t - \tau', \tau')$$
$$+ \theta(t)\theta(\tau' - t)R_3^{\mathrm{H}}(t, \tau' - t, t)]\}. \quad (10.35)$$

Note that the first term in Eq. (10.34), associated with the Liouville pathway R_1, does not contribute to the PLSLE signal in this case, since $R_1^{\text{H}}(-\tau', t+\tau', \tau') = 0$ when $\tau' > 0$.

Equation (10.35) can be interpreted as follows. If the locked frequencies of the two fields are identical, the phase shift, ϕ, determines whether the detected signal is produced by the real (i.e., $\phi = 0$; in-phase) or imaginary (i.e., $\phi = \pi/2$; inquadrature) parts of the nonlinear response functions. Furthermore, time-gated detection of the spontaneous emission makes it possible to selectively measure each of the two contributions, $R_2^{\text{H}}(\tau', t-\tau', \tau')$ and $R_3^{\text{H}}(t, \tau'-t, t)$. The θ factors in Eqs. (10.34) and (10.35) make sure that causality is maintained [see Eq. (5.17)]. Consequently, for $t \le \tau'$, only the Raman term $R_3^{\text{H}}(t, \tau'-t, t)$ makes a contribution to the spontaneous emission, whereas for $t \ge \tau'$ only the fluorescence term $R_2^{\text{H}}(\tau', t-\tau', \tau')$ contributes.

The possible measurements can be summarized as

$$S_{\text{SLE}}^{\text{PL}}(\omega_1 = \omega_2, \phi, \tau', t)$$
$$= \left| \begin{array}{lll} \left\{ \begin{array}{ll} 2\,\text{Re}[R_3^{\text{H}}(t, \tau'-t, t)] & \text{for } \phi = 0 \\ 2\,\text{Im}[R_3^{\text{H}}(t, \tau'-t, t)] & \text{for } \phi = \pi/2 \end{array} \right\} & \text{when } t \le \tau' \\ \left\{ \begin{array}{ll} 2\,\text{Re}[R_2^{\text{H}}(\tau', t-\tau', \tau')] & \text{for } \phi = 0 \\ 2\,\text{Im}[R_2^{\text{H}}(\tau', t-\tau', \tau')] & \text{for } \phi = \pi/2 \end{array} \right\} & \text{when } t > \tau' \end{array} \right| . \qquad (10.36)$$

The phase factor ϕ can be used to selectively measure the real and the imaginary parts of each complex Liouville-space path. Since R_2 and R_3 represent the Liouville pathways associated with the excited and the ground state dynamics during the t_2 time period, respectively, we can use this technique to look separately at excited state and ground state vibrational dynamics. An example of a phase-locked spontaneous emission of Iodine in the gas phase is shown in Figure 10.13.

To compare the phase-locked spontaneous light emission (PLSLE) measurement with photon echoes, we rewrite the heterodyne-detected stimulated photon echo (HSPE) signal as

$$S_{\text{HSPE}}(t', t'', \psi) = S_{\text{SLE}}^{\text{PL}}(\omega_1 = \omega_2, \phi = \psi, \tau' = t', t = t' + t'') + S_{\text{SLE}}^{\text{PL}}(\omega_1 = \omega_2, \phi = \psi, \tau' = t' + t'', t = t')]. \qquad (10.37)$$

Equations (10.23) and (10.37) connect the three-pulse and the phase-locked photon echoes to the phase-locked spontaneous light emission signal. Phase locking in pump probe spectroscopy and its connection to hole-burning spectroscopy will be discussed in Chapter 11.

In summary, we presented correlation function expressions for the two-pulse, the three-pulse, and the accumulated photon echo. Inhomogeneous broadening may be eliminated by rephasing processes that completely reverse the inhomogeneous dephasing. Conditions for the total elimination of inhomogeneous broadening were specified. We then considered the impulsive limit where the excitation field is short compared with the nuclear dynamic timescale. In this limit, all the echo signals [Eqs. (10.19)–(10.23)] are directly related to the correlation function $\mathscr{R}$, and no time integrations are necessary. The three-pulse echo then appears as a sharply peaked signal whose temporal width is inversely

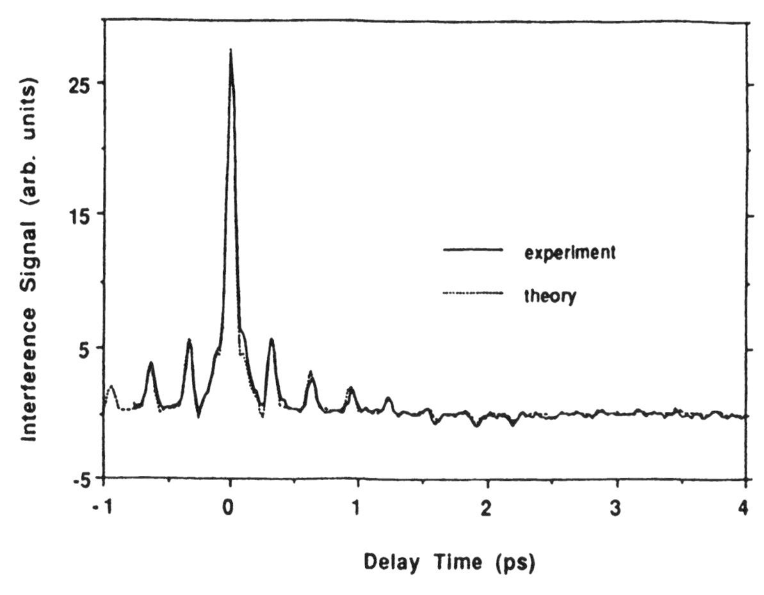

FIG. 10.13 Phase-locked spontaneous emission in iodine. Fluorescence is detected as a result of single interactions with two 50 fs pulses whose phases are set equal. The time-delay between the two pulses is varied and the total fluorescence intensity recorded is perpendicular to the laser beams. The signal is symmetric around zero when the two pulses are identical. It shows positive peaks at delays corresponding to the excited state vibrational period that result from constructive interference. If the phase of the second pulse is set at π with respect to the first pulse, negative-going peaks are observed. The form of the fluorescence interferogram can be accurately calculated from the vibrational and rotational constants of I_2 as shown by the curve labeled "theory." Rotational dephasing is the main contribution to the fast initial decay. [From N. F. Scherer, A. Matro, L. D. Ziegler, M. Du, R. J. Carlson, J. A. Cina, and G. R. Fleming, "Fluorescence-detected wave packet interferometry. II. Role of rotations and determination of the susceptibility," *J. Chem. Phys.* **96**, 4180 (1992).]

proportional to the inhomogeneous linewidth [Eq. (10.24b)]. Thus the echo works best when it is most badly needed, i.e. when the inhomogeneous broadening is very large. For nonimpulsive measurements, the echo will depend on the field profiles. These can be calculated using Eqs. (10.16) through (10.22). The multimode Brownian oscillator model allows us to incorporate different modes with a multitude of timescales. The calculation of a specific nonlinear measurement requires only a few Liouville space pathways, which are complex (i.e., have real and imaginary parts) rather than the entire response function $S^{(3)}$, which is real. Our calculation of photon echo spectroscopy employing phase-locked pulses demonstrates how the real and imaginary contributions to the response function can be measured separately by controlling the relative phase of the two pulses.

APPENDIX 10A

Accumulated Photon Echoes

In the accumulated photon echo technique [14] the system interacts with two noncollinear trains of coherent laser pulses. The first will be denoted as the pump train whereas the second, the probe train, is delayed with respect to the pump by time τ'. The pulse spacing in both trains is $\tau + \tau'$. Each pulse in the pump (probe) train has a wavevector $\mathbf{k}_1$ ($\mathbf{k}_2$), frequency ω_1 (ω_2), and a temporal envelope $E_1(t)\,[E_2(t)]$. The echoes stimulated by the pump train appear at the probe train wavevector $\mathbf{k}_S = \mathbf{k}_1 + \mathbf{k}_2 - \mathbf{k}_1 = \mathbf{k}_2$, are coincident temporally, and oscillate in phase with the probe train (Figure 10.7). Consider a four-pulse cycle, two from the pump train and two from the probe train. The echo field is in this case identical to the stimulated echo generated from the first three (pump–probe–pump) pulses. Furthermore, the echo field interferes in phase with the fourth (probe) pulse, and the signal is combined with a carrier beam with the same frequency. The accumulated echo corresponds therefore to the heterodyne detection mode. Setting $\mathbf{k}_3 = \mathbf{k}_1$, $E_3 = E_1$, and $\omega_3 = \omega_1$ in Eq. (10.17) we get for the signal

$$
\begin{aligned}
S_{\mathrm{APE}}(\tau', \tau) &= -2\,\mathrm{Im} \int_{-\infty}^{\infty} dt\, E_2^*(t - \tau') P_{\mathrm{SPE}}^{(3)}(\mathbf{k}_S = \mathbf{k}_1 + \mathbf{k}_2 - \mathbf{k}_1, t) \\
&= 2\,\mathrm{Re} \int_{-\infty}^{\infty} dt \int_0^{\infty} dt_3 \int_0^{\infty} dt_2 \int_0^{\infty} dt_1\, \mathcal{R}(t_3, t_2, t_1)\chi(t_3 - t_1) \\
&\quad \times E_2^*(t - \tau')E_1(t - t_3)E_2(t + \tau - t_3 - t_2)E_1^*(t + \tau + \tau' - t_3 - t_2 - t_1) \\
&\quad \times \exp[i\omega_2 t_3 + i(\omega_2 - \omega_1)t_2 - i\omega_1 t_1].
\end{aligned}
\tag{10.A.1}
$$

In the impulsive limit the accumulated echo becomes

$$
S_{\mathrm{APE}}(\tau', \tau) = \mathrm{Re}[\mathcal{R}(\tau', \tau, \tau')],
\tag{10.A.2}
$$

which can thus be related to the heterodyne detected echo:

$$
S_{\mathrm{APE}}(\tau', \tau) = S_{\mathrm{HSPE}}(\tau', \tau, \psi = 0).
\tag{10.A.3}
$$

The multi-pulse accumulation and the heterodyne detection make this technique particularly sensitive, and it has been widely used to probe vibronic dephasing and relaxation dynamics in crystals and liquids. This technique is usually applied to systems with a long absorption recovery time. Fast lifetime will cause the signal to decay, and the dephasing information will be lost. To avoid this problem, the technique is usually applied to systems in which the excited state relaxes to a third level rather than to the ground state. The ground state grating information is thus preserved and accumulates, and is not limited by the excited state lifetime.

In ordinary applications of the technique, the train repetition of the four-pulse sequence simply accumulates and amplifies the echo signal which is given by Eq. (10.A.1). Complications may occur when the system is characterized by a broad distribution of relaxation timescales including very long timescales as is the case, e.g., in glasses. In this case, slow spectral diffusion processes can take place, and their incorporation requires a more elaborate theory that depends on higher order response functions. As a result, the four-pulse description of the accumulated photon echo is adequate only when the timescale τ can be made longer than any other relaxation timescale of the system.

APPENDIX 10B

Accumulated Photon Echoes with Incoherent Light Sources

Equation (10.A.1) gives the accumulated photon echo generated using two trains of coherent pulses. Alternatively, such echoes can be created using incoherent broad band light sources [27–30]. The beam configuration is similar to the ordinary APE, and the electric field is given by

$$E(\mathbf{r}, t) = E(t + \tau')\exp(i\mathbf{k}_1\mathbf{r} - i\omega t) + E(t)\exp(i\mathbf{k}_2\mathbf{r} - i\omega t) + c.c. \tag{10.B.1}$$

We have a pump beam and a delayed probe beam with wavevectors $\mathbf{k}_1$ and $\mathbf{k}_2$, respectively. The two beams are generated by splitting a single, incoherent beam and passing one of the resulting beams through a delay line, which determines the delay τ'. The temporal envelope of the incoherent light is assumed to be a complex stochastic stationary Gaussian process with

$$\langle E^*(t + \tau)E(t)\rangle = \Phi(\tau) = \Phi^*(-\tau), \tag{10.B.2a}$$

$$\langle E(t + \tau)E(t)\rangle = \langle E(t)\rangle = \langle E^*(t)\rangle = 0. \tag{10.B.2b}$$

Here $\langle \cdots \rangle$ denotes an average over the stochastic fluctuations and $\Phi(t)$ denotes the correlation function of the incoherent laser field. For a Gaussian process, higher order correlation functions can be calculated as well: All odd correlation functions vanish. The even correlation functions with order $2n$ are given by sums of all possible products of two time correlation functions obtained by pairing the E factors [there are $(2n!)/(2^n n!)$ distinct terms]. The signal generated in the probe direction $\mathbf{k}_2$ and detected as a heterodyne beat with the probe beam is

$$\begin{aligned} S(\mathbf{k}_2; \tau') = &-2\,\mathrm{Im}\langle E^*(t)P^{(3)}(\mathbf{k}_S = \mathbf{k}_1 + \mathbf{k}_2 - \mathbf{k}_1, t)\rangle \\ = &\ 2\,\mathrm{Re}\int_0^\infty dt_3 \int_0^\infty dt_2 \int_0^\infty dt_1\, \{\exp[i\omega(t_3 - t_1)][\mathcal{F}_2 + \mathcal{F}_3] \\ &\times [R_2(t_3, t_2, t_1) + R_3(t_3, t_2, t_1)] \\ &+ 2\,\mathrm{Re}\int_0^\infty dt_3 \int_0^\infty dt_2 \int_0^\infty dt_1\, \{\exp[i\omega(t_3 + t_1)][\mathcal{F}_1 + \mathcal{F}_4] \\ &\times [R_1(t_3, t_2, t_1) + R_4(t_3, t_2, t_1)], \end{aligned} \tag{10.B.3}$$

where

$$\left.\begin{aligned}
\mathscr{F}_1 &= \langle E^*(t)E(t-t_3)E^*(t+\tau'-t_3-t_2)E(t+\tau'-t_3-t_2-t_1)\rangle,\\
\mathscr{F}_2 &= \langle E^*(t)E(t+\tau'-t_3)E(t-t_3-t_2)E^*(t+\tau'-t_3-t_2-t_1)\rangle,\\
\mathscr{F}_3 &= \langle E^*(t)E(t-t_3)E(t+\tau'-t_3-t_2)E^*(t+\tau'-t_3-t_2-t_1)\rangle,\\
\mathscr{F}_4 &= \langle E^*(t)E(t+\tau'-t_3)E^*(t+\tau'-t_3-t_2)E(t-t_3-t_2-t_1)\rangle.
\end{aligned}\right\} \qquad (10.B.4)$$

In the derivation of Eq. (10.B.3) we invoked the rotating wave approximation but we did not assume temporal separation of the pump and the probe, since this is not the case in the incoherent experiment. The four-time-correlation functions $\mathscr{F}_\alpha$ [Eqs. (10.B.4)] can be evaluated by assuming Gaussian statistics for the field [Eqs. (10.B.2)]. We then get,

$$\left.\begin{aligned}
\mathscr{F}_1 &= \Phi(t_3+t_2+t_1-\tau')\Phi(\tau'-t_2)+\Phi(t_3)\Phi(t_1),\\
\mathscr{F}_2 &= \Phi^*(\tau'-t_3)\Phi(\tau'-t_1)+\Phi(t_3+t_2)\Phi^*(t_2+t_1),\\
\mathscr{F}_3 &= \Phi(t_3+t_2-\tau')\Phi^*(t_1+t_2-\tau')+\Phi(t_3)\Phi^*(t_1),\\
\mathscr{F}_4 &= \Phi^*(\tau'-t_3)\Phi(\tau'+t_1)+\Phi(t_3+t_2+t_1)\Phi^*(t_2).
\end{aligned}\right\} \qquad (10.B.5)$$

Equations (10.B.3) can be further simplified. First, each $\mathscr{F}_\alpha$ with $\alpha = 1, \ldots, 4$ contains two terms. The second terms do not depend on the delay time τ' and thus contribute a dc background, which can be eliminated. Furthermore, the echo signal is generated only when τ_c, the timescale of the field correlation function $\Phi(t)$, is short compared with the dephasing time of the matter, and the delay time τ'. Both $\mathscr{F}_1$ and $\mathscr{F}_4$ are negligible under these conditions, and we obtain the final expression for the incoherent accumulated photon echo signal

$$\begin{aligned}
S_{\mathrm{IAPE}}(\tau') = 2\,\mathrm{Re}\int_0^\infty dt_3\int_0^\infty dt_2\int_0^\infty dt_1\\
\times\{\exp[i\omega(t_3-t_1)][\Phi^*(\tau'-t_3)\Phi(\tau'-t_1)\\
+\Phi(t_3+t_2-\tau')\Phi^*(t_1+t_2-\tau')]\mathscr{R}(t_3,t_2,t_1)\chi(t_3-t_1)\}.
\end{aligned} \qquad (10.B.6)$$

In the impulsive limit we get

$$S_{\mathrm{IAPE}}(\tau') = \int_0^\infty d\tau\, S_{\mathrm{APE}}(\tau',\tau) + \int_0^{\tau'} d\tau\, S_{\mathrm{APE}}(\tau',\tau'-\tau). \qquad (10.B.7)$$

The temporal resolution of this technique is determined by the correlation time τ_c, rather than by the pulse durations. Strangely enough, the resolution is improved when the quality of the laser field is deteriorated, rendering it more incoherent. This is the ultimate poor man's femtosecond spectroscopy. Similar techniques have long been used in microwave and magnetic resonance, and they were pioneered in the optical regime by Hartmann and Yajima and co-workers [27, 28]. It should also be noted that these ideas are not limited to photon echoes, and incoherent pulses can be used to perform other types of measurements, such as coherent Raman.

NOTES

1. W. E. Moerner, Ed., *Persistent Spectral Hole Burning: Science and Applications* (Springer-Verlag, Berlin, 1988).

2. A. M. Stoneham, *Rev. Mod. Phys.* **41**, 82 (1969).

3. W. R. Lambert, P. M. Felker, and A. M. Zewail, *J. Chem. Phys.* **75**, 5958 (1981).

4. U. P. Wild and A. Renn, *J. Mol. Elec.* **7**, 1 (1991); M. Pirotta, F. Güttler, H. Gygax, A. Renn, J. Sepiol, and U. P. Wild, *Chem. Phys. Lett.* **208**, 379 (1993).

5. W. E. Moerner and T. Basche, *Angew. Chem.* **105**, 537 (1993).

6. M. Orrit and J. Bernard, *Phys. Rev. Lett.* **65**, 2716 (1990); M. Orrit, J. Bernard, and R. I. Personov, *J. Phys. Chem.* **97**, 10256 (1993).

7. E. Betzig, R. J. Chichester, *Science* **262**, 1422 (1993); R. C. Dunn, G. R. Holtom, L. Mets, and X. S. Xie, *J. Phys. Chem.* **98**, 3094 (1994).

8. V. S. Letokhov and V. P. Chebotayev, *Nonlinear Laser Spectroscopy* (Springer-Verlag, Berlin, 1977).

9. Y. J. Yan and S. Mukamel, *J. Chem. Phys.* **94**, 179 (1991).

10. R. I. Personov, E. I. Al'Shits, and L. A. Bykovskaya, *Opt. Commun.* **6**, 169 (1972).

11. A. Mokhtari, J. Chesnoy, and A. Laubereau, *Chem. Phys. Lett.* **155**, 593 (1989).

12. N. A. Kurnit, I. D. Abella, and S. R. Hartmann, *Phys. Rev. Lett.* **13**, 567 (1964); I. D. Abella, N. A. Kurnit, and S. R. Hartmann, *Phys. Rev.* **141**, 391 (1966); S. R. Hartmann, *IEEE J. Quant. Electron.* **4**, 802 (1968).

13. T. Mossberg, A. Flusberg, R. Kachru, and S. R. Hartmann, *Phys. Rev. Lett.* **39**, 1523 (1977); T. W. Mossberg, R. Kachru, A. M. Flusberg, and S. R. Hartmann, *Phys. Rev. A* **20**, 1976 (1979).

14. W. H. Hesselink and D. A. Wiersma, *J. Chem. Phys.* **73**, 648 (1980); in *Modern Problems in Condensed Matter Sciences*, Vol. 4, V. M. Agranovich and A. A. Maradudin, Eds. (North-Holland, Amsterdam, 1983), p. 249.

15. Y. S. Bai and M. D. Fayer, *Chem. Phys.* **128**, 135 (1988); L. R. Narasimhan, K. A. Littau, D. W. Pack, Y. S. Bai, A. Elschner, and M. D. Fayer, *Chem. Rev.* **90**, 439 (1990).

16. P. C. Becker, H. L. Fragnito, J. Y. Bigot, C. H. Brito-Cruz, R. L. Fork, and C. V. Shank, *Phys. Rev. Lett.* **63**, 505 (1989).

17. E. T. J. Nibbering, D. A. Wiersma, and K. Duppen, *Phys. Rev. Lett.* **66**, 2464 (1991); **68**, 514 (1992).

18. M. Koch, D. Weber, J. Feldman, E. O. Göbel, T. Meier, A. Schulze, P. Thomas, S. Schmitt-Rink, and K. Ploog, *Phys. Rev. B* **47**, 1532 (1993); U. Siegner, D. Weber, E. O. Göbel, D. Bennhardt, V. Heuckeroth, R. Saleh, S. D. Baranovskii, P. Thomas, H. Schwab, C. Klingshirn, J. M. Hvam, and V. G. Lyssenko, *Phys. Rev. B* **46**, 4564 (1992).

19. E. L. Hahn, *Phys. Rev.* **80**, 580 (1950).

20. R. R. Ernst, G. Bodenhausen, and A. Wokaun, *Principles of Nuclear Magnetic Resonance in One and Two Dimensions* (Clarendon, Oxford, 1987).

21. A. Pines, in *Proceedings of the 100th School of Physics "Enrico Fermi"* (North-Holland, Amsterdam, 1988), p. 43.

22. S. Mukamel and K. Shan, *Chem. Phys. Lett.* **117**, 489 (1985).

23. A. M. Weiner, S. DeSilvestri, and E. P. Ippen, *J. Opt. Soc. Am. B* **2**, 654 (1985); A. M. Weiner and E. P. Ippen, *Chem. Phys. Lett.* **114**, 456 (1985).

24. M. Cho, N. F. Scherer, G. R. Fleming, and S. Mukamel, *J. Chem. Phys.* **96**, 5618 (1992).

25. J. L. Black and B. I. Halperin, *Phys. Rev. B* **16**, 2879 (1977).

26. N. F. Scherer, R. J. Carlson, A. Matro, M. Du, A. J. Ruggiero, V. Romero-Rochin, J. A. Cina, G. R. Fleming, and S. A. Rice, *J. Chem. Phys.* **95**, 1487 (1991).

27. N. Morita and T. Yajima, *Phys. Rev. A* **30**, 2525 (1984).

28. R. Beach and S. R. Hartmann, *Phys. Rev. Lett.* **53**, 663 (1984).
29. S. Asaka, N. Nakatsuka, M. Fujiwara, and M. Matsuoka, *Phys. Rev. A* **29**, 2286 (1984).
30. T. Kobayashi, A. Tersaki, T. Hattori, and K. Kurokawa, *Appl. Phys. B* **47**, 107 (1988).

CHAPTER 11

Resonant Gratings, Pump-Probe, and Hole-Burning Spectroscopy

The calculation of the signal in the most general experiment related to the third-order polarization $P^{(3)}$ may require a three- or even a four-fold time integration involving the response function and the light pulses. For sufficiently simple models, these integrations may be carried out analytically. In ideal time-domain techniques, the number of integrations is reduced, and the nonlinear response function provides the simplest and most natural method for calculating the signal. The photon echoes discussed in Chapter 10 constitute a beautiful example of time-domain spectroscopy. These experiments can be performed with very short (impulsive) pulses that allow three of the time integrations to be eliminated, so that the signal is directly related to the response function. A single time integration (over the observation time) still remains, but even this can be eliminated in the presence of a large inhomogeneous broadening or by using heterodyne detection. The "secret" of echo spectroscopies is that they are genuinely time-domain techniques, which depend on two carefully designed evolution periods, and require no frequency resolution.

An additional advantage of time-domain techniques is the absolute control they offer over the relative time ordering of the interactions with the various fields. This generally reduces the number of terms in the calculation of the signal and makes it easier to focus on a specific process of interest. For a measurement involving n fields, the number of terms is usually reduced by a factor of $n!$ (the number of permutations of the fields). This is why the nonlinear susceptibility $\chi^{(n)}$, which represents a frequency-domain measurement with no control over time ordering, has $n!$ times more terms than the corresponding nonlinear response function $S^{(n)}$.

For techniques requiring some kind of frequency resolution, and for experiments involving complex systems, such as a polyatomic solute in a solvent, the multiple time integrations are often so tedious that it becomes desirable to abandon the response function and reformulate the problem in terms of a few relevant degrees of freedom that are kept alive. This is the basis for the semiclassical wavepacket and the doorway-window pictures that will be developed in Chapters 12 and 13. The issue is whether to carry out all averagings over the matter degrees of freedom first, obtain the response functions, and then worry about the convolution with the fields, or to include the fields from the start and calculate the dynamics in terms of wavepackets, performing the trace over the relevant material coordinates only at the end of the calculation. The method of choice depends on the nature of the system and the type of experiment.

In this chapter we consider a few additional "almost ideal" time-domain resonant four-wave mixing techniques whereby the number of integrations is reduced to one or two, and the nonlinear response function still provides a practical approach. These techniques include transient grating and pump-probe spectroscopy, which can be carried out either in an impulsive or in a hole-burning mode. We shall first develop a general expression for the polarization in a resonant four-wave mixing measurement performed on a two level system, invoking the RWA, but allowing for arbitrary time ordering of the fields. We next consider a *partial* control of time ordering obtained in a *sequential* process whereby the third pulse is well separated from the first two. This is the situation for ideal transient gratings. Specific applications will then be made to hole-burning and impulsive pump-probe spectroscopies using the Brownian oscillator model.

An important question that frequently comes up in the discussion of nonlinear spectroscopies is the interrelationships among frequency-domain and time-domain techniques, and their relative merits. In linear spectroscopy, one can either measure the response function in the frequency domain, or the free induction decay generated by exciting the system with an impulsive pulse and monitoring the time-dependent emitted field. The two observations are related by a simple Fourier transform. The real and the imaginary parts of the response function are connected by the Kramers–Kronig relation, which is a direct consequence of causality. Which technique to adopt then becomes a matter of convenience depending on the desired spectral and temporal resolution, signal to noise, sampling time, and other considerations. In nonlinear four-wave mixing spectroscopy the question is more subtle. Here time-domain and frequency-domain observations are connected by a triple Fourier transform, relating t_1, t_2, and t_3 to ω_1, ω_2, and ω_3. Such a relation, though important in principle, cannot be usually used in practice in its most general form, since the entire response function at all times or frequencies is rarely available. An important practical question is then under what conditions does a single Fourier transform relation exist between given techniques. The question can be stated differently: when can we focus on one particular combination of frequencies that is conjugate to a single time variable in the response function? There is no simple general answer to this question; it must be studied on a case by case basis. We shall address this issue as well in this chapter by establishing several Fourier transform relations between pump-probe, hole-burning, and photon echo spectroscopies. Additional Fourier transform relationships will be discussed in the coming chapters.

RESONANT, NON-TIME-ORDERED, FOUR-WAVE MIXING

Consider a general four-wave mixing experiment involving a two-level system and three laser pulses:

$$E(\mathbf{r}, t) = E_1(t + \tau + \tau') \exp[i\mathbf{k}_1\mathbf{r} - i\omega_1 t] + E_3(t + \tau) \exp[i\mathbf{k}_3\mathbf{r} - i\omega_3 t] + E_2(t) \exp[i\mathbf{k}_2\mathbf{r} - i\omega_2 t] + c.c. \quad (11.1)$$

The system and the pulse configuration are identical to those discussed in the previous chapter. However, there we focused on ideal time-domain measurements, in which absolute control over time ordering can be maintained by taking the delay periods τ and τ' to be long compared with the pulse durations (limit of well-separated pulses). We also addressed questions such as what are the possible outgoing wavevectors $\mathbf{k}_s$ and which of the Liouville space pathways contribute to each of them under the RWA. In contrast, at this point we do not attempt to maintain any control over time ordering, i.e., there are no restrictions on the delay periods τ and τ', so the pulse envelopes may overlap.* As a result, all permutations of $\mathbf{k}_1$, $\mathbf{k}_2$, and $\mathbf{k}_3$ are equivalent, and do not constitute distinguishable techniques. Therefore, unlike Chapter 10, we need consider here only a single outgoing wavevector, $\mathbf{k}_s \equiv \mathbf{k}_1 - \mathbf{k}_3 + \mathbf{k}_2$ and frequency $\omega_s \equiv \omega_1 - \omega_3 + \omega_2$. Making use of Eq. (5.45) with the present pulses and choice of $\mathbf{k}_s$ we get

$$P^{(3)}(\mathbf{k}_s, t) = P_{\mathrm{I}}^{(3)}(\mathbf{k}_s, t) + P_{\mathrm{II}}^{(3)}(\mathbf{k}_s, t), \quad (11.2)$$

with

$$\begin{aligned} P_{\mathrm{I}}^{(3)}(\mathbf{k}_s, t) = \left(\frac{i}{\hbar}\right)^3 \int_0^\infty dt_3 \int_0^\infty dt_2 \int_0^\infty dt_1\, [R_1^{\mathrm{H}}(t_3, t_2, t_1) + R_4^{\mathrm{H}}(t_3, t_2, t_1)]\chi(t_3 + t_1) \\ \times \{E_2(t - t_3)E_3^*(t + \tau - t_3 - t_2)E_1(t + \tau + \tau' - t_3 - t_2 - t_1) \\ \times \exp[i\omega_s t_3 + i(\omega_1 - \omega_3)t_2 + i\omega_1 t_1] \\ + E_1(t + \tau + \tau' - t_3)E_3^*(t - \tau - t_3 - t_2)E_2(t - t_3 - t_2 - t_1) \\ \times \exp[i\omega_s t_3 + i(\omega_2 - \omega_3)t_2 + i\omega_2 t_1]\} \end{aligned} \quad (11.3a)$$

and

$$\begin{aligned} P_{\mathrm{II}}^{(3)}(\mathbf{k}_s, t) = \left(\frac{i}{\hbar}\right)^3 \int_0^\infty dt_3 \int_0^\infty dt_2 \int_0^\infty dt_1\, [R_2^{\mathrm{H}}(t_3, t_2, t_1) + R_3^{\mathrm{H}}(t_3, t_2, t_1)]\chi(t_3 - t_1) \\ \times \{E_2(t - t_3)E_1(t + \tau + \tau' - t_3 - t_2)E_3^*(t + \tau - t_3 - t_2 - t_1) \\ \times \exp[i\omega_s t_3 + i(\omega_1 - \omega_3)t_2 - i\omega_3 t_1] \\ + E_1(t + \tau + \tau' - t_3)E_2(t - t_3 - t_2)E_3^*(t + \tau - t_3 - t_2 - t_1) \\ \times \exp[i\omega_s t_3 + i(\omega_2 - \omega_3)t_2 - i\omega_3 t_1]\}. \end{aligned} \quad (11.3b)$$

* Note the interchange of the pulse labels 2 and 3 in Eq. (11.1) compared with Eq. (10.11). This is a better choice for presenting the pump-probe results later in this chapter.

Equation (11.2) has eight terms; recall that only two terms survive when a complete time ordering is imposed [Eq. (10.13) or (10.14)]. A specific intermediate case, where time ordering is only partially imposed, will be considered next.

PARTIAL CONTROL OVER TIME ORDERING: PHASE-LOCKED TRANSIENT GRATINGS

We now consider an important case in which, by taking the second delay, τ, to be sufficiently large, we arrange for the interaction with the E_2 pulse to be the last. We thus attain partial time ordering since pulses 1 and 3 can still interact in any order (Figure 11.1). This class of techniques will be labeled transient gratings [1–3], and the generated signal can be interpreted as follows: Pulses 1 and 3 interfere in the medium, creating a grating of matter response with a wavevector $\mathbf{k}_1 - \mathbf{k}_3$. The E_2 beam then undergoes a Bragg diffraction off this grating, resulting in the scattered beam at $\mathbf{k}_s = \mathbf{k}_1 + \mathbf{k}_2 - \mathbf{k}_3$. Coherent Raman and pump-probe spectroscopy are special cases of this configuration. The pioneering work of Weiner, Desilvestri, and Ippen displayed in Figure 11.2 had demonstrated the utilization of this grating configuration.

In addition to the three pulses of Eq. (11.1) we include a fourth (local oscillator) field that will be used for heterodyne detection as described in Chapter 4. The electric field is now taken to be

$$E(\mathbf{r}, t) = E_1(t + \tau + \tau')\exp(i\mathbf{k}_1\mathbf{r} - i\omega_1 t) + E_3(t + \tau)\exp(i\mathbf{k}_3\mathbf{r} - i\omega_3 t - i\phi) + E_2(t)\exp(i\mathbf{k}_2\mathbf{r} - i\omega_2 t) + E_{LO}(t)\exp(i\mathbf{k}_s\mathbf{r} - i\omega_{LO} t - i\psi) + c.c. \quad (11.4)$$

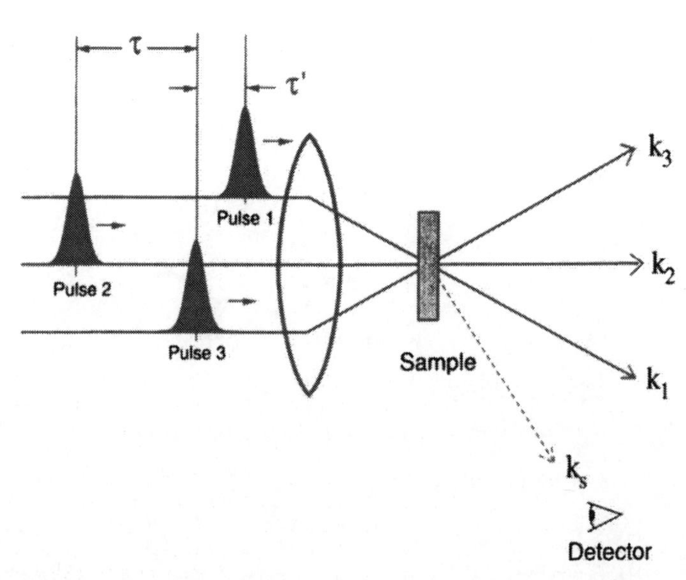

FIG. 11.1 Schematic representation of the timing in a sequential three-pulse measurement.

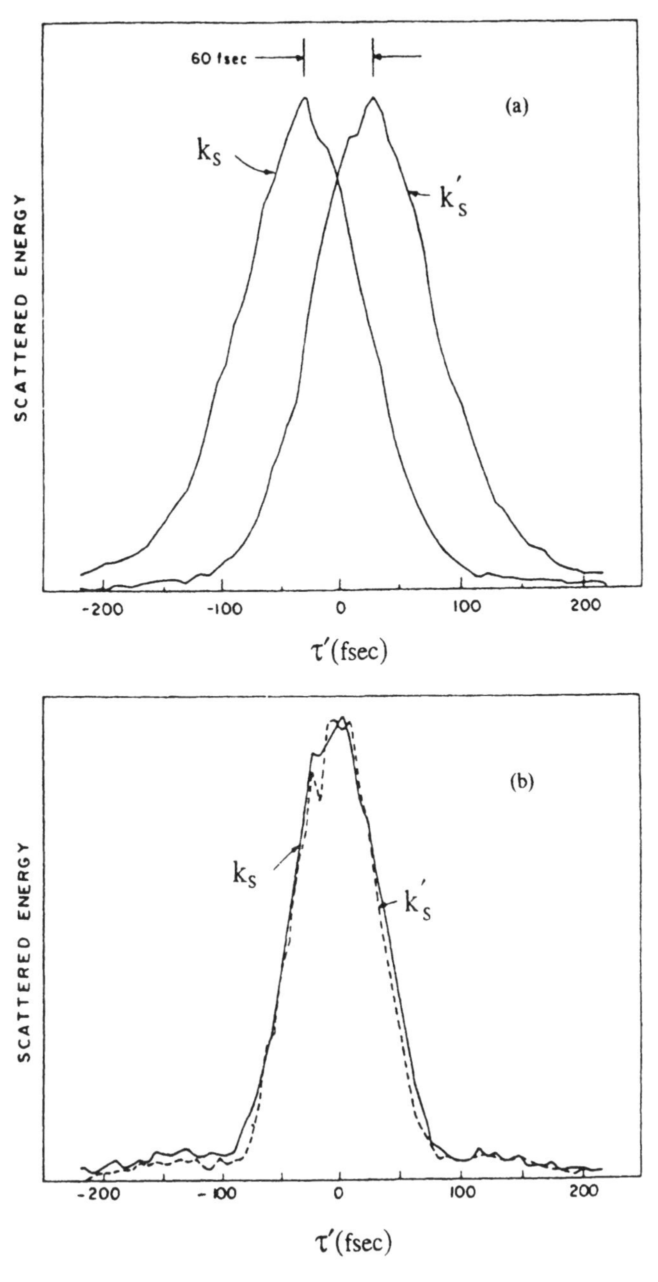

FIG. 11.2 Time-resolved four-wave mixing signal for the dye cresyl violet in a polymer thin film PMMA as a function of delay τ' between pulses1 and 3. The temperatures are (a) 15 K, (b) 290 K. The delay of pulse 2 was set to 1.3 ps. [A. M. Weiner, S. DeSilvestri, and E. P. Ippen, "Three-pulse scattering for femtosecond dephasing studies: Theory and experiment," *J. Opt. Soc. Am. B* **2**, 654 (1985).] Unlike the notation of Chapter 10, here the delay τ' variable can be either positive (pulse 1 first) or negative (pulse 3 first). Thus, $\mathbf{k}_s = \mathbf{k}_1 - \mathbf{k}_3 + \mathbf{k}_2$ implies paths $R_2 + R_3$ for $\tau' > 0$ and paths $R_1 + R_4$ for $\tau' < 0$. The converse is true for $\mathbf{k}_3' = \mathbf{k}_1 - \mathbf{k}_3 - \mathbf{k}_2$. The curves thus display both the echo and the nonecho contributions.

where we have assumed a phase control of the pulses. ϕ is the relative phase of E_3 with respect to E_1, and ψ is the phase of the local oscillator with respect to E_2. For the present applications these are the only two phases that need to be controlled (the phase between E_2 and E_1 is arbitrary).

The heterodyne signal is given by Eq. (4.80), integrated over all times t

$$S_{\mathrm{HTG}}(\mathbf{k}_s) = -2\omega_s \,\mathrm{Im} \int_{-\infty}^{\infty} dt \,\{E_{\mathrm{LO}}^*(t) P^{(3)}(\mathbf{k}_s, t) \exp[i(\omega_{\mathrm{LO}} - \omega_s)t + i(\psi + \phi)]\}. \tag{11.5}$$

By selecting the terms where E_2 interacts last [i.e., the first field permutation in P_{I} and the first field permutation in P_{II} in Eq. (11.3)], we get

$$\begin{aligned} S_{\mathrm{HTG}}(\mathbf{k}_s) = {} & 2\left(\frac{1}{\hbar}\right)^3 \omega_s \,\mathrm{Re} \int_{-\infty}^{\infty} dt \int_0^{\infty} dt_3 \int_0^{\infty} dt_2 \int_0^{\infty} dt_1 \\ & \times \{[R_1^{\mathrm{H}}(t_3, t_2, t_1) + R_4^{\mathrm{H}}(t_3, t_2, t_1)]\chi(t_3 + t_1) \\ & \times E_{\mathrm{LO}}^*(t) E_2(t - t_3) E_3^*(t + \tau - t_3 - t_2) E_1(t + \tau + \tau' - t_3 - t_2 - t_1) \\ & \times \exp[i(\omega_{\mathrm{LO}} - \omega_s)t + i\omega_s t_3 + i(\omega_1 - \omega_3)t_2 + i\omega_1 t_1 + i\psi + i\phi] \\ & + [R_2^{\mathrm{H}}(t_3, t_2, t_1) + R_3^{\mathrm{H}}(t_3, t_2, t_1)]\chi(t_3 - t_1) \\ & \times E_{\mathrm{LO}}^*(t) E_2(t - t_3) E_1(t + \tau + \tau' - t_3 - t_2) E_3^*(t + \tau - t_3 - t_2 - t_1) \\ & \times \exp[i(\omega_{\mathrm{LO}} - \omega_s)t + i\omega_s t_3 + i(\omega_1 - \omega_3)t_2 - i\omega_3 t_1 + i\psi + i\phi]\}. \end{aligned} \tag{11.6}$$

The double-sided Feynman diagrams for the response function in the RWA were displayed in Figure 10.6. Specific applications of this expression will be made in the remainder of this chapter.

SEQUENTIAL PUMP-PROBE SPECTROSCOPY

In a pump-probe experiment the system is subjected to two light pulses: the pump and the probe, whose frequencies are centered around ω_1 and ω_2, respectively. The incoming field is

$$E(\mathbf{r}, t) = E_1(t + \tau) \exp(i\mathbf{k}_1\mathbf{r} - i\omega_1 t) + E_2(t) \exp(i\mathbf{k}_2\mathbf{r} - i\omega_2 t) + c.c. \tag{11.7}$$

Here $E_1(t + \tau)$ is the temporal envelope of the pump field peaked at time $t = -\tau$, while $E_2(t)$ is the probe field, peaked at $t = 0$. The time delay τ is assumed to be sufficiently large to enforce a sequential (pump-first) time ordering. The relevant double-sided Feynman diagrams are given in Figure 11.3.

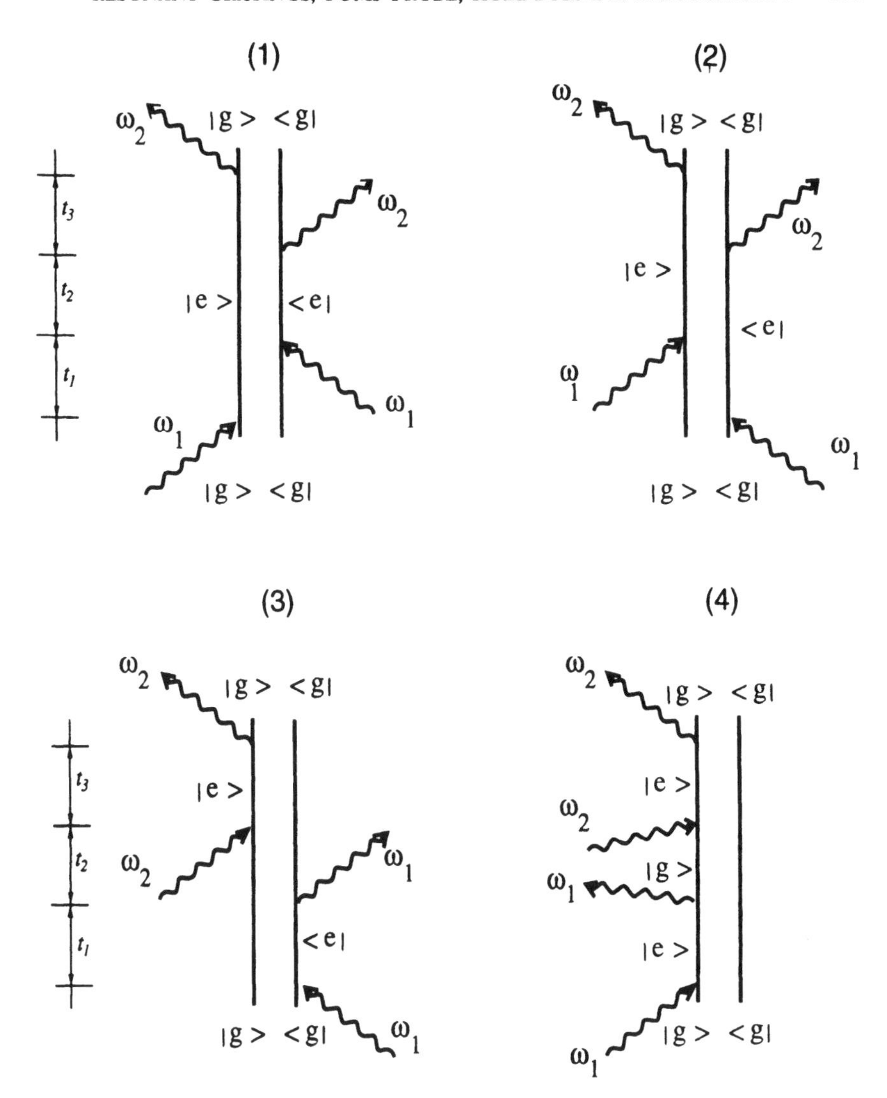

FIG. 11.3 Double-sided Feynman diagrams representing the probe difference absorption in the RWA. The complex conjugate contribution can be obtained by exchanging the ket and the bra. Diagram (1) and its complex conjugate represent the "particle" dynamics in the excited electronic state, whereas diagram (2) and its complex conjugate represent the "hole" dynamics in the ground electronic state. t_1, t_2, and t_3 represent the time intervals between the successive matter-field interactions.

We shall calculate the probe difference absorption, defined as the total probe absorption in the presence of the pump minus the probe absorption in the absence of the pump. In Chapter 4 we provided a formal definition for the probe absorption; both integrated [Eq. (4.85)] and dispersed [Eq. (4.90)] detection schemes were considered. Hereafter we use the integrated scheme. The observed spectrum is a special case of the heterodyne signal Eq. (11.6), obtained with the choice $E_1 = E_3$ and $\mathbf{k}_s = \mathbf{k}_1 - \mathbf{k}_1 + \mathbf{k}_2 = \mathbf{k}_2$. Furthermore, the E_2 field serves as its own local oscillator $E_{LO} = E_2$, and we have $\tau' = 0$ and $\phi = \psi = 0$. We thus get

$$
\begin{aligned}
&S_{PP}(\omega_1, \omega_2; \tau) \\
&= \left(\frac{1}{\hbar}\right)^3 2\omega_2 \, \text{Re} \int_{-\infty}^{\infty} dt \int_0^{\infty} dt_3 \int_0^{\infty} dt_2 \int_0^{\infty} dt_1 \\
&\quad \times \{E_2^*(t - \tau + t_3)E_2(t - \tau)E_1^*(t - t_2)E_1(t - t_2 - t_1) \exp[i\omega_2 t_3 + i\omega_1 t_1] \\
&\quad \times [R_1^H(t_3, t_2, t_1) + R_4^H(t_3, t_2, t_1)]\chi(t_3 + t_1) \\
&\quad + E_2^*(t - \tau + t_3)E_2(t - \tau)E_1(t - t_2)E_1^*(t - t_2 - t_1) \exp[i\omega_2 t_3 - i\omega_1 t_1] \\
&\quad \times [R_2^H(t_3, t_2, t_1) + R_3^H(t_3, t_2, t_1)]\chi(t_3 - t_1)\}. \qquad (11.8)
\end{aligned}
$$

We have made the change of variables $t \to t - \tau + t_3$ in order to simplify the subsequent analysis of limiting cases. Obviously, the phases of the fields are not important in this configuration, since the signal depends on products of each field amplitude E_j and its complex conjugate E_j^*, $j = 1, 2$. We have considered only contributions to S_{PP} that are to second order in the pump field and the probe. There are, of course, additional terms that are zeroth order in the pump and fourth order in the probe, which contribute to the ω_2 Fourier component of the polarization $\mathbf{k}_2 = \mathbf{k}_2 - \mathbf{k}_2 + \mathbf{k}_2$. Those terms, however, represent saturation by the probe and do not contribute to the difference signal calculated here. They can be easily eliminated experimentally by subtracting the results of two measurements, with and without the pump.

Equation (11.8) is the complete formal expression for the probe difference absorption signal, which may be studied as a function of the delay time of the probe with respect to the pump (τ), the pump frequency (ω_1), and the probe frequency (ω_2). As the pulse durations increase, the technique varies from the impulsive to the hole-burning mode. These cases will be discussed below.

IMPULSIVE PUMP-PROBE SPECTROSCOPY AND QUANTUM BEATS

The development of femtosecond laser sources has made it possible to probe coherent nuclear motions in real time. A particular mode can be coherently excited only if the laser pulse is sufficiently short compared with the characteristic period of the mode. Thus, a 10-fs pulse can excite coherent modes with frequencies as high as 3000 cm^{-1}. Nuclear motions on the femtosecond timescale have been observed in neat liquids, in semiconductors, in solution, in dye-doped polymer films, in proteins, and in isolated molecules in supersonic beams. Coherent vibrations may appear in femtosecond spectroscopy as quantum beats, which constitute the simplest example of a quantum mechanical interference. When two levels are excited coherently and emit to a common final level, the emission spectrum will oscillate with the two-level frequency. Microsecond to nanosecond quantum beats are well known in atomic [4] and ultracold molecular spectroscopy (see Figure 11.4) [5]. Femtosecond excitation can produce beats in large dye molecules at room temperature and in condensed phases [6–12]. Using an eigenstate description, a short pulse is needed to prepare a wavepacket in which different eigenstates (separated by the inverse oscillator period) are excited in

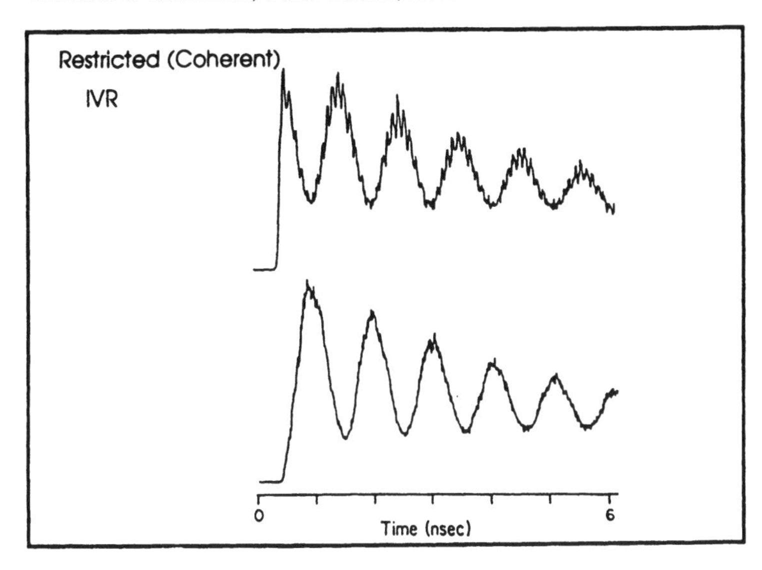

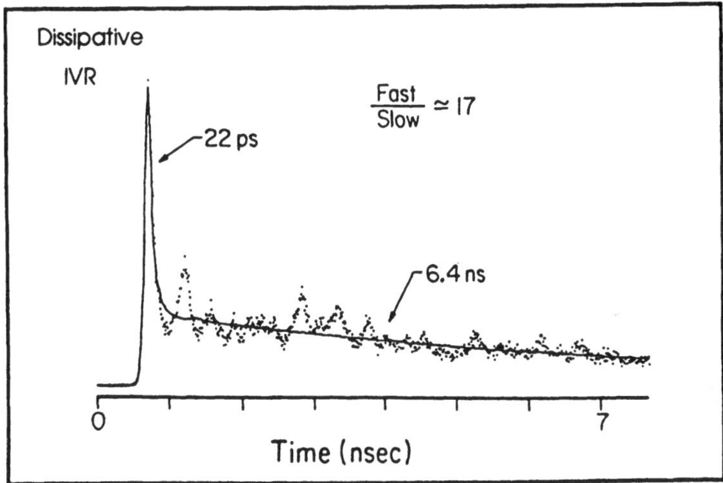

FIG. 11.4 Intramolecular vibrational redistribution (IVR) in large molecules. The in-phase and out-of-phase quantum beats observed in the emission spectrum for an anthracene molecular beam with picosecond time resolution. At higher energy, this restricted behavior of IVR changes to more of a dissipative behavior, still with residual coherence. [From P. M. Felker and A. H. Zewail, "Direct observation of nonchaotic multilevel vibrational energy flow in isolated polyatomic molecules," *Phys. Rev. Lett.* **53**, 501 (1984); see also W. R. Lambert, P. M. Felker, and A. H. Zewail, *J. Chem. Phys.* **75**, 5958 (1981); ibid. **82**, 2975 (1985).]

phase. Adopting a time-domain classical picture, a short pulse gives a "kick" to the nuclear mode, thus initiating a coherent oscillation. When the pulse is not short enough, it excites many systems with different phases, and the coherent oscillation disappears upon averaging.

The first terahertz oscillations in saturated resonant absorption using two pulses were observed by Tang et al. in malachite green (MG) and similar dyes in solution using 40-fs optical pulses [6]. The combined transmitted intensity of the pump and the probe was measured as a function of the delay time τ. Shank et al. reported the probe absorption of the same dye in solution using 6-fs optical pulses, showing coherent vibrational motions in the 1200–1600 cm^{-1} frequency range [7]. Quantum beats observed in another dye (Nile blue) showing the 590 cm^{-1} ring distortion mode are shown in Figure 11.5.

The vibrational quantum beats observed in these experiments contain simple damped harmonic oscillations related to only a few molecular vibrations (even though complex dye molecules have a multitude of vibrational modes), and the multimode Brownian oscillator model introduced in Chapter 8 is adequate for their description. We shall now apply this model to the analysis of the probe absorption signal [Eq. (11.8)] in a time-domain (impulsive) experiment. In the impulsive limit, the pump and probe pulses are short compared to the dynamic timescales of the solvent and solute nuclear degrees of freedom. We can therefore make the following assumption in Eq. (11.8).

1. The field E_1 is very short

$$E_1^*(t - t_2)E_1(t - t_2 - t_1) \sim \delta(t - t_2)\delta(t_1) \tag{11.9}$$

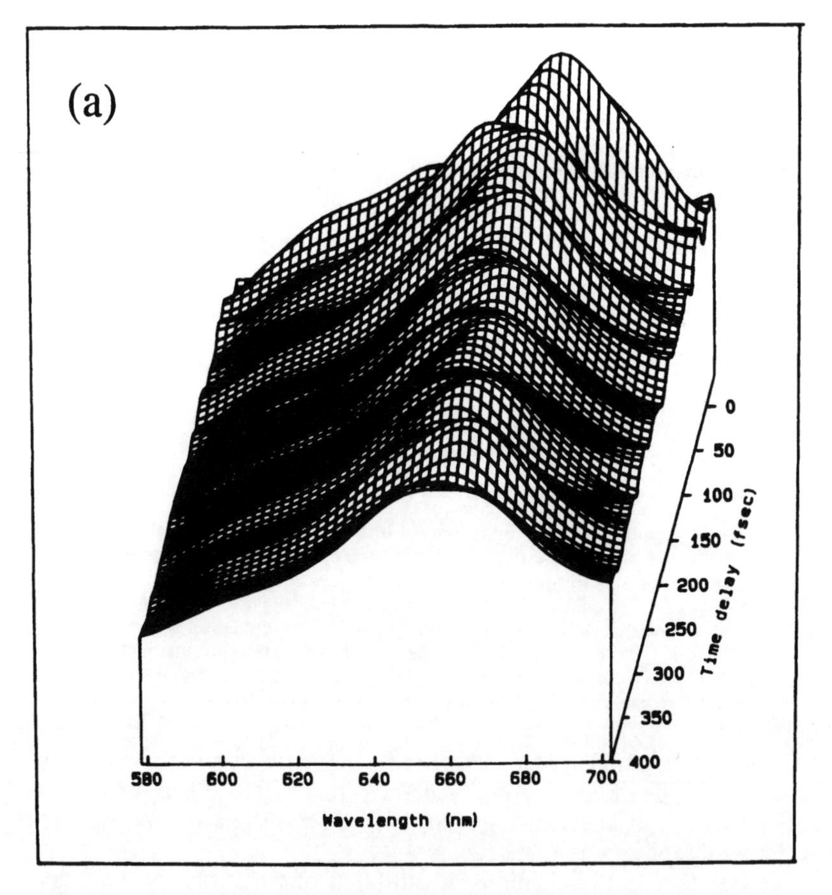

FIG. 11.5 (a) Three-dimensional plots of the differential absorption spectrum as a function of time delay and wavelength for the dye nile blue.

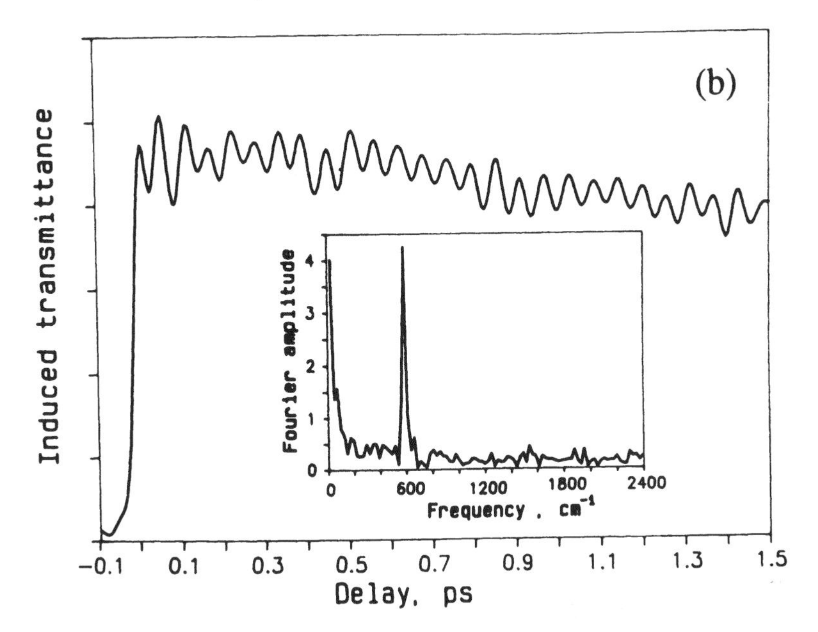

FIG. 11.5 (*continued*) (b) Pump-probe signal as a function of time delay at a selected wavelength and its Fourier transform (inset) for the dyes nile blue, signal at 666 nm. [Adapted from H. L. Fragnito, J. Y. Bigot, P. C. Becker, and C. V. Shank, "Evolution of the vibronic absorption spectrum in a molecule following impulsive excitation with a 6 fs optical pulse," *Chem. Phys. Lett.* **160**, 101 (1989).]

This allows us to perform the t_1 and the t integrations. Note that when $t_1 = 0$ then $R_1 = R_2$ and $R_3 = R_4$.

2. The E_2 field is long compared to the absorption linewidth.

$$E_2^*(t - \tau + t_3) \sim E_2^*(t - \tau).$$

We thus have

$$|E_2(t - \tau)|^2 \cong |E_2(t_2 - \tau)|^2. \tag{11.10}$$

3. The E_2 field is short compared with the timescale of nuclear motions in the excited state or the ground state

$$|E_2(t_2 - \tau)|^2 \sim \delta(t_2 - \tau). \tag{11.11}$$

The impulsive pump-probe signal [Eq. (11.8)] finally becomes (omitting some prefactors)

$$S_{\text{IPP}}(\omega_2; \tau) = \text{Re} \int_0^\infty dt_3 \exp(i\omega_2 t_3) \chi(t_3) \mathscr{R}(t_3, \tau, 0), \tag{11.12a}$$

$$\mathscr{R}(t_3, \tau, 0) = R_2^{\text{H}}(t_3, \tau, 0) + R_3^{\text{H}}(t_3, \tau, 0). \tag{11.12b}$$

The $\mathscr{R}$ correlation function was introduced previously in Eq. (10.15) in the analysis of photon echoes. The impulsive optical response is thus determined by the two-time response functions R_2 and R_3, which, when invoking the Condon approximation, are given by

$$R_2^{\mathrm{H}}(t_3, \tau, 0) = \langle \mathscr{G}_{eg}(t_3)\mathscr{G}_{ee}(\tau)\rho_g \rangle, \tag{11.13}$$

$$R_3^{\mathrm{H}}(t_3, \tau_2, 0) = \langle \mathscr{G}_{eg}(t_3)\mathscr{G}_{gg}(\tau)\rho_g \rangle. \tag{11.14}$$

These two terms can be interpreted as follows: The pump transfers a small fraction of the molecules from the ground to the excited state, thereby creating a "particle" in the excited state and a "hole" in the ground state. The particle and the hole then evolve during the delay period τ, followed by an electronic coherence $|e\rangle\langle g|$ during the t_3 period. R_2 and R_3 represent the particle and the hole contributions, respectively. Note that within the Condon approximation the hole contribution $R_3^{\mathrm{H}}(t_3, \tau, 0) = R_3^{\mathrm{H}}(t_3, 0, 0)$ does not depend on the time delay τ. We further note that $R_2^{\mathrm{H}}(t_3, 0, 0) = R_3^{\mathrm{H}}(t_3, 0, 0)$. Consequently, at short delay times the two contributions to the probe absorption in Eq. (11.12) are identical.

Using the multimode Brownian oscillator model [Eq. (8.15)], we have [13]

$$R_2^{\mathrm{H}}(t_3, \tau, 0) = \exp(-i\omega_{eg}t_3)\exp\left\{-\sum_j [g_j^*(t_3) + f_j(t_3; \tau)]\right\}, \tag{11.15}$$

and

$$R_3^{\mathrm{H}}(t_3, \tau, 0) = \exp(-i\omega_{eg}t_3)\exp\left[-\sum_j g_j(t_3)\right], \tag{11.16}$$

where using Eq. (8.26) we have

$$g_j(t_3) = \Delta_j^2 \int_0^{t_3} d\tau_2 \int_0^{\tau_2} d\tau_1 M_j'(\tau_1) - i\lambda_j \int_0^{t_3} d\tau_1 [1 - M_j''(\tau_1)], \tag{11.17}$$

$$f_j(t_3; \tau) = 2ig_j''(\tau + t_3) - 2ig_j''(\tau) = -2i\lambda_j \int_{\tau}^{\tau + t_3} d\tau_1 [1 - M_j''(\tau_1)]. \tag{11.18}$$

Here $g_j''(t)$ denotes the imaginary part of $g_j(t)$, i.e., $g_j(t) \equiv g_j'(t) + ig_j''(t)$ and $g_j(t)$ is given in Eqs. (8.26) or (8.67).

Impulsive spectroscopy of solvated dyes typically shows simple harmonic beats with only a few fundamental vibrational frequencies and a very small contributions of overtones. Note that the contributions of the various modes to the signal are not additive. Each Brownian oscillator is characterized by its frequency ω_j (not to be confused with the field frequencies ω_1 and ω_2), the dissipative friction constant γ_j, and the temperature T. Its coupling to the electronic transition is given by the parameters λ_j and Δ_j. In Chapter 8 we discussed the various limiting cases of this model, which can describe both high frequency vibrations undergoing coherent motions for many periods and low frequency overdamped vibrations. We shall now analyze the impulsive response in these limiting cases.

1. Coherent Oscillations. For high frequency (underdamped) modes we have $\gamma_j < \omega_j$. When $\gamma_j = 0$, the oscillator experiences a coherent motion with $M_j'(\tau_1) = M_j''(\tau_1) = \cos \omega_j \tau_1$. Using Eq. (8.39), Eqs. (11.17) and (11.18) then assume the form

$$g_j(t_3) = S_j\{\coth[\beta\hbar\omega_j/2)[1 - \cos(\omega_j t_3)] + i[\sin(\omega_j t_3) - \omega_j t_3]\}, \quad (11.19a)$$

$$f_j(t_3, \tau) = 2iS_j[\sin \omega_j(\tau + t_3) - \sin(\omega_j \tau) - \omega_j t_3]. \quad (11.19b)$$

Here S_j is the Huang Rhys factor. Calculated impulsive pump probe spectra for a single underdamped Brownian oscillator are displayed in Figure 11.6. For comparison we also show the three pulse echo for the same system.

2. Strongly Overdamped Motions. We next consider the effect of low frequency modes with $\gamma_j \gg \omega_j$ on the impulsive signal. This case has a few additional limits, which will be considered below.

2a. Spectral Diffusion Limit. In this case we neglect the nuclear dynamics during the optical transition, and assume that M_j' and M_j'' do not vary during the t_3 period. We thus set in Eq. (11.17) $M_j'(\tau_1) = M_j''(\tau_1) = 1$ and in Eq. (11.18) we set $M_j''(\tau_1) = M_j''(\tau)$, resulting in

$$f_j(t_3; \tau) = -2i\lambda_j t_3[1 - M_j''(\tau)], \quad (11.20a)$$

$$\hbar g_j(t_3) = \lambda_j k_B T t_3^2. \quad (11.20b)$$

We note that the phase of R_2^H at short delay times τ is $\omega_{eg} t_3$ and for long time delays its changes to $[(\omega_{eg} - 2\sum_j \lambda_j)t_3]$. The phase of R_3^H is independent of the delay and is given by $\omega_{eg} t_3$. Hence, for long delay times the signal splits into two components with reflection symmetry: the hole contribution (R_3^H) centered at $\omega_2 = \omega_{eg}$ and the particle contribution of R_2^H centered at $\omega_2 = \omega_{eg} - 2\sum_j \lambda_j$. This is known as the Stokes shift. The probe absorption in this case will be discussed further in Chapter 13 using a wavepacket representation. As an example, the time-dependent Stokes shift calculated for a single overdamped mode is displayed in Figure 11.7. The two peaks separated by the Stokes shift correspond to the particle and the hole contributions. The absorption and the long time (relaxed) fluorescence spectra are given in this case by Eqs. (8.52).

2b. Inhomogeneous Broadening. This is an extreme case of 2a obtained by further setting $M_j''(\tau) = 1$. In this case, the mode is completely static and using Eq. (8.51) we obtain

$$f_j(t_3; \tau) = 0, \quad (11.21a)$$

$$\hbar g_j(t_3) = \lambda_j k_B T t_3^2. \quad (11.21b)$$

No time-dependent Stokes shift is observed in this case.

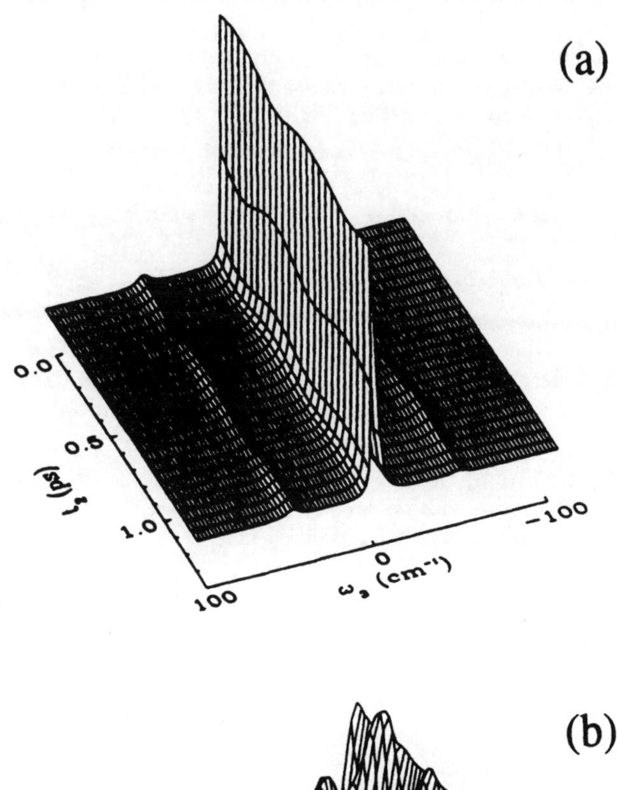

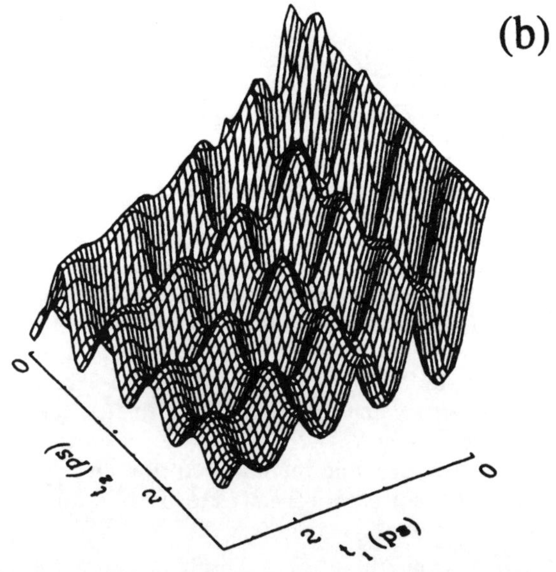

FIG. 11.6 (a) The impulsive pump probe signal S_{IPP} for a single underdamped Brownian oscillator is shown as a function of the detection frequency ω_3 and the delay time t_2. $\omega = 50\ \text{cm}^{-1}$, $d = 0.34$, $\gamma = 6.25\ \text{cm}^{-1}$, $\Delta = 12.5\ \text{cm}^{-1}$, and $T = 20$ K. (b) The three-pulse photon echo signal S_{SPE} is shown as a function of the two time delays t_1 and t_2 for the same system. [From L. E. Fried and S. Mukamel, "Simulation of nonlinear electronic spectroscopy in the condensed phase," *Adv. Chem. Phys.* **84**, 435 (1993).]

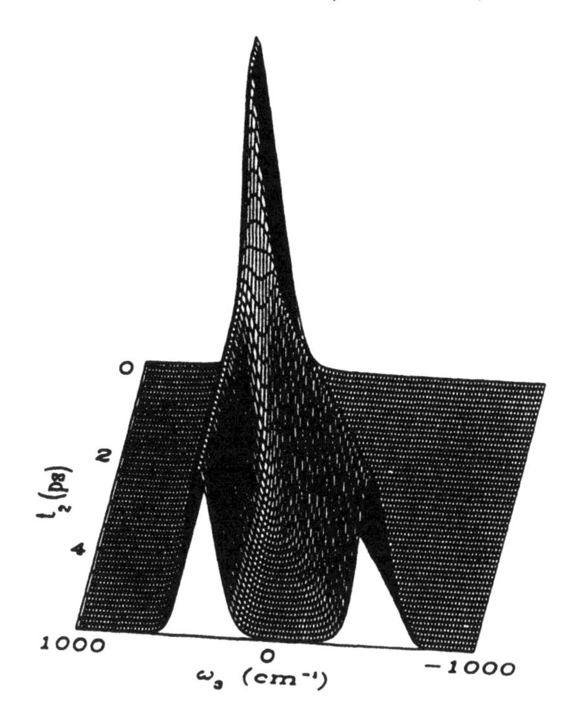

FIG. 11.7 The impulsive pump-probe signal S_{pp} for an overdamped slow modulation limit Brownian oscillator shown as a function of the detection frequency ω_3 and the delay time t_2. The two peaks splitting corresponds to the time-dependent Stokes shift. $\omega = 1.6 \text{ cm}^{-1}$, $d = 22.1$, $\gamma = 2.5 \text{ cm}^{-1}$, $\Delta = 50 \text{ cm}^{-1}$, and $T = 5$ K. [From L. E. Fried and S. Mukamel, "Simulation of nonlinear electronic spectroscopy in the condensed phase," *Adv. Chem. Phys.* **84**, 435 (1993).]

The contribution of inhomogeneous broadening to the signal may be incorporated by a frequency domain convolution of the lineshape with a Gaussian inhomogeneous distribution function with width $(2\lambda_j k_B T)^{1/2}$.

2c. Homogeneous Broadening. This is the opposite extreme to case 2b. Here, the solvent nuclear relaxation is very fast [Eq. (8.53)] so that

$$M_j''(t) = 0, \tag{11.22a}$$

$$M_j'(t) = \frac{1}{\Lambda}\,\delta(t), \tag{11.22b}$$

and

$$f_j(t_3; \tau) = -2i\lambda_j t_3, \tag{11.23a}$$

$$g_j(t_3) = -i\lambda_j t_3 + \hat{\Gamma}_j t_3. \tag{11.23b}$$

The relevant oscillator parameters reduce to two: λ_j and $\hat{\Gamma}_j \equiv 2\lambda_j k_B T/\Lambda_j$. In this limit, the oscillator coordinates' contribution to the signal may be modeled by a frequency domain convolution with a Lorentzian lineshape [Eq. (8.54)]) centered at $\omega_{eg}^{\circ} = \omega_{eg} - \lambda$. The impulsive signal in both the inhomogeneous and the homogeneous broadening limits does not vary with the delay τ. This is due to different reasons. In the former case there are no dynamics at all, whereas in the latter the dynamics are so fast that they are completed instantly and cannot be resolved on the experimental timescale.

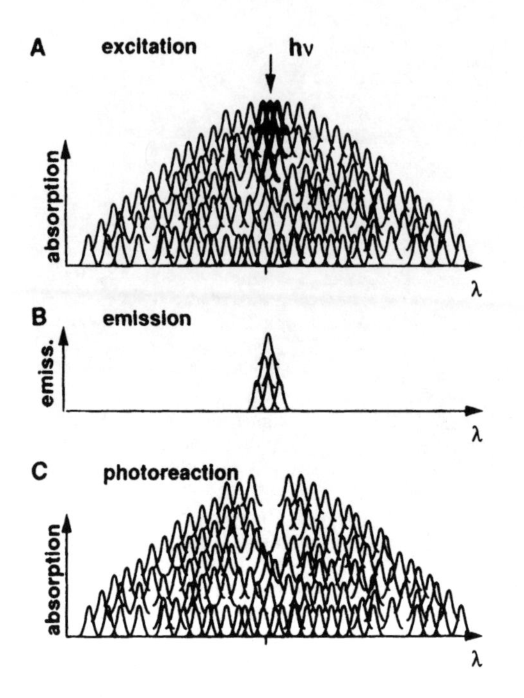

FIG. 11.8 Irradiation of an inhomogeneously broadened absorption band by a narrow-band light source selects a specific set of molecules by their transition energy. This energy selection leads to narrow emission lines or if a photoreaction takes place a narrow dip in the absorption spectrum is created. [After U. P. Wild and A. Renn, "Molecular computing: A review," *J. Mol. Electronics* **7**, 1 (1991).].

HOLE-BURNING SPECTROSCOPY

We now turn our attention to hole burning [14], which, together with photon echo spectroscopy, constitutes the most common nonlinear techniques used for the selective elimination of inhomogeneous broadening. The discussion in this chapter will be brief and is primarily intended to establish the relation between hole burning and the nonlinear response function and thereby connect it to other techniques. In Chapter 13 we present a more thorough analysis of hole burning based on a semiclassical wavepacket representation.

The experimental configuration (pulses and detection mode) for hole burning is identical to pump-probe spectroscopy. However, the experiment is carried out under different conditions. In the impulsive limit, the pump pulse is very short and offers no spectral resolution. Conversely, in the hole-burning mode, spectral resolution is crucial, so a longer (and thus spectrally narrower) pump pulse is used. The experiment is based on spectrally selecting a group of molecules within a broad inhomogeneous line, and measuring the change in the absorption of a probe induced by the photoexcitation. A narrow bandwidth is essential for the selective excitation. This is illustrated schematically in Figure 11.8.

We start our analysis with the expression for sequential pump-probe spectroscopy [Eq. (11.8)] and make the following additional assumptions:

1. Both pulses are long compared with the electronic dephasing timescale (inverse linewidth). We thus neglect the variation of the fields on the t_1 and t_3

time variables, resulting in the following field factors

$$|E_2(t-\tau)|^2|E_1(t-t_2)|^2. \tag{11.24a}$$

2. The pulses are short compared with nuclear dynamics during the t_2 period, so that we can safely set

$$|E_2(t-\tau)|^2|E_1(t-t_2)|^2 \rightarrow \delta(t-\tau)\delta(t_2-\tau). \tag{11.24b}$$

Performing the t_2 and the t integrations, finally yields the ideal "snapshot" spectrum (the origin of this terminology will be clarified in Chapter 13),

$$S_{HB}(\omega_1, \omega_2, \tau) = \left(\frac{1}{\hbar}\right)^3 2\omega_2 \operatorname{Re} \int_0^\infty dt_1 \int_0^\infty dt_3$$

$$\{\exp(i\omega_2 t_3 + i\omega_1 t_1)\chi(t_3+t_1)[R_1^{\mathrm{H}}(t_3,\tau,t_1) + R_4^{\mathrm{H}}(t_3,\tau,t_1)]$$

$$+ \exp(i\omega_2 t_3 - i\omega_1 t_1)\chi(t_3-t_1)[R_2^{\mathrm{H}}(t_3,\tau,t_1) + R_3^{\mathrm{H}}(t_3,\tau,t_1)]\}. \tag{11.26}$$

The probe absorption signal is now independent of the temporal profiles of the applied fields.

3. Inhomogeneous broadening is much larger than the contribution of any other electronic dephasing. In this case we may approximate the inhomogeneous dephasing function $\chi(t)$ by a delta function; the $R_1 + R_4$ term then vanishes since the t_1 and t_3 integration variables are both positive whereas $\chi(t_3+t_1) = \delta(t_3+t_1)$ requires that $t_1 = -t_3$. We thus get (omitting some prefactors)

$$S_{HB}(\omega_1, \omega_2, \tau) = \operatorname{Re} \int_0^\infty dt_1 \exp[i(\omega_2-\omega_1)t_1]\mathscr{R}(t_1,\tau,t_1), \tag{11.27a}$$

$$\mathscr{R}(t_1,\tau,t_1) = R_2^{\mathrm{H}}(t_1,\tau,t_1) + R_3^{\mathrm{H}}(t_1,\tau,t_1). \tag{11.27b}$$

In the Condon approximation we have

$$R_2^{\mathrm{H}}(t_1,\tau,t_1) = \langle \mathscr{G}_{eg}(t_1)\mathscr{G}_{ee}(\tau)\mathscr{G}_{ge}(t_1)\rho_g \rangle, \tag{11.28a}$$

$$R_3^{\mathrm{H}}(t_1,\tau,t_1) = \langle \mathscr{G}_{eg}(t_1)\mathscr{G}_{gg}(\tau)\mathscr{G}_{ge}(t_1)\rho_g \rangle. \tag{11.28b}$$

Inhomogeneous broadening contributes a multiplicative factor of $\chi(t_3 - t_1)$ to R_2^{H} and R_3^{H}, and as with the photon echo it is completely eliminated from the signal, which depends only on $t_1 = t_3$. The hole profile, measured by tuning ω_2 with respect to ω_1, will thus contain narrow resonances, free of inhomogeneous broadening.

As an example, consider a system with a Gaussian inhomogeneous and a Lorentzian homogeneous broadening. The linear absorption is then given by the Voigt profile [Eqs. (10.7) and (10.8)]. For the nonlinear response function we get

$$R_2^{\mathrm{H}}(t_1, \tau, t_1) = R_3^{\mathrm{H}}(t_1, \tau, t_1) = \exp(-2\hat{\Gamma} t_1) \tag{11.29a}$$

and

$$S_{\mathrm{HB}}(\omega_1, \omega_2; \tau) = \frac{2\hat{\Gamma}}{(\omega_1 - \omega_2)^2 + 4\hat{\Gamma}^2}. \tag{11.29b}$$

Hole-burning lineshapes of solvated dyes show vibronic (Franck–Condon) progressions, broadened by the homogeneous width [15]. These can be easily described using Eqs. (6.19) for R_2^{H} and R_3^{H}. If the timescale of nuclear motions is very slow but finite, the effect of the bath can be considered inhomogeneous for short delay times τ, but as τ increases, nuclear motions will result in shift and broadening of the hole. The variation of a spectral line with time, induced by slow nuclear motions is known as spectral diffusion.

Fluorescence line narrowing [16] is a technique closely related to hole burning. The pump excitation process is identical, however, the spontaneous light emission is detected. Using the results of Chapter 9, the signal is given by the particle contribution to Eq. (11.27) (R_3 and not R_2). Fluorescence line narrowing thus probes exclusively excited-state dynamics, whereas hole-burning probes both ground and excited-state dynamics. A wavepacket analysis of hole burning and spectral diffusion, including the additional nonsequential contributions which show up when the pulses overlap temporally, will be given in Chapter 13.

THREE-PULSE PHASE-LOCKED PUMP-PROBE ABSORPTION

We next consider a three-pulse pump-probe absorption measurement carried out with a pair of phase-locked pump pulses and optical heterodyne detection [17]. The pulses are given by Eq. (11.4) and the pulse configuration is shown in Figure 10.7. The second pump pulse (E_3) is delayed by τ' with respect to the first (E_1), and its phase factor, ϕ, is controlled relative to that of the first pump pulse. The probe pulse (E_2) follows the second pump with time delay τ, and the local oscillator pulse has a phase ψ relative to the probe. For simplicity we assume that the envelopes, frequencies, and the wavevectors of the two pump pulses are identical ($E_1 = E_3$, $\omega_1 = \omega_3$, $\mathbf{k}_1 = \mathbf{k}_3$) except for the phase ϕ. This is a pump-probe configuration since the signal and probe wavevectors coincide, $\mathbf{k}_s = \mathbf{k}_1 - \mathbf{k}_3 + \mathbf{k}_2 = \mathbf{k}_2$. The signal, which measures the probe transmission, is given by Eq. (11.6) except that now we do not distinguish between $\mathbf{k}_1 - \mathbf{k}_3 + \mathbf{k}_2$ and $\mathbf{k}_3 - \mathbf{k}_1 + \mathbf{k}_2$ and we need to add both contributions. We also enforce the time ordering of $\mathbf{k}_1$ first. A potential complication is that additional terms with wavevector combinations $\mathbf{k}_1 - \mathbf{k}_1 + \mathbf{k}_2$ and $\mathbf{k}_3 - \mathbf{k}_3 + \mathbf{k}_2$ also contribute to the signal. These represent processes where the system interacts twice with either the first or with the second pump pulse (instead of once with each). These contributions, however, can be easily eliminated by subtracting the signal obtained with each pump pulse individually from the total signal obtained using both phase-locked pump pulses. Since these additional terms are independent of the phase factor ϕ, they can alternatively be eliminated by subtracting two measurements with two different choices of ϕ.

The signal for this experiment is thus given by

$$
\begin{aligned}
S_{\mathrm{PLPP}}(\mathbf{k}_2) = 2\left(\frac{1}{\hbar}\right)^3 \omega_2 \operatorname{Re} \int_{-\infty}^{\infty} dt \int_0^{\infty} dt_3 \int_0^{\infty} dt_2 \int_0^{\infty} dt_1 \\
\times \{[R_1^{\mathrm{H}}(t_3, t_2, t_1) + R_4^{\mathrm{H}}(t_3, t_2, t_1)]\chi(t_3 + t_1) \\
\times E_{\mathrm{LO}}^*(t)E_2(t - t_3)E_1^*(t + \tau - t_3 - t_2)E_1(t + \tau + \tau' - t_3 - t_2 - t_1) \\
\times \exp[i(\omega_{\mathrm{LO}} - \omega_2)t + i\omega_2 t_3 + i\omega_1 t_1 + i\psi + i\phi] \\
+ [R_2^{\mathrm{H}}(t_3, t_2, t_1) + R_3^{\mathrm{H}}(t_3, t_2, t_1)]\chi(t_3 - t_1) \\
\times E_{\mathrm{LO}}^*(t)E_2(t - t_3)E_1(t + \tau + \tau' - t_3 - t_2)E_1^*(t + \tau - t_3 - t_2 - t_1) \\
\times \exp[i(\omega_{\mathrm{LO}} - \omega_2)t + i\omega_2 t_3 - i\omega_1 t_1 + i\psi + i\phi]\}. \qquad (11.30)
\end{aligned}
$$

When the two pump pulses are impulsive, and the inhomogeneous broadening is large so that $\chi(t) \cong \delta(t)$, we get

$$
\begin{aligned}
S_{\mathrm{PLPP}}(\omega_1, \omega_2, \tau', \tau, \phi, \omega_{\mathrm{LO}}, \psi) \propto \operatorname{Re}\{\exp[i(\omega_{\mathrm{LO}} - \omega_1)\tau' + i(\psi + \phi)] \\
\times \int_{-\infty}^{\infty} dt\, \mathscr{R}(\tau', t + \tau, \tau')E_{\mathrm{LO}}^*(t + \tau')E_2(t) \\
\times \exp[i(\omega_{\mathrm{LO}} - \omega_2)t]\}. \qquad (11.31)
\end{aligned}
$$

Finally, when the probe pulse is short compared with the molecular nuclear dynamics, $E_2(t) \cong E_2\delta(t)$, and $\omega_1 = \omega_2 = \omega_{\mathrm{LO}}$ we obtain

$$
S_{\mathrm{PLPP}}(\tau', \tau, \phi, \psi) \propto \operatorname{Re}\{\exp[i(\psi + \phi)]\mathscr{R}(\tau', \tau, \tau')E_{\mathrm{LO}}^*(\tau')\}. \qquad (11.32)
$$

If the local oscillator pulse is longer than the optical dephasing time, we can set $E_{\mathrm{LO}}^*(\tau') = 1$ in Eq. (11.31). It then becomes identical to the heterodyne-detected stimulated photon echo (Eq. (10.20)), with the phase ψ replaced by $\psi - \phi$.

PLPP, like the photon echo, constitutes an ideal time-domain technique, which can be carried out in a totally impulsive mode and requires no frequency resolution. The variation of the signal with τ', the separation of the two pump interactions, provides a direct look on optical dephasing. This is clearly in contrast with the conventional two-pulse pump probe absorption. Since the latter is given by the integration over the first (and third) optical coherence time period, τ', it does not give any information on the optical dephasing process, unless the probe frequency spectrum is resolved. Equation (11.32) shows that the PLPP technique is capable of measuring separately the real and the imaginary parts of $\mathscr{R}(\tau', \tau, \tau')$, by a proper change of the relative phase $\psi - \phi$. We can then relate it to photon echo as well as to ordinary pump-probe spectroscopy (in the limit of large inhomogeneous broadening), as discussed in Chapter 10 and earlier in this chapter.

MACROSCOPIC VERSUS MICROSCOPIC INTERFERENCE

So far we considered quantum beats that reflect a microscopic interference, namely a coherent superposition of vibrational states. However, this is not the only possible mechanism for the generation of quantum beats. Often such beats are macroscopic in nature. Consider, for example, an experiment performed on a mixture of various species so that the nonlinear polarization $P^{(3)}$ is the superposition of their contributions. If two molecules have a different frequency (e.g., two isotopes), then when the sum of contributions is squared we get beats with the difference frequency. These beats are macroscopic in nature and are formed only after $P^{(3)}$ is calculated, by a superposition of the fields generated by both types of molecules. This is different from the microscopic beats discussed here and in Chapter 10. In general, interference within a given Liouville space pathway occurs in a single molecule and is, therefore, microscopic, whereas interference between different paths can be either microscopic or macroscopic, since it can occur between different molecules. Microscopic interference is naturally more interesting than its macroscopic counterpart, since it may reveal additional information about the matter. Another manifestation of macroscopic interference has been demonstrated in a three-pulse pump-probe experiment carried out using pulses with no phase control [18]. It has been shown that the quantum beat signal can be manipulated by controlling the time interval τ'. As τ' is varied, the vibrational beat signal can be periodically canceled or enhanced. When the optical phase is not controlled, the contribution of Eq. (11.31) vanishes, and the signal is given by the contributions of the $\mathbf{k}_1 - \mathbf{k}_1 + \mathbf{k}_2$ and the $\mathbf{k}_3 - \mathbf{k}_3 + \mathbf{k}_2$ terms, which do not depend on the phase. These measurements can be interpreted by noting that the quantum beat amplitude resulting from one pulse can be either in-phase or out-of-phase with that created by the other pulse. Unlike S_{PLPP}, the manipulation of the beat signal is in this case purely macroscopic [17, 19].

FOURIER TRANSFORM RELATIONSHIPS

It is of great fundamental as well as practical interest to identify conditions whereby a single frequency variable in a frequency-domain technique is related by a Fourier transform to a particular time variable in a time-domain technique. The existence of a general triple transform relationship for four-wave mixing is a direct consequence of the nonlinear response function. However, it does not provide a practical connection since the complete response function is seldom known. We have already established earlier some general connections among different techniques, for example, the relations among the various echo spectroscopies and the connection between PLSPE and spontaneous light emission, established in Chapter 10.

We shall turn now to consider the possibility of a direct Fourier transform relation between pump-probe and hole-burning spectroscopies, which are frequency-domain experiments, and photon echoes. All of these spectroscopies depend on the same correlation function $\mathscr{R} \equiv R_2^{\mathrm{H}} + R_3^{\mathrm{H}}$ [Eq. (10.15)]. In general, this correlation function is complex and has two independent components (i.e., the real and imaginary parts). The hole-burning spectrum is the frequency-domain analogue of the photon echo, since it can be calculated using a one-sided Fourier transform of the echo correlation function [Eq. (11.27)]. The PLSPE technique allows us to measure these two components independently. A particularly simple general relation holds in the hole-burning mode, provided we assume (1) large inhomogeneous broadening, (2) the pulses are short compared with the molecular nuclear dynamics, and (3) pulses are long compared to the electronic dephasing time. The signal is then directly related to the in-phase and in-quadrature HSPE signals

$$S_{HB}(\omega_2 - \omega_1; \tau) = \text{Re} \int_0^\infty dt \exp[i(\omega_2 - \omega_1)t] \times \{S_{HSPE}(t, \tau, \psi = 0) - iS_{HSPE}(t, \tau, \psi = \pi/2)], \quad (11.33)$$

which can also be written in the form

$$S_{HB}(\omega_2 - \omega_1; \tau) = \int_0^\infty dt \cos[(\omega_2 - \omega_1)t] S_{HSPE}(t, \tau, \psi = 0) + \int_0^\infty dt \sin[(\omega_2 - \omega_1)t] S_{HSPE}(t, \tau, \psi = \pi/2). \quad (11.34)$$

We next turn to the inverse relation, i.e., expressing photon echoes in terms of the pump-probe signal. The pump-probe spectrum does not generally yield the complete complex function $\mathscr{R}(t, \tau, t)$; therefore, there is no general relationship between the pump-probe absorption signal and echo measurements. However, such transform relations do exist when the delay time is long compared with the timescales of the molecular nuclear relaxation processes, $\tau > \tau_R$. We can then factorize R_2 and R_3 as

$$\left.\begin{aligned} R_2(t, \tau > \tau_R, t) &= J_e(t) \exp(-\gamma\tau) J_g^*(t), \\ R_3(t, \tau > \tau_R, t) &= J_g(t) \exp(-\gamma'\tau) J_g^*(t) \approx J_g(\tau) J_g^*(t), \end{aligned}\right\} \quad (11.35)$$

with

$$J_n(t) = \text{Tr}[\mathscr{G}_{eg}(t)\rho_n], \qquad n = \text{e or g}. \quad (11.36)$$

Here, γ and $\gamma'(\approx 0)$ represent the inverse lifetimes of the electronic excited and ground states, respectively. $J_n(t)$ is a molecular correlation function with respect to the thermal equilibrium distribution in the $|n\rangle$th electronic state. The Fourier transforms of $J_g(t)$ and $J_e(t)$ give, respectively, the stationary absorption and fluorescence line shapes. These functions satisfy the symmetry relations: $J_n(-t) = J_n^*(t)$ and therefore $R_\alpha(-t, \tau > \tau_R, -t) = R_\alpha^*(t, \tau > \tau_R, t)$. In this case we have

$$\mathscr{R}(t, \tau > \tau_R, t) = \frac{1}{\pi} \int_{-\infty}^\infty d\omega_{21} \exp(-i\omega_{21}t) S_{HB}(\omega_{21}; \tau > \tau_R), \quad (11.37)$$

with $\omega_{21} \equiv \omega_2 - \omega_1$. Upon the substitution of this in Eqs. (10.20) and (10.23), we can immediately connect the hole-burning spectrum to any of the photon echo spectroscopies. For example, the stimulated photon echo signal is given by

$$S_{HSPE}(\tau', \tau > \tau_R, \psi) = \frac{2}{\pi} \text{Re} \int_{-\infty}^\infty d\omega_{21} \exp(-i\omega_{21}t) \exp(i\psi) S_{HB}(\tau > \tau_R; \omega_{21}). \quad (11.38)$$

This connection, which has been verified experimentally [20], does not always hold. If the system has many timescales τ_R then experiments performed on different timescales will show a different "homogeneous" width. In Figure 11.9 we compare the hole-burning measurement with the Fourier transform of a picosecond two-pulse echo signal for a dye in ethanol glass. The fwhm of the hole is 1.54 GHz while width of the echo curve is only 326 MHz. The factor of $\sim$5 increase in the width measured by photon echoes is due to the contribution of spectral diffusion in the hole-burning experiment. Glass dynamics (spectral diffusion) as slow as the time scale of hole-burning experiment ($\sim$100 sec) contributes to the holewidth. These very slow dynamics are rephased in the echo experiment, which operates on a $\sim$100 ps timescale. As shown in Eq. (11.38), the response function that describes the hole-burning experiment is the Fourier transform of the correlation function for the stimulated photon echo (three-pulse echo), not that of the two-pulse echo; comparison of photon echo and hole-burning data permits the extent of spectral diffusion to be assessed. A beautiful example of the variation of the apparent "homogeneous" width observed in a three- pulse echo of a chromophone in a protein is shown in Figure 10.12. The various points represent measurements with a different value of the delay time τ from 10 ns to 10 ms. Again, the different dephasing rates refect spectral diffusion processes.

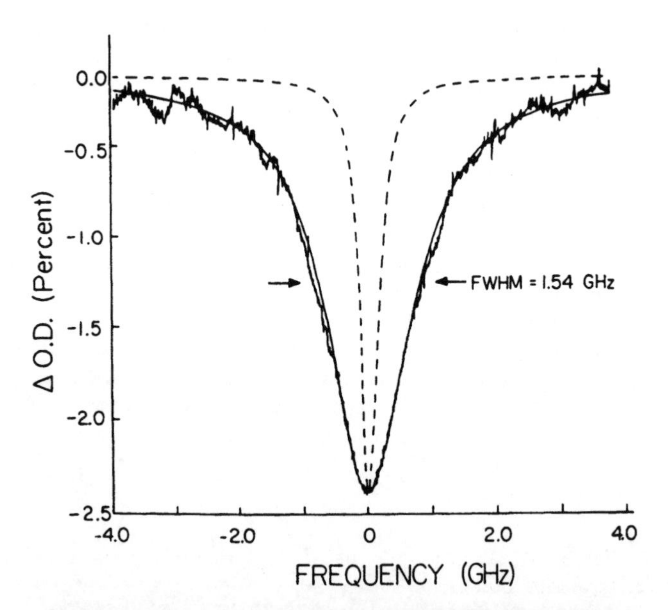

FIG. 11.9 Nonphotochemical hole burning data and a curve obtained from the Fourier transform of picosecond photon echo exponential decay data (dashed curve), both taken at 582 nm in the spectrum of resorufin in ethanol glass at 1.5 K. The solid line through the hole burning data is a Lorentzian fit. [From C. A. Walsh, M. Berg, L. R. Narasimhan, and M. D. Fayer, "Optical dephasing of chromophores in an organic glass: Picosecond photon echo and hole burning experiments," *Chem. Phys. Lett.* **130**, 6 (1986).]

In conclusion, we have established a general Fourier transform relationship between hole-burning and photon-echo spectroscopies. Conditions for the existence of such relationships were specified using the response function and without alluding to any particular model for the material system. We note that although both techniques are related to the third-order polarization, they provide complementary information. The hole-burning experiment gives dynamic information on the system when its density operator is in an electronic population, either in the ground or in the excited electronic state. Homogeneous and inhomogeneous dephasing processes due to the solvent appear as line broadening in the frequency domain, as is the case in the ordinary (linear) absorption and fluorescence lineshapes; however, these dephasing processes do not affect the temporal profile of the signal. Photon echoes, on the other hand, probe the system, while the density operator is in an electronic coherence. As such, they reveal directly the timescale of electronic dephasing processes. Furthermore, since electronic coherence depends on superpositions of vibrational states in both the ground and in the excited electronic manifolds, the signal may show combinations of ground-state and excited-state frequencies. Quantum beats observed in impulsive off resonant experiments will be discussed in Chapter 14. What we do see in the pump-probe signal are signatures of nuclear motions in the form of quantum beats, and the dynamic Stokes shift, characteristic of the bath relaxation, which stabilizes the excited electronic state. The present analysis allows us to address key questions, such as how do nuclear motions couple to the signal? Do the beats represent ground state or excited state evolution? and what is the role of coupling to the medium (homogeneous and inhomogeneous broadening, and spectral diffusion processes) in these measurements?

The Fourier transform of the hole-burning signal in the large inhomogeneous broadening limit [Eq. (11.37)] gives the correlation function $\mathscr{R}(\tau', \tau, \tau')$, whose real part is the accumulated photon echo signal [Eq. (10.A.1) with $\omega_1 = \omega_2$]. However, the reverse procedure of obtaining the hole-burning signal from the accumulated echo signal is generally not feasible, unless the correlation function $\mathscr{R}(\tau', \tau, \tau')$ is real so that a cosine transform relation holds. It should be emphasized that an exact transform relation does not exist when either the large inhomogeneous broadening assumption or the impulse limit in photon echoes is relaxed. Furthermore, in general, the pump-probe absorption signal depends on the pulse envelopes, and the ideal spectrum [Eq. (11.12)] should be replaced by more general expressions.

NOTES

1. H. J. Eichler, P. Gunter, and D. W. Pohl, *Laser-Induced Dynamic Gratings* (Springer-Verlag, Berlin, 1986).

2. M. D. Fayer, *Annu. Rev. Phys. Chem.* **33**, 63 (1982).

3. J. Knoester and S. Mukamel, *Phys. Rep.* **205**, 1 (1991).

4. J. N. Dodd, R. D. Kaul, and D. Warrington, *Proc. Phys. Soc. London* **89**, 176 (1964).

5. J. Manz and A. W. Castleman, Jr., Eds., Proceedings of the Berlin Conference on Femtosecond Chemistry, *J. Phys. Chem.* **97**, 48 (1993), special issue.

6. M. J. Rosker, F. W. Wise, and C. L. Tang, *Phys. Rev. Lett.* **57**, 321 (1986); F. W. Wise, M. J. Rosker, and C. L. Tang, *J. Chem. Phys.* **86**, 2827 (1987).

7. R. L. Fork, B. I. Greene, and C. V. Shank, *Opt. Lett.* **38**, 671 (1981); R. L. Fork, C. H. Brito-Cruz, P. C. Becker, and C. V. Shank, *Opt. Lett.* **12**, 483 (1987); P. C. Becker, R. L. Fork, C. H. Brito-Cruz, J. P. Gordon, and C. V. Shank, *Phys. Rev. Lett.* **60**, 2462 (1988); S. L. Dexheimer, Q. Wang, L. A. Peteanu, W. T. Pollard, R. A. Mathies, and C. V. Shank, *Chem. Phys. Lett.* **188**, 61 (1992).

8. J. Chesnoy and A. Mokhtari, *Phys. Rev. A* **38**, 3566 (1988); A. Mokhtari and J. Chesnoy, *Europhys. Lett.* **5**, 523 (1988).

9. K. A. Nelson and E. P. Ippen, *Adv. Chem. Phys.* **75**, 1 (1989).

10. M. H. Vos, F. Rappaport, J. C. Lambry, J. Breton, and J. L. Martin, *Nature* (*London*) **363**, 320 (1993).

11. K. Duppen, F. de Haan, E. T. J. Nibbering, and D. A. Wiersma, *Phys. Rev. A* **47**, 5120 (1993).

12. K. Leo, E. O. Göbel, T. C. Damen, J. Shah, S. Schmitt-Rink, W. Schäfer, J. F. Müller, K. Köhler, and P. Ganser, *Phys. Rev. B* **44**, 5726 (1991).

13. W. B. Bosma, Y. J. Yan, and S. Mukamel, *Phys. Rev. A* **42**, 6920 (1990); Y. J. Yan and S. Mukamel, *Phys. Rev. A* **41**, 6485 (1990).

14. D. Haarer and R. J. Silbey, *Phys. Today* **43**, 58 (1990).

15. R. Jankowiak and G. J. Small, *Science* **237**, 618 (1987).

16. R. I. Personov, E. I. Al'Shits, and L. A. Bykovskaya, *Opt. Commun.* **6**, 169 (1972).

17. M. Cho, N. F. Scherer, G. R. Fleming, and S. Mukamel, *J. Chem. Phys.* **96**, 5618 (1992).

18. J. J. Gerdy, M. Dantus, R. M. Bowman, and A. H. Zewail, *Chem. Phys. Lett.* **171**, 1 (1990).

19. M. Koch, J. Feldman, G. Von Plessen, E. O. Göbel, P. Thomas, and K. Köhler, *Phys. Rev. Lett.* **69**, 3631 (1992).

20. S. Saikan, T. Kishda, Y. Kanematsu, H. Aota, A. Harada, and M. Kamachi, *Chem. Phys. Lett.* **166**, 358, 1990; S. Saikan, A. Imaoka, Y. Kanematsu, K. Sakoda, K. Kominami, and Iwamoto, *Phys. Rev. B* **41**, 3185 (1990); S. Saikan, T. Nakabayashi, Y. Kanematsu, and N. Tato, *Phys. Rev. B* **38**, 7777 (1987).

CHAPTER 12

Wavepacket Dynamics in Liouville Space: The Wigner Representation

In previous chapters we have presented several applications of the nonlinear response function in the time domain. The strategy for computing the nonlinear polarization then involves the calculation of the four-time correlation functions, followed by a triple Fourier transform. A complete knowledge of these four-time correlation functions for all values of t_1, t_2, and t_3 allows the calculation of the signal for arbitrary field profiles $E_j(t)$ and time delays. This approach is particularly useful for spectroscopies involving ultrashort pulses where the necessary time integrations can be trivially performed and the signal is related directly to the time-domain response function.

The multiple time integrations may also be calculated for simple models involving only a few levels, even for complex pulse shapes. However, these integrations become a formidable task for complicated material systems and arbitrary laser pulses. In this case, it may prove desirable to keep one or a few relevant nuclear degrees of freedom "alive" in the theoretical description and to integrate over them only at the end of the calculation. This approach, which will be developed in the present chapter, allows us to calculate the time integrals as we go along, and suggests the interpretation of four-wave mixing spectroscopies in terms of the dynamics of nuclear wavepackets in Liouville space [1, 2]. These wavepackets can be calculated using either the vibronic eigenstate, or the phase space (Wigner) representation. The latter can be used to develop a path integral representation in phase space for the nonlinear response [3–5]. It is also most suitable for developing semiclassical approximations.* Since nuclear motions are essentially classical, it makes sense to develop classical or semiclassical procedures for the evaluation of the response function. Classical dynamics are much simpler

* A beautiful example of localized wavepackets of electrons in configuration space is provided by time-domain spectroscopy of highly excited atoms in Rydberg states. In this case, the density of electronic states is very high and a wavepacket in real space is the simplest and the clearest way to describe the optically prepared system. A semiclassical treatment of the electron is then appropriate [J. A. Yeazell and C. R. Stroud, *Phys. Rev. A* **43**, 5153 (1991)].

than the full quantum dynamics. Consequently, it is possible to model classically complex systems, with many more degrees of freedom than is possible quantum mechanically. Classical simulations of systems with $\sim 10^4$ degrees of freedom are not uncommon [6]. The wavepacket representation further yields the intuitively appealing doorway–window picture, which will be introduced in this chapter and developed further in Chapter 13. The problem of nonadiabatic curve crossing in the condensed phase is formally identical to the calculation of optical nonlinearities and there is tremendous insight to be gained by comparison of the two [7–10].

LIOUVILLE SPACE-GENERATING FUNCTION FOR THE LINEAR RESPONSE

The present approach is based on the propagation of a Liouville space-generating function (LGF), which is a wavepacket in the nuclear phase space. We shall introduce it by considering the linear response function for the two electronic level system. Equation (7.4b) can be recast in the form

$$J(t_1) = \langle\langle V_{eg}|\mathcal{G}_{eg}(t_1)\mathcal{V}_{eg,gg}|\rho_g\rangle\rangle. \tag{12.1}$$

Equation (12.1) can be evaluated as follows: Using Eqs. (3.31) and (7.5) we calculate the operator

$$\rho_1(t_1) \equiv \mathcal{G}_{eg}(t_1)\mathcal{V}_{eg,gg}\rho_g = \exp\left(-\frac{i}{\hbar}H_e t_1\right)V_{eg}\rho_g \exp\left(\frac{i}{\hbar}H_g t_1\right). \tag{12.2}$$

(The significance of the subscript 1 will be clarified in the next section.) This operator represents a wavepacket in Liouville space. Using the nuclear eigenstates we have

$$\rho_1(t_1) = \sum_{a,b} P(a)\mu_{ba}\exp(-i\omega_{ba}t_1)|b\rangle\langle a|. \tag{12.3a}$$

$J(t_1)$ is then given by multiplying $\rho_1(t_1)$ by V_{ge} from the left and taking a trace

$$J(t_1) = \mathrm{Tr}[V_{ge}\rho_1(t_1)]. \tag{12.3b}$$

$\rho_1(t_1)$ represents the time-dependent nuclear wavepacket when the electronic system is in an optical coherence ρ_{eg}. Its time evolution from the left (ket) and the right (bra) sides is determined by different nuclear Hamiltonians. $\rho_1(t)$ should be interpreted as a *Liouville space-generating function* (LGF) that allows the calculation of the two-time correlation function of the dipole operator using Eq. (12.3). The totál electronic and nuclear density operator of the system, to first order in V_{ge}, can also be expressed in terms of the LGF

$$\rho^{(1)}(t) = \int_0^\infty dt_1\, E(t-t_1)\bar{\rho}^{(1)}(t_1) \tag{12.4a}$$

with

$$\bar{\rho}_1^{(1)}(t_1) = |e\rangle\rho_1(t_1)\langle g| - |g\rangle\rho_1^\dagger(t_1)\langle e|, \tag{12.4b}$$

$\rho^\dagger(t_1)$ being the Hermitian conjugate of $\rho(t_1)$.

GENERATING FUNCTIONS FOR NONLINEAR RESPONSE: LIOUVILLE SPACE PATHWAYS

The above picture can be extended to the nonlinear response function. In this case each of the Liouville space pathways corresponding to the various contributions R_α to the nonlinear response function has its own generating function. We need to calculate four generating functions ρ_α that depend on the three evolution periods [1, 2]. We shall review in detail the calculation of the generating function ρ_1 corresponding R_1. The other generating functions can be calculated in a similar fashion. We start with the expression

$$R_1(t_3, t_2, t_1) = \langle\langle V_{eg}|\underbrace{\mathscr{G}_{eg}(t_3)}_{\uparrow\ \rho_1(t_1+t_2+t_3)}\mathscr{V}_{eg,ee}\underbrace{\mathscr{G}_{ee}(t_2)}_{\uparrow\ \rho_1(t_1+t_2)}\mathscr{V}_{ee,eg}\underbrace{\mathscr{G}_{eg}(t_1)}_{\uparrow\ \rho_1(t_1)}\mathscr{V}_{eg,gg}|\rho_g\rangle\rangle. \tag{12.5}$$

Equation (12.5) can be evaluated by starting with ρ_g in the far right and acting successively in a time-ordered fashion with the various $\mathscr{V}$ and $\mathscr{G}$ operators from right to left. The generating function will be evaluated during each of the propagation periods [i.e., t_1, t_2, and t_3, as indicated by the arrows in Eq. (12.5)]. For clarity, we also express the generating function using ordinary Hilbert space notation. We first act with one $\mathscr{V}$ and one $\mathscr{G}$ operator:

$$\rho_1(\tau) = \mathscr{G}_{eg}(\tau)\mathscr{V}_{eg,gg}\rho_g, \qquad 0 < \tau < t_1$$

where the subscript in ρ_1 indicates that this is the generating function relevant for pathway 1 (R_1). In this case $\mathscr{V}$ acts from the left (i.e., on the "ket" side of ρ_1), and we recover Eq. (12.2)

$$\rho_1(t_1) = \exp\left(-\frac{i}{\hbar}H_e t_1\right)V_{eg}\rho_g \exp\left(\frac{i}{\hbar}H_g t_1\right).$$

We next act with the second pair of $\mathscr{V}$ and $\mathscr{G}$ factors:

$$\rho_1(t_1 + \tau) = \mathscr{G}_{ee}(\tau)\mathscr{V}_{ee,gg}\rho_1(t_1), \qquad 0 < \tau < t_2$$

i.e.,

$$\rho_1(t_1 + t_2) = \exp\left(-\frac{i}{\hbar}H_e t_2\right)\rho_1(t_1)V_{ge}\exp\left(\frac{i}{\hbar}H_e t_2\right).$$

Finally, we act with the third $\mathscr{V}$ and $\mathscr{G}$ pair

$$\rho_1(t_1 + t_2 + \tau) = \mathscr{G}_{eg}(\tau)\mathscr{V}_{eg,ee}\rho_1(t_1 + t_2), \qquad 0 < \tau < t_3$$

i.e.,

$$\rho_1(t_1 + t_2 + t_3) = \exp\left(-\frac{i}{\hbar}H_e t_3\right)\rho_1(t_1 + t_2)V_{eg}\exp\left(\frac{i}{\hbar}H_g t_3\right),$$

and we have

$$R_1(t_3, t_2, t_1) = \mathrm{Tr}[V_{ge}\rho_1(t_1 + t_2 + t_3)].$$

While calculation of the linear response function Eq. (12.3) required propagating ρ_g, for a single time interval (t_1), the calculation of the nonlinear response function requires propagating the proper LGF ρ_1 for three time intervals t_1, t_2, and t_3 successively. *It should be emphasized that* $\rho_1(t_1 + t_2 + t_3)$, *which denotes the LGF at time* $(t_1 + t_2 + t_3)$, *depends separately on all three time arguments* t_1, t_2, *and* t_3, *and not only on their sum* $t_1 + t_2 + t_3$. The reason is that in each time interval there is a different propagation [i.e., $\mathcal{G}_{eg}(t_1)$, $\mathcal{G}_{ee}(t_2)$, and $\mathcal{G}_{eg}(t_3)$]. Consequently, the LGF $\rho_\alpha(t_1 + t_2 + t_3)$ entering into the calculations of R_α $\alpha = 1, \ldots, 4$ are different, since they correspond to different choices of e and g, as shown by the Liouville space pathways in Figure 7.1. The generating functions ρ_2, ρ_3, and ρ_4 corresponding to the other Liouville space pathways can be evaluated similarly, by simply changing the propagation operators $\mathcal{G}_{ee}$, $\mathcal{G}_{gg}$, $\mathcal{G}_{eg}$, and $\mathcal{G}_{ge}$.

In summary, the complete expression for the nonlinear response function in terms of the LGF is

$$R_\alpha(t_3, t_2, t_1) = \mathrm{Tr}[V_{ge}\rho_\alpha(t_1 + t_2 + t_3)] \qquad \alpha = 1, \ldots, 4 \tag{12.6}$$

where

$$\rho_1(t_1) = \exp\left(-\frac{i}{\hbar}H_e t_1\right)V_{eg}\rho_g \exp\left(\frac{i}{\hbar}H_g t_1\right), \tag{12.7a}$$

$$\rho_1(t_1 + t_2) = \exp\left(-\frac{i}{\hbar}H_e t_2\right)\rho_1(t_1)V_{ge} \exp\left(\frac{i}{\hbar}H_e t_2\right), \tag{12.7b}$$

$$\rho_1(t_1 + t_2 + t_3) = \exp\left(-\frac{i}{\hbar}H_e t_3\right)\rho_1(t_1 + t_2)V_{eg} \exp\left(\frac{i}{\hbar}H_g t_3\right). \tag{12.7c}$$

$$\begin{aligned}
\rho_2(t_1) &= \exp\left(-\frac{i}{\hbar}H_g t_1\right)\rho_g V_{ge} \exp\left(\frac{i}{\hbar}H_e t_1\right), \\
\rho_2(t_1 + t_2) &= \exp\left(-\frac{i}{\hbar}H_e t_2\right)V_{eg}\rho_2(t_1) \exp\left(\frac{i}{\hbar}H_e t_2\right), \\
\rho_2(t_1 + t_2 + t_3) &= \exp\left(-\frac{i}{\hbar}H_e t_3\right)\rho_2(t_1 + t_2)V_{eg} \exp\left(\frac{i}{\hbar}H_g t_3\right).
\end{aligned} \tag{12.8}$$

$$\begin{aligned}
\rho_3(t_1) &= \exp\left(-\frac{i}{\hbar}H_g t_1\right)\rho_g V_{ge} \exp\left(\frac{i}{\hbar}H_e t_1\right), \\
\rho_3(t_1 + t_2) &= \exp\left(-\frac{i}{\hbar}H_g t_2\right)\rho_3(t_1)V_{eg} \exp\left(\frac{i}{\hbar}H_g t_2\right), \\
\rho_3(t_1 + t_2 + t_3) &= \exp\left(-\frac{i}{\hbar}H_e t_3\right)V_{eg}\rho_3(t_1 + t_2) \exp\left(\frac{i}{\hbar}H_g t_3\right).
\end{aligned} \tag{12.9}$$

$$\rho_4(t_1) = \exp\left(-\frac{i}{\hbar} H_e t_1\right) V_{eg} \rho_g \exp\left(\frac{i}{\hbar} H_g t_1\right),$$

$$\rho_4(t_1 + t_2) = \exp\left(-\frac{i}{\hbar} H_g t_2\right) V_{ge} \rho_4(t_1) \exp\left(\frac{i}{\hbar} H_g t_2\right),$$

$$\rho_4(t_1 + t_2 + t_3) = \exp\left(-\frac{i}{\hbar} H_e t_3\right) V_{eg} \rho_4(t_1 + t_2) \exp\left(\frac{i}{\hbar} H_g t_3\right). \tag{12.10}$$

This representation constitutes a *path integral in Liouville space* and gives a clear picture regarding the interplay of population and coherence evolution periods in the nonlinear response functions.

In complete analogy with Eq. (12.4) derived for the linear response, the present LGF represents the total (electronic and nuclear) density operator to third order in V_{ge}. We thus write

$$\rho^{(3)}(t) = \int_0^\infty dt_1 \int_0^\infty dt_2 \int_0^\infty dt_3\, E(t - t_1 - t_2 - t_3) E(t - t_2 - t_3) \times E(t - t_3) \bar{\rho}^{(3)}(t_1 + t_2 + t_3), \tag{12.11a}$$

where

$$\bar{\rho}^{(3)}(t_1 + t_2 + t_3) = \sum_{\alpha=1}^{4} |e\rangle \rho_\alpha(t_1 + t_2 + t_3) \langle g| - \sum_{\alpha=1}^{4} |g\rangle \rho_\alpha^\dagger(t_1 + t_2 + t_3) \langle e|. \tag{12.11b}$$

Equations (12.7)–(12.10) can be expanded in the nuclear eigenstates, explicitly showing their wavepacket character. This is done in Appendix 12A. In addition, the LGF lends itself readily to the development of useful semiclassical approximations for the response function, as will be shown next.

CLASSICAL SIMULATION OF NUCLEAR WAVEPACKETS: PHASE-AVERAGING REVISITED

In Chapter 7 we presented a simulation procedure for the response functions based on phase averaging. The approach reduces the calculation to running classical trajectories on a reference potential (which is different for each Liouville space path). A phase is accumulated along the trajectory and an ensemble average over the phase finally yields the response functions. We shall now show how this result may be obtained by making a classical approximation for the LGF. This will provide an additional insight and will allow us to explore various generalizations such as the semiclassical wavepacket procedure to be presented in the next section.

The adiabatic Hamiltonian of our two-level system is given by Eq. (7.1). Let us consider the time evolution of ρ_α in a period when the electronic state is $|n\rangle\langle m|$

$$\frac{d\rho_\alpha^{nm}(t)}{dt} = -\frac{i}{\hbar}[H_n\rho_\alpha^{nm}(t) - \rho_\alpha^{nm}(t)H_m], \tag{12.12}$$

where the choice $n, m = \text{e, g}$ depends on the path (α) as well as on the specific interval (t_1, t_2, or t_3). We next choose a *reference Hamiltonian* that is taken to be some weighted average of H_n and H_m, i.e.,

$$H_\alpha \equiv \eta H_n + (1 - \eta)H_m, \tag{12.13}$$

where $0 \leq \eta \leq 1$. Some obvious choices that will be adopted below are $\eta = 1$ ($H_\alpha = H_n$), $\eta = 0$ ($H_\alpha = H_m$), or $\eta = 1/2$ $H_\alpha = (H_m + H_n)/2$. Using this definition, we can recast Eq. (12.12) in the form

$$\frac{d\rho_\alpha^{nm}}{dt} = -\frac{i}{\hbar}[H_\alpha, \rho_\alpha^{nm}] - \frac{i}{\hbar}[(1 - \eta)U_{nm}\rho_\alpha^{nm} + \eta\rho_\alpha^{nm}U_{nm}], \tag{12.14}$$

where $U_{nm} \equiv W_n - W_m$ is the difference of the two adiabatic potentials.

A classical approximation can now be obtained by replacing the commutator in the right-hand side with the classical Liouville equation and assuming that ρ_α^{nm} and U_{nm} commute. We then have

$$\frac{d\rho_\alpha^{nm}}{dt} = \left[\frac{\partial H_\alpha}{\partial \mathbf{q}}\frac{\partial \rho_\alpha^{nm}}{\partial \mathbf{p}} - \frac{\partial H_\alpha}{\partial \mathbf{p}}\frac{\partial \rho_\alpha^{nm}}{\partial \mathbf{q}}\right] - \frac{i}{\hbar}\rho_\alpha^{nm}U_{nm}. \tag{12.15}$$

A more systematic approach involves a switch to the Wigner representation. Making use of Eq. (3.110) and taking the classical ($\hbar \to 0$) limit we get Eq. (12.15) with the choice $\eta = 1/2$, i.e., $H_\alpha = (H_n + H_m)/2$. For physical reasons, as explained in Chapter 7, we may wish to retain the more general form (12.15) with some flexibility in the choice of η. The solution to this equation is

$$\rho_\alpha^{nm}(\mathbf{pq}; t) = \iint d\mathbf{p}'\, d\mathbf{q}'\, \mathcal{G}_{nm}(\mathbf{pq}t; \mathbf{p}'\mathbf{q}')\rho_\alpha^{nm}(\mathbf{p}'\mathbf{q}'; 0),$$

which gives for the Liouville space Green function $\mathcal{G}_{nm}(t)$ [Eq. (7.5)]

$$\mathcal{G}_{nm}(\mathbf{pq}t; \mathbf{p}'\mathbf{q}') = \delta[\mathbf{q} - \mathbf{q}_c(t)]\delta[\mathbf{p} - \mathbf{p}_c(t)] \exp\left\{-\frac{i}{\hbar}\int_{t_0}^{t} d\tau\, U_{nm}[\mathbf{q}_c(\tau)]\right\}, \tag{12.16}$$

with the classical trajectory calculated using the reference potential

$$W_\alpha \equiv \eta W_n + (1 - \eta)W_m, \tag{12.17}$$

i.e.,

$$\left.\begin{aligned} \dot{\mathbf{q}}_c &= \mathbf{p}_c, \\ \dot{\mathbf{p}}_c &= -\frac{\partial W_\alpha}{\partial \mathbf{q}_c}, \end{aligned}\right\} \tag{12.18}$$

and the initial condition $\mathbf{p}_c(0) = \mathbf{p}'$, $\mathbf{q}_c(0) = \mathbf{q}'$.

Combining the classical approximation with Eqs. (7.11) we finally get

$$\rho_\alpha(\mathbf{p}, \mathbf{q}, t_1 + t_2 + t_3) = V(0)V(t_1)V(t_1 + t_2)V(t_1 + t_2 + t_3)$$
$$\times \delta[\mathbf{p} - \mathbf{p}_\alpha(t_1 + t_2 + t_3]\delta[\mathbf{q} - \mathbf{q}_\alpha(t_1 + t_2 + t_3)]$$
$$\times \exp\left[-\frac{i}{\hbar}\int_0^{t_1} d\tau\, U_\alpha(\tau) - \frac{i}{\hbar}\int_{t_1+t_2}^{t_1+t_2+t_3} d\tau\, U_\alpha(\tau) \right], \quad (12.19)$$

where $V_{ge} = V_{eg} \equiv V$, and

$$V(t) \equiv V[\mathbf{q}_c(t)], \quad (12.20a)$$

$$U_\alpha(t) \equiv \pm U[\mathbf{q}_c(t)]. \quad (12.20b)$$

The $\pm$ sign depends on the path. For an "eg" or a "ge" evolution period we have $U_\alpha \equiv U$ and $U_\alpha \equiv -U$, respectively.

The choice of a reference potential W_α for each of the Liouville paths and time intervals was discussed in Chapter 7 where we made the following choice: For R_3 and R_4 we took $W_\alpha = W_g$ whereas for R_1 and R_2 we used W_g during t_1 and W_e during t_2 and t_3. Equation (12.19) recovers our classical phase averaging procedure developed in Chapter 7 [Eq. (7.19)], provided we adopt the same reference potential. The nonuniqueness of the classical approximation and its dependence on the reference potential is a signature of the nonanalytical dependence of the response function on $\hbar$, as discussed in Chapter 7.

SEMICLASSICAL EQUATIONS OF MOTION FOR THE LIOUVILLE SPACE GENERATING FUNCTIONS

We are now in a position to develop the *semiclassical* procedure for the evaluation of the response functions, which goes beyond the fully classical approximation of the previous section. For simplicity we shall invoke the Condon approximation and neglect the nuclear dependence of the electronic dipole V_{eg}. This approximation, which follows from the weak dependence of the adiabatic electronic wavefunctions on the nuclear coordinates, is usually adequate for resonant spectroscopies (for off-resonant techniques, the Condon approximation must be relaxed).

We shall consider first the generating function for the linear optical response. In the Condon approximation it becomes

$$\rho_1(t_1) = \mathcal{G}_{eg}(t_1)\rho_g = \exp\left(-\frac{i}{\hbar}H_e t_1\right)\rho_g \exp\left(\frac{i}{\hbar}H_g t_1\right). \quad (12.21)$$

The generating function is an operator. If $H_g = H_e$, the trace of this operator is time independent and is equal to 1; otherwise, ρ_1 varies with time. This variation reflects the effect of *dephasing processes*, namely, loss of coherence by the accumulation of phase in the off diagonal elements of the density operator. The linear response function is then given by

$$J(t_1) \equiv \langle \mathcal{G}_{eg}(t_1)\rho_g \rangle = \mathrm{Tr}[\rho_1(t_1)] = \iint d\mathbf{p}\, d\mathbf{q}\, \rho_1(\mathbf{pq}; t_1). \quad (12.22)$$

We next define the *normalized* Liouville space-generating function:

$$\sigma_1(t_1) \equiv \rho_1(t_1)/\mathrm{Tr}[\rho_1(t_1)], \tag{12.23a}$$

i.e.

$$\sigma_1(t_1) \equiv \left[\exp\left(-\frac{i}{\hbar}H_e t_1\right)\rho_g \exp\left(\frac{i}{\hbar}H_g t_1\right)\right]\Big/\mathrm{Tr}\left[\exp\left(-\frac{i}{\hbar}H_e t_1\right)\rho_g \exp\left(\frac{i}{\hbar}H_g t_1\right)\right].$$

Using these definitions we can now recast the LGF in the form of a product of $J(t_1)$ times the normalized wavepacket,

$$\rho_1(t_1) = J(t_1)\sigma_1(t_1). \tag{12.23b}$$

To compute the LGF we further define

$$\langle U(t_1)\rangle \equiv \mathrm{Tr}[\mathrm{U}\sigma_1(t_1)] = \iint d\mathbf{p}\, d\mathbf{q}\; U(\mathbf{q})\sigma_1(\mathbf{pq}; t_1), \tag{12.24a}$$

where $U(\mathbf{q})$ was defined in Eq. (7.6),

$$U(\mathbf{q}) \equiv W_e(\mathbf{q}) - W_g(\mathbf{q}) - \langle [W_e(\mathbf{q}) - W_g(\mathbf{q})]\rho_g\rangle. \tag{12.24b}$$

We recall that U is a collective coordinate, which depends, in principle, on all the nuclear degrees of freedom. When the ground-state and the excited-state potentials are identical ($W_g = W_e$), $U = 0$, and the bath is insensitive to the electronic state of the system. Consequently, $J(t) = \exp(-i\omega_{eg}t)$, and the nuclear degrees of freedom are decoupled from the optical process. More generally, U introduces a time-dependent phase into the evolution of ρ_g. That phase results in a variation of the trace and induces line broadening. The magnitude of U and its dynamics are therefore crucial for determining the response function. We reiterate that additional time-dependent terms resulting from the evolution of $V_{ge}(t)$ must be added to go beyond the Condon approximation, as was shown in Chapter 7.

Using these definitions it is straightforward to show that $J(t)$ satisfies the following equation of motion

$$\frac{dJ}{dt} = -i\omega_{eg}J - \frac{i}{\hbar}\langle U(t)\rangle J, \tag{12.25a}$$

whose solution with the initial condition $J(0) = 1$ is

$$J(t_1) = \exp\left[-i\omega_{eg}t_1 - \frac{i}{\hbar}\int_0^{t_1} d\tau \langle U(\tau)\rangle\right]. \tag{12.25b}$$

Note that $\langle U(\tau)\rangle$ is a somewhat unusual object. It is evaluated with respect to the normalized non-Hermitian generating function $\sigma_1(t)$, which propagates from the left (ket side) and from the right (bra side) with different Hamiltonians. As a consequence of this unusual propagation, $\langle U(t)\rangle$ is a complex quantity. Its imaginary part will result in the decay of $J(t_1)$ at long times. The decay is a manifestation of the distribution of phases (loss of overall phase) and is thus termed *dephasing*. This decay is readily observable experimentally since it induces spectral line broadening.

We shall now extend this procedure for the semiclassical evaluation of the nonlinear response function within the Condon approximation. To that end we define the normalized, path-dependent, wavepackets

$$\sigma_\alpha(t) \equiv \rho_\alpha(t)/\text{Tr}\,\rho_\alpha(t), \qquad \alpha = 1, \ldots, 4 \tag{12.26a}$$

where ρ_α were introduced in Eqs. (12.7)–(12.10), and

$$\langle U_\alpha(t)\rangle \equiv \text{Tr}[\sigma_\alpha(t)U]. \tag{12.26b}$$

In analogy with Eq. (12.23b) we have

$$\rho_\alpha(t_1 + t_2 + t_3) = R_\alpha(t_3, t_2, t_1)\sigma_\alpha(t_1 + t_2 + t_3),$$

where

$$R_\alpha(t_3, t_2, t_1) = \text{Tr}\,\rho_\alpha(t_1 + t_2 + t_3).$$

R_α satisfy equations of motion analogous to Eqs. (12.25a), i.e.,

$$\frac{dR_\alpha}{dt} = -i\omega_\alpha R_\alpha - \frac{i}{\hbar}\langle U_\alpha(t)\rangle R_\alpha, \tag{12.27}$$

where $\omega_\alpha = \omega_{eg}$, ω_{ge}, or 0 depending on the path and the time interval (t_1, t_2, or t_3). The solution of Eqs. (12.27) results in the following expression for the response function:

$$R_1(t_3, t_2, t_1) = \exp(-i\omega_{eg}t_1 - i\omega_{eg}t_3) \times \exp\left[-\frac{i}{\hbar}\int_{t_1+t_2}^{t_1+t_2+t_3} d\tau\langle U_1(\tau)\rangle - \frac{i}{\hbar}\int_0^{t_1} d\tau\langle U_1(\tau)\rangle\right], \tag{12.28a}$$

$$R_2(t_3, t_2, t_1) = \exp(i\omega_{eg}t_1 - i\omega_{eg}t_3) \times \exp\left[-\frac{i}{\hbar}\int_{t_1+t_2}^{t_1+t_2+t_3} d\tau\langle U_2(\tau)\rangle + \frac{i}{\hbar}\int_0^{t_1} d\tau\langle U_2(\tau)\rangle\right], \tag{12.28b}$$

$$R_3(t_3, t_2, t_1) = \exp(i\omega_{eg}t_1 - i\omega_{eg}t_3) \times \exp\left[-\frac{i}{\hbar}\int_{t_1+t_2}^{t_1+t_2+t_3} d\tau\langle U_3(\tau)\rangle + \frac{i}{\hbar}\int_0^{t_1} d\tau\langle U_3(\tau)\rangle\right], \tag{12.28c}$$

$$R_4(t_3, t_2, t_1) = \exp(-i\omega_{eg}t_1 - i\omega_{eg}t_3) \times \exp\left[-\frac{i}{\hbar}\int_{t_1+t_2}^{t_1+t_2+t_3} d\tau\langle U_4(\tau)\rangle - \frac{i}{\hbar}\int_0^{t_1} d\tau\langle U_4(\tau)\rangle\right]. \tag{12.28d}$$

So far the only approximation made in the derivation of Eqs. (12.25b) and (12.28) is the Condon approximation for the dipole moment. Our expression thus retains the full quantum nature of the system. However, the main reason for recasting the response function in this form is that it suggests a useful semiclassical procedure for computing the LGF $\sigma_\alpha(t)$ and the nonlinear response function. For simplicity we shall consider a single nuclear coordinate; the generalization to multidimensional systems is straightforward. We start by assuming that $\sigma_\alpha(t)$ in the Wigner representation is a complex normalized Gaussian wavepacket in phase space. We thus make the following ansatz [1]:

$$\sigma_\alpha(p, q; t) = \frac{1}{(2\pi)[\det W(t)]^{1/2}} \times \exp\left\{-\tfrac{1}{2}[q - q_\alpha(t), p - p_\alpha(t)]W^{-1}(t)\begin{bmatrix} q - q_\alpha(t) \\ p - p_\alpha(t)\end{bmatrix}\right\}, \quad (12.29a)$$

with

$$W(t) = \begin{bmatrix} \sigma^\alpha_{qq}(t) & \sigma^\alpha_{pq}(t) \\ \sigma^\alpha_{pq}(t) & \sigma^\alpha_{pp}(t)\end{bmatrix}. \quad (12.29b)$$

This form is exact for harmonic systems, even when the frequencies in the ground and in the excited states are different. For anharmonic systems it provides a semiclassical approximation that should hold at short times. Hereafter in this section, angular brackets will denote an "expectation value" with respect to the normalized generating function σ_α [see Eq. (12.26b)], i.e.,

$$\langle A\rangle \equiv \mathrm{Tr}[A\sigma_\alpha(t)] = \int A(p, q)\sigma_\alpha(p, q, t)\, dp\, dq. \quad (12.30)$$

In Eqs. (12.29), q and p represents the coordinate and momentum; q_α and p_α represent their "expectation values" for the α path

$$q_\alpha(t) = \langle q\rangle \quad (12.31a)$$

$$p_\alpha(t) = \langle p\rangle \quad (12.31b)$$

σ^α_{qq}, σ^α_{pq}, and σ^α_{qp} represent the "second moments" of our dynamic variables

$$\sigma^\alpha_{qq}(t) = \langle q^2\rangle - \langle q\rangle^2 \quad (12.31c)$$

$$\sigma^\alpha_{pp}(t) = \langle p^2\rangle - \langle p\rangle^2 \quad (12.31d)$$

$$\sigma^\alpha_{qp}(t) = \tfrac{1}{2}[\langle pq\rangle + \langle qp\rangle] - \langle p\rangle\langle q\rangle \quad (12.31e)$$

W is a 2×2 matrix and [det $W(t)$] denotes its determinant. The parameters [Eqs. (12.31)] constitute a complete set of first- and second-order "moments" of the LGF $\rho(p, q; t)$ with respect to the coordinates and momenta. Strictly speaking, these are not moments, since the LGF σ_α is not a probability distribution. Equation (12.29a) implies that σ_α has a Gaussian form at all times. If initially it is not a Gaussian, we may still represent it as a superposition of Gaussians.

The time evolution of the LGF [Eq. (12.29)] is given by the Green function

$$\sigma_\alpha(p, q, t + t_0) = \iint dp\, dq\, \mathscr{G}_{nm}(p, q, t; p_0, q_0)\sigma_\alpha(p_0, q_0, t_0), \tag{12.32}$$

where $n, m = \mathrm{e}$ or g depending on the path α and the specific time interval t_1, t_2, or t_3. We shall further adopt the average reference potential

$$W_\alpha(q) \equiv \tfrac{1}{2}[W_m(q) + W_n(q)], \tag{12.33}$$

and define

$$U_\alpha(q) \equiv W_n(q) - W_m(q) - \langle [W_n(q) - W_m(q)]\rho_g \rangle. \tag{12.34}$$

By substituting the ansatz Eq. (12.29) in the Liouville equation, and taking the first two moments of both sides, we obtain the following equations of motion for the parameters characterizing the LGF [1]:

$$\dot{q}_\alpha = p_\alpha/m - \frac{i}{\hbar}\langle \partial U_\alpha/\partial q \rangle \sigma^\alpha_{qq} \tag{12.35a}$$

$$\dot{p}_\alpha = -\langle \partial W_\alpha/\partial q \rangle - \frac{i}{\hbar}\langle \partial U_\alpha/\partial q \rangle \sigma^\alpha_{pq} \tag{12.35b}$$

$$\dot{\sigma}^\alpha_{qq} = 2\sigma^\alpha_{qp}/m - \frac{i}{\hbar}\langle \partial^2 U_\alpha/\partial q^2 \rangle (\sigma^\alpha_{qq})^2 \tag{12.35c}$$

$$\dot{\sigma}^\alpha_{pp} = -2\langle \partial^2 W_\alpha/\partial q^2 \rangle \sigma^\alpha_{qp} - \frac{i}{\hbar}[(\sigma^\alpha_{qp})^2 - 1/4]\langle \partial^2 U_\alpha/\partial q^2 \rangle \tag{12.35d}$$

$$\dot{\sigma}^\alpha_{qp} = -\langle \partial^2 W_\alpha/\partial q^2 \rangle \sigma^\alpha_{qq} + \frac{1}{m}\sigma^\alpha_{pp} - \frac{i}{\hbar}\langle \partial^2 U_\alpha/\partial q^2 \rangle \sigma^\alpha_{qq}\sigma^\alpha_{qp}. \tag{12.35e}$$

These equations are closed, since the expectation values in the right-hand side are also functions of the LGF parameters. The appearance of $\hbar$ in these semiclassical equations reflects the quantum nature of our system. When the two potential surfaces W_g and W_e are identical $U_\alpha = 0$, and assuming that the second moments are vanishingly small $\sigma^\alpha_{\nu\mu} = 0$, then Eqs. (12.35) reduce to the purely classical equations of motion [Eqs. (12.18)]. Note that the classical limit of Eqs. (12.35) represents a different choice of the reference potential than was used in Chapter 7. During the t_1 and t_3 time intervals the evolution here is with respect to the average potential $(W_g + W_e)/2$. During the t_2 interval, the system is either in the excited state (evolves with W_e) or in the ground state (evolves with W_g).

In the classical approach developed in Chapters 7 and 8 we solve a classical trajectory many times in order to sample the distribution σ_α. In each trajectory we keep only the first moments q_α and p_α. In the semiclassical procedure we assume a Gaussian form for the phase space wavepacket, and solve the coupled equations for its first two moments. The calculation is done only once, but includes more variables, namely the second moments [Eqs. (12.31)].

REDUCED DYNAMICS OF LIOUVILLE SPACE PATHS: THE MULTIMODE BROWNIAN OSCILLATOR MODEL

In Chapter 8 we calculated the nonlinear response function for the multimode Brownian oscillator model. The model focuses on a few relevant nuclear coordinates such as optically active intramolecular vibrations, phonons, as well as collective solvent modes. The effects of dissipation were introduced by coupling these primary relevant coordinates to a thermal bath represented by a continuous distribution of oscillators. The optical response functions were calculated using the second-order cumulant expansion (which is exact for this model), and adopting the ground state as a reference. In this section we return to this model with the goal of developing closed form expressions for the Liouville space-generating functions (LGF) in the Condon approximation in the reduced phase space of the relevant coordinates. The reduced description is a convenient computation tool, in which only the necessary information is kept. The corresponding LGF wavepackets provide a deeper physical insight on the connection between nuclear dynamics and the optical response, and will enable us to develop useful approximations for time-domain spectroscopies. Since we are not interested in the dynamics of the bath, we trace over its coordinates. We thus introduce the reduced density operator

$$\rho(t) \equiv \mathrm{Tr}_{\mathrm{B}}\{\rho_{\mathrm{tot}}(t)\}. \tag{12.36}$$

Here, $\mathrm{Tr}_{\mathrm{B}}\{\ \}$ represents the trace over the bath degrees of freedom and $\rho_{\mathrm{tot}}(t)$ is the total (system + bath) density operator. We first express the LGF in the coordinate representation $\langle \mathbf{q}|\rho_\alpha(t)|\mathbf{q}'\rangle$. It will be convenient to introduce the center of mass and the difference coordinates

$$\mathbf{q} \equiv (\mathbf{q} + \mathbf{q}')/2; \qquad x \equiv \mathbf{q} - \mathbf{q}', \tag{12.37}$$

and define

$$\sigma_\alpha(\mathbf{x}, \mathbf{q}, t) = \prod_j \sigma_\alpha(x_j, q_j, t), \tag{12.38a}$$

with

$$\sigma_\alpha(x_j, q_j, t) \equiv \langle q_j - x_j/2|\rho_\alpha(t)|q_j + x_j/2\rangle. \tag{12.38b}$$

For this model, the wavepacket σ_α may be calculated exactly, e.g. by using the real time path integral approach where the bath degrees of freedom are eliminated using the Feynman–Vernon influence functional [3, 11, 12]. We then obtain

$$\sigma_\alpha(x_j, q_j, t) = (2\pi\langle q_j^2\rangle_g)^{-1/2} \exp\left\{-\frac{1}{2\langle q_j^2\rangle_g}[q_j - \bar{q}_\alpha^j(t_3, t_2, t_1)]^2 - \frac{1}{2\hbar^2}\langle p_j^2\rangle_g x_j^2 + \frac{i}{\hbar}\bar{p}_\alpha^j(t_3, t_2, t_1)x_j\right\}. \tag{12.39}$$

In the Wigner representation Eq. (12.39) becomes

$$\sigma_\alpha(p_j, q_j, t) \equiv \frac{1}{2\pi}[\langle p_j^2\rangle_g\langle q_j^2\rangle_g]^{-1/2} \exp\left\{-\frac{1}{2\langle q_j^2\rangle_g}[q_j - \bar{q}_\alpha^j(t_3, t_2, t_1)]^2 - \frac{1}{2\langle p_j^2\rangle_g}[p_j - \bar{p}_\alpha^j(t_3, t_2, t_1)]^2\right\}. \tag{12.40}$$

The position and the momentum "expectation values" $\bar{q}_\alpha^j$ and $\bar{p}_\alpha^j$ are given by

$$\left.\begin{aligned}
\bar{q}_1^j(t_3, t_2, t_1) &= -i\xi_j^{-1}[\dot{g}_j^*(t_1 + t_2 + t_3) - \dot{g}_j(t_2 + t_3) + \dot{g}_j(t_3)],\\
\bar{q}_2^j(t_3, t_2, t_1) &= -i\xi_j^{-1}[-\dot{g}_j(t_1 + t_2 + t_3) + \dot{g}_j^*(t_2 + t_3) + \dot{g}_j(t_3)],\\
\bar{q}_3^j(t_3, t_2, t_1) &= -i\xi_j^{-1}[-\dot{g}_j(t_1 + t_2 + t_3) + \dot{g}_j(t_2 + t_3) + \dot{g}_j^*(t_3)],\\
\bar{q}_4^j(t_3, t_2, t_1) &= -i\xi_j^{-1}[\dot{g}_j^*(t_1 + t_2 + t_3) - \dot{g}_j^*(t_2 + t_3) + \dot{g}_j^*(t_3)],
\end{aligned}\right\} \tag{12.41}$$

and

$$\left.\begin{aligned}
\bar{p}_1^j(t_3, t_2, t_1) &= -im_j\xi_j^{-1}[\ddot{g}_j^*(t_1 + t_2 + t_3) - \ddot{g}_j(t_2 + t_3) + \ddot{g}_j(t_3)],\\
\bar{p}_2^j(t_3, t_2, t_1) &= -im_j\xi_j^{-1}[-\ddot{g}_j^*(t_1 + t_2 + t_3) + \ddot{g}_j^*(t_2 + t_3) + \ddot{g}_j(t_3)],\\
\bar{p}_3^j(t_3, t_2, t_1) &= -im_j\xi_j^{-1}[-\ddot{g}_j(t_1 + t_2 + t_3) + \ddot{g}_j(t_2 + t_3) + \ddot{g}_j^*(t_3)],\\
\bar{p}_4^j(t_3, t_2, t_1) &= -im_j\xi_j^{-1}[\ddot{g}_j^*(t_1 + t_2 + t_3) - \ddot{g}_j^*(t_2 + t_3) + \ddot{g}_j^*(t_3)],
\end{aligned}\right\} \tag{12.42}$$

The equilibrium second moments of the coordinates and momenta are given by

$$\langle q_j^2 \rangle_g = \int_{-\infty}^{\infty} d\omega\, \tilde{C}_j''(\omega) \coth(\beta\hbar\omega/2), \tag{12.43a}$$

$$\langle p_j^2 \rangle_g = m_j^2 \int_{-\infty}^{\infty} d\omega\, \omega^2 \tilde{C}_j''(\omega) \coth(\beta\hbar\omega/2), \tag{12.43b}$$

where $\tilde{C}_j''(\omega)$ is given by Eq. (8.64b), and ξ_j was defined in Eq. (8.62c).

We reiterate that although not positive definite, the Wigner function is extremely useful for comparison with the classical density operator in phase space and for developing semiclassical approximations for anharmonic systems. We further note that the actual expectation values of the coordinate and momentum with respect to the complete density operator $\langle q_j \rangle = \sum_\alpha \bar{q}_\alpha^j$ and $\langle p_j \rangle = \sum_\alpha \bar{p}_\alpha^j$ are real, as they should be! However, $\bar{q}_\alpha^j$ and $\bar{p}_\alpha^j$, which represent the contributions of individual Liouville space paths to these expectation values, are complex.

The present reduced wavepacket analysis provides a microscopic insight on the role of dissipation. The use of an exactly solvable model is most valuable since, unlike the Langevin equations [Eqs. (8.65)], the results hold at any temperature. These results can be extended to higher order optical processes ($P^{(5)}$ etc.). Another possible extension, which is particularly important in off resonant grating techniques such as coherent Raman scattering, is to include the dependence of the dipole moment on coordinates, i.e., non-Condon corrections [2].

REDUCED EQUATIONS OF MOTION FOR LIOUVILLE SPACE WAVEPACKETS

In the previous section we have demonstrated the utility of the reduced Liouville space wavepackets. We shall now explore the possibility of deriving reduced equations of motion for the LGF [13–15]. This will provide an additional perspective on the multimode Brownian oscillator model, and connect the optical response with standard formultions such as the Fokker–Planck and the Smoluchowski equations.

We start our analysis by considering the phase space representation of the LGF. Taking R_1 as an example, we can recast Eq. (12.5) in the form

$$\rho_1(\Gamma, t_1 + t_2 + t_3) = \iiint d\Gamma_1 d\Gamma_2\, d\Gamma_3\, \mathscr{G}_{eg}(\Gamma, t_3; \Gamma_3)\mathscr{V}_{eg,ee}\mathscr{G}_{ee}(\Gamma_3, t_2; \Gamma_2)\mathscr{V}_{ee,eg} \times \mathscr{G}_{eg}(\Gamma_2, t_1; \Gamma_1)\mathscr{V}_{ee,gg}\rho_g(\Gamma_1) \tag{12.44}$$

$$R_1(t_3, t_2, t_1) = \int d\Gamma\, V_{ge}(\Gamma)\rho_1(\Gamma, t_1 + t_2 + t_3). \tag{12.45}$$

Here Γ represents the phase space $\{p_j, q_j\}$ of all active nuclear coordinates. $\mathscr{G}_{nm}(\Gamma, t; \Gamma')$ represents the nuclear propagation from Γ' to Γ when the electronic system density operator is in the nm state ($n, m =$ e, g). As long as we work in the complete phase space, and include all nuclear degrees of freedom Γ, Eq. (12.45) is exact. The question we wish to address here is under what conditions can we still use this equation in a reduced description, when Γ represents only a partial set of coordinates. The problem is not simple since the representation (12.44) implies a factorization of the propagation into segments with $\mathscr{G}_{nm}$ serving as a kind of conditional probability. We have shown earlier in this chapter how propagation of the Green function $\mathscr{G}_{nm}(t)$ may be interpreted in terms of a motion with respect to a reference potential W_α accompanied by an accumulated phase. The motion with respect to the reference potential has a well-defined classical analogue and will be described using the Langevin equation where the effects of the remaining degrees of freedom are introduced through a Langevin random force and friction. The q_j coordinates satisfy the Langevin equation [Eq. (8.65a)]:

$$m_j \ddot{q}_j + m_j\omega_j^2 q_j + m_j\gamma_j\dot{q}_j = f_j(t). \tag{12.46}$$

This equation is for the ground-state evolution. For the excited state we simply replace q_j by $q_j + d_j$. Using standard techniques it is then possible to derive a Fokker–Planck equation for the corresponding generating function in the Wigner representation [16]

$$\left[\frac{\partial}{\partial t} - m_j\omega_j^2(q_j + \eta d_j) + \frac{p_j}{m^j}\frac{\partial}{\partial q_j}\right]\rho_\alpha^{nm}(p_j q_j, t)$$
$$= \gamma_j\left[\frac{\partial}{\partial p_j}p_j + K_B T m_j \frac{\partial^2}{\partial p_j^2}\right]\rho_\alpha^{nm}(p_j q_j, t) - \frac{i}{\hbar}[(1-\eta)U_{nm} + \eta\rho_\alpha^{nm}(p_j q_j, t)U_{nm}], \tag{12.47}$$

where the parameter η introduced in Eq. (12.13) is related to the choice of a reference potential. The left-hand side is simply the Liouville equation for the harmonic oscillator with the reference potential W_α [Eq. (12.17)]. The first term in the square brackets in the right-hand side represents Brownian motion in momentum space [13, 14]. The second term in the right-hand side carries the phase information for $n \neq m$ [1, 13]. The evaluation of the nonlinear response function requires the propagation of our equations for three consecutive time intervals, t_1, t_2, and t_3. The choice of the reference potential (i.e., η) depends on the specific pathway R_α and the time interval, as discussed previously.

Low frequency nuclear modes and solvent modes are usually strongly overdamped (i.e., the friction is much larger than the relevant frequency $\gamma_j \gg \omega_j$). Consequently, they do not show up in the spectra as progressions of well-solved lines, but rather contribute to line broadening. Since, in this case, the momentum rapidly attains its equilibrium value, it need not be considered an independent dynamic variable. The oscillator distribution function then satisfies a Smoluchowski equation in configuration space $\Gamma = q$ (rather than a Fokker–Planck equation in phase space $\Gamma = p, q$).

$$\frac{\partial \rho_\alpha^{nm}(q_j, t)}{\partial t} - \frac{\omega_j^2}{\gamma_j}\left[\frac{\partial}{\partial q_j}(q_j + \eta d_j) + \frac{k_N T}{m_j \omega_j^2}\frac{\partial^2}{\partial q_j^2}\right]\rho_\alpha^{nm}(q_j, t) = -\frac{i}{\hbar} U_{nm}\rho_\alpha^{nm}(q_j, t). \quad (12.48)$$

In this limit, the correlation function $M_j(t)$ is equal to $\exp(-\Lambda_j t)$ with $\Lambda_j \equiv \omega_j^2/\gamma_j$.

In summary, using the present reduced equation of motion, the response functions may be calculated by solving Eqs. (12.47) or (12.48) for the wavepackets $\rho_\alpha(\Gamma, t_1 + t_2 + t_3)$ and substituting in Eq. (12.45) to obtain the LGF.

We shall now compare these results with those obtained in Chapter 8 using an exact solution of a microscopic model for the bath [Eq. (8.61)], without factorizing the propagation into segments. The response functions R_α obtained from the above equations of motion [Eqs. (12.47) or (12.48)] are exact and coincide with Eq. (8.15) together with Eqs. (8.31) and (8.64b). This is, however, not the case for the resulting LGF. Although the system was treated quantum mechanically, the inclusion of a classical Langevin force necessarily limits the resulting wavepackets to high temperatures ($k_B T \gg \hbar\omega_j$). By comparing the complete wavepacket, rather than just the response function with the exact one, we find that the classical Langevin equation holds only when the following three conditions are simultaneously met: (1) high temperature, (2) Ohmic dissipation $\gamma(\omega)$ independent of ω, and (3) strongly overdamped motion $\gamma_j \gg \omega_j$. When any of these conditions is not satisfied, then the solution of the Langevin equation is different from the exact result. Because of the correlation between the system and the bath, it is not generally possible to factorize the propagator in the form of Eq. (12.44). Such factorization implies the existence of a reduced description whereby the bath degrees of freedom are always in equilibrium, which allows us to follow only the relevant motions. When such a level of description can be adopted, we can describe each segment of the Liouville space path in terms of a conditional probability and an accumulating phase factor calculated using the classical Langevin equation.

The Langevin approach can further be extended to include general anharmonic potentials,

$$m_j \ddot{q}_j + m_j \gamma_j \dot{q}_j + \partial W_\alpha / \partial q_j = f_j(t). \quad (12.49)$$

Corresponding Fokker–Planck equations that generalize Eq. (12.46) may then be derived and solved numerically [13–15].

THE DOORWAY–WINDOW REPRESENTATION OF THE NONLINEAR RESPONSE FUNCTION

The wavepacket approach can be used to recast the response function in a new representation that provides some additional physical insight. We start by recasting Eq. (7.11) in the form

$$\left.\begin{aligned} R_1(t_3, t_2, t_1) &= \langle \rho_w(t_3)\mathcal{G}_{ee}(t_2)\rho_D(t_1)\rangle, \\ R_2(t_3, t_2, t_1) &= \langle \rho_w(t_3)\mathcal{G}_{ee}(t_2)\rho_D^\dagger(t_1)\rangle, \\ R_3(t_3, t_2, t_1) &= \langle \rho'_w(t_3)\mathcal{G}_{gg}(t_2)\rho_D'^\dagger(t_1)\rangle, \\ R_4(t_3, t_2, t_1) &= \langle \rho'_w(t_3)\mathcal{G}_{gg}(t_2)\rho'_D(t_1)\rangle, \end{aligned}\right\} \tag{12.50}$$

where

$$\left.\begin{aligned} \rho_D(t_1) &= \exp\left(-\frac{i}{\hbar}H_e t_1\right)V_{eg}\rho_g \exp\left(\frac{i}{\hbar}H_g t_1\right)V_{ge}, \\ \rho_W(t_3) &= \exp\left(\frac{i}{\hbar}H_e t_3\right)V_{eg} \exp\left(-\frac{i}{\hbar}H_g t_3\right)V_{ge}, \\ \rho'_D(t_1) &= V_{ge}\exp\left(-\frac{i}{\hbar}H_e t_1\right)V_{eg}\rho_g \exp\left(\frac{i}{\hbar}H_g t_1\right), \\ \rho'_W(t_3) &= V_{ge}\exp\left(\frac{i}{\hbar}H_e t_3\right)V_{eg} \exp\left(-\frac{i}{\hbar}H_g t_3\right). \end{aligned}\right\} \tag{12.51}$$

Equations (12.50) and (5.27) can be combined to yield the following compact expression for the nonlinear response function

$$\begin{aligned} S^{(3)}(t_3, t_2, t_1) &= \left(\frac{i}{\hbar}\right)^3 \theta(t_1)\theta(t_2)\theta(t_3)\langle[\rho_W(t_3) - \rho_W^\dagger(t_3)]\mathcal{G}_{ee}(t_2)[\rho_D(t_1) + \rho_D^\dagger(t_1)]\rangle \\ &+ \left(\frac{i}{\hbar}\right)^3 \theta(t_1)\theta(t_2)\theta(t_3)\langle[\rho'_W(t_3) - \rho_W'^\dagger(t_3)]\mathcal{G}_{gg}(t_2)[\rho'_D(t_1) + \rho_D'^\dagger(t_1)]\rangle \end{aligned} \tag{12.52}$$

Unlike Eq. (12.6) where we propagated the LGF for three time periods to obtain the wavepacket at time $t_1 + t_2 + t_3$, here we propagate it first for time t_1, resulting in the doorway wavepacket $\rho_D + \rho_D^\dagger$. We then construct a window wavepacket $\rho_W - \rho_W^\dagger$ obtained by a propagation of a different wavepacket for the t_3 period. The response function is finally calculated by propagation of the doorway for an additional t_2 period [as given by the $\mathcal{G}_{ee}(t_2)$ or $\mathcal{G}_{gg}(t_2)$ propagators in Eq. (12.52)], and computing its overlap with the window. This representation has significant advantages for some optical techniques. A specific application of this representation for pump-probe experiments conducted using pulses with finite spectral and temporal width will be given in Chapter 13. The two terms in Eq. (12.52) reflect the contribution of excited state and ground state propagation, respectively, to the nonlinear response function. Equation (12.52) clearly illustrates that $S^{(3)}$ is real, since all factors are Hermitian. This factorized form can also be maintained for ideal frequency-domain measurements, and the nonlinear susceptibility is given by

$$\begin{aligned} \chi^{(3)}(-\omega_s; \omega_1, \omega_2, \omega_3) &= \frac{1}{3!}\left(\frac{1}{\hbar}\right)^3 \rho_0 \sum_p \\ &[\langle \rho_W(\omega_s) - \rho_W^\dagger(\omega_s)|\, \mathcal{G}_{ee}(\omega_1 + \omega_2)|\rho_D(\omega_1) + \rho_D^\dagger(\omega_1)\rangle \\ &+ \langle \rho'_W(\omega_s) - \rho_W'^\dagger(\omega_s)|\mathcal{G}_{gg}(\omega_1 + \omega_2)|\rho'_D(\omega_1) + \rho_D'^\dagger(\omega_1)\rangle]. \end{aligned} \tag{12.53}$$

Here $\rho_j(\omega)$ are the one-sided Fourier transforms of $\rho_j(t)$ where ρ_j stands for ρ_D, ρ_W, ρ'_D, ρ'_W [see Eq. (5.30b)]

$$\rho_j(\omega) \equiv -i \int_0^\infty d\tau\, \rho_j(\tau) \exp(i\omega\tau),$$

$$\rho_j^\dagger(\omega) \equiv -i \int_0^\infty d\tau\, \rho_j^\dagger(\tau) \exp(i\omega\tau).$$

For arbitrary pulse durations, it is possible to adopt the doorway window representation only when certain requirements are met. This will be shown in the coming chapters for pump-probe (Chapter 13) and for Raman spectroscopies (Chapter 14).

When invoking the Condon approximation, $\rho_W = \rho'_W$, $\rho_D = \rho'_D$, and we have

$$\left.\begin{aligned} \rho_D(t_1) &= \exp\left(-\frac{i}{\hbar} H_e t_1\right)\rho_g \exp\left(\frac{i}{\hbar} H_g t_1\right), \\ \rho_W(t_3) &= \exp\left(\frac{i}{\hbar} H_g t_3\right)\exp\left(-\frac{i}{\hbar} H_e t_3\right). \end{aligned}\right\} \tag{12.54}$$

For the multimode Brownian oscillator system we can solve Eqs. (12.35), to obtain

$$\mathscr{G}_{gg}(t_2)\rho_D(t_1) = \prod_j F_{gj}(t_1, t_2; p_j q_j) \tag{12.55a}$$

$$\mathscr{G}_{ee}(t_2)\rho_D(t_1) = \prod_j F_{ej}(t_1, t_2; p_j q_j) \tag{12.55b}$$

$$\rho_W(t_3) = \prod_j G_{gj}(t_3; p_j q_j) \tag{12.55c}$$

and

$$\langle \cdots \rangle = \iint dp_j\, dq_j. \tag{12.56}$$

The function $F_{gj}(t_1; p_j q_j)$ is simply the doorway wavepacket propagated for the first two time intervals. The system starts with the equilibrium ground-state density operator distribution. It then propagates for the time period t_1 while the system is in the electronic coherence, followed by propagation for the time period t_2 in the electronic ground ($m = g$) state. The function F_{ej} can be obtained in a similar way, except that the second propagation for the t_2 period takes place in the electronic excited ($m = e$) state. The window function is calculated by starting with a uniform distribution in phase space (this corresponds to the density operator at infinite temperature), which then propagates for the time period t_3 in the electronic coherence. Note that within the Condon approximation the ground state and the excited state window functions are identical, i.e., $G_{gj} = G_{ej}$.

For the displaced oscillator model (i.e., the Brownian oscillator model with no friction $\gamma_j = 0$), the ground state doorway and window functions are given by

$$F_{gj}(t_1, t_2; p_j q_j) = \frac{1}{2\pi} [\langle p_j^2 \rangle_g \langle q_j^2 \rangle_g]^{-1/2} \exp[-i\omega_{eg} t_1 - g_j(t_1)]$$

$$\times \exp\left\{ -\frac{1}{2\langle q_j^2 \rangle_g} [q_j - \bar{q}_j(t_1, t_2)]^2 - \frac{1}{2\langle p_j^2 \rangle_g} [p_j - \bar{p}_j(t_1, t_2)]^2 \right\}, \tag{12.57a}$$

where

$$\langle q_j^2 \rangle_g = \frac{\hbar}{\omega_j m_j} \coth(\beta\hbar\omega_j/2) \tag{12.57b}$$

$$\langle p_j^2 \rangle_g = \hbar\omega_j m_j \coth(\beta\hbar\omega_j/2) \tag{12.57c}$$

$$\bar{q}_j(t_1, t_2) = -\frac{i}{\xi_j} [\dot{g}_j(t_1 + t_2) - \dot{g}_j(t_2)] \tag{12.57d}$$

$$\bar{p}_j(t_1, t_2) = -\frac{im_j}{\xi_j} [\ddot{g}_j(t_1 + t_2) - \ddot{g}_j(t_2)] \tag{12.57e}$$

$$g_j(t) = S_j \coth(\beta\hbar\omega_j/2)(1 - \cos\omega_j t) + i(\sin\omega_j t - \omega_j t). \tag{12.57f}$$

The window wavepacket is given by

$$G_{gj}(t_3; p_j q_j) = \exp[i\omega_{eg} t_3 - g_j^*(t_3)] \exp[\eta_0(t_3) + i\eta_1(t_3) q_j + i\eta_2(t_3) p_j], \tag{12.58a}$$

where

$$\eta_0(t_3) = \frac{\xi_j^2 \coth(\beta\hbar\omega_j/2)}{2\omega_j^2} \left\{ \omega_j^2 \left[\int_0^{t_3} d\tau\, M_j(\tau) \right]^2 - [1 - M_j(t_3)]^2 \right\}, \tag{12.58b}$$

$$\eta_1(t_3) = -\xi_j \int_0^{t_3} d\tau\, M_j(\tau), \tag{12.58c}$$

$$\eta_2(t_3) = \frac{\xi_j}{m_j \omega_j^2} [1 - M_j(t_3)], \tag{12.58d}$$

$$M_j(t_3) = \cos\omega_j t. \tag{12.58e}$$

The excited state wavepackets are given by similar expressions. $F_{ej}(t_1, t_2)$ is given by Eq. (12.57a) where the only difference is replacing $\dot{g}_j(t_2)$ and $\ddot{g}_j(t_2)$ in Eqs. (12.57d) and (12.57e) by $\dot{g}_j^*(t_2)$ and $\ddot{g}_j^*(t_2)$, respectively.

Equations (12.57) and (12.58) do not hold for the general Brownian oscillator model with an arbitrary friction. They require the factorization of the response function [Eq. (12.45)]. As pointed out earlier, this can be justified only for the overdamped case, where the Fokker–Planck equation reduces to the Smoluchowski equation. We shall change variables and define $U_j \equiv \xi_j q_j/\hbar$, which is the contribution of the q_j coordinate to the electronic energy gap. The phase space doorway and the window functions reduce to functions, $F_{mj}(t_1, \tau; U_j)$ and $G_{mj}(t_3; U_j)$, in the coordinate space with $M_j' = M_j'' \equiv M_j$. Furthermore, the phase space integration over $p_j q_j$ in Eq. (12.56) is replaced by an integration over the coordinate U_j. We thus obtain

$$F_{gj}(t_1, t_2; U_j) = (2\pi\Delta_j^2)^{-1/2} \exp\{-\tfrac{1}{2}[U_j - \bar{U}_j(t_1, t_2)]^2/\Delta_j^2 - i\omega_{eg}t_1 - g_j(t_1)\}, \tag{12.59a}$$

$$G_{gj}(t_3) = \exp[i\omega_{eg}t_3 - g_j^*(t_3)] \exp\{\eta_0(t_3) + i\eta_1(t_3)U_j\}, \tag{12.59b}$$

with

$$\bar{U}_j(t_1, t_2) = \xi_j \bar{q}_j(t_1, t_2) = -\frac{i\Delta_j^2}{\Lambda_j} \exp(-\Lambda_j t_2)[1 - \exp(-\Lambda_j t_1)] + (1 - 2\eta)\lambda_j$$
$$+ \lambda_j \exp(-\Lambda_j t_2)\{1 - (1 - 2\eta)\exp(-\Lambda_j t_1)\}, \tag{12.60a}$$

$$\Delta_j^2 = \xi_j^2 \langle q_j^2 \rangle_g = 2k_B T\lambda_j, \tag{12.60b}$$

$$\left.\begin{aligned} M_j(t) &= \exp(-\Lambda_j t), \\ \eta_0(t_3) &= -\frac{2\xi^2}{\beta\hbar\Lambda_j^3}[1 - \exp(-\Lambda_j t_3)]^2, \\ \eta_1(t_3) &= -\frac{\hbar}{\Lambda_j}[1 - \exp(-\Lambda_j t_3)], \end{aligned}\right\} \tag{12.60c}$$

and $g_j(t)$ is given by Eq. (8.49).

The doorway–window form of the response function is particularly useful for the description of certain time-domain techniques such as pump-probe spectroscopy. This will be demonstrated in the next chapter, where we further present calculations showing the wavepackets and the corresponding spectra for the various limiting cases of the Brownian oscillator model. However, this form also gives a compact and useful form in the frequency domain. In Appendix 12B we present an expression for $\chi^{(3)}$ obtained using a model system consisting of a vibronic manifold (Chapter 7) plus a single overdamped mode. This expression is identical to Eq. (8.105), but recast in a form that reflects the underlying wavepacket dynamics.

APPENDIX 12A

Eigenstate Expansion of the Liouville Space-Generating Functions

Expanding Eqs. (12.7)–(12.10) in nuclear eigenstates (Figure 7.1), we get

$$\left.\begin{aligned}
\rho_1(t_1) &= \sum_{a,b} P(a)\mu_{ba}\exp(-i\omega_{ba}t_1)|b\rangle\langle a| \\
\rho_1(t_1+t_2) &= \sum_{a,b,d} P(a)\mu_{ba}\mu_{ad}\exp(-i\omega_{ba}t_1 - i\omega_{bd}t_2)|b\rangle\langle d| \\
\rho_1(t_1+t_2+t_3) &= \sum_{a,b,c,d} P(a)\mu_{ba}\mu_{ad}\mu_{dc}\exp(-i\omega_{ab}t_1 - i\omega_{bd}t_2 - i\omega_{bc}t_3)|b\rangle\langle c|.
\end{aligned}\right\} \tag{12.A.1}$$

$$\left.\begin{aligned}
\rho_2(t_1) &= \sum_{a,b} P(a)\mu_{ab}\exp(-i\omega_{ab}t_1)|a\rangle\langle b| \\
\rho_2(t_1+t_2) &= \sum_{a,b,d} P(a)\mu_{ba}\mu_{ab}\exp(-i\omega_{ab}t_1 - i\omega_{db}t_2)|d\rangle\langle b| \\
\rho_2(t_1+t_2+t_3) &= \sum_{a,b,c,d} P(a)\mu_{da}\mu_{ab}\mu_{bc}\exp(-i\omega_{ab}t_1 - i\omega_{db}t_2 - i\omega_{dc}t_3)|d\rangle\langle c|.
\end{aligned}\right\} \tag{12.A.2}$$

$$\left.\begin{aligned}
\rho_3(t_1) &= \sum_{a,b} P(a)\mu_{ab}\exp(-i\omega_{ab}t_1)|a\rangle\langle b| \\
\rho_3(t_1+t_2) &= \sum_{a,b,c} P(a)\mu_{ab}\mu_{bc}\exp(-i\omega_{ab}t_1 - i\omega_{ac}t_2)|a\rangle\langle c| \\
\rho_3(t_1+t_2+t_3) &= \sum_{a,b,c,d} P(a)\mu_{da}\mu_{ab}\mu_{bc}\exp(-i\omega_{ab}t_1 - i\omega_{ac}t_2 - i\omega_{dc}t_3)|d\rangle\langle c|.
\end{aligned}\right\} \tag{12.A.3}$$

$$\left.\begin{aligned}
\rho_4(t_1) &= \sum_{a,b} P(a)\mu_{ba}\exp(-i\omega_{ba}t_1)|b\rangle\langle a| \\
\rho_4(t_1+t_2) &= \sum_{a,b,c} P(a)\mu_{cb}\mu_{ba}\exp(-i\omega_{ba}t_1 - i\omega_{ca}t_2)|c\rangle\langle a| \\
\rho_4(t_1+t_2+t_3) &= \sum_{a,b,c,d} P(a)\mu_{dc}\mu_{cb}\mu_{ba}\exp(-i\omega_{ba}t_1 - i\omega_{ca}t_2 - i\omega_{da}t_3)|d\rangle\langle a|.
\end{aligned}\right\} \tag{12.A.4}$$

APPENDIX 12B

The Doorway–Window Picture in the Frequency Domain: Vibronic State Representation

The doorway–window form of the response function can be used to recast the optical susceptibilities in a form that has fewer summation indexes. Starting with Eqs. (8.100) and (8.105), we can combine terms corresponding to the doorway and the window wavepackets and obtain [17]

$$J(\omega_j) = \sum_a P(a) K_{aa}^{(0)}(\varepsilon_a + \omega_j), \tag{12.B.1}$$

$$R_1(\omega_j + \omega_k + \omega_q, \omega_j + \omega_k, \omega_j) = \sum_{n=0}^{\infty} \sum_{b,d} \frac{(-z)^n}{n!} \frac{1}{\omega_j + \omega_k - \omega_{db} + i(\gamma + n\Lambda)} \times \bar{K}_{db}^{(n)*}(\varepsilon_d - \omega_j - \omega_k - \omega_q) \tilde{K}_{bd}^{(n)*}(\varepsilon_d - \omega_j), \tag{12.B.2a}$$

$$R_2(\omega_j + \omega_k + \omega_q, \omega_j + \omega_k, \omega_j) = -\sum_{n=0}^{\infty} \sum_{b,d} \frac{z^{*n}}{n!} \frac{1}{\omega_j + \omega_k - \omega_{db} + i(\gamma + n\Lambda)} \times \bar{K}_{db}^{(n)*}(\varepsilon_d - \omega_j - \omega_k - \omega_q) \tilde{K}_{db}^{(n)}(\varepsilon_b + \omega_j), \tag{12.B.2b}$$

$$R_3(\omega_j + \omega_k + \omega_q, \omega_j + \omega_k, \omega_j) = -\sum_{n=0}^{\infty} \sum_{a,c} P(a) \frac{z^{*n}}{n!} \frac{1}{\omega_j + \omega_k - \omega_{ac} + i(\gamma' + n\Lambda)} \times K_{ca}^{(n)}(\varepsilon_c + \omega_j + \omega_k + \omega_q) K_{ca}^{(n)*}(\varepsilon_a - \omega_j), \tag{12.B.2c}$$

$$R_4(\omega_j + \omega_k + \omega_q, \omega_j + \omega_k, \omega_j) = \sum_{n=0}^{\infty} \sum_{a,c} P(a) \frac{(-z)^n}{n!} \frac{1}{\omega_j + \omega_k - \omega_{ca} + i(\gamma' + n\Lambda)} \times \sum_{n=0}^{\infty} \sum_{a,c} P(a) \frac{(-z)^n}{n!} \frac{1}{\omega_j + \omega_k - \omega_{ca} + i(\gamma' + n\Lambda)} \times K_{ac}^{(n)}(\varepsilon_a + \omega_j + \omega_k + \omega_q) K_{ca}^{(n)}(\varepsilon_a + \omega_j). \tag{12.B.2d}$$

Here ε_v is the energy of state v [Eq. (6.1)], and

$$T(t) \equiv V_{ge} \exp\left[-\frac{i}{\hbar} H_e t - (\gamma + \gamma')t/2 \right] V_{eg}, \tag{12.B.3a}$$

$$\bar{T}(t) \equiv V_{eg} \exp\left[-\frac{i}{\hbar} H_g t - (\gamma + \gamma')t/2 \right] V_{ge}, \tag{12.B.3b}$$

$$\tilde{T}(t) \equiv V_{eg}\rho_g \exp\left[-\frac{i}{\hbar} H_g t - (\gamma + \gamma')t/2 \right] V_{ge}. \tag{12.B.3c}$$

γ and γ' denote the inverse lifetimes of the vibronic levels of the excited and ground electronic states, respectively. ρ_g is the density matrix of the system in thermal equilibrium in the ground electronic state.

$$K^{(n)}(\omega) \equiv -i \int_0^\infty dt \exp(i\omega t) T(t) J_n(t), \tag{12.B.4a}$$

$$\bar{K}^{(n)}(\omega) \equiv -i \int_0^\infty dt \exp(i\omega t) \bar{T}(t) J_n(t), \tag{12.B.4b}$$

$$\tilde{K}^{(n)}(\omega) \equiv -i \int_0^\infty dt \exp(i\omega t) \tilde{T}(t) \tilde{J}_n(t). \tag{12.B.4c}$$

These forms are formally identical to Eqs. (8.107). For the linearly displaced harmonic model it is possible to evaluate Eqs. (12.B.4) in the coordinate representation without any summation over eigenstates.

NOTES

1. S. Mukamel, S. Abe, Y. J. Yan, and R. Islampour, *J. Phys. Chem.* **89**, 201 (1985); S. Mukamel and Y. J. Yan, *Adv. Chem. Phys.* **73**, 579 (1988); *J. Chem. Phys.* **89**, 5160 (1988).
2. Y. Tanimura and S. Mukamel, *Phys. Rev. E* **47**, 118 (1993); Y. Tanimura and S. Mukamel, *J. Opt. Soc. Am. B* **10**, 2263 (1993); *J. Phys. Soc. Jpn.* **63**, 66 (1994).
3. A. Schmid, *J. Low Temp. Phys.* **49**, 609 (1982).
4. G. W. Ford and M. Kac, *J. Stat. Phys.* **46**, 803 (1987); G. W. Ford, J. T. Lewis, and R. F. O'Connell, *Phys. Rev. A* **37**, 4419 (1988).
5. N. Hashitsume, M. Mori, and T. Takahashi, *J. Phys. Soc. Jpn.* **55**, 1887 (1986).
6. M. P. Allen and D. J. Tildesley, *Computer Simulations of Liquids* (Oxford Science Publications, Oxford, 1987); M. P. Allen and D. J. Tildesley, Eds., *Computer Simulations in Chemical Physics*, NATO ASI Series, Vol. 397 (Kluwer, 1993).
7. A. Garg, J. N. Onuchic, and V. Ambegaoker, *J. Chem. Phys.* **83**, 4491 (1985); L. D. Zusman, *Chem. Phys.* **49**, 295 (1980).
8. J. C. Tully and R. K. Preston, *J. Chem. Phys.* **55**, 562 (1971); J. Tully, *J. Chem. Phys.* **93**, 1061 (1990); H. Nakamura, *Int. Rev. Phys. Chem.* **10**, 123 (1991).
9. F. Webster, P. J. Rossky, and R. A. Friesner, *Comp. Phys. Commun.* **63**, 494 (1991).
10. S. Mukamel and Y. J. Yan, *Acc. Chem. Res.* **22**, 301 (1989).
11. R. P. Feynman and F. L. Vernon, *Annu. Phys.* **24**, 118 (1963).

12. A. O. Caldeira and A. J. Leggett, *Physica* **121A**, 587 (1983); A. J. Leggett, S. Chakravarty, et al., *Rev. Mod. Phys.* **59**, 1 (1987); H. Grabert, P. Schramm, and G.-L. Ingold, *Phys. Rep.* **168**, 115 (1988).
13. N. Wax, Ed., *Selected Papers on Noise and Stochastic Processes* (Dover, New York, 1954).
14. N. G. Van Kampen, *Stochastic Processes in Physics and Chemistry* (North-Holland, New York, 1981).
15. H. Risken, *The Fokker–Planck Equation*, 2nd ed. (Springer-Verlag, Berlin, 1989).
16. Y. J. Yan and S. Mukamel, *J. Chem. Phys.* **89**, 5160 (1988).
17. S. Mukamel and Y. J. Yan, in *Recent Trends in Raman Spectroscopy*, S. B. Banerjee and S. S. Jha, Eds. (World Scientific, Singapore, 1989), p. 160.

CHAPTER 13

Wavepacket Analysis of Nonimpulsive Measurements

In Chapter 12 we developed a Liouville space wavepacket representation for the nonlinear response function. By rearranging it into the doorway–window form, we showed that the role of the three time variables of the response function in an impulsive measurement may be viewed as follows: during t_1 the density operator evolves, creating a *doorway wavepacket*. During t_3, a *window wavepacket* is formed. The nonlinear response function [Eq. (12.52)] is given by the overlap of the doorway wavepacket, propagated for the period t_2, with the window wavepacket. The wavepacket dynamics may be described using either the Wigner phase space representation or the vibronic eigenstate representation for nuclear motions.

In this chapter we carry out an in-depth analysis of time resolved pump-probe spectroscopy conducted using pulses of finite duration [1–11]. The wavepacket representation provides a transparent physical picture of time-resolved pump probe experiments which is valid for arbitrary excitation pulse durations compared with nuclear dynamics timescales. It can, therefore, describe homogeneous, inhomogeneous, as well as intermediate broadening mechanisms. In the impulsive pump limit, in which the duration of the pump is much shorter than the nuclear dynamics timescale, vibronic motions in phase space may result in coherent oscillations (quantum beats). When the pump is long compared to the inverse linewidth but short compared to the timescales of overdamped solvent motions, the technique is known as time-resolved hole-burning spectroscopy. It can then probe spectral diffusion processes and the time-dependent Stokes shift. When a large number of nuclear eigenstates are involved in the process (which is the case for short pulses and low frequency motions) the wavepacket can easily be computed semiclassically in phase space; this way we avoid the tedious multiple summations over eigenstates. In the reverse situation (long pulses and high frequency modes), only a few vibronic levels are coherently excited, and the vibronic level representation of the wavepacket is more appropriate. Pump-probe spectroscopy has been applied to study a variety of processes, including carrier and phonon dynamics in semiconductor quantum wells (Figure 13.1), photochemical processes such as isomerization and chemical bond breaking (Figure 13.2), nonadiabatic transitions (Fig. 1.10), and coherent

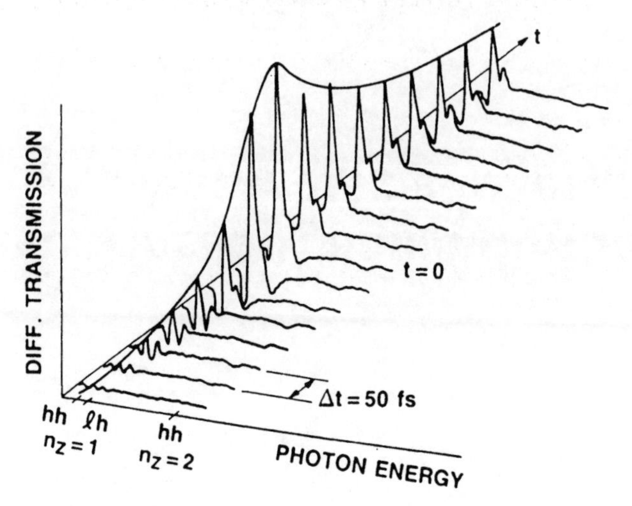

FIG. 13.1 Pump-probe spectroscopy of a GaAs quantum well. The figure shows the resonant excitonic nonlinearity. The fast overshoot of the excitonic bleaching signal is primarily a result of exciton ionization from absorption of LO phonons. The signal rises rapidly, then slowly drops as the carriers are spread out in energy by the relatively slow 300–500 fs absorption of LO phonons. [W. H. Knox, "Optical Studies in GaAs" in *Hot Carriers in Semiconductor Nanostructures*, J. Shah, Ed. (Academic Press, New York, 1992), p.333.]

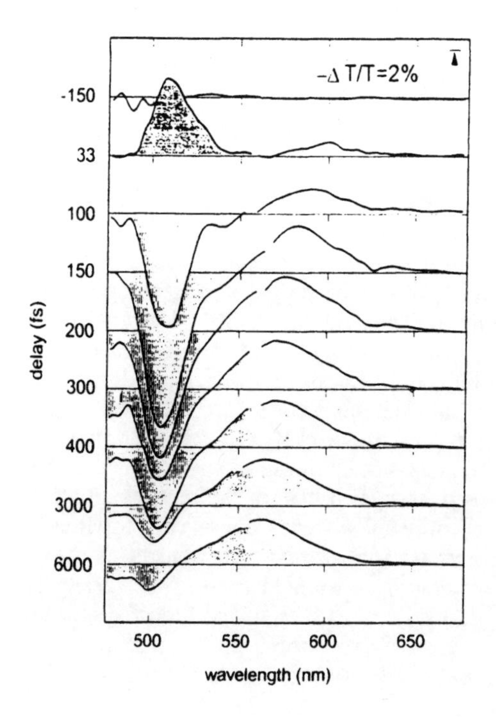

FIG. 13.2 Differential absorption spectra of rhodopsin probed with 10 fs blue (490–570 nm) and red (570–670 nm) pulses after 35 fs excitation pulses. The molecule undergoes photoisomerization. The spectral feature near 550 nm is attributed to photoproduct absorption. The progression of this spectral feature indicates that the first step of vision is completed in 200 fs. [Reprinted from L. A. Peteanu et al., "Femtosecond photoisomerization of rhodopsin as the primary event in vision," *Proc. Natl. Acad. Sci. U.S.A.* **90**, 11762 (1993).]

vibrational motions of isolated molecules in supersonic beams and in solution (Figures 1.6, 11.5, and 13.3). Typical hole-burning spectra are shown in Figures 13.4 and 13.5.

This chapter is organized as follows. We first consider an ideal situation in which the pump and the probe are well separated in time (i.e., their delay τ is much longer than their durations). We show that the doorway-window picture of Chapter 12 applies in this case as well, provided the doorway and window wavepackets are modified to incorporate the pulse envelopes. The optical process is then *sequential* and can be described by the following picture: The pump

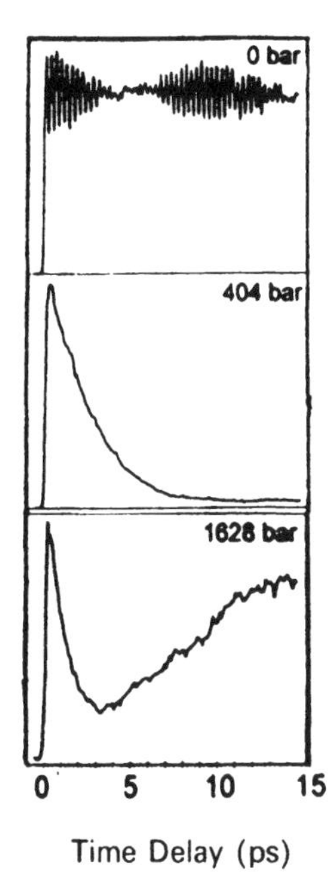

FIG. 13.3 Wave packets in a bound molecular system. The oscillatory behavior (top) reflects the change in internuclear separation of the iodine molecule as a function of time. Femtosecond pump-probe transients observed for excitation of the iodine in compressed supercritical argon at 295 K and various pressures. Note the increase in the decay rate with increase in the pressure and the rise due to caging at 1628 bar. [From Ch. Lienau, J. C. Wiliamson, and A. H. Zewail, "Femtochemistry at high pressures. The dynamics of an elementary reaction in the gas–liquid transition region," *Chem. Phys. Lett.* **213**, 289 (1993).]

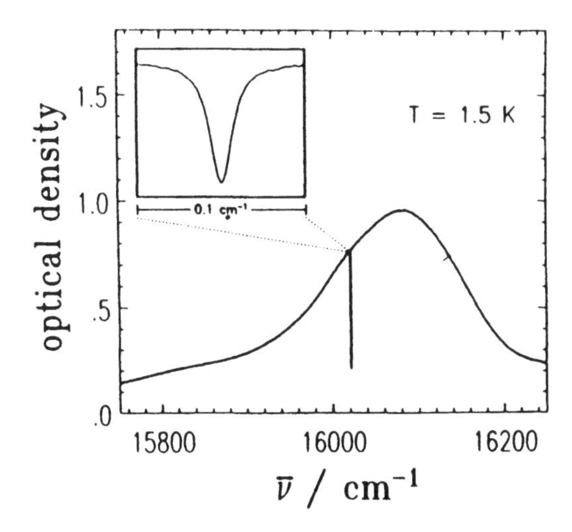

FIG. 13.4 Hole-burning spectrum of protoporphyrin-IX substituted myoglobin in a glass. [Courtesy of J. Friedrich.]

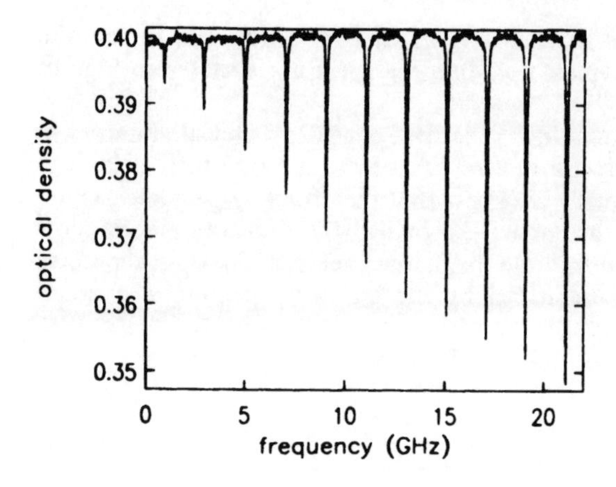

FIG. 13.5 Hole spectra and the holewidth as a function of temperature and time of the dye molecule free base phthalocyanine in a polymer glass. A series of holes burned with decreasing laser fluence are shown (from right to left) at $T = 50$ mK. The total range is less than one wave number. [From K.-P. Müller and D. Haarer, "Spectral diffusion of optical transitions in doped polymer glasses below 1 K," *Phys. Rev. Lett.* **66**, 2344 (1991).]

interacts with the system and creates a nonequilibrium *doorway wavepacket* in Liouville space. This wavepacket then evolves in time for a period τ (the delay between the pump and the probe). The probe field creates a *window wavepacket*, and the probe absorption is calculated by evaluating the overlap of the doorway and window in Liouville space. The signature of a sequential process is that the doorway wavepacket depends on the pump pulse shape and is independent of the probe, whereas the window wavepacket depends only on the probe pulse shape. We then consider the *snapshot spectrum*, a useful limiting case in which the doorway and the window depend on the pulse frequencies but not on their temporal profiles. Formal expressions for the snapshot doorway and window wavepackets are given, and expanded using the vibronic state representation. General conditions are specified whereby the observed spectrum may be calculated by convoluting the snapshot spectrum with either the temporal or the spectral intensity profiles of the incoming pulses. We further consider the classical limit as well as off-resonant impulsive spectroscopy, where non-Condon effects become significant.

The nuclear wavepackets are subsequently expressed using the Wigner phase space representation. We focus on the contribution of low frequency modes and solvation dynamics by considering strongly overdamped (diffusive) nuclear motions. These coordinates may describe dielectric relaxation, spectral diffusion, as well as homogeneous and inhomogeneous broadening. Their spectroscopic signatures such as the time-dependent Stokes shift in hole-burning and fluorescence line narrowing measurements, are analyzed.

To be specific, we restrict the discussion to pump-probe spectroscopy with the integrated detection scheme. However, the present picture applies to a general class of techniques invoking two well-separated pulses, including pump probe with dispersed detection and heterodyne-detected transient gratings. These generalizations are straightforward and the necessary changes are outlined in Appendices 13A and 13B. Wavepacket analysis of off-resonant measurements will be discussed briefly in this chapter and will be developed further in Chapter 14.

When the pump and the probe do overlap in time, the optical process is no longer sequential, and we need to consider an additional contribution to the signal. This contains valuable dynamic information since it depends in a delicate way on the interplay of the field and the material coherences. It also offers some interesting possibilities of manipulating nuclear motions using phase-locked pulses. Since this contribution does not have a simple classical analogue and is harder to interpret, it is sometimes given the unfortunate label of "coherent artifact." We shall attempt to unveil the mystery of this term by considering the complete (coherent as well as sequential) response of a three-level model system. The three-level case is interesting in many real applications to dye molecules as well as to crystals, semiconductors, and aggregates. Strangely enough, the interpretation of the nonlinear response of a three-level system is often simpler than that of a two-level system, since the pump and the probe may interact with transitions of different frequencies, reducing the number of terms to be considered. Semiclassical approximations for the coherent and the sequential components of the pump probe spectrum will be presented.

THE DOORWAY–WINDOW PICTURE FOR WELL SEPARATED PULSES

The formal expression for pump-probe spectroscopy was developed in Chapter 11. It contains a sum over permutations representing the different possible time orderings of the excitation pulses interacting with the system. We shall first focus on sequential processes in which the delayed probe pulse is well separated from the pump. In this case the system is forced to interact first with the pump, and then with the probe. This allows us to retain only the first field combination in Eqs. (11.3b) and (11.3c). The other, "coherent artifact," combinations, which contribute only when the pulses overlap in time, will be considered later in this chapter.

Using the pulse configuration specified by Eq. (11.7), the sequential pump-probe signal is given by

$$S_{PP}(\omega_1, \omega_2; \tau) = 2\omega_2\left(\frac{1}{\hbar}\right)^3 \mathrm{Re}\int_{-\infty}^{\infty} dt \int_0^{\infty} dt_3 \int_0^{\infty} dt_2 \int_0^{\infty} dt_1$$

$$[E_2^*(t+t_3)E_2(t)E_1^*(t+\tau-t_2)E_1(t+\tau-t_2-t_1)\exp(i\omega_2 t_3 + i\omega_1 t_1)$$

$$+ E_2^*(t+t_3)E_2(t)E_1(t+\tau-t_2)E_1^*(t+\tau-t_2-t_1)\exp(i\omega_2 t_3 - i\omega_1 t_1)]$$

$$\times \langle\langle V|\mathcal{G}(t_3)\mathcal{V}\mathcal{G}(t_2)\mathcal{V}\mathcal{G}(t_1)\mathcal{V}|\rho(-\infty)\rangle\rangle. \tag{13.1}$$

Equation (13.1) is similar to Eq. (11.8) except that we have not yet made the RWA, so that we have more terms here (the correlation function has eight terms when expanded in Liouville space paths), and we have also changed the integration variable t to $t + \tau$. We next change the t_2 integration variable to $t' = t + \tau - t_2$, and use the identity $\mathcal{G}(t_2) = \mathcal{G}(t)\mathcal{G}(\tau)\mathcal{G}^\dagger(t')$ to obtain

$$S_{PP}(\omega_1, \omega_2; \tau) = 2\omega_2\left(\frac{1}{\hbar}\right)^3 \mathrm{Re}\int_{-\infty}^{\infty} dt \int_0^{\infty} dt_3 \int_{-\infty}^{\tau+t} dt' \int_0^{\infty} dt_1$$

$$\times [E_2^*(t+t_3)E_2(t)E_1^*(t')E_1(t'-t_1)\exp(i\omega_2 t_3 + i\omega_1 t_1)$$

$$+ E_2^*(t+t_3)E_2(t)E_1(t')E_1^*(t'-t_1)\exp(i\omega_2 t_3 - i\omega_1 t_1)]$$

$$\times \langle\langle V|\mathcal{G}(t_3)\mathcal{V}\mathcal{G}(t)\mathcal{G}(\tau)\mathcal{G}^\dagger(t')\mathcal{V}\mathcal{G}(t_1)\mathcal{V}|\rho(-\infty)\rangle\rangle. \tag{13.2}$$

When the probe is well separated from the pump (i.e., the duration of both pulses is much shorter than their relative delay time τ), we can extend the upper limit of the t' integration from $\tau + t$ to ∞. This approximation is justified in that the integral contains an $E_1(t')$ factor, which by assumption vanishes for $t' \geq \tau$. Invoking the rotating wave approximation, we can then recast the signal in the form [11]:

$$S_{PP}(\omega_1, \omega_2; \tau) = \frac{2\omega_2}{\hbar} \mathrm{Tr}[W_e(\omega_2)\mathcal{G}_{ee}(\tau)D_e(\omega_1)] + \frac{2\omega_2}{\hbar} \mathrm{Tr}[W_g(\omega_2)\mathcal{G}_{gg}(\tau)D_g(\omega_1)], \tag{13.3}$$

where

$$D_e(\omega_1) = \frac{i}{\hbar}\int_{-\infty}^{\infty} dt' \int_0^{\infty} dt_1\, E_1^*(t')E_1(t' - t_1)\exp(i\omega_1 t_1)$$
$$\times \left[\exp\left(\frac{i}{\hbar}H_e t'\right)\exp\left(-\frac{i}{\hbar}H_e t_1\right)V_{eg}\rho_g \exp\left(\frac{i}{\hbar}H_g t_1\right)V_{ge}\exp\left(-\frac{i}{\hbar}H_e t'\right)\right] - h.c., \tag{13.4a}$$

$$D_g(\omega_1) = \frac{i}{\hbar}\int_{-\infty}^{\infty} dt' \int_0^{\infty} dt_1\, E_1^*(t')E_1(t' - t_1)\exp(i\omega_1 t_1)$$
$$\times \left[\exp\left(\frac{i}{\hbar}H_g t'\right)V_{ge}\exp\left(-\frac{i}{\hbar}H_e t_1\right)V_{eg}\rho_g \exp\left(\frac{i}{\hbar}H_g t_1\right)\exp\left(-\frac{i}{\hbar}H_g t'\right)\right] - h.c., \tag{13.4b}$$

$$W_e(\omega_2) = \frac{i}{\hbar}\int_{-\infty}^{\infty} dt \int_0^{\infty} dt_3\, E_2^*(t + t_3)E_2(t)\exp(i\omega_2 t_3)$$
$$\times \left[\exp\left(\frac{i}{\hbar}H_e t\right)V_{eg}\exp\left(\frac{i}{\hbar}H_g t_3\right)V_{ge}\exp\left(-\frac{i}{\hbar}H_e t_3\right)\exp\left(-\frac{i}{\hbar}H_e t\right)\right] - h.c., \tag{13.5a}$$

$$W_g(\omega_2) = \frac{i}{\hbar}\int_{-\infty}^{\infty} dt \int_0^{\infty} dt_3\, E_2^*(t + t_3)E_2(t)\exp(i\omega_2 t_3)$$
$$\times \left[\exp\left(\frac{i}{\hbar}H_g t\right)\exp\left(\frac{i}{\hbar}H_g t_3\right)V_{ge}\exp\left(-\frac{i}{\hbar}H_e t_3\right)V_{eg}\exp\left(-\frac{i}{\hbar}H_g t\right)\right] - h.c. \tag{13.5b}$$

Here h.c. denotes the Hermitian conjugate. Note that Eq. (13.3) is identical to Eq. (11.8). The various time variables appearing in Eqs. (13.4) are displayed in Figure 13.6. Double-sided Feynman diagrams representing the probe absorption were given in Figure 11.3.

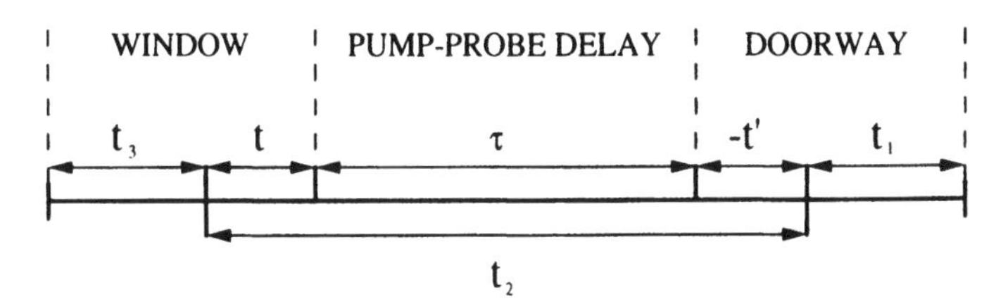

FIG. 13.6 Time variables for the doorway–window representation of pump-probe spectroscopy. The doorway is created during t_1 and t'. The window is created during t_3 and t. τ is the pump-probe delay.

Equations (13.3)–(13.5) constitute our final formal doorway–window expression of pump-probe spectroscopy. In this picture the fourfold temporal integrations implied in Eqs. (11.8) are separated into a twofold integration for the doorway and a twofold integration for the window. We reiterate that the only approximations made in the present derivation are the assumption that the pulses do not overlap, and the rotating wave approximation. Equation (13.3) provides a three-step physical picture of the pump-probe experiment: preparation, propagation, and detection. First, the pump pulse prepares an initial doorway wavepacket by exciting a "particle" into the electronic excited level, leaving a "hole" in the electronic ground level. The density operators representing the particle and the hole are given by doorway operators D_e and D_g, respectively. Similarly, the probe field creates the window wavepackets, W_e for the particle and W_g for the hole. The first term in Eq. (13.3) represents the particle contribution to the probe absorption, which is simply given by the overlap of the particle doorway, propagated for the delay time τ, with the particle window. The propagation takes place on the excited potential surface and is represented by $\mathcal{G}_{ee}(\tau)$. The hole makes a similar contribution with its own doorway and window wavepackets, with the propagation $\mathcal{G}_{gg}(\tau)$ taking place on the ground state surface. This contribution is given by the second term in Eq. (13.3). Note that the doorway wavepackets [Eq. (13.4)] contain the equilibrium density operator ρ_g whereas the window wavepackets [Eq. (13.5)] can be obtained from the doorway wavepackets by interchanging the fields E_1 and E_2 and by replacing the ground state density operator ρ_g with the unit operator. For that reason the windows may be formally interpreted as infinite temperature wavepackets.

To get a better feel for these wavepackets, we shall expand them in a complete basis set of nuclear eignestates (see Figure 7.1). Using this representation and taking the field amplitudes to be real we have

$$D_e(\omega_1) = \frac{2}{\hbar}\int_{-\infty}^{\infty} dt' \int_0^{\infty} dt_1\, E_1(t')E_1(t'-t_1)$$

$$\times \sum_{a,b,d} P(a)\mu_{ba}\mu_{ad}\cos[(\omega_1+\omega_{ab})t_1+\omega_{bd}t']|b\rangle\langle d|, \qquad (13.6a)$$

$$D_g(\omega_1) = \frac{2}{\hbar}\int_{-\infty}^{\infty} dt' \int_0^{\infty} dt_1\, E_1(t')E_1(t'-t_1)$$

$$\times \sum_{a,b,c} P(a)\mu_{cb}\mu_{ba}\cos[(\omega_1+\omega_{ab})t_1+\omega_{ca}t']|c\rangle\langle a|, \qquad (13.6b)$$

$$W_e(\omega_2) = \frac{2}{\hbar}\int_{-\infty}^{\infty} dt \int_0^{\infty} dt_3\, E_2(t)E_2(t+t_3) \times \sum_{c,b,d} \mu_{bc}\mu_{cd} \cos[(\omega_2 + \omega_{cb})t_3 + \omega_{db}t]|b\rangle\langle d|, \tag{13.7a}$$

$$W_g(\omega_2) = \frac{2}{\hbar}\int_{-\infty}^{\infty} dt \int_0^{\infty} dt_3\, E_2(t)E_2(t+t_3) \times \sum_{a,b,c} \mu_{cb}\mu_{cd} \cos[(\omega_2 + \omega_{cb})t_3 + \omega_{ca}t]|c\rangle\langle a|. \tag{13.7b}$$

Here $\hbar\omega_{vv'} \equiv \varepsilon_v - \varepsilon_{v'}$ is the energy difference between the vibronic levels v and v', and $P(v)$ is the thermal equilibrium population of the vth vibronic level.

With the help of Figure 13.6, we shall now have a closer look at our results. The doorway contains an integration over two time variables t_1 and t', and similarly the window depends on the time variables t_3 and t. The pump absorption requires two interactions with the radiation field. The transition amplitude is first order in the pump field but the signal is related to the amplitude squared and involves two interactions. The field amplitudes $E_1(t' - t_1)$ and $E_1(t')$ indicate that these two interactions take place at times $t' - t_1$ and t'. t_1 is the time delay between these two interactions; accordingly, t_1 may be interpreted as the time it takes for the ω_1 photon to be absorbed. When the absorption process is completed, the system finds itself in the excited state and propagates for a time t' on the excited state potential. The doorway state is obtained by integrating over all possible values of t' and t_1. Similarly, the probe absorption involves two interactions with the probe pulse. The time arguments in $E_2(t)$ and $E_2(t + t_3)$ indicate that these two interactions take place at times t and $t + t_3$ so that t_3 may be interpreted as the time interval over which the ω_2 probe photon is absorbed. During the t period the system is in the excited state. The window wavepacket is obtained by integrating over all values of t_3 and t. When Eqs. (13.6) and (13.7) are substituted in Eq. (13.3), we finally obtain the spectrum. This requires propagating the doorway for the delay period τ and calculating its overlap with the window wavepacket. The nonlinear polarization is determined by the time evolution during the three time intervals t_1, t_2, and t_3. During t_1 and t_3 the system is in an electronic coherence, whereas during the $t_2 \equiv \tau + t + t'$ period the system is in an electronic population (either in the ground state on in the excited state). The partitioning of t_2 into τ, t and t' is required in order to derive the doorway–window expression.

Having defined precisely these time intervals, let us consider their typical magnitudes. Using Eq. (13.4) we note that t_1 may be controlled by two factors: Since the function $E_1(t')$ is peaked around $t' = 0$, it is clear that t_1 cannot be longer than the pump duration. In addition, during the t_1 interval the system is in an optical coherence (ρ_{eg}). The time evolution of the coherence includes a time-dependent phase caused by the nuclear degrees of freedom (i.e., the difference between H_g and H_e [see Eqs. (7.19)]. This phase depends on the initial configuration of nuclei. When the signal is averaged over the distribution of phases, contributions with different initial conditions interfere destructively; this interference causes an irreversible decay in the response function, which is denoted pure dephasing. Dephasing is brought about by the $\cos[(\omega_1 + \omega_{ab})t_1]$ factors in Eq. (13.6), which describe the evolution of optical coherence. Optical dephasing is also responsible for the linear absorption linewidth. For a long pump pulse, the relevant t_1 period contributing to the doorway state is given by $\hbar$ divided by the electronic absorption linewidth. To summarize, t_1 is controlled by the shorter of the two timescales: the pump field duration and the decay of optical coherence (dephasing time). Similar arguments apply to the window state, where t_3 is restricted by either the duration of the probe pulse E_2 or by the optical dephasing timescale [as reflected in the $\cos[(\omega_2 + \omega_{cb})t_3]$ factors].* The relevant t' and t timescales are determined by the durations of the pump and the probe pulses, respectively, and by nuclear dynamics on the excited state (ω_{bd}) or on the ground state (ω_{ca}). Assuming that the pump (probe) absorption takes place at the center of the t_1 (t_3) intervals, the delay between these events is $\sim\tau + t + t' + 1/2(t_1 + t_3)$. Thus, only when all time variables t, t', t_1, and t_3 are much smaller than τ will the time delay between pump and probe interactions be equal to τ.

Depending on the relative magnitudes of t_1, t_3, t, t', and τ we can identify several limiting cases in which the theoretical description simplifies considerably. These will be discussed in the next section. Finally, the present picture can be extended to other sequential measurements involving two nonoverlapping pulses; we simply need to modify the definitions of the doorway and the window wavepackets. The generalization to pump-probe spectroscopy using the frequency-dispersed detection scheme is given in Appendix 13A and to the transient grating configuration is given in Appendix 13B.

THE SNAPSHOT SPECTRUM AND RELATED LIMITING CASES

In this section we discuss some typical experimental situations in which the calculation of the probe absorption is simplified considerably.

The snapshot spectrum: pulses short compared with nuclear dynamics and long compared with electronic dephasing

Equation (13.3) contains four time integrations. As illustrated in Figure 11.3, during the time periods t_1 and t_3 the system is in an *optical coherence* (ρ_{eg} or ρ_{ge}), whereas during the time periods t, t', and τ it is in a population (ρ_{ee} or ρ_{gg}). The t_1 and t_3 timescales are dominated either by pure dephasing processes, or by pulse durations, and are typically very short. During the τ, t, and t' periods the system is in an electronic population (ρ_{ee} for the particle and ρ_{gg} for the hole). The relevant timescale of nuclear motion is usually much longer than the dephasing timescale. Typically in solvated dyes, the absorption bandwidth is $\sim 500\ \text{cm}^{-1}$ which restricts t_1 and t_3 to about 50 fs, whereas the solvent nuclear relaxation, reflected, e.g., in the time-dependent Stokes shift, is in the picosecond range. The simplified picture given below holds when the dephasing timescale is much shorter than the nuclear dynamics, and the pulse durations are adjusted

* It is shown in Appendix 13A that by employing a frequency dispersed detection scheme, the probe pulse can be made arbitrarily short. In this case, t_3 is solely controlled by the optical dephasing time of matter.

to be intermediate between these two timescales. We then obtain the ideal snapshot spectrum, for which the pulses can be considered both monochromatic (with respect to the relevant linewidth related to the dephasing rate) and infinitely short (with respect to nuclear dynamics). Starting with Eqs. (13.4) and (13.5), the snapshot spectrum can be calculated by making the following two approximations:

1. Since the pulses are short compared with nuclear dynamics, we may neglect the nuclear motions during the time t' and t periods in the integrands by approximating

$$\exp\left(\pm\frac{i}{\hbar}H_m t'\right) \approx \exp\left(\pm\frac{i}{\hbar}H_m t\right) \approx 1, \qquad m = g, e. \tag{13.8}$$

2. Since the dephasing timescale is much shorter than the pulse durations, we neglect the variation of the external pulses on the t_1 and t_3 timescales, resulting in

$$E_1^*(t')E_1(t' - t_1) \approx |E_1(t')|^2, \tag{13.9a}$$

$$E_2^*(t + t_3)E_2(t) \approx |E_2(t)|^2. \tag{13.9b}$$

We can then define the snapshot doorway (D^0) and window (W^0) wavepackets, which do not depend on the laser pulse envelopes; i.e., they are purely material quantities. The resulting *snapshot spectrum*, denoted S_0, is (up to a proportionality factor)

$$S_0(\omega_1, \omega_2;\tau) \equiv \frac{2\omega_2}{\hbar}\,\mathrm{Tr}[W_e^0(\omega_2)\mathcal{G}_{ee}(\tau)D_e^0(\omega_1)] + \frac{2\omega_2}{\hbar}\,\mathrm{Tr}[W_g^0(\omega_2)\mathcal{G}_{gg}(\tau)D_g^0(\omega_1)], \tag{13.10}$$

with

$$D_e^0(\omega_1) \equiv \frac{i}{\hbar}\int_0^\infty dt_1\, \exp(i\omega_1 t_1)\left[\exp\left(-\frac{i}{\hbar}H_e t_1\right)V_{eg}\rho_g \exp\left(\frac{i}{\hbar}H_g t_1\right)V_{ge}\right] - h.c., \tag{13.11a}$$

$$D_g^0(\omega_1) \equiv \frac{i}{\hbar}\int_0^\infty dt_1\, \exp(i\omega_1 t_1)\left[V_{ge}\exp\left(-\frac{i}{\hbar}H_e t_1\right)V_{eg}\rho_g \exp\left(\frac{i}{\hbar}H_g t_1\right)\right] - h.c., \tag{13.11b}$$

$$W_e^0(\omega_2) \equiv \frac{i}{\hbar}\int_0^\infty dt_3\, \exp(i\omega_2 t_3)\left[V_{eg}\exp\left(\frac{i}{\hbar}H_g t_3\right)V_{ge}\exp\left(-\frac{i}{\hbar}H_e t_3\right)\right] - h.c., \tag{13.12a}$$

$$W_g^0(\omega_2) \equiv \frac{i}{\hbar}\int_0^\infty dt_3\, \exp(i\omega_2 t_3)\left[\exp\left(\frac{i}{\hbar}H_g t_3\right)V_{ge}\exp\left(-\frac{i}{\hbar}H_e t_3\right)V_{eg}\right] - h.c. \tag{13.12b}$$

Equations (13.11) and (13.12) define the *snapshot doorway* and the *snapshot window* wavepackets, respectively. They are simpler than their more general counterparts [Eqs. (13.4) and (13.5)] since each contains a single (rather than a double) time integration. In addition, they do not depend on the light pulse shapes, but only on their frequencies ω_1 and ω_2, and thus constitute an intrinsic property of the material system (although proper light pulses are essential in order to prepare these wavepackets and observe the snapshot spectrum).

Using the basis set of nuclear eigenstates, Eqs. (13.11) and (13.12) become

$$D_{\mathrm{e}}^{0}(\omega_1) = \frac{1}{\hbar}\sum_{d,b} |d\rangle\langle b| \rho_{\mathrm{D,e}}^{0}(db; \omega_1), \tag{13.13a}$$

$$D_{\mathrm{g}}^{0}(\omega_1) = \frac{1}{\hbar}\sum_{c,a} |c\rangle\langle a| \rho_{\mathrm{D,g}}^{0}(ca; \omega_1), \tag{13.13b}$$

$$W_{\mathrm{e}}^{0}(\omega_2) = \frac{1}{\hbar}\sum_{d,b} |d\rangle\langle b| \rho_{\mathrm{W,e}}^{0}(db; \omega_2), \tag{13.14a}$$

$$W_{\mathrm{g}}^{0}(\omega_2) = \frac{1}{\hbar}\sum_{c,a} |c\rangle\langle a| \rho_{\mathrm{W,g}}^{0}(ca; \omega_2), \tag{13.14b}$$

where

$$\rho_{\mathrm{D,e}}^{0}(db; \omega_1) = \sum_a \mu_{da}\mu_{ab}\left[\frac{P(a)}{\omega_1 - \omega_{da} + i\gamma/2} - \frac{P(a)}{\omega_1 - \omega_{ba} - i\gamma/2}\right], \tag{13.15a}$$

$$\rho_{\mathrm{D,g}}^{0}(ca; \omega_1) = \sum_d \mu_{cd}\mu_{da}\left[\frac{P(a)}{\omega_1 - \omega_{da} + i\gamma/2} - \frac{P(c)}{\omega_1 - \omega_{dc} - i\gamma/2}\right], \tag{13.15b}$$

$$\rho_{\mathrm{W,e}}^{0}(db; \omega_2) = \sum_c \mu_{dc}\mu_{cb}\left[\frac{1}{\omega_2 - \omega_{bc} + i\gamma/2} - \frac{1}{\omega_2 - \omega_{dc} - i\gamma/2}\right], \tag{13.16a}$$

$$\rho_{\mathrm{W,g}}^{0}(ca; \omega_2) = \sum_b \mu_{cb}\mu_{ba}\left[\frac{1}{\omega_2 - \omega_{bc} + i\gamma/2} - \frac{1}{\omega_2 - \omega_{ba} - i\gamma/2}\right]. \tag{13.16b}$$

In Eqs. (13.13)–(13.16) the doorway and the window wavepackets are expanded in the vibronic representation and $\rho_{\mathrm{D}}^{0}(vv'; \omega_1)$ and $\rho_{\mathrm{W}}^{0}(vv'; \omega_2)$ are their vv' matrix elements. γ is the inverse excited electronic state lifetime. The probe absorption is then given by

$$\begin{aligned} S_0(\omega_1, \omega_2; \tau) = &\sum_{b,d} \rho_{\mathrm{W,e}}^{0}(db; \omega_2)\exp(-i\omega_{db}\tau - \gamma_{db}\tau)\rho_{\mathrm{D,e}}^{0}(db; \omega_1) \\ &+ \sum_{a,c} \rho_{\mathrm{W,g}}^{0}(ca; \omega_2)\exp(-i\omega_{ca}\tau - \gamma_{ca}\tau)\rho_{\mathrm{D,g}}^{0}(ca; \omega_1). \end{aligned} \tag{13.17}$$

Equation (13.17) may be interpreted as follows: The pump field creates a wavepacket in the excited electronic state. For $v = v'$, $\rho_{\mathrm{D}}^{0}(vv'; \omega_1)$ represents the population of state v, whereas for $v \neq v'$ this is the coherence between the two vibrational states. The denominator $\omega_1 - \omega_{vv'}$ selects preferentially near resonance transitions with $\omega_1 \cong \omega_{vv'}$. Note also that creating a coherence between v and v' requires both states to be coupled to a common ground state level a. $\rho_W^0(vv'; \omega_2)$ is a wavepacket prepared by the probe. It tells us how important the coherence between v and v' is to the absorption of an ω_2 photon. Equations (13.15) and (13.16) are very similar except that Eq. (13.15) contains an average over the a states with a weight given by the thermal populations $P(a)$ and $P(c)$, whereas Eq. (13.16) has a sum over the b or c states with no corresponding weight $P(b)$ or $P(c)$ (since the window is an infinite temperature wavepacket). Equation (13.17) implies that the doorway propagates for a period τ and the spectrum is given by its overlap with the window.

The vibronic eigenstate representation used here for the doorway and the window wavepackets is convenient when the relevant nuclear levels are so sparse that only a few eigenstates dominate. For experiments involving ultrashort pulses and high densities of eigenstates, the eigenstate representation becomes impractical since we need to include a large number of states. It may then become advantageous to adopt the Wigner phase space representation introduced in Chapter 12, which is most suitable for developing semiclassical approximations. We shall denote the Wigner representation of $\rho_D^0(\omega_1)$ and $\rho_W^0(\omega_2)$ by $\rho_D^0(\mathbf{pq}; \omega_1)$ and $\rho_W^0(\mathbf{pq}; \omega_2)$. Equation (13.17) thus assumes the form

$$
\begin{aligned}
S_0(\omega_1, \omega_2; \tau) = & \iiiint d\mathbf{p}\, d\mathbf{q}\, d\mathbf{p}'\, d\mathbf{q}' \\
& \rho_{W,e}^0(\mathbf{pq}; \omega_2)\mathcal{G}_{ee}(\mathbf{pq}\tau; \mathbf{p}'\mathbf{q}')\rho_{D,e}^0(\mathbf{p}'\mathbf{q}'; \omega_1) \\
& + \iiiint d\mathbf{p}\, d\mathbf{q}\, d\mathbf{p}'\, d\mathbf{q}' \\
& \rho_{W,g}^0(\mathbf{pq}; \omega_2)\mathcal{G}_{gg}(\mathbf{pq}\tau; \mathbf{p}'\mathbf{q}')\rho_{D,g}^0(\mathbf{p}'\mathbf{q}'; \omega_1). \quad (13.18)
\end{aligned}
$$

$\rho_{D,m}^0$ is now the density operator (in phase space) prepared by the pump. $\rho_{W,m}^0$ is the corresponding window and $\mathcal{G}_{mm}(\mathbf{pq}\tau; \mathbf{p}'\mathbf{q}')$ is the quantum mechanical propagator from phase space point $\mathbf{p}'\mathbf{q}'$ to $\mathbf{pq}$ at time τ. $m = $ e or g represents the particle or the hole contribution, respectively. Although they are in different representations, Eqs. (13.17) and (13.18) are identical. It is also possible to adopt a mixed representation whereby some nuclear degrees of freedom are described in phase space and others in the state representation.

Femtosecond experiments are most meaningful in the snapshot limit where the condition $t_1, t_3, t, t' \ll \tau$ holds, and they can be interpreted as probing the dynamics during the time τ. The ability to control the delay τ is the great advantage of time-domain pump-probe techniques. In steady-state measurements we integrate over all values of τ, and the signal is mostly determined by the long time behavior, rendering it insensitive to short time dynamics. By selecting τ we may probe the dynamics on a desired timescale. The snapshot limit represents an ideal pump-probe experiment, in which the finite temporal and spectral widths of the laser pulses do not limit the information gained on the system dynamics, and the pulses can be considered both monochromatic and infinitely short in time.*

We shall derive exact expressions for the snapshot doorway and window wavepackets using the multimode displaced harmonic model system introduced in Chapter 12 [Eqs. (12.57) and (12.58) with no friction $\gamma_j = 0$]. To that end we first recast Eq. (13.18) in the form

$$S_0(\omega_1, \omega_2; \tau) = \iint d\mathbf{p}\, d\mathbf{q}\, \rho^0_{\mathrm{W,e}}(\omega_2; \mathbf{pq})\rho^0_{\mathrm{D,e}}(\omega_1, \tau; \mathbf{pq}) + \iint d\mathbf{p}\, d\mathbf{q}\, \rho^0_{\mathrm{W,g}}(\omega_2; \mathbf{pq})\rho^0_{\mathrm{D,g}}(\omega_1, \tau; \mathbf{pq}). \tag{13.19}$$

The function $\rho^0_{\mathrm{D},m}(\omega_1, \tau; \mathbf{pq})$ is the Wigner representation of the snapshot doorway propagated for the delay time τ, i.e., $\mathscr{G}_{mm}(\tau)\rho^0_{\mathrm{D},m}(\omega_1)$, $m = \mathrm{g, e}$. Using the expressions for the wavepackets [Eqs. (12.57) and (12.58)] we obtain

$$\rho^0_{\mathrm{D},m}(\omega_1, \tau; \mathbf{pq}) = 2\,\mathrm{Re} \int_0^\infty dt_1 \exp[i(\omega_1 - \omega_{\mathrm{eg}})t_1 \prod_j F_{mj}(t_1, \tau; p_j q_j), \tag{13.20a}$$

$$\rho^0_{\mathrm{W},m}(\omega_2; \mathbf{pq}) = 2\,\mathrm{Re} \int_0^\infty dt_3 \exp[i(\omega_2 - \omega_{\mathrm{eg}})t_3 \prod_j G_{mj}(t_3; p_j q_j), \qquad m = \mathrm{g, e}. \tag{13.20a}$$

Numerical methods for quantum propagation of systems with a few degrees of freedom [12, 13] may be used in the exact calculation of the doorway and the window wavepackets for more general systems with anharmonic potentials.

The present picture contains an apparent contradiction. The temporal resolution ($\Delta\tau$) and the spectral resolution of ω_2, $\Delta\omega_2$ may be controlled independently (when using the frequency dispersed detection, they are even controlled by different instruments!). Consequently, we can make both $\Delta\omega_2$ and $\Delta\tau$ as small as we desire. This may imply that we know both the time of the absorption event and the frequency of the ω_2 photon with infinite accuracy, in violation of the Heisenberg uncertainty relation $\Delta\omega\Delta\tau \geq 1$. Our previous discussion of the relevant timescales underlying the process resolves this apparent contradiction. Although τ may be fixed accurately, the actual time between the pump and the probe absorption events is $\sim\tau + t + t' + (t_1 + t_3)/2$. If the pulses are short ($t, t' \ll \tau$) and the lineshapes are broad ($t_1, t_3 \ll \tau$), then the approximate time is τ, and the temporal resolution is high. However, since the lineshapes are broad, we do not know ω_2 to a better accuracy than the absorption linewidth. Conversely, if the lines are narrow, we lose the temporal resolution (since t_1 and t_3 are long) but can maintain a good spectral resolution. It makes no sense to perform femtosecond experiments on systems with narrow spectral lines, since no information is gained beyond a cw measurement; the setup then becomes an expensive spectrometer! It is important to recognize that the genuine temporal

* The combined role of both spectal and temporal properties of the field can be visualized by adopting a Wigner representation for the field as well [J. Paye, *IEEE J. Quant. Electr.* **28**, 2262 (1992)]. This way the coherence properties of the field and matter may be represented in a more similar fashion.

resolution of the experiment is fixed by the system being probed as well as by the pulses. One cannot necessarily assume that if the pulses are short and their delay τ is controlled with infinite accuracy, then the actual delay between the pump and the probe absorption events is τ. Quantum mechanical principles will determine the actual delay, and only when the system's dephasing is fast will it be equal to τ.

Pulses short compared with nuclear dynamics

We shall now consider the limiting case when the durations of the both pulses are short compared with the nuclear dynamics timescales. We thus make only approximation (1) [Eq. (13.8)] but not (2) [Eq. (13.9)]. In this case we can neglect the excited state dynamics during the pulse durations. The doorway function, D_m is simply the spectral convolution of snapshot doorway, D_m^0 [Eqs. (13.11)] with the spectral profile of pump field, $|E_1|^2$. Similarly, the window function, W_m, is simply the spectral convolution of snapshot window, W_m^0 [Eqs. (13.12)], with the spectral profile of pump field, $|E_2|^2$. The signal [Eq. (13.3)] is then given by

$$S_{PP}(\omega_1, \omega_2; \tau) = \int_{-\infty}^{\infty} d\omega_2' \int_{-\infty}^{\infty} d\omega_1' \, I_1(\omega_1' - \omega_1) I_2(\omega_2' - \omega_2) S_0(\omega_1', \omega_2'; \tau), \tag{13.21a}$$

where

$$I_j(\omega - \omega_j) = \frac{1}{2\pi} \left| \int_{-\infty}^{\infty} dt \, E_j(t) \exp[-i(\omega - \omega_j)t] \right|^2, \tag{13.21b}$$

is the power spectrum of the jth pulse. The interpretation of this result is straightforward: Since the pulses are short, their power spectrum is broad and their bandwidth cannot be ignored. The actual signal is then given by a spectral convolution of the snapshot spectrum S_0 with the pulse intensity profiles.

Pulses long compared with the electronic dephasing timescale

When the pulses are long compared with the dephasing timescale, we make only approximation (2) [Eq. (13.9)] but not (1) [Eq. (13.8)]. In this case the signal may be expressed as the *temporal convolution* of the total intensity of the external fields with the snapshot spectrum:

$$S_{PP}(\omega_1, \omega_2; \tau) = \int_{-\infty}^{\infty} dt \, I(\tau - t) S_0(\omega_1, \omega_2; t), \tag{13.22a}$$

with

$$I(\tau - t) = \int_{-\infty}^{\infty} dt' \, |E_2(t)|^2 |E_1(t' + \tau - t)|^2. \tag{13.22b}$$

The time of observation, t, is controlled by the durations of the pump and the probe, and we need to average the calculated spectrum over the distribution of t around τ, as given by $I(\tau - t)$. For sufficiently short pulses, we can set in Eq. (13.22a) $I(\tau - t) = \delta(\tau - t_2)$. t is then simply equal to the time delay τ, and we recover the snapshot limit.

In this case, as well as in the previous case, the signal depends on the light pulses only through their intensities (either spectral or temporal). The field amplitudes and phases do not enter into the picture.

Resonant impulsive pump

We now assume that the pump is short compared with both the electronic dephasing and the nuclear dynamics of the system. The signal [Eq. (13.21)] in this impulsive pump limit can be expressed as

$$S_{PP}(\omega_2;\tau) = \int_{-\infty}^{\infty} d\omega_2\, I_2(\omega_2' - \omega_2) \times \{\mathrm{Tr}[W_e^0(\omega_2')D_e(\tau)] + \mathrm{Tr}[W_g^0(\omega_2')D_g(\tau)]\}. \tag{13.23}$$

$W_m^0(\omega_2')$ with $m = g, e$ is the snapshot window function [Eqs. (13.12)] at frequency ω_2'.

The formal expression of the doorway D_m in impulsive pump excitation is different for resonant and off-resonant conditions. In the *resonant* impulsive pump limit, we may represent the pump pulse by a delta function: $E_1(t) \sim \delta(t)$. The excited-state and ground-state doorway wavepackets [Eq. (13.4)], propagated over the delay period τ, are in this case

$$D_e(\tau) \equiv 2\mathcal{G}_{ee}(\tau)V_{eg}\rho_g V_{ge}, \tag{13.24a}$$

$$D_g(\tau) \equiv \mathcal{G}_{gg}(\tau)[V_{ge}V_{eg}\rho_g + \rho_g V_{ge}V_{eg}], \qquad |\omega_1 - \omega_{eg}| \approx \omega_{vv'}. \tag{13.24b}$$

The present results are identical to those of Chapter 11 where impulsive excitation was analyzed starting with the nonlinear response function (rather than the wavepacket dynamics). We can then write

$$R_2(t_3, \tau, 0) = \mathrm{Tr}[V_{eg}(t_3)V_{ge}(0)D_e(\tau)],$$

$$R_3(t_3, \tau, 0) = \mathrm{Tr}[V_{ge}(t_3)V_{eg}(0)D_g(\tau)].$$

$R_2(t_3, \tau, 0)$ and $R_3(t_3, \tau, 0)$ evaluated in Chapter 11 can thus be interpreted as two-time correlation functions of the dipole operator calculated using the non-equilibrium distribution functions $D_e(\tau)$ and $D_g(\tau)$ [Eqs. (13.24)], respectively.

We next consider the effect of the Condon approximation on the probe absorption. In general, the electronic dipole V_{eg} depends on nuclear coordinates. Consequently, both the excited doorway $D_e(\tau)$ ("particle") and the ground doorway $D_g(\tau)$ ("hole") vary with the delay time τ. However, in the Condon approximation in which V_{eg} is assumed to be independent of nuclear coordinates, the impulsive hole wavepacket (D_g) will no longer vary with τ. In this case only excited-state dynamics, as given by $D_e(\tau)$, are probed; this was already pointed out in Chapter 11. The doorway wavepacket shows why this is so: In the Condon approximation the doorway wavepacket is equal to the ground state equilibrium density operator ρ_g, which, of course, does not vary in time when subjected to the ground-state Hamiltonian H_g. Another case where the dynamics do not show up in the spectrum is when we invoke the Condon approximation and use a detection mode that lacks any frequency resolution. This is realized when the probe is impulsive or when the signal is integrated over ω_2. The absence of any signature of dynamics can be rationalized as follows: Time-dependent features such as quantum beats, reflect coherent motions, causing the system to emit at different frequencies at different times. In the Condon approximation, the impulsive window function is equal to the unit operator $W = 1$. At the detection stage we therefore integrate over all frequencies, thereby losing the spectral information and the beat pattern.

Off-resonance spectroscopy

When the pump excitation is tuned far off resonance from the electronic transition, we expect the particle contribution to the spectrum to vanish, and the signal should solely reflect the hole dynamics in the ground state. To see how this limit is obtained, we shall examine the behavior of the doorway states created by off-resonant pump excitation, in which the detuning, $|\omega_1 - \omega_{eg}|$, is much larger than the relevant vibrational level spacings in the same electronic state, ω_{db} or ω_{ca}. Equations (13.13) then reduce to

$$D_e^0(\omega_1) \approx \sum_{d,b} |d\rangle\langle b| \left\{ \sum_a P(a)\mu_{da}\mu_{ab} \frac{\gamma}{(\omega_1 - \omega_{eg})^2} \right\}, \tag{13.25a}$$

$$D_g^0(\omega_1) \approx \sum_{c,a} |c\rangle\langle a| \left\{ \sum_d \mu_{cd}\mu_{da} \left[i\frac{P(a) - P(c)}{\omega_1 - \omega_{eg}} + \frac{\gamma}{2}\frac{P(a) + P(c)}{(\omega_1 - \omega_{eg})^2} \right] \right\},$$
$$|\omega_1 - \omega_{eg}| \gg \omega_{vv'}. \tag{13.25b}$$

Both the excited-state and the ground-state doorway wavepackets contain a real contribution that varies as $\gamma/(\omega_1 - \omega_{eg})^2$. The ground doorway contains in addition an imaginary contribution that scales as $i/(\omega_1 - \omega_{eg})$. As the detuning is increased, the real part vanishes much more rapidly than the imaginary part; this is not only due to the higher power $[(\omega_1 - \omega_{eg})^{-2}$ vs. $(\omega_1 - \omega_{eg})^{-1}]$, but also to the relaxation constant γ, which is usually frequency dependent $[\gamma(\omega_1)]$, vanishing for large off-resonance detunings. The absence of a $i/(\omega_1 - \omega_{eg})$ contribution to D_e^0 is a result of an interference between two Liouville space pathways. It is the same interference that leads to the absence of excited-state frequencies in coherent anti-Stokes Raman (CARS) spectroscopy of isolated systems with no dephasing, as shown in Chapter 9. Consequently, the excited-state (particle) doorway state can be neglected for off-resonance excitation and we need consider only the ground-state (hole) doorway whose contribution is purely imaginary. We then have

$$D_e(\tau) = 0 \tag{13.26a}$$

and

$$D_g(\tau) = i\mathcal{G}_{gg}(\tau)[\alpha, \rho_g], \tag{13.26b}$$

with the ground-state electronic polarizability

$$\alpha \equiv \frac{1}{\hbar}\frac{|V_{eg}|^2}{\omega_1 - \omega_{eg}}.$$

Only the ground-state dynamics are retained in this case. Nuclear degrees of freedom enter through their ͜ffect on the electronic polarizability α. When the dependence of α on nuclear coordinates is neglected, α becomes a number and Eq. (13.26b) vanishes. Non-Condon effects, therefore, dominate the off-resonant response. The off-resonant doorway [Eqs. (13.26)] is purely imaginary, whereas the resonant doorway [Eqs. (13.24)] is purely real.

The present analysis contains an internal inconsistency since the RWA used in Eqs. (13.25) is no longer justified for off-resonant detunings. A more careful analysis of this case will be presented in Chapter 14. We shall show that Eq. (13.26b) holds even if we do not make the RWA, provided the polarizability α is modified to include non RWA terms [Eq. (14.4b)].

The classical limit

In the classical limit, we assume that the optical transition is instantaneous, and that the transition dipole does not depend on nuclear coordinates so that we can set $V_{eg} = 1$ (the Condon approximation). It is preferable in this case to start with the formal expression [Eqs. (13.11) and (13.12)], rather than with the eigenstate expansion. In the following we shall use the same notation introduced in Chapter 7. Assuming that H_g and H_e commute, we write [see Eq. (7.21)]

$$\exp\left(-\frac{i}{\hbar}H_e t\right)\exp\left(\frac{i}{\hbar}H_g t\right) \cong \exp[-i\omega_{eg}(\Gamma)t]; \qquad t = t_1, t_3. \tag{13.27}$$

Substituting this in Eqs. (13.11) and (13.12) we get

$$D_e^0(\omega_1) = D_g^0(\omega_1) = 2\pi\,\delta[\omega_1 - \omega_{eg}(\Gamma)]\rho_g(\Gamma), \tag{13.28a}$$

$$W_e^0(\omega_2) = W_g^0(\omega_2) = 2\pi\,\delta[\omega_2 - \omega_{eg}(\Gamma)]. \tag{13.28b}$$

These results have a simple classical interpretation; they reflect the classical Condon approximation, which states that in the classical limit a photon ω can be absorbed or emitted only if the system is in nuclear configurations Γ with resonant $\omega = \omega_{eg}(\Gamma)$. The snapshot spectrum [Eq. (13.10)] then becomes [12]

$$\begin{aligned} S_0(\omega_1, \omega_2; \tau) = {} & 4\pi^2\,\mathrm{Tr}[\delta(\omega_2 - \omega_{eg}(\Gamma))\mathcal{G}_{ee}(\tau)\delta(\omega_1 - \omega_{eg}(\Gamma))\rho_g(\Gamma)] \\ & + 4\pi^2\,\mathrm{Tr}[\delta(\omega_2 - \omega_{eg}(\Gamma))\mathcal{G}_{gg}(\tau)\delta(\omega_1 - \omega_{eg}(\Gamma))\rho_g(\Gamma)]. \end{aligned} \tag{13.29}$$

NUCLEAR WAVEPACKETS FOR THE OVERDAMPED BROWNIAN OSCILLATOR

Spectral diffusion and hole burning

An interesting question is whether the doorway–window picture holds even if we work in a *reduced space* where some degrees of freedom have been eliminated. In Chapter 12 we showed that for the multimode Brownian oscillator model, once the bath degrees of freedom are eliminated, we can no longer break up the propagation into segments for the t_1, t_2, and the t_3 periods, and we have to treat the entire nonlinear response as a single event. However, such factorization is still justified for the strongly overdamped Brownian oscillator, whose motion is described by the Smoluchowski equation.

The doorway and the window wavepackets for this important case, which may represent low frequency and collective nuclear modes, can be obtained by substituting Eqs. (12.59) in Eqs. (13.20). When the lineshapes are broad (fast dephasing) and the oscillator motion is much slower than the dephasing timescale, we can further neglect its nuclear dynamics during the electronic transition in the time intervals t_1 and t_3. This is the static limit of line broadening discussed in Chapter 7. We then set $M_j(t) \cong 1 - \Lambda_j t$ for $t \leq t_1, t_3$ in Eqs. (12.59) and (12.60), resulting in

$$F_{mj}(t_1, \tau; U_j) = (2\pi\Delta_j^2)^{-1/2} \exp[-\tfrac{1}{2}[U - \bar{U}_j(t_1, \tau)]^2/\Delta_j^2 - i\omega_{eg}t_1 - g_j(t_1)], \quad m = g, e \tag{13.30a}$$

$$G_{mj}(t_3; U_j) = \exp\left(i\omega_{eg}t_3 + \frac{i}{\hbar}U_j t_3\right), \qquad m = g, e \tag{13.30b}$$

with

$$\Delta_j^2 = 2\lambda_j k_B T, \tag{13.30c}$$

$$g_j(t_1) = \tfrac{1}{2}\Delta_j^2 t_1^2, \tag{13.30d}$$

$$\bar{U}_j = (1 - 2\eta)\lambda_j[1 - M_j(\tau)] + \lambda_j M_j(\tau) - i\Delta_j^2 M_j(\tau)t_1. \tag{13.30e}$$

Here we set $\eta = 0$ for F_{gj} and G_{gj} and $\eta = 1$ for F_{ej} and G_{ej}. The absorption and the fluorescence lineshapes are inhomogeneously broadened in this case [see Eqs. (8.52)]

$$\sigma_a(\omega) = (2\pi\Delta_j^2)^{-1/2} \exp[-(\omega - \omega_{eg}^0 - \lambda_j)^2/2\Delta_j^2], \tag{13.31a}$$

$$\sigma_f(\omega) = (2\pi\Delta_j^2)^{-1/2} \exp[-(\omega - \omega_{eg}^0 + \lambda_j)^2/2\Delta_j^2]. \tag{13.31b}$$

Upon the substitution of Eqs. (13.30) in (13.20) we can carry out all the integrations, and calculate the snapshot spectrum. However, for this model we can go beyond the snapshot limit and calculate the spectrum for pulses that are short compared with nuclear dynamics [Eqs. (13.21)]. The pump and probe will be assumed to have a Gaussian profile with widths w_1 and w_2, respectively, i.e.,

$$I_j(\omega) = \frac{1}{\sqrt{2\pi}w_j} \exp(-\omega^2/2w_j^2), \qquad j = 1, 2.$$

Combining all these expressions we finally obtain [14]

$$S_{PP}(\omega_1, \omega_2; \tau) = 2\pi[(\Delta_j^2 + w_1^2)\alpha^2(\tau)]^{-1/2} \exp[-(\omega_1 - \omega_{eg}^0 - \lambda_j)^2/2(\Delta_j^2 + w_1^2)] \times \{\exp[-(\omega_2 - \omega_e(\tau))^2/2\alpha^2(\tau)] + \exp[-(\omega_2 - \omega_g(\tau))^2/2\alpha^2(\tau)]\}, \tag{13.32}$$

with

$$\omega_e(\tau) \equiv \omega_{eg}^0 - \lambda_j + M_j(\tau)(\omega_0 - \omega_{eg}^0 + \lambda_j), \tag{13.33a}$$

$$\omega_g(\tau) \equiv \omega_{eg}^0 + \lambda_j + M_j(\tau)(\omega_0 - \omega_{eg}^0 - \lambda_j), \tag{13.33b}$$

$$\omega_0 \equiv \omega_1 \frac{\Delta_j^2}{\Delta_j^2 + w_1^2} + (\omega_{eg}^0 + \lambda_j)\frac{w_1^2}{\Delta_j^2 + w_1^2}, \tag{13.33c}$$

$$\alpha^2(\tau) \equiv \Delta_j^2\left[1 - \frac{\Delta_j^2}{\Delta_j^2 + w_1^2} M_j^2(\tau)\right] + w_2^2. \tag{13.33d}$$

The first term in the curly brackets in Eq. (13.32) represents the "particle" dynamics, whereas the second term represents the "hole" dynamics. At $\tau = 0$ both terms are identical; they are Gaussians centered at $\omega_2 = \omega_0$, with a spectral width $\alpha(0)$. As the delay time is increased, the particle and the hole undergo diffusive dynamics on their respective potential surfaces (the excited state for the particle and the ground state for the hole). The probe absorption then consists of two contributions, centered at $\omega_e(\tau)$ for the particle and at $\omega_g(\tau)$ for the hole. This τ-dependent splitting represents the dynamic Stokes shift. At long times we have $\omega_e(\infty) = \omega_{eg}^0 - \lambda_j$ and $\omega_g(\infty) = \omega_{eg}^0 + \lambda_j$. These terms then reflect the stationary fluorescence and absorption [Eq. (13.31)] and their splitting $2\lambda_j$ is the Stokes shift [15–17]. The magnitude and sign of the particle and the hole spectral shifts depend on the pump frequency ω_1. For resonant pump excitation $\omega_1 = \omega_{eg}^0 + \lambda_j$, the "particle" experiences a time-dependent red shift from $\omega_2 = \omega_{eg}^0 + \lambda_j$ to $\omega_2 = \omega_{eg}^0 - \lambda_j$. The "hole" position, in this case, does not evolve in time. When $\omega_1 = \omega_{eg}^0 - \lambda_j(1 + 2w_1^2/\Delta_j^2)$, the "particle" does not shift in time while the "hole" undergoes a blue shift from $\omega_2 = \omega_{eg}^0 - \lambda_j$ to $\omega_2 = \omega_{eg}^0 + \lambda_j$. In general both particle and hole contributions undergo time dependent shifts, as illustrated in Figure 13.7.

$\alpha(\tau)$ (not to be confused with the polarizability) represents the time-dependent spectral width of the particle and the hole. This width increases with time, with the initial value

$$\alpha(0) = [\Delta_j^2 w_1^2/(\Delta_j^2 + w_1^2) + w_2^2]^{1/2}, \tag{13.34a}$$

and the long time value

$$\alpha(\infty) = (\Delta_j^2 + w_2^2)^{1/2}. \tag{13.34b}$$

In the impulsive pump limit where $w_1 \gg \Delta_j$, we have $\omega_0 = \omega_{eg}^0 + \lambda_j$ and $\alpha(0) = (\Delta_j^2 + w_2^2)^{1/2} = \alpha(\infty)$. In this case only the particle experiences a red shift; the spectral width as well as the hole position do not evolve in time. In the other extreme, where w_1 and $w_2 \ll \Delta_j$, we have $\alpha(0) \approx (w_1^2 + w_2^2)^{1/2} \ll \alpha(\infty)$. In

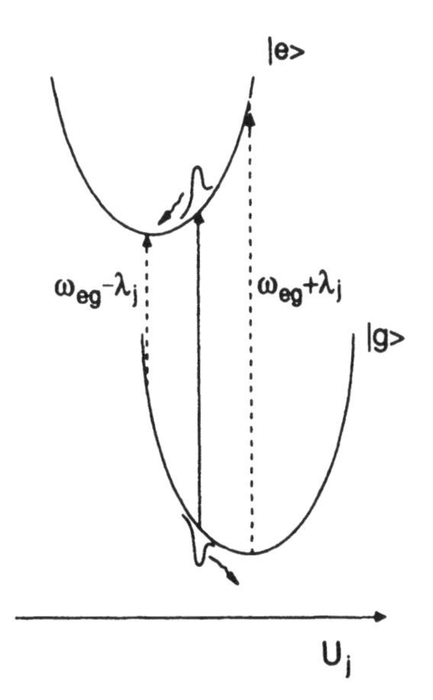

FIG. 13.7 Schematic representation of pump-probe measurement and the dynamic Stokes shift for a strongly overdamped mode in the spectral diffusion limit. The potential function has a displaced equilibrium position in the ground and in the excited electronic states. Shown is the excited state "particle" and the ground-state "hole" relaxation. For the excitation wavelength shown (solid arrow) $\omega_{eg} - \lambda < \omega_1 < \omega_{eg} + \lambda$, where ω_1 is the pump frequency, the "particle" will show a time-dependent red shift toward the emission maximum $\omega_{eg} - \lambda$, and the "hole" makes a blue shift to the absorption maximum at $\omega_{eg} + \lambda$.

this limit, the spectral width shows a diffusive broadening and the pump-probe technique is referred to as "hole burning," as discussed in Chapter 11. The name implies that the pump pulse creates (burns) a narrow hole in the absorption lineshape, since it interacts only with a small subgroup of the ensemble. As time evolves, the hole fills, eventually acquiring the equilibrium value $\alpha(\infty) \approx \Delta_j$. Finally, we note that the signal is sensitive to the spectral bandwidth w_2 of the probe pulse which determines the spectral width of the window wavepacket. For impulsive probe, $w_2 \to \infty$ and $\alpha(\tau) \to \infty$, and the signal loses its time dependence, i.e., the probe absorption no longer depends on τ. The probe pulse must, therefore, be sufficiently long to allow for spectral selectivity in the detection process. For that reason, measurements performed with an impulsive probe must adopt the dispersed [Eq. (4.92)], rather than the integrated [Eq. (4.85)] detection mode.

The dynamic processes described here are known as spectral diffusion [6–8, 18], which was already discussed in Chapter 11 using the nonlinear response functions. In this case, the line broadening is inhomogeneous, and nuclear dynamics can be ignored during the coherence (t_1 and t_3) periods. However, during the delay period τ, the spectrum changes (both shifts and broadens) due to nuclear diffusive motions that are slow compared with the inverse linewidth, and that change the inhomogeneous distribution of nuclear configurations. The absence of nuclear dynamics during t_1 and t_3 makes it possible to adopt a simple classical picture of these processes, which can then be modeled phenomenologically. Spectroscopy of glasses and polymers is usually described adequately using the spectral diffusion model [6–8, 19, 20]. In the extreme static limit, where nuclear motions are frozen, we should set $M(\tau) = 1$ for all times. In this case the hole is permanent and the spectrum does not change with the delay time. Permanent hole-burning spectra, known as "persistent hole-burning," are commonly obtained using molecules that undergo a photochemical change such as isomerization upon optical excitation, or when a physical change affects the surrounding of an impurity in a crystal or glass. These are denoted photochemical and nonphotochemical hole burning, respectively. Hole burning has potentially important practical applications in, e.g., memory storage. A remarkable look at the spectral diffusion process can be obtained using fluorescence excitation spectroscopy of a single molecule [21], as displayed in Figure 13.8. The figure clearly shows how the molecular frequency undergoes a kind of random walk, which is the origin of spectral diffusion.

In many physical situations, inhomogeneous broadening results from a very large number of small contributions of various molecules in the medium. The Gaussian inhomogeneous profile is then a direct consequence of the central limit theorem. The Brownian oscillator model used here does indeed yield a Gaussian lineshape in the inhomogeneous limit. As was pointed out in Chapter 10, however, inhomogeneous line profiles are not necessarily Gaussian; other forms appear in various situations. The spectral diffusion model can be easily extended to represent more general line profiles, e.g., the two state jump model where we assume that the frequency ω_{eg}^0 changes its magnitude randomly between two values on a certain timescale. A popular model, used in the interpretation of spectral lineshapes of impurities in glasses, assumes that each absorber interacts with a large number of perturbers, which in turn are described by the two-state jump model [19].

Finally, fluorescence line narrowing experiments [22] may also be described by the present theory. In these experiments, we monitor the time and frequency

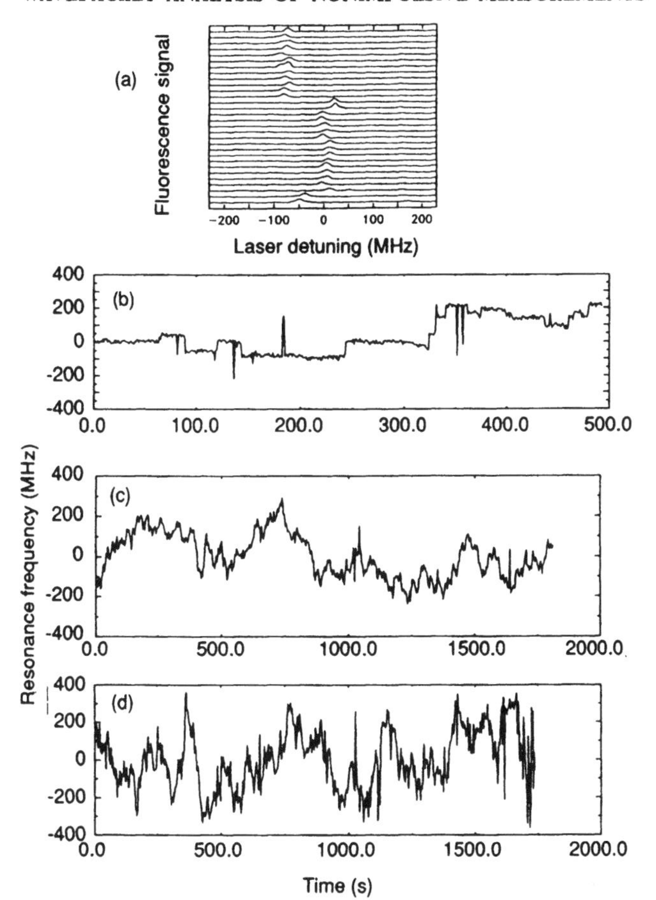

FIG. 13.8 Examples of single molecule spectral diffusion for pentacene in *p*-terphenyl at 1.5 K. (a) A series of fluorescence excitation spectra each 2.5 s long spaced by 2.75 s showing discontinuous shifts in resonance frequency, with zero detuning at 592.546 mm. (b) Trend or trajectory of the resonance frequency over a long time scale for the molecule in (a). (c) Resonance frequency trend for a different molecule at 592.582 nm at 1.5 K and (d) at 4.0 K [W. E. Moerner, "Examining nanoenvironments in solids on the scale of a single, isolated impurity molecule," *Science* **265**, 46 (1994)].

resolved spontaneous emission following the excitation by the pump. The emitted spectrum is initially narrow, reflecting the spectral selectivity of the pump, which can select a subgroup of molecules out of the inhomogeneous ensemble. Using the results of Chapter 9, it immediately follows that this spectrum is given by Eq. (13.32), provided we keep only the particle contribution (the first term in the curly brackets) and neglect the hole contribution. w_2 then represents the spectral bandwidth of the detection process.

Homogeneous broadening

The strongly overdamped Brownian oscillator model has two important limiting behaviors, depending on the value of $\kappa_j \equiv \Lambda_j/\Delta_j$ [Eq. (8.50)], which represents the bath relaxation rate compared with the static linewidth. The spectral diffusion considered above represents the slow motion limit ($\kappa_j \ll 1$). We shall now consider the opposite extreme $\kappa_j \gg 1$, where nuclear motions are very fast. We then have [see Eq. (8.53a)]

$$g_j(t) = \hat{\Gamma}_j t - i\lambda_j t, \tag{13.35}$$

with $\hat{\Gamma}_j \equiv \Delta_j^2/\Lambda_j$. The contributions of the overdamped coordinate to the hole doorway and window functions are in this case

$$F_{gj}(t_1, \tau; U_j) = \exp(-\hat{\Gamma}_j t_1)(2\pi\Delta_j^2)^{-1/2} \exp[(U_j - \lambda_j)^2/2\Delta_j^2], \tag{13.36a}$$

$$G_{gj}(t_3; U_j) = \exp(-\hat{\Gamma}_j t_3). \tag{13.36b}$$

The particle contribution F_{ej} and G_{ej} may be obtained using the same expressions by simply replacing $U_j - \lambda_j$ by $U_j + \lambda_j$.

Since the window function [Eq. (13.36b)] does not depend on the U_j coordinate, we can perform the U_j integration on F_{gj} and eliminate U_j altogether, while maintaining the simple doorway–window picture (which could depend on other nuclear degrees of freedom). We do not need to consider the oscillator coordinate U_j explicitly at all in this case; all we need is to simply replace ω_{eg} by $\omega_{eg} - i\hat{\Gamma}_j$ in the doorway and the window functions. This simplification reflects the fact that nuclear relaxation is so fast that the U_j coordinate retains its equilibrium distribution at all times during the optical process, and we do not need to keep track of its evolution. The absorption and the fluorescence lineshapes assume a Lorentzian form with a width equal to the homogeneous dephasing rate $\hat{\Gamma}_j$ and no Stokes shift, as shown in Eqs. (8.54). This is not the case for slower modes, which are not in the homogeneous limit.

We have limited the present discussion to the strongly overdamped Brownian oscillator where the factorization of the response function using a reduced description is rigorously justified. It may be possible to derive approximate expressions for other cases too. The semiclassical equations of motions [Eq. (12.35)] allow the numerical computation of these functions for more general systems with anharmonic potentials.

A model calculation for a multimode solute which includes the 29 Raman active vibrations of bacteriorhodopsin is shown in Figure 13.9a. In this calculation the particle contribution is red shifted compared to ω_{eg}, whereas the hole contribution is shifted to the blue. The two contributions are therefore clearly separated. For comparison we present in Figure 13.9b the calculated fluorescence line narrowing measurement of the same system. The doorway and its evolution are identical to the pump-probe measurement. The only difference is in the detection stage, which observes only the particle contribution (since only the excited state can emit light).

We shall now examine the hole-burning spectra for a model system with a single strongly overdamped mode (hereafter denoted the solvent mode). In Figure 13.10 we display spectra calculated using a stochastic model (which corresponds to a single overdamped Brownian oscillator mode with the neglect of the Stokes shift). In all calculations the fwhm of the absorption line is fixed $\Gamma_0 = 3000\ \text{cm}^{-1}$ but the underlying solvent timescale Λ^{-1} is varied from 100 ps to 100 fs. The linear absorption profile (indicated by the outermost dashed line) is virtually unchanged as the timescale is varied in the frequency range plotted. Hole burning, however, is much more sensitive to the solvent timescale. The pump pulse is Gaussian, with fwhm of 150 fs. The innermost dashed line in the figure is

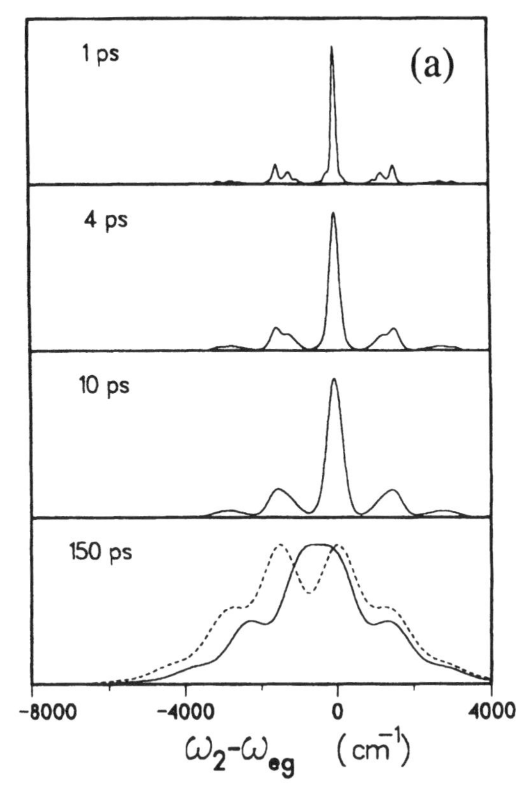

FIG. 13.9 (a) Hole-burning line shapes of a polyatomic solute in ethanol at 247 K, following a 1 ps excitation pulse. The model solute has the 29 Raman active vibrational modes of the retinal chromophore in bacteriorhodopsin, and undergoes rapid vibrational relaxation. $\omega_1 = \omega_{eg}$.

(continued)

the pump-pulse power spectrum. The solid lines, from the inside out, are the $\tau = 200$ fs differential absorption lineshapes for $\Lambda^{-1} = 100$ ps, 10 ps, 1 ps, and 100 fs. The lineshapes have fwhms of 260, 610, 1700, and 2950 cm^{-1}, respectively [as predicted by Eq. (13.33d)]. The variation of the hole linewidth with the solvent timescale can be rationalized as follows: Because we are considering a delay of $\tau = 200$ fs, and because the doorway state is prepared over a finite time, we see that the linewidths in Figure 13.10 vary with the solvent timescale. If Δ^{-1} is very large (slow solvent), the initial hole linewidth is determined by the pulse spectral width. [In Eq. (13.33d) we set $M(\tau) = 1$ throughout the experiment, and we have $\alpha(\tau) = w$.] In the other extreme, for small values of Λ^{-1}, the solvent equilibrates completely during the preparation of the doorway state and the first 200 fs of propagation; in this case, the pulse "burns" the whole absorption line, and the solvent broadening can be thought of as purely homogeneous. [In Eq. (13.33d) we set $M(\tau) = 0$; hence, $\alpha(\tau) = \Delta$.] As can be seen from Figure 13.10, the pump-pulse envelope and the absorption linewidth give lower bounds for Λ^{-1}, should we wish to view the evolution of the spectral diffusion process.

In all calculations presented so far we considered only the sequential contributions to the signal whereby the two pulses do not overlap. Using the response function it is possible to calculate the entire signal including all terms (sequential and nonsequential). As an example, in Figure 13.11 we display an experimental pump-probe spectrum obtained using a 70 fs pump, and conducted on a single crystal of a conjugated polymer (PTS). The fit to a three Brownian oscillator model with two high frequency modes and one overdamped mode is shown as well.

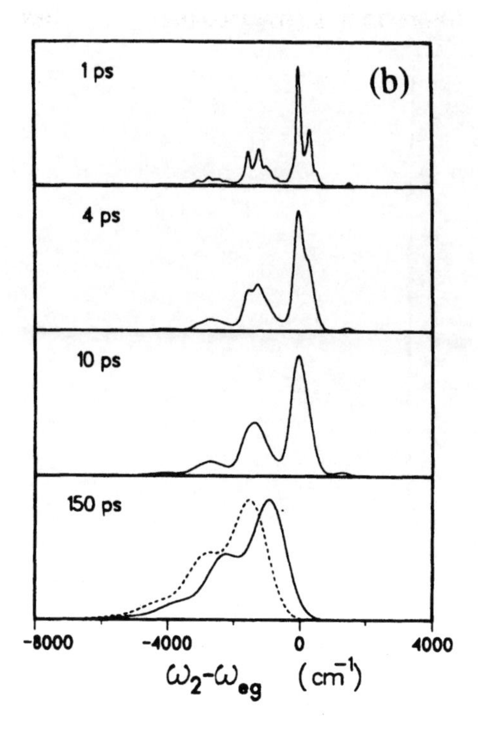

FIG. 13.9 (*continued*) (b) Time-resolved fluorescence line narrowing spectra for the same system, following a 1 ps pump pulse. [After R. F. Loring, Y. J. Yan, and S. Mukamel, "Time-resolved fluorescence and hole-burning line shapes of solvated molecules: Longitudinal dielectric relaxation and vibrational dynamics," *J. Chem. Phys.* **87**, 5840 (1987).]

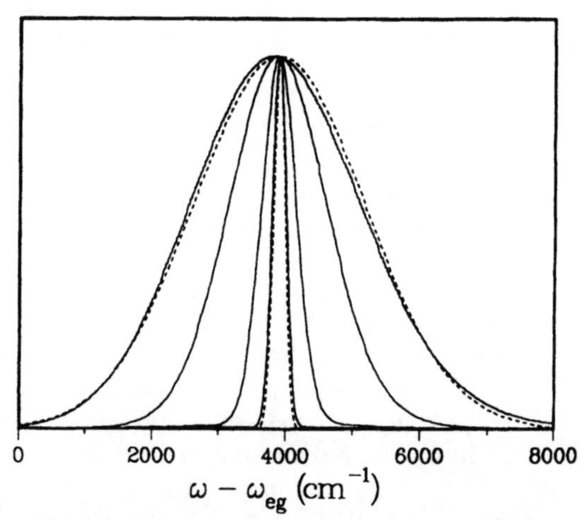

FIG. 13.10 Hole-burning lineshapes at the delay $\tau = 200$ fs for a single solvent Brownian oscillator mode, with different values of the solvent timescale Λ^{-1}, $\Delta = 1300\ \mathrm{cm}^{-1}$ for all curves. The absorption lineshape ($\Gamma_0 = 3000\ \mathrm{cm}^{-1}$) is given by the outer dashed line; the inner dashed line gives the pump-pulse envelope. The pump fwhm is 150 fs, and tuned to the center of the absorption line. From the inside out, $\Lambda^{-1} = 100$ ps, 10 ps, 1 ps, and 100 fs. All curves are normalized to the same maximum height. [From W. B. Bosma, Y. J. Yan, and S. Mukamel, "Intramolecular and solvent dynamics in femtosecond pump-probe spectroscopy," *J. Chem. Phys.* **93**, 3863 (1990).]

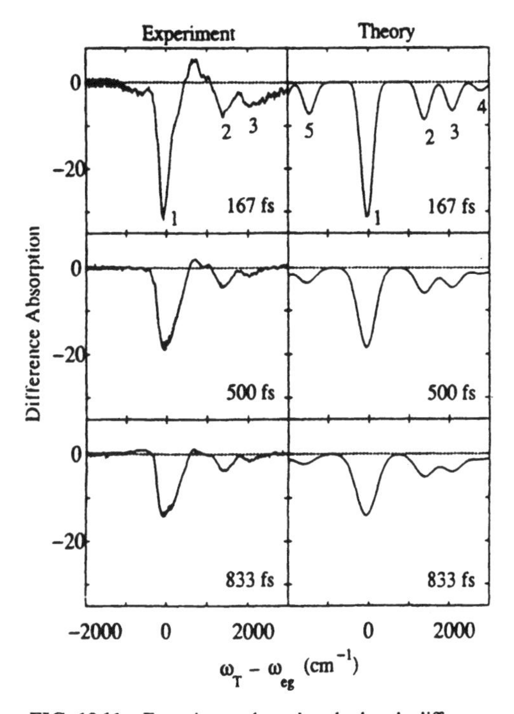

FIG. 13.11 Experimental and calculated difference absorption lineshapes for a conjugated polymer single crystal PTS, at three different τ values, with a resonant pump pulse ($\Omega_p = \omega_{eg}$) The pump duration is 70 fs. The experimental signals have been scaled to match the calculated 0-phonon line peak heights. The experimental (calculated) fwhms for the 0–0 transition are 308 cm^{-1} (382 cm^{-1}), 543 cm^{-1} (524 cm^{-1}), and 558 cm^{-1} (615 cm^{-1}), for τ = 167, 500, and 833 fs, respectively. These reflect a spectral diffusion process with time scale Λ^{-1} = 1.3 ps. [From W. B. Bosma, S. Mukamel, B. I. Greene, and S. Schmitt-Rink, "Femtosecond pump-probe spectroscopy of conjugated polymers: Coherent and sequential contributions," *Phys. Rev. Lett.* **68**, 2456 (1992).]

SEMICLASSICAL PICTURE OF PUMP-PROBE SPECTROSCOPY IN A THREE-ELECTRONIC-LEVEL SYSTEM

The semiclassical applications made so far were restricted to a two-electronic-level system. The generalization to other level schemes is straightforward. We shall illustrate this by considering pump-probe spectroscopy in a three-electronic-level [23], which provides beautiful insight into excited-state dynamics. This model applies for dye molecules and for molecular assemblies as well as semiconductors, where the third level represents two exciton states (see Chapter 17). The present calculation of pump-probe spectroscopy in a three-level system is simpler than for the two-level model, since we assume that the pump and the probe interact with different electronic transitions, thus reducing the number of terms.

We consider a system with three electronic states: a ground state $|g\rangle$ and two excited states, $|e\rangle$ (an intermediate electronic state), and $|f\rangle$ (the final excited electronic state). The molecular Hamiltonian is given by (see Figure 13.13)

$$H = |g\rangle H_g \langle g| + |e\rangle H_e \langle e| + |f\rangle H_f \langle f|, \tag{13.37a}$$

with

$$H_n = \hbar\omega_{ng}^0 + T(\mathbf{q}) + W_n(\mathbf{q}), \tag{13.37b}$$

and the dipole operator

$$V = V_{ge}(\mathbf{q})|g\rangle\langle e| + V_{eg}(\mathbf{q})|e\rangle\langle g| + V_{ef}(\mathbf{q})|e\rangle\langle f| + V_{fe}(\mathbf{q})|f\rangle\langle e|. \tag{13.38}$$

Similar to the two-level case [Eqs. (7.1)], the adiabatic Hamiltonian H_n (n = g, e, or f) describes the nuclear degrees of freedom $\mathbf{q}$ of the system in its electronic state $|n\rangle$. $W_n(\mathbf{q})$ is the potential and the 0–0 electronic transition energies will be denoted ω_{eg}^0 and $\omega_{fe}^0 \equiv \omega_{fg}^0 - \omega_{eg}^0$. We further assume that the system is initially in thermal equilibrium, occupying the vibronic manifold belonging to the electronic level $|g\rangle$, with the density operator $\rho(-\infty) = |g\rangle\rho_g\langle g|$.

For this model, the four Liouville space pathways for the nonlinear response function can be written in the form

$$\left.\begin{aligned}
R_1(t_3, t_2, t_1) &\equiv \langle\langle V_{ef}|\mathcal{G}_{ef}(t_3)\mathcal{V}_{ef,ee}\mathcal{G}_{ee}(t_2)\mathcal{V}_{ee,eg}\mathcal{G}_{eg}(t_1)\mathcal{V}_{eg,gg}|\rho_g\rangle\rangle,\\
R_2(t_3, t_2, t_1) &\equiv \langle\langle V_{ef}|\mathcal{G}_{ef}(t_3)\mathcal{V}_{ef,ef}\mathcal{G}_{ee}(t_2)\mathcal{V}_{ee,ge}\mathcal{G}_{ge}(t_1)\mathcal{V}_{ge,gg}|\rho_g\rangle\rangle,\\
R_3(t_3, t_2, t_1) &\equiv \langle\langle V_{ef}|\mathcal{G}_{ef}(t_3)\mathcal{V}_{ef,gf}\mathcal{G}_{gf}(t_2)\mathcal{V}_{gf,ge}\mathcal{G}_{ge}(t_1)\mathcal{V}_{ge,gg}|\rho_g\rangle\rangle,\\
R_4(t_3, t_2, t_1) &\equiv \langle\langle V_{ef}|\mathcal{G}_{ef}(t_3)\mathcal{V}_{eg,fg}\mathcal{G}_{fg}(t_2)\mathcal{V}_{fg,eg}\mathcal{G}_{eg}(t_1)\mathcal{V}_{eg,gg}|\rho_g\rangle\rangle.
\end{aligned}\right\} \tag{13.39}$$

A pictorial representation of these pathways is given in Figure 13.12. We have written down only the terms that include the final state f, since they will be the only relevant terms for the application considered below. There are four additional terms obtained by changing all f indexes to g. These are the four

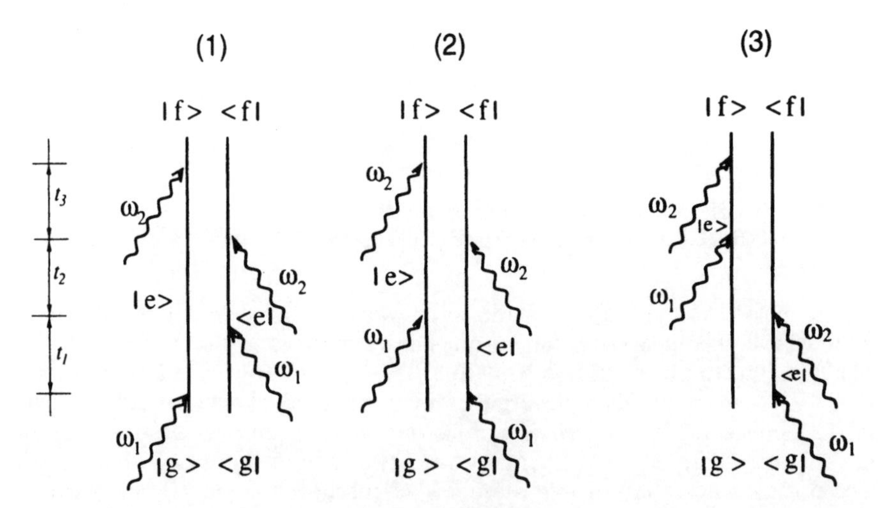

FIG. 13.12 The three double-sided Feynman diagrams and Liouville space paths contributing to the pump-probe spectrum in a three-level system in the RWA.

"two-level" terms, which do not depend on f at all, and which were discussed earlier in this chapter.

Coherent versus sequential components

The pulse configuration for a pump-probe experiment was given in Eqs. (11.7). We consider an ideal three-level pump-probe experiment whereby the frequency ω_{eg}^0 of the $|g\rangle$ to $|e\rangle$ electronic transition is very different from the frequency ω_{fe}^0 of the $|e\rangle$ to $|f\rangle$ electronic transition. The pump frequency ω_1 is tuned near resonance with ω_{eg}^0, while the probe frequency ω_2 is tuned near resonance with ω_{fe}^0. Consequently, we neglect all terms in which ω_1 induces an $|e\rangle$ to $|f\rangle$ transition or ω_2 induces an $|g\rangle$ to $|e\rangle$ transition. This allows us to neglect the "two-level" contributions to the response function, and we can focus only on the given four terms. Were the ω_{eg}^0 and the ω_{fe}^0 frequencies close, we should have included the additional terms. We shall further invoke the RWA and retain only near-resonant terms. When both of these approximations are made, we are left with only three terms (and their complex conjugates). Proceeding along the same steps that lead to Eq. (13.3) we now get

$$S_{PP}(\omega_1, \omega_2; \tau) = 2\omega_2[S_S + S_C], \tag{13.40a}$$

with a *sequential* contribution

$$\begin{aligned} S_S = 2\,\mathrm{Re} \int_{-\infty}^{\infty} dt \int_0^{\infty} dt_3 \int_0^{\infty} dt_2 \int_0^{\infty} dt_1 \\ \times \{\exp(-i\omega_2 t_3 - i\omega_1 t_1) E_2(t + t_3 - \tau) E_2^*(t - \tau) E_1^*(t - t_2) \\ \times E_1(t - t_2 - t_1) R_1(t_3, t_2, t_1) \\ + \exp(-i\omega_2 t_3 - i\omega_1 t_1) E_2(t + t_3 - \tau) E_2^*(t - \tau) E_1(t - t_2) \\ \times E_1^*(t - t_2 - t_1) R_2(t_3, t_2, t_1)\}, \end{aligned} \tag{13.40b}$$

and a *coherent* contribution

$$\begin{aligned} S_C = 2\,\mathrm{Re} \int_{-\infty}^{\infty} dt \int_0^{\infty} dt_3 \int_0^{\infty} dt_2 \int_0^{\infty} dt_1 \\ \times \exp[-i\omega_2 t_3 - i(\omega_2 + \omega_1) t_2 - i\omega_1 t_1] E_2(t + t_3 - \tau) E_1(t) \\ \times E_2^*(t - t_2 - \tau) E_1^*(t - t_2 - t_1) R_3(t_3, t_2, t_1). \end{aligned} \tag{13.40c}$$

Equations (13.40) can alternatively be derived by assuming the following simplified interaction with the electromagnetic field:

$$\begin{aligned} H_{\text{int}} = & - V_{eg} |e\rangle\langle g| E_1(t + \tau) \exp(i\mathbf{k}_1 \cdot \mathbf{r} - i\omega_1 t) \\ & - V_{fe} |f\rangle\langle e| E_2(t) \exp(i\mathbf{k}_2 \cdot \mathbf{r} - i\omega_2 t) + c.c. \end{aligned} \tag{13.41}$$

where *c.c.* denotes the complex conjugate. In this form we build in the RWA, as well as the assumption that each pulse interacts only with its own electronic transition. If we use this interaction instead of Eq. (5.1a) in Eq. (5.5), we immediately obtain Eqs. (13.40). The double-sided Feynman diagrams are given in Figure 13.12.

Equations (13.40) are compact formal expressions for the probe absorption. S_S, which contains two terms, represented by (1) and (2) of Figure 13.12, describes a sequential process in which the system interacts first with the pump and then with the probe field. Its wavepacket analysis can be carried out in complete analogy with the particle contribution to the two level case [Eqs. (13.3)], by simply replacing the ge coherence during t_3 with ef. S_C, which contains the term (3) of Figure 13.12, describes a coherent process in which one of the interactions with the probe field E_2 takes place between two interactions with the pump. If the delay time τ between the pump and the probe is large compared with their temporal widths, we can enforce the time ordering of "pump first and probe second." In this case the contribution of S_C will be negligible compared with S_S. In steady-state pump-probe experiments where the pump and the probe act simultaneously, all time orderings of the fields are equally important; S_C is then as relevant as S_S and may give rise to interesting extra resonances.

We further note the close analogy between the present calculation and the spontaneous light emission considered in Chapter 9. The S_S and S_C terms are formally analogous to the fluorescence and to the Raman components of light emission, which also represent a sequential and a coherent process, respectively. The only difference is that there, the second transition (ω_2) represents photon emission rather than absorption.

The snapshot limit

We next introduce the doorway and the window wavepackets in the snapshot limit [since there is only one (particle) doorway wavepacket here we shall omit the label e]

$$\rho_D^0(\omega_1) = \int_0^\infty dt_1 \exp(i\omega_1 t_1)\left[\exp\left(-\frac{i}{\hbar}H_e t_1\right)V_{eg}\rho_g \exp\left(\frac{i}{\hbar}H_g t_1\right)V_{ge}\right] + h.c., \tag{13.42a}$$

$$\rho_W^0(\omega_2) \equiv \int_0^\infty dt_3\left[\exp\left(\frac{i}{\hbar}H_e t_3\right)V_{ef}\exp\left(-\frac{i}{\hbar}H_f t_3\right)V_{fe}\right]\exp(i\omega_2 t_3) + h.c. \tag{13.42b}$$

The sequential contribution thus becomes [see Eq. (13.28)]

$$\begin{aligned} S_S \equiv M_S(\omega_2, \tau, \omega_1) &= \langle \rho_W(\omega_2)\mathcal{G}_{ee}(\tau)\rho_D(\omega_1)\rangle \\ &= \int d\mathbf{p}\, d\mathbf{q}\, \rho_W(\omega_2; \mathbf{p}, \mathbf{q})\mathcal{G}_{ee}(\tau)\rho_D(\omega_1; \mathbf{p}, \mathbf{q}). \end{aligned} \tag{13.43}$$

The doorway wavepacket $\rho_D(\omega_1; \mathbf{p}, \mathbf{q})$ is the Wigner distribution representing the state of the system prepared by the pump with frequency ω_1. The doorway function then evolves in time for a period τ, where its evolution is determined by the Hamiltonian H_e. At the end of that period we calculate the overlap of the resulting wavepacket with the window wavepacket $\rho_W(\omega_2; \mathbf{p}, \mathbf{q})$, which represents the region in phase space selected by the probe. In short, the system enters the excited state through the doorway, propagates, and is finally observed through the window.

The coherent term S_C [Eq. (13.40c)] does not offer such a simple classical picture. During the t_2 period it represents a coherent evolution of $|g\rangle$ and $|f\rangle$, very much analogous to Raman scattering or two photon absorption. The time ordering of the interactions is pump, probe, pump, and we cannot interpret it simply in terms of probing the wavepacket prepared by the pump. All time arguments t_1, t_2, and t_3 in S_C represent the evolution of coherences, which may be subject to fast electronic dephasing; this allows us to invoke a short time approximation:

$$E_1^*(t)E_1(t - t_2 - t_1) \cong |E_1(t)|^2 \equiv I_1(t), \tag{13.44a}$$

$$E_2^*(t + t_3 - \tau)E_2(t - t_2 - \tau) \cong |E_2(t - \tau)|^2 \equiv I_2(t - \tau). \tag{13.44b}$$

When Eqs. (13.44) are substituted in Eq. (13.40c) we get

$$S_C = 2I_1(t)I_2(t - \tau)\,\mathrm{Re}\int_0^\infty dt_2\, M_C(\omega_2, t_2, \omega_1), \tag{13.45}$$

$$M_C(\omega_2, t_2, \omega_1) = \exp[i(\omega_1 + \omega_2)t_2]\,\mathrm{Tr}[\Phi_W^\dagger(\omega_2)\mathcal{G}_{gf}(t_2)\Phi_D(\omega_1)]. \tag{13.46}$$

The associated doorway amplitude $\Phi_D(\omega_1)$ and the window amplitude $\Phi_W(\omega_2)$ are defined by

$$\Phi_D(\omega_1) = \int_0^\infty dt_1 \exp(-i\omega_1 t_1)\exp\left(-\frac{i}{\hbar}H_g t_1\right)\rho_g V_{ge}\exp\left(\frac{i}{\hbar}H_e t_1\right)V_{ef}, \tag{13.47a}$$

$$\Phi_W(\omega_2) = \int_0^\infty dt_3 \exp(-i\omega_2 t_3)\exp\left(-\frac{i}{\hbar}H_f t_3\right)V_{fe}\exp\left(-\frac{i}{\hbar}H_e t_3\right)V_{eg}, \tag{13.47b}$$

We emphasize the fundamental difference between the doorway and the window *amplitudes* (Φ_D, Φ_W), which appear in M_C and the doorway and the window *wavepackets* (ρ_D^0, ρ_W^0), which appear in M_S. The doorway amplitude represents the system in an electronic coherence $|f\rangle\langle g|$ and $|g\rangle\langle f|$ and its evolution during the t_2 period is dominated by electronic dephasing arising from the difference between the Hamiltonians H_g and H_f. This evolution does not have a simple classical analogue. The doorway wavepacket represents the system in the electronic state $|e\rangle\langle e|$, and its evolution during the t_2 period is the ordinary dynamics given by H_e, which has a well-defined classical analogue.

Since the operators Φ_W and Φ_D are not Hermitian, the corresponding Wigner functions $\Phi_W(\mathbf{p}, \mathbf{q})$ and $\Phi_D(\mathbf{p}, \mathbf{q})$ are complex. In the Wigner representation, Eq. (13.46) reads

$$M_C(\omega_2, t_2, \omega_1) = \exp[i(\omega_1 + \omega_2)t_2]\int d\mathbf{p}\, d\mathbf{q}\, \Phi_W^*(\omega_2; \mathbf{p}, \mathbf{q})\Phi_D(\omega_1; \mathbf{p}, \mathbf{q}, t_2), \tag{13.48}$$

where the time evolution of the Wigner function is given by the coherent propagator from point $\mathbf{p}'\mathbf{q}'$ to point $\mathbf{pq}$ in phase space $\mathcal{G}_{fg}(\mathbf{p}, \mathbf{q}, \mathbf{p}', \mathbf{q}'; t_2)$.

The pump probe measurement of ICN photodissociation (Figure 13.13) allows the direct observation of the breaking of a chemical bond [3, 23, 24]. The doorway

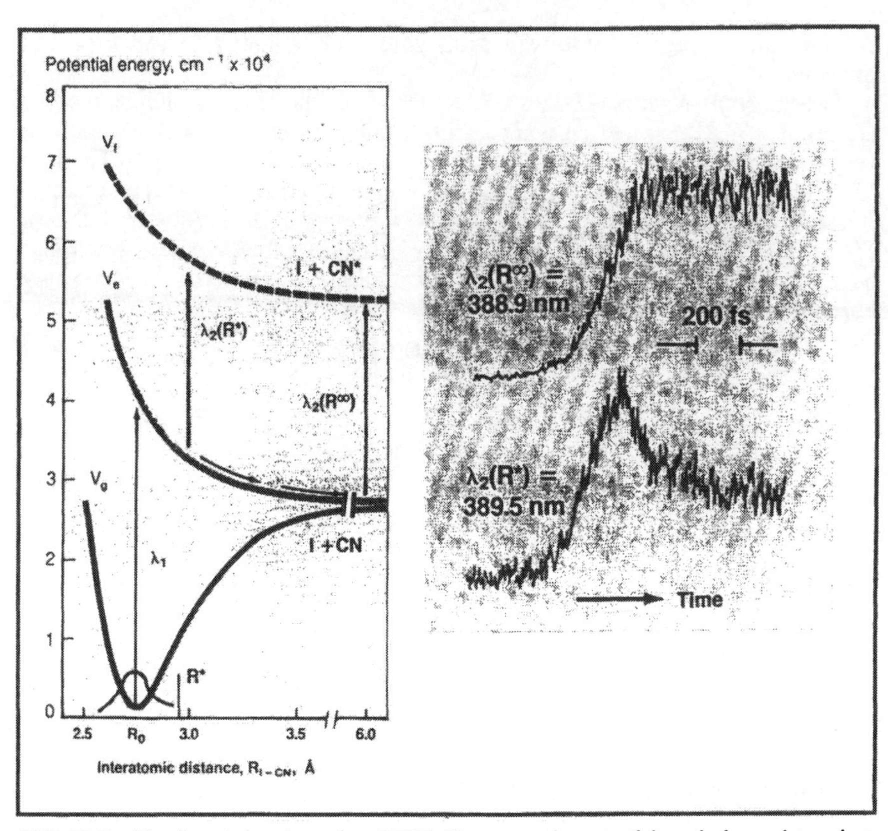

FIG. 13.13 The dissociation dynamics of ICN. Shown are the potentials and observed transients at two different internuclear separation. The excitation pulse is absorbed into the V_1 potential, which is unstable with respect to dissociation as indicated on the right. Coherent motion along $V_1(R)$ is monitored by absorption of variably delayed, tunable probe pulses into $V_2(R)$. The excited-state absorption progressively shifts to the blue. [From M. Dantus, M. J. Rosker, and A. H. Zewail, "Femtosecond real-time probing of reactions. II. The dissociation reaction of ICN," *J. Chem. Phys.* **87**, 2395 (1987); **89**, 6128 (1988).]

and the window wavepackets calculated using model potentials are shown in Figures 13.14 and 13.15. For off resonant pump, the doorway wavepacket is identical to the ground state equilibrium wavepacket since its preparation is impulsive. The doorway shown in Figure 13.14 corresponds to resonant excitation near the absorption maximum. The wavepacket now spreads and assumes a complex oscillatory pattern. Nevertheless, if we trace it over momentum and look at it in coordinate space we find it to be peaked around the classical value q_c predicted by the Frank–Condon principle ($q = 0$ here is the minimum of the ground state potential and the chosen pump wavelength is resonant at this position). The window wavepacket is more delocalized, particularly in momentum space. The reason is that it does not have an equilibrium density operator ρ_g and it thus corresponds to an infinite temperature. Again, if we trace it over momentum, we find that the distribution in coordinate space is peaked at the classical Condon coordinate and that the peak shifts to larger coordinates as the probe wavelength is tuned to the blue, as can be predicted from the potential functions displayed in Figure 13.13.

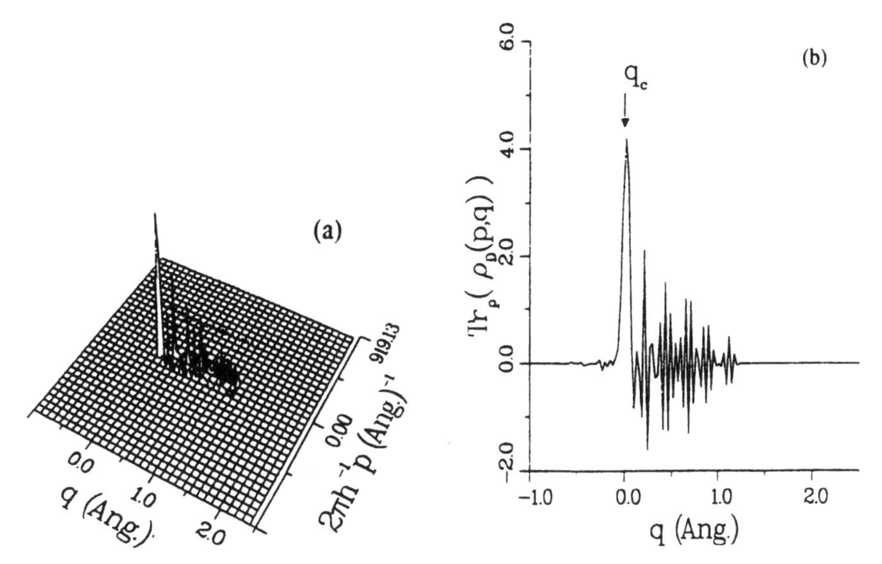

FIG. 13.14 (a) The doorway Wigner function for the ICN photodissociation. The potential surfaces are shown schematically in Figure 13.13. The excitation frequency is at the absorption maximum. This resonant doorway is spread over a large range of q values. (b) The trace over momentum p of the doorway Wigner function of (a). The trace is peaked near the classical value q_c, with oscillations occurring around the peak. [Y. J. Yan, L. E. Fried, and S. Mukamel, "Ultrafast pump-probe spectroscopy: Femtosecond dynamics in Liouville space," *J. Chem. Phys.* **93**, 8149 (1989)]

We next show model calculations that mimic the coherent vibrational motions observed in isolated I_2 molecule (Figure 13.3). The relevant potential surfaces are shown in Figure 13.16a. Figure 13.16b shows the time evolution of the doorway wavepacket in the Wigner representation, and its projection onto coordinate space is given in Figure 13.16c. We have chosen one particular probe wavelength, which corresponds to a classical Condon transition at 4.25 Å (see probe arrow in Figure 13.16a). The time resolved probe absorption signal at this wavelength is shown in Figure 13.16d.

Interplay of field and matter coherence: beyond the snapshot limit

To highlight the dual role of the spectral and the temporal profiles of the radiation fields in pump-probe spectroscopy, we shall recast our results using a mixed time- and frequency-domain representation. Since the material response at time t depends only on the fields at earlier times, we shall represent the pump field in a mixed time-frequency domain that maintains causality [25]. The sequential and coherent contributions represent two distinct physical processes in the pump-probe measurement, and the effect of the field on these two contributions should be represented differently. In the sequential process, the effect of the pump and the probe fields can be described by the following temporal-spectral density:

$$F_S(t, \omega'', \omega') = \int_{-\infty}^{\infty} dt' \, I_{22}^{+}(t + t', \omega'') I_{11}^{-}(t', \omega'), \qquad (13.49a)$$

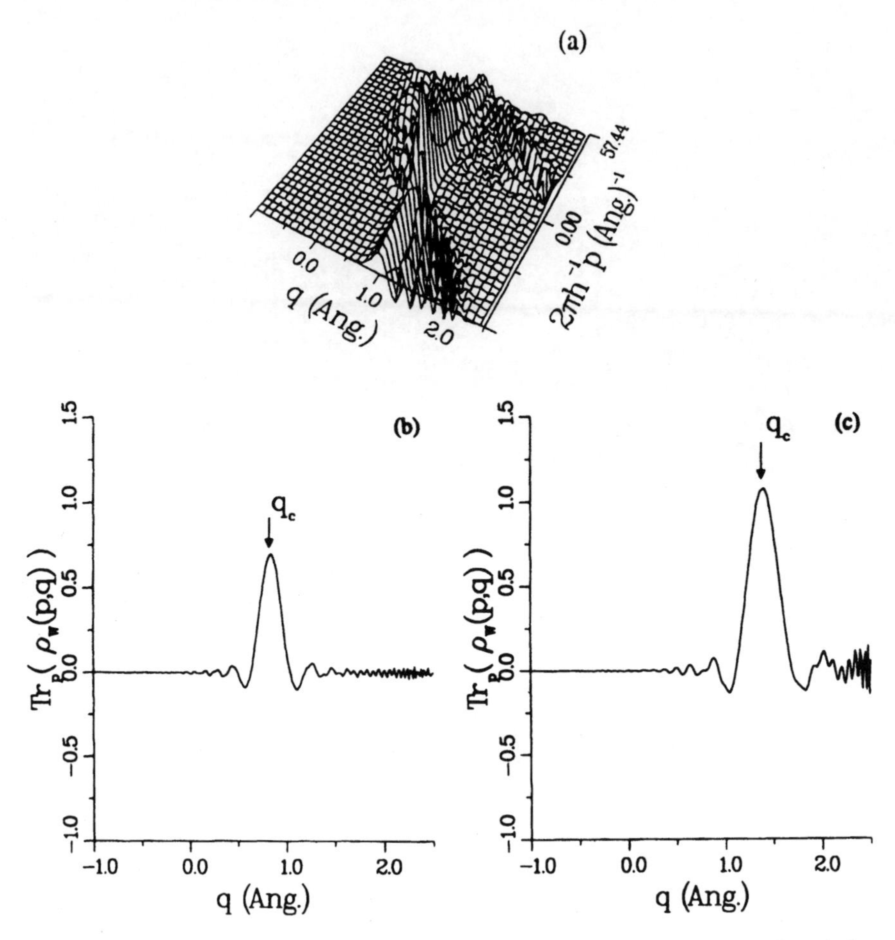

FIG. 13.15 (a) The window Wigner function for the system of Figure 13.14 is displayed. The window shows a ridge in q, surrounded by highly oscillatory regions. (b) The trace over momentum p of the window Wigner functions of (a). It is peaked at the coordinate q_c predicted by the classical Condon approximation, for the given probe frequency. (c) Same as (b) but when the probe is detuned to the blue. The trace shows a clear shift to a larger internuclear separation as the frequency is increased. This Wigner function too is peaked closely to the classical prediction q_c.

with

$$I_{jj}^{\pm}(t,\omega) = 2\,\mathrm{Re}\left[E_j^*(t)\int_0^\infty dt'\,\exp[\pm i(\omega-\omega_j)t']E_j(t\pm t')\right]. \tag{13.49b}$$

Note that only a one-sided Fourier transform enters in Eq. (13.49a). This reflects causality: in the sequential term (S_S) the system interacts first with the pump and then with the probe; therefore, F_S for any given time t contains only the contribution from the pump field prior to that time (I_{11}^-) and from the probe after that time (I_{22}^+).

The effect of the field in the coherent process can be represented by a second joint temporal–spectral density:

$$F_C(t, \omega'', \omega'; \tau) = \int_{-\infty}^{\infty} dt' [I_{21}^{+}(t + t', \omega''; \tau)]^{*} I_{12}^{-}(t, \omega'; \tau), \tag{13.50a}$$

with

$$I_{12}^{-}(t, \omega; \tau) = E_2(t - \tau) \int_0^{\infty} dt' \exp[-i(\omega - \omega_1)t'] E_1(t - t'), \tag{13.50b}$$

$$I_{21}^{+}(t, \omega; \tau) = E_1(t) \int_0^{\infty} dt' \exp[i(\omega - \omega_2)t'] E_2(t - \tau + t'). \tag{13.50c}$$

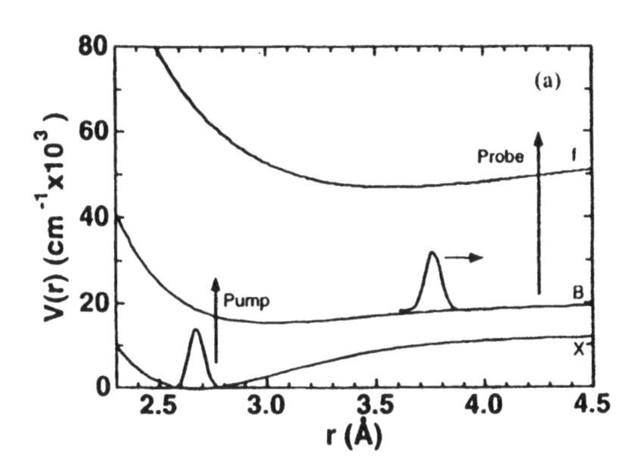

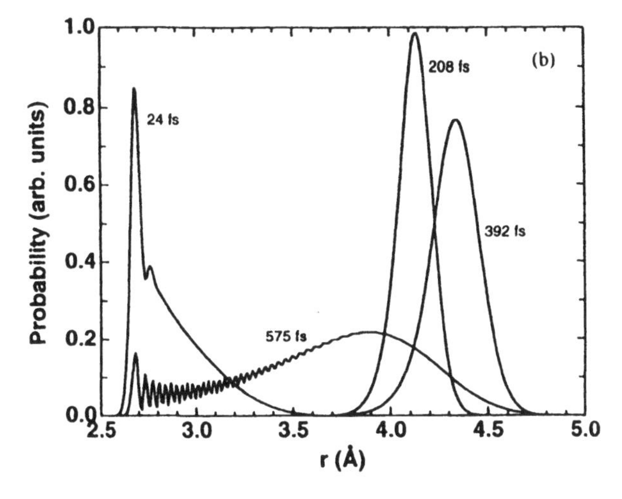

FIG. 13.16 (a) Schematic for iodine pump-probe experiment. Calculations show a wave packet excited on the B state with a 20 fs pulse, with a center wavelength of 570 nm. The wave packet is probed with a second pulse as a function of the delay time between the pump and probe pulses. The probe pulse in this example is also a 20 fs pulse, with a center wave length of 329 nm. At this wavelength, the probe pulse is resonant at 4.25 Å (denoted by the arrow in the schematic), and also at 3.4 Å. The inner resonance is responsible for the small peak at early delay times in (d). (b) Configuration space wave packets $|\psi(r)|^2$ created on the Iodine B state for the conditions described in figure (a). The time labels refer to the delay time between the pump and the probe pulses.

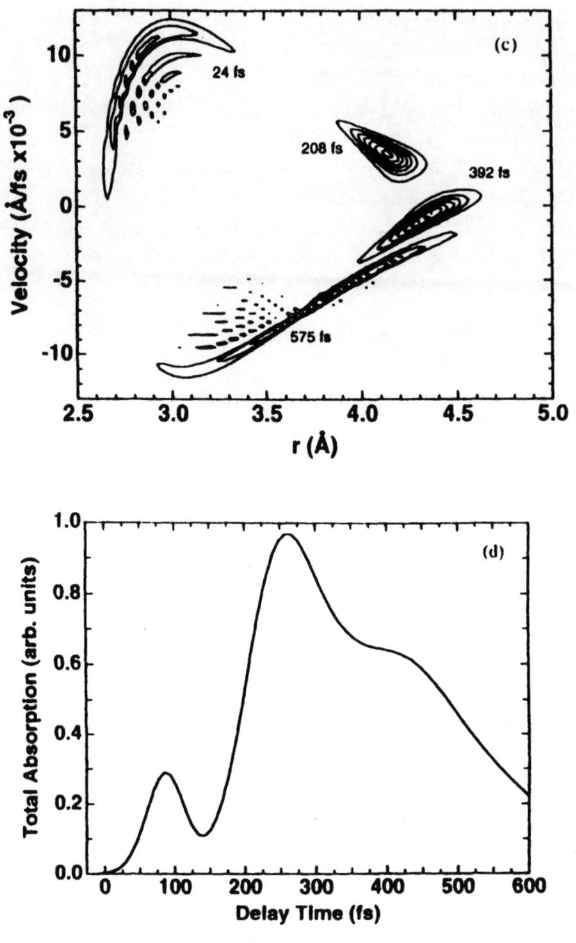

FIG. 13.16 (*continued*) (c) Wigner phase space wave packets $F_w(p, q)$ created on the iodine B state for the conditions described in (a). The time labels refer to the delay time between the pump and the probe pulses. (d) Total absorption spectrum for the conditions described in (a). The x-axis corresponds to the time difference between the peak of the probe and pump pulses. [For details of calculation see J. L. Krause, R. M. Whitnell, K. R. Wilson, and Y. Yan, "Light packet control of wave packet dynamics," in *Femtosecond Chemistry*, J. Manz and L. Wöste, Eds., p. 743 (Verlag Chemie, Weinheim, 1994).]

Using these definitions, the sequential and coherent contributions can then be recast in the form

$$S_S = \int_{-\infty}^{\infty} d\omega'' \int_{-\infty}^{\infty} d\omega' \int_{0}^{\infty} dt\, F_S(t - \tau, \omega'', \omega')M_S(\omega'', t, \omega'), \tag{13.51}$$

$$S_C = 2\,\mathrm{Re} \int_{-\infty}^{\infty} d\omega'' \int_{-\infty}^{\infty} d\omega' \int_{0}^{\infty} dt\, F_C(t, \omega'', \omega'; \tau)M_C(\omega'', t, \omega'). \tag{13.52}$$

Here the double frequency convolution runs over the frequencies of the pump and the probe fields, while the temporal convolution represents the uncertainties in the probe delay due to finite pulse durations.

M_S and M_C are intrinsic material properties (independent of the nature of the pump and probe fields). They represent the material response functions relevant for the sequential and for the coherent processes, respectively. As an example, consider an ideal case in which the pump and the probe are well separated ($S_C = 0$), and both pulses can be approximated as delta functions in time as well as in frequency: $F_S(t-\tau, \omega'', \omega') \propto \delta(t-\tau)\delta(\omega''-\omega_2)\delta(\omega'-\omega_1)$. The probe absorption spectrum [Eq. (13.51)] then reduces to the snapshot spectrum M_S.

Semiclassical picture beyond the snapshot limit: the classical Franck–Condon principle

Finally, we consider the limiting case of classical nuclear dynamics. For simplicity, we restrict our discussion to well-separated pulses, so that we can neglect the coherent term S_C and focus on the sequential S_S contribution. We recall that the relevant t_1 timescale is the inverse linewidth of the $|g\rangle$ to $|e\rangle$ electronic absorption profile. If this linewidth is broader than the vibrational frequencies, we can neglect the changes in the molecular momentum during the $|g\rangle$ to $|e\rangle$ transition. This is the classical Franck–Condon principle introduced in Chapter 7. We further assume that the classical Condon approximation holds also for the $|e\rangle$ to $|f|$ transition. In this case, H_g and H_e, and H_e and H_f commute, and we can approximate the coherence Green functions by [see Eq. (7.21)]

$$\mathcal{G}_{eg}(t_1) \sim \exp\left[-\frac{i}{\hbar}(H_e - H_g)t_1\right] \equiv \exp[-i\omega_{eg}(\mathbf{q})t_1], \tag{13.53a}$$

$$\mathcal{G}_{fe}(t_3) \sim \exp\left[-\frac{i}{\hbar}(H_f - H_e)t_3\right] \equiv \exp[-i\omega_{fe}(\mathbf{q})t_3], \tag{13.53b}$$

where

$$\hbar\omega_{eg}(\mathbf{q}) \equiv H_e - H_g = \hbar\omega_{eg}^0 + W_e - W_g, \tag{13.54a}$$

$$\hbar\omega_{fe}(\mathbf{q}) \equiv H_f - H_e = \hbar\omega_{fe}^0 + W_f - W_e. \tag{13.54b}$$

Substituting these in Eqs. (13.42), setting $V_{ge} = V_{ef} = 1$, and using the Wigner representation we obtain [10]

$$\rho_D(\omega_1, \mathbf{pq}) = i\int_{-\infty}^{\infty} dt' \int_{-\infty}^{\infty} d\omega\, E_1^*(t')\tilde{E}_1(\omega)\exp(-i\omega t')$$
$$\times\, \mathcal{G}_{ee}(t')\frac{1}{\omega_1 + \omega - \omega_{eg}(\mathbf{q}) + i\varepsilon}\rho_g(\mathbf{p},\mathbf{q}) + c.c., \tag{13.55a}$$

$$\rho_W(\omega_2, \mathbf{pq}) = i\int_{-\infty}^{\infty} dt \int_{-\infty}^{\infty} d\omega\, E_2(t)\tilde{E}_2(\omega)\exp(i\omega t)$$
$$\times\, \mathcal{G}_{ee}(t)\frac{1}{\omega_2 + \omega - \omega_{fe}(\mathbf{q}) + i\varepsilon} + c.c. \tag{13.55b}$$

The Lorentzian factors $1/[\omega_1 + \omega - \omega_{eg}(\mathbf{q}) + i\varepsilon]$ and $1/[\omega_2 + \omega - \omega_{fe}(\mathbf{q}) + i\varepsilon]$ result from Fourier transforming the classical propagators [Eqs. (13.53)]; ε is a small positive number that is set to 0 at the end of the calculation.

These expressions hold in the classical limit where the doorway, the window, and the equilibrium density operator ρ_g become classical functions in phase space. In addition, the evolution of the doorway state during the delay time τ, represented by $\mathcal{G}_{ee}(\tau)$, is classical. In physical terms, we expect our formulas to be valid for doorway and window wavepackets that include many nuclear states. This could be either a consequence of short pulses or fast dephasing processes leading to broad spectral lines. We interpret Eqs. (13.55) as follows: The double integral over time and frequency reflects the dual role the fields play in pump-probe spectroscopy. The frequency integral selects a distribution of phase space points weighted by the laser bandwidth. A spectrally broad laser will select a broad distribution in phase space. During the time integration, which is controlled by pulse durations, the distribution is broadened by nuclear dynamics. Hence, a very long pulse leads to doorway and window wavepackets that are broad functions of $\mathbf{q}$ and $\mathbf{p}$. It is also important to note that the fields enter at the amplitude level in Eqs. (13.55), and their phases affect the nature of the wavepacket. This is to be contrasted with Eqs. (13.21) and (13.22) where the fields enter through their intensity profiles.

Additional simplifications can be made if certain relations between pulse durations and the system time scales are obeyed, which lead to the snapshot limit. In complete analogy with our previous analysis of the two level case [see Eqs. (13.8) and (13.9)], we assume the following two simplifying conditions.

1. The duration of the pump and probe pulses is short compared to the time scale of nuclear motion (i.e., $W_e - W_g$ or $W_f - W_e$ is slowly varying on the time scale of the pulse). In this case, we set $\mathcal{G}_{ee}(t)$ and $\mathcal{G}_{ee}(t')$ to 1 in Eqs. (13.55).
2. The pump (probe) pulse is long compared to the inverse linewidth of the pump (probe) absorption (i.e., pulse long compared to dephasing). We can then set $\omega_1 + \omega$ to ω_1 and set $\omega_2 + \omega$ to ω_2.

As indicated earlier, these two conditions are not mutually exclusive, and when both conditions hold we obtain the snapshot limit, in which the doorway and window wavepackets (and hence the spectrum) no longer depend on the pulse envelopes $E_1(t)$ and $E_2(t)$. We then have

$$\rho_D^0(\omega_1) \propto \delta[\omega_1 - \omega_{eg}(\mathbf{q})]\rho_g, \tag{13.56a}$$

$$\rho_W^0(\omega_2) \propto \delta[\omega_2 - \omega_{fe}(\mathbf{q})]. \tag{13.56b}$$

Equation (13.43) then becomes [see Eq. (13.27)]

$$\begin{aligned} S_0(\omega_1, \omega_2; \tau) &= \mathrm{Tr}\{\rho_W^0(\omega_2)\mathcal{G}_{ee}(\tau)\rho_D^0(\omega_1)\} \\ &= \mathrm{Tr}\{\delta[\omega_2 - \omega_{fe}(\mathbf{q})]\mathcal{G}_{ee}(\tau)\delta[\omega_1 - \omega_{eg}(\mathbf{q})]\rho_g\}. \end{aligned} \tag{13.57}$$

Equation (13.57) corresponds to the classical Franck–Condon picture, where the absorption of ω_1 and ω_2 takes place only at nuclear configurations where $\omega_{eg}(\mathbf{q}) = \omega_1$ and $\omega_{fe}(\mathbf{q}) = \omega_2$, respectively. Other limiting cases discussed earlier for the two-level model apply here as well. If only condition (1) holds, the doorway and window function are broadened by the spectral width of the laser pulses and we can use Eq. (13.19). This equation can be alternatively recast in that form

$$S_{\mathrm{PP}}(\omega_1, \omega_2, \tau) = \int d\mathbf{p}\, d\mathbf{q}\, |\tilde{E}_2[-\Delta_2(\mathbf{q}[\tau])]|^2 |\tilde{E}_1[-\Delta_1(\mathbf{q})]|^2 \rho_{\mathrm{g}}(\mathbf{p}, \mathbf{q}), \quad (13.58\mathrm{a})$$

where

$$\Delta_1(\mathbf{q}) \equiv \omega_1 - \omega_{\mathrm{eg}}^0 + W_{\mathrm{g}}(\mathbf{q}) - W_{\mathrm{e}}(\mathbf{q}), \quad (13.58\mathrm{b})$$

$$\Delta_2(\mathbf{q}) \equiv \omega_2 - \omega_{\mathrm{fe}}^0 + W_{\mathrm{e}}(\mathbf{q}) - W_{\mathrm{f}}(\mathbf{q}), \quad (13.58\mathrm{c})$$

are the detunings of point $\mathbf{q}$ from the g $\to$ e and e $\to$ f resonances, respectively. $\mathbf{q}[\tau]$ is the coordinate of the phase point $(\mathbf{q}, \mathbf{p})$ at time τ, propagated by classical mechanics.

Additional insight may be obtained by considering a single trajectory approximation to Eq. (13.58). This may be derived by replacing $\rho_{\mathrm{g}}(\mathbf{p}, \mathbf{q})$ with a delta function: $\rho_{\mathrm{g}}(\mathbf{p}, \mathbf{q}) \to \delta(\mathbf{q} - \mathbf{q}_0)\delta(\mathbf{p})$. We then obtain

$$S_{\mathrm{PP}}(\omega_1, \omega_2, \tau) \propto |\tilde{E}_2(-\Delta_2(\mathbf{q}[\tau]))|^2. \quad (13.59)$$

This approximation is valid for classical processes where the pulse is short compared to nuclear dynamics, and where the finite width of the ground-state distribution in phase space may be ignored.

Equation (13.59) was found to accurately reproduce the ICN photodissociation experiment [23], where the linewidth of a few thousand cm^{-1} corresponds to t_1, t_3 timescales of $\sim$10 fs. The relevant molecular dynamics (in this case the bond breaking) takes place in $\sim$200 fs, and the pulse is indeed short compared with nuclear dynamics.

If only condition (2) holds, the spectrum depends on the temporal profile of the laser pulses, and the probe absorption can once again be expressed as a convolution involving the snapshot spectrum [Eq. (13.20)]. Alternatively, we can write the probe absorption in this case as

$$S(\omega_1, \omega_2, \tau) = \iint d\mathbf{p}\, d\mathbf{q} \frac{I(t^* - \tau)\rho_{\mathrm{g}}(\mathbf{p}, \mathbf{q})}{|W_{\mathrm{f}}'(\mathbf{q}[t^*]) - W_{\mathrm{e}}'(\mathbf{q}[t^*])||\dot{\mathbf{q}}[t^*]||W_{\mathrm{e}}'(\mathbf{q}) - W_{\mathrm{g}}'(\mathbf{q})|} \quad (13.60)$$

where $\mathbf{q}$ is the point in resonance with the g $\to$ e transition, $W_n' \equiv \partial W_n/\partial \mathbf{q}$, and t^* is the time at which a trajectory starting at $(\mathbf{q}, \mathbf{p})$ becomes resonant with the e $\to$ f transition: $\Delta_1(\mathbf{q}) = 0$, $\Delta_2(\mathbf{q}[t^*]) = 0$. It should be noted that Eq. (13.60) becomes singular whenever the probe absorption is resonant with a classical turning point. If we replace $\rho_{\mathrm{g}}(\mathbf{p}, \mathbf{q})$ by a delta function, we obtain $\sigma(\omega_1, \omega_2, \tau) \propto I(t^* - \tau)$. In other words, the spectral profile as a function of τ is given by the combined pump-probe intensity I. Changing the probe frequency ω_2 has the effect of changing the t^* value, thus modifying the peak of the spectrum as a function of τ, but leaving the shape invariant.

The present analysis illustrates the physical insight provided by the Liouville space formulation into the classical description of nuclear dynamics, even in the case of a single degree of freedom at zero temperature. It would be a tedious task to recover these results, that appear here so naturally, using a Hilbert space (wavefunction) representation. The Liouville space representation shows that in addition to the sequential contribution we need to consider the coherent contribution S_C, which is important when the pump and the probe pulses overlap. This contribution does not have a simple classical analogue. By using the wavefunction, both contributions $S_S + S_C$ are lumped together, and the derivation and significance of the classical result are much less transparent. The Liouville-space analysis shows the range of validity of the classical picture and the approximations made. One of the most attractive aspects of the classical Franck–Condon principle is that it allows the direct inversion of experimental spectra to obtain the intermolecular potentials. This may be done since in this approximation at every nuclear configuration the system absorbs only one frequency. The exact expressions for the doorway and the window function show the limitations of this procedure, and point out practical ways for improvements by a more rigorous calculation of the doorway and the window functions.

APPENDIX 13A

Wavepacket Representation of Pump-Probe Spectroscopy with Frequency Dispersed Detection

The frequency-dispersed detection scheme [Eq. (4.92)] leads to an expression similar to Eqs. (13.3)–(13.5) with the window operators $W_e(\omega_2)$ and $W_g(\omega_2)$ replaced by $W'_e(\omega_2; \omega_2)$ and $W'_g(\omega_2; \omega_2)$, respectively,

$$W'_e(\omega_2; \omega'_2) = \tilde{E}_2^*(\omega'_2) \int_{-\infty}^{\infty} dt \int_0^{\infty} dt_3\, E_2(t) \exp[i(\omega'_2 - \omega_2)t + i\omega_2 t_3] \times \left[\exp\left(\frac{i}{\hbar} H_e t\right) \mu \exp\left(\frac{i}{\hbar} H_g t_3\right) \mu \exp\left(-\frac{i}{\hbar} H_e t_3\right) \exp\left(-\frac{i}{\hbar} H_e t\right) \right] + h.c. \tag{13.A.1}$$

$$W'_g(\omega_2; \omega'_2) = \tilde{E}_2^*(\omega'_2) \int_{-\infty}^{\infty} dt \int_0^{\infty} dt_3\, E_2(t) \exp[i(\omega'_2 - \omega_2)t + i\omega_2 t_3] \times \left[\exp\left(\frac{i}{\hbar} H_g t\right) \exp\left(\frac{i}{\hbar} H_g t_3\right) \mu \exp\left(-\frac{i}{\hbar} H_e t_3\right) \mu \exp\left(-\frac{i}{\hbar} H_g t\right) \right] + h.c. \tag{13.A.2}$$

APPENDIX 13B

Wavepacket Representation of Transient Grating with Heterodyne Detection

The wavepacket formulation also applies for a general sequential experiment with heterodyne detection; to that end we assume that

1. E_1 can represent a sum of two pulses, and
2. the factor $E_2^*(t + t_3)$ in Eqs. (13.4) can be replaced by $E_3^*(t + t_3)$, where E_3 represents the heterodyne field. We then have

$$S_{PP}(\omega_1, \omega_2; \tau) = \mathrm{Tr}[W_e(\omega_2)\mathcal{G}_{ee}(\tau)D_e(\omega_1)] + \mathrm{Tr}[W_g(\omega_2)\mathcal{G}_{gg}(\tau)D_g(\omega_1)], \tag{13.B.1}$$

where

$$D_e(\omega_1) = \int_{-\infty}^{\infty} dt' \int_0^{\infty} dt_1\, E_2^*(t')E_1(t' - t_1)\exp(i\omega_1 t_1) \times \left[\exp\left(\frac{i}{\hbar}H_e t'\right)\exp\left(-\frac{i}{\hbar}H_e t_1\right)V_{eg}\rho_g \exp\left(\frac{i}{\hbar}H_g t_1\right)V_{ge}\exp\left(-\frac{i}{\hbar}H_e t'\right)\right] + h.c., \tag{13.B.2a}$$

$$D_g(\omega_1) = \int_{-\infty}^{\infty} dt' \int_0^{\infty} dt_1\, E_2^*(t')E_1(t' - t_1)\exp(i\omega_1 t_1) \times \left[\exp\left(\frac{i}{\hbar}H_g t'\right)V_{ge}\exp\left(-\frac{i}{\hbar}H_e t_1\right)V_{eg}\rho_g \exp\left(\frac{i}{\hbar}H_g t_1\right)\exp\left(-\frac{i}{\hbar}H_g t'\right)\right] + h.c., \tag{13.B.2b}$$

$$W_e(\omega_2) = \int_{-\infty}^{\infty} dt \int_0^{\infty} dt_3\, E_{LO}^*(t + t_3)E_3(t)\exp(i\omega_2 t_3) \times \left[\exp\left(\frac{i}{\hbar}H_e t\right)V_{eg}\exp\left(\frac{i}{\hbar}H_g t_3\right)V_{ge}\exp\left(-\frac{i}{\hbar}H_e t_3\right)\exp\left(-\frac{i}{\hbar}H_e t\right)\right] + h.c., \tag{13.B.3a}$$

$$W_g(\omega_2) = \int_{-\infty}^{\infty} dt \int_0^{\infty} dt_3\, E_{LO}^*(t + t_3)E_3(t)\exp(i\omega_2 t_3) \times \left[\exp\left(\frac{i}{\hbar}H_g t\right)\exp\left(\frac{i}{\hbar}H_g t_3\right)V_{ge}\exp\left(-\frac{i}{\hbar}H_e t_3\right)V_{eg}\exp\left(-\frac{i}{\hbar}H_g t\right)\right] + h.c. \tag{13.B.3b}$$

NOTES

1. C. H. Brito-Curz, R. L. Fork, W. H. Knox, and C. V. Shank, *Chem. Phys. Lett.* **132**, 341 (1986); R. A. Mathies, C. H. Brito Cruz, W. T. Pollard, and C. V. Shank, *Science* **240**, 777 (1988); H. L. Fragnito, J. Y. Bigot, P. C. Becker, and C. V. Shank, *Chem. Phys. Lett.* **160**, 101 (1989).

2. M. Mitsunaga and C. L. Tang, *Phys. Rev. A* **35**, 1720 (1987); I. A. Walmsley, M. Mitsunaga, and C. L. Tang, *Phys. Rev. A* **38**, 4681 (1988).

3. M. J. Rosker, M. Dantus, and A. H. Zewail, *Science* **241**, 1200 (1988); M. Dantus, M. J. Rosker, and A. H. Zewail, *J. Chem. Phys.* **89**, 6113 (1988); L. R. Khundkar and A. H. Zewail, *Ann. Rev. Phys. Phys. Chem.* **41**, 15 (1990); A. H. Zewail, *Faraday Discuss. Chem. Soc.* **91**, 207 (1991); A. H. Zewail, *J. Phys. Chem.* **97**, 12427 (1993).

4. T. Tokizaki, Y. Ishida, and T. Yajima, *Opt. Commun.* **71**, 355 (1989).

5. J. H. Glownia, J. A. Misewich, and P. P. Sorokin, *J. Chem. Phys.* **92**, 3335 (1990); R. E. Walkup, J. A. Misewich, J. H. Glownia, and P. P. Sorokin, *J. Chem. Phys.* **94**, 3389 (1991).

6. J. Friedrich and D. Haarer, *J. Chem. Phys.* **76**, 61 (1982); J. Friedrich, J. D. Swalen, and D. Haarer, *J. Chem. Phys.* **73**, 705 (1980); G. Shulta, W. Grond, D. Haarer, and R. J. Silbey, *J. Chem. Phys.* **88**, 679 (1988).

7. S. Volker, *Annu. Rev. Phys. Chem.* **40**, 499 (1989).

8. M. Berg, C. A. Walsh, L. R. Narasimhare, K. A. Littau, and M. D. Fayer, *J. Chem. Phys.* **88**, 1564 (1988); L. R. Narasimhan, K. A. Littau, D. W. Pack, Y. S. Bai, A. Elschner, and M. D. Fayer, *Chem. Rev.* **90**, 439 (1990).

9. S. Kinoshita, N. Nishi, and T. Kushida, *Chem. Phys. Lett.* **124**, 605 (1987).

10. J. Manz and L. Wöste, Eds., *Femtosecond Chemistry* (Verlag Chemie, Weinheim, 1994).

11. Y. J. Yan and S. Mukamel, *Phys. Rev. A* **41**, 6485 (1990).

12. R. Kosloff, S. A. Rice, P. Gaspard, S. Tersigni, and D. J. Tannor, *Chem. Phys.* **139**, 201 (1989); C. Leforestier, R. H. Bisseling, C. Cerjan, M. D. Feit, R. Friesner, A. Guldberg, A. Hammerich, G. Jolicard, W. Karrlein, H.-D. Meyer, N. Lipkin, O. Roncero, and R. Kosloff, *J. Comput. Phys.* **94**, 59 (1991).

13. S. E. Choi and J. C. Light, *J. Chem. Phys.* **90**, 2593 (1989); V. Engel and H. Metiu, *J. Chem. Phys.* **90**, 6116 (1989); R. Kosloff, A. D. Hammerich, D. Tannor, *Phys. Rev. Lett.* **69**, 2172 (1992).

14. R. F. Loring, Y. J. Yan, and S. Mukamel, *J. Chem. Phys.* **87**, 5840 (1987).

15. Yu T. Mazurenko and V. S. Udaltsov, *Opt. Spectrosc.* **44**, 417 (1977); Mataga.

16. M. Maroncelli and G. R. Fleming, *J. Chem. Phys.* **89**, 5044 (1988); M. Maroncelli, *J. Mol. Liq.* **57**, 1 (1993).

17. P. F. Barbara and W. Jarzeba, *Adv. Photochem.* **15**, 1 (1990).

18. W. B. Mims, *Phys. Rev.* **168**, 370 (1968); J. R. Klauder and P. W. Anderson, *Phys. Rev.* **125**, 912 (1962).

19. P. W. Anderson, B. I. Halperin, and C. M. Varma, *Phil. Mag.* **25**, 1 (1972); W. A. Phillips, *J. Low. Temp. Phys.* **7**, 351 (1972).

20. S. Saikan, J. W-I. Lin, and H. Nemoto, *Phys. Rev. B* **46**, 7123 (1992).

21. W. E. Moerner and T. Basche, *Angew. Chem.* **105**, 537 (1993); W. E. Moerner, *Science*, **265**, 46 (1994).

22. B. M. Kharlamov, R. I. Personov, and L. A. Bykovskaya, *Opt. Commun.* **12**, 191 (1974); M. Orrit, J. Bernard and R. I. Personov, *J. Phys. Chem.* **97**, 10256 (1993).

23. Y. J. Yan, L. E. Fried, and S. Mukamel, *J. Phys. Chem.* **93**, 8149 (1989); S. Mukamel, *Annu. Rev. Phys. Chem.* **41**, 647 (1990); L. E. Fried and S. Mukamel, *J. Chem. Phys.* **93**, 3063 (1990).

24. I. Benjamin and K. R. Wilson, *J. Chem. Phys.* **90**, 4176 (1989).

25. Y. J. Yan, *Chem. Phys. Lett.* **198**, 43 (1992); Y. J. Yan, R. M. Wintell, K. R. Wilson, and A. H. Zewail, *Chem. Phys. Lett.* **193**, 402 (1992).

CHAPTER 14

Off-Resonance Raman Scattering

Off-resonant Raman spectroscopies constitute a simple class of nonlinear techniques related to the third-order polarization $P^{(3)}$ [1–6]. We showed in Chapter 5 that the third-order response may contain single-photon, two-photon, and three-photon resonances. In the Raman process we are looking at two-photon resonances, whereby a difference of two field frequencies is resonant with a vibrational transition, $\omega_1 - \omega_2 = \omega_{vv'}$. The Raman resonance is thus generated during the t_2 evolution period, as demonstrated in Chapter 9. The technique is "off-resonant" when all the laser frequencies involved (ω_j) are detuned far from any electronic transition (ω_{eg}).

Off-resonant measurements are attractive for the following reasons: (1) Excited-state populations are limited by the Heisenberg relation to very short times $\Delta t \approx 1/\Delta\omega$, $\Delta\omega$ being the detuning from resonance. As $\Delta\omega$ is increased, these populations become negligible and the technique probes only ground-state dynamics. (2) The time the system spends in an electronic coherence is also limited by the Heisenberg relation. Consequently, nuclear dynamics can be neglected during the coherence periods t_1 and t_3, and all the interesting dynamics takes place during t_2 [see Eq. (7.11)]. The ability to focus on a single (ground-state) propagation period greatly simplifies the interpretation of off-resonant measurements. (3) Raman resonances are not affected by electronic dephasing since the t_2 propagation period corresponds to an electronic population ($\mathcal{G}_{gg}$) rather than a coherence ($\mathcal{G}_{eg}$). Consequently, these resonances are usually very narrow and contain most valuable information regarding vibrational dynamics, including frequencies, vibrational relaxation, and dephasing rates. (Electronic dephasing does enter in resonant Raman excitation profiles as discussed in Chapter 9, but not in the Raman resonances themselves.) On the other hand, simplifying assumptions such as the rotating wave approximation and the approximation that the dipole moment is independent of nuclear coordinates (the Condon approximation), which usually apply in resonant measurements, cannot be justified under off-resonant conditions. Near resonance, excited state dynamics represented by $\mathcal{G}_{ee}(t_2)$ contributes in addition to the ground state evolution $\mathcal{G}_{gg}(t_2)$. The Raman signal is then usually accompanied by fluorescence, and its interpretation becomes more complicated. This situation was discussed in Chapter 9.

Raman measurements can be carried out on a broad range of timescales, ranging from the continuous wave mode to the femtosecond regime. In the frequency domain, spontaneous Raman and its coherent analogues (CARS and CSRS) [5] show up as specific resonances. All the information regarding vibrational frequencies and relaxation processes is then contained in the positions and widths of these resonances. In the other extreme, Raman experiments are performed using ultrafast (femtosecond) laser pulses that are short compared with the relevant vibrational periods and relaxation times (i.e., inverse line broadening). In these impulsive techniques, developed in the 1980s, all the information is in the time domain: vibrational frequencies show up as coherent oscillations, and relaxation processes appear as a decay of the time-dependent signal [7–9]. Impulsive Raman techniques were applied to study a variety of systems including acoustic phonons in glasses, optical phonons in molecular crystals, and intramolecular, orientational, and intermolecular motions in neat molecular liquids. Figure 14.1 shows coherent vibrational motions in liquid CH_2Br_2, observed using an off-resonant Raman experiment. The transmitted probe intensity shows a 173 cm^{-1} oscillation corresponding to a bending mode fundamental. This reflects a periodic spectral shift in the transmitted probe field as the delay time τ is varied. The figure clearly demonstrates that the phase of the beats depends strongly on the optical frequency of observation. Figure 14.2a shows the nuclear part of the heterodyne detected Raman transient for iodobenzene liquid. The nuclear and electronic contributions to the transient were separated using Fourier transform methods. Figure 14.2b shows the imaginary part of the

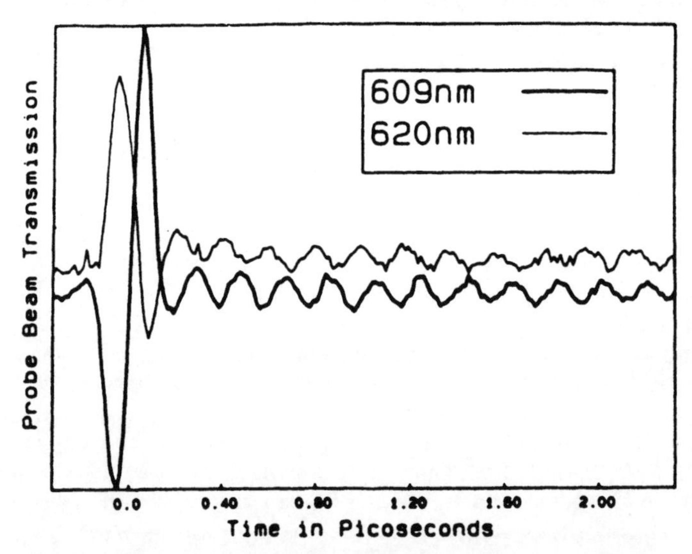

FIG. 14.1 Impulsive heterodyne detected Raman signal from the 173 cm^{-1} Br–C–Br "bending" vibrational mode in CH_2Br_2. A single excitation pulse was used, and the time-dependent intensities of red (620 nm) and blue (609 nm) frequency components of the transmitted probe pulse were measured. The spectrum of the transmitted probe alternates from red to blue shifting at the vibrational frequency. This causes the intensities of the two spectral components to undergo antiphased oscillations. [From K. A. Nelson and E. P. Ippen, "Femtosecond coherent spectroscopy," *Adv. Chem. Phys.* **75**, 1 (1989).]

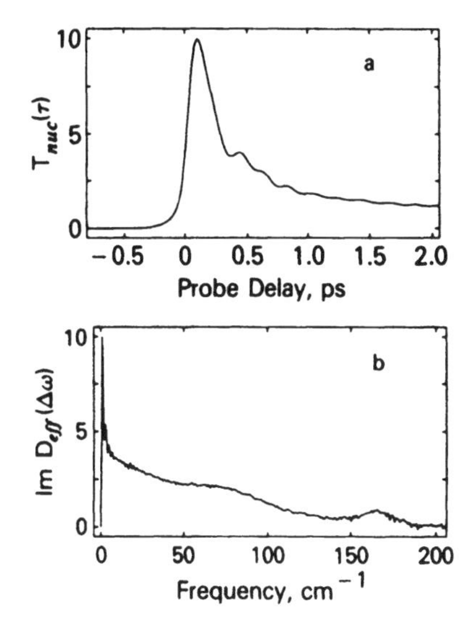

FIG. 14.2 (a) Nuclear part of the heterodyne detected Raman (optical Kerr) transient for iodobenzene liquid measured at 298°K, and (b) its corresponding spectral density representation. The nuclear and electronic contributions to the transient were separated using Fourier transform methods. Generation of the spectral density includes a deconvolution step that eliminates the spectral-filter effects of the finite-duration excitation and probing pulses. [D. McMorrow, *Opt. Commun.* **86**, 236 (1991); D. McMorrow and W. T. Lotshaw, "The frequency response of condensed-phase media to femtosecond optical pulse: Spectral filter effects," *Chem. Phys. Lett.* **174**, 85 (1990).]

impulsive frequency response function for the data of Figure 14.2a. This spectrum contains all of the information on the Raman-active nuclear dynamics of the liquid for the frequency range shown. Evident is a sharp, low-frequency resonance that is associated with diffusive reorientational dynamics, a broad band with maxima at 15 and 70 cm^{-1} extending to $\sim$150 cm^{-1} that can be identified with nondiffusive intermolecular degrees of freedom, and a narrow resonance at 170 cm^{-1} that arises from the intramolecular iodine bending mode (the oscillations in Figure 14.2a can be identified with the 170 cm^{-1} mode of Figure 14.2b). Picosecond Raman techniques are intermediate between continuous wave and femtosecond techniques [10, 11]. Picosecond pulses are typically long compared with vibrational periods but short compared with inverse vibrational linewidths. In picosecond Raman spectroscopy, pioneered by Kaiser and co-workers, the vibrational modes are selected in the frequency domain, but their relaxation appears in the time domain.*

In this chapter we apply the semiclassical Liouville space wavepacket picture of the nonlinear response function developed in Chapter 13 to the description of off-resonant Raman scattering spectroscopies. Raman techniques can be performed also using either homodyne or heterodyne detection. For clarity we restrict the discussion in this chapter to the quadratic (homodyne) detection. However, the Raman polarization calculated in this chapter can be used to calculate the heterodyne signal as well [see Eq. (14.51)]. A discussion of the various detection modes will be given in the next chapter. We adopt a dynamic approach and obtain two alternative expressions for coherent off-resonant Raman scattering. The signal is first expressed as the expectation value of the molecular polarizability, calculated using the nonequilibrium time-dependent doorway state prepared by the pump field. This expression can then be rearranged in a different form, whereby the molecular information is recast in terms of equilibrium correlation functions. We show that the continuous wave

* A more precise terminology would be to denote these as "intermediate" Raman techniques (between the impulsive and frequency domain limits). Depending on the particular vibrational frequency, a picosecond Raman can be impulsive and a femtosecond Raman can be intermediate! For historical reasons we retain the terminology of picosecond and femtosecond Raman.

(frequency-domain) spontaneous and the coherent Raman lineshapes can be expressed in terms of the correlation function and the linear response function, respectively, of the electronic polarizability. Comparison is made with our previous treatment of Chapter 9. We thus have at our disposal two basic viewpoints of the scattering process. We either express the signal in terms of an *expectation value with respect to a nonequilibrium doorway wavepacket* or as an *equilibrium two-time correlation function* of the polarizability. In the former approach, we break the process into preparation, evolution, and detection stages, as implied by the doorway–window wavepacket picture. In the latter form the entire process is treated as a single scattering event. The expressions derived in this chapter hold for an arbitrary time profile of the incoming field, and an arbitrary coordinate dependence of the electronic transition dipole; this serves to provide a unified description of stationary, femtosecond, and picosecond Raman techniques. We then make use of the weak dependence of electronic wavefunctions on nuclear coordinates, and expand the electronic polarizability in a Taylor series in these coordinates. To lowest order, the polarizability depends linearly on nuclear coordinates, and we recover the linearly driven Brownian harmonic oscillator model that has been widely used in the analysis of impulsive Raman scattering. Pulse shaping is shown to result in frequency filtering of the Raman active modes, and the temporal profile of the scattered field is studied. Next we show that in an off-resonant Raman scattering process, the molecular dynamics are controlled by an effective Hamiltonian [Eq. (14.34)], and demonstrate how, using the effective Hamiltonian, the doorway wavepacket can be calculated to an arbitrary order in the incoming fields. We can thus generalize the previous expressions to include strong field effects. When the doorway state is expanded to second order in the pump, we recover the correlation function expression obtained using $P^{(3)}$. We further examine the doorway wavepacket obtained using the effective Hamiltonian, and discuss the role of pulse shaping and Fourier transform relationships among Raman techniques and impulsive pump-probe spectroscopy. Finally we consider the extension of these ideas to off-resonant multipulse measurements. These constitute an N-dimensional vibrational spectroscopy, which can be applied, e.g., as a direct probe of intermolecular modes in liquids.

DYNAMIC APPROACH TO COHERENT RAMAN SCATTERING

Coherent Raman scattering is often carried out in a *transient grating* configuration, in which two simultaneous pump pulses with center frequencies ω_1 and ω_2 and wavevectors $\mathbf{k}_1$ and $\mathbf{k}_2$, first interact with the system, creating a dynamic grating with wavevector $\mathbf{k}_1 - \mathbf{k}_2$. After a delay period τ, a third pulse with frequency ω_3 and wavevector $\mathbf{k}_3$ is scattered from the sample. The coherent Raman scattering signal with wavevector $\mathbf{k}_s = \mathbf{k}_3 + \mathbf{k}_1 - \mathbf{k}_2$ corresponding to the Bragg diffraction off the grating is detected.* The formulation of this section [12] applies to the time-domain as well as frequency-domain Raman techniques.

* In many applications $\mathbf{k}_3 = \mathbf{k}_1$ and the signal is generated at $2\mathbf{k}_1 - \mathbf{k}_2$. For the sake of clarity, and to highlight the role of the various pulses, we retain the more general wavevector.

The electric field in a coherent Raman experiment is given by

$$E(\mathbf{r}, t) = E_1(t + \tau)\exp(i\mathbf{k}_1\mathbf{r} - i\omega_1 t) + E_2(t + \tau)\exp(i\mathbf{k}_2\mathbf{r} - i\omega_2 t) + E_3(t)\exp(i\mathbf{k}_3\mathbf{r} - i\omega_3 t) + c.c. \tag{14.1}$$

This is similar to Eq. (11.4) with a few differences. We interchange the labels E_2 and E_3 in order to keep the conventional notation. Also pulses 1 and 2 are coincident in time and centered around $t = -\tau$ so that $\tau' = 0$, whereas pulse 3 is centered around $t = 0$. We further set the phase $\phi = 0$.

We shall calculate the transient Raman scattering process when the pulses are well separated. The first pair of pulses prepares the system in a nonequilibrium state, which the third pulse then detects. This situation is analogous to the doorway–window picture of pump-probe spectroscopy, described in Chapter 13. The following derivation proceeds along the same steps used in Chapter 11 for a general sequential process, except that we adopt the homodyne detection scheme (rather than the heterodyne detection relevant for pump-probe). Starting with Eq. (11.3) and making the necessary changes, we have

$$S_{\text{CRS}}(\tau) = \int |P^{(3)}(\mathbf{k}_s, t)|^2\, dt, \tag{14.2a}$$

with

$$\begin{aligned} P^{(3)}(\mathbf{k}_s, t) = \int_0^\infty dt_1 \int_0^\infty dt_2 \int_0^\infty dt_3\, S^{(3)}(t_3, t_2, t_1) \\ \times \{E_3(t - t_3)E_2^*(t + \tau - t_2 - t_3)E_1(t + \tau - t_1 - t_2 - t_3) \\ \times \exp[i\omega_s t_3 + i\omega_1 t_1 + i(\omega_1 - \omega_2)t_2] \\ + E_3(t - t_3)E_2^*(t + \tau - t_1 - t_2 - t_3)E_1(t + \tau - t_2 - t_3) \\ \times \exp[i\omega_s t_3 - i\omega_2 t_1 + i(\omega_1 - \omega_2)t_2]\}. \end{aligned} \tag{14.2b}$$

Equations (14.2) contains 16 terms arising from the 8 terms of $S^{(3)}$ and the two field permutations. We need only two permutations of the fields, since we maintain a partial control over time ordering, as described in Chapter 11. We next introduce the following simplifications: (1) We keep only ground state contributions to $S^{(3)}$. (2) For off-resonant processes, t_1 and t_3 are very short. We assume that they are short compared with the field durations, so that we can neglect these arguments in the fields. This situation corresponds to pulses long compared with dephasing but not necessarily short compared with nuclear dynamics [see Eq. (13.22)]. In addition, we shall use the doorway–window form of the nonlinear response function [Eqs. (12.50)].* Putting all of these together,

* Strictly speaking, under off-resonant conditions we need to consider many excited electronic states and the present two-level model is not realistic. However, all our results apply to a multilevel system as well. All we need to introduce a summation in the polarizability α [Eq. (14.4b)] over all excited states e, $\sum_e$. For clarity, we, therefore, keep the two-level model.

we get

$$P^{(3)}(\mathbf{k}_s, t) = \frac{i}{\hbar^3} E_3(t) \int_0^\infty dt_2 \exp[i(\omega_1 - \omega_2)t_2] E_1(t + \tau - t_2) E_2^*(t + \tau - t_2)$$
$$\times \langle\langle \rho'_W(\omega_s) - \rho'^{\dagger}_W(\omega_s) | \mathscr{G}_{gg}(t_2) | \rho'_D(\omega_1) + \rho'_D(-\omega_2) + \rho'^{\dagger}_D(\omega_1) + \rho'^{\dagger}_D(-\omega_2) \rangle\rangle. \tag{14.3}$$

The frequency-dependent wavepackets were defined in Eq. (12.53), and $\omega_s = \omega_1 + \omega_3 - \omega_2$. Equation (14.3) provides a sequential description of the Raman process and separates it into a preparation of a doorway state, followed by a ground-state evolution period, and a detection through the overlap with the window wavepacket. This equation depends on the doorway and the window wavepackets at the frequencies ω_1, ω_2, and ω_s. Under off-resonant conditions, this dependence is very weak and we can safely replace all of these frequencies by some average frequency say $\omega_p \equiv (\omega_1 + \omega_2)/2$. In impulsive (femtosecond) Raman techniques, $\omega_1 = \omega_2$ so that this substitution is exact. Note that we make this approximation only during the t_1 and the t_3 periods and we still retain the $\exp[-i(\omega_1 - \omega_2)t_2]$ term in Eq. (14.3), since it represents two-photon resonances. Equations (14.3) can now be expressed using the snapshot wavepackets introduced in Eqs. (13.11) and (13.12), i.e.,

$$P^{(3)}(\mathbf{k}_s, t) = \frac{i}{\hbar} E_3(t) \int_0^\infty dt_2 \exp[i(\omega_1 - \omega_2)t_2] E_1(t + \tau - t_2) E_2^*(t + \tau - t_2)$$
$$\times \langle\langle W_g^0(\omega_p) | \mathscr{G}_{gg}(t_2) | D_g^0(\omega_p) + D_g^0(-\omega_p) \rangle\rangle,$$

where

$$\hbar D_g^0(\omega_p) = \rho'_D(\omega_p) + \rho'^{\dagger}_D(\omega_p)$$

and

$$\hbar W_g^0(\omega_p) = \rho'_W(\omega_p) - \rho'^{\dagger}_W(\omega_p).$$

Equation (14.3) can also be recast in an alternative form that treats the entire Raman process as a single event. To that end, we introduce the polarizability operator at the average frequency*

$$\hbar\alpha \equiv V_{ge} \mathscr{G}_{eg}(\omega_p) V_{eg} + V_{ge} \mathscr{G}_{eg}(-\omega_p) V_{eg}, \tag{14.4a}$$

i.e.,

$$\alpha = \frac{|V_{ge}|^2}{\hbar} \left[\frac{1}{\omega_p - \omega_{eg}} - \frac{1}{\omega_p + \omega_{eg}} \right]. \tag{14.4b}$$

We thus have

$$\rho'_W(\omega_p) - \rho'^{\dagger}_W(\omega_p) = \hbar\alpha, \tag{14.5a}$$

$$\rho'_D(\omega_p) + \rho'_D(-\omega_p) = \hbar\alpha\rho_g, \tag{14.5b}$$

$$\rho'^{\dagger}_D(\omega_p) + \rho'^{\dagger}_D(-\omega_p) = \hbar\rho_g\alpha, \tag{14.5c}$$

* The polarizability operator is obtained from Eq. (5.33) by expanding it in the electronic basis set but keeping V_{ge} as an operator in the nuclear space.

We next introduce the two-time correlation function of the polarizability

$$C_{\alpha\alpha}(t) \equiv \langle \alpha(t)\alpha(0)\rho_g \rangle, \tag{14.6}$$

where we have defined the Heisenberg operator

$$\alpha(t) \equiv \exp\left(\frac{i}{\hbar} H_g t\right)\alpha \exp\left(-\frac{i}{\hbar} H_g t\right) = \alpha\mathcal{G}_{gg}(t). \tag{14.7}$$

Using these definitions we get

$$P^{(3)}(\mathbf{k}_s, t) = -E_3(t) \int_0^\infty dt_2 \exp[i(\omega_1 - \omega_2)t_2]E_1(t + \tau - t_2)E_2^*(t + \tau - t_2)$$
$$\times \langle (-i/\hbar)[\alpha(t_2), \alpha(0)]\rho_g \rangle, \tag{14.8}$$

where we have used the cyclic permutation symmetry of the trace. The signal is obtained by combining Eqs. (14.2a) and (14.8)

$$S_{\text{CRS}}(\tau) =$$
$$\int_{-\infty}^{\infty} dt\, |E_3(t)|^2 \left| \int_0^\infty dt_2\, E_2^*(t + \tau - t_2)E_1(t + \tau - t_2) \exp[i(\omega_1 - \omega_2)t_2]\chi_{\alpha\alpha}(t_2) \right|^2, \tag{14.9}$$

$$\chi_{\alpha\alpha}(t_2) \equiv -\frac{i}{\hbar} \langle [\alpha(t_2), \alpha(0)]\rho_g \rangle. \tag{14.10}$$

Here $\chi_{\alpha\alpha}$ is the linear response function associated with the polarizability. Using the notation of Chapter 8 we can separate $C_{\alpha\alpha}$ into its real and imaginary parts $C_{\alpha\alpha}(t) = C'_{\alpha\alpha}(t) + iC''_{\alpha\alpha}(t)$. We then have $\chi_{\alpha\alpha}(t) = (2/\hbar)C''_{\alpha\alpha}(t)$. All the symmetries and relationships among $C'(t)$ and $C''(t)$ discussed in Chapter 8 apply here as well.

We have thus shown that the coherent Raman process is controlled by the response function associated with the electronic polarizability $\chi_{\alpha\alpha}$. This general result can be used to compare the various Raman spectroscopies, which differ only by the pulse envelopes $E_j(t)$. Raman measurements contain two types of nuclear dynamical information. The positions of the Raman resonances $\omega_1 - \omega_2 = \omega_{vv'}$ reveal the frequencies of the Raman active modes, and their linewidths carry information on vibrational relaxation (both population relaxation and dephasing). Equation (14.9) allows us to define precisely how these motions may be probed in the time and in the frequency domain by different Raman techniques, as will be shown below.

Frequency-domain coherent and spontaneous Raman scattering

In Chapter 9 we discussed frequency-domain spontaneous and coherent Raman spectroscopy. The semiclassical picture of Raman processes becomes most transparent by recasting these results using nuclear correlation functions and response functions. In this form, the relationships to time-domain (whether picosecond or femtosecond) Raman techniques are readily established. We shall now analyze the off-resonant limit of stationary Raman measurements. In a frequency-domain measurement, all field envelopes are independent of time, $E_j(t) = E_j$, $j = 1, 2, 3$. For comparison, we also derive the spontaneous Raman scattering cross section introduced in Chapter 9. To maintain a notation consistent with that used for coherent Raman techniques, we label the fields here as $\omega_2 = \omega_S$, and $\omega_1 \equiv \omega_L$.

In Chapter 9 we focused on resonant conditions. We therefore invoked the rotating wave approximation and considered both ground state and excited state resonances. Here, on the other hand, we are interested in off-resonant conditions where we need consider only the ground-state contributions. Furthermore, we can no longer invoke the RWA, since the terms neglected under the RWA are comparable to those retained when the frequencies are tuned off resonance. We therefore have to start with Eq. (9.9), rather than simply take the off-resonance limit of Eq. (9.12). The coherent and spontaneous Raman signals are then given by

$$
\begin{aligned}
S_{\mathrm{CRS}}(\omega_1, \omega_2) &= |P^{(3)}(\mathbf{k}_s, t)|^2 \\
&= \frac{1}{\hbar^6}\Big|\langle\langle \rho'_{\mathrm{W}}(\omega_s) - \rho'^{\dagger}_{\mathrm{W}}(\omega_s)|\mathscr{G}_{gg}(\omega_1 - \omega_2)|\rho'_{\mathrm{D}}(\omega_1) \\
&\quad + \rho'_{\mathrm{D}}(-\omega_2) + \rho'^{\dagger}_{\mathrm{D}}(\omega_1) + \rho'^{\dagger}_{\mathrm{D}}(-\omega_2)\rangle\rangle\Big|^2,
\end{aligned}
\tag{14.11a}
$$

$$
\begin{aligned}
&S_{\mathrm{SRS}}(\omega_1, \omega_2) \\
&= 2\,\mathrm{Im}\langle\langle \rho'_{\mathrm{w}}(\omega_2) - \rho'^{\dagger}_{\mathrm{w}}(\omega_2)|\mathscr{G}_{gg}(\omega_1 - \omega_2)|\rho'_{\mathrm{D}}(\omega_1) + \rho'_{\mathrm{D}}(-\omega_2)\rangle\rangle.
\end{aligned}
\tag{14.11b}
$$

The frequency dependence of the signal contains all the microscopic information about the system: The positions of the Raman lines give the nuclear frequencies, and their lineshapes provide information on vibrational relaxation and dephasing. The response function [Eq. (14.10)] has 8 terms (each α has two terms and the commutator multiplies the number of terms by 2). Only two of these terms survive if we further invoke the RWA for α. Equation (14.11a) then becomes identical to the expression derived in Chapter 9, where a detailed analysis of this case for off-resonant as well as resonant conditions was carried out. Equation (14.11b) generalizes the results of Chapter 9, since we did not make the RWA here [which amounts to neglecting the second term in the electronic polarizability Eq. (14.4b)]. By replacing all frequency arguments in $\rho_j(\omega)$ by the average frequency ω_p we obtain

$$
S_{\mathrm{CRS}}(\omega_1 - \omega_2) = \left|\int_0^{\infty} dt \exp[i(\omega_1 - \omega_2)t]\langle -\frac{i}{\hbar}[\alpha(t), \alpha(0)]\rho_g\rangle\right|^2,
\tag{14.12a}
$$

$$
S_{\mathrm{SRS}}(\omega_1 - \omega_2) = \int_{-\infty}^{\infty} dt \exp[i(\omega_1 - \omega_2)t]\,\langle \alpha(t)\,\alpha(0)\rho_g\rangle.
\tag{14.12b}
$$

In Chapter 9 we calculated stationary Raman processes starting with the response function. The formal expression for coherent Raman is given by the first two terms of Eq. (9.22), and for spontaneous Raman by [Eq. (9.12b)]

$$S_{\text{CRS}}(\omega_1 - \omega_2) = |S^{(3)}(2\omega_1 - \omega_2, \omega_1 - \omega_2, \omega_1) + S^{(3)}(2\omega_1 - \omega_2, \omega_1 - \omega_2, -\omega_2)|^2, \tag{14.13a}$$

$$S_{\text{SRS}}(\omega_1 - \omega_2) = 2\,\text{Im}\,R_3(\omega_2, \omega_2 - \omega_1, -\omega_1). \tag{14.13b}$$

Upon replacing the field frequencies during the coherent periods (i.e., the first and the third arguments) by the average frequency, we obtain the following off-resonant form of Eqs. (14.13):

$$S_{\text{CRS}}(\omega_1 - \omega_2) = |S^{(3)}(\omega_{\text{p}}, \omega_1 - \omega_2, \omega_{\text{p}}) + S^{(3)}(\omega_{\text{p}}, \omega_1 - \omega_2, -\omega_{\text{p}})|^2, \tag{14.14a}$$

$$S_{\text{SRS}}(\omega_1 - \omega_2) = 2\,\text{Im}\,R_3(\omega_{\text{p}}, \omega_2 - \omega_1, -\omega_{\text{p}}). \tag{14.14b}$$

Equations (14.14) coincide with Eq. (14.12). This illustrates the connection between the response function and the wavepacket approach. By expanding Eqs. (14.14) in eigenstates we recover Eqs. (9.19) and (9.23).

Let us now consider a system with a few Raman active high frequency modes in addition to many bath modes. We can then expand all quantities in a basis set of the high frequency modes, keeping them as operators in the remaining phase space. We thus have

$$C_{\alpha\alpha}(t) = \sum_{a,c} \langle \alpha_{ac}(t)\alpha_{ca}(0)\rho_{aa}\rangle \exp(-i\omega_{ca}t), \tag{14.15}$$

and

$$\chi_{\alpha\alpha}(t) = \sum_{a,c} \chi_{ac}(t) \exp(-i\omega_{ca}t), \tag{14.16a}$$

with

$$\chi_{ac}(t) \equiv -\frac{i}{\hbar}[\langle \alpha_{ac}(t)\alpha_{ca}(0)\rho_{aa}\rangle - \langle \alpha_{ca}(0)\alpha_{ac}(t)\rho_{cc}\rangle], \tag{14.16b}$$

with

$$\alpha_{vv'} \equiv \langle v|\alpha|v'\rangle, \tag{14.17a}$$

and

$$\rho_{vv} \equiv \langle v|\rho_g|v\rangle. \tag{14.17b}$$

Here the time evolution of $\alpha(t)$ is given by Eq. (14.7) except that the Hamiltonian H_g now excludes the higher frequency modes.

As a special case, let us assume homogeneous vibrational dephasing, i.e.,

$$\langle \alpha_{ac}(t)\alpha_{ca}(0)\rho_{aa}\rangle = \langle |\alpha_{ac}(0)|^2 \rho_{aa}\rangle \exp(-\Gamma_{ac}t), \tag{14.18a}$$

$$\alpha_{ca} = \frac{i}{\hbar}\sum_b \left[\frac{\mu_{cb}\mu_{ba}}{\omega_{\text{p}} - \omega_{ba}} + \frac{\mu_{cb}\mu_{ba}}{\omega_{\text{p}} + \omega_{ba}}\right]. \tag{14.18b}$$

We then get

$$S_{\text{CRS}}(\omega_1 - \omega_2) = \left|\sum_{a,c} [P(a) - P(c)] \frac{|\alpha_{ac}|^2}{\omega_1 - \omega_2 - \omega_{ca} + i\Gamma_{ac}}\right|^2, \tag{14.18c}$$

and Eq. (14.12b) then becomes

$$S_{SRS}(\omega_1 - \omega_2) = \sum_{a,c} P(a)|\alpha_{ca}|^2 \frac{\Gamma_{ac}}{(\omega_1 - \omega_2 - \omega_{ca})^2 + \Gamma_{ac}^2}, \tag{14.18d}$$

which is equivalent to the Kramers–Heisenberg formula derived in Chapter 9, with the addition of vibrational dephasing.

Impulsive (femtosecond) CARS

Impulsive (very short) pulses have a sufficiently broad bandwidth to excite high frequency vibrations coherently. The E_1 and E_2 pump pulses may be taken to have identical envelopes, which will be denoted E_P, and the same carrier frequency (but a different wavevector). Typical examples of femtosecond Raman measurements in molecular liquids are shown in Figures 14.1 and 14.2.

To calculate the polarization induced in an impulsive measurement we set $E_1 = E_2 = E_P$ and $\omega_1 = \omega_2$ in Eq. (14.8) and obtain

$$P(\mathbf{k}_s, t) = -E_3(t) \int_0^\infty dt_2\, |E_P(t + \tau - t_2)|^2 \chi_{\alpha\alpha}(t_2). \tag{14.19}$$

The effect of the pump field bandwidth on the Raman signal can be illustrated by recasting Eq. (14.9) in the frequency domain

$$S_{ICRS}(\tau) \approx \int_{-\infty}^{\infty} dt\, |E_3(t)|^2 \left| \frac{1}{2\pi} \int_{-\infty}^{\infty} d\omega \exp[-i\omega(t + \tau)] I_P(\omega) \chi_{\alpha\alpha}(\omega) \right|^2, \tag{14.20a}$$

where

$$I_P(\omega) \equiv \int_{-\infty}^{\infty} dt \exp(i\omega t) |E_P(t)|^2, \tag{14.20b}$$

and

$$\chi_{\alpha\alpha}(\omega) \equiv \int_{-\infty}^{\infty} dt \exp(i\omega t) \langle (-i/\hbar)[\alpha(t), \alpha(0)] \rangle. \tag{14.20c}$$

This shows that the pump field merely affects the Raman signal through frequency filtering, i.e., by selecting only the Raman active modes whose frequencies lie within its spectral profile $I_P(\omega)$. Figure 14.3 shows the time- and frequency-domain representation for the Raman response of pyridine liquid. The figure illustrates the spectral filter effects of finite-duration/finite-bandwidth optical pulses. This filtering effect will be discussed further in the following sections. For very short pulses (impulsive pump), $I_P(\omega)$ is independent of frequency and all the Raman modes show up.

$$S_{ICRS}(\tau) = |\chi_{\alpha\alpha}(\tau)|^2. \tag{14.21}$$

Here the entire microscopic information appears in the time-domain. When the delay time τ is varied, vibrational frequencies show up as oscillations (quantum beats) whereas relaxation enters as damping [12, 13].

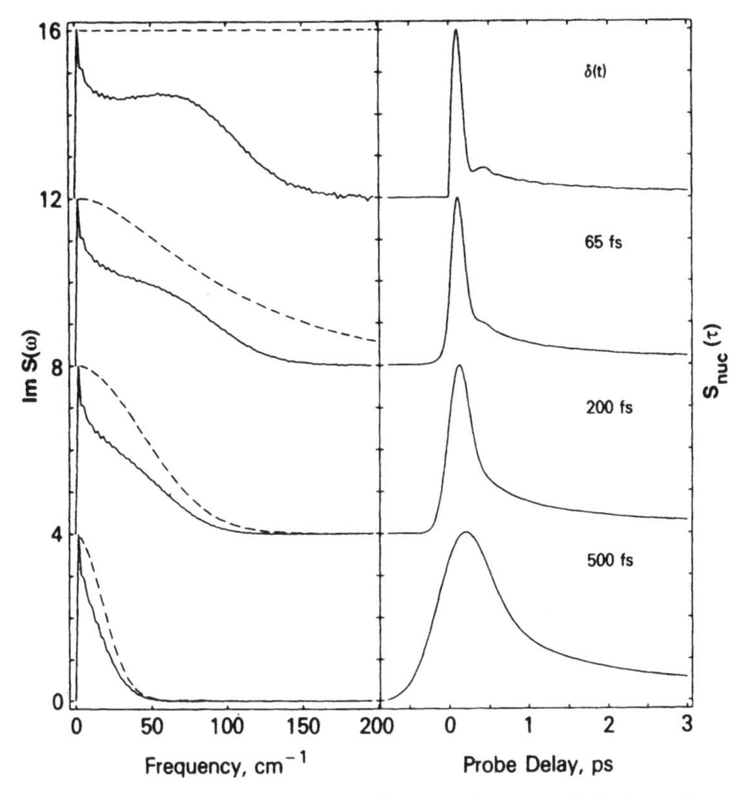

FIG. 14.3 Shown in the right panel is the nuclear part of the heterodyne-detected Raman response of pyridine liquid as a function of laser pulse width. The left panel gives the corresponding frequency response functions (solid). The effective filter functions, which for optical heterodyne detection with transform limited laser pulses are given by the Fourier transform of the laser pulse autocorrelation function, are given by dashed lines. The 65 fs curves are experimental data. The other three sets are calculated. The top set of curves, which corresponds to the impulsive limit, was generated by deconvolution of the experimental data (65 fs). The lower two sets of curves are generated from the impulsive response (upper curves) for Gaussian pump and probe pulses of the durations indicated. [D. McMorrow and W. T. Lotshaw, "The frequency response of condensed-phase media to femtosecond optical pulse: Spectral-filter effects," *Chem. Phys. Lett.* **174**, 85 (1990).]

Picosecond CARS

Picosecond pulses are usually long compared with vibrational frequencies but short compared with vibrational dephasing times. Such pulses are sufficiently spectrally narrow to select a given vibrational transition. However, the lineshape of that transition is seen in the time domain. This is therefore an *intermediate* situation whereby the information is obtained in both the time and the frequency domain. When Eqs. (14.16a) are substituted in Eq. (14.8) we get

$$P(\mathbf{k}_s, t) = E_3(t) \sum_{a,c} \chi_{ac}(t+\tau) \int_0^\infty dt_2 \, |E_1(t+\tau-t_2)|^2 \exp[i(\omega_1 - \omega_2 + \omega_{ac})t_2].$$

In a given experiment, the integral selects a particular vibronic transition, since $|E_1(\tau_1)|^2$ is sufficiently spectrally narrow to cover a single transition. For that transition we have

$$P(\mathbf{k}_s, t) \cong E_3(t)\chi_{ac}(t + \tau), \tag{14.22}$$

$$S_{\text{PCRS}}(\tau) = |\chi_{ac}(\tau)|^2. \tag{14.23}$$

In summary, we note that the precise manner in which nuclear dynamic information is reflected in the signal is intimately related to the uncertainty in the t_2 interval. In the impulsive techniques there is no uncertainty and the entire information is obtained in the time domain. In frequency-domain CARS, there is no external control over t_2, and the uncertainty is large (of the order of the vibrational dephasing timescale). Consequently, all the information is in the frequency domain. In picosecond techniques, the uncertainty is large compared with the vibrational period but short compared with the dephasing time. The spectral resolution is then used to select a given vibration, but the dephasing process is still observed in the time domain.

THE MULTIMODE BROWNIAN OSCILLATOR MODEL

Off-resonant Raman scattering is induced by the dependence of the electronic polarizability on nuclear coordinates $\alpha = \alpha(\mathbf{q})$. Usually this dependence is weak, and for resonant excitation it is often neglected (the Condon approximation). Making use of this weak dependence, we expand $\alpha(\mathbf{q})$ to first order around some equilibrium configuration $\mathbf{q}^0$ [2, 6]:

$$\alpha(\mathbf{q}) = \alpha(\mathbf{q}^0) + \sum_j \kappa_j q_j + \cdots. \tag{14.24}$$

Here $\kappa_j \equiv \partial\alpha_{\text{P}}/\partial q_j$ evaluated at $\mathbf{q}^0$. This expansion is usually justified for intramolecular modes or for phonons in crystals where the relevant values of $\mathbf{q}$ are confined to a small region around an equilibrium configuration. For nonbounded nuclear motions (e.g., intermolecular coordinates in liquids) it should be applicable only for very short times. Truncating the expansion to linear order in $\mathbf{q}$, and assuming that the various modes q_j can be modeled as uncorrelated harmonic Brownian oscillators, we have

$$\chi_{\alpha\alpha}(t) = -\frac{i}{\hbar}\sum_j \kappa_j^2 \langle [q_j(t), q_j(0)]\rho_g \rangle, \tag{14.25}$$

with

$$q_j(t_2) \equiv \exp\left(\frac{i}{\hbar}H_g t_2\right) q_j \exp\left(-\frac{i}{\hbar}H_g t_2\right). \tag{14.26}$$

Note that the leading (Condon) term in the expansion (14.24) does not contribute to the signal since it is simply a number, and thus commutes with all the operators. Using the notation introduced in Chapter 8 [Eq. (8.63b)] we have $\langle -(i/\hbar)[q_j(t), q_j(0)]\rho_g \rangle = (2/\hbar)C_j''(t)$. The coherent Raman scattering signal in this case assumes the form

$$S_{\text{CRS}}(\tau) = \int_{-\infty}^{\infty} dt\, |E_3(t)|^2 \left[\int_0^{\infty} dt_2\, |E_{\text{P}}(t + \tau - t_2)|^2 \sum_j \frac{2}{\hbar}\kappa_j^2 C_j''(t_2)\right]^2. \tag{14.27}$$

For impulsive excitation, where the durations of the pump and probe pulses are short compared with the molecular nuclear dynamics, we may approximate $E_3(t)$ and $E_P(t + \tau - t_2)$ by delta functions. In this case, Eq. (14.27) reduces to

$$S_{ICRS}(\tau) \propto \left[\sum_j \frac{2}{\hbar} \kappa_j^2 C_j''(\tau) \right]^2, \tag{14.28}$$

where $C_j''(\tau)$ is given by Eq. (8.67b). Various limiting forms for $C_j''(\tau)$ for underdamped and overdamped motions, as well as high temperature approximations were given in Chapter 8. It is also possible to compute $C_j''(\tau)$ using classical simulations. The procedure involves simulating the correlation function $C_j(\tau) \equiv \langle q_j(\tau) q_j(0) \rangle$ and then using the fluctuation dissipation theorem to get the response function [see Eqs. (8.89)]

$$\frac{2}{\hbar} C_j''(\tau) = \langle (-i/\hbar)[q_j(\tau), q_j(0)] \rho_g \rangle \cong \frac{1}{k_B T} \frac{d}{d\tau} \langle q_j(\tau) q_j \rho_g \rangle. \tag{14.29}$$

The development of picosecond and femtosecond pulse shaping techniques makes it possible to create "Taylor-made" pulses with control on their timing, shapes, and phases [14]. It is also possible to measure the absolute electric field using nonlinear correlation optical gating methods, which are based on off-resonant nonlinearities [15, 16]. Some examples of shaped pulses are displayed in Figure 14.4. Potentially these developments offer some interesting possibilities for manipulating elementary nuclear motions by optical means [17]. It is also possible to describe the field using a Wigner representation by considering a product of two field amplitudes in complete analogy with the density matrix. This elegant representation highlights the role of temporal,

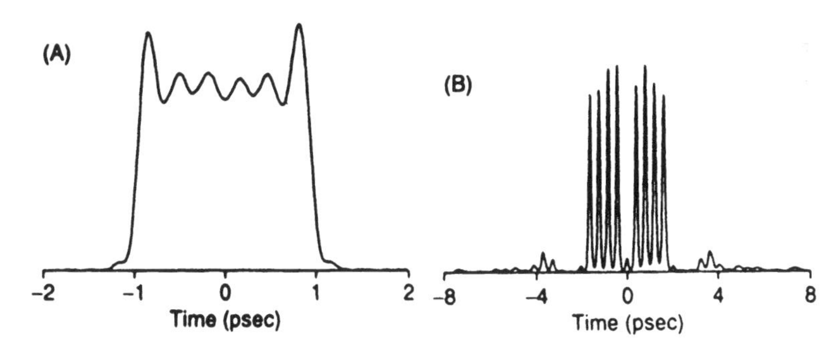

FIG. 14.4 Specially shaped ultrafast optical waveforms can be generated by spatial filtering within a gratings and lens femtosecond pulse shaping apparatus. [A. M. Weiner, J. P. Heritage, and E. M. Kirschner, "High-resolution femtosecond pulse shaping," *J. Opt. Soc. Am. B* **5**, 1563 (1988).] Shaped waveforms can include phase as well as amplitude modulation, and nearly arbitrary pulse shapes are possible. Computer control of the optical waveform has been demonstrated. [A. M. Weiner, D. E. Leaird, J. S. Patel, and J. R. Wullert, "Programmable shaping of femtosecond pulses by use of a 128-element liquid-crystal phase modulator," *IEEE J. Quantum Electron.* **28**, 908 (1992).] Some examples of shaped waveforms are shown above. (A) A 2-ps duration "square" pulse with 100 fs rise and fall times. (B) A 2.5-THz sequence of femtosecond pulses with the center pulse missing. [A. M. Weiner and D. E. Leaird, "Generation of terahertz-rate trains of femtosecond pulses by phase-only filtering," *Optics Lett.* **15**, 51 (1990).]

spectral, and spatial features of the field and may be used for visualization as well as for for treating the field and the matter on the same footing [18].

We shall now present model calculations which demonstrate the filtering effect and mode-selectivity achieved by pulse shaping on a model system with three Raman active modes [12]. For simplicity we assume that the probe pulse is impulsive, whereas the pump field consists of a train of N identical ultrashort pulses with

$$|E_{\mathrm{P}}(t)|^2 = \frac{I_{\mathrm{P}}(0)}{N} \sum_{n=1}^{N} \delta[t + (n-1)T]. \tag{14.30}$$

Here T denotes the train period and $I_{\mathrm{P}}(0)$ is the pulse train energy. The train is normalized such that its total energy does not change as N is varied. The spectral profile of the pump field is given by

$$I_{\mathrm{P}}(\omega) = I_{\mathrm{P}}(0) \exp[-i(N-1)\omega T/2] \frac{\sin(N\omega T/2)}{N \sin(\omega T/2)}. \tag{14.31}$$

In single pulse experiments, we have $N = 1$ and $I_{\mathrm{P}}(\omega) = I_{\mathrm{P}}(0)$. Multiple femtosecond pulse sequences improve the mode selectivity of these experiments. If a train of femtosecond pulses is used, $I_{\mathrm{P}}(\omega)$ assumes the form of a series of peaks around a selected set of frequencies $\omega_n = 2\pi n/T$. As the number of pulses N increases, these peaks narrow, and in the ideal cases where $N \to \infty$, the spectral profile of the pump field becomes a series of delta functions. This behavior is illustrated in Figure 14.5. The figure demonstrates the filtering capacity of pulse shaping, which allows only selected Fourier components of the optical response $\chi_\alpha(\omega)$ to be observed and discriminates against other frequency components. Optical phonons in α-perylene crystal were excited in this way [14]. For an ordinary impulsive Raman experiment using a single pulse [Eq. (14.30) with $N = 1$], all the Raman modes are equally excited. Consider now a Raman experiment conducted using a shaped train of ultrashort pulses with a fixed total energy [Eq. (14.30)]. If the train period is tuned to match a particular Raman mode j, the pump profile $I_{\mathrm{P}}(\omega_j)$, which acts as a frequency filter function, excites only the jth Raman mode, whose amplitude is unchanged by the pulse shaping, but discriminates against the other modes: $I_{\mathrm{P}}(\omega_k) \approx 0$ for $k \neq j$ (cf. Figure 14.5). Figure 14.6 demonstrates this filtering effect. The figures show the calculated Raman sighal of a three-mode system. Figure 14.6a displays the impulsive (nonselective) limit ($N = 1$). Figure 14.6b shows the signal generated by a train of 30 pulses, with repetition time $T = 334$ fs (corresponding to a vibrational frequency of $100\ \mathrm{cm}^{-1}$). In Figure 14.6c we present the Raman signal for the same model system with $N = 30$ versus the time delay τ_{D} and the train period T. The 67, 100, and $117\ \mathrm{cm}^{-1}$ modes are now clearly separated and show up at different values of the train period.

EFFECTIVE HAMILTONIAN FOR RAMAN SCATTERING

So far, our calculation of the off-resonant Raman process was based on the nonlinear response function. This way the connection with the previous treatment of resonant Raman spectroscopy (Chapter 9), as well as other nonlinear techniques, is clearly established. There is an alternative method for deriving

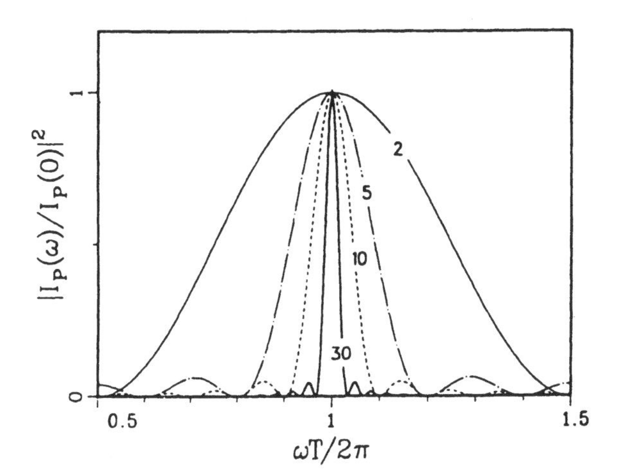

FIG. 14.5 The spectrum $|I_p(\omega)/I_p(0)|^2$ [Eq. (14.31)] of the pump train of N identical ultrashort pulses vs. $\omega T/2\pi$. T is the train period and $I_p(\omega) = I_p(\omega + 2\pi/T)$ is a periodic function. The various curves correspond to different values of N as indicated. As the number of pulses N increases, $I_p(\omega)$ becomes more sharply peaked around $\omega = 2\pi n/T$, with $n = 1, 2, \ldots$, and the frequency filtering of the pump is improved. [From Y. J. Yan and S. Mukamel, "Pulse shaping and coherent Raman spectroscopy in condensed phases," *J. Chem. Phys.* **94**, 997 (1991).]

these results, which exploits from the start the off-resonant nature of the excitation. This greatly simplifies the derivation, and provides a clear physical insight in terms of a driven oscillator picture. Furthermore, we are no longer limited to the weak field $P^{(3)}$ limit and can explore strong field effects.

We start with the preparation process and define the doorway state. Following the interaction with the pump [the E_1 and the E_2 fields in Eq. (14.1)], the density matrix of a molecule located at $\mathbf{r}$, may be represented in the form

$$\rho(\mathbf{r}, t) = |g\rangle\rho_{gg}(\mathbf{r}, t)\langle g| + |e\rangle\rho_{ee}(\mathbf{r}, t)\langle e| + |e\rangle\rho_{eg}(\mathbf{r}, t)\langle g| + |g\rangle\rho_{ge}(\mathbf{r}, t)\langle e|, \quad (14.32)$$

where each matrix element ρ_{mn}, with $m, n = \mathrm{e}$ or g, is an operator in the nuclear space. ρ_{gg} or ρ_{ee} describes the nuclear dynamics when the system is in the electronic ground or excited state, whereas ρ_{eg} or ρ_{ge} represents nuclear dynamics when the system is in an electronic (optical) coherence. The four coupled Heisenberg equations of motion of $\rho_{mn}(\mathbf{r}, t)$ are given in Eqs. (14.A.2). The density matrix depends on the position $\mathbf{r}$ through the phase of the pump field. We shall focus now on off-resonant Raman spectroscopy, in which all electronic detunings, $\omega_{eg} \pm \omega_j$, $j = 1, 2, 3$, are large compared with the inverse timescales of the excitation pulses and the nuclear dynamics. In this case the equations of motion for ρ_{gg} and ρ_{ee} are decoupled. Since the system is initially in the ground electronic state, the state prepared by the pump (the doorway state) is characterized only by $\rho_{gg}(\mathbf{r}, t)$, which satisfies the equation of motion (Appendix 14A):

$$\dot{\rho}_{gg}(\mathbf{r}, t) = -\frac{i}{\hbar}[H_{eff}(\mathbf{r}, t), \rho_{gg}(\mathbf{r}, t)]. \tag{14.33}$$

Here the effective Hamiltonian is

$$H_{eff}(\mathbf{r}, t) \equiv H_g - \alpha|\tilde{E}_p(\mathbf{r}, t)|^2, \tag{14.34a}$$

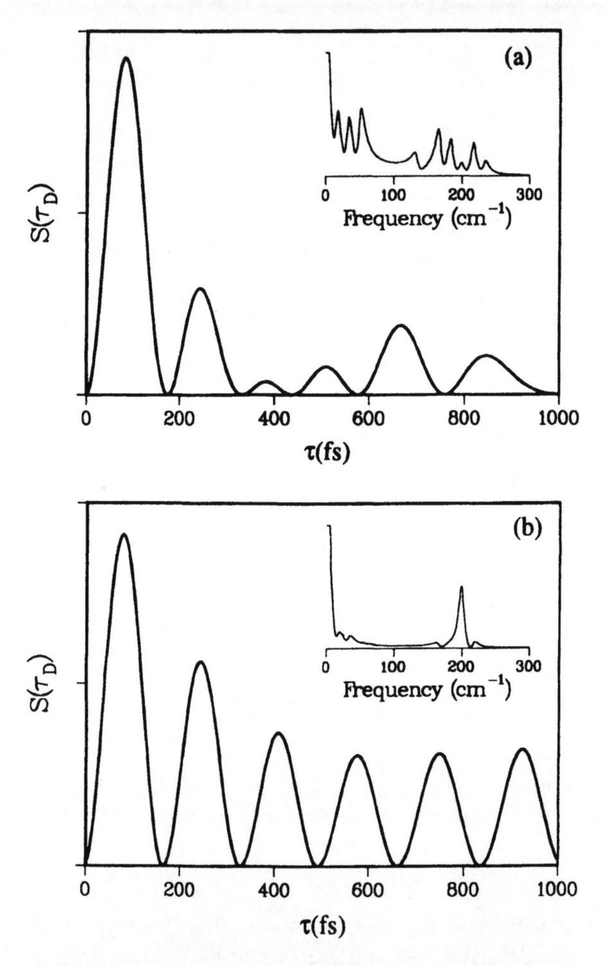

FIG. 14.6 (a) The Raman signal [Eq. (14.27)] of a molecule with three Brownian harmonic modes: $\omega_1 = 100\,\text{cm}^{-1}$, $\omega_2 = 67\,\text{cm}^{-1}$, and $\omega_3 = 117\,\text{cm}^{-1}$, the friction parameters $\gamma_1 = \gamma_2 = \gamma_3 = 5\,\text{cm}^{-1}$. The calculation is performed in the impulsive limit where both the probe and the pump fields are a single peaked ($N = 1$) ultrashort pulses. The power spectrum of the Raman signal is shown in the inset. No mode selectivity is observed. (b) The same as (a) but for the shaped pump train of $N = 30$ pulses, and $T = 334$ fs (or $1/cT = 100\,\text{cm}^{-1}$). The optical selectivity for mode 1 ($\omega_1 = 100\,\text{cm}^{-1}$) is evident.

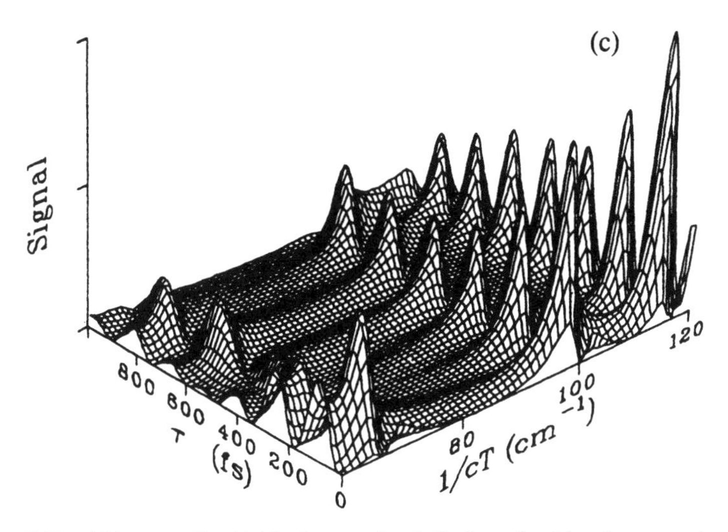

FIG. 14.6 (*continued*) (c) The Raman signal $S(\tau_D)$ vs. the delay time τ_D and the train period T for the same model of Figure 14.5b. The three modes are well resolved at different values of the train period T. [From Y. Yan and S. Mukamel, "Pulse shaping and coherent Raman spectroscopy in condensed phases," *J. Chem. Phys.* **94**, 997 (1991).]

with the complex pump amplitude

$$\tilde{E}_p(\mathbf{r}, t) \equiv E_1(t+\tau)\exp(i\mathbf{k}_1\mathbf{r} - i\omega_1 t) + E_2(t+\tau)\exp(i\mathbf{k}_2\cdot\mathbf{r} - i\omega_2 t), \quad (14.34b)$$

and α, defined in Eq. (14.4b), is the electronic polarizability at frequency $\omega_p \equiv (\omega_1 + \omega_2)/2$. These equations are valid for an arbitrary intensity and shape of the pump. For resonant excitation, the nuclear dynamics depend also on the electronic excited state Hamiltonian H_e. In this case, the equations of motion for ρ_{gg} and ρ_{ee} are coupled (cf. Appendix 14A) and the effective Hamiltonian is inapplicable.

We shall be interested in the $\mathbf{k}_1 - \mathbf{k}_2$ spatial Fourier component of the doorway wavepacket, which constitutes the relevant grating for the Raman process:

$$\rho_{gg}(\mathbf{r}, t) = \rho_D(t)\exp[-i(\mathbf{k}_1 - \mathbf{k}_2)\mathbf{r} - i(\omega_1 - \omega_2)t] + \text{other terms} \quad (14.35)$$

In a coherent Raman process, a weak probe is scattered and the signal in the direction $\mathbf{k}_s = \mathbf{k}_3 + \mathbf{k}_1 - \mathbf{k}_2$ is measured as the delay time τ of the probe with respect to the pump fields is varied. The detection process with a weak probe can be described using the window function introduced earlier. We thus get

$$P(\mathbf{k}_s; t) = E_3(t)\,\mathrm{Tr}[\alpha\rho_D(t)]. \quad (14.36)$$

Substituting Eq. (14.36) in (14.2a), we obtain

$$S_{CRS}(\tau) = \int_{-\infty}^{\infty} dt\,|E_3(\tau)|^2\,\{\mathrm{Tr}[\alpha\rho_D(t)]\}^2. \quad (14.37)$$

Equation (14.37), together with (14.33) and (14.35), constitutes the main formal result of the effective Hamiltonian approach. These equations are valid for arbitrary (anharmonic) nuclear adiabatic Hamiltonians H_g or H_e, arbitrary shapes of the pulse envelopes $E_P(t)$ and $E_3(t)$, and for any intensity of the pump. The probe field is assumed to be weak. The key quantity in the calculation of the signal is $\mathrm{Tr}[\alpha\rho_D(t)]$ (i.e., the expectation value of the molecular electronic polarizability calculated using the time dependent doorway wavepacket). The electronic polarizability depends parametrically on nuclear coordinates. The Raman active modes are characterized by a relatively strong coordinate dependence, and the Raman signal reflects their coherent motions. Obviously, nuclear motions that do not affect the polarizability will not show up in the spectrum.

When the pump and the probe are sufficiently weak, we can expand the dynamics of ρ_{gg} [Eq. (14.33)] to lowest order in $E_1E_2^*$. We then have

$$\mathrm{Tr}[\alpha\rho_D(t)] = \int_0^\infty dt_2\, E_1(t+\tau-t_2)E_2^*(t+\tau-t_2)\exp[i(\omega_1-\omega_2)t_2]\chi_{\alpha\alpha}(t_2). \tag{14.38}$$

The nuclear dynamics is in this case described by the response function of the electronic polarizability in the ground electronic state. Equations (14.37) and (14.38) are identical to Eq. (14.9), and establish the connection between the wavepacket and the correlation function formulations of off-resonant Raman signals.

Off-resonant impulsive stimulated experiments may thus be interpreted using a driven oscillator model for nuclear dynamics where the driving force is induced by the coupling of the pump to the electronic polarizability. The probe field then interacts with the driven oscillator. The exchange of energy between the oscillator and the driving field will depend on their relative phases. This model can be readily applied to interpret the coherent vibrational motions of phonons in crystals and dye molecules in glasses and in solvent environments. By applying the multimode Brownian oscillator model to Eq. (14.33) we recover Eqs. (14.28) and (14.29). However, the effective Hamiltonian applies for anharmonic vibrational motions with an arbitrary dependence of the electronic polarizability on nuclear coordinates, and the classical limit of the theory can be readily obtained. In addition, the present derivation shows why the effective Hamiltonian approach does not apply for the interpretation of impulsive resonant experiments, where the electronic excited state is populated, and nuclear dynamics in both the ground and the excited electronic states becomes important.

We have previously analyzed the filtering effect of Raman excitation using pulse shaping, which allows the observation of a few selected Raman modes. We shall now analyze this process by examining the doorway wavepacket prepared by the pump when applied to the multimode Brownian oscillator model. For simplicity we assume that the E_1 and the E_2 pulses have an identical time profile $E_P(t)$ (but different propagation directions). This is typically the case in femtosecond Raman measurements. By combining Eqs. (14.24) and (14.34) we obtain the effective Hamiltonian

$$H_{\mathrm{eff}}(\mathbf{r},t) = H_0 - \sum_j F_j(t)q_j, \tag{14.39a}$$

where $H_0 = H_g + H'$ is the multimode Brownian oscillator model in the ground state [see Eqs. (8.61)], and

$$F_j(t) \equiv 2[1 + \cos(\mathbf{k}_1 - \mathbf{k}_2)\mathbf{r}]\kappa_j|E_P(t)|^2, \tag{14.39b}$$

is the driving force for the jth mode. Since the driving field forms a grating with a wavevector $\mathbf{k}_1 - \mathbf{k}_2$, the force depends on the position $\mathbf{r}$ of the molecule. An exact solution of Eq. (14.33) with the effective Hamiltonian (14.39) can be obtained using a variety of methods, including expansion in coherent states and path integral techniques [see, e.g., Eqs. (12.55)]. The reduced doorway wavepacket where the bath degrees of freedom have been traced over, in the Wigner representation, is given by

$$\rho_D(\mathbf{p}, \mathbf{q}; t) = \prod_j \rho_D(p_j, q_j; t), \tag{14.40a}$$

with

$$\rho_D(p_j, q_j; t) = \frac{1}{2\pi}[\langle p_j^2\rangle_g \langle q_j^2\rangle_g]^{-1/2}$$
$$\times \exp\left\{-\frac{1}{2\langle q_j^2\rangle_g}[q_j - \langle q_j(t)\rangle]^2 - \frac{1}{2\langle p_j^2\rangle_g}[p_j - \langle p_j(t)\rangle]^2\right\}. \tag{14.40b}$$

Here $\langle q_j(t)\rangle$ and $\langle p_j(t)\rangle$ are the expectation values of the coordinate and momentum of the jth mode in the doorway state:

$$\langle q_j(t)\rangle \equiv \text{Tr}[q_j \rho_D(t)] = \frac{2}{\hbar}\int_{-\infty}^{t} d\tau\, F_j(\tau) C_j''(t - \tau), \tag{14.41a}$$

$$\langle p_j(t)\rangle \equiv \text{Tr}[p_j \rho_D(t)] = \frac{-2m_j}{\hbar}\int_{-\infty}^{t} d\tau\, F_j(\tau) \dot{C}_j''(t - \tau). \tag{14.41b}$$

The correlation function $C_j''(t - \tau)$ was introduced and analyzed in Chapter 8.

At zero temperature, $\coth(\beta\hbar\omega_j/2) = 1$ and the doorway state [Eq. (14.40)] represents a coherent state. Furthermore, $\langle q_j(t)\rangle$ is directly related to the Raman signal via the relation: $\text{Tr}[\alpha\rho_D(t)] = \sum \kappa_j \langle q_j(t)\rangle$. The total energy of the jth mode is

$$\langle E_j(t)\rangle = \langle E_j\rangle_g + \langle \Delta E_j(t)\rangle. \tag{14.42a}$$

$\langle E_j\rangle_g$ is the thermal equilibrium energy, which, using Eqs. (12.43), is given by

$$\langle E_j\rangle_g = \tfrac{1}{2}\hbar\omega_j \coth(\beta\hbar\omega_j/2).$$

$\langle \Delta E_j(t)\rangle$ is the excess energy acquired by the oscillator from the driving field

$$\langle \Delta E_j(t)\rangle = \frac{1}{2m_j}\langle p_j(t)\rangle^2 + \tfrac{1}{2}m_j\omega_j^2\langle q_j(t)\rangle^2. \tag{14.42b}$$

Making use of Eqs. (14.41), and substituting Eq. (8.68) for $C_j''(\tau)$ we finally have

$$\langle \Delta E_j(t)\rangle = \frac{1}{2m_j}\left|\int_{-\infty}^{t} d\tau\, F(\tau)\exp(i\omega_j\tau)\right|^2. \tag{14.42c}$$

When the pump field is over, $t \to \infty$, $\langle \Delta E_j(t) \rangle$ is proportional to $|I_P(\omega_j)|^2$, the pump field spectral density at frequency ω_j:

$$\langle \Delta E_j(\infty) \rangle = 4[1 + \cos(\mathbf{k}_1 - \mathbf{k}_2)\cdot\mathbf{r}]^2 \kappa_j^2 |I_P(\omega_j)|^2. \tag{14.43}$$

The distribution of the driven oscillator energy can be seen by recasting the doorway wavepacket in the vibronic state representation. The probability of the oscillator to be in its v_j state is given by the diagonal element $\rho(v_j, v_j; t)$.

$$\rho_D(\mathbf{v}, \mathbf{v}; t) = \prod_j \langle v_j | \rho_j(t) | v_j \rangle \equiv \prod_j \rho_D(v_j, v_j; t), \tag{14.44a}$$

where

$$\rho_D(v_j, v_j, t) = \exp[-S_j(t)/(\bar{n}_j + 1)] \times \sum_{l=0}^{v_j} \frac{v_j!}{(v_j - l)!(l!)^2} (\bar{n}_j + 1)^{-2l} [S_j(t)]^l \rho_{eq}(v_j - l, v_j - l). \tag{14.44b}$$

Here $\bar{n}_j$ is the thermal occupation number of mode j [Eq. (8.35)], and $\rho_{eq}(v_j', v_j')$ is the initial thermal equilibrium population in the vibronic level $|v_j'\rangle$, and $S_j(t) \equiv \langle \Delta E_j(t) \rangle / \hbar\omega_j$ is the dimensionless excess energy. At zero temperature this becomes a Poisson distribution

$$\rho_D(v_j, v_j, t) = \exp[-S_j(t)] \frac{[S_j(t)]^{v_j}}{v_j!}. \tag{14.45}$$

Typical value of S_j in off resonant Raman experiments is $\sim 10^{-4}$, which implies that a very small amount of energy is acquired by the driven oscillators.

FOURIER TRANSFORM RELATIONSHIPS

The present formulation establishes very clearly the relationships among various Raman spectroscopies. Upon comparing the spontaneous (frequency-domain) with the coherent Raman signals, we note that the former is related to the correlation function of the polarizability α and the latter to the response function of α. The two are related by the fluctuation dissipation theorem, as shown in Chapter 8

$$\chi_{\alpha\alpha}(t) = \int_{-\infty}^{\infty} d\omega\, S_{SRS}(\omega)[1 - \exp(-\beta\hbar\omega)] \exp(i\omega t). \tag{14.46}$$

Using Eqs. (14.9) or (14.12a) together with Eq. (14.46) we can then calculate the coherent Raman signal using a Fourier transform of the spontaneous Raman spectrum.

Similarly, the picosecond Raman signal is closely related to spontaneous Raman. This may best be seen by expanding the spontaneous correlation function in vibrational states. Combining Eqs. (14.12b) and (14.15) we obtain

$$S_{SRS}(\omega_1 - \omega_2) = \sum_{a,c} \int_{-\infty}^{\infty} d\tau \exp[i(\omega_1 - \omega_2 - \omega_{ca})\tau] \langle \alpha_{ac}(\tau) \alpha_{ca}(0) \rho_{aa} \rangle. \tag{14.47}$$

The picosecond Raman technique selects a particular ac pair and the signal may be related to the same correlation function. To see that, we start with Eq. (14.16b). We assume that the Hamiltonian of the bath modes depends only weakly on the state of the high frequency mode (whether a or c). We then have $\rho_{cc} \cong \exp(-\beta\omega_{ca})\rho_{aa}$. When the temperature is low compared with the vibrational frequency, $k_B T \ll \hbar\omega_{ca}$ (we take $\omega_{ca} > 0$), we can neglect the second term in Eq. (14.16b) and obtain

$$\chi_{ac}(\tau) = -\frac{i}{\hbar}\langle\alpha_{ac}(\tau)\alpha_{ca}(0)\rho_{aa}\rangle.$$

Combining with Eq. (14.23) we finally have

$$S_{\text{PCRS}}(\tau) = \frac{1}{\hbar^2}|\langle\alpha_{ac}(\tau)\alpha_{ca}(0)\rho_{aa}\rangle|^2. \tag{14.48}$$

The picosecond coherent Raman signal can be obtained using the Fourier transform of the spontaneous Raman signal.* Similarly, impulsive Raman scattering is related to the Fourier transform of the stationary CARS.

We shall now discuss the heterodyne detection mode (which is equivalent to pump probe spectroscopy) [7]. For off-resonant pump we need consider only the ground-state contribution to Eq. (13.3) since the excited-state doorway vanishes. It should be noted that the ground-state doorway D_g (and the signal) would also vanish had we invoked the Condon approximation [see Eqs. (13.24)], leaving no signal. Relaxing the Condon approximation and using the notation of this chapter, the ground doorway state D_g is

$$D_g(\omega_p) = \alpha\rho_g - \rho_g\alpha. \tag{14.49}$$

D_g, which is different from the equilibrium ground-state density matrix, evolves in time and shows signatures of nuclear motions, such as quantum beats. Using Eq. (12.52) we can thus write

$$P^{(3)}(\mathbf{k}_S, t) = \left(\frac{i}{\hbar}\right)^3 \int_0^\infty dt_3 \int_0^\infty dt_2\, E_2(t - t_3)|E_1(t + \tau - t_2 - t_3)|^2 \exp(i\omega_2 t_3)$$
$$\times \langle\langle \rho'_w(t_3) - \rho'^{\dagger}_w(t_3)|\mathcal{G}_{gg}(t_2)|\alpha\rho_g - \rho_g\alpha\rangle\rangle. \tag{14.50}$$

Note that in this case $\mathbf{k}_S = \mathbf{k}_2 + \mathbf{k}_1 - \mathbf{k}_1 = \mathbf{k}_2$ and $\omega_S = \omega_2$. When the probe is either off resonant or long compared with electronic dephasing, the time t_3 is short and we can safely assume that the pulses do not vary on that timescale. The signal [Eq. (4.84)] then becomes

$$S(\mathbf{k}_s, t) = -2\omega_S|E_2(t)|^2 \int_0^\infty dt_2 |E_1(t + \tau - t_2)|^2 \,\text{Im}\,\chi_{\alpha\alpha}(t_2) \tag{14.51}$$

Eq. (14.51) also agrees with Eq. (11.6) provided we set $E_{\text{LO}}(t) = E_2(t - t_3)$ and $E_3 = E_1$.

---

* Early theoretical treatments of picosecond CARS had incorrectly argued that the spatial phase matching condition in these experiments makes them selective, i.e., they can distinguish between homogeneous and inhomogeneous vibrational line broadening and, therefore, contain more information than the spontaneous Raman spectrum. This Fourier transform relationship shows however that the two techniques are equivalent [R. F. Loring and S. Mukamel, "Selectivity in coherent transient Raman measurements of vibrational dephasing in liquids," *J. Chem. Phys.* **83**, 2116 (1985)].

If both pulses are further short compared with the vibrational dephasing timescale, the technique is impulsive and we get

$$S(\mathbf{k}_s, \tau) \propto \operatorname{Im} \chi_{\alpha\alpha}(\tau). \tag{14.52}$$

Off-resonance pump-probe spectroscopy is thus related to the same response function that dominates Raman spectroscopies.

BEYOND RAMAN SCATTERING: MULTIDIMENSIONAL OFF-RESONANT SPECTROSCOPY OF LIQUIDS

Nuclear motions in liquids typically span a broad range of timescales, and vibrational lineshapes cannot be simply classified as either homogeneous or inhomogeneous. Even when such classification is possible by virtue of separation of timescales, it is not easy to firmly establish it experimentally. Picosecond CARS experiments conducted on isolated intramolecular high frequency vibrations [10] do not have the time resolution to directly observe the vibrational motions, since the light pulses are longer than the vibrational periods. The decay of the signal with the delay between the excitation and the probe pulses then merely reflects vibrational dephasing. Femtosecond techniques made it possible to probe intermolecular vibrations in the frequency range 0–1000 [cm^{-1}] using an impulsive excitation with pulses short compared with the vibrational periods [9, 19–21]. Under these conditions the time-resolved signal can show the coherent vibrations as well as their dephasing.

Femtosecond four-wave mixing measurements in liquids and crystals (including impulsive Raman, optical Kerr, and pump-probe spectroscopy) yield spectral densities that provide characteristic signatures of intermolecular nuclear degrees of freedom, both local and collective [22]. Our present analysis shows that off-resonant $P^{(3)}$ measurements are related to the Raman response function [Eq. (14.20c)] and depend on a two-time correlation function of the electronic polarization. As far as nuclear dynamics are concerned, we are looking at a single propagation period, and the information gained is formally equivalent to linear infrared absorption, which is related to the two time correlation function of the dipole operator.* It then follows from our analysis of photon echoes (Chapter 10) that it is impossible to deconvolute the inhomogeneous contributions to these spectral densities using off-resonant four-wave mixing techniques, and only higher order techniques related to $P^{(5)}$, $P^{(7)}$, etc. could provide this additional information. By looking at the joint dynamics of N evolution periods we have in effect an N-dimensional spectroscopy. Two-dimensional NMR spectroscopy has proven extremely valuable in the analysis of complex systems such as proteins [23]. A two-dimensional analysis of infrared spectra provides a direct information regarding correlations among vibrational modes [24]. The following analysis has some connection with its NMR counterpart, although the type of information

* This is not to imply that the infrared and Raman spectra are the same. In general, different nuclear modes will have different coupling strengths to the electronic polarizability and to the dipolar operator [W. B. Bosma, L. E. Fried, and S. Mukamel, "Simulation of the intermolecular vibrational spectra of liquid water and water clusters," *J. Chem. Phys.* **98**, 4413 (1993)].

obtained here is very different, and nuclear vibrational Hamiltonians are more complicated than spin Hamiltonians.

In this section we present a closed form expression for the nuclear response function corresponding to $P^{(3)}$ and $P^{(5)}$ for the inhomogeneous multimode Brownian oscillator model (see Chapter 8) with a nonlinear coupling to the radiation field (i.e., through the nonlinear dependence of the electronic polarizability on nuclear coordinates). We analyze the application of a five-pulse ($P^{(5)}$) measurement to molecular liquids [25].

Consider a pulse configuration consisting of a train of N pairs of simultaneous pulses, followed by a final (probe) pulse (Figure 14.7)

$$E(\mathbf{r}, t) = \sum_{j=1}^{N} E_j(\mathbf{r}, t) + E_T(\mathbf{r}, t),$$

where

$$E_j(\mathbf{r}, t) = E_j(t)[\exp(i\Omega_j t - i\mathbf{k}_j\mathbf{r}) + \exp(i\Omega_j' t - i\mathbf{k}_j'\mathbf{r})] + c.c.$$

and

$$E_T(\mathbf{r}, t) = E_T(t)\exp[i(\Omega_T t - \mathbf{k}_T\mathbf{r})] + c.c.$$

Here $E_j(t)$ denotes the temporal profile of the jth pulse. We assume that the pulse pairs are well separated in time, and that the system is initially in thermal equilibrium in the ground electronic state. In an off-resonant experiment conducted with $2N + 1$ laser pulses, related to the polarizations $P^{(2N+1)}$, we need consider only N (rather than $2N$) time evolution periods, in which the system is in the ground state. The third-order polarization for the present pulse configurations is given by

$$P^{(3)}(t) = \int_0^\infty d\tau_1\, E_T(t)E_1^2(t - \tau_1)$$
$$\times \exp(i\Omega_T t - i\mathbf{k}_T\mathbf{r})\{\cos[\Delta\Omega_1(t - \tau_1) - \Delta\mathbf{k}_1\mathbf{r}] + 1\}R^{(3)}(\tau_1), \quad (14.53a)$$

with

$$R^{(3)}(\tau_1) = -\frac{i}{\hbar}\langle[\alpha(0), \alpha(\tau_1)]\rho_g\rangle. \quad (14.53b)$$

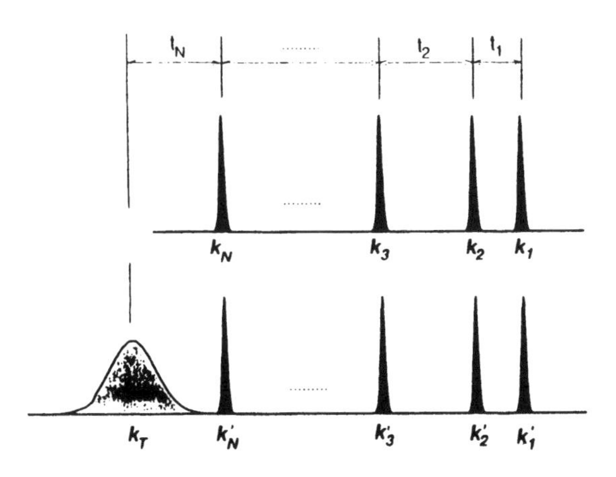

FIG. 14.7 Pulse configuration for a $2N + 1$th-order impulsive experiment. The system first interacts with N pairs of pulses, which have the same time profile $E_j(t)$, but different wavevectors $\mathbf{k}_j$ and $\mathbf{k}_j'$ and frequencies Ω_j and Ω_j' for the jth pair of pulses. The last pulse (k_T, Ω_T) is the probe that generates the signal.

The fifth-order polarization is given by

$$P^{(5)}(t) = \int_0^\infty d\tau_2 \int_0^\infty d\tau_1\, E_T(t)E_2^2(t-\tau_2)E_1^2(t-\tau_2-\tau_1)\exp(i\Omega_T t - i\mathbf{k}_T\mathbf{r})$$
$$\times \{\cos[\Delta\Omega_2(t-\tau_2) - \Delta\mathbf{k}_2\mathbf{r}] + 1\}$$
$$\times \{\cos[\Delta\Omega_1(t-\tau_2-\tau_1) - \Delta\mathbf{k}_1\mathbf{r}] + 1\}R^{(5)}(\tau_2, \tau_1), \tag{14.54a}$$

with

$$R^{(5)}(\tau_2, \tau_1) = -\frac{1}{\hbar^2}\langle[[\alpha(0), \alpha(\tau_2)], \alpha(\tau_1 + \tau_2)]\rho_g\rangle, \tag{14.54b}$$

where $\alpha(t_j)$ represents the operator $\alpha(\mathbf{q})$ in the interaction picture at time t_j [Eq. (14.7)].

We next turn to the multimode Brownian oscillator Hamiltonian [Eqs. (8.61)]. We assume that the electronic polarizability has an exponential dependence on the nuclear coordinates, i.e.,*

$$\alpha(\mathbf{q}) = \alpha_0 \exp\left(a \sum_j A_j q_j\right). \tag{14.55}$$

Here q_j are the optically active modes that couple to the electronic polarizability, and A_j is the coupling strength for the jth oscillator. We further include an inhomogeneous distribution of the oscillator parameters [see Eqs. (8.84)]. In analogy with Eq. (8.81) we define

$$C''(\omega; \Gamma) = \sum_j C_j''(\omega; \Gamma_j). \tag{14.56}$$

$C_j''(\omega)$ are given by Eqs. (8.64b) with $\lambda_j = \hbar A_j^2/(2m_j\omega_j^2)$. $\Gamma_j \equiv \{\lambda_j, \omega_j, \gamma_j\}$ represents the parameters of the model, namely the strength of the dipole interaction (λ_j), the frequency (ω_j), and the relaxation rate (γ_j) of the jth mode. As pointed out in Chapter 8, this model can be used to represent specific coordinates, whether local (e.g., intramolecular) or collective in nature. Even if we do not have a clear idea of the nature of the modes of the system, it can be used as a convenient parameterization. In the condensed phase, the distribution of the values of $\{\lambda_j, \omega_j, \gamma_j\}$ may reflect different slowly interconverting local environments. The Γ dependence is taken into account by averaging over the distribution function of the parameters $S(\Gamma)$.

Expanding in powers of a^2, the lowest two terms in the third- and the fifth-order response functions are given by

$$R^{(3)}(\tau_1) = \frac{2}{\hbar}\alpha_0^2 a^2 \int d\Gamma\, S(\Gamma)C''(\tau_1; \Gamma) \tag{14.57a}$$

* If $\alpha(\mathbf{q})$ were to depend linearly on the nuclear coordinates, the present model would have been linear, and all nonlinear response functions should vanish identically. This can be easily seen from the Heisenberg equation for q, which is given by Eq. (8.65a). In Liouville space this linearity is a result of a destructive interference of various nonlinear paths. The absence of an a^3 term in $R^{(5)}$ Eq. (14.57a) reflects the destructive interference that eliminates the nonlinear nuclear response for the linear model with $\alpha = \alpha_0 a q$.

and

$$R^{(5)}(\tau_2, \tau_1) = \frac{4\alpha_0^4 a^4}{\hbar^2} \int d\Gamma\, S(\Gamma) C''(\tau_2; \Gamma)[C''(\tau_1; \Gamma) + C''(\tau_1 + \tau_2; \Gamma)]. \quad (14.57b)$$

For the multimode Brownian oscillator model we can represent C'' in the form

$$C''(t; \Gamma) = \int d\omega\, J(\omega; \Gamma) \sin(\omega t). \quad (14.58a)$$

Here we have introduced the spectral distribution function

$$J(\omega; \Gamma) = \sum_j \eta_j f(\omega; \omega_j, \gamma_j), \quad (14.58b)$$

where

$$f(\omega; \omega_j, \gamma_j) = \frac{1}{2\pi} \frac{\omega\gamma_j}{(\omega_j^2 - \omega^2)^2 + \omega^2\gamma_j^2}. \quad (14.58c)$$

The effect of the heat bath is expressed by the friction $\gamma_j(\omega)$. The coupling strength is given by

$$\eta_j \equiv \frac{\hbar A_j^2}{m_j}.$$

Using this representation, we obtain the following form for the response function

$$R^{(3)}(\tau_1) = \frac{2}{\hbar} \alpha_0^2 a^2 \int d\omega\, \mathscr{S}_1(\omega) \sin(\omega\tau_1), \quad (14.59a)$$

$$R^{(5)}(\tau_2, \tau_1) = \frac{4}{\hbar^2} \alpha_0^4 a^4 \left\{ \int d\omega\, \mathscr{S}_1(\omega) \sin(\omega\tau_2) \sin[\omega(\tau_1 + \tau_2)] + \int d\omega \int d\omega'\, \mathscr{S}_2(\omega, \omega') \sin(\omega\tau_1) \sin(\omega'\tau_2) \right\}, \quad (14.59b)$$

with

$$\mathscr{S}_1(\omega) \equiv \int d\Gamma\, S(\Gamma) J(\omega; \Gamma)$$

and

$$\mathscr{S}_2(\omega, \omega') \equiv \int d\Gamma\, S(\Gamma) J(\omega; \Gamma) J(\omega'; \Gamma).$$

These expressions illustrate the additional information gained from multidimensional spectroscopy. Higher order measurements are related to a generalized multi-point spectral density which reveals the homogeneous or inhomogeneous nature of the single-frequency spectral density obtained from conventional Raman measurements. $\mathscr{S}_1(\omega)$ depends on the average of the homogeneous spectral density J, whereas $\mathscr{S}_2(\omega, \omega')$ is a kind of two-point correlation function of that spectral density. We next present calculations of femtosecond two-dimensional Raman spectroscopy that probes the low frequency Raman active modes of liquid water in the range between 0 and 1000 cm^{-1}. The impulsive

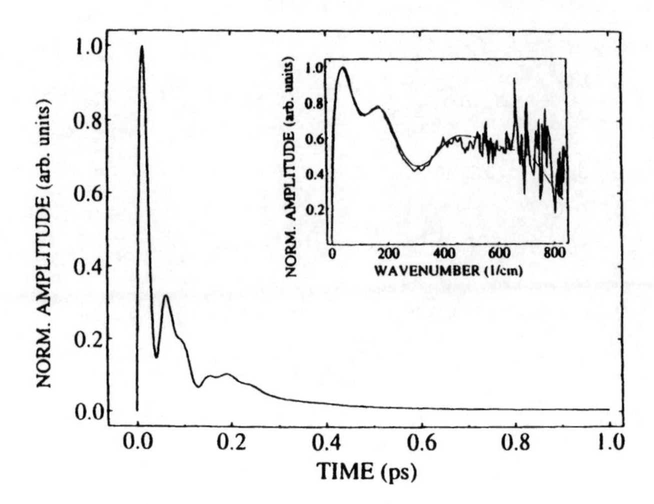

FIG. 14.8 The third-order calculated deconvolved response for water. The homogeneous and inhomogeneous limits are equivalent in this order. The parameters for the calculation have been obtained by fitting the imaginary part of the FFT for the deconvolved, heterodyne, out-of-phase OKE response of water. The inset shows this frequency domain spectrum and the parameterized fit to six Brownian oscillator modes. [S. Palese, J. T. Buontempo, L. Schilling, W. T. Lotshaw, Y. Tanimura, S. Mukamel, and R. J. D. Miller, "Femtosecond two dimensional Raman spectroscopy of liquid water," *J. Phys. Chem.* **98**, 12466 (1994).]

off-resonant Raman spectrum of liquid water and its fit to a 6 mode Brownian oscillator model are displayed in Figure 14.8. Shown is the third-order signal, and the corresponding $R^{(3)}(\omega)$ is given in the insert. $R^{(3)}$ depends on the homogeneous J and inhomogeneous S components only through the combination $\mathscr{S}_1(\omega)$. There are therefore infinite numbers of choices of inhomogeneous distribution $S(\Gamma_s)$ and homogeneous spectral density $J(\omega; \Gamma_s)$, that give the same third-order signal $R^{(3)}$ (Eq. 14.59a). Hereafter we adopt the following choices:

1. A purely homogeneous model, where spectral density is attributed to the six oscillator modes. We then assume that S is a δ function, and the measured spectral density is identified with $J(\omega)$.
2. A pure inhomogeneous model, where the entire distribution is attributed to S.

As can be seen from Eq. (14.59a), the third-order signal is identical for both cases. This is in agreement with our previous analysis to the effect that we cannot distinguish between homogeneous and inhomogeneous contributions to the spectral density using experiments based on the third-order response function under off-resonant excitation. However, the fifth-order off-resonant experiment may be used to distinguish between the two contributions.

For impulsive pump experiments, we set

$$E_T(t) = \delta(t - \tau_1 - \tau_2), \quad E_1(t) = \delta(t), \quad \text{and} \quad E_2(t) = \delta(t - \tau_1);$$

the signal, related to the square of the polarization, is given by (up to a proportionality constant)

$$I^{(3)}(\tau_1) = |R^{(3)}(\tau_1)|^2, \tag{14.60a}$$

$$I^{(5)}(\tau_1, \tau_2) = |R^{(5)}(\tau_2, \tau_1)|^2. \tag{14.60b}$$

Since $J(\omega)$ enters in a different way in Eq. (14.59a) than in Eq. (14.59b), we can observe the difference between the homogeneous and the inhomogeneous models using the fifth-order signal. This is illustrated in the numerical calculations for the pure homogeneous and the pure inhomogeneous cases, shown in Figure 14.9. The fifth-order (three-pulse) signal is very different for the two cases. An example of a seventh-order response $I^{(7)}$ of a single Brownian oscillator in the inhomogeneous limit is shown in Figure 14.10. Several echo signals are clearly seen.

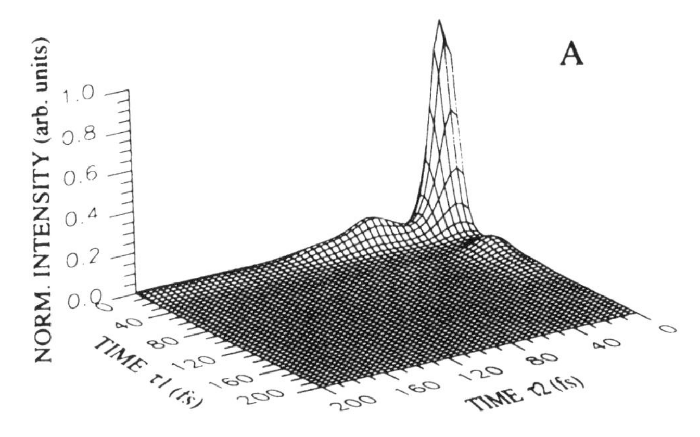

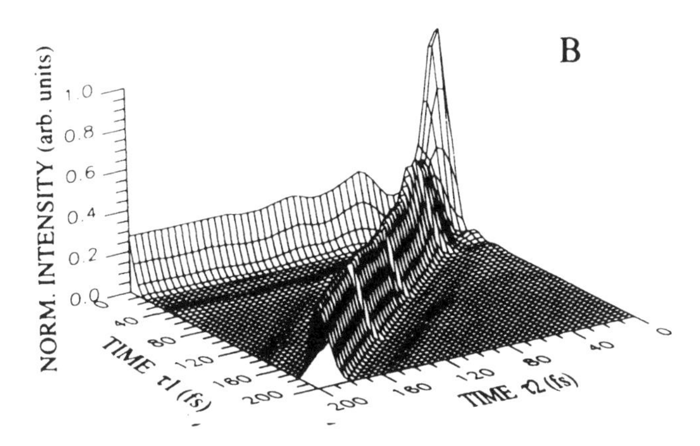

FIG. 14.9 The fifth-order $I^{(5)}(\tau_1, \tau_2)$ calculated response for the deconvolved water spectrum in the (A) homogeneous limit, and (B) inhomogeneous limit. The delay between the excitations is τ_1 and the probe delay from the final excitation is τ_2.

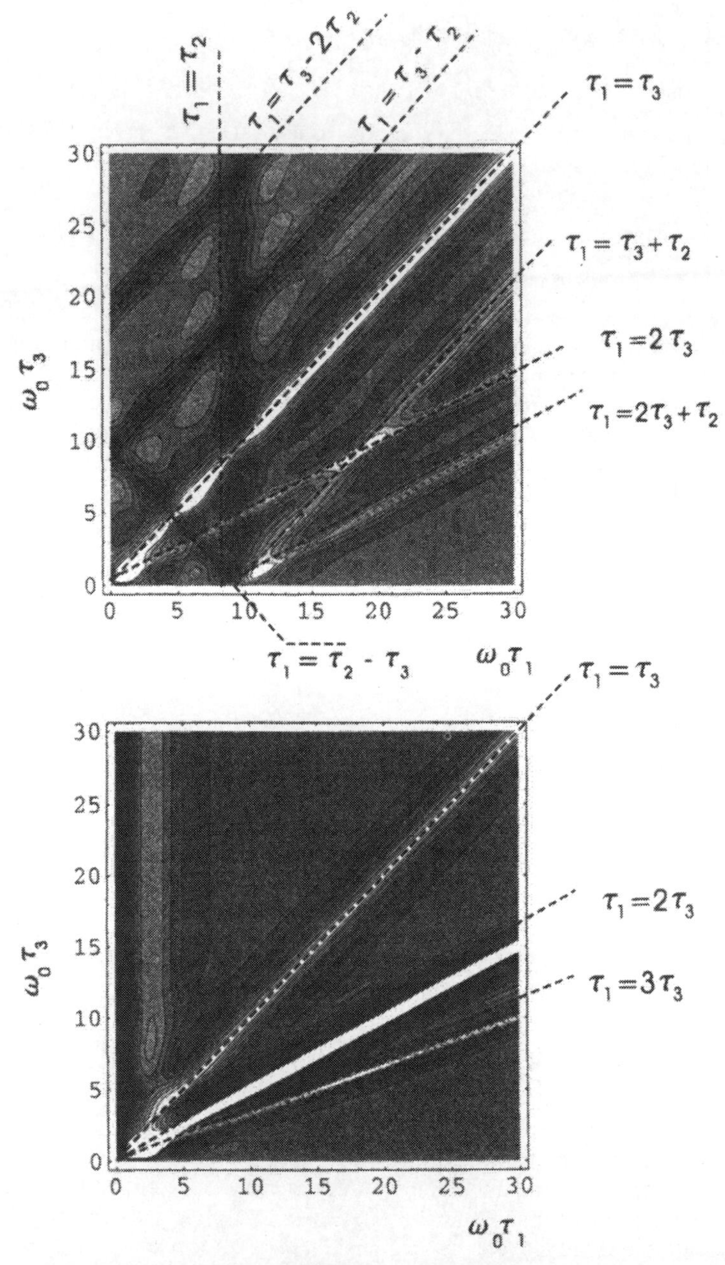

Fig. 14.10 Response function $R^{(7)}(\tau_3,\tau_2,\tau_1)$ of a single Browian oscillator in the inhomogeneous limit. The time arguments are scaled by ω_0, the central frequency of the oscillator's mode distribution. The width of the distribution is $\gamma = \omega_0/2$. The grey background indicates the magnitude of $R^{(7)}$. Dark (light) regions correspond to positive (negative) values. Upper Panel: τ_1 and τ_3 are varied for a fixed $\tau_2(\omega_0\tau_2 = 3\pi)$. Eight echo signals are clearly seen in the directions specified. Lower Panel: $\tau_2 = \tau_3$. Three echo signals can be seen (From: V. Khidekel and S. Mukamel, "High-Order Echoes in Vibrational Spectroscopy of Liquids," *Chem. Phys. Lett.* **240**, 313 (1995)).

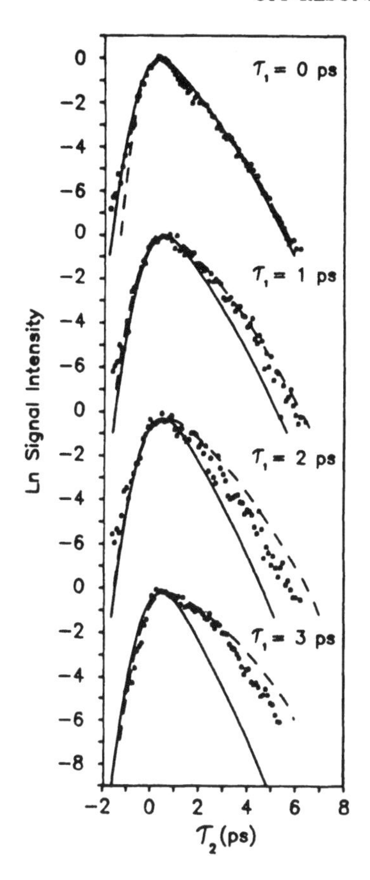

FIG. 14.11 Raman echo data of the 2960 cm^{-1} ν_1 vibration of CH_3I in the solvent $CDCl_3$. (Points) Natural logarithm of the signal intensity vs. τ_2 for various values of τ_1. (Solid curves) Signal decay in the absence of rephasing as predicted from picosecond two-pulse Raman data. The slow decay of the echo data indicates inhomogeneous broadening of the Raman line. (Dashed curves) Signal decay with a static inhomogeneity. The discrepancy with the data is attributed to spectral diffusion within the inhomogeneous distribution. [Adapted from L. J. Muller, D. Vanden Bout, and M. Berg, "Broadening of vibrational lines by attractive forces: Ultrafast Raman echo experiments in a CH_3I:$CDCl_3$ mixture," *J. Chem. Phys.* **99**, 810 (1993).]

The $I^{(5)}$ signal constitutes a two-dimensional spectroscopy with two independent time periods during which the nuclear coherence evolves. A different perspective on these results can be obtained by performing two-dimensional (2D) Fourier transformation, as follows:

$$I^{(5)}(\omega_2, \omega_1) = \left| \int_0^\infty d\tau_1 \int_0^\infty d\tau_2 \exp(i\omega_1\tau_1 + i\omega_2\tau_2) I^{(5)}(\tau_2, \tau_1) \right|^2. \quad (14.61)$$

The homogeneous and inhomogeneous cases, which have an identical 1D spectrum, clearly show very different 2D spectra. Realistic situations of the liquid spectral density are expected to be intermediate between these purely homogeneous and inhomogeneous cases. The form of $C''(\omega)$ and $S(\Gamma\alpha)$ may thus be probed by performing higher order measurements [25, 26].

An analogous technique is related to the vibrational photon echo induced by off-resonant measurements. This is an $R^{(7)}$ technique with

$$\mathbf{k}_{PE} = 2(\mathbf{k}_2' - \mathbf{k}_2) - (\mathbf{k}_1' - \mathbf{k}_1) + \mathbf{k}_T.$$

The theory is identical to the stimulated echo, except that each transition now involves two interactions with the field. The Raman echo was suggested by

Loring and Mukamel as a way of overcoming the limitations of off-resonant CARS and obtaining homogeneous vibrational linewidths in liquids by eliminating inhomogeneous vibrational dephasing [27]. This was demonstrated experimentally (Figure 14.11) [28]. It is, of course, possible to use pulsed infrared lasers [29] to generate vibrational echoes, which will then be simply related to $R^{(3)}$, as discussed in Chapter 10.

APPENDIX 14A

The Doorway State and the Effective Hamiltonian H_{eff}: Derivation of Eqs. (14.33)

The density operator of our system driven by the pump satisfies the Liouville equation:

$$\dot{\rho}_D(\mathbf{r}, t) = -(i/\hbar)[H, \rho_D(\mathbf{r}, t)] + (i/\hbar)E_P(\mathbf{r}, t)[V, \rho_D(\mathbf{r}, t)]. \qquad (14.A.1)$$

Here, H is the total Hamiltonian of the material system [Eq. (7.1)]. $E_P(\mathbf{r}, t) \equiv \tilde{E}_P(\mathbf{r}, t) + c.c.$ is the electric field of the pump pulse [where $\tilde{E}_p$ was defined in Eq. (14.34b)]. Using Eq. (14.32), we shall expand the density operator as a 2×2 matrix in the electronic space and keep each matrix element as an operator in the nuclear space. The Liouville equation thus reads [12]:

$$\dot{\rho}_{gg}(\mathbf{r}, t) = -\frac{i}{\hbar}\mathscr{L}_{gg}\rho_{gg}(\mathbf{r}, t) + \frac{i}{\hbar}E_P(\mathbf{r}, t)[\mathscr{V}_{gg,eg}\rho_{eg}(\mathbf{r}, t) + \mathscr{V}_{gg,ge}\rho_{ge}(\mathbf{r}, t)], \qquad (14.A2a)$$

$$\dot{\rho}_{ee}(\mathbf{r}, t) = -\frac{i}{\hbar}\mathscr{L}_{ee}\rho_{ee}(\mathbf{r}, t) + \frac{i}{\hbar}E_P(\mathbf{r}, t)[\mathscr{V}_{ee,ge}\rho_{ge}(\mathbf{r}, t) + \mathscr{V}_{ee,eg}\rho_{eg}(\mathbf{r}, t)], \qquad (14.A2b)$$

$$\dot{\rho}_{eg}(\mathbf{r}, t) = -\frac{i}{\hbar}(\mathscr{L}_{eg} + \omega_{eg})\rho_{eg}(\mathbf{r}, t) + \frac{i}{\hbar}E_P(\mathbf{r}, t)[\mathscr{V}_{eg,gg}\rho_{gg}(\mathbf{r}, t) + \mathscr{V}_{eg,ee}\rho_{ee}(\mathbf{r}, t)], \qquad (14.A.2c)$$

$$\dot{\rho}_{ge}(\mathbf{r}, t) = -\frac{i}{\hbar}(\mathscr{L}_{ge} - \omega_{eg})\rho_{ge}(\mathbf{r}, t) + \frac{i}{\hbar}E_P(\mathbf{r}, t)[\mathscr{V}_{ge,ee}\rho_{ee}(\mathbf{r}, t) + \mathscr{V}_{ge,gg}\rho_{gg}(\mathbf{r}, t)], \qquad (14.A.2d)$$

where we adopt the notation of Chapter 7, and

$$\mathscr{L}_{mn}A \equiv \hbar^{-1}[H_m A - AH_n]. \qquad (14.A.3)$$

The formal solution of Eqs. (14.A.2) can be obtained as follows: The electronic coherences, ρ_{mn} with $m \neq n$, are first expressed in terms of the electronic populations, ρ_{gg} and ρ_{ee}, by a direct integration of Eqs. (14.A.2c) and (14.A.2d), with the initial conditions $\rho_{eg}(-\infty) = \rho_{ge}(-\infty) = 0$. We then have

$$\rho_{eg}(\mathbf{r}, t) = \frac{i}{\hbar}\int_{-\infty}^{t} d\tau\, E_P(\mathbf{r}, \tau)\exp[-i\omega_{eg}(t - \tau)]\mathscr{G}_{eg}(t - \tau)$$
$$\times [\mathscr{V}_{eg,gg}\rho_{gg}(\mathbf{r}, \tau) + \mathscr{V}_{eg,ee}\rho_{ee}(\mathbf{r}, \tau)]. \qquad (14.A.4)$$

The Hermitian conjugate $\rho_{ge} = [\rho_{eg}]^\dagger$ can be obtained from Eq. (14.A.4) by simply interchanging the indexes g and e throughout. Closed equations of motion for the electronic populations can now be obtained by substituting ρ_{eg} and ρ_{ge} in Eqs. (14.A.2a) and (14.A.2b). We thus have,

$$\dot{\rho}_{gg}(\mathbf{r}, t) = -\frac{i}{\hbar}\mathscr{L}_{gg}\rho_{gg}(\mathbf{r}, t) - \int_{-\infty}^{t} d\tau\, K(t-\tau)\rho_{gg}(\mathbf{r}, \tau) + \int_{-\infty}^{t} d\tau\, K'(t-\tau)\rho_{ee}(\mathbf{r}, \tau), \tag{14.A.5}$$

with

$$K(t-\tau) = E_P(\mathbf{r}, t)E_P(\mathbf{r}, \tau)\exp[-i\omega_{eg}(t-\tau)]\mathscr{V}_{gg,eg}\mathscr{G}_{eg}(t-\tau)\mathscr{V}_{eg,gg} + E_P(\mathbf{r}, t)E_P(\mathbf{r}, \tau)\exp[i\omega_{eg}(t-\tau)]\mathscr{V}_{gg,ge}\mathscr{G}_{ge}(t-\tau)\mathscr{V}_{ge,gg} \tag{14.A.6}$$

$$K'(t-\tau) = -E_P(\mathbf{r}, t)E_P(\mathbf{r}, \tau)\exp[-i\omega_{eg}(t-\tau)]\mathscr{V}_{gg,eg}\mathscr{G}_{eg}(t-\tau)\mathscr{V}_{eg,ee} - E_P(\mathbf{r}, t)E_P(\mathbf{r}, \tau)\exp[i\omega_{eg}(t-\tau)]\mathscr{V}_{gg,ge}\mathscr{G}_{ge}(t-\tau)\mathscr{V}_{ge,ee}. \tag{14.A.7}$$

ρ_{ee} satisfies a similar equation of motion, which can be obtained from Eqs. (14.A.5)–(14.A.7) by simply interchanging the indexes g and e throughout. The first term in the right-hand side of Eq. (14.A.5) describes the adiabatic motions of the system in the electronic ground state in the absence of the external field. $K(t-\tau)$ is a Liouville space operator representing the nuclear relaxation processes following the electronic excitation, both relaxation within the g state and the loss of population from g to e. $K'(t-\tau)$ represents processes in which the system transfers from e to g. Equations of motion for the diagonal elements $\rho_{gg}(\mathbf{r}, t)$ [Eq. (14.A.5)] and $\rho_{ee}(\mathbf{r}, t)$ are called the generalized master equations. Eq. (14.A.5) can be recast more explicitly as

$$\dot{\rho}_{gg}(\Gamma, t) = -i\int d\Gamma'\,\mathscr{L}_{gg}(\Gamma; \Gamma')\rho_{gg}(\Gamma', t) - \int_{-\infty}^{t} d\tau \int d\Gamma'\, K(\Gamma, t-\tau; \Gamma')\rho_{gg}(\Gamma', \tau) + \int_{-\infty}^{t} d\tau \int d\Gamma'\, K'(\Gamma, t-\tau; \Gamma')\rho_{ee}(\Gamma', \tau). \tag{14.A.8}$$

Here, Γ stands for the nuclear degrees of freedom. In the Wigner representation, $\Gamma = \mathbf{pq}$ and $\int d\Gamma' = \iint d\mathbf{p}'\, d\mathbf{q}'$ represents the integration over the entire nuclear phase space. In the vibronic eigenstate representation, $\Gamma = v, v'$ and $\int d\Gamma' = \sum_{vv'}$.

The spatial direction and the optical phase factors associated with the kernels K and K' are determined by the product of the pump fields. From Eqs. (14.34b) and (14.A.1), we have

$$\begin{aligned} E_P(\mathbf{r}, t)E_P(\mathbf{r}, \tau) &= 2[1+\cos(\mathbf{k}_1-\mathbf{k}_2)\mathbf{r}]E_P^*(t)E_P(\tau)\exp[i\omega_p(t-\tau)] \\ &+ 2[1+\cos(\mathbf{k}_1-\mathbf{k}_2)\mathbf{r}]E_P(t)E_P^*(\tau)\exp[-i\omega_p(t-\tau)] \\ &+ [\exp(i\mathbf{k}_1\mathbf{r})+\exp(i\mathbf{k}_2\mathbf{r})]^2 E_P(t)E_P(\tau)\exp(-i2\omega_p t)\exp[i\omega_p(t-\tau)] \\ &+ [\exp(-i\mathbf{k}_1\mathbf{r})+\exp(-i\mathbf{k}_2\mathbf{r})]^2 E_P^*(t)E_P^*(\tau) \\ &\times \exp(i2\omega_p t)\exp[-i\omega_p(t-\tau)]. \end{aligned} \tag{14.A.9}$$

Consider an off-resonant excitation optical process in which $|\omega_p \pm \omega_{eg}|$ are large compared with the inverse timescales of the pulse duration and the nuclear dynamics. In this case, the optical phase contributions to the integrand in the right-hand side of Eq. (14.A.5) are fast varying quantities and we can consider the slowly varying amplitude of the integrand only at $\tau = t$. In evaluating the last two terms of Eq. (14.A.5), we can thus assume

$$E_{\text{P}}^*(t)E_{\text{P}}(\tau) \approx |E_{\text{P}}(t)|^2 \tag{14.A.10a}$$

and

$$\rho_{mm}(\mathbf{r}, \tau) \approx \rho_{mm}(\mathbf{r}, t), \qquad m = \text{e or g}. \tag{14.A.10b}$$

Furthermore, for off-resonant optical processes, we may neglect the molecular nuclear dynamics in the optical coherence Green function and set $\mathscr{G}_{\text{eg}}(t - \tau) \sim \exp[-i\omega_{\text{eg}}(t - \tau)]$. Finally, the last two terms in Eq. (14.A.9) make a negligibly small contribution when we make an optical cycle average and integrate Eq. (14.A.6) in the off-resonant configuration, since they contain the highly oscillatory phase factor $\exp(\pm i2\omega_{\text{p}}t)$. We then find that the kernel K' is negligibly small, whereas the kernel K is pure imaginary and is proportional to $1/(\omega_{\text{p}} \pm \omega_{\text{eg}})$. The excited state population ρ_{ee} is negligible in this limit, and Equation (14.A.6) reduces to Eqs. (14.33).

NOTES

1. C. V. Raman and K. A. Krishnan, *Nature (London)* **121**, 501 (1928); G. Landsberg and L. I. Mandel'stamm, *Naturwissenschaften* **16**, 557 (1928).
2. G. Venkataraman, *Journal into Light: Life and Science of C. V. Raman* (Indian Academy of Sciences, Bangalore, 1988); G. Placzek, *The Rayleigh and Raman Scattering* (1934), UCRL Transl. 256(L) (Washington, D.C.: U.S. Department of Commerce, 1962).
3. B. J. Berne and R. Pecora, *Dynamic Light Scattering* (Wiley, Toronto, 1976).
4. Proceedings of the Fourteenth International Conference on Raman Spectroscopy, N.-T. Yu and X.-Y. Li, Eds. (Wiley, New York, 1994).
5. N. Bloembergen, *Am. J. Phys.* **35**, 989 (1967).
6. R. W. Hellwarth, *Progr. Quant. Electron.* **5**, 2 (1977).
7. M. J. Rosker, F. W. Wise and C. L. Tang, *Phys. Rev. Lett.* **57**, 321 (1986); F. W. Wise, M. J. Rosker, and C. L. Tang, *J. Chem. Phys.* **86**, 2827 (1987).
8. K. A. Nelson and E. P. Ippen, *Adv. Chem. Phys.* **75**, 1 (1989); L. Dhar, J. A. Rogers, and K. A. Nelson, *Chem. Rev.* **94**, 157 (1994).
9. C. Kalpouzos, D. McMorrow, W. T. Lotshaw, and G. A. Kenney-Wallace, *Chem. Phys. Lett.* **155**, 240 (1989); **150**, 138 (1988); D. McMorrow, W. T. Lotshaw, G. A. Kenney-Wallace, *IEEE J. Quant. Electron.* **QE-24**, 443 (1988); D. McMorrow and W. T. Lotshaw, *J. Phys. Chem.* **95**, 10395 (1991).
10. A. Laubereau and W. Kaiser, *Rev. Mod. Phys.* **50**, 607 (1978); W. Zinth, M. C. Nussn and W. Kaiser, *Phys. Rev. A* **30**, 1139 (1984); W. Zinth, H.-J. Polland, A. Laubereau, and W. Kaiser, *Appl. Phys.* **B26**, 77 (1981); W. Zinth, R. Leonhardt, H. Holzapfel, and W. Kaiser, *IEEE J. Quant. Electron.* **24**, 455 (1988).
11. S. M. George, A. M. Harris, M. Berg, and C. B. Harris, *J. Chem. Phys.* **73**, 5573 (1984); R. Inaba, H. Okamoto, K. Yoshihara, and M. Tasumi, *J. Phys. Chem.* **96**, 8385 (1992).
12. Y. J. Yan and S. Mukamel, *J. Chem. Phys.* **94**, 997 (1991).
13. A. M. Walsh and R. F. Loring, *J. Chem. Phys.* **93**, 7566 (1990); R. F. Loring, *J. Phys. Chem.* **94**, 513 (1990).
14. A. M. Weiner and J. P. Heritage, *Rev. Phys. Appl.* **22**, 1619 (1987); A. M. Weiner, D. E. Leaird, G. P. Wiederecht, and K. A. Nelson, *Science* **247**, 1317 (1990).

15. J. Paye, M. Ramaswsamy, J. G. Fujimoto, and E. P. Ippen, *Opt. Lett.* **18**, 1946 (1993).

16. D. J. Kane and R. Trebino, *IEEE J. Quant. Electr.* **29**, 571 (1993).

17. W. S. Warren, H. Rabitz, and M. Dahleh, *Science* **259**, 1581 (1993); H. Rabitz and S. Shi, *Adv. Mol. Vib. Coll. Dyn.* **1A**, 187 (1991).

18. J. Paye, *IEEE J. Quant. Electr.* **28**, 2262 (1992).

19. T. Hatori and T. Kobayashi, *J. Chem. Phys.* **94**, 3332 (1991).

20. N. F. Scherer, L. D. Ziegler, and G. R. Fleming, *J. Chem. Phys.* **96**, 5544 (1992); M. Cho, G. R. Fleming, and S. Mukamel, *J. Chem. Phys.* **98**, 5314 (1993); M. Cho, M. Du, N. F. Scherer, G. R. Fleming, and S. Mukamel, *J. Chem. Phys.* **99**, 2410 (1993).

21. C. E. Barker, R. Trebino, A. G. Kostenbauder, and A. E. Seigman, *J. Chem. Phys.* **92**, 4740 (1990);

22. M. Buchner, B. M. Ladanyi, and R. M. Straat, *J. Chem. Phys.* **97**, 8522 (1992); T. Keyes and B. M. Ladanyi, *Adv. Chem. Phys.* **56**, 411 (1984); Z. Chen and R. M. Stratt, *J. Chem. Phys.* **95**, 2669 (1991).

23. R. R. Ernst, G. Bodenhausen, and A. Wokaun, *Principles of Nuclear Magnetic Resonance in One and Two Dimensions* (Clarendon Press, Oxford, 1987).

24. I. Noda, *J. Am. Chem. Soc.* **111**, 8116 (1989); *Appl. Spectrosc.* **444**, 550 (1990).

25. Y. Tanimura and S. Mukamel, *J. Chem. Phys.* **99**, 9496 (1993).

26. K. Tominaga, Y. Naitoh, T. Kang, and K. Yoshihara, in *Ultrafast Phenomena IX*, G. A. Mourou, A. H. Zewail, P. F. Barbara, and W. H. Knox, Eds. (Springer-Verlag, Berlin, 1994), p. 143; K. Tominaga, G. P. Keogh, Y. Naitoh, and K. Yoshihara, *J. Raman Spectrosc.*, **26** (1995).

27. R. F. Loring and S. Mukamel, *J. Chem. Phys.* **83**, 2116 (1985).

28. D. V. Bout, L. J. Muller, and M. Berg, *Phys. Rev. Lett.* **67**, 3700 (1991).

29. D. Zimdars, A. Tokmakoff, S. Chen, S. R. Greenfield, M. D. Fayer, T. I. Smith, and H. A. Schwettmann, *Phys. Rev. Lett.* **70**, 2718 (1993).

BIBLIOGRAPHY

J. P. Hansen and I. R. MacDonald, *Theory of Simple Liquids* (Academic, New York, 1976).

CHAPTER 15

Polarization Spectroscopy: Birefringence and Dichroism

The optical susceptibilities (and response functions) are high-rank tensors. The nth-order response function connects n vectors (the electric fields) to the polarization vector and is, therefore, a tensor of rank $n + 1$ [e.g., $\chi^{(3)}$ is a fourth-rank tensor]. As such it requires $n + 1$ indices to denote its components [1–6]. The tensor notation was introduced in Chapter 5 [Eqs. (5.45)] but in all the applications made so far we did not consider it explicitly.

In this chapter we focus on four-wave mixing techniques that depend on the polarization of light and reveal the tensor nature of the response functions. The simplest polarization-sensitive techniques include the depolarization of spontaneous Raman spectra, and time-resolved fluorescence anisotropy measurements [7, 8]. Polarization spectroscopy can be performed, however, using many other configurations. For example, off-resonant picosecond Raman measurements with homodyne detection [9, 10], described in Chapter 14. Heterodyne detection introduced in Chapter 4 has many advantages [11] since it provides an improved sensitivity as well as information about the phase of the signal field. The first femtosecond heterodyne measurements were carried out in liquid CS_2 [12, 13]. There is a multitude of terminologies and four letter acronyms associated with polarization techniques, and this is often a source of confusion; some of these terms refer to the operational aspects of the techniques whereas others are connected with the fundamental level, i.e., their information content. Heterodyne detected Raman is obviously identical to pump-probe spectroscopy (it is a matter of taste whether one views the second pulse as a probe or as a local oscillator). The dependence of the probe absorption on its relative polarization direction with respect to the pump is known as the optical Kerr effect (OKE)[14–19]. The difference of probe transmission with two orthogonal polarizations (usually $\pm 45°$) with respect to the pump is known as the optical Kerr spectrum. The same type of measurements can thus be denoted stimulated Raman, heterodyne-detected optical Kerr, impulsive stimulated light scattering, or pump probe. This is not yet the end of the story, and other names are also used to describe the same techniques. Since the phase of the signal field can be measured in heterodyne detection, it is possible to classify the techniques by the relative phase of the signal with respect to the pump. The dispersive in-phase contribution to the optical polarization (which gives an out-of-phase signal field) is known as the birefringence component [20]. The absorptive out-of-phase polarization (which gives an in-phase signal field) is denoted the dichroism component [21, 22]. These components dominate the off resonant and the resonant response, respectively. The term optical Stark effect (or ac Stark effect) is often used to describe the spectral shift induced in the probe absorption due to a near-resonant pump [23]. This is another feature of the same class of techniques.

The tensorial properties of the susceptibility have been extensively studied in crystals, and general rules regarding the number of independent components of the tensor for various crystal symmetries have been developed and tabulated [2, 3]. In this chapter we discuss the tensorial aspects of four-wave mixing spectroscopies in liquids, which provide a direct probe for molecular rotational motions. Resonant fluorescence anisotropy [8] and off-resonant light scattering [7] measurements are widely used to study rotational motions of large molecules with strong fluorescence emission and large scattering cross section, respectively. For small molecules, time-dependent coherent Raman spectroscopy using polarized light pulses was shown to be an effective tool [24]. Figure 15.1 illustrates how the reorientation relaxation of an aromatic molecule (biphenyl) in solution can be probed in an off-resonant optical Kerr measurement. The symmetry of the $\chi^{(3)}$ tensor makes it possible to select certain properties of a system for observation by choosing the appropriate polarizations of the three input optical pulses and the polarization of the signal. Signals 1 and 2 in Figure 15.1a are recorded with the electronic Kerr effect and nuclear Kerr effect (molecular orientation) polarization configurations, respectively. Signal 1 is the instrument response which is identical to the contribution of the instantaneous electronic polarization within the Born-Oppenheimer (adiabatic) approximation. Signal 2 reflects the short time scale orientational dynamics (the nuclear Kerr response). The initial ultrafast response ($\sim$100 fs) arises from liberational (hindered rotational) dynamics of biphenyl. The displacement of curve 2 relative to 1 and the rise time reflect the short-time inertial motions associated with the librations following excitation. The $\sim$picosecond decay is due to librational damping and dephasing. Following that, a residual orientational anisotropy remains, which decays on a much longer time scale by rotational diffusion, as shown in Figure 15.1b.

To elucidate the rotational dynamics of a chromophore embedded in a liquid, we shall consider a general three-pulse transient grating experiment with heterodyne detection mode [14, 20]. This technique, which allows the independent control of all polarization directions, is conceptually the simplest and is best suited for defining precisely the information gained in polarization spectroscopy. Experimentally it is very demanding, and many ingenious ways have been designed to get away with using fewer laser fields. Heterodyne Raman measurements, discussed briefly in Chapter 14, are a special case of this general configuration. We first formulate the third-order polarization using a fourth-rank tensorial nonlinear response function. The contribution of each Liouville space pathway is factorized into two terms: (1) the electronic and vibrational parts described by the tetradic Green function in Liouville space and (2) the reorientational contribution that is incorporated through its conditional probability in phase space. The vibrational contribution will be treated using the doorway-window representation. Since we shall be interested in resonant as well as off-resonant excitation, we generalize the results of Chapter 13 by not invoking the RWA. Rotational dynamics will be treated using the Fokker–Planck equation for the rotational diffusive motion of the transition dipole moment vector [25–27]. Dichroism and birefringence measurements and their dependence on detuning will be analyzed.

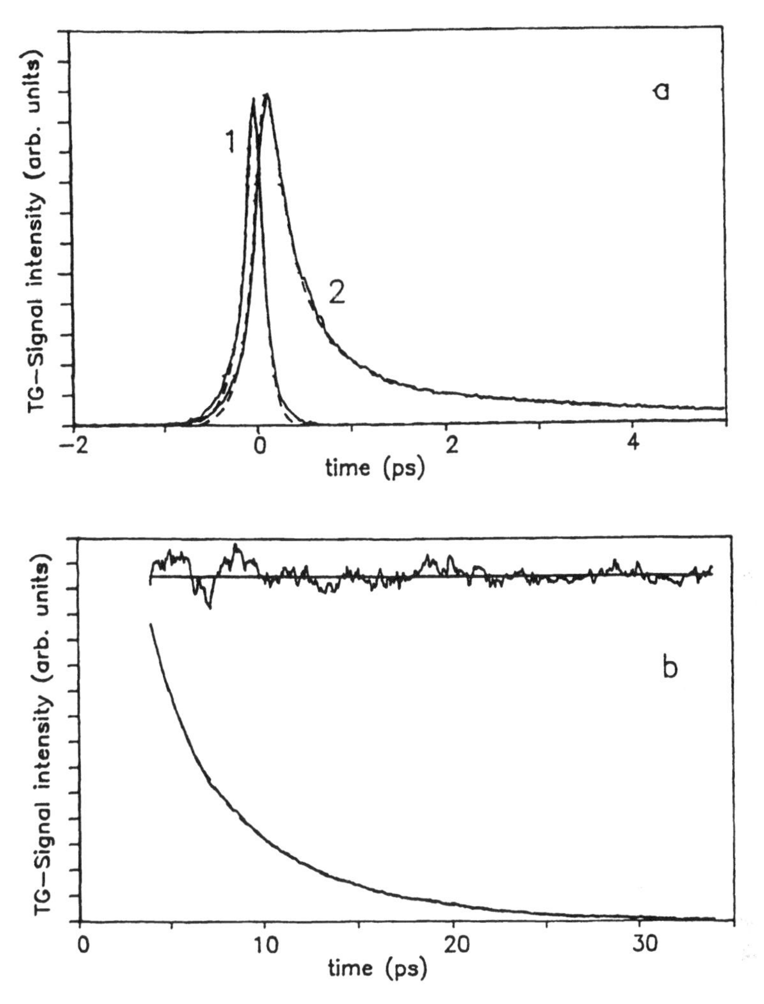

FIG. 15.1 Transient grating optical Kerr effect (TG-OKE) signals $S(t)$ for a 0.77 mol/liter biphenyl/n-heptane solution at 23. (a) Fast time scale behavior. (b) Slow time scale behavior of the signal. The dashed line is a fit using a biexponential decay function. The dashed lines through the signals are calculated through the evaluation of the proper convolution integrals. The upper trace shows the residuals of the fit on a 10× enlarged vertical scale. A biexponential is appropriate for orientational relaxation when the moment of inertia polarizability tensors coincide, or almost coincide, as in the case of biphenyl. [From F. W. Deeg, J. J. Stankus, S. R. Greenfield, V. J. Newell, and M. D. Fayer, "Anistropic reorientational relaxation of biphenyl: Transient grating optical Kerr effect measurements," *J. Chem. Phys.* **90**, 6893 (1989).].

ROTATIONAL CONTRIBUTION TO THE NONLINEAR RESPONSE FUNCTION

The third-order polarization vector, $\mathbf{P}^{(3)}(\mathbf{k}_s, t)$, can be described by using the nonlinear response function, $S^{(3)}(t_3, t_2, t_1)$, which is a fourth-rank tensor. The

η-component of $\mathbf{P}^{(3)}(\mathbf{k}_s, t)$ in Cartesian coordinates is given by

$$P_\eta^{(3)}(\mathbf{k}_s, t) = \sum_p \int_0^\infty dt_3 \int_0^\infty dt_2 \int_0^\infty dt_1\, S_{\eta\alpha\beta\gamma}^{(3)}(t_3, t_2, t_1)$$
$$\times \exp[i(\omega_1 + \omega_2 + \omega_3)t_3 + i(\omega_1 + \omega_2)t_2 + i\omega_1 t_1]$$
$$\times E_{3\alpha}(t - t_3)E_{2\beta}(t - t_3 - t_2)E_{1\gamma}(t - t_3 - t_2 - t_1). \qquad (15.1)$$

Denoting the initial field polarization directions in the laboratory frame as γ, β, and α, the η-component of the polarization vector is associated with $S_{\eta\alpha\beta\gamma}^{(3)}(t_3, t_2, t_1)$. We use the conventional summation notation for the repeated Greek letters α, β, and γ. The Greek letters take the values X, Y, and Z in the laboratory frame. (To avoid confusion with the Euler angles ν_j defined below we use $\alpha, \beta, \gamma, \eta$ rather than ν_1, ν_2. ν_3, ν_s used in Chapter 5 [28].

We consider a chromophore in a solvent. Neglecting the coupling between the vibronic and rotational degrees of freedom, the molecular Hamiltonian is partitioned as follows:

$$H = H_{\text{vib}} + H_{\text{rot}}. \qquad (15.2)$$

We shall adopt the two-level Hamiltonian [Eq. (7.1)] for the vibronic Hamiltonian H_{vib}, which describes the electronic and vibrational motions. H_{rot} is the rotational part of the Hamiltonian, and the transition dipole moment is given by

$$V = V^0 \hat{\mu}, \qquad (15.3)$$

V^0 is the vibronic part and $\hat{\mu}$ is the unit vector defining the orientation of the transition dipole moment of the chromophore in the laboratory frame.

Since the vibrational and rotational motions are completely decoupled in our model, we have

$$\rho(-\infty) = \rho_V(-\infty)\rho_R(-\infty).$$

Consequently, the contribution of each Liouville space path to the nonlinear response functions may be factorized into a product of rotational and vibrational terms

$$S_{\eta\alpha\beta\gamma}^{(3)}(t_3, t_2, t_1) = Y_{\eta\alpha\beta\gamma}(t_3, t_2, t_1) \sum_{n=1}^{4} [R_n(t_3, t_2, t_1) - R_n^*(t_3, t_2, t_1)], \qquad (15.4)$$

where R_n are the vibronic contributions discussed in Chapters 6 and 13, and Y is the rotational contribution. The four Liouville space pathways differ by the various actions of $\hat{\mu}$ from the left and from the right. Had we treated the rotations quantum mechanically, we would need four different terms, Y_n in analogy with R_n. However, here we treat the rotational motion classically in phase space and assume that it is identical in both electronic states so that the rotational contribution to all Liouville space paths is identical. The Cartesian components (X, Y, Z) in the space-fixed (laboratory) frame L can be related to the Cartesian components (x, y, z) in the molecule-fixed (body) frame M by a unitary transformation Φ that is parameterized by the Euler angles ϕ, θ, and χ [28]:

$$\begin{bmatrix} x \\ y \\ z \end{bmatrix} = \Phi_{LM} \begin{bmatrix} X \\ Y \\ Z \end{bmatrix} \tag{15.5a}$$

$$\Phi_{LM} = \begin{pmatrix} \cos\chi & \sin\chi & 0 \\ -\sin\chi & \cos\chi & 0 \\ 0 & 0 & 1 \end{pmatrix} \begin{pmatrix} \cos\theta & 0 & -\sin\theta \\ 0 & 1 & 0 \\ \sin\theta & 0 & \cos\theta \end{pmatrix} \begin{pmatrix} \cos\phi & \sin\phi & 0 \\ -\sin\phi & \cos\phi & 0 \\ 0 & 0 & 1 \end{pmatrix}. \tag{15.5b}$$

Since Φ_{LM} are the real elements of a unitary transformation, $\Phi_{LM}^{-1} = \Phi_{ML}$. The fourth-rank tensor Y represents the rotational contribution to the response function and is given by the four-time correlation function of the unit vector defining the orientation of the optical transition dipole moment, which is taken to be on the z-axis in the molecule-fixed frame, $(\hat{x}, \hat{y}, \hat{z}) = (0, 0, 1)$. Using the direction cosine matrix, $\hat{\mu}$ is completely specified by the Euler angles,

$$\hat{\mu}(\phi, \theta, \chi) = (\hat{\mu}_X, \hat{\mu}_Y, \hat{\mu}_Z) = (\cos\phi \sin\theta, \sin\phi \sin\theta, \cos\theta). \tag{15.6}$$

The rotational phase space coordinates are the Euler angles $\nu \equiv (\phi, \theta, \chi)$, and the angular velocities $\bar{\omega}_1$, $\bar{\omega}_2$, $\bar{\omega}_3$. The components of the angular velocity, $\bar{\omega}_1$, $\bar{\omega}_2$, and $\bar{\omega}_3$, are defined in the molecular frame M.

Let $P(\bar{\omega}, \nu, t \,|\, \bar{\omega}_0, \nu_0)\, d\bar{\omega}\, d\nu$ represent the conditional rotational phase-space probability density that the molecule has angular velocity $\bar{\omega} = (\bar{\omega}_1, \bar{\omega}_2, \bar{\omega}_3)$ in $d\bar{\omega}$ and orientation $\nu = (\phi, \theta, \chi)$ in $d\nu \equiv \sin\theta\, d\phi\, d\theta\, d\chi$ at time $t \geq 0$. The initial condition for $P(\bar{\omega}, \nu, t \,|\, \bar{\omega}_0, \nu_0)$ is

$$P(\bar{\omega}, \nu, t = 0 \,|\, \bar{\omega}_0, \nu_0) = \delta(\bar{\omega} - \bar{\omega}_0)\, \delta(\nu - \nu_0). \tag{15.7}$$

It is useful to define the conditional probability density in angular configuration space by tracing over the angular velocity, i.e.,

$$W(\nu, t \,|\, \nu_0) \equiv \int d\bar{\omega} \int d\bar{\omega}_0\, P(\bar{\omega}, \nu, t \,|\, \bar{\omega}_0, \nu_0) P_0(\bar{\omega}_0), \tag{15.8}$$

where $P_0(\bar{\omega}_0)$ is the initial distribution of angular velocities, given by the Maxwell–Boltzmann expression

$$P_0(\bar{\omega}_0) = (I_1 I_2 I_3)^{1/2}/(2\pi k_B T)^{3/2} \exp\left[-\sum_{i=1}^{3} I_i \bar{\omega}_i^2 / 2k_B T \right]. \tag{15.9}$$

Here I_j are three principal moments of inertia. The rotational tensorial contribution to the response function thus becomes

$$Y_{\eta\alpha\beta\gamma}(t_3, t_2, t_1) \equiv \int d\nu_3 \int d\nu_2 \int d\nu_1 \int d\nu_0\, (\hat{\mu}_3 \cdot \hat{e}_\eta) W(\nu_3, t_3 \,|\, \nu_2) \times (\hat{\mu}_2 \cdot \hat{e}_\alpha) W(\nu_2, t_2 \,|\, \nu_1)(\hat{\mu}_1 \cdot \hat{e}_\beta) W(\nu_1, t_1 \,|\, \nu_0)(\hat{\mu}_0 \cdot \hat{e}_\gamma) P_0(\nu_0). \tag{15.10}$$

Here $\hat{e}_\varepsilon$ ($\varepsilon = \eta, \alpha, \beta, \gamma$) are the unit vectors defining the polarization direction of the external fields, and $\hat{\mu}$ is related to the Euler angles ν, as given in Eq. (15.6). We define

$$\hat{\mu}_j \equiv \hat{\mu}(\nu_j), \qquad j = 0, 1, 2, 3 \tag{15.11}$$

where ν_0, ν_1, ν_2, and ν_3 denote the Euler angles defining the molecular orientation at times 0, t_1, $t_1 + t_2$, and $t_1 + t_2 + t_3$, respectively.

In Eq. (15.10), $W(\nu_1, t_1 \mid \nu_0)$ describes the evolution of the rotational degrees of freedom from ν_0 to ν_1 during the first propagation period t_1, and so on. The conditional probability satisfies the initial condition, $W(\nu_i, t = 0 \mid \nu_{i-1}) = \delta(\nu_i - \nu_{i-1})$. The scalar product $\hat{\mu} \cdot \hat{e}$ is related to the amplitude that the molecule having the transition dipole moment along $\hat{\mu}$ will be excited by interaction with the external field polarized along $\hat{e}$. For example, if the field is linearly polarized along the Z-axis, $(\hat{\mu} \cdot \hat{e}_Z)^2 = \cos^2 \theta$, and molecules oriented parallel to the Z-axis are excited preferentially. If all initial orientations are distributed uniformly (i.e., isotropic system) the initial distribution function, $P_0(\nu)$, is simply $1/8\pi^2$.

VIBRATIONAL CONTRIBUTION TO HETERODYNE-DETECTED TRANSIENT GRATING

The phase-locked transient grating was introduced in Chapter 11, and the signal is given by Eq. (11.6). For now we shall neglect the rotational motions and focus on the vibrational contribution.

We expand the doorway and the window wavepackets in the complete basis of molecular vibronic eigenstates. We then express the electronic polarizability in terms of matrix elements of the doorway and the window wavepackets.

We shall calculate the signal in the snapshot limit discussed in Chapter 13. We can then use Eqs. (13.17). However, since we are interested in resonant as well as off-resonant contributions we have to extend these equations to include non-RWA terms. The derivation follows the same steps that led to Eq. (13.17) and results in the snapshot signal given by

$$\begin{aligned} S_0(\omega_1, \omega_2, \tau; \omega_{\mathrm{LO}}, \psi) &= \sum_{db} \rho^0_{\mathrm{W}}(db; \omega_2, \omega_{\mathrm{LO}}, \psi) \exp(-i\omega_{db}\tau - \Gamma_{db}\tau) \rho^0_{\mathrm{D}}(db; \omega_1) \\ &\quad + \sum_{a,c} \rho^0_{\mathrm{W}}(ca; \omega_2, \omega_{\mathrm{LO}}, \psi) \exp(-i\omega_{ca}\tau - \Gamma_{ca}\tau) \rho^0_{\mathrm{D}}(ca; \omega_1), \end{aligned} \tag{15.12}$$

with

$$\begin{aligned} \rho^0_{\mathrm{D}}(ca; \omega_1) = \sum_d \mu_{ad}\mu_{dc} \Bigg[&\frac{P(a)}{\omega_1 - \omega_{da} + i\Gamma_{da}} - \frac{P(a)}{\omega_1 + \omega_{da} - i\Gamma_{da}} \\ &- \frac{P(c)}{\omega_1 - \omega_{dc} - i\Gamma_{dc}} + \frac{P(c)}{\omega_1 + \omega_{dc} + i\Gamma_{dc}} \Bigg], \end{aligned} \tag{15.13a}$$

$$\begin{aligned} \rho^0_{\mathrm{D}}(db; \omega_1) = \sum_a P(a) \mu_{ba}\mu_{ad} \Bigg[&\frac{1}{\omega_1 - \omega_{da} + i\Gamma_{da}} - \frac{1}{\omega_1 + \omega_{da} - i\Gamma_{da}} \\ \times &- \frac{1}{\omega_1 - \omega_{ba} - i\Gamma_{ba}} + \frac{1}{\omega_1 + \omega_{ba} + i\Gamma_{ba}} \Bigg], \end{aligned} \tag{15.13b}$$

$$\rho_{\rm w}^{0}(ca; \omega_2, \omega_{\rm LO}, \psi) = \sum_b \mu_{cb}\mu_{ba}$$

$$\times \left[\frac{\exp(i\psi)}{\omega_{\rm LO} - \omega_{bc} + i\Gamma_{bc}} - \frac{\exp(-i\psi)}{\omega_{\rm LO} + \omega_{bc} - i\Gamma_{bc}} - \frac{\exp(-i\psi)}{\omega_{\rm LO} - \omega_{ba} - i\Gamma_{ba}} + \frac{\exp(i\psi)}{\omega_{\rm LO} + \omega_{ba} + i\Gamma_{ba}} \right], \tag{15.13c}$$

$$\rho_{\rm w}^{0}(db; \omega_2, \omega_{\rm LO}, \psi) = \sum_c \mu_{dc}\mu_{cb}$$

$$\times \left| \frac{\exp(i\psi)}{\omega_{\rm LO} - \omega_{bc} + i\Gamma_{bc}} - \frac{\exp(-i\psi)}{\omega_{\rm LO} + \omega_{bc} - i\Gamma_{bc}} - \frac{\exp(-i\psi)}{\omega_{\rm LO} - \omega_{dc} - i\Gamma_{dc}} + \frac{\exp(i\psi)}{\omega_{\rm LO} + \omega_{dc} + i\Gamma_{dc}} \right|. \tag{15.13d}$$

These expressions generalize Eqs. (13.13)–(13.16) to include non-RWA terms. The relaxation parameters $\Gamma_{\nu\nu'}$ were defined in Eq. (6.15).

THE POLARIZATION-DEPENDENT GRATING SIGNAL

We now put all our results together, and calculate the grating signal including the vibrational and the rotational contributions [20]. We first introduce a key approximation, which is usually justified for typical dyes in the condensed phases. This approximation greatly simplifies the treatment of rotational motions and will allow us to consider a more general Hamiltonian than was used in Chapter 3. The time periods t_1 and t_3 are limited by the ultrafast optical dephasing processes as well as pulse durations. Both of these are usually short compared with the timescale of the reorientational motion. We shall therefore neglect the molecular rotation during these time periods (this is analogous to the snapshot limit discussed in Chapter 13), and set

$$W(\nu_1, t_1 \,|\, \nu_0) \cong \delta(\nu_1 - \nu_0),$$

$$W(\nu_3, t_3 \,|\, \nu_2) \cong \delta(\nu_3 - \nu_2),$$

and

$$W(\nu_2, t' + \tau - t'' \,|\, \nu_1) \cong W(\nu_2, \tau \,|\, \nu_1). \tag{15.14}$$

In general we should consider the rotational motion when the system is either in an electronic population (ρ_{gg}, ρ_{ee}) or a coherence (ρ_{ge}, ρ_{eg}). The former has a well-defined classical analogue, whereas the latter is less obvious and requires specifying a more detailed model. Once approximations [Eq. (15.14)] are made, we no longer need the rotational propagation during electronic coherence. The optical heterodyne-detected pump-probe signal can be written as*

$$\bar{\bar{S}}(\omega_1, \omega_2, \tau; \omega_{LO}, \psi) = S_0(\omega_1, \omega_2, \tau; \omega_{LO}, \psi)\bar{\bar{Y}}_0(\tau), \tag{15.15}$$

where S_0 denotes the scalar (orientation independent) part, and the rotational contribution is given by the tensor

$$\bar{\bar{Y}}_0(\tau) \equiv \int dv \int dv_0\, \bar{Y}_w(\hat{\mu}) W(v, \tau \,|\, v_0) \otimes \bar{Y}_D(\hat{\mu}_0). \tag{15.16}$$

Here $\bar{Y}_D(\hat{\mu}_0)$ and $\bar{Y}_w(\hat{\mu})$ are second-rank tensor associated with the doorway and window wavepackets, defined as

$$\bar{Y}_D(\hat{\mu}_0) \equiv \hat{\mu}_0 \otimes \hat{\mu}_0 P_0(v_0) \tag{15.17a}$$

and

$$\bar{Y}_W(\hat{\mu}) \equiv \hat{\mu} \otimes \hat{\mu}. \tag{15.17b}$$

$\bar{Y}_D(\hat{\mu}_0)$ contains the distribution function for initial molecular orientations, $P_0(v_0)$, while $\bar{Y}_W(\hat{\mu})$ does not. The propagation of the dipole operator is described by the conditional probability function, $W(v, \tau; v_0)$.

We shall now calculate two components of the signal. We assume that the pump pulse is linearly polarized along the Z-axis and that the probe pulse and the local oscillator are polarized either along X or Z. The corresponding signals are given by

$$\left.\begin{aligned} S^{\parallel}(\omega_1, \omega_2, \tau; \omega_{LO}, \psi) &= S_0(\omega_1, \omega_2, \tau; \omega_{LO}, \psi) \times [\bar{\bar{Y}}_0(\tau)]_{ZZZZ}, \\ S^{\perp}(\omega_1, \omega_2, \tau; \omega_{LO}, \psi) &= S_0(\omega_1, \omega_2, \tau; \omega_{LO}, \psi) \times [\bar{\bar{Y}}_0(\tau)]_{YYZZ}. \end{aligned}\right\} \tag{15.18}$$

Using the direction cosine matrix elements, the tensor elements in Eq. (15.21) are

$$\left.\begin{aligned} [\bar{\bar{Y}}_0(\tau)]_{ZZZZ} &= (8\pi^2)^{-1} \int \sin\theta\, d\phi\, d\theta\, d\chi \\ &\quad \times \int \sin\theta_0\, d\phi_0\, d\theta_0\, d\chi_0 \cos^2\theta\; W(v, \tau; v_0) \cos^2\theta_0, \\ [\bar{\bar{Y}}_0(\tau)]_{YYZZ} &= (8\pi^2)^{-1} \int \sin\theta\, d\phi\, d\theta\, d\chi \\ &\quad \times \int \sin\theta_0\, d\phi_0\, d\theta_0\, d\chi_0 \sin^2\phi \sin^2\theta\; W(v, \tau; v_0) \cos^2\theta_0. \end{aligned}\right\} \tag{15.19}$$

The difference between the parallel and perpendicular grating signals is obtained using Eqs. (15.18) and (15.19)

$$\begin{aligned} S_D(\omega_1, \omega_2, \tau; \omega_{LO}, \psi) &\equiv S^{\parallel}(\omega_1, \omega_2, \tau; \omega_{LO}, \psi) - S^{\perp}(\omega_1, \omega_2, \tau; \omega_{LO}, \psi) \\ &= 2S_0(\omega_1, \omega_2, \tau; \omega_{LO}, \psi)\langle \Omega^2_{00}(v)\Omega^{2*}_{00}(v_0)\rangle, \end{aligned}$$

* In the snapshot limit it is easy to generalize Eq. (15.15) by incorporating a different rotational dynamics in both electronic states, described by $\bar{\bar{Y}}_0^e$ and $\bar{\bar{Y}}_0^g$. These will enter in the particle and in the hole contribution, respectively.

which finally gives [20]

$$S_D(\omega_1, \omega_2, \tau; \omega_{LO}, \psi) \propto S_0(\omega_1, \omega_2, \tau; \omega_{LO}, \psi)\omega^2_{00}(t). \tag{15.20}$$

Ω^J_{MN} are the normalized rotation matrices, and W^J_{KN} are defined in Appendix 15A, where we present the conditional probability functions governing the time evolution of the dipole moment vector in the laboratory coordinate system for a spherical particle undergoing rotational Brownian motion, and described by a rotational Fokker–Planck equation [26]. For a free particle we obtain

$$\omega^2_{00}(t) = \frac{1}{5}\sum_{L=-2}^{2}(1 - \varepsilon L^2 t^2)\exp[-\varepsilon L^2 t^2/2],$$

where $\varepsilon = k_B T/I$ and I is the moment of inertia. In the small friction limit

$$\omega^2_{00}(t) = \frac{1}{5}\Bigg[(1 + 2\sqrt{6\zeta t})\exp(-2\sqrt{6\zeta t}) + 2\sum_{L=-2}^{2}(1 - \varepsilon L^2 t^2)\exp(-\varepsilon L^2 t^2/2)\Bigg],$$

where ζ is the rotational friction. Finally, in the strong friction limit we have

$$\omega^2_{00}(t) = \exp\{-6D_R[t + \zeta^{-1}\exp(-\zeta t) - \zeta^{-1}]\},$$

where $D_R = k_B T/I\zeta$ denotes the rotational diffusion coefficient. If $6D_R/\zeta \ll 1$, we obtain the Debye rotational diffusion model:

$$\omega^2_{00}(t) = \exp[-6D_R t].$$

In general, fluorescence anisotropy experiments [7] measure the spatially averaged second-order Legendre polynomial of $\mu^2_E(t)\cdot\mu^2_A(0)$, where $\mu_E(t)$ and $\mu_A(0)$ denote the propagated emission dipole moment and the initial absorption dipole moment, respectively. This is the same rotational property given by Eq. (15.20). However, fluorescence anisotropy measurements can be performed only when the molecule has a strong fluorescence emission, and probe only its rotational dynamics in the excited electronic state, whereas pump-probe measurements utilizing polarized fields can probe the rotational dynamics in both the excited (particle) and ground (hole) electronic states.

Since the window wavepacket depends on the phase factor ψ, it is possible to manipulate the signal by controlling ψ with respect to the probe field in heterodyne-detected grating or Raman spectroscopy. The *optically induced molecular dichroism and birefringence signals are defined as the in-phase* ($\psi = 0$) *and in-quadrature* ($\psi = \pi/2$) components of Eq. (15.20)

$$\left.\begin{aligned} \bar{S}_{dic}(\tau) &\equiv S_0(\omega_1, \omega_2, \tau; \omega_{LO}, \psi = 0) \times \bar{Y}_0(\tau) \\ \bar{S}_{bir}(\tau) &\equiv S_0(\omega_1, \omega_2, \tau; \omega_{LO}, \psi = \pi/2) \times \bar{Y}_0(\tau). \end{aligned}\right\} \tag{15.21}$$

Example of a dichroism measurement of I_2 in solution is shown in Figure 15.2. The birefringence spectrum of liquid $CHCl_3$ is shown in Figure 15.3. The spectra show vibrational quantum beats despite the fact that the absorption spectrum

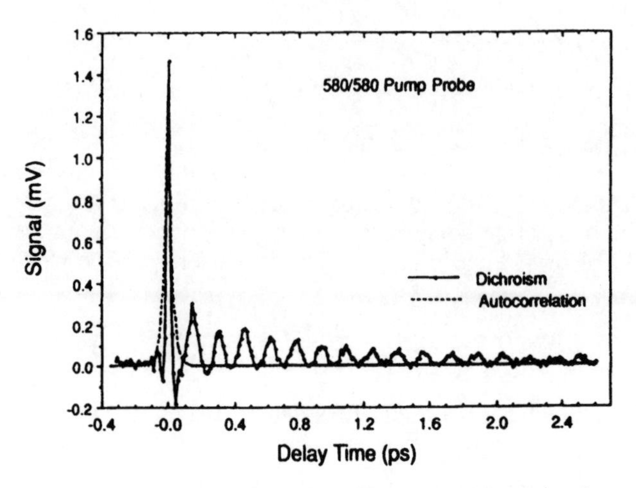

FIG. 15.2 Transient dichroism (difference in transmission for two orthogonal polarizations) observed for I_2 in hexane solution at room temperature. Excitation and probing is with 30 fs pulses centered at 580 nm, which corresponds to probing near the outer turning point of the ground state potential and near the inner turning point of the B state potential. Both ground and excited state contributions are present in the signal. The signal can be decomposed into the sum of a positive dichroic response from B-X absorption (420 and 220 cm^{-1} damped cosines and 1500 fs exponential decay) and a negative response from absorption by the B state wavepacket (112 cm^{-1} damped cosine with 200 fs decay damping constant, and 180 fs exponential decay). The excited state time constants correspond to curve crossing to a dissociative potential surface leading to bond breaking. Experiments at different probe wavelengths have followed the dissociating wavepacket out to a bond length of nearly 4 Å. [From N. Scherer, D. M. Jonas, and G. R. Fleming, "Femtosecond wave packet and chemical reaction dynamics of iodine in solution: Tunable probe study of motion along the reaction coordinate," *J. Chem. Phys.* **99**, 153 (1993).]

is broad and featureless. This is characteristic of Raman resonances which probe vibrational coherences in the ground and excited state and are not affected by the broad electronic dephasing.

When the optical pulses are resonant with the excited state of the molecule, both ground- and excited-state dynamics contribute to the optically induced dichroism and birefringence signals. Thus both measurements are useful tools for studying electronic dephasing, vibrational dephasing processes in the ground and excited electronic manifolds, and the lifetime of the excited electronic states. It should be noted that the rotational contribution enters as a product of a tensor multiplying the vibronic (scalar) part of the response function. Thus, the anisotropic measurement can be used to obtain the conditional probability function of the reorientational motion. If the local oscillator frequency is equal to that of the probe field, the phase factor is adjusted to be 0 compared to that of the probe field, and both pump and probe pulses are depolarized; we obtain

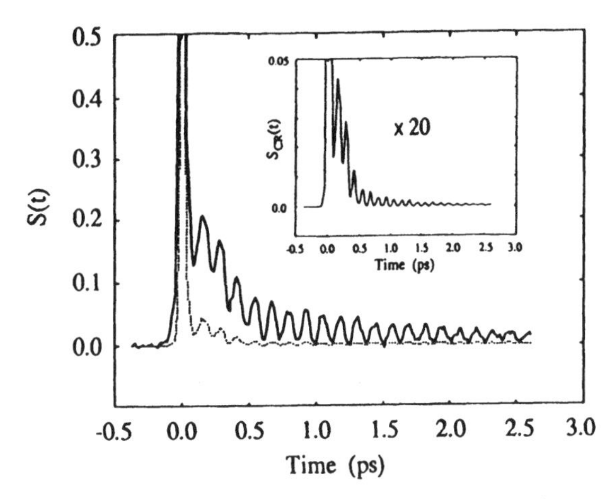

FIG. 15.3 Experimental off-resonance transient birefringence (solid line) and simulated coherent Raman scattering (dotted line) signals of $CHCl_3$. The inset shows the coherent Raman scattering signal enlarged by 20 times. The maxima of the two signals at $t = 0$ are matched. [From M. Cho, M. Du, N. F. Scherer, G. R. Fleming, and S. Mukamel, "Off-resonant transient birefringence in liquids," *J. Chem. Phys.* **99**, 2410 (1993).]

the scalar part of the ordinary two-pulse pump-probe absorption signal, discussed in Chapter 13.

OFF-RESONANT BIREFRINGENCE AND DICHROISM

The doorway and the window wavepackets given in Eqs. (15.13) can be recast in the form [9]

$$\left.\begin{aligned} \rho_{\mathrm{D}}^{0}(jj'; \omega_1) &\equiv A_{jj'}^{\mathrm{D}} \exp[i\phi_{jj'}^{\mathrm{D}}] \\ \rho_{\mathrm{w}}^{0}(jj'; \omega_2, \omega_{\mathrm{LO}}, \psi) &\equiv A_{jj'}^{\mathrm{W}}(\psi) \exp[i\phi_{jj'}^{\mathrm{W}}(\psi)] \end{aligned}\right\} \tag{15.22}$$

where $A_{\mu\nu}$ and $\phi_{\mu\nu}$ denote the absolute magnitude and the phase of the corresponding electronic polarizability.

The vibronic coherences created in the excited and in the ground states can be observed as an exponentially decaying cosinusoidal curve, whose frequencies are ω_{db} and ω_{ca}, respectively. In the case of resonant excitation, the antirotating wave terms in Eqs. (15.16) are negligibly small compared with the resonant terms. Thus, the phase of the vibrational coherences is zero in resonant dichroism and birefringence measurements. Vibrational coherence is observed as quantum beats in the signal.

For off-resonant excitation, the existence of the antirotating wave terms in Eqs. (15.13) affects the phase of the vibrational coherences in both the ground and the excited states. Using Eq. (15.12) and (15.22), the off-resonant snapshot signal can be written as

$$S^0(\omega_1, \omega_2, \tau; \omega_{\text{LO}}, \psi) = \sum_{db} A_{db}^{\text{W}}(\psi) A_{db}^{\text{W}} \{ \cos[\phi_{db}^{\text{W}}(\psi) + \phi_{db}^{\text{D}}] \cos \omega_{db}\tau$$

$$- \sin[\phi_{db}^{\text{W}}(\psi) + \phi_{db}^{\text{D}}] \sin \omega_{db}\tau \} \exp(-\Gamma_{db}\tau)$$

$$+ \sum_{ca} A_{ca}^{\text{W}}(\psi) A_{ca}^{\text{W}} \{ \cos[\phi_{ca}^{\text{W}}(\psi) + \phi_{ca}^{\text{D}}] \cos \omega_{ca}\tau$$

$$- \sin[\phi_{ca}^{\text{W}}(\psi) + \phi_{ca}^{\text{D}}] \sin \omega_{ca}\tau \} \exp(-\Gamma_{ca}\tau). \qquad (15.23)$$

The phases of the vibrational coherences are determined by the combination of the phase factors $\phi_{vv'}^{\text{D}}$ and $\phi_{vv'}^{\text{W}}$. Since the birefringence and dichroism signals are related to real and imaginary parts of the macroscopic polarization, the interferences between the two Liouville space pathways are different for each case. We shall denote the average detuning of all laser frequencies from the electronic transitions by $\Delta \equiv (\omega_{eg} - \omega)$, and the electronic dephasing rate by Γ. For off-resonance detunings we have $\Delta \gg \Gamma$; let us consider the variation of the doorway and the window wavepackets as the field frequencies are tuned. Using Eq. (15.13), we obtain for the doorway states under these conditions

$$\rho_{\text{D}}^0(ca; \omega_1) \sim 1/\Delta, \qquad (15.24a)$$

$$\rho_{\text{D}}^0(db; \omega_1) \sim i\Gamma/\Delta^2. \qquad (15.24b)$$

Comparing these two limiting forms shows that due to interference, the ground-state doorway wavepacket is always dominant for large detuning. We can thus neglect the first term in Eq. (15.23). In practice, for large detuning the dephasing rate always depends on the detuning $\Gamma(\Delta)$ and drops very rapidly as Δ is increased, since the bath does not have Fourier components at frequencies as high as Δ. Therefore, the ground state term wins not only by a factor of Δ (i.e., Δ^{-1} vs. Δ^{-2}) but by a much stronger factor $\Gamma(\Delta)/\Delta$.

A similar type of interference takes place for the window wavepacket. Starting with Eq. (15.13c) we have for dichroism,

$$\rho_{\text{w}}^0(ca; \omega_2, \omega_{\text{LO}}, \psi = 0) \cong 2 \sum_b \mu_{cb}\mu_{ba} \left[\frac{\omega_{bc} - i\Gamma_{bc}(\Delta)}{\omega_{\text{LO}}^2 - \omega_{bc}^2} - \frac{\omega_{ba} + i\Gamma_{ba}(\Delta)}{\omega_{\text{LO}}^2 - \omega_{ba}^2} \right]; \qquad (15.25a)$$

and for birefringence,

$$\rho_{\text{w}}^0\left(ca; \omega_2, \omega_{\text{LO}}, \psi = \frac{\pi}{2} \right) \cong 2i \sum_b \mu_{cb}\mu_{ba} \left[\frac{\omega_{\text{LO}}}{\omega_{\text{LO}}^2 - \omega_{bc}^2} + \frac{\omega_{\text{LO}}}{\omega_{\text{LO}}^2 - \omega_{ba}^2} \right]. \qquad (15.25b)$$

The two terms in square brackets in these expressions come from two Liouville space pathways that go from $|g\rangle\langle g|$ to $|g\rangle\langle g|$ either via $|e\rangle\langle g|$ or via $|g\rangle\langle e|$. For off-resonance dichroism the two pathways interfere destructively yielding

$$\rho_{\text{w}}^0(jj'; \omega_2, \omega_{\text{LO}}, \psi = 0) \sim i\Gamma/\Delta^2, \qquad jj' = ca, bd. \qquad (15.26a)$$

For birefringence the two pathways add rather than interfere and we have

$$\rho_{\mathrm{w}}^{0}(jj';\omega^{2},\omega_{\mathrm{LO}},\psi=\pi/2)\sim i/\Delta. \tag{15.26b}$$

Because of this interference at the window wavepacket, the off-resonant birefringence term always wins over the dichroism component by an additional factor of $\Gamma(\Delta)/\Delta$.

APPENDIX 15A

Solution of the Rotational Fokker–Planck Equations

The Fokker–Planck equation for rotational motion is written by [25–27]

$$\frac{\partial P}{\partial t} = -\sum_{j=1}^{3} i\omega_j J_j P - \frac{\partial}{\partial \omega_j}[(\zeta_j \omega_j - \lambda_j \omega_i \omega_k)P] - \frac{k_B T \zeta_j}{I_j}\frac{\partial^2 P}{\partial \omega_j^2}, \qquad (15.A.1)$$

where I_j are the principal moments of inertia along the jth principal axis. ζ_j are the rotational damping coefficients, and $\lambda_j = (I_j - I_k)/I_j$. The components J_j of the quantum mechanical angular-momentum operator are defined in the molecular reference frame **M**. The Euler equations and rotational Langevin equations were used to obtain Eq. (15.A.1):

$$\dot{\omega}_j = \lambda_j \omega_i \omega_k - \zeta_j \omega_j + f_j(t) \qquad (15.A.2)$$

where i, j, and k are a cyclic permutation of 1, 2, and 3 and $I_j f_j(t)$ is a thermal random fluctuating torque assumed to be Markovian process.

The normalized rotation matrices are defined as

$$\Omega^J_{MN}(\nu) \equiv \left(\frac{2J+1}{8\pi^2}\right)^{1/2} [D^J_{MN}(\nu)]^* \qquad (15.A.3)$$

where the rotation matrices are

$$D^J_{MK}(\nu) \equiv \langle JM | \exp(-i\phi J_Z)\exp(-i\theta J_Y)\exp(-iJ_Z) | JK \rangle,$$
$$J = 0, 1, \ldots; \qquad -J \le M, K \le J. \qquad (15.A.4)$$

Using the normalized rotation matrices, $\Omega^J_{MN}(\nu)$, the conditional probability function can be expanded as

$$W(\nu, t | \nu_0) \equiv \sum_{J=0}^{\infty} \sum_{K,M,N=-J}^{J} \omega^J_{KN}(t)[\Omega^J_{MN}(\nu_0)]^* \Omega^J_{MK}(\nu) \qquad (15.A.5)$$

Hubbard had derived analytic expressions for $\omega^J_{KN}(t)$ in several limiting cases for a spherical particle with $I_1 = I_2 = I_3 = I$. For a free particle (no friction)

$$\omega^J_{KN}(t) = \delta_{KN}\frac{1}{2J+1}\sum_{L=-J}^{J}(1 - \varepsilon L^2 t^2)\exp[-\varepsilon L^2 t^2/2], \qquad (15.A.6)$$

where $\varepsilon = K_B T/I$. In the small friction limit we have

$$\omega^J_{KN}(t) = \delta_{KN} \frac{1}{2J+1}\left[\{1 + 2\sqrt{\zeta J(J+1)t}\} \exp\{-2\sqrt{\zeta J(J+1)t}\} + 2 \sum_{L=-J}^{J} (1 - \varepsilon L^2 t^2) \exp(-\varepsilon L^2 t^2/2)\right], \quad (15.A.7)$$

and, finally, in the large friction limit

$$\omega^J_{KN}(t) = \delta_{KN} \exp[-D_R J(J+1)\{t + \zeta^{-1} \exp(-\zeta t) - \zeta^{-1}\}]. \quad (15.A.8)$$

Here $\tilde{w}(t)$ is the square matrix of dimension $2J + 1$ whose KNth element is $w^J_{KN}(t)$ for a given J. $\tilde{I}$ denotes the unit matrix of dimension $2J + 1$.

$$\omega^2_{00}(t) = \exp\{-6D_R[t + \zeta^{-1} \exp(-\zeta t) - \zeta^{-1}]\}. \quad (15.A.9)$$

D_R denotes the rotational diffusion coefficient. If $J(J + 1)D_R/\zeta \ll 1$, we obtain the Debye rotational diffusion model:

$$\omega^2_{00}(t) = \exp[-D_R J(J+1)t]. \quad (15.A.10)$$

NOTES

1. N. Bloembergen, *Nonlinear Optics* (Benjamin, New York, 1964).
2. P. N. Butcher, *Nonlinear Optical Phenomena* (Ohio State University, 1965).
3. R. W. Hellwarth, *Prog. Quant. Electron.* **5**, 1 (1977).
4. D. A. Kleinman, *Phys. Rev.* **126**, 1977 (1962); A. D. Buckingham, *Adv. Chem. Phys.* **12**, 107 (1967).
5. S. F. Mason, *Molecular Optical Activity and the Chiral Discriminations* (Cambridge University Press, Cambridge, Massachusetts, 1982); D. J. Caldwell and H. Eyring, *The Theory of Optical Activity* (Wiley, New York, 1971).
6. D. S. Kliger, J. W. Lewis, and C. E. Randall, *Polarized Light in Optics and Spectroscopy* (Academic Press, New York, 1990).
7. B. J. Berne and R. Pecora, *Dynamic Light Scattering* (Wiley, Toronto, 1976); G. R. Fleming, *Chemical Applications of Ultrafast Spectroscopy* (Oxford University Press, New York, 1986).
8. A. B. Myers and R. M. Hochstrasser, *J. Chem. Phys.* **87**, 2116 (1987).
9. K. Sala and M. C. Richardson, *Phys. Rev. A* **12**, 1036 (1975).
10. P. P. Ho and R. R. Alfano, *Phys. Rev. A* **20**, 2170 (1979).
11. M. D. Levenson and G. L. Eesley, *Appl. Phys.* **19**, 1 (1979); A. Owyoung, *IEEE J. Quan. Elect.* **QE14**, 192 (1978).
12. E. P. Ippen and C. V. Shank, *Appl. Phys. Lett.* **25**, 92 (1976).
13. B. E. Green and R. C. Farrow, *J. Chem. Phys.* **77**, 4779 (1982).
14. M. D. Levenson and S. S. Kano, *Introduction to Nonlinear Laser Spectroscopy* (Academic Press, New York, 1988).
15. J. Chesnoy and A. Mokhtari, *Phys. Rev. A* **38**, 3566 (1988).
16. J. A. Walmsley, M. Mitsunaga, and C. L. Tang, *Phys. Rev. A* **38**, 4681 (1988).
17. M. Fayer, *J. Chem. Phys.* **91**, 2269 (1989); J. T. Fourkas and M. D. Fayer, *Acc. Chem. Res.* **25**, 227 (1992).
18. T. Hattori, A. Terasaki, T. Kobayashi, T. Wada, A. Yamada, and H. Sasabe, *J. Chem. Phys.* **85**, 937 (1991).

19. D. McMorrow, W. T. Lotshaw, and G. A. Kenney-Wallace, *IEEE J. Quant. Electronics* **QE24**, 443 (1988); D. McMorrow and W. T. Lotshaw, *J. Phys. Chem.* **95**, 10395 (1991).

20. M. Cho, G. R. Fleming, and S. Mukamel, *J. Chem. Phys.* **98**, 5314 (1993); N. F. Scherer, L. D. Ziegler, and G. R. Fleming, *J. Chem. Phys.* **96**, 5544 (1992).

21. X. Xie and J. D. Simon, *Rev. Sci. Instrum.* **60**, 2614 (1989).

22. D. H. Waldeck, A. J. Cross, D. B. McDonald, and G. R. Fleming, *J. Chem. Phys.* **74**, 3381 (1981); G. R. Fleming, W. T. Lotshaw, R. J. Gulotty, M. C. Chang, and J. W. Petrich, *Laser. Chem.* **3**, 181 (1983); D. S. Alavi, R. S. Hartman, and D. H. Waldeck, *J. Chem. Phys.* **94**, 4509 (1991).

23. M. Joffre, D. Hulin, A. Migus, and A. Antonetti, *J. Mod. Optics* **35**, 1951 (1988); M. Combescot, *Phys. Rep.* **221**, 167 (1992).

24. L. D. Ziegler, Y. C. Chung, P. Wang, and Y. P. Zhang, *J. Chem. Phys.* **90**, 4125 (1989); O. S. Mortensen and S. Hassing, in *Advances in Infrared and Raman Spectroscopy*, Vol. 6, R. J. H. Clark and R. E. Hester, Eds. (Heyden, London, 1980), p. 1.

25. P. Debye, *Polar Molecules* (Dover, New York, 1929).

26. P. S. Hubbard, *Phys. Rev.* **46**, 2421 (1972).

27. A. Polimeno and J. H. Freed, *Adv. Chem. Phys.* **83**, 89 (1993).

28. R. N. Zare, *Angular Momentum: Understanding Spatial Aspects in Chemistry and Physics* (Wiley, New York, 1988).

BIBLIOGRAPHY

M. Abramovitz and I. A. Stegun, *Handbook of Mathematical Functions* (Dover, New York, 1980).

L. D. Barron, *Molecular Light Scattering and Optical Activity* (Cambridge, New York, 1982).

D. M. Brink and G. R. Satchler, *Angular Momentum* (Clarendon Press, Oxford, 1962).

A. R. Edmonds, *Angular Momentum in Quantum Mechanics* (Princeton University Press, Princeton, New Jersey, 1957).

U. Fano, *Rendiconti Dell' Accademia Nazionale Dei Lincei* (1995).

M. E. Rose, *Elementary Theory of Angular Momentum* (Wiley, New York, 1957).

CHAPTER 16

Nonlinear Response of Molecular Assemblies: The Local-Field Approximation

In Chapter 5 we derived exact formal expressions for optical susceptibilities in terms of correlation functions that are nonlocal in time and in space. However, in all applications made so far, we have analyzed only the nonlinear response of a collection of noninteracting particles, where the response is spatially local. This model applies when the density of particles is sufficiently low (e.g., gas phase, dilute solutions, doped crystals and glasses). Under these conditions, the average electric field in the medium is equal to the external field, and the optical susceptibility is simply given by the hyperpolarizability of a single particle, multiplied by the particle density [Eqs. (5.51)–(5.54)]. However, optically dense media often consist of many interacting particles (e.g., solutions at finite concentration, molecular crystals, and semiconductors), and calculating the response becomes an intractable many-body problem. The systematic calculation of nonlinear susceptibilities and the precise relationship between individual particle hyperpolarizabilities and the macroscopic optical response have drawn considerable theoretical attention. This problem is of particular significance for the interpretation of nonlinear optical measurements in terms of microscopic properties and intermolecular forces as well as for the design of new optical materials with desirable characteristics.

The local-field approximation (LFA) provides a simple phenomenological way to relate the polarizabilities of isolated particles to the macroscopic susceptibilities. In this approach, the effect of interparticle forces is introduced through an effective local electric field. The problem of calculating the response of an interacting ensemble of particles to the electromagnetic field is then reduced to the response of isolated particles interacting with the local field $\mathbf{E}_l$, through the coupling $-\mathbf{E}_l \cdot V$, where V denotes the particle's dipole operator [see Eq. (5.1)]. The Lorentz relation between the local field and the Maxwell field $\mathbf{E}$ can then be used to calculate the dielectric function. This procedure, which reduces the complex many-body problem to a single-body problem, has long been used for calculating the linear response (i.e., the dielectric function) [1–7], and was subsequently extended and applied also to the calculation of nonlinear susceptibilities [8–12]. The nonlinear susceptibilities at a given order are then expressed by sums of products of polarizabilities of that order and lower orders. The remarkable simplicity and elegance of this approach are the main reasons for its popularity; however, this procedure is not rigorous, and, as will be shown in the next chapter, may be justified only for off-resonant frequencies. Nevertheless, a detailed discussion of the LFA is desirable since it is so widely used and, moreover, the LFA results provide a reference for identifying and analyzing the genuine many-body characteristics of the problem, which are responsible for deviations from the LFA.

In this chapter we calculate the nonlinear optics of atomic or molecular systems with localized electronic states and multipolar intermolecular forces using the LFA. We first derive the LFA using the conventional phenomenological procedure, which applies to a collection of polarizable particles with non-overlapping charge distributions [8]. We calculate the nonlocal response functions, and relate the macroscopic susceptibilities to microscopic hyperpolarizabilities, including local-field corrections and cascading effects. We then present a microscopic derivation of the LFA based on equations of motion [11, 13], which pinpoints the approximations involved. In the next chapter we shall extend the microscopic approach beyond the LFA and explore a hierarchy of improved approximations for the time-domain and frequency-domain optical response. We shall then incorporate phenomena such as cooperative effects, two-exciton resonances, and exciton transport, which are missed by the LFA.

PHENOMENOLOGICAL APPROACH TO THE NONLINEAR RESPONSE IN REAL SPACE: LOCAL-FIELD AND CASCADING CORRECTIONS

The local-field approximation is based on a self-consistent mean-field calculation of the optical polarization. It is usually formulated for homogeneous infinite media in momentum (**k**) space. However, deriving it in real space is advantageous since it provides a clearer physical insight and allows us to analyze the results in terms of properties of individual particles; this is somewhat less transparent in **k** space. In addition, the real space procedure applies also to finite systems such as molecular aggregates. In this section we therefore derive all the results in **r** space. **k** space will be discussed in the next section.

Consider an assembly of particles with an arbitrary geometry. We neglect translational motions and assume that the particles have fixed positions and orientations. Denoting the electric local field acting on the nth particle by $E_{l,n}$, we can expand its polarization in terms of its hyperpolarizabilities (all quantities are in ω space, as will be specified later)

$$P_n = \alpha E_{l,n} + \beta E_{l,n} E_{l,n} + \gamma E_{l,n} E_{l,n} E_{l,n} + \cdots. \tag{16.1}$$

The optical susceptibilities are defined on the other hand in terms of the average transverse Maxwell field

$$P_n = \sum_m \chi^{(1)}_{n,m} E_m + \sum_{m,r} \chi^{(2)}_{n,mr} E_m E_r + \sum_{m,r,s} \chi^{(3)}_{n,mrs} E_m E_r E_s + \cdots. \tag{16.2}$$

An important feature to note is that the susceptibilities are now *nonlocal* in space.

$\chi^{(3)}_{n,mrs}$, for example, represents the contribution of three fields interacting at points m, r, and s to the polarization at point n. This is fundamentally different from the susceptibilities of noninteracting particles that are local.

Our goal is to relate the susceptibilities $\chi^{(n)}$ to the given set of hyperpolarizabilities α, β, γ, etc. This requires connecting the Maxwell and the local fields. Adopting the procedure of Lorentz we write*

$$E_{l,n}(\omega) = E_n(\omega) + \sum_m{}' T_{nm} \cdot P_m(\omega). \tag{16.3a}$$

Here E_n is the transverse Maxwell field, which is given by the external field plus the contribution of retarded interactions among particles through the Maxwell equation (an exact Green function expression that relates E_n to the external applied field was given in Chapter 4). The second term in the right-hand-side represents the electrostatic longitudinal field created by all other particles, i.e., the instantaneous dipole– dipole interactions between the particles, and the prime excludes terms with $m = n$ from the summation. T_{nm} is the dipole–dipole tensor [Eq. (4.A.11)]

$$T_{nm} = \frac{1 - 3\hat{\mathbf{r}}_{nm}\hat{\mathbf{r}}_{nm}}{r_{nm}^3}. \tag{16.3b}$$

(Note that here n, m denote the sites and not the tensor components as used in Appendix 4A) $\mathbf{r}_{mn} \equiv \mathbf{r}_m - \mathbf{r}_n$, where $\mathbf{r}_m$ denotes the position of the mth particle.

We next partition the polarization into a linear $P_n^{(1)}$ and a nonlinear P_n^{NL} part with respect to the Maxwell field,

$$P_n(\omega) = P_n^{(1)}(\omega) + P_n^{\mathrm{NL}}(\omega). \tag{16.4a}$$

Similarly we have for the local field

$$E_{l,n}(\omega) = E_{l,n}^{(1)}(\omega) + E_{l,n}^{\mathrm{NL}}(\omega). \tag{16.4b}$$

Using Eq. (16.3a), and retaining only terms to linear order in the Maxwell field we have

$$E_{l,n}^{(1)}(\omega) = E_n(\omega) + \sum_m{}' T_{nm} \cdot P_m^{(1)}(\omega). \tag{16.5}$$

$P_m^{(1)}$ can now be written in two ways:

$$P_m^{(1)}(\omega) = \sum_{n'} \chi_{m,n'}^{(1)}(\omega) E_{n'}(\omega) \tag{16.6a}$$

or

$$P_m^{(1)}(\omega) = \alpha(\omega) E_{l,m}^{(1)}(\omega). \tag{16.6b}$$

We can use these expressions to relate $E_l^{(1)}$ to E,

$$E_{l,n}^{(1)}(\omega) = \sum_m S_{nm}(\omega) E_m(\omega). \tag{16.7}$$

* Eq. (16.3a) will be derived later in this chapter [Eq. (16.41)] using a microscopic model for the system and the Heisenberg equation of motion approach.

The matrix S can be expressed in three equivalent forms:

Combining Eqs. (16.5) and (16.6b) we have

$$[S^{-1}(\omega)]_{nm} \equiv \delta_{nm} - \alpha(\omega)T_{nm}. \tag{16.8a}$$

From Eqs. (16.5) and (16.6a) we obtain

$$S_{nm}(\omega) = \delta_{nm} + \sum_{m'} T_{nm'}\chi^{(1)}_{m',m}(\omega) \tag{16.8b}$$

Finally, comparing Eqs. (16.6a) and (16.6b) we get

$$S_{nm}(\omega) = \alpha^{-1}(\omega)\chi^{(1)}_{n,m}(\omega). \tag{16.8c}$$

Adopting a matrix notation $\mathbf{T}$, $\chi^{(1)}$, etc., we have

$$\mathbf{S}(\omega) = [1 - \alpha(\omega)\mathbf{T}]^{-1}, \tag{16.9a}$$

or

$$\mathbf{S}(\omega) = \alpha^{-1}(\omega)\chi^{(1)}(\omega). \tag{16.9b}$$

Combining Eqs. (16.9a) and (16.9b) we finally get for the linear response

$$\chi^{(1)}(\omega) = \alpha(\omega)[1 - \alpha(\omega)\mathbf{T}]^{-1} \tag{16.10a}$$

or the inverse relation

$$\alpha(\omega) = [1 + \mathbf{T}\chi^{(1)}(\omega)]^{-1}\chi^{(1)}(\omega). \tag{16.10b}$$

We next turn to the nonlinear response. Expanding the polarization order by order in the Maxwell field we get

$$\begin{aligned} P_n^{\mathrm{NL}}(\omega) = {} & \frac{1}{2!}\frac{1}{2\pi}\sum_{m,r}\iint d\omega_1\, d\omega_2\, \chi^{(2)}_{n,mr}(-\omega;\omega_1,\omega_2)E_m(\omega_1)E_r(\omega_2)\delta(\omega-\omega_1-\omega_2) \\ & + \frac{1}{3!}\frac{1}{(2\pi)^2}\sum_{m,r,s}\iiint d\omega_1\, d\omega_2\, d\omega_3\, \chi^{(3)}_{n,mrs}(-\omega;\omega_1,\omega_2,\omega_3) \\ & E_m(\omega_1)E_r(\omega_2)E_s(\omega_3)\delta(\omega-\omega_1-\omega_2-\omega_3) + \cdots \end{aligned} \tag{16.11a}$$

or, in terms of the local field

$$\begin{aligned} P_n^{\mathrm{NL}}(\omega) = {} & \alpha(\omega)\sum_m{}' T_{nm}P_m^{\mathrm{NL}}(\omega) \\ & + \frac{1}{2!}\frac{1}{2\pi}\iint d\omega_1\, d\omega_2\, \beta(\omega_1,\omega_2)E_{l,n}(\omega_1)E_{l,n}(\omega_2)\delta(\omega-\omega_1-\omega_2) \\ & + \frac{1}{3!}\frac{1}{(2\pi)^2}\iiint d\omega_1\, d\omega_2\, d\omega_3\, \gamma(\omega_1,\omega_2,\omega_3)E_{l,n}(\omega_1)E_{l,n}(\omega_2)E_{l,n}(\omega_3) \\ & \times \delta(\omega-\omega_1-\omega_2-\omega_3) + \cdots. \end{aligned} \tag{16.11b}$$

These equations can be combined in the form

$$P_n^{\mathrm{NL}}(\omega) = \frac{1}{2!}\frac{1}{2\pi}\sum_m \iint d\omega_1\, d\omega_2\, \beta(\omega_1, \omega_2) S_{nm}(\omega) E_{l,m}(\omega_1) E_{l,m}(\omega_2)\delta(\omega - \omega_1 - \omega_2)$$

$$+ \frac{1}{3!}\frac{1}{(2\pi)^2}\sum_m \iiint d\omega_1\, d\omega_2\, d\omega_3\, \gamma(\omega_1, \omega_2, \omega_3) S_{nm}(\omega)$$

$$\times E_{l,m}(\omega_1) E_{l,m}(\omega_2) E_{l,m}(\omega_3)\delta(\omega - \omega_1 - \omega_2 - \omega_3) + \cdots, \qquad (16.12)$$

with

$$E_{l,n} = \sum_m S_{nm}(\omega) E_m(\omega) + \sum_m{}' T_{nm} P_m^{\mathrm{NL}}(\omega). \qquad (16.13)$$

Upon the substitution of Eq. (16.13) in (16.12), and collecting terms in orders of the Maxwell field E, we can express the nonlinear susceptibilities in terms of the hyperpolarizabilities. For the second and third order we get

$$\chi^{(2)}_{n,mr}(-\omega; \omega_1, \omega_2) = \beta(\omega_1, \omega_2) \sum_{n'} S_{nn'}(\omega) S_{n'm}(\omega_1) S_{n'r}(\omega_2), \qquad (16.14)$$

$$\chi^{(3)}_{n,m_1m_2m_3}(-\omega; \omega_1, \omega_2, \omega_3) = \gamma(\omega_1, \omega_2, \omega_3) \sum_{n'} S_{nn'}(\omega) S_{n'm_1}(\omega_1) S_{n'm_2}(\omega_2) S_{n'm_3}(\omega_3)$$

$$+ \frac{1}{2} \sum_{\substack{\text{perm} \\ (\omega_j m_j)}} \sum_{n',m} \beta(\omega_1, \omega_2 + \omega_3)$$

$$\times S_{nn'}(\omega_s) S_{n'm_1}(\omega_1) T_{n'm} \chi^{(2)}_{m,m_2m_3}(-\omega_2 - \omega_3; \omega_2, \omega_3) \qquad (16.15)$$

The following points should now be noted:

1. If we neglect the second term in Eq. (16.13) then the nth-order susceptibility $\chi^{(n)}$ will have $n + 1$ local field correction S factors (3 for $\chi^{(2)}$, 4 for $\chi^{(3)}$, etc.).
2. The second term in Eq. (16.13) representing the nonlinear local field is responsible for additional *cascading* (sequential) contributions to the nonlinear response [9–11]. The second term in $\chi^{(3)}$ [Eq. (16.15)] provides an example for a cascading contribution; two fields mix at sites r and s via $\chi^{(2)}$ resulting in a polarization at q, which then propagates to site n'. The resulting field mixes with another field at m via β resulting in the final field. $\chi^{(3)}$ has thus a contribution from two consecutive $\chi^{(2)}$ events.
3. Cascading contributions can be neglected to lowest (first)-order in particle density since they arise from interparticle interactions. As an example, the first term in Eq. (16.15) is proportional to the number density ρ_0 (plus corrections in higher order) whereas the second term scales at least as ρ_0^2.

4. Within the present approximation, the nonlocal character of the nonlinear response is contained entirely in the S_{nm} factors [Eq. (16.7)] that relate the local field at point n to the Maxwell field at point m.

THE LOCAL-FIELD APPROXIMATION IN k SPACE

The response of an infinite and homogeneous medium to optical fields is usually formulated in terms of the wavevector and frequency-dependent optical susceptibilities, introduced in Chapter 5. Consider a frequency-domain experiment involving a few relevant modes of the radiation field. We shall expand the local field, the Maxwell field, and the polarization in a discrete Fourier series, i.e.,

$$E_l(\mathbf{r}, t) = \sum_j [E_l(\mathbf{k}_j, \omega_j) \exp(i\mathbf{k}_j \cdot \mathbf{r} - i\omega_j t) + E_l^*(\mathbf{k}_j, \omega_j) \exp(-i\mathbf{k}_j \cdot \mathbf{r} + i\omega_j t)], \tag{16.16a}$$

$$E(\mathbf{r}, t) = \sum_j [E(\mathbf{k}_j, \omega_j) \exp(i\mathbf{k}_j \cdot \mathbf{r} - i\omega_j t) + E^*(\mathbf{k}_j, \omega_j) \exp(-i\mathbf{k}_j \cdot \mathbf{r} + i\omega_j t)], \tag{16.16b}$$

$$P(\mathbf{r}, t) = \sum_j [P(\mathbf{k}_j, \omega_j) \exp(i\mathbf{k}_j \cdot \mathbf{r} - i\omega_j t) + P^*(\mathbf{k}_j, \omega_j) \exp(-i\mathbf{k}_j \cdot \mathbf{r} + i\omega_j t)], \tag{16.16c}$$

where j labels the few relevant modes for a given experiment. The frequencies ω_j are by definition positive, and are related to $\mathbf{k}_j$ by the system's dispersion relation [Eq. (4.55)]. Hereafter, $P(\mathbf{r}, t)$ denotes the polarization per unit volume at point $\mathbf{r}$ (rather than the polarization of a particle). We further note that optical wavevectors are usually small compared with the length scale of microscopic fluctuations, so that the system may be safely assumed spatially homogeneous. Equation (16.1) thus becomes

$$\begin{aligned} P(\mathbf{r}, t) = {} & \rho_0 \sum_j \alpha(\omega_j) E_l(\mathbf{k}_j, \omega_j) \exp(i\mathbf{k}_j \cdot \mathbf{r} - i\omega_j t) \\ & + \rho_0 \sum_{j \geqq n} \beta(\omega_j, \omega_n) E_l(\mathbf{k}_j, \omega_j) E_l \mathbf{k}_n, \omega_n) \exp[i(\mathbf{k}_j + \mathbf{k}_n) \cdot \mathbf{r} - i(\omega_j + \omega_n)t] \\ & + \rho_0 \sum_{j \geqq n \geqq m} \gamma(\omega_j, \omega_n, \omega_m) E_l(\mathbf{k}_j, \omega_j) E_l(\mathbf{k}_n, \omega_n) E_l(\mathbf{k}_m, \omega_m) \\ & \times \exp[i(\mathbf{k}_j + \mathbf{k}_n + \mathbf{k}_m) \cdot \mathbf{r} - i(\omega_j + \omega_n + \omega_m)t] + \cdots . \end{aligned} \tag{16.17}$$

We have included here only the $\mathbf{k}_j + \mathbf{k}_n$ or $\mathbf{k}_j + \mathbf{k}_n + \mathbf{k}_m$ Fourier components of $P(\mathbf{r}, t)$. In fact, all choices of signs in $\pm\mathbf{k}_j \pm \mathbf{k}_n$ or $\pm\mathbf{k}_j \pm \mathbf{k}_n \pm \mathbf{k}_m$ represent possible Fourier components. As indicated in Chapter 5, these Fourier components may be obtained from Eq. (16.17) by changing one (or more) $\mathbf{k}_j$ and ω_j to $-\mathbf{k}_j$ and $-\omega_j$ and replacing $E_l(\mathbf{k}_j, \omega_j)$ by $E_l^*(\mathbf{k}_j, \omega_j)$.

At this point we introduce the optical susceptibilities that relate the polarization $P(\mathbf{r}, t)$ to the Fourier components of the Maxwell field

$$P(\mathbf{r}, t) = P^{(1)}(\mathbf{r}, t) + P^{(2)}(\mathbf{r}, t) + P^{(3)}(\mathbf{r}, t) + \cdots, \tag{16.18}$$

$$P^{(1)}(\mathbf{r}, t) = \sum_j \chi^{(1)}(-\mathbf{k}_j - \omega_j, \mathbf{k}_j\omega_j)E(\mathbf{k}_j, \omega)\exp(i\mathbf{k}_j\cdot\mathbf{r} - i\omega_j t), \tag{16.19a}$$

$$P^{(2)}(\mathbf{r}, t) = \sum_{j\geqq n} \chi^{(2)}(-\mathbf{k}_j - \mathbf{k}_n - \omega_j - \omega_n; \mathbf{k}_j\omega_j, \mathbf{k}_n\omega_n)$$
$$\times E(\mathbf{k}_j, \omega_j)E(\mathbf{k}_n, \omega_n)\exp[i(\mathbf{k}_j + \mathbf{k}_n)\cdot\mathbf{r} - i(\omega_j + \omega_n)t], \tag{16.19b}$$

$$P^{(3)}(\mathbf{r}, t) = \sum_{j\geqq n\geqq m} \chi^{(3)}(-\mathbf{k}_j - \mathbf{k}_n - \mathbf{k}_m - \omega_j - \omega_n - \omega_m; \mathbf{k}_j\omega_j, \mathbf{k}_n\omega_n, \mathbf{k}_m\omega_m)$$
$$\times E(\mathbf{k}_j, \omega_j)E(\mathbf{k}_n, \omega_n)E(\mathbf{k}_m, \omega_m)$$
$$\times \exp[i(\mathbf{k}_j + \mathbf{k}_n + \mathbf{k}_m)\cdot\mathbf{r} - i(\omega_j + \omega_n + \omega_m)]. \tag{16.19c}$$

To express the susceptibilities $\chi^{(n)}$ in terms of the microscopic susceptibilities α, β, γ, we first transform Eq. (16.3) to **k** space and recast the local field in the form

$$E_l(\mathbf{k}, t) \equiv \mathbf{E}(\mathbf{k}, t) + \frac{4\pi}{3}\eta(\mathbf{k})\cdot P(\mathbf{k}, t), \tag{16.20a}$$

with

$$\eta(\mathbf{k}) \equiv \frac{3T(\mathbf{k})}{4\pi\rho_0}. \tag{16.20b}$$

The linear-optical properties of a medium are determined by its frequency- and wavevector-dependent dielectric function $\varepsilon(\mathbf{k}, \omega)$, which generalizes Eq. (5.52a), i.e.,

$$\varepsilon(\mathbf{k}, \omega) \equiv 1 + 4\pi\chi^{(1)}(-\mathbf{k} - \omega; \mathbf{k}\omega). \tag{16.21}$$

Using Eq. (16.8b) we then have

$$S(\mathbf{k}, \omega) = \frac{3 + \eta(\mathbf{k})[\varepsilon(\mathbf{k}, \omega) - 1]}{3}. \tag{16.22}$$

By splitting the polarization into its linear and nonlinear parts we then get

$$E_l(\mathbf{k}, \omega) = S(\mathbf{k}, \omega)E(\mathbf{k}, \omega) + \frac{4\pi}{3}\eta(\mathbf{k})P_{\mathrm{NL}}(\mathbf{k}, \omega), \tag{16.23}$$

and $P_{\mathrm{NL}}(\mathbf{k}, \omega)$ is obtained by transforming Eq. (16.4a) to **k** space. The linear polarization is given by

$$P^{(1)}(\mathbf{k}, \omega) = \chi^{(1)}(-\mathbf{k} - \omega; \mathbf{k}\omega)E(\mathbf{k}, \omega), \tag{16.24a}$$

where

$$\chi^{(1)}(-\mathbf{k} - \omega; \mathbf{k}\omega) = \rho_0\alpha(\omega)S(\mathbf{k}, \omega). \tag{16.24b}$$

The local-field correction factor in **k** space becomes

$$S(\mathbf{k}, \omega) \equiv E_l^{(1)}(\mathbf{k}, \omega)/E(\mathbf{k}, \omega). \tag{16.25}$$

For optical wavevectors and dipolar interactions we have [14, 15]

$$T(\mathbf{k}) = -\frac{4\pi}{3}\rho_0(3\hat{\mathbf{k}}\hat{\mathbf{k}} - 1), \qquad \mathbf{k} \to 0$$

so that for a transverse field $\eta(\mathbf{k}) = 1$. This results in the *Lorentz expression for the local field*

$$E_l(\mathbf{k}, t) \equiv E(\mathbf{k}, t) + \frac{4\pi}{3} P(\mathbf{k}, t). \tag{16.26}$$

Equation (16.22) then becomes

$$S(\mathbf{k}, \omega) = \frac{\varepsilon(\mathbf{k}, \omega) + 2}{3}. \tag{16.27}$$

In this limit, Eqs. (16.24b) and (16.27) result in the *Clausius–Mossotti* expression for the dielectric function

$$\frac{\varepsilon(\omega) - 1}{\varepsilon(\omega) + 2} = \frac{4\pi}{3}\rho_0\alpha(\omega). \tag{16.28}$$

For second-order nonlinear processes we assume two incoming fields: $\mathbf{k}_1\omega_1$ and $\mathbf{k}_2\omega_2$. By substituting Eq. (16.23) in Eq. (16.17) and comparing with Eq. (16.19b), we obtain

$$P^{(2)}(\mathbf{k}, \omega) = \chi^{(2)}(-\mathbf{k} - \omega; \mathbf{k}_1\omega_1; \mathbf{k}_2\omega_2)E(\mathbf{k}_1, \omega_1)E(\mathbf{k}_2, \omega_2), \tag{16.29a}$$

where

$$\chi^{(2)}(-\mathbf{k} - \omega; \mathbf{k}_1\omega_1; \mathbf{k}_2\omega_2) = \rho_0\beta(\omega_1; \omega_2)S(\mathbf{k}_1, \omega_1)S(\mathbf{k}_2, \omega_2)S(\mathbf{k}, \omega), \tag{16.29b}$$

and $\mathbf{k} = \mathbf{k}_1 + \mathbf{k}_2$; $\omega = \omega_1 + \omega_2$.

For third-order nonlinear processes we introduce three incoming fields: $\mathbf{k}_1\omega_1$, $\mathbf{k}_2\omega_2$, and $\mathbf{k}_3\omega_3$. Using the same procedure, we then get for the polarization at $\mathbf{k} = \mathbf{k}_1 + \mathbf{k}_2 + \mathbf{k}_3$, $\omega = \omega_1 + \omega_2 + \omega_3$,

$$\begin{aligned} P^{(3)}(\mathbf{k}, \omega) = {} & \rho_0\gamma(\omega_1, \omega_2, \omega_3)E_l(\mathbf{k}_1\omega_1)E_l(\mathbf{k}_2\omega_2)E_l(\mathbf{k}_3, \omega_3) \\ & + \rho_0\beta(\omega_1 + \omega_2, \omega_3)E_l(\mathbf{k}_1 + \mathbf{k}_2, \omega_1 + \omega_2)E_l(\mathbf{k}_3, \omega_3) \\ & + \rho_0\beta(\omega_1 + \omega_3, \omega_2)E_l(\mathbf{k}_1 + \mathbf{k}_3, \omega_1 + \omega_3)E_l(\mathbf{k}_2, \omega_2) \\ & + \rho_0\beta(\omega_2 + \omega_3, \omega_1)E_l(\mathbf{k}_2 + \mathbf{k}_3, \omega_2 + \omega_3)E_l(\mathbf{k}_1, \omega_1) \\ & + \rho_0\alpha(\omega)E_l(\mathbf{k}, \omega). \end{aligned} \tag{16.30}$$

By substituting Eq. (16.23) into Eq. (16.30) and collecting terms to third order in E, we get

$$P^{(3)}(\mathbf{k}, \omega) = \chi^{(3)}(-\mathbf{k} - \omega; \mathbf{k}_1\omega_1, \mathbf{k}_2\omega_2, \mathbf{k}_3\omega_3)E(\mathbf{k}_1, \omega_1)E(\mathbf{k}_2, \omega_2)E(\mathbf{k}_3, \omega_3), \tag{16.31a}$$

where

$$\chi^{(3)}(-\mathbf{k}-\omega; \mathbf{k}_1\omega_1, \mathbf{k}_2\omega_2, \mathbf{k}_3\omega_3)$$
$$= \rho_0[\gamma(\omega_1, \omega_2, \omega_3)$$
$$+ \beta(\omega_1, \omega_2)\beta(\omega_1+\omega_2, \omega_3)Q(\mathbf{k}_1+\mathbf{k}_2, \omega_1+\omega_2)$$
$$+ \beta(\omega_1, \omega_3)\beta(\omega_1+\omega_3, \omega_2)Q(\mathbf{k}_1+\mathbf{k}_3, \omega_1+\omega_3)$$
$$+ \beta(\omega_2, \omega_3)\beta(\omega_2+\omega_3, \omega_1)Q(\mathbf{k}_2+\mathbf{k}_3, \omega_2+\omega_3)]$$
$$\times S(\mathbf{k}_1, \omega_1)S(\mathbf{k}_2, \omega_2)S(\mathbf{k}_3, \omega_3)S(\mathbf{k}, \omega), \quad (16.31b)$$

with

$$Q(\mathbf{k}, \omega) \equiv \frac{4\pi}{3}\rho_0\eta(\mathbf{k})S(\mathbf{k}, \omega). \quad (16.31c)$$

The first term in the square brackets in Eq. (16.31b) represents the nonlinear process whereby three waves, $\mathbf{k}_1\omega_1$, $\mathbf{k}_2\omega_2$, and $\mathbf{k}_3\omega_3$, mix to generate a new wave, $\mathbf{k}\omega$. The second term represents a cascading process [see Eq. (16.15)] whereby a low-order nonlinear process (β) contributes to the higher-order nonlinear process $\chi^{(3)}$. In this process two waves, $\mathbf{k}_1\omega_1$ and $\mathbf{k}_2\omega_2$, mix to generate a wave $\mathbf{k}' = \mathbf{k}_1 + \mathbf{k}_2, \omega' = \omega_1 + \omega_2$. This wave mixes further with the third wave, $\mathbf{k}_3\omega_3$, to generate the final wave with wavevector $\mathbf{k} = \mathbf{k}' + \mathbf{k}_3$ and frequency $\omega = \omega' + \omega_3$. The last two terms represent the same sequence of events with all possible permutations of the three fields.

MICROSCOPIC DERIVATION OF THE LFA: THE DRIVEN ANHARMONIC OSCILLATOR

Calculating the optical response using the density operator is straightforward for a system of dilute and noninteracting particles. Under these conditions it is possible to evaluate the complete density operator and the corresponding optical polarization to any desired order in the electric field. The susceptibilities are then given in terms of multiple summations over eigenstates, as shown in Chapter 6. When an interacting many-body system is considered, the number of relevant global states of the system is prohibitively large. For N two-level systems we have 2^N levels. It may then be desirable to adopt a different approach, which is based on the Heisenberg equations, as outlined in Chapter 6. We recall that in the Heisenberg picture the expectation value of any operator A is given by

$$\langle A(t)\rangle = \mathrm{Tr}[A(t)\rho],$$

where the time evolution is now incorporated in the operators rather than in the density operator [see Eqs. (3.61c) and (3.61d)], and we write

$$\dot{A} = \frac{i}{\hbar}[H, A] = \frac{i}{\hbar}LA, \quad (16.32a)$$

$$\dot{\rho} = 0. \quad (16.32b)$$

The polarization operator is our primary physical quantity of interest. By writing its Heisenberg equation of motion we find that it is coupled to other more complex operators. We can then proceed by writing the Heisenberg equations for these operators, thus creating a hierarchy of coupled equations. To obtain a practical procedure, we need to close the hierarchy by some approximate truncation. The proper truncation is crucial for the success of this procedure.

The Heisenberg picture has several advantages:

1. The number of equations is typically smaller than when working with the full density operator. The operators A are usually simpler than the complete density operator. In addition, many states (and density operator elements) do not contribute to the response at a given order. By calculating the complete density operator we inevitably calculate much more information than is necessary. In contrast, irrelevant states will not be calculated in the equation of motion approach.
2. The polarization is calculated by solving a set of coupled nonlinear equations of motion. The equations are obtained by identifying the relevant dynamic variables and deriving equations of motion for their expectation values. This provides a rigorous way of modeling the matter as a collection of coupled anharmonic oscillators. The resulting intuitive picture generalizes the anharmonic oscillator picture developed in Chapter 6 and suggests new physically motivated approximations.
3. The local field approximation can be derived microscopically by using the simplest possible truncation of the hierarchy. This clarifies its limitations and immediately suggests possible systematic generalizations.

To explore the role of many-body effects in the macroscopic nonlinear optical response, we consider a model system consisting of an assembly of two-level molecules. The system can have an arbitrary geometry and dimensionality. In this chapter we derive the LFA and obtain explicit expressions for $\chi^{(1)}$ and $\chi^{(3)}$ for this model. In the next chapter we extend this procedure beyond the LFA. We shall denote the creation and the annihilation operators for an excitation on particle m by $B_m^\dagger$ and B_m, respectively, (i.e., $B_m \equiv |0\rangle\langle m|$, where $|0\rangle$ is the ground state, and $|m\rangle$ denotes a state where excitation resides on particle m) [16]. The index m runs over all the particles in the system, and may also denote other internal (e.g., nuclear) degrees of freedom. The Bs satisfy the following commutation rules

$$[B_n, B_m] = [B_n^\dagger, B_m^\dagger] = B_n^2 = (B_n^\dagger)^2 = 0 \tag{16.33a}$$

and

$$[B_n, B_m^\dagger] = \delta_{nm}(1 - W_m), \tag{16.33b}$$

where δ_{nm} is the Kroenecker delta, and $W_m \equiv 2B_m^\dagger B_m$ is known as the exciton population operator. Equations (16.33) are equivalent to the Pauli anticommutation rule

$$B_m^\dagger B_m + B_m B_m^\dagger = 1. \tag{16.34}$$

For harmonic oscillators (Bosons), we simply have $[B_m, B_m^\dagger] = 1$. Therefore, neglecting the W_m operator in Eq. (16.33b) is usually denoted as the Bose approximation. The Bs representing different particles thus have the commutation rules of Bosons, and when corresponding to the same particle behave as fermions. We note that in the present model the Pauli exclusion (i.e., the fact that each particle can carry only one excitation) is the only source of nonlinearities. In systems with multilevel (or polar two level) molecules, other sources arise from intermolecular interaction terms that are cubic and quartic in the molecular exciton creation and annihilation operators. Such terms give rise to nonlinearities even if a Bose approximation is applied [17, 18].

Using these definitions, the semiclassical Hamiltonian for our system in the dipole approximation [see Eq. (4.47b)] is

$$H = H_{\text{mat}} - \sum_n P_n \cdot E(\mathbf{r}_n, t), \tag{16.35}$$

where H_{mat} denotes the material Hamiltonian [16]

$$H_{\text{mat}} = \sum_n \hbar(\Omega_n - i\Gamma_n)B_n^\dagger B_n + \frac{\hbar}{2}\sum_{m,n}{}' J_{nm}(B_m^\dagger + B_m)(B_n^\dagger + B_n). \tag{16.36a}$$

This is known as the *Frenkel exciton* Hamiltonian. It represents the dynamics of electronic excitations that can migrate among the particles, and constitutes the basic model in the theory of molecular crystals. The first term represents noninteracting particles with transition frequency Ω_n. It also includes a phenomenological exciton damping rate Γ_n that may represent, e.g., exciton scattering by phonons. More realistic microscopic models for exciton relaxation will be considered in Chapter 17. The second term in H_{mat} accounts for the instantaneous dipole–dipole interactions between the molecules in their equilibrium positions and orientations (the prime excludes terms with $m = n$ from the summation).

$$\hbar J_{nm} \equiv \vec{\mu}_n \cdot T_{nm} \cdot \vec{\mu}_m = |\mu|^2 \, \frac{1 - 3(\hat{\mu}_n \cdot \hat{r}_{nm})(\hat{\mu}_m \cdot \hat{r}_{nm})}{r_{nm}^3}. \tag{16.36b}$$

In the present model the excitons therefore have two types of interaction: the dipole–dipole interaction and a repulsion indicating that two excitons cannot occupy the same site. This repulsion allows us to consider the excitons as hard core bosons (similar to liquid ^{4}He).* The second term in H represents the coupling to a classical transverse electric field $E(\mathbf{r})$.

We assume that each molecule has a transition (off diagonal) but not a permanent (diagonal) dipole matrix element. This implies that the molecules are polarizable but not polar. The optical polarization operator of the nth molecule is thus given by [16]

$$P_n = \vec{\mu}_n(B_n + B_n^\dagger), \tag{16.37}$$

where $\vec{\mu}_n$ denotes its transition dipole matrix element.

* This exclusion is reminiscent of the Pauli exclusion of fermions and is most commonly referred to as "Pauli exclusion." However, in this case it does not result from a fundamental symmetry but is merely a consequence of our two-level model. A "hard core boson" is thus a better terminology.

Since the polarization operator is expressed in terms of B_n and $B_n^\dagger$, it is natural to start with the equations of motion for these operators. Using the Hamiltonian Eq. (16.35) and the commutation relation (16.34), we obtain the following Heisenberg equations of motion (all operators taken at time t):

$$\frac{1}{i}\frac{d}{dt}B_n = (-\Omega_n + i\Gamma_n)B_n - \sum_m{}' J_{nm}(B_m^\dagger + B_m)$$

$$+ 2\sum_m{}' J_{nm}(B_n^\dagger B_n B_m + B_n^\dagger B_n B_m^\dagger) + \frac{1}{\hbar}\vec{\mu}_n \cdot \vec{E}(r_n, t)[1 - 2B_n^\dagger B_n]. \quad (16.38)$$

The first two terms in the right-hand side of this equation represent linear dynamics whereby B_n is coupled to B_m and $B_m^\dagger$. The other two terms are more complex and reflect the nonlinear coupling to higher order operators. When the expectation value of Eq. (16.38) is taken, we find that the single-exciton variables $\langle B_n\rangle$ are coupled to the two-operator variables $\langle B_n^\dagger B_n\rangle$ and to the three operator (two-exciton) variables $\langle B_n^\dagger B_n B_m\rangle$ and $\langle B_n^\dagger B_n B_m^\dagger\rangle$. Equation (16.38) is therefore not closed. We may proceed by taking $A = B_m^\dagger B_n$, $B_n^\dagger B_n B_m$, and $B_n^\dagger B_n B_m^\dagger$, and writing the Heisenberg equation (16.32) for these operators. We shall then couple them to even more complex operators involving products of four B operators. In general, Eq. (16.38) will therefore result in an infinite hierarchy of coupled dynamic equations whereby single-body operators are successively coupled to more complex operators. Fortunately, the optical response to electromagnetic fields that are not too strong requires the explicit introduction of only few-particle states. This allows us to truncate the hierarchy at a very early stage. The main problem in the many-body formulation of the nonlinear response is the development of truncation schemes that yield a hierarchy of approximations for the nonlinear optical response. This situation is typical for zero temperature many-body theory where a few quasiparticles dominate the dynamic behavior. In the next chapter we shall demonstrate how a truncation at the two-exciton level may be adequate for the calculation of a variety of optical measurements. At this point, however, we shall invoke the simplest approximation by adopting the single-exciton factorization

$$\langle B_n^\dagger B_n B_m\rangle = \langle B_n^\dagger\rangle\langle B_n\rangle\langle B_m\rangle,$$

$$\langle B_n^\dagger B_n B_m^\dagger\rangle = \langle B_n^\dagger\rangle\langle B_n\rangle\langle B_m^\dagger\rangle,$$

and

$$\langle B_n^\dagger B_n\rangle = \langle B_n^\dagger\rangle\langle B_n\rangle. \quad (16.39)$$

Taking the expectation value of Eq. (16.38) and using these factorizations we get

$$\frac{1}{i}\frac{d}{dt}\langle B_n\rangle = -\left(\Omega_n - i\frac{\Gamma_n}{2}\right)\langle B_n\rangle - \frac{1}{\hbar}\vec{\mu}_n \cdot \vec{E}_{l,n}[2\langle B_n^\dagger\rangle\langle B_n\rangle - 1]. \quad (16.40)$$

Here we have introduced the local field at site n generated by the intermolecular interactions. This field is related to the average Maxwell field $E(r_n)$ by

$$\vec{\mu}_n \cdot \vec{E}_{l,n} \equiv \vec{\mu}_n \cdot \vec{E}(\mathbf{r}_n) - \hbar\sum_m{}' J_{nm}[\langle B_m\rangle + \langle B_m^\dagger\rangle], \quad (16.41)$$

which is identical to Eq. (16.3a). The equation of motion for $\langle B_n^\dagger \rangle$ can be obtained by taking the complex conjugate of Eq. (16.40). Since the Maxwell field and the local field enter here only through their scalar products with the dipole operator, it will be useful to introduce the notation $E_{l,n} \equiv \hat{\mu}_n \cdot \vec{E}_{l,n}$ and $E_n \equiv \hat{\mu}_n \cdot E(\mathbf{r}_n)$, where $\hat{\mu}_n$ is a unit vector in the direction of $\vec{\mu}_n$. We then have

$$\vec{\mu}_n \cdot \vec{E}(\mathbf{r}_n) \equiv \mu_n E_n,$$

$$\vec{\mu}_n \cdot \vec{E}_{l,n} \equiv \mu_n E_{l,n}.$$

This way, we can avoid using vector notation in much of the following derivations.

Combining Eq. (16.40) with its complex conjugate, and using Eq. (16.37), we finally get a closed equation of motion for the polarization [13]

$$\ddot{P}_n(t) + 2\Gamma_n \dot{P}_n(t) + \Omega_n^2 P_n(t)$$

$$= 2\Omega_n \hbar^{-1} |\vec{\mu}_n|^2 E_{l,n}(t) - \frac{1}{\hbar \Omega_n} \cdot E_{l,n} |(\Omega_n + i\Gamma_n) P_n(t) + i\dot{P}_n(t)|^2. \quad (16.42)$$

Here the polarization at every site behaves as an oscillator driven by the local electric field and by anharmonic (nonlinear) "forces." All intermolecular interactions were formally eliminated by introducing the local electric field, $E_{l,n}$. This result is identical to the equations derived in Chapter 6 [Eq. (6.44) together with (6.50)] except that the Maxwell field E is replaced by the local field $E_{l,n}$. Therefore, the discussion following these equations in Chapter 6, in particular, regarding the differences with the phenomenological cubic-oscillator model, applies here as well.

In concluding this discussion let us consider a homogeneous excitation of an assembly of identical molecules so that $E_n(t) = E(t)$, $\mu_n = \mu$ and $\Omega_n = \Omega$. In this case we have

$$\psi(t) \equiv \langle B_n(t) \rangle,$$

independent on n. Equation (6.40) then assumes the form

$$\frac{1}{i}\frac{d\psi}{dt} = \left(-\Omega - J(k=0) + i\frac{\Gamma}{2}\right)\psi$$

$$+ \frac{1}{\hbar}\mu E(t)[1 - 2|\psi|^2] + 2J(k=0)|\psi|^2\psi. \quad (16.43)$$

This is identical to the Landau–Ginzburg equation of second-order phase transitions [12]. It has been used to interpret femtosecond four-wave mixing measurements in semiconductor nanostructures, which show some clear signatures of local-field corrections. This will be shown next.

The absolute magnitudes of optical susceptibilities do not usually provide an unambiguous test for the validity of the LFA. Typically, many unknowns in the structure and parameters of the system prohibit a quantitative test. However, direct evidence for the contribution of local field corrections can be obtained through the following time-domain four-wave mixing measurement. The system interacts with two light beams $\mathbf{k}_1$ and $\mathbf{k}_2$, and the signal at $\mathbf{k}_s = 2\mathbf{k}_2 - \mathbf{k}_1$ is detected. Let us denote the delay between the pulses as τ. For $\tau > 0$ (pulse $\mathbf{k}_1$ first) this is the two pulse echo (see Chapter 10). For $\tau < 0$ this is a different measurement (which does not have a name) [see Eqs. (10.13) and (10.14)]. The interesting point is that for a two-level model, and when the light beams are nearly resonant with the electronic transition, there is no pathway that is completely resonant, and by invoking the RWA the signal vanishes identically for $\tau < 0$. However, Eq. (16.43), which includes a local field correction, does predict a signal for $\tau < 0$. When Eq. (16.43) is solved, one finds that the signal rises exponentially as $\exp(-2\Gamma\tau)$ for $\tau < 0$ and then decays as $\exp(-\Gamma\tau)$ for $\tau > 0$ (photon echo). Figure 16.1a shows an experimental time-integrated signal for a semiconductor multiple quantum well. In qualitative agreement with the predictions of Eq. (16.43), the signals are asymmetric in time, with exponentially rising and decaying tails at long time delays. As the lattice temperature is raised, both the decay and the rise times get shorter, due to increasing dephasing of the excitons by thermal phonons. Figure 16.1b shows calculations based on a generalization of Eq. (16.43). Instead of adding a phenomenological dephasing rate (Γ) we used a stochastic model for the exciton frequency Ω [see Eq. (8.60)] and set $\Gamma = 0$. In the homogeneous limit $\kappa \gg 1$, we recover the behavior described above. As κ decreases toward the inhomogeneous limit $\kappa \ll 1$, the rise and decay become faster and the signal for $|\tau| < \Lambda^{-1}$ is suppressed. However, for time delays much longer than the correlation time, $\tau \gg \Lambda^{-1}$, the rise and decay remain exponential, in agreement with the homogeneous limit. The fact that intermolecular interactions induce this signal can be rationalized by recognizing that the presence of two exciton states (see Figure 17.4) makes the system effectively a three-level, rather than a two-level, system. In a three-level system one can easily satisfy the RWA with this pulse sequence. The local field correction

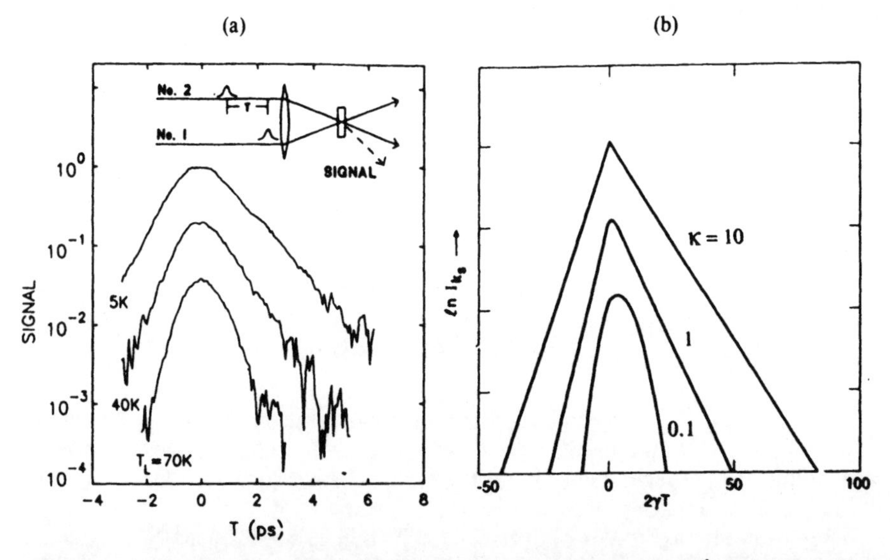

FIG. 16.1 (a) Experimental time-integrated four-wave mixing signal in (a) 170 Å $GaAs/Ga_{1-x}Al_xAs$ multiple-quantum well structure vs. time delay τ, taken with 500 fs laser pulses for different lattice temperatures T as indicated. The inset shows the geometry of the experiment. [From K. Leo, M. Wegener, J. Shah, D. S. Chemla, E. O. Göbel, T. C. Damen, S. Schmitt-Rink, and W. Schaefer, *Phys. Rev. Lett.* **65**, 1340 (1990).] (b) Time-integrated four-wave mixing signal vs. time delay τ for different line-shape parameters κ [Eq. (8.50)], as indicated. [From S. Schmitt-Rink, S. Mukamel, K. Leo, J. Shah, and D. S. Chemla, "Stochastic theory of time-resolved four-wave mixing in interacting media," *Phys. Rev. A* **44**, 2124 (1991).]

contains (in some approximate way, as will be shown in Chapter 17) the effects of these additional levels. When put in this manner, the new contribution induced by the local field becomes less mysterious.

LOCAL FIELD EXPRESSIONS FOR OPTICAL SUSCEPTIBILITIES OF HOMOGENEOUS SYSTEMS

The linearized part of Eq. (16.42) is obtained by neglecting the second term on the right-hand side. Introducing the particle's polarizability, we then get,

$$\alpha(\omega) = \frac{1}{\hbar}\frac{2\Omega\bar{\mu}\bar{\mu}}{\Omega^2 - (\omega + i\Gamma)^2}. \tag{16.44}$$

This is known as the Drude oscillator formula [see Eq. (6.21)].

The harmonic oscillator representation of the linear optical response of a collection of two-level systems has long been recognized [19]. The dielectric function of a cubic crystal [7] can be obtained by substituting $\alpha(\omega)$ in the Clausius–Mossotti expression [Eq. (16.28)]. This result can alternatively be obtained by working directly in **k** space. To that end we introduce the material polarization field in momentum space

$$P(\mathbf{k}) \equiv \sqrt{N}\,\bar{\mu}(B_{\mathbf{k}} + B^{\dagger}_{-\mathbf{k}}), \tag{16.45}$$

where we have defined the exciton annihilation $B_{\mathbf{k}}$ and creation $B^{\dagger}_{-\mathbf{k}}$ operators in the momentum representation:

$$B_{\mathbf{k}} \equiv \frac{1}{\sqrt{N}}\sum_m B_m \exp(-i\mathbf{k}\cdot\mathbf{r}_m), \tag{16.46a}$$

$$B^{\dagger}_{\mathbf{k}} \equiv \frac{1}{\sqrt{N}}\sum_m B^{\dagger}_m \exp(i\mathbf{k}\cdot\mathbf{r}_m). \tag{16.46b}$$

We shall consider the linearized part of Eq. (16.38) obtained by neglecting all terms with more than one B or B^+ factor in the right-hand side. This will be justified in Chapter 17. Assuming that the particles are arranged in a periodic crystal, we can switch to **k**-space and the equation reads

$$\frac{1}{i}\frac{d}{dt}B_{\mathbf{k}} = [-\Omega - J(\mathbf{k}) + i\Gamma]B_{\mathbf{k}} - J(\mathbf{k})B^{\dagger}_{-\mathbf{k}} + \frac{1}{\hbar}\bar{\mu}E(\mathbf{k}, t), \tag{16.47}$$

where $J(\mathbf{k})$ is the lattice Fourier transform of the intermolecular interaction

$$J(\mathbf{k}) = \sum_m{}' J_{mn}\exp(-i\mathbf{k}\cdot\mathbf{r}_m).$$

For a centrosymmetric lattice, we have $J(\mathbf{k}) = J(-\mathbf{k})$. By combining Eq. (16.47) with its Hermitian conjugate equation for $B^{\dagger}_{-\mathbf{k}}$, and performing a Fourier transform to the frequency domain, we get

$$[-(\omega + i\Gamma)^2 + \Omega_{\mathbf{k}}^2]P(\mathbf{k}, \omega) = 2\Omega\rho\hbar^{-1}|\bar{\mu}|\cdot E^{\perp}(\mathbf{k}, \omega), \tag{16.48}$$

where

$$\Omega_{\mathbf{k}} = [\Omega(\Omega + 2J(\mathbf{k}))]^{1/2}. \tag{16.49}$$

The system behaves as a collection of oscillators, denoted *Coulomb excitons*. Equations (16.47) and its Hermitian conjugate for $B^{\dagger}_{-\mathbf{k}}$ define an eigenstate problem whose solutions are the creation and annihilation operators for the Coulomb excitons. The Coulomb exciton frequency $\Omega_{\mathbf{k}} - i\Gamma$ is determined by the secular equation of the problem. For $|J(\mathbf{k})| \ll \Omega$, which is, by definition, the case in molecular crystals, this yields $\Omega_{\mathbf{k}} \approx \Omega + J(\mathbf{k})$. This is known as the Heitler–London approximation in which the Coulomb excitons are simply created (annihilated) by $B^{\dagger}_{-\mathbf{k}}(B_{\mathbf{k}})$.

Equation (16.48) immediately yields the following expression for the frequency and wavevector dependent dielectric tensor:

$$\varepsilon(\mathbf{k}, \omega) = 1 + \frac{8\pi\Omega\rho_0\hbar^{-1}\vec{\mu}\vec{\mu}}{\Omega_{\mathbf{k}}^2 - (\omega + i\Gamma)^2}. \tag{16.50}$$

We next turn to the calculation of the nonlinear optical response. We note that $\chi^{(3)}$ is the lowest nonlinearity allowed by the present model, since $\chi^{(2)}$ vanishes for a centrosymmetric medium. When Eqs. (16.38) are solved directly in powers of the Maxwell field, we obtain for $\chi^{(3)}$

$$\chi^{(3)}(-\mathbf{k}_s - \omega_s; \mathbf{k}_1\omega_1, -\mathbf{k}_2 - \omega_2, \mathbf{k}_3\omega_3)$$
$$= \frac{1}{3!} 4\Omega\rho_0 \frac{\vec{\mu}\vec{\mu}\vec{\mu}\vec{\mu}}{\hbar^3} \sum_P \frac{(\Omega + \omega_2 - i\Gamma)(\Omega + \omega_1 + i\Gamma)}{\Delta(\mathbf{k}_s, \omega_s)\Delta(\mathbf{k}_1, \omega_1)\Delta(-\mathbf{k}_2, -\omega_2)} \left[\frac{2\Omega J(\mathbf{k}_3)}{\Delta(\mathbf{k}_3, \omega_3)} - 1\right], \tag{16.51a}$$

where $\sum_P$ denotes a sum over permutations of the fields [see Eq. (5.48) and

$$\Delta(\mathbf{k}, \omega) \equiv \Omega_{\mathbf{k}}^2 - (\omega + i\Gamma)^2. \tag{16.51b}$$

Alternatively, since the single-particle factorization used here leads to a local-field description, it is possible to write the susceptibility as the third-order polarizability of a single molecule $\gamma(-\omega_s; \omega_1, -\omega_2, \omega_3)$ multiplied by local-field correction factors

$$\chi^{(3)}(-\mathbf{k}_s - \omega_s; \mathbf{k}_1\omega_1, -\mathbf{k}_2 - \omega_2, \mathbf{k}_3\omega_3)$$
$$= \rho_0\gamma(-\omega_s; \omega_1, -\omega_2, \omega_3)S(\mathbf{k}_1\omega_1)S(-\mathbf{k}_2 - \omega_2)S(\mathbf{k}_3\omega_3)S(\mathbf{k}_s\omega_s), \tag{16.52a}$$

with

$$\gamma(-\omega_s; \omega_1, -\omega_2, \omega_3)$$
$$= \frac{4}{3!}\Omega\frac{\vec{\mu}\vec{\mu}\vec{\mu}\vec{\mu}}{\hbar^3}\sum_P \frac{(-\Omega - \omega_2 + i\Gamma)(\Omega + \omega_1 + i\Gamma)}{[-(\omega_s + i\Gamma)^2 + \Omega^2][-(\omega_1 + i\Gamma)^2 + \Omega^2][-(-\omega_2 + i\Gamma)^2 + \Omega^2]},$$
$$(+\text{permutations of } j = 1, 2, 3) \tag{16.52b}$$

and

$$S(\mathbf{k}, \omega) \equiv \frac{\Omega^2 - (\omega + i\Gamma)^2}{\Omega_{\mathbf{k}}^2 - (\omega + i\Gamma)^2}. \tag{16.53}$$

The numerators in the $S(\mathbf{k}, \omega)$ factors cancel the three molecular (Ω) resonances in γ and introduce new exciton ($\Omega_{\mathbf{k}}$) resonances. The permutations of $(\mathbf{k}_1, \omega_1)$, $(-\mathbf{k}_2, -\omega_2)$, and $(\mathbf{k}_3, \omega_3)$ account for the six different time orderings with which the electric fields can interact with the system.

NOTES

1. H. A. Lorentz, *The Theory of Electrons* (Dover, New York, 1952).
2. P. Mazur, *Adv. Chem. Phys.* **1**, 309 (1958); J. deGoede and P. Mazur, *Physica* **58**, 568 (1972); S. R. De Groot, *The Maxwell Equations* (North-Holland, Amsterdam, 1969).
3. A. D. Buckingham, *Adv. Chem. Phys.* **12**, 107 (1967); B. Linder, *Adv. Chem. Phys.* **12**, 225 (1967); K. L. C. Hunt, Y. Q. Liang, R. Nimalakirthi, and R. A. Harris, *J. Chem. Phys.* **91**, 5251 (1989).
4. J. Van Kranendonk and J. E. Sipe, in *Progress in Optics*, Vol. 15, E. Wolf, Ed. (North-Holland, Amsterdam, 1969), p. 245.
5. P. Madden and D. Kivelson, *Adv. Chem. Phys.* **56**, 467 (1984).
6. C. J. F. Bottcher, *Theory of Electric Polarization*, Vol. I (Elsevier, Amsterdam, 1973); C. J. F. Bottcher and P. Bordewijk, *Theory of Electric Polarization*, Vol. II (Elsevier, Amsterdam, 1978).
7. L. V. Keldysh, D. A. Kirzhnitz, and A. A. Maradudin, *The Dielectric Function of Condensed Systems* (North-Holland, Amsterdam, 1989).
8. D. Bedeaux and N. Bloembergen, *Physica* **69**, 67 (1973).
9. G. R. Meredith, *J. Chem. Phys.* **75**, 4317 (1981); **77**, 5863 (1982); *Phys. Rev. B* **24**, 5522 (1981); M. Hurst and R. W. Munn, *J. Mol. Electr.* **3**, 75 (1987).
10. W. E. Torruellas, R. Zanoni, G. I. Stegeman, G. R. Mohlmann, E. W. P. Erdhuisen, and W. H. G. Horsthuis, *J. Chem. Phys.* **94**, 6851 (1991); J. H. Andrews, K. L. Kowalski, and K. D. Singer, *Phys. Rev. A* **46**, 4712 (1992); W. E. Torruellas, D. Y. Kim, M. Jaegger, G. Krijnen, R. Schief, G. I. Stegeman, P. Vidakovic, and J. Zyss, in *ACS Tutorial Series*, G. Lindsay and K. Singer, Eds. (1995).
11. S. Mukamel, Z. Deng, and J. Grad, *J. Opt. Soc. Am. B* **5**, 804 (1988).
12. W. Wegener, D. S. Chemla, S. Schmitt-Rink, and W. Schäfer, *Phys. Rev. A* **42**, 5675 (1990); S. Schmitt-Rink, S. Mukamel, K. Leo, J. Shah, and D. S. Chemla, *Phys. Rev. A* **44**, 2124 (1991); W. Wegener, D. S. Chemla, S. Schmitt-Rink, and W. Schäfer, *Phys. Rev. A* **44**, 2124 (1991).
13. J. Knoester and S. Mukamel, *Phys. Rev. A* **41**, 3812 (1990).
14. W. R. Heller and A. Marcus, *Phys. Rev.* **84**, 809 (1951).
15. M. H. Cohen and K. Keffer, *Phys. Rev.* **99**, 1128 (1955).
16. A. S. Davydov, *Theory of Molecular Excitons* (Plenum, New York, 1971).
17. V. M. Agranovich and M. D. Galanin, *Electronic Exciton Energy Transfer in Condensed Matter* (North-Holland, Amsterdam, 1982).
18. L. N. Ovander, *Usp. Fiz. Nauk* **86**, 3 (1965) [*Sov. Phys.-Usp.* **8**, 337 (1965)].
19. M. Born and K. Huang, *Dynamical Theory of Crystal Lattices* (Oxford, London, 1954).

CHAPTER 17

Many-Body and Cooperative Effects in the Nonlinear Response

Calculations of nonlinear optical response in condensed phases are usually based on a mean-field ansatz: the local-field approximation. As was shown in the previous chapter, underlying this approximation is the implicit assumption that the only relevant dynamic variables are *local* exciton variables. The system may then be viewed as a collection of localized anharmonic oscillators with intermolecular harmonic dipole–dipole coupling that can be lumped into an effective local field. In this picture, the nonlinear susceptibilities assume the form of a product of a single molecule hyperpolarizability, times local-field corrections, which account in a simple way for intermolecular interactions.

This calculation of macroscopic susceptibilities is, however, not rigorous. It fails to take properly into account the correlated dynamics of the interacting many-body system, i.e., correlations among the particles, as well as correlations between the particles and the radiation field. Short-range (e.g., exchange) forces are totally neglected in this procedure, and even the dipole–dipole forces are not fully incorporated. In addition, the wavevector dependence of the resulting susceptibilities fails to account for processes such as exciton migration and energy transfer and transport (e.g., the Forster transfer) [1–3]. These processes are often added phenomenologically in order to interpret transient grating spectroscopy, which is a four-wave mixing technique that measures transport processes by following the wavevector dependence of the susceptibilities [4]. Moreover, the local-field approximation totally misses many-body effects such as cooperative radiative decay (superradiance) resulting from cooperative interactions with the radiation field [6] as well as collective effects on the magnitude of off-resonant optical nonlinearities and the nature of their resonant structure. Such effects are particularly important for the linear and nonlinear optical properties of confined excitons in small clusters and in molecular [7–14] and semiconductor [15–19] nanostructures. The significant progress made in the fabrication of artificial structures with a nanometer scale control, has created great interest in the effects of geometric confinement on their optical properties. Nanostructures are classified by the number of confined dimensions. Two-dimensional quantum

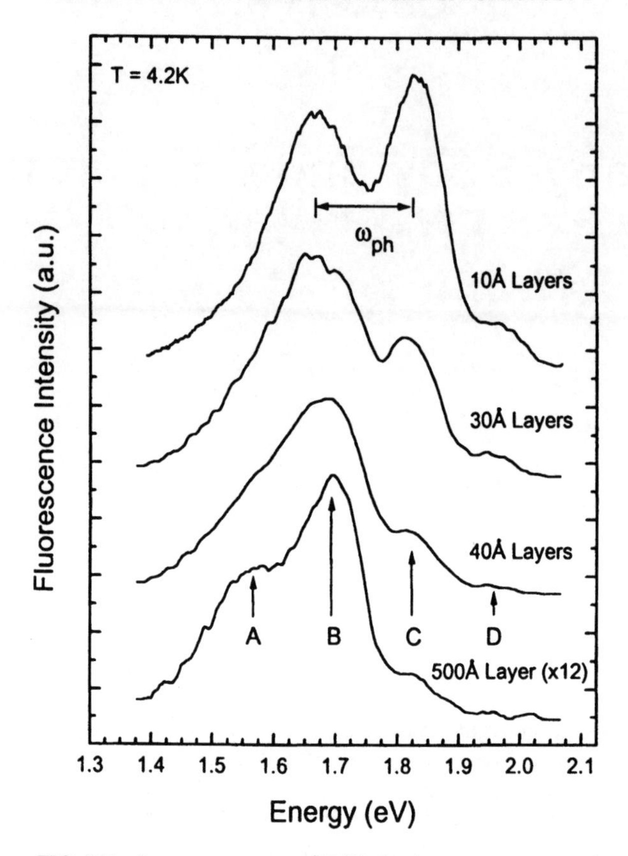

FIG. 17.1 Low temperature (4.2 K) luminescence spectra of stacks of alternating layers of the crystalline organic compounds 3,4,9,10-perylenetetracarboxylic dianhydride (PTCDA) and 3,4,7,8-naphthalenetetracarboxylic dianhydride (NTCDA). The layer thickness of PTCDA ranging from 10 to 500 Å as indicated. The spectra are displaced vertically for clarity. [From E. I. Haskal, Y. Zhang, P. E. Burrows, and S. R. Forrest, "Finite size effects observed in the fluorescence of ultrathin crystalline organic thin films grown by organic molecular beam deposition," *Chem. Phys. Lett.* **219**, 325 (1994).]

wells, one-dimensional quantum wires, and zero-dimensional small particle quantum dots have been studied. The low temperature (4.2°K) fluorescence spectra of superlattice structures consisting of alternating layers of crystalline organic thin films grown by the ultrahigh vacuum process of organic molecular beam deposition are displayed in Figure 17.1. They clearly show that as the layer thickness of the organic films is decreased (from 500 to ~10 Å), the phonon vibrational frequency increases, and that there is a redistribution of fluorescence intensity from lower to higher energy vibronic transitions. These observations may be interpreted using exciton confinement and a systematic change in the exciton–phonon coupling in the restricted environment of ultrathin molecular films. A clear signature of exciton confinement is provided by the sharp exciton resonances observed in semiconductor quantum well structures. This is illustrated

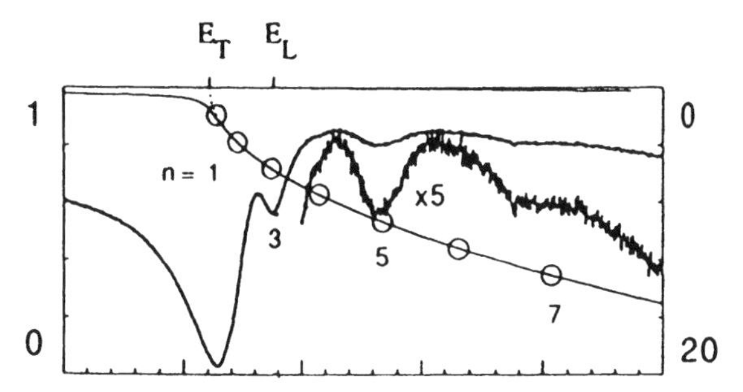

FIG. 17.2 Transmission spectra in Z_3 exciton region of CuCl thin films at 2 K with thickness $L = 157$ Å. Polariton and exciton dispersions are also shown, by thin solid lines and dotted lines, respectively. Open circles show the exciton positions of $K_n = n\pi/L$ with $n = 1, 2, 3, \ldots$. [From Z. K. Tang, A. Yanase, T. Yasui, Y. Segawa, and K. Cho, "Optical selection rule and oscillator strength of confined exciton system in CuCl thin films," *Phys. Rev. Lett.* **71**, 1431 (1993).]

in the transmission spectra of CuCl epitaxial thin films displayed in Figure 17.2. The various lines represent quantized exciton states. The value of the oscillator strength for the nth exciton is found to be proportional to L/n^2 for odd n and zero for even n, in accordance with theoretical predictions.

One of the most fascinating open questions raised by these studies is the possibility of maintaining a large coherence size [19–21], which may give rise to enhanced nonlinear optical susceptibilities and ultrafast radiative decay rates, resulting from collective interactions with the radiation field. As an example, in Figure 17.3a we show cooperative radiative decay in molecular J-aggregates. These are strongly coupled quasi-one-dimensional aggregates of the dye pseudo-isocyanine bromide (PIC) [7]. The coherence size N_{eff} is defined as the ratio of the actual radiative decay rate to the decay rate of a single molecule, and reflects the number of molecules which interact in phase with the radiation field. Figure 17.3a shows the calculated N_{eff} of a J-aggregate coupled to a bath of optical phonons vs. the temperature and physical size N. Obviously at low temperatures $N_{\text{eff}} = N$ reflecting a complete cooperativity. At high temperatures phonon-induced dephasing destroys the cooperativity, and N_{eff} reaches eventually the value of 1. The figure establishes the existence of a temperature-dependent coherence size, provided the exciton dephasing timescale is much shorter than the fluorescence lifetime. A comparison of these calculations with experiment is shown in Figure 17.3b.

The main problem in predicting the nonlinear optical response in the condensed phase is not how to better calculate the local field, but rather the fundamental failure of the local-field approximation to describe and predict some very important effects related to the many-body nature of the problem. A new theoretical approach that does not invoke the local-field approximating is called for. In this chapter we present a microscopic theoretical framework for calculating the nonlinear optical response of assemblies of molecules with localized electronic states and an arbitrary geometry. The present approach is based on the Heisenberg equations of motion for the matter variables which determine the

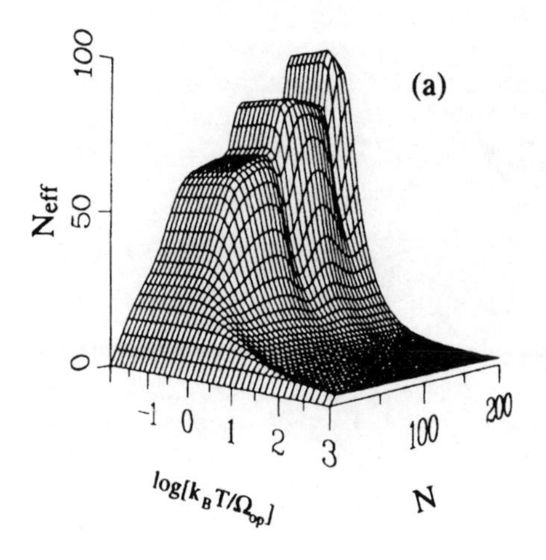

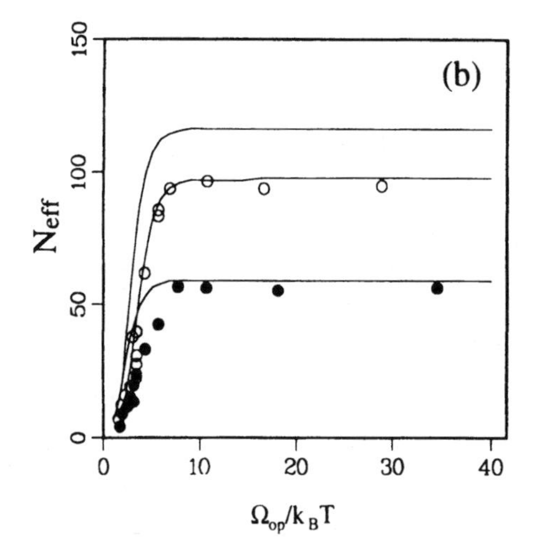

FIG. 17.3 Temperature variation of exciton coherence size in a molecular aggregate (a) calculating N_{eff} as a function of aggregate size N and $\log_{10}(kT/\Phi_{op})$, for optical phonons with frequency Ω_{op}. (b) N_{eff} vs. Ω_{op}/kT for aggregates of size $N = 75$, 125, and 150 (bottom to top) (solid curves). The circles represent the experimental measurements of S. DeBoer and D. A. Wiersma [13]; the open circles correspond to the blue site and the solid circles correspond to the red site. [F. C. Spano, J. R. Kuklinski, and S. Mukamel, "Temperature- dependent superradiant decay of excitons in small aggregates," *Phys. Rev. Lett.* **65**, 211 (1990).]

optical response [see Eq. (16.39)].* It accounts for intermolecular interactions and dynamic correlations and offers a close look at the limitations of the local-field picture and a systematic way for its generalization. We find that in addition to the single-particle variables used in the derivation of LFA, we need consider the contribution of all quadratic (bilinear) and cubic exciton variables that are responsible for intermolecular coherences. The material system can thus be modeled as a collection of coupled nonlocal anharmonic oscillators [20]. An anharmonic oscillator picture for the nonlinear response has been suggested as a qualitative model since the early days of nonlinear optics [see Eqs. (6.44) and (6.45)]. The equation of motion approach shows how such a picture can be rigorously established and applied toward the fully microscopic calculation of the nonlinear response. The equations are solved using Green function techniques,

* The phenomenological derivation of the local-field approximation as reviewed in the beginning of Chapter 16 cannot be extended to include many-body effects, since it is intrinsically a mean-field single-molecule theory.

resulting in closed expressions for the optical response and susceptibilities. An explicit expression for $\chi^{(3)}$ is derived by solving the equations in the absence of relaxation, and numerical calculations for one-, two-, and three-dimensional assemblies are presented.

GREEN FUNCTION EXPRESSION FOR THE OPTICAL RESPONSE OF MOLECULAR NANOSTRUCTURES WITH ARBITRARY GEOMETRY

Nonlinear susceptibilities of simple quantum systems are usually calculated using the time-dependent density operator in the Schrödinger picture. As demonstrated in Chapter 6, the resulting susceptibilities are then given in terms of multiple summations over eigenstates. In the previous chapter we showed that in order to map the problem into coupled anharmonic oscillators, we should adopt a different approach, which is based on the Heisenberg picture. The polarization is then calculated by solving a set of coupled nonlinear equations of motion. The equations are obtained by identifying the relevant dynamic variables and deriving equations of motion for their expectation values. We further pointed out the advantages of the oscillator picture and showed how the simplest truncation of the Heisenberg equations leads to the LFA. We shall now generalize these results and treat the many-body aspects of the problem.

As in Chapter 16, we consider here a molecular assembly containing N coupled two-level molecules with nonoverlapping charge distributions and with an arbitrary geometry and dimensionality. We shall adopt a different Frenkel exciton Hamiltonian containing a simplified dipole coupling [22, 23].

$$H_{\mathrm{mat}} = \hbar \sum_n \Omega_n B_n^\dagger B_n + \frac{\hbar}{2} \sum_{n,m}' J_{nm}(B_m^\dagger B_n + B_n^\dagger B_m). \tag{17.1}$$

The second term in Eq. (17.1) is the dipole–dipole interaction in the Heitler–London approximation. This coupling allows for exciton migration among molecules but, unlike Eq. (16.36), it conserves the number of excitations. J_{nm} can represent dipole–dipole interactions as well as short-range exchange exciton couplings. Ω_n is the isolated particle transition frequency. The total Hamiltonian of the system is given by Eq. (16.35) together with Eq. (17.1), and the polarization operator is given by Eq. (16.37). The global level scheme of the assembly described by the present material Hamiltonian is shown in Figure 17.4. The levels are divided into bands. The global ground state is the state in which all molecules are in their ground state. We then have a single-exciton band with N states, a two-exciton band with $N(N-1)/2$ states, etc. for a total of 2^N states. Had we included the full dipole coupling [Eq. (16.36)], these bands will be coupled, and the ground state (as well as all other states) will change. These corrections are of the order of J/Ω_n, which is a very small parameter for molecular systems ($\sim 10^{-2}$ for anthracene), and will be neglected. Invoking the Heitler– London approximation, we thus work in first-order perturbation theory in this parameter. As we shall shortly see, this allows us to classify the dynamic variables by their nonlinear order with respect to the field and truncate the hierarchy of Heisenberg equations generated from our Hamiltonian, commensurate with the order of the optical process of interest. This is the crucial step in a successful

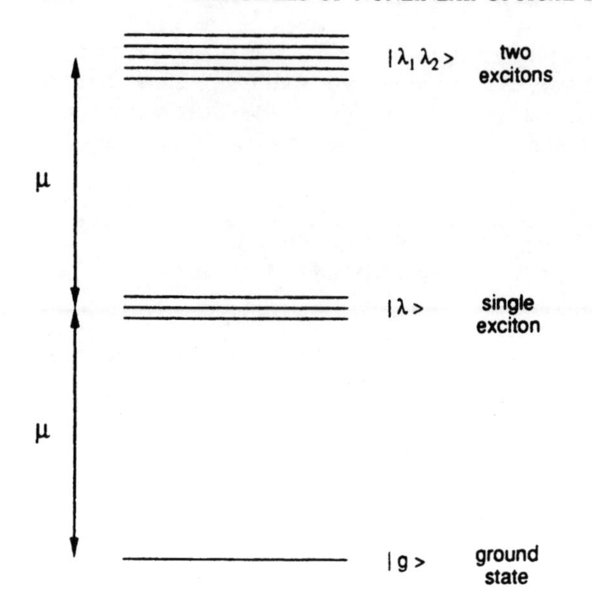

FIG. 17.4 Energy level diagram for a molecular aggregate, showing the single-photon allowed exciton states and the two-photon allowed two-exciton states. The exciton coupling J is taken to be negative (attractive interaction, as in J aggregates) [7] and the exciton-two-exciton splitting $\omega_q - \omega_e$ is therefore positive. In this case strong two-photon absorption will occur to the blue of the exciton resonance.

implementation of the equation of motion procedure. If necessary, it is possible to calculate corrections of higher order in J/Ω_n perturbatively.

To connect the dynamic variables to the exciton level scheme, it will be useful to start with the Heisenberg equation, take the expectation values of both sides, and switch back to the Schrödinger picture. The expectation value of an arbitrary operator A then satisfies the following equation of motion

$$\frac{\partial}{\partial t}\langle\psi(t)|A|\psi(t)\rangle = \frac{i}{\hbar}\langle\psi(t)|[H, A]|\psi(t)\rangle,$$

with A being time independent (since we are using the Schrödinger picture). For our model, the wavefunction can be expanded in a complete basis set of exciton states

$$|\psi(t)\rangle = |0\rangle + \sum_n c_n^{(1)}(t)B_n^\dagger|0\rangle + \sum_{n,m} c_{n,m}^{(2)}(t)B_n^\dagger B_m^\dagger|0\rangle + \cdots$$

where $|0\rangle$ is the ground state (vacuum) of the system and $c^{(p)}$ is the expansion coefficient of states in the p-exciton band. We shall write this expression in an abbreviated notation

$$|\psi(t)\rangle = \sum_{p=0}^{\infty} c^{(p)}(t)(B^\dagger)^p|0\rangle. \tag{17.2a}$$

In the Schrödinger picture, calculating the dynamics thus involves finding the time-dependent coefficients $c^{(p)}(t)$. Since the dipole coupling [Eq. (16.37)] can create (or annihilate) only one exciton at a time, it is clear that $c^{(p)}$ contains only contributions to order p and higher in the incoming fields. This important observation will form the basis for our classification and systematic truncation scheme. The one-particle nature of the transition dipole results in an additional simplification. For calculating the third-order response we need in principle to truncate the sum in Eq. (17.2a) at $p = 3$. However, the three-exciton ($p = 3$)

states will not contribute to the expectation value of the polarization to third order since the matrix element $\langle 0|(B + B^\dagger)(B^\dagger)^3|0\rangle$ vanishes identically. It is therefore sufficient to truncate the expression at the two-exciton $p = 2$ level for calculating the third-order response. The system behaves effectively as a three-level system!

We next introduce the notion of a normally ordered product of operators where all $B^\dagger$ are to the left and all B to the right

$$B_n^\dagger B_m^\dagger \cdots B_{n'} B_{m'} \cdots$$

We shall, hereafter, work only with normally ordered products. This does not restrict the generality of our treatment since, by using the commutation rules, we can bring any product to a normally ordered form. We next note that the expectation value of any normally ordered product of order $p + q$ (pth order in $B^\dagger$ and qth order in B) must be at least to $p + q$th order in the field. This can be shown as follows:

$$\begin{aligned}\langle (B^\dagger)^p (B)^q \rangle &= \langle \psi(t)|(B^\dagger)^p (B)^q|\psi(t)\rangle \\ &= \langle \psi(t)|(B^\dagger)^p|0\rangle\langle 0|(B)^q|\psi(t)\rangle + \cdots. \end{aligned} \tag{17.2b}$$

Here we have introduced a complete basis set between the $B^\dagger$ and B. The first term involving the vacuum state is written explicitly, and the higher terms (not written) include single-exciton, two-exciton states, etc. The lowest term in the expansion of the wavefunction that contributes to this matrix element is thus $c^{(p)*}c^{(q)}$, which implies that this term is at least of order $p + q$ in the field. The higher terms in the expansion can contribute only to higher orders in this field. For the sake of evaluating the third-order response it is thus sufficient to consider only products with up to three B operators. The Heisenberg equations of motion then truncate naturally and yield,

$$\begin{aligned}\frac{1}{i}\frac{d}{dt}B_n = &-\Omega_n B_n - {\sum_m}' J_{nm} B_m + 2{\sum_m}' J_{nm} B_n^\dagger B_n B_m \\ &+ \frac{1}{\hbar}\mu_n E_n(t)(1 - 2B_n^\dagger B_n). \end{aligned} \tag{17.3}$$

We adopt here the scalar notation introduced in Chapter 16 where E_n denotes the component of the Maxwell field along the (fixed) direction of the transition dipole $\vec{\mu}_n$. For weak fields, the polarization, which is related to the expectation values of B_n and $B_n^\dagger$, is small, so to find the nonlinear response we can make a perturbative expansion in the number of Bs, provided they are normally ordered. We are interested in the response of the system to the Maxwell electric fields, in particular in the expectation value of the dipole operator $\langle P_n(t)\rangle$, as this is what is observed in optical measurements. In the Heisenberg picture the expectation value of an operator is taken with respect to the ground (vacuum) state, so that $\langle \cdots \rangle = \langle 0|1 \cdots |0\rangle$. In Eq. (17.3) we then need the expectation value of a product of two and three Bs.

It is possible to derive entirely general Green function formulas for the nonlinear response. Before doing so, however, we shall incorporate dephasing and relaxation induced by coupling of our electronic degrees of freedom to nuclear motions. We have outlined in Appendix 17.A a simple model for relaxation that results in a prescription of how to incorporate relaxation in the equations of motion once expectation values are taken [24]. The model depends on two real parameters $\hat{\Gamma}$ and γ. $\hat{\Gamma}$ is the pure dephasing rate and γ represents the population relaxation rate.

$$\left[\frac{d}{dt}\langle B_m^\dagger(t)B_n(t)\rangle\right]_{\text{ph}} = -[\hat{\Gamma}(1-\delta_{nm})+\gamma]\langle B_m^\dagger(t)B_n(t)\rangle, \tag{17.4a}$$

and

$$\left[\frac{d}{dt}\langle B_n^\dagger(t)B_m(t)B_{n'}(t)\rangle\right]_{\text{ph}} = -[\hat{\Gamma}(\tfrac{3}{2}-\delta_{n,m}-\delta_{nn'})+\tfrac{3}{2}\gamma]\langle B_n^\dagger(t)B_m(t)B_{n'}(t)\rangle. \tag{17.4b}$$

We further introduce the notation

$$\zeta_{nm} \equiv 1-\delta_{nm}, \tag{17.4c}$$

and

$$F_{nm} \equiv -\Omega_n\delta_{nm} - J_{nm} + i\frac{\Gamma}{2}\delta_{nm}, \tag{17.4d}$$

with the total dephasing rate.

$$\Gamma \equiv \hat{\Gamma}+\gamma,$$

The equations of motion now read [19]

$$\frac{1}{i}\frac{d}{dt}\langle B_n\rangle = \sum_m F_{nm}\langle B_m\rangle + 2\sum_m{}' J_{nm}\langle B_n^\dagger B_n B_m\rangle + \frac{1}{\hbar}\mu_n E_n(t)(1-2\langle B_n^\dagger B_n\rangle), \tag{17.5a}$$

$$\begin{aligned}\frac{1}{i}\frac{d}{dt}\langle B_n B_m\rangle = {} & \zeta_{nm}\sum_l (F_{n,l}\langle B_l B_m\rangle + F_{m,l}\langle B_n B_l\rangle) \\ & + \mu_n\zeta_{nm}\frac{E_n(t)}{\hbar}\langle B_m\rangle + \frac{1}{\hbar}\mu_m\zeta_{nm}E_m(t)\langle B_n\rangle,\end{aligned} \tag{17.5b}$$

$$\begin{aligned}\frac{1}{i}\frac{d}{dt}\langle B_n^\dagger B_m\rangle = {} & \sum_l (F_{m,l}\langle B_n^\dagger B_l\rangle - F_{n,l}^*\langle B_l^\dagger B_m\rangle) \\ & - i\hat{\Gamma}\delta_{nm}\langle B_n^\dagger B_m\rangle - \frac{1}{\hbar}\mu_n E_n^*(t)\langle B_m\rangle + \frac{1}{\hbar}\mu_m E_m(t)\langle B_n^\dagger\rangle,\end{aligned} \tag{17.5c}$$

$$\begin{aligned}\frac{1}{i}\frac{d}{dt}\langle B_n^\dagger B_m B_l\rangle = {} & \zeta_{ml}\sum_i [F_{m,i}\langle B_n^\dagger B_i B_l\rangle + F_{l,i}\langle B_n^\dagger B_m B_i\rangle - F_{n,i}^*\langle B_i^\dagger B_m B_l\rangle] \\ & - i\hat{\Gamma}\zeta_{ml}(\delta_{nm}+\delta_{nl})\langle B_n^\dagger B_m B_l\rangle - \frac{1}{\hbar}\mu_n E_n^*(t)\zeta_{ml}\langle B_m B_l\rangle \\ & + \frac{1}{\hbar}\mu_m E_m(t)\zeta_{ml}\langle B_n^\dagger B_l\rangle + \frac{1}{\hbar}\mu_l E_l(t)\zeta_{ml}\langle B_n^\dagger B_m\rangle.\end{aligned} \tag{17.5d}$$

When $\langle B_n\rangle$ obtained by solving these equations is substituted in Eq. (16.37) we obtain the polarization. This should allow us to calculate the optical response.

The ζ_{nm} factors in these equations result from the Pauli exclusion and guarantee that $\langle B_n B_n \rangle = 0$ at all times (only a single exciton is allowed at a given site). Their presence complicates the solution. It is possible to use a different way of accounting for the on-site exclusion in the dynamics. We eliminate the ζ_{nm} factors, and to account for Pauli exclusion we add a repulsive exciton–exciton interaction and change the frequency of $\langle B_n B_m \rangle$ in Eq. (17.5b) from $\Omega_n + \Omega_m$ to $\Omega_n + \Omega_m + g\delta_{n,m}$. This way we are allowing two excitons to reside on the same site, but we add a penalty: The energy of the state is increased by g. For a finite g we have a soft-core model of excitons. By taking the $g \to \infty$ limit in the end, we exclude the two exciton states and obtain the solution of Eqs. (17.5).

The present procedure maps the calculation of optical nonlinearities onto solving the dynamics of coupled nonlinear oscillators that correspond to the single-site variables $\langle B_n \rangle$, as well as to nonlocal variables $\langle B_n B_m \rangle$, $\langle B_n^\dagger B_m \rangle$, and $\langle B_n^\dagger B_m B_l \rangle$. The formal solution of these equations can be most conveniently expressed in terms of a few Green functions that describe the free (no field) evolution of our dynamic variables. We shall now introduce these Green functions, starting with the evolution of single excitons. Setting the field $E(t) = 0$ in Eq. (17.5a), and neglecting the $\langle B_n^\dagger B_n B_m \rangle$ terms that contribute only to third (and higher) orders in the field, we can then solve for $\langle B_n(t) \rangle$

$$\langle B_n(t) \rangle = \sum_m G_{nm}(t) \langle B_m(0) \rangle, \tag{17.6}$$

where the one-exciton Green function G is given by

$$G_{nm}(t_1) = [\exp(iFt_1)]_{nm}, \tag{17.7a}$$

with

$$F_{nm} \equiv -\Omega_n \delta_{nm} - J_{nm} + i\frac{\Gamma}{2}\delta_{nm}. \tag{17.7b}$$

For the adjoint variables we have

$$\langle B_n^\dagger(t) \rangle = \langle B_n(t) \rangle^*. \tag{17.8}$$

Setting $E(t) = 0$ in Eq. (17.5b) we recast it in the form

$$\frac{1}{i}\frac{d}{dt}\langle B_n B_m \rangle = \sum_{n'm'} F^{(2)}_{nm,n'm'} \langle B_{n'} B_{m'} \rangle, \tag{17.9}$$

where the $F^{(2)}$ tetradic matrix is defined by comparing Eqs. (17.9) and (17.5b). The time evolution of the two-exciton variables is thus described by the two-exciton Green function which is the formal solution of Eq. (17.9)

$$\langle B_n(t) B_m(t) \rangle = \sum_{n'm'} G^{(2)}_{nm,n'm'}(t) \langle B_{n'}(0) B_{m'}(0) \rangle, \tag{17.10}$$

i.e.,

$$G^{(2)}_{nm,n'm'}(t) = [\exp(iF^{(2)}t)]_{nm,n'm'}. \tag{17.11}$$

The zero field $[E(t) = 0]$ evolution of the exciton population variables $\langle B^\dagger B \rangle$

[Eq. (17.5c)] is similarly given by

$$\langle B_n^\dagger(t)B_m(t)\rangle = \sum_{n'm'} G^{(2')}_{nm,n'm'}(t)\langle B_{n'}^\dagger(0)B_{m'}(0)\rangle. \tag{17.12}$$

In analogy with Eq. (17.9) we can recast Eq. (17.5c) with $E_n(t) = 0$ in the form

$$\frac{1}{i}\frac{d}{dt}\langle B_n^\dagger B_m\rangle = \sum_{n'm'} F^{(2')}_{nm,n'm'}\langle B_{n'}^\dagger B_{m'}\rangle, \tag{17.13}$$

where the tetradic matrix $F^{(2')}$ is defined upon comparison with Eq. (17.5c). The exciton population Green function, which is the formal solution of Eq. (17.13) is thus given by

$$G^{(2')}_{nm,n'm'}(t_2) = [\exp(iF^{(2')}t_2)]_{nm,n'm'}. \tag{17.14}$$

Finally, by solving Eq. (17.5d) with $E(t) = 0$ we obtain the Green function for the third-order variables

$$\langle B_n^\dagger(t)B_m(t)B_l(t)\rangle = \sum_{n'm'l'} G^{(3)}_{nml,n'm'l'}(t)\langle B_{n'}^\dagger(0)B_{m'}(0)B_{l'}(0)\rangle. \tag{17.15}$$

Using these Green functions we can now solve our coupled oscillator equations perturbatively in the applied field. The complete expression for $P^{(3)}$, which depends on four Green functions describing the evolution of excitons, two-excitons, exciton population, and three exciton variables, is given in Appendix 17C. In the coming sections we develop some simplified decoupling schemes that provide a clear insight into the role of each of these Green functions.

FACTORIZED APPROXIMATIONS FOR THE GREEN FUNCTION SOLUTION

In many cases it is sufficient to adopt one of the approximate factorization schemes outlined below, which simplify the exact Green function expression for $P^{(3)}$ derived in Appendix 17C. Hereafter, we explore systematically the hierarchy of possible factorizations.

The local field approximation

In the simplest possible approximation we keep only single-particle variables. We thus take expectation values of all operators in Eq. (17.5a) and factorize

$$\langle B_n^\dagger B_n B_m\rangle = \langle B_n^\dagger\rangle\langle B_n\rangle\langle B_m\rangle \tag{17.16a}$$

and

$$\langle B_n^\dagger B_n\rangle = \langle B_n^\dagger\rangle\langle B_n\rangle. \tag{17.16b}$$

This truncation retains only the exciton amplitude variables. Equation (17.5a) can then be recast in the form of Eq. (16.40) with the local field at site n

$$E_{l,n}(t) \equiv E_n(t) + \sum_{m\neq n} J_{nm}\langle B_m(t)\rangle. \tag{17.17}$$

We have thus recovered the LFA, and converted the dynamics to that of a single molecule in a local field. The resulting equation is closed and can be solved for the nonlinear response. The nonlinear response in this approximation was calculated in Chapter 16. The only difference is that here we have made the Heitler–London approximation and neglected $\sim(J/\Omega)$ terms, resulting in the local field correction factor

$$S(\mathbf{k}, \omega) = \frac{\Omega - \omega - i\Gamma}{\Omega_k - \omega - i\Gamma}, \tag{17.18}$$

instead of Eq. (16.53).

This result can be also obtained using our general expression [Eq. (17.C.5)], provided we factorize the Green functions

$$\left.\begin{aligned} G^{(2)}_{nm,n'm'}(t) &= G_{nn'}(t)G_{mm'}(t), \\ G^{(2')}_{nm,n'm'}(t) &= G^{*}_{nn'}(t)G_{mm'}(t), \\ G^{(3)}_{nml,n'm'l'}(t) &= G^{*}_{nn'}(t)G_{mm'}(t)G_{ll'}(t). \end{aligned}\right\} \tag{17.19}$$

Pure state factorization: two-exciton variables

An improved level of the hierarchy can be obtained by invoking a factorization into two-exciton variables. If we set the pure-dephasing rate $\hat{\Gamma}$ to zero, the only dephasing mechanism is population relaxation, and the system is in a pure state at all times. The density matrix in this case assumes the form $\rho(t) = |\psi(t)\rangle\langle\psi(t)|$, where $\psi(t)$ is the wavefunction. Consider now the expectation value of the normally ordered product $\langle(B^\dagger)^p(B)^q\rangle$. It follows from Eq. (17.2b) that its expectation value has contributions to order $p + q$ and higher in the incoming fields. It can easily be verified that to lowest order $(p + q)$ we can rigorously factorize the normally ordered product of any number of exciton creation and annihilation operators,

$$\langle(B^\dagger)^p B^q\rangle = \langle(B^\dagger)^p\rangle\langle B^q\rangle. \tag{17.20}$$

In particular, we have

$$\langle B_n^\dagger B_n B_m\rangle = \langle B_n^\dagger\rangle\langle B_n B_m\rangle, \tag{17.21a}$$

$$\langle B_n^\dagger B_n\rangle = \langle B_n^\dagger\rangle\langle B_n\rangle. \tag{17.21b}$$

Were the system not in a pure state, we would have to sum the right-hand side over the statistical mixture [see Eq. (3.11)] and the above factorization would no longer hold. The absence of dephasing is therefore essential here. The $\langle B_n B_m\rangle$ variables represent two-exciton dynamics. This follows directly from the wavefunction Eq. (17.2a), which yields $\langle B_n(t)\rangle = c_n^{(1)}(t)$, and $\langle B_n(t)B_m(t)\rangle = c_{n,m}^{(2)}(t)$. We further note that

$$\langle B_n(t)B_m(t)\rangle \neq \langle B_n(t)\rangle\langle B_m(t)\rangle.$$

This can be seen by taking for instance $n = m$; the left-hand side then necessarily vanishes (by the Pauli exclusion we do not allow two excitons to reside on the same molecule), whereas in the factorized form we lose this correlation and $\langle B_n\rangle^2$

does not necessarily vanish. The Green functions now factorize in the form

$$G^{(2')}_{nm,n'm'}(t) = G^{*}_{nn'}(t)G_{mm'}(t), \tag{17.22a}$$

$$G^{(3)}_{nml,n'm'l'}(t) = G^{*}_{nn'}(t)G^{(2)}_{ml,m'l'}(t). \tag{17.22b}$$

This level of the hierarchy will be discussed in detail in the next section.

Exciton population variables

We next consider a factorization into exciton population variables. If pure dephasing is incorporated, the factorization Eq. (17.21a) no longer holds. For large dephasing rates, however, it makes sense to neglect all correlations among different sites, and set for $n \neq m$

$$\begin{aligned} \langle B_n B_m \rangle &= \langle B_n \rangle \langle B_m \rangle, \\ \langle B_n^{\dagger} B_n B_m \rangle &= \langle B_n^{\dagger} B_n \rangle \langle B_m \rangle. \end{aligned} \tag{17.23}$$

The $\langle B_n^{\dagger} B_n \rangle$ variables represent exciton populations and they are the source of nonlinearity in the single-particle formulation of nonlinear optics, which is based on the Bloch equations (see Chapter 6). The dynamics of $\langle B_n^{\dagger} B_n \rangle$ may be related to standard transport equations (the Boltzmann and the diffusion equations), as will be demonstrated later in this chapter. In terms of the Green functions, the factorization (17.23) implies

$$\left.\begin{aligned} G^{(2)}_{nm,n'm'}(t) &= G_{nn'}(t)G_{mm'}(t), \\ G^{(3)}_{nml,n'm'l'}(t) &= G^{(2')}_{nm,n'm'}(t)G_{ll'}(t). \end{aligned}\right\} \tag{17.24}$$

We note that Eqs. (17.5a) and (17.5b) depend only on the total dephasing rate Γ and it makes no difference whether it comes from pure-dephasing or finite lifetime. In contrast, Eqs. (17.5c) and (17.5d) depend specifically on the pure dephasing rate. If $\hat{\Gamma} = 0$ we can factorize $\langle B_n^{\dagger} B_m \rangle = \langle B_n^{\dagger} \rangle \langle B_m \rangle$ and $\langle B_n^{\dagger} B_m B_l \rangle = \langle B_n^{\dagger} \rangle \langle B_m B_l \rangle$, and these equations become redundant. Pure dephasing thus introduces a new class of relevant variables, $\langle B_n^{\dagger} B_m \rangle$. This agrees with the previous arguments made in the pure state case [Eqs. (17.22)].

Complete factorization into two-particle variables

A more general factorization that includes the previous factorizations as special cases is

$$\langle B_n^{\dagger} B_n B_m \rangle = \langle B_n^{\dagger} B_n \rangle \langle B_m \rangle + \langle B_n^{\dagger} B_m \rangle \langle B_n \rangle + \langle B_n^{\dagger} \rangle \langle B_n B_m \rangle - 2\langle B_n^{\dagger} \rangle \langle B_n \rangle \langle B_m \rangle. \tag{17.25}$$

This may be derived by postulating an approximate maximum-entropy form for the density matrix $\rho(t)$. (In general, if we make any ansatz regarding the time-dependent density matrix, which may depend on a few parameters, we can then close the hierarchy of equations of motion for these parameters.) When Eq. (17.25) is used, we incorporate all exciton populations ($\langle B_n^{\dagger} B_n \rangle$) exciton-coherence ($\langle B_n^{\dagger} B_m \rangle$) and two-exciton ($\langle B_n B_n \rangle$) variables. These binary variables, together with the single-exciton ($\langle B_n \rangle$) variables and their Hermitian conjugates, constitute the relevant set of dynamic variables in the present level of description [25].

The three-particle Green function is now given by

$$G^{(3)}_{nml,n'm'l'} = G^{(2')}_{nm,n'm'}(t)G_{ll'}(t) + G^{(2')}_{nl,n'l'}(t)G_{mm'}(t) + G^{(2)}_{ml,m'l'}(t)G^{*}_{nn'}(t) - 2G^{*}_{nn'}(t)G_{mm'}(t)G_{ll'}(t). \tag{17.26}$$

The first three cases are special cases of the fourth case. Starting with Eq. (17.26), if we further factorize all operators, or creation and annihilation operators, or operators belonging to different sites, we recover Eqs. (17.19), (17.22), and (17.24), respectively.

ANHARMONIC OSCILLATOR REAL SPACE EXPRESSION FOR THE OPTICAL RESPONSE BEYOND THE LFA

We now consider the simple case where relaxation is negligible $\hat{\Gamma} = \gamma = 0$, so that the system is in a pure state at all times, and the factorization Eq. (17.21) holds. Equations (17.5) then become

$$\frac{1}{i}\frac{d}{dt}\langle B_n\rangle = -\Omega_n\langle B_n\rangle - \sum_m J_{nm}\langle B_m\rangle + 2\sum_{m\neq n} J_{nm}\langle B_n^\dagger\rangle\langle B_nB_m\rangle + \frac{1}{\hbar}\mu_n E_n(t)(1 - 2\langle B_n^\dagger\rangle\langle B_n\rangle), \tag{17.27a}$$

$$\frac{1}{i}\frac{d}{dt}\langle B_nB_m\rangle = -(\Omega_n + \Omega_m)\langle B_nB_m\rangle - \sum_{n'\neq m} J_{nn'}\langle B_{n'}B_m\rangle - \sum_{m'\neq n} J_{mm'}\langle B_nB_{m'}\rangle + \frac{1}{\hbar}\mu_nE_n(t)\langle B_m\rangle + \frac{1}{\hbar}\mu_mE_m(t)\langle B_n\rangle, \qquad n\neq m. \tag{17.27b}$$

There are two nonlinear terms in Eq. (17.27a). The origin of nonlinearities is that two-level molecules are not harmonic oscillators, and the nonlinear terms correct for the difference. The last term in Eq. (17.27a) implies that an excited molecule cannot be excited again, and, therefore, constitutes a local, or intramolecular nonlinearity. The Bloch equations commonly used in the calculation of the nonlinear response contain only this nonlinearity, which represents Pauli exclusion of excitons. In the semiconductor literature this contribution is known as "phase space filling" [22]. The $\langle B_n^\dagger\rangle\langle B_nB_m\rangle$ term in Eq. (17.27a) represents an additional nonlinearity; the coupling *between* sites is also nonlinear due to the on-site exclusion. The interaction between molecules n and m depends on whether the n molecule is excited or not, since the molecule can be excited only once. We will refer to this term as the nonlocal, or intermolecular nonlinearity.

To calculate the linear response we neglect the nonlinear terms in Eq. (17.5a) that contribute only to third (and higher) orders in the field, and obtain

$$P_n^{(1)}(\omega) = \sum_m \alpha_{nm}(\omega) E_m(\omega), \tag{17.28a}$$

with the linear polarizability

$$\alpha_{nm}(\omega) = \vec{\mu}_n \vec{\mu}_m [G_{nm}(\omega) + G_{nm}^*(-\omega)]. \tag{17.28b}$$

Here G_{nm} is the single-exciton Green function defined as follows

$$G(\omega) = (\omega - F + i\varepsilon)^{-1}, \tag{17.29}$$

where ε is a small positive number that is set to zero at the end of the calculation. The single-exciton Green function can be evaluated explicitly by introducing the one-exciton wavefunction $\psi_\alpha(n)$:

$$\Psi_\alpha = \sum_n \psi_\alpha(n) B_n^\dagger |0\rangle,$$

which satisfy the equation

$$H_{\text{mat}} \Psi_\alpha = \varepsilon_\alpha \Psi_\alpha,$$

and H_{mat} is given by Eq. (17.1). In the frequency domain we then have

$$G_{nm}(\omega) = \sum_\alpha \frac{\psi_\alpha(n)\psi_\alpha^*(m)}{\omega - \varepsilon_\alpha + i\varepsilon}. \tag{17.30}$$

We next turn to the calculation of the third-order nonlinear polarization $P^{(3)}$. This is done by the iterative procedure outlined below: We calculate $\langle B_n \rangle$ to first order, then solve for the second order term $\langle B_n B_m \rangle$, and finally substitute these back in Eq. (17.27a) and solve for $\langle B_n \rangle$ to third order. We finally get [26]

$$P_n^{(3)}(\omega_s) = -\frac{1}{8\pi^2} \int d\omega_1\, d\omega_2\, d\omega_3\, \delta(\omega_s - \omega_1 - \omega_2 - \omega_3)$$
$$\times \sum_{m_1, m_2, m_3} \gamma_{nm_1m_2m_3}(-\omega_s; \omega_1, \omega_2, \omega_3) E_{m_1}(\omega_1) E_{m_2}(\omega_2) E_{m_3}(\omega_3), \tag{17.31}$$

with the third-order polarizability

$$\gamma_{nm_1m_2m_3}(-\omega_s; \omega_1, \omega_2, \omega_3) = \frac{1}{6} \sum_p \sum_{n', n''} \vec{\mu}_n \vec{\mu}_{m_1} \vec{\mu}_{m_2} \vec{\mu}_{m_3}$$
$$\times \{G_{nn'}(\omega_s) G_{n'm_3}^*(-\omega_3) G_{n''m_2}(\omega_2) G_{n''m_1}(\omega_1) \bar{\Gamma}_{n'n''}(\omega_1 + \omega_2)$$
$$+ G_{nn'}^*(-\omega_s) G_{n'm_3}(\omega_3) G_{n''m_2}^*(-\omega_2) G_{n''m_1}^*(-\omega_1) \bar{\Gamma}_{n'n''}^*(-\omega_1 - \omega_2)\}. \tag{17.32}$$

Here $\bar{\Gamma}$ is a two-exciton scattering matrix given by

$$\bar{\Gamma}_{n'n''}(\omega) = -2[\mathscr{F}^{-1}(\omega)]_{n'n''}, \tag{17.33a}$$

where the $\mathscr{F}$ matrix is

$$\mathscr{F}_{n'n''}(\omega) \equiv \int \frac{d\omega'}{2\pi i} G_{n''n'}(\omega')G_{n''n'}(\omega - \omega'). \tag{17.33b}$$

Equation (17.33b) can be expanded in the single exciton eigenstates, resulting in

$$\mathscr{F}_{n'n''}(\omega) = \sum_{\alpha',\alpha} \frac{\psi_{\alpha'}(n')\psi_{\alpha}(n')\psi^*_{\alpha'}(n'')\psi^*_{\alpha}(n'')}{\omega - \varepsilon_{\alpha'} - \varepsilon_{\alpha} + i\varepsilon}. \tag{17.34}$$

The p summation in Eq. (17.32) is over all six permutations of three pairs (m_1, ω_1), (m_2, ω_2), (m_3, ω_3).

In the notation of Eq. (5.45) we get

$$\chi^{(3)}(\mathbf{r}, \mathbf{r}_1\mathbf{r}_2\mathbf{r}_3; -\omega_s; \omega_1\omega_2\omega_3) = -\frac{1}{2 \times 3!} \sum_p \sum_{nm_1m_2m_3} \gamma_{nm_1m_2m_3}(-\omega_s; \omega_1\omega_2\omega_3)$$
$$\times\, \delta(\mathbf{r} - \mathbf{r}_n)\, \delta(\mathbf{r}_1 - \mathbf{r}_{m_1})\, \delta(\mathbf{r}_2 - \mathbf{r}_{m_2})\, \delta(\mathbf{r}_3 - \mathbf{r}_{m_3}),$$

where $\sum_p$ now denotes the permutation over $(\mathbf{r}_1\omega_1, \mathbf{r}_2\omega_2, \mathbf{r}_3\omega_3)$.

A key ingredient that entered the derivation of Eq. (17.32) is the two-exciton Green function which is evaluated in Appendix 17B. It is both convenient and instructive to express this Green function in terms of the two-exciton scattering ($\mathscr{T}$) matrix. $\bar{\Gamma}$ is given by a particular matrix element of the $\mathscr{T}$ matrix that is inversely proportional to the amplitude of returning to the origin for a particle that is hopping over the lattice. We can, therefore, establish a general formal connection between enhanced nonlinearities and the probability of returning to the origin in a random walk [27]. Apart from being an interesting theoretical observation, this connection immediately shows that the dimensionality is going to be of prime importance, as the total number of returns to the origin in a random walk diverges in one and two dimensions, and is finite in three dimensions.

GREEN FUNCTION EXPRESSION FOR THE FOUR-WAVE MIXING SIGNAL INCLUDING RADIATIVE DECAY

So far we have expanded the response functions in the Maxwell field. When the expansion is substituted in the Maxwell equations, we can calculate the signal. Ultimately, we need to relate the signal to the external field, since this is the field that can be controlled. To that end we can use the Green function solution of the Maxwell equation given in Appendix 4C, which relates the Maxwell field E to the external field E^{ext}. We shall now relate the polarization directly to the external field. The total (radiation field and matter) Hamiltonian of our assembly of two level molecules with arbitrary geometry, and within the Heitler–London approximation is obtained by starting with Eq. (4.29) and substituting $\hat{H}_{\text{mat}}$ [Eq. (17.1)] for the first two terms $\hat{H}_{\text{mol}} + \hat{V}_{\text{inter}}$. We then get

$$H = \hbar \sum_m \Omega_m B_m^\dagger B_m + \hbar \sum_{m,n}' J_{mn} B_n^\dagger B_n - \int d\mathbf{r}\, \hat{P}(\mathbf{r}) \cdot \hat{D}^\perp(\mathbf{r})$$
$$+ H_{\text{rad}} + 2\pi \int d\mathbf{r}\, |\hat{P}^\perp(\mathbf{r})|^2. \tag{17.35}$$

H_{rad} is the radiation field Hamiltonian. $\hat{P}^\perp(\mathbf{r})$ and $\hat{\mathbf{D}}^\perp(\mathbf{r})$ are the transverse parts of the polarization and electric displacement. The interaction with an external field can be incorporated by adding a term,

$$-\int d\mathbf{r}\, \mathbf{P}(\mathbf{r}, t) \cdot \mathbf{E}^{\text{ext}}(\mathbf{r}, t), \tag{17.36}$$

to the Hamiltonian, and expanding the density operator in powers of E^{ext}. We then get, in complete analogy with Eq. (5.40),

$$P_n^{(j)}(t) = \sum_{m_1 m_2 \cdots m_j} \int_0^\infty dt_1 \int_0^\infty dt_2 \cdots \int_0^\infty dt_j\, \hat{S}^{(j)}_{nm_1 m_2 \cdots m_j}(t_j, t_{j-1}, \ldots, t_1)$$
$$\times E^{\text{ext}}_{m_j}(t - t_j) E^{\text{ext}}_{m_{j-1}}(t - t_j - t_{j-1}) \cdots E^{\text{ext}}_{m_1}(t - t_j - \cdots - t_1), \tag{17.37a}$$

where

$$\hat{S}^{(j)}_{nm_1 m_2 \cdots m_j}(t_j, t_{j-1}, \ldots, t_1)$$
$$= \left(\frac{i}{\hbar}\right)^j \langle \hat{P}_n(t_1 + \cdots + t_j)[\hat{P}_{m_j}(t_1 + \cdots + t_{j-1}) \ldots, [\hat{P}_{m_2}(t_1), [\hat{P}_{m_1}(0), \rho_{\text{eq}}]]]\rangle, \tag{17.37b}$$

$$\hat{P}_n(\tau) \equiv \exp\left(\frac{i}{\hbar} H\tau\right) \hat{P}_n \exp\left(-\frac{i}{\hbar} H\tau\right), \tag{17.38}$$

and the Hamiltonian H is given by Eq. (17.35). This correlation function is defined in the joint matter + field space (but in the absence of the external field), ρ_{eq} is the equilibrium material + field density matrix, and $\langle \cdots \rangle$ denotes a quantum mechanical average.

We reiterate that $E_m^{\text{ext}}(t)$ is the external field. Consequently, $\hat{S}^{(j)}$ are directly related to the signal, and are different from the response functions $S^{(j)}$ introduced in Chapter 5, which are defined with respect to the transverse Maxwell field. By expressing the Maxwell field in terms of the external field, Eq. (17.3) assumes the form

$$\frac{1}{i}\frac{d}{dt} B_n = -\Omega_n B_n - \sum_m' J_{nm} B_m + 2 \sum_m' J_{nm} B_n^\dagger B_n B_m + \frac{1}{\hbar} \mu_n E_n^{\text{ext}}(t)(1 - 2B_n^\dagger B_n)$$
$$-\frac{1}{2} \sum_m \int dt'\, \phi_{nm}(t - t')[B_m(t') + B_m^\dagger(t'), 1 - 2B_n^\dagger(t) B_n(t)]_+, \tag{17.39}$$

where $[\,,\,]_+$ is an anticommutator.

$\phi_{mn}(t)$ is the material memory kernel resulting from the interaction with the transverse electric field. Hereafter, we consider its Fourier transform $\tilde{\phi}_{nm}(\omega)$ (see Eqs. (5.29)). Its real part Δ_{mn} represents a level shift (for $n = m$ this is the Lamb shift and for $n \neq m$ this is the radiative correction to intermolecular interactions). Its imaginary part Γ_{mn} is responsible for radiative decay which, for confined excitons, may be cooperative. Cooperative radiative decay has been reported in molecular and semiconductor materials [12–14, 28]. The diagonal elements ϕ_{nn} diverge for the point dipole model. Their evaluation thus requires using a more realistic model in which the nth molecule is characterized by a polarization density $\rho_n(\mathbf{r} - \mathbf{R}_n)$ with $\int d^3r\, \rho(\mathbf{r}) = 1$ [the point dipole limit is recovered when $\rho_n(\mathbf{r} - \mathbf{R}_n) = \delta(\mathbf{r} - \mathbf{R}_n)$]. We then get

$$\begin{aligned}\tilde{\phi}_{nm}(\omega) &\equiv \Delta_{mn}(\omega) - i\Gamma_{mn}(\omega) \\ &= \int d\mathbf{r} \int d\mathbf{r}'\, \rho_m(\mathbf{r} - \mathbf{R}_m)\rho_n(\mathbf{r}' - \mathbf{R}_n)\vec{\mu}_m \cdot \mathscr{G}^{\perp}(\mathbf{r} - \mathbf{r}', \omega)\cdot \vec{\mu}_n \qquad (17.40)\end{aligned}$$

$\mathscr{G}^{\perp}$ is the Green function of the transverse electromagnetic field in vacuum (see Appendix 4C).

$$\begin{aligned}\mathscr{G}^{\perp}(\mathbf{r}, \omega) &= \int \frac{d^3q}{2\pi^2} \frac{\omega^2}{\omega^2 - q^2c^2 + i\varepsilon}\left(1 - \frac{\hat{q}\hat{q}}{q^2}\right)\exp(i\mathbf{q}\cdot\mathbf{r}) \\ &= -\left[\left(\frac{\omega}{c}\right)^2 + \nabla\nabla\right]\frac{\exp[i(\omega/c)r]}{\mathbf{r}} + \nabla\nabla\frac{1}{\mathbf{r}}. \qquad (17.41)\end{aligned}$$

It is possible to use the polarization density $\rho_n(\mathbf{r} - \mathbf{R}_n)$ (rather than the point dipole approximation) throughout. In the present calculation, we shall use it only for evaluating the self-energy. All other quantities will be evaluated using the point dipole approximation.

The linear polarization in the frequency domain is now given by

$$P_n^{(1)}(\omega) = -\sum_m \alpha_{nm}^{\mathrm{ext}}(\omega)\cdot E_m^{\mathrm{ext}}(\omega), \qquad (17.42a)$$

with the external linear polarizability

$$\alpha_{nm}^{\mathrm{ext}}(\omega) = \vec{\mu}_n\vec{\mu}_m[\hat{G}_{nm}(\omega) + \hat{G}_{nm}^*(-\omega)]. \qquad (17.42b)$$

The $\hat{G}_{mn}(\omega)$ matrix represents the single particle Green function,

$$\hat{G}_{mn} = (G(\omega)\cdot\{1 - \tilde{\phi}(\omega)\cdot[G(\omega) + G^*(-\omega)]\}^{-1})_{mn}, \qquad (17.43)$$

where $G(\omega)$ is the material Green's function [Eq. (17.33)]. In the rotating wave approximation (RWA), Eq. (17.41) yields

$$\hat{G}_{mn}(\omega) = [\omega - H^{\mathrm{eff}}(\omega) + i\varepsilon]_{mn}^{-1}. \qquad (17.44a)$$

with $H^{\mathrm{eff}}(\omega)$ being the effective Hamiltonian matrix

$$H_{mn}^{\mathrm{eff}}(\omega) \equiv \Omega_m\delta_{mn} + J_{mn} + \tilde{\phi}_{mn}(\omega). \qquad (17.44b)$$

For the third-order polarization, we have [26]

$$P_n^{(3)}(\omega_s) = -\frac{1}{2(2\pi)^2}\int d\omega_1 \int d\omega_2 \int d\omega_3\, \delta(\omega_s - \omega_1 - \omega_2 - \omega_3)$$
$$\times \sum_{m_1, m_2, m_3} \gamma^{\text{ext}}_{nm_1m_2m_3}(\omega_s; \omega_1, \omega_2, \omega_3)$$
$$\times E^{\text{ext}}_{m_1}(\omega_1)E^{\text{ext}}_{m_2}(\omega_2)E^{\text{ext}}_{m_3}(\omega_3), \quad (17.45)$$

with the external third-order polarizability

$$\gamma^{\text{ext}}_{nm_1m_2m_3}(\omega_s; \omega_1, \omega_2, \omega_3) = \frac{1}{6}\sum_p \sum_{n'n''} \vec{\mu}_n\vec{\mu}_{m_1}\vec{\mu}_{m_2}\vec{\mu}_{m_3}$$
$$\times [\hat{G}'_{nn'}(\omega_s)\hat{G}^*_{n'm_3}(-\omega_3)G_{n''m_2}(\omega_2)\hat{G}_{n''m_1}(\omega_1)\bar{\Gamma}_{n'n''}(\omega_1 + \omega_2)$$
$$+ \hat{G}'^*_{nn'}(-\omega_s)\hat{G}_{n'm_3}(\omega_3)\hat{G}^*_{n''m_2}(-\omega_2)\hat{G}^*_{n''m_1}(-\omega_1)\bar{\Gamma}^*_{n'n''}(-\omega_1 - \omega_2)]. \quad (17.46)$$

Here

$$\bar{\Gamma}_{n'n''}(\omega_1 + \omega_2) = -2[\mathscr{F}^{-1}(\omega_1 + \omega_2)]_{n'n''}, \quad (17.47a)$$

is the two-exciton scattering matrix, with

$$\mathscr{F}_{n'n''}(\omega_1 + \omega_2) = \frac{i}{2\pi}\int d\omega'\, \hat{G}_{n'n''}(\omega')\hat{G}_{n'n''}(\omega_1 + \omega_2 - \omega'). \quad (17.47b)$$

Note that in this equation, the first single particle Green's function is $\hat{G}'$ instead of $\hat{G}$,

$$\hat{G}'_{nn'} \equiv (\{1 - [G(\omega) + G^*(-\omega)]\cdot\tilde{\phi}(\omega)\}^{-1}\cdot G(\omega))_{nn'}, \quad (17.48)$$

and in the RWA we have $\hat{G}' = \hat{G}$.

The local field approximation (LFA) is equivalent to the factorization approximation [Eq. (16.39)]. The LFA expression is also given by Eq. (17.45) except that the two-exciton scattering matrix should be replaced by

$$\bar{\Gamma}_{n'n''}(\omega_1 + \omega_2) \cong -2(\omega_1 + \omega_2 - 2\Omega)\delta_{n'n''}. \quad (17.49)$$

By comparing this with Eq. (17.47a) we note that the LFA holds only when $\omega_1 + \omega_2$ is tuned off all two-exciton states.

OPTICAL SUSCEPTIBILITIES OF PERIODIC STRUCTURES IN *k* SPACE

The Green function expression derived above holds for a molecular assembly with arbitrary geometry. Nanostructures and superlattices often have a periodic structure, where these expressions can be simplified further. Consider a periodic system in which the molecules occupy a d-dimensional lattice with constant a.

Each lattice site is occupied by a two-level molecule with a transition dipole moment μ. All dipole moments are assumed parallel.

To make use of the translational symmetry of the problem, we perform a d-dimensional Fourier transform and recast the Green functions in momentum ($\mathbf{k}$) space

$$G(\mathbf{k}, \omega) \equiv \sum_{\mathbf{m}} G_{\mathbf{mn}}(\omega) \exp[-i\mathbf{k}\cdot(\mathbf{r_m} - \mathbf{r_n})].$$

In a periodic geometry with $\Omega_n = \Omega$ independent of n, the single-exciton states have a well-defined momentum $\mathbf{k}$, i.e., $\alpha = \mathbf{k}$ and energy $\Omega_{\mathbf{k}}$

$$\left.\begin{aligned} \psi_{\mathbf{k}}(\mathbf{n}) &= \exp(i\mathbf{k}\cdot\mathbf{r_n}) \\ \Omega_{\mathbf{k}} &= \Omega + J(\mathbf{k}), \end{aligned}\right\} \tag{17.50}$$

with

$$J(\mathbf{k}) = \sum_{\mathbf{n}} \exp(-i\mathbf{k}\cdot\mathbf{r_n}) J_{\mathbf{no}}.$$

These wavefunctions are normalized as

$$\sum_{n} \psi_{\mathbf{k}}^{*}(n)\psi_{\mathbf{k}'}(n) = \left(\frac{2\pi}{a}\right)^{d} \delta(\mathbf{k} - \mathbf{k}').$$

We then have:

$$G(\mathbf{k}, \omega) = \frac{1}{\omega - \Omega_k + i\varepsilon}. \tag{17.51}$$

Equations (17.28) then yield

$$P^{(1)}(\mathbf{k}, \omega) = \chi^{(1)}(\mathbf{k}, \omega) E(\mathbf{k}, \omega), \tag{17.52a}$$

with

$$\chi^{(1)}(\mathbf{k}, \omega) = \frac{1}{\hbar} \frac{2\Omega_{\mathbf{k}} \rho_0 |\mu|^2}{\Omega_{\mathbf{k}}^2 - (\omega + i\varepsilon)^2}. \tag{17.52b}$$

This agrees with Eq. (16.50). Equations (17.31)–(17.33) then assume the form [26]

$$\begin{aligned} P^{(3)}(\mathbf{k}_s, \omega_s) = (2\pi)^{-2(d+1)} \int & d\omega_1\, d\omega_2\, d\omega_3\, d\mathbf{k}_1\, d\mathbf{k}_2\, d\mathbf{k}_3 \\ & \times \delta(\omega_s - \omega_1 - \omega_2 - \omega_3)\, \delta(\mathbf{k}_s - \mathbf{k}_1 - \mathbf{k}_2 - \mathbf{k}_3) E(\mathbf{k}_1, \omega_1) \\ & \times E(\omega_2, \mathbf{k}_2) E(\omega_3, \mathbf{k}_3) \chi^{(3)}(-\mathbf{k}_s, -\omega_s; \mathbf{k}_1\omega_1, \mathbf{k}_2\omega_2, \mathbf{k}_3\omega_3), \end{aligned} \tag{17.53}$$

with the susceptibility

$$\chi^{(3)}(-\mathbf{k}_s, -\omega_s; \mathbf{k}_1\omega_1, \mathbf{k}_2\omega_2, \mathbf{k}_3\omega_3)$$
$$= -\frac{\rho_0}{2}\frac{1}{\hbar^3}|\mu|^4 \frac{1}{6}\sum_p \{G(\mathbf{k}_s, \omega_s)G^*(-\mathbf{k}_3, -\omega_3)G(\mathbf{k}_1, \omega_1)G(\mathbf{k}_2, \omega_2)$$
$$\times \bar{\Gamma}(\mathbf{k}_1 + \mathbf{k}_2, \omega_1 + \omega_2) + G^*(-\mathbf{k}_s, -\omega_s)G(\mathbf{k}_3, \omega_3)G^*(-\mathbf{k}_1, -\omega_1)$$
$$\times G^*(-\mathbf{k}_2, -\omega_2)\bar{\Gamma}^*(-\mathbf{k}_1 - \mathbf{k}_2, -\omega_1 - \omega_2)\}, \quad (17.54)$$

and

$$\bar{\Gamma}(\omega, \mathbf{k}) = -2\left(\frac{2\pi}{a}\right)^d \left[\int d\mathbf{k}_1(\omega - \Omega_{\mathbf{k}_1} - \Omega_{\mathbf{k}-\mathbf{k}_1} + i\varepsilon)^{-1}\right]^{-1}. \quad (17.55)$$

Here we have replaced in Eq. (17.34)

$$\sum_\alpha \to \left(\frac{a}{2\pi}\right)^d \int d\mathbf{k}.$$

The expressions for the optical susceptibilities simplify if we consider an aggregate with a size small compared to optical wavelength. We can then set $\mathbf{k}_j = 0$. A Green function solution of Eqs. (17.5) may be derived using the factorization (17.25) [25]. The resulting $\chi^{(3)}$ has three types of contributions originating from the $\langle B_n^\dagger\rangle\langle B_n\rangle\langle B_m\rangle$, $\langle B_n^\dagger B_m\rangle$, and the $\langle B_n B_m\rangle$ nonlinearities. When expanded in powers of molecular number density ρ_0, we find that the first contribution scales as $\sim\rho_0$, whereas the second and the third contributions scale as $\sim\rho_0^2$. This is to be expected since the first represents the contribution of *local nonlinearities* whereas the other two are induced by *nonlocal* interactions. Nonlocal interactions, therefore, enter the nonlinear response in two ways: they modify the local term and induce additional terms. The appearance and the form of these new terms provide a direct probe for nonlocal interactions.

SIGNATURES OF COOPERATIVITY: TWO-EXCITON RESONANCES AND ENHANCED NONLINEAR SUSCEPTIBILITIES IN MOLECULAR AGGREGATES

In this section we explore the role of the two-exciton variables $\langle B_n B_m\rangle$ in the cooperative enhancement of optical nonlinearities. These variables are particularly significant in finite-size systems such as molecular aggregates or monolayers. We shall therefore analyze a specific model with geometric confinement.

Consider a linear aggregate [7, 8, 12–14] consisting of N coupled two-level molecules with transition frequency Ω and nearest neighbor dipole–dipole coupling V. The nth molecule is located at $\mathbf{r}_n = n\mathbf{a}$ where $\mathbf{a}$ is the lattice vector, and $\mathbf{r}_{N+1} = \mathbf{r}_1$ (periodic boundary conditions). We further neglect exciton–phonon coupling and simply add phenomenologically a relaxation rate which represents finite exciton lifetime.

We shall first present the single-exciton and two-exciton eigenstates that can be exactly calculated for the present model. Using these eigenstates, we can calculate the single- and the two-exciton Green functions and $\chi^{(3)}$. It will be instructive, however, to adopt a different route and evaluate $\chi^{(3)}$ through the conventional sum-over-states expression. This demonstrates how cooperative effects show up in a different manner using the Green function and the sum over state expressions. We explore the roles of intra- and intermolecular nonlinearities, show the limitations of the local-field approximation, and discuss the factors affecting cooperative enhancement.

The relevant eigenstates of the aggregate shown in Figure 17.4 are grouped in three levels. The lowest is the ground state $|0\rangle$. The next level includes N single-exciton states

$$|\lambda\rangle = \frac{1}{\sqrt{N}} \sum_n \exp(i\lambda n) B_n^\dagger |0\rangle,$$

where

$$\lambda = \frac{2\pi}{N} k; \qquad k = 0, 1, \ldots, N-1.$$

The eigenvalues are

$$\Omega_\lambda \equiv \Omega + 2V \cos\left(\frac{2\pi k}{N}\right). \qquad (17.56b)$$

We next consider the two-exciton states. They are given by

$$|\lambda_1 \lambda_2\rangle = \frac{1}{N} \sum_{n,m} \exp(i\lambda_1 n + i\lambda_2 m)\, \mathrm{sgn}(n-m) B_n^\dagger B_m^\dagger |0\rangle,$$

with eigenvalues

$$\Omega_{\lambda_1 \lambda_2} = 2\Omega + 4V(\cos\lambda_1 + \cos\lambda_2).$$

Here λ are determined by the condition $\exp(i\lambda N) = -1$ which gives

$$\lambda = \frac{\pi q}{N}, \qquad q = 1, 3, \ldots, 2N-1,$$

and

$$\mathrm{sgn}\, n = \begin{cases} 1 & n > 0 \\ -1 & n < 0 \\ 0 & n = 0. \end{cases}$$

Note that the state $|\lambda_1 \lambda_2\rangle = -|\lambda_2 \lambda_1\rangle$, which implies that the state $|\lambda\lambda\rangle$ vanishes. There are altogether $N(N-1)/2$ distinct two-exciton states. This can be easily seen from their definition since by the Pauli exclusion we have N possibilities for the first excitation and only $N-1$ for the second. (The factor 2 results from the permutation of these two excitations.) Altogether we need consider $1 + N(N+1)/2$ eigenstates.

For simplicity we hereafter assume a uniform excitation induced by an external electric field with a wavevector oriented normal to the aggregate axis, which excites only the single-exciton $k = 0$ state. We denote this state by

$$|e\rangle = \frac{1}{\sqrt{N}} \sum_n B_n^\dagger |0\rangle. \tag{17.56a}$$

Its energy is

$$\omega_e \equiv \Omega + 2V$$

and transition dipole moment

$$\mu_{ge} = \sqrt{N}\,\mu, \tag{17.56c}$$

μ being the transition dipole of a single molecule. A general analysis for $k \neq 0$ excitons, excited when the aggregate axis is not normal to the laser beam wavevector, is straighforward; however, the essential physics is already contained in the $k = 0$ case considered here.

The only two-exciton states with finite transition dipole are those with zero momentum, i.e., $\exp[i(\lambda_1 + \lambda_2)] = 1$. We then get

$$|\lambda, -\lambda\rangle = \frac{1}{N} \sum_{n,m} \exp[i\lambda(n - m)]\, \mathrm{sgn}(n - m) B_n^\dagger B_m^\dagger |0\rangle, \tag{17.57a}$$

with energy

$$2\omega_q = 2\left[\Omega + 2V \cos\left(\frac{\pi q}{N}\right)\right], \tag{17.57b}$$

and transition dipole moment

$$\mu_{eq} = (2\mu/\sqrt{N}) \cot\left(\frac{\pi q}{N}\right). \tag{17.57c}$$

We thus need consider only one single-exciton state $|e\rangle \equiv |k = 0\rangle$ and only $(N^* + 1)/2$ two-exciton states, where $N^* = N - 2$ for N odd and $N^* = N - 1$ for N even. We further specialize to a particular $\chi^{(3)}$, which depends only on a single frequency and is responsible for nonlinear propagation and two-photon absorption of a single beam. Using the standard sum over states expression for $\chi^{(3)}$ [Eq. (6.23)] together with Eqs. (17.56) and (17.57) we then get [20],

$$\chi^{(3)}(-\omega; \omega, -\omega, \omega) = \frac{4}{3!}\left(\frac{1}{\hbar}\right)^3 \rho_0 \frac{1}{(\omega - \omega_e)^2 + (\gamma_e/2)^2} \times \left[\frac{|\mu_{ge}|^4}{\omega - \omega_e + i\gamma_e/2} - \sum_q \frac{|\mu_{ge}|^2 |\mu_{eq}|^2}{2\omega - 2\omega_q + i\gamma_q}\right], \tag{17.58}$$

Note that $\omega_q \equiv \omega_e + 2V[\cos(\pi q/N) - 1]$; $q = 1, 3, \ldots, N^*$. γ_e and γ_q are phenomenological relaxation rates of the single-exciton and two-exciton states, respectively. ρ_0 is the number of aggregates per unit volume. The transition dipole moments satisfy the sum rule

$$\sum_{q=1,3}^{N^*} |\mu_{eq}|^2 = 2|\mu_{ge}|^2 \frac{N - 1}{N}.$$

$\chi^{(3)}$ has two terms (Liouville space pathways). The first term in the square brackets represents the contribution of the single-exciton level and scales as $|\mu_{eg}|^4 \sim N^2$. The second term consists of a series of two-exciton resonances. Using the above sum rule, the integrated area of these resonance scales as

$$(1/2)|\mu_{ge}|^2 \sum_q |\mu_{eq}|^2 = |\mu|^4 N(N-1). \tag{17.59}$$

When the laser beams are tuned far from an excitonic or two-excitonic resonances, the two terms interfere destructively and the N^2 parts exactly cancel out, leaving an overall linear dependence of $\chi^{(3)}$ on size

$$\chi^{(3)}(-\omega; \omega, -\omega, \omega) = \frac{4}{3!}\left(\frac{1}{\hbar}\right)^3 \rho_0 N|\mu|^4 \frac{1}{(\omega-\omega_e)^2 + (\gamma_e/2)^2} \frac{1}{\omega - \omega_e + i\gamma_e/2}. \tag{17.60}$$

This behavior can be rationalized by the following simple argument: When all frequencies are tuned far off resonance, the optical response is instantaneous (by the Heisenberg uncertainty, the process is completed in a timescale $\tau \sim \hbar/\Delta E$ with ΔE being an average detuning). Intermolecular interactions then do not have enough time to be effective, and may be neglected. The off-resonant optical response of molecular aggregates thus reduces to essentially that of a single molecule.

In small aggregates the single-exciton and the two-exciton contributions may be well separated spectrally (i.e., $\omega_e - \omega_q$ is larger than the dephasing rates γ_e and γ_q). By carefully tuning the frequency near resonance, it may become possible to spectrally select either the first or the second term of Eq. (17.58), resulting in an $\sim N^2$ scaling of the nonlinear response due to cooperative enhancement. An increase in aggregate size N reduces the exciton-two-exciton splitting of the lowest q states which contain most of the oscillator strength. It is impossible to spectrally select one of these terms when that splitting becomes comparable to the exciton linewidth γ. The interference will thus cancel the enhancement, resulting in an $\sim N$ scaling of $\chi^{(3)}$. The crossover size N_c, whereby the magnitude of the aggregate response changes from $\sim N^2$ to $N_c \sim N$, can be obtained from the solution to $8 \sin^2(\pi/2N_c) = \gamma/V$, which gives

$$N_c = \pi[2V/\gamma]^{1/2}. \tag{17.61}$$

Nonradiative damping, due, e.g., to coupling with phonons, dominates the exciton relaxation γ and reduces the coherence size.

The present expression for the nonlinear optical response of aggregates clarifies the origin of the crossover from the small aggregate to the bulk limit. The nonlinear susceptibility is expressed as the sum of Liouville pathways, and it has two contributions, both of order N^2. These contributions interfere destructively for moderate to rapid dephasing rates, resulting in a susceptibility proportional to the number of molecules N. Cooperatively enhanced optical nonlinearities ($\sim N^2$) are possible only in small aggregates with $N < N_c$, where the two-level excitonic resonance is spectrally well separated from the two-photon resonances. The present calculation provides a beautiful demonstration for the danger in relying on "typical" terms of $\chi^{(3)}$ (which are all $\sim N^2$), since $\chi^{(3)}$

may be strongly affected by interference (which, in this case, is between single-exciton and two-exciton contributions).

The same results can be obtained using the equation of motion approach that rigorously maps the system onto a collection of coupled anharmonic oscillators, which represent one or more excitons. In the oscillator picture the interference is built in naturally and we obtain a starting harmonic reference system in which the excitons are bosons: cooperative enhancement is then obtained through exciton–exciton interactions.

Using the Green function expression for $\chi^{(3)}$ we have [26]

$$\chi^{(3)}(-\omega; \omega, -\omega, \omega) = \frac{4}{3!}\left(\frac{1}{\hbar}\right)^3 N\rho_0|\mu|^4 \frac{1}{(\omega-\omega_e)^2 + (\gamma_e/2)^2} \frac{1}{[\omega - \omega_e + i(\gamma_e/2)]^2}$$
$$\times \left\{\frac{1}{N}\sum_{k=0}^{N-1} \frac{1}{\omega - \Omega - 2V\cos(2\pi k/N) + i(\gamma_k/2)}\right\}^{-1}. \quad (17.62)$$

The anharmonic-oscillator picture offers an alternative interpretation of the cooperative $\sim N^2$ enhancement and the crossover to $\sim N$ scaling discussed earlier. In accordance with the Pauli exclusion principle, a single site cannot be doubly excited and as a result excitons are not true bosons (or fermions). Linear optical properties can be exactly evaluated using the boson approximation for single excitons, but the approximation breaks down for the dynamics of two excitons and higher excitons. These are vital for the nonlinear optical response and can therefore be proved via second and higher order nonlinear optical techniques. Spatially confining the exciton enhances the nonbosonic nature and hence the nonlinear susceptibilities (which would vanish if Frenkel excitons were bosons). The nonboson nature of the elementary excitations is amplified in systems with restricted geometries due to exciton–exciton scattering. The combined influence of the Pauli exclusion and excitonic confinement on the third-order nonlinear hyperpolarizability could result in a cooperative $\sim N^2$ scaling for small sizes and an $\sim N$ scaling for larger sizes where the excitons become more Boson-like.

The spectroscopic manifestations of the two contributions to Eq. (17.58) can be illustrated by considering the two-photon absorption signal

$$W_{\text{TPA}} = \text{Im}\, \chi^{(3)}_{\text{TPA}},$$

where

$$\chi^{(3)}_{\text{TPA}} \equiv \chi^{(3)}(-\mathbf{k} - \omega; \mathbf{k}\omega, -\mathbf{k} - \omega, \mathbf{k}\omega).$$

In Figure 17.5 we show the two photon absorption signal as a function of ω for several size aggregates (as indicated in each panel).* In the small aggregate region, $\chi^{(3)}(-\omega; \omega, -\omega, \omega)$ reduces, near resonance, to that of a single excitonic two-level system with a transition dipole $\sqrt{N}\mu$, with additional two-photon resonances at $2\omega = 2\omega_q$ on the blue side of the spectrum (we have assumed $V < 0$). Positive values of the signal correspond to two-photon absorption, whereas negative

* This calculation also took into account radiative decay, which does not show up in $\chi^{(3)}$ but is present in the signal [see Eq. (17.44a)].

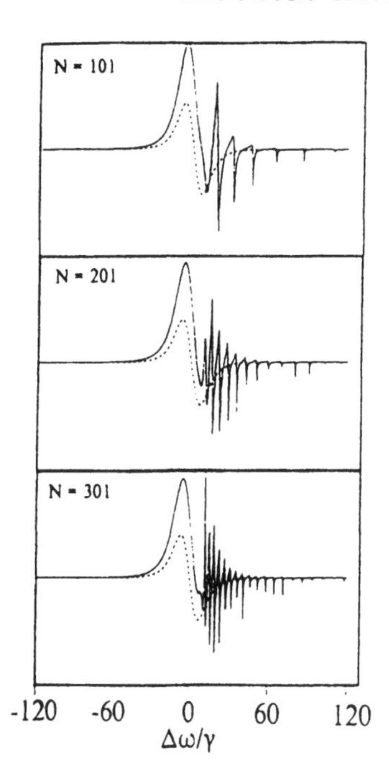

FIG. 17.5 The nonlinear two-photon absorption of one-dimensional aggregates (W_{TPA}) calculated using the imaginary part of Eq. (17.58). The various panels are for a succession of aggregate sizes. The dashed curves show the local-field approximation and have been reduced by a factor of 10. The two-photon resonances are absent in this case. Note the reduction of the two-exciton blueshift as the aggregate size is increased. [From F. C. Spano and S. Mukamel, "Cooperative nonlinear optical response of molecular aggregates: Crossover to bulk behavior," *Phys. Rev. Lett.* **66**, 1197 (1991).]

values represent bleaching of the excitonic line (saturated absorption). As the aggregate size increases, the $q = 1$ resonance moves toward the exciton line and eventually interferes destructively with it. For larger sizes, the destructive interference is complete, resulting in the cancellation of the $\sim N^2$ prefactor in $\chi^{(3)}(-\omega; \omega, -\omega, \omega)$. At this point, the signal scales linearly with N; the red side of the spectrum has converged completely, while the blue side still shows resolved two-photon resonances. These, however, become more congested as N increases and eventually the finite experimental spectral resolution will exceed the line spacing, making the blue side a smooth function of ω, with no N dependence. In general, when ω is close to the exciton resonance (ω_e), the local-field approximation is inadequate, completely missing the intermolecular nonlinearities; the LFA (dashed curves) does not reproduce the correct magnitude of $\chi^{(3)}$. In addition, within this approximation the small aggregate region is completely missed, the N^2 prefactor never appears, and the two-photon resonances never show up.

J-aggregates provide a clear demonstration of cooperative exciton behavior. Their cooperative radiative decay was illustrated in Figure 17.3. The monomer dye absorption and fluorescence spectra are very broad and show a large Stokes shift (Figure 17.6a). Upon aggregation there is a drastic narrowing of both absorption and fluorescence, and the spectra show no Stokes shift (Figure 17.6b). These effects can be attributed to the fast exciton motion due to the strong itermolecu-lar coupling. Effects of inhomogeneous broadening then average out, due to motional narrowing, as discussed in Chapter 8 [8]. Figure 17.6c displays the differential absorption spectrum (difference of the probe absorption spectra with and without pumping) for zero delay and a pump wavelength of 576.0 nm, which

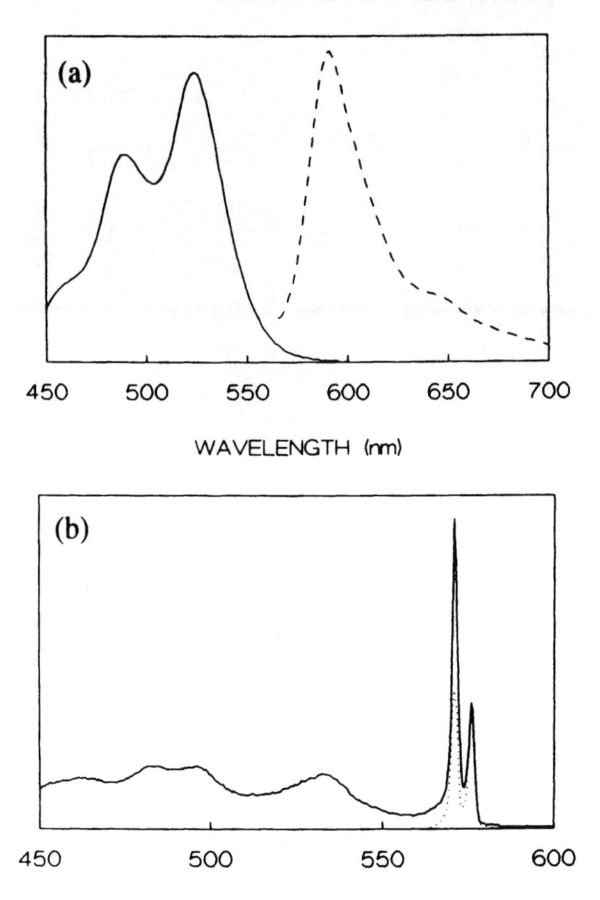

FIG. 17.6 (a) Absorption (——) and fluorescence (---) spectrum of the PIC-Br monomer at room temperature. The fluorescence was excited at 550 nm. (b) PIC-Br aggregate absorption (——) and fluorescence (····) at 1.5 K. Fluorescence after excitation at 514.5 nm.

is near the peak of the red J band. The strength of this signal strongly depends on the pump wavelength and correlates very well with the aggregate absorption. The spectrum shows a strong bleach at the excitation wavelength, which represents the saturation of the Frenkel exciton line, and an equally intense, but slightly blue-shifted, increased absorption which can be assigned to a one-exciton to two-exciton transition.

In Figures (17.7)–(17.9) we present numerical calculations for periodic two- and three-dimensional clusters with dipolar interactions. Figure 17.7 shows the TPA signal for a three-dimensional $79 \times 79 \times 79$ cubic aggregate with periodic boundary conditions. This should mimic an infinite lattice. We see that the cooperativity gives a large enhancement of the signal, compared with the local field approximation even for the quite strong damping, and essentially no

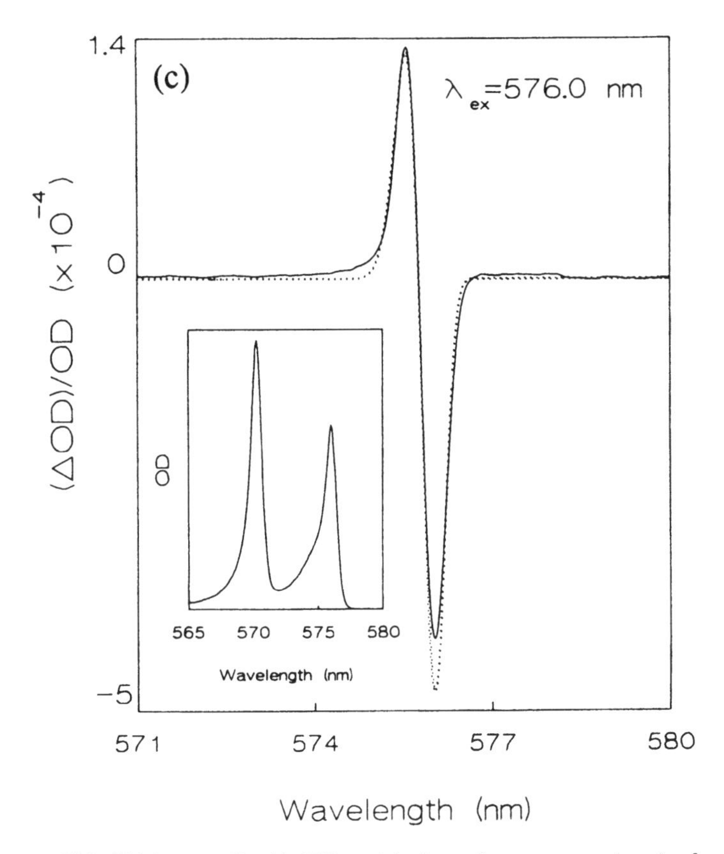

FIG. 17.6 (*continued*) (c) Differential absorption spectrum (——) of PIC-Br aggregates at 1.5 K, when pumping at 576.0 nm and at zero delay between the pump and the probe. The dotted line is the calculated spectrum. The inset displays the (linear) absorption spectrum at 1.5 K. [From H. Fidder, J. Knoester, and D. A. Wiersma, "Observation of the one-exciton to two-exciton transition in a J aggregate," *J. Chem. Phys.* **98**, 6564 (1993).].

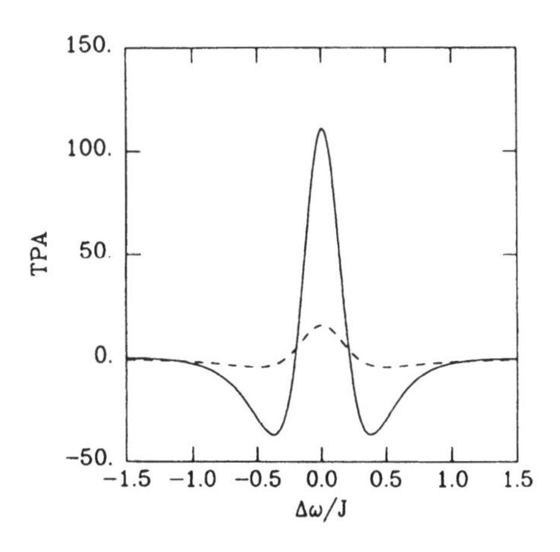

FIG. 17.7 Calculated two-photon absorption W_{TPA} for a 79 × 79 × 79 lattice of dipoles in three dimensions. $\Delta\omega = \omega - \Omega_s$. The frequency is expressed in units of $J \equiv \mu^2/a^3$, where a is the lattice constant. Solid line, numerical results. Dashed line, local-field approximation. The absorption is enhanced by a factor 7 compared with the local field approximation. [J. A. Leegwater and S. Mukamel, "Exciton scattering mechanism for enhanced nonlinear response of molecular nanostructures," *Phys. Rev. A* **46**, 452 (1992).]

shift of the resonance. The local-field approximation misses the enhancement entirely.

Equations (17.5) were solved approximately for a one-dimensional aggregate with N molecules, using a somewhat different relaxation matrix (this is not essential for the present discussion). The resulting expression for $\chi^{(3)}$ contains the contributions of local as well as nonlocal nonlinearities. The latter may be responsible for an enhanced (cooperative) nonlinear optical response. The enhancement can most conveniently be described in terms of an exciton coherence size N_c, which represents the separation of two sites that can still respond coherently to the applied fields. In order to define the coherence size more precisely, gain a clear insight into its role, and make the connection with the optical response of aggregates, we can proceed in the following way. We adopt the complete factorization into single operators [Eqs. (17.16)] if sites n and m are separated by M bonds or more, and a more general factorization if they are separated by less than M bonds. M is a cutoff size that can be varied at will. Nonlocal coherences are important only as long as $|n - m| < N_c$, N_c being the coherence size. We expect $\chi^{(3)}$ to vary with the cutoff M as long as M is smaller than the coherence size N_c. As M exceeds N_c, the factorization (17.16) should hold, and $\chi^{(3)}$ should become independent on M. Observing the convergence of $\chi^{(3)}$ as M is varied provides an operational definition of the nonlinear coherence size N_c [20]. The local-field approximation is obtained by taking $M = 1$. In this case all intermolecular interactions enter via the local field. As M is increased, we treat the intermolecular interactions more rigorously, and the local field becomes closer to the average field. For $M = N$ we treat all intermolecular interactions explicitly.

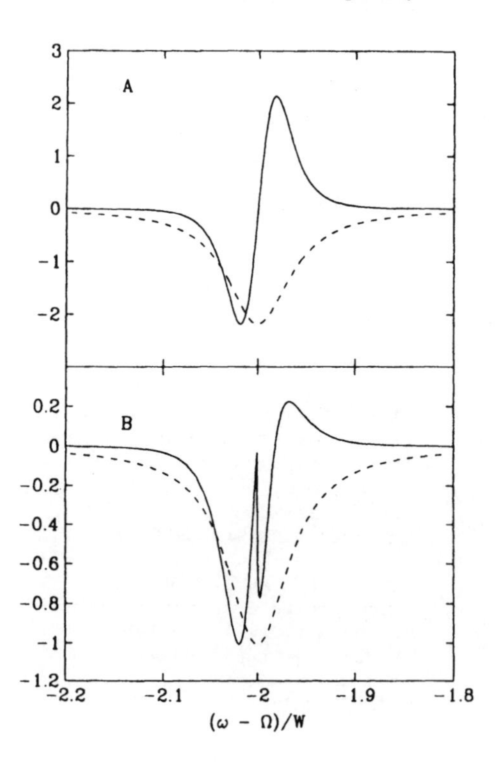

FIG. 17.8 The calculated frequency-dependent nonlinear reflection signal (solid line) from a molecular monolayer. For comparison we also show the negative linear reflection (dashed line). Both signals are normalized to the same height. W is the exciton bandwidth. (A) is the result of LFA. (B) is the result of Green function expression. The vertical scale shows the relative magnitude of the signal. [From N. Wang, V. Chernyak, and S. Mukamel, "Cooperative ultrafast nonlinear optical response of molecular nanostructures," *J. Chem. Phys.* **100**, 2465 (1994).]

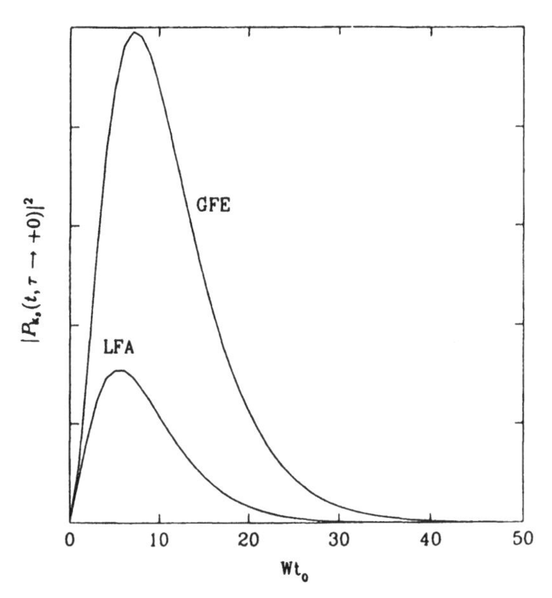

FIG. 17.9 The time-resolved impulsive two pulse echo signal at $\mathbf{k}_s = 2\mathbf{k}_2 - \mathbf{k}_1$ from a molecular monolayer. Shown is the Green function calculation (GFE) and the local-field (LFA) result. The rise time reflects exciton–exciton scattering, and decay is radiative. [From N. Wang, V. Chernyak, and S. Mukamel, "Cooperative ultrafast nonlinear optical response of molecular nanostructures," *J. Chem. Phys.* **100**, 2465 (1994).]

Linear and nonlinear reflection spectra for a molecular monolayer are shown in Figure 17.8. These correspond to the lowest order nonlinear correction to the reflection spectrum, that scales as the fourth power of the incident field, and is related to $\chi^{(3)}(-\omega; \omega, -\omega, \omega)$. We consider first the local-field approximation. The linear $S_{LR}(0, \omega)$ (dashed line) and the nonlinear $S_{NR}(0, \omega)$ (solid line) reflection spectra are shown in Figure 17.8. This is a typical spectrum for a three-level system shown in Figure 17.2. The negative peak represents bleaching of the ground state to one exciton state transition. The positive component represents induced reflection caused by the one-exciton to two-exciton transitions. A calculation of an impulsive two pulse echo measurement (signal at $\mathbf{k}_s = 2\mathbf{k}_2 - \mathbf{k}_1$ with pulse 1 first) is shown in Figure 17.9. The signal is displayed as a function of observation time for a fixed small delay between pulses. The rise of the signal reflects exciton scattering processes that are necessary to generate the polarization at the desired wavevector, and the subsequent decay is radiative. The local field approximation predicts a similar signal but its magnitude is ~ 3 times weaker.

EXCITON-POPULATION VARIABLES AND EXCITON TRANSPORT

In this section, we pursue further our study of the nonlinear susceptibilities by adopting the factorization [Eq. (17.23)] of the three-operator variables,

$$\langle B_n^\dagger B_n B_m \rangle = \langle W_n \rangle \langle B_m \rangle, \tag{17.63}$$

with $W_n \equiv B_n^\dagger B_n$. This factorization is expected to hold in the presence of sufficiently fast phonon-induced dephasing processes, which destroy intermolecular coherences. Using the present factorization we therefore neglect intermolecular coherences and miss the cooperative enhancement associated with the two-exciton variables. However, the $\langle W_n \rangle$ variables introduce exciton transport into the nonlinear response.

Exciton transport in phase space: the Wigner representation

Once the factorization [Eq. (17.63)] is made, we need to derive an equation of motion for the exciton population variables $\langle W_n \rangle$. Instead, we consider the more general two-particle variables $\langle B_n^\dagger B_m \rangle$. To make the connection to macroscopic transport equations, such as the Boltzmann equation and the diffusion equation, we shall work with the Wigner phase-space representation $\phi(\mathbf{r}, \mathbf{p}, t)$ of these variables. To that end we introduce the relative $\mathbf{s} \equiv \mathbf{r}_m - \mathbf{r}_n$ and the center of mass $\mathbf{r} \equiv (\mathbf{r}_m + \mathbf{r}_n)/2$ coordinates, and write

$$\langle B_m^\dagger B_n \rangle \equiv \langle B_{\mathbf{r}+\mathbf{s}/2}^\dagger B_{\mathbf{r}-\mathbf{s}/2} \rangle .$$

We then define the Wigner distribution [see Eq. (3.100a)]:

$$\phi(\mathbf{r}, \mathbf{p}, t) \equiv \frac{1}{\sqrt{N}} \sum_{\mathbf{s}} \langle B_{\mathbf{r}+\mathbf{s}/2}^\dagger B_{\mathbf{r}-\mathbf{s}/2} \rangle \exp(i\mathbf{p}\cdot\mathbf{s}). \tag{17.64}$$

Using the Wigner distribution, our equations of motion [Eq. (17.5)] read [29]

$$\frac{1}{i}\frac{d}{dt}\langle B_n \rangle = \left(-\Omega_n + i\frac{\Gamma}{2}\right)\langle B_n \rangle - \sum_m J_{nm}\langle B_m \rangle + 2\sum_m J_{nm}\langle W_n \rangle \langle B_m \rangle + \frac{1}{\hbar}\mu_n E_n(t)(1 - 2\langle W_n \rangle), \tag{17.65a}$$

with

$$\frac{d}{dt}\phi(\mathbf{r}, \mathbf{p}, t) = 2\sum_{\mathbf{a}} J(\mathbf{a}) \sin(\mathbf{p}\cdot\mathbf{a})\phi(\mathbf{r} + \mathbf{a}/2, \mathbf{p}, t) - \hat{\Gamma}[\phi(\mathbf{r}, \mathbf{p}, t) + W(\mathbf{r}, t)] - \gamma\phi(\mathbf{r}, \mathbf{p}, t) + \frac{i}{\hbar}\frac{1}{\sqrt{N}}\sum_{\mathbf{s}} \exp(i\mathbf{p}\cdot\mathbf{s})[\mu_{\mathbf{r}-\mathbf{s}/2}E_{\mathbf{r}-\mathbf{s}/2}(t)\langle B_{\mathbf{r}+\mathbf{s}/2}^\dagger \rangle - \mu_{\mathbf{r}+\mathbf{s}/2}E_{\mathbf{r}+\mathbf{s}/2}^*(t)\langle B_{\mathbf{r}-\mathbf{s}/2} \rangle], \tag{17.65b}$$

and

$$\langle W_n(t) \rangle = \frac{2}{N}\sum_{\mathbf{p}} \phi(\mathbf{r}_n, \mathbf{p}, t). \tag{17.65c}$$

The first term in Eq. (17.65b) describes coherent exciton motion on the lattice. When the interaction is short-range and centrosymmetric [$J(\mathbf{r}) = J(-\mathbf{r})$] it reduces to $(\mathbf{p}/m^*)\cdot\vec{\nabla}_{\mathbf{r}}\phi(\mathbf{r}, \mathbf{p}, t)$ with the effective mass given by $(m^*)^{-1} = (1/d)\sum_n J(r_n)r_n^2$, d being the dimensionality of the system. Here, n runs over the lattice and $r_n \equiv |\mathbf{r}_n|$.

The second term in this equation has the form of the BGK strong collision operator in the Boltzmann equation [24], in which collisions occur with rate $\hat{\Gamma}$ and the momentum after each collision is distributed according to $g(\mathbf{p})$, where $g(\mathbf{p})$ is the equilibrium momentum distribution. The strong collision operator conserves the number of particles (population); a population loss with rate γ is described by the last term. If we take $g(\mathbf{p}) = 1/N$ where N is the number of lattice sites, then Eq. (17.65b) is identical to the Haken–Strobl model [Eq. (17.A.7)]. This is, therefore, a high-temperature model, as the equilibrium distribution is then uniform over all momenta. Equation (17.65b) can also be connected to other common transport equations (the diffusion and the master equation), using standard limiting cases.

When Eqs. (17.65) are transformed to the frequency domain and solved iteratively we obtain the nonlinear susceptibility $\chi^{(3)}$. For the sake of clarity we shall focus our attention on a specific four-wave mixing technique: degenerate four-wave mixing and its time-domain analogue, the transient grating. These techniques provide a direct probe for the exciton-population variables, $\langle W_n \rangle$. They use two incoming fields $\mathbf{k}_1$, $\mathbf{k}_2$ and the signal is observed at $\mathbf{k}_s = 2\mathbf{k}_1 - \mathbf{k}_2$ and $\omega_s = 2\omega_1 - \omega_2$. In the degenerate four-wave mixing (D4WM) technique we study the signal in the vicinity of $\omega_1 = \omega_2$, looking for a sharp resonance. The signal intensity is, within the slowly varying amplitude approximation,

$$S_{\mathrm{D4WM}}(\mathbf{k}_s, \omega_s) \propto |\chi^{(3)}(\mathbf{k}_s - \omega_s; \mathbf{k}_1\omega_1, -\mathbf{k}_2 - \omega_2, \mathbf{k}_1\omega_1)|^2.$$

The solution of Eqs. (17.65) results in the following expression for $\chi^{(3)}$

$$\begin{aligned}
&\chi^{(3)}(-\mathbf{k}_s - \omega_s; \mathbf{k}_1\omega_1, -\mathbf{k}_2 - \omega_2, \mathbf{k}_1\omega_1) \\
&= 2\rho_0 \frac{\mu\mu\mu\mu}{\hbar^3} \frac{1}{\omega_s - \Omega_{\mathbf{k}_s} + i(\hat{\Gamma} + \gamma)/2} \frac{1}{\omega_1 - \Omega_{\mathbf{k}_1} + i(\hat{\Gamma} + \gamma)/2} \frac{1}{\omega_2 - \Omega_{\mathbf{k}_2} - i(\hat{\Gamma} + \gamma)/2} \\
&\times \left\{ 1 - \frac{\hat{\Gamma}}{N} \sum_{\mathbf{p}} [-i\omega_{12} - iJ(\mathbf{p} - \mathbf{k}_g/2) + iJ(\mathbf{p} + \mathbf{k}_g/2) + \hat{\Gamma} + \gamma]^{-1} \right\}^{-1}, \qquad (17.66)
\end{aligned}$$

where $\omega_{12} \equiv \omega_1 - \omega_2$. This result is the product of the rotating-wave (resonant) part of Eq. (16.51), obtained using the single-particle factorization, and a dephasing-induced correction factor given by the curly brackets, which is equal to one if $\hat{\Gamma} = 0$.

We shall now examine a few limiting cases:

1. We first consider a system in a pure state (i.e., in the absence of pure dephasing $\hat{\Gamma} = 0$). We note that the first three denominators in Eq. (17.66) depend on the frequencies ω_1, ω_2, and $\omega_s \equiv 2\omega_1 - \omega_2$ (single and three-photon resonances) but not on $\omega_{12} \equiv \omega_1 - \omega_2$, which is a two-photon resonance. The only $\omega_1 - \omega_2$ frequency dependence in $S_{\mathrm{D4WM}}(\mathbf{k}_s, \omega_s)$ that may give a sharp resonance is then contained in the last (dephasing-induced) factor in this expression, which is completely missed by the single-particle factorization. In this limit the model describes coherent exciton motion on the lattice, and the D4WM signal exhibits no resonance as a function of ω_{12}.

2. We next consider noninteracting molecules [$J(\mathbf{k}) = 0$] and find

$$S_{\mathrm{D4WM}}(\mathbf{k}_s, \omega_s) \propto 1 + \frac{\hat{\Gamma}(\hat{\Gamma} + 2\gamma)}{\omega_{12}^2 + \gamma^2}. \tag{17.67}$$

This signal shows a Lorentzian resonance at $\omega_{12} = 0$ whose width is the inverse of the excited state lifetime. The resonance vanishes in the absence of dephasing. Dephasing-induced resonances have been observed by Bloembergen et al. in the gas phase and were discussed in Chapter 9. The counterintuitive aspect of these resonances is that most commonly dephasing results in line-broadening associated with the loss of coherence, whereas here it induces new sharp resonances as ω_{12} is varied. This is the result of a delicate interference of various terms contributing to $\chi^{(3)}$, which exactly cancel in the absence of dephasing. The addition of dephasing affects different Liouville paths differently, eliminates this cancellation, and results in the new resonance.

3. We finally discuss the strong dephasing (incoherent or diffusive) limit, defined by $\hat{\Gamma} \gg |J(\mathbf{p} - \mathbf{k}_g/2) - J(\mathbf{p} + \mathbf{k}_g/2)|$ (for all $\mathbf{p}$) and $\hat{\Gamma} \gg \gamma$. In this limit, the Haken–Strobl model describes diffusive exciton motion, and we have

$$S_{\mathrm{D4WM}}(\mathbf{k}_s, \omega_s) \propto 1 + \frac{\hat{\Gamma}(\hat{\Gamma} + 2\gamma)}{\omega_{12}^2 + [\gamma + (\mathbf{k}_1 - \mathbf{k}_2)^2 D_e]^2}, \tag{17.68a}$$

with the exciton diffusion constant given by

$$\begin{aligned} D_e &\equiv \frac{1}{|\mathbf{k}_g|^2} \frac{1}{\hat{\Gamma} N} \sum_{\mathbf{p}} [J(\mathbf{p} - \mathbf{k}_g/2) - J(\mathbf{p} + \mathbf{k}_g/2)]^2 \\ &= \frac{4}{|\mathbf{k}_1 - \mathbf{k}_2|^2} \frac{1}{\hat{\Gamma}} \sum_m J^2(\mathbf{r}^m) \sin^2\left[\frac{(\mathbf{k}_1 - \mathbf{k}_2)\cdot \mathbf{r}_m}{2}\right]. \end{aligned} \tag{17.68b}$$

In the incoherent limit the D4WM signal shows a Lorentzian resonance with a width that is the sum of the inverse excited state lifetime (γ) and a wavevector-dependent contribution from exciton diffusion [$D_e(\mathbf{k}_1 - \mathbf{k}_2)^2$].

Transient grating: the time-domain analogue of degenerate four-wave mixing

The exciton-population $\langle B_n^\dagger B_n \rangle$ variables can also be probed by a time-domain technique, transient grating (TG) spectroscopy, which may be related to the Fourier transform of the degenerate four wave mixing. The following typical setup is considered (Figure 17.10). At time $t = 0$ two short excitation pulses, $(\mathbf{k}_1, \omega_1)$ and $(\mathbf{k}_2, \omega_2)$, crossed under an angle θ, interfere in the sample and create an excitonic grating. The decay of the grating as a result of dephasing and population relaxation is monitored by applying a probe pulse, $(\mathbf{k}_3, \omega_3)$, at $t = \tau$. The observable $S_{\mathrm{TG}}(\tau)$ is the time-integrated intensity of the nonlinear ("diffracted") signal with wavevector $\mathbf{k}_s = \mathbf{k}_1 - \mathbf{k}_2 + \mathbf{k}_3$ and frequency $\omega_s = \omega_1 - \omega_2 + \omega_3$ as a function of the pump-probe delay τ [4, 29, 30]. We shall now consider again the three cases discussed above.

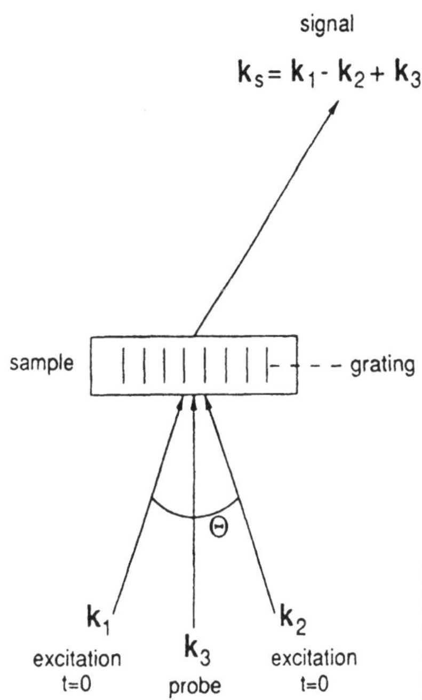

FIG. 17.10 Transient grating setup. Two excitation beams crossed under angle θ create a grating in the sample with wavevector $\mathbf{k}_1 - \mathbf{k}_2$. After a variable delay τ, the grating is probed by a third pulse $\mathbf{k}$, resulting in a nonlinear ("diffracted") signal at $\mathbf{k}_s = \mathbf{k}_1 - \mathbf{k}_2 + \mathbf{k}_3$.

1. In the absence of dephasing $\hat{\Gamma} = 0$, no grating decay is observed, apart from the trivial population decay $\exp(-2\gamma t)$. This result can be easily understood since for $\hat{\Gamma} = 0$ the initial state created by the incoming pulses is an exact eigenstate of the crystal.
2. For noninteracting molecules we get

$$S_{\mathrm{TG}}(\tau) = \exp[-2(\gamma + \hat{\Gamma})\tau]. \tag{17.69}$$

   Here, dephasing destroys the grating and contributes to the signal decay.
3. We now turn to the opposite limit of strong scattering, $\hat{\Gamma} \gg \gamma$ and $\hat{\Gamma} \gg J$, where the exciton scattering length is much shorter than the grating length scale (diffusive or incoherent motion). The signal intensity is then given by

$$S_{\mathrm{TG}}(\tau) = \exp[-2\gamma\tau - 2D_e|\mathbf{k}_1 - \mathbf{k}_2|^2\tau]. \tag{17.70}$$

The signal decay rate consists of a wavevector-independent contribution reflecting population relaxation (2γ), and a contribution from exciton motion, which is proportional to $|\mathbf{k}_1 - \mathbf{k}_2|^2$. This is characteristic for diffusive motion and leads, for small cross-angles θ between the two pump pulses, to a linear relation between the observed decay rate and θ^2. This relation provides a useful means to distinguish experimentally between diffusive and coherent exciton motion. In the incoherent limit, an interesting relation exists between the D4WM and TG signals, namely, the amplitude of the D4WM signal [Eq. (17.68a)] can be obtained by evaluating the Fourier transform of the amplitude of the TG signal [Eq. (17.70)] at the frequency ω_{12}. An example of a TG signal of a molecular crystal (anthracene) which probes exciton diffusion is given in Figure 17.11.

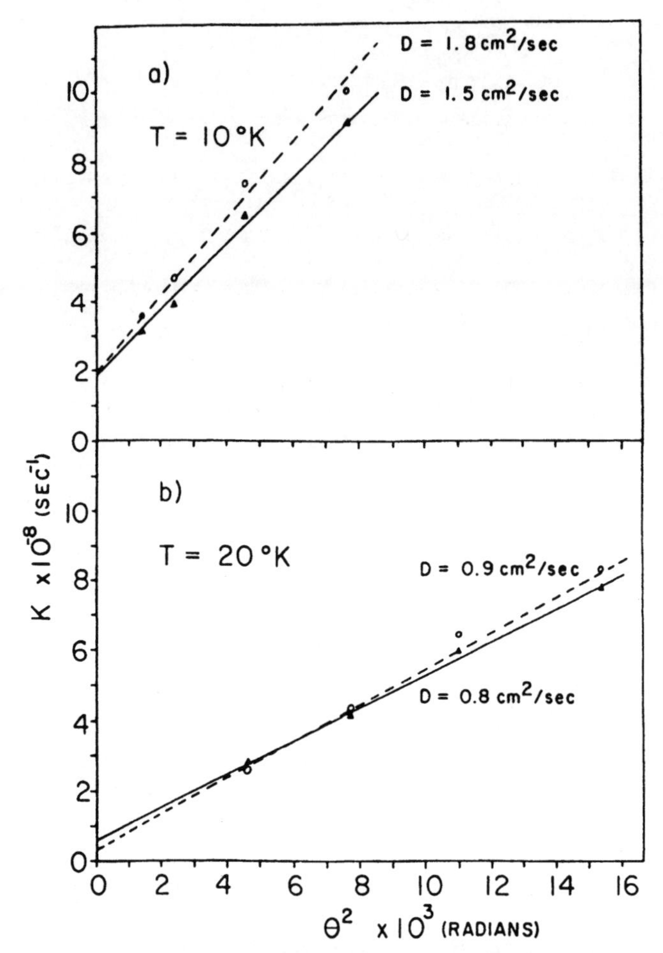

FIG. 17.11 (a) The decay rate of the transient grating signal versus θ^2 for two anthracene crystals at 10 K, along the *a* axis. The magnitude of the slope equals $8\pi^2 D/\lambda^2$, with λ the wavelength of the pump pulses, and thus yields directly the exciton diffusion constant *D*. The value of the intercept equals 2γ, where γ is the population relaxation rate. (b) The decay rate of the transient grating signal for the same two crystals at 20 K. The difference in slopes is due to the temperature dependence of the exciton diffusion constant. As the temperature increases, a decrease in the diffusion constant is observed. The average diffusion constant obtained from these measurements is roughly 10 times larger than the expected value for incoherent exciton diffusion. [T. S. Rose, R. Righini, and M. D. Fayer, "Picosecond transient grating measurements of singlet exciton tránsport in anthracene single crystals," *Chem. Phys. Lett.* **106**, 13 (1984).]

Finally, we note that it is essential to include the population variables in order to reproduce the wavevector dependence of both the degenerate four-wave mixing and the transient grating signals. This dependence is a signature of exciton diffusion which is missed by the local field factorization.

DISCUSSION

The equations of motion (Green function) approach establishes an anharmonic-oscillator picture for calculating and interpreting the nonlinear optical response of Frenkel exciton systems such as molecular assemblies or weakly confined semiconductor nanostructures where the confinement size is large compared with the Wannier exciton radius. It has long been recognized that the linear optical response of such systems can be calculated using a harmonic (Drude) oscillator picture. The optical polarization of a crystal of N two-level atoms is rigorously given by a sum of contributions of N coupled harmonic oscillators representing the individual molecular polarizations [31]. A natural extension of these ideas to nonlinear spectroscopy suggests the use of an anharmonic oscillator picture for the material polarization. The key step in the implementation of this approach is the microscopic derivation of equations of motion for the optical polarization, including other relevant dynamic variables that are nonlocal in space and represent intermolecular coherences. The complex many-body problem thus reduces to the coupled nonlinear dynamics of relatively few oscillators related to the exciton amplitudes $\langle B_n \rangle$, the two-exciton $\langle B_n B_m \rangle$, exciton-population and coherence $\langle B_n^\dagger B_m \rangle$, and three-exciton $\langle B_n^\dagger B_n B_m \rangle$ variables. These variables together with their Hermitian conjugates constitute the relevant set of anharmonic oscillators, which are the natural elementary excitation (quasiparticle) of the system [32]. In the coupled oscillator picture, obtained using the equations of motion, cooperativity arises from the existence of nonlocal intermolecular coherences, and is the source of new resonances, possible enhancements (unusually large nonlinearities) as well as cooperative radiative decay (superradiance), and exciton transport. The local-field approximation fails completely to take into account any contribution of nonlocal nonlinearities and interactions among excitons and, therefore, misses all cooperative effects. Studies of multiquantum NMR [33] have some common features to the present two exciton cooperativity (although they are treated using a very different terminology). In the design of optical materials it is important to optimize the detunings such that they are kept sufficiently large to avoid absorptive losses, but sufficiently small to make use of cooperative effects.

Comparison of the equation of motion Green function approach and the sum over states reveals several advantages for the former, when applied to molecular assemblies and nanostructures. To start with, the resulting expressions contain fewer terms and lend themselves more easily to numerical computations. Moreover, since interference between single-exciton and two-exciton transitions is naturally built in, the susceptibilities no longer contain two large contributions that almost cancel and need to be carefully added. This interference makes it hard to develop physical intuition based on the sum over states expression. Simple necessary approximations such as truncation of the summation over states then become dangerous since they may have a large effect by destroying the delicate balance between large terms.

The Frenkel exciton model considered here includes only exciton repulsion through Pauli exclusion. Attractive interactions may arise from different mechanisms, either direct dipole–dipole interactions, or phonon mediated (very much like the Cooper pairs in superconductivity [32]). Attractive exciton–exciton interactions can create new bound states of two excitons (denoted biexcitons). Their effect can be incorporated in the present treatment in a straightforward way, e.g., by simply changing the sign of the parameter g [see discussion following Eqs. (17.5)]. Biexcitons usually show up as strong two-photon resonances, that are red shifted with respect to the exciton frequency ($\omega < \Omega$). The red shift reflects the biexciton binding energy [34–36].

The equation of motion approach and the resulting anharmonic oscillator representation can be applied to materials other than molecular crystals. Systems with delocalized electronic states such as semiconductors and conjugated polymers constitute an important class of materials with interesting nonlinear optical properties. In semiconductors one writes equations of motion for carriers (electrons and holes) as well as for bound pairs of carriers (Wannier excitons) [15, 16]. This requires working with different types of creation and annihilation operators for Fermions and invoking a Hartree–Fock approximation for the decoupling of higher variables. The same approach can also be applied to conjugated polymers such as polydiacetylenes, polyacetylenes, or polysilanes [37–39]. These systems are intermediate between the Frenkel and the Wannier limits and resemble more closely charge transfer excitons. Nevertheless, the description of off-resonant as well as resonant $\chi^{(3)}$ spectroscopies of conjugated polymers can be effectively developed using an equation of motion, Green function procedure [40]. The differences between various types of materials can be very clearly identified and investigated by analyzing the nature of the nonlinear oscillators in each case. The origin of these differences is much less transparent when using the sum-over-states expression, which is formally identical for all materials and whereby the differences enter through the complete set of eigenstates [17]. The modeling of optical nonlinearities in various types of materials in terms of coupled anharmonic oscillators offers a simple and physically transparent unifying picture that constitutes a natural extension of the harmonic oscillator picture of linear optics. It is possible to incorporate additional (e.g., phonon) degrees of freedom in this picture. In the applications presented here, phonons simply result in relaxation and damping of the other degrees of freedom. Phonons can also show up as explicit new resonances (rather than broadening). In measurements, such as Raman and the optical Stark effect, these can be described by adding more oscillator variables corresponding to phonons in the equations of motion [17].

To account for strong field effects we can either consider higher order terms in the perturbative expansion and discuss $\chi^{(5)}$, etc., or may use a canonical transformation of the boson operators. For example, in one-dimensional systems with nearest-neighbor interactions, hard core bosons can be transformed to noninteracting fermions using the Wigner–Jordan transformation [41]. At high exciton densities created by intense laser fields [42], excitons may undergo a Bose condensation. Excitons form a hard core bose fluid, pretty much like superfluid helium where short range repulsive forces play the role of the Pauli exclusion. Experimental attempts to observe Bose condensation of Wannier exciton systems in semiconductors have been made [43]. Frenkel exciton systems could also be promising candidates for observing Bose condensation, as much higher exciton densities can be attained without destroying their boson character. Finally, in ordered low temperature materials with a large oscillator strength per unit volume, the radiation field combines with the material polarization to form new elementary excitations: polaritons (see Figure 17.12) [44–49]. The signatures of polaritons in optical nonlinearities include a significant change in damping and unusually fast transport rates. Polariton effects result from the fact that the

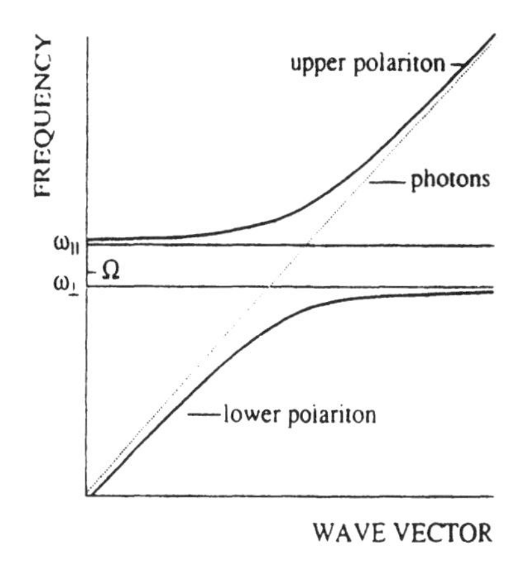

FIG. 17.12 Typical polariton dispersion curves in the optical region for an atomic crystal (thick solid curves). The diagonal line represents the pure photon dispersion curve ($\omega_k = kc$). The shaded region between the transverse ($\omega_\perp$) and longitudinal ($\omega_\parallel$) crystal exciton frequencies is the stopgap, where no polariton modes exist. Ω indicates the atomic transition frequency. For a crystal of two-level molecules, the stopgap position depends on the direction of propagation.

anharmonic oscillators are in general combined material and field degrees of freedom (rather than purely material variables). Studies of polariton effects in nonlinear optical optics are still in their infancy [29]. The frequency-dependent self energy $\phi_{nm}(\omega)$ introduced here allows us to incorporate polariton effects which show up in extended systems, such as waveguide configurations, and discuss their signatures in nonlinear optics.

APPENDIX 17A

Exciton Dephasing and Relaxation Processes

Equation (17.3) contains no exciton relaxation mechanism. To incorporate relaxation microscopically we need to incorporate nuclear degrees of freedom (e.g., phonons) in the Hamiltonian. The exciton–phonon interaction arises from the dependence of the intermolecular interactions on the displacements of the nuclear coordinates from their equilibrium values. The coupling with phonons induces relaxation once the nuclear degrees of freedom are eliminated, and plays an important role in determining the coherence size and the magnitude of optical nonlinearities. We shall introduce here some simple models for relaxation. A more rigorous treatment may be carried out using the projection operator methods discussed in Chapter 3.

We shall incorporate exciton population relaxation and pure-dephasing mechanisms into the equations of motion. We first introduce a simple model for exciton dephasing corresponding to relaxation of population resulting in a finite exciton lifetime. The model assumes a coupling to a heat bath $\mathbf{q}_n$ with a coupling $\sum_n [V_n^\dagger(\mathbf{q}_n)B_n^\dagger + V_n(\mathbf{q}_n)B_n]$. The coupling operators V_n at different sites are assumed to be uncorrelated, so that each molecule has its own heat bath. Using standard procedures [19], we can then incorporate the effects of the bath by adding a damping term to the equations of motion. To second order in V_n, and assuming a very short correlation time of the bath, we find that the following relaxation term has to be added to the equations of motion

$$\left.\frac{d}{dt}\right|_{\text{lifetime}} B_{n_1}^\dagger \cdots B_{n_i}^\dagger B_{n_i+1} \cdots B_{n_m} = \frac{m}{2}\gamma B_{n_1}^\dagger \cdots B_{n_i}^\dagger B_{n_i+1} \cdots B_{n_m} \qquad (17.A.1)$$

where the (site-independent) population relaxation rate γ is given by

$$\gamma = \int_0^\infty d\tau \langle V_n(\tau)V_n(0)\rangle_{\text{bath}} \exp(i\Omega_n\tau). \qquad (17.A.2)$$

It should be emphasized that this relaxation term may be added only to the time evolution of a *normally ordered product* of operators, that is, all the $B^\dagger$ are left of the Bs. As we can bring any product of operators to a normally ordered form using the commutation relations, this is not a limitation. It can easily be verified that this relaxation model guarantees conservation of probability, i.e. $(d/dt)(B_n^\dagger B_n + B_n B_n^\dagger) = 0$.

The second relaxation mechanism corresponds to *pure dephasing* resulting from fluctuations of the isolated molecule frequency. For this we assume a coupling to another heat bath, $(\mathbf{q}'_n)$ with a coupling $\sum_n (V'_n(\mathbf{q}'_n) B_n^\dagger B_n)$. Again each molecule has its own heat bath, which has a very short correlation time. This will introduce an additional damping term, such as Eq. (17.A.1), but now the right-hand side depends on whether different indices are equal. The pure-dephasing rate is given by

$$\hat{\Gamma} = 2 \int_0^\infty d\tau \langle V'_n(\tau) V'_n(0) \rangle_{\text{bath}}. \tag{17.A.3}$$

The combined (population relaxation and pure dephasing) contributions to the equations of motion are

$$\left[\frac{d}{dt} \langle B_m(t) \rangle \right]_{\text{ph}} = -\tfrac{1}{2}(\hat{\Gamma} + \gamma) \langle B_m(t) \rangle, \tag{17.A.4}$$

$$\left[\frac{d}{dt} \langle B_m^\dagger(t) B_n(t) \rangle \right]_{\text{ph}} = -[\hat{\Gamma}(1 - \delta_{nm}) + \gamma] \langle B_m^\dagger(t) B_n(t) \rangle, \tag{17.A.5}$$

$$\frac{d}{dt}\bigg|_{\text{ph}} \langle B_n^\dagger B_m B_{n'} \rangle(t) = -[\hat{\Gamma}(\tfrac{3}{2} - \delta_{n,m} - \delta_{nn'}) + \tfrac{3}{2}\gamma] \langle B_n^\dagger B_m B_{n'} \rangle(t), \tag{17.A.6}$$

where $[\cdots]_{\text{ph}}$ denotes the phonon contribution.

The parameters $\hat{\Gamma}$ and γ are taken to be real. $\hat{\Gamma}$ is the pure dephasing rate and γ represents the population relaxation rate. We use the notation $\Gamma \equiv \hat{\Gamma} + \gamma$ for the total dephasing rate. The relaxation model introduced in Eqs. (17.A.4)–(17.A.6) allows for analytical results while still preserving the essential physical aspects related to pure dephasing.

The pure dephasing contribution can alternatively be derived by adopting the following stochastic model for the exciton-phonon coupling, known as the Haken–Strobl model [24].

$$H_{\text{ex-phon}} = \hbar \sum_n \delta\Omega_n(t) B_n^\dagger B_n. \tag{17.A.7}$$

$\delta\Omega_n(t)$ is a stochastic Gaussian random variable with the following properties:

$$\langle \delta\Omega_n(t) \rangle = 0, \tag{17.A.8a}$$

$$\langle \delta\Omega_n(t)\, \delta\Omega_m(t) \rangle = \frac{\hat{\Gamma}}{2} \delta(t - t') \delta_{m,n}. \tag{17.A.8b}$$

When the equations of motion are averaged over the stochastic part, we obtain the relaxation Eqs. (17.A.4)–(17.A.6).

APPENDIX 17B

Scattering ($\mathscr{T}$) Matrix Expression for the Two-Exciton Green Function

We are interested in the Green function solution to Eq. (17.9). We shall use the model discussed following Eqs. (17.5) where we add a repulsive energy g and set $g \to \infty$ at the end [19].

We start by defining a zero-order Green function $\bar{G}^{(2)}$ for the reference (noninteracting boson) system with $g = 0$.

$$\bar{G}^{(2)}(\omega_1 + \omega_2) = \left[\frac{1}{\omega_1 + \omega_2 + \bar{F}^{(2)}}\right]. \tag{17.B.1}$$

The Green function for noninteracting Boson $\bar{G}^{(2)}$ can be expressed as a convolution of one-exciton Green functions. Alternatively, if we compute the single-exciton eigenstates $|\psi_j\rangle$ we have

$$G(\omega) = \sum_j \frac{|\psi_j\rangle\langle\psi_j|}{\omega - \lambda_j + i\Gamma/2} \tag{17.B.2}$$

and

$$\bar{G}^{(2)}(\omega) = \sum_{j,j'} \frac{|\psi_j\rangle|\psi_{j'}\rangle\langle\psi_j|\langle\psi_{j'}|}{\omega - \lambda_j - \lambda_{j'} + i\Gamma}. \tag{17.B.3}$$

Using standard techniques, we can then express the actual Green function using the reference one and a scattering ($\mathscr{T}$) matrix. In this way we can pin down the contribution caused by the on-site repulsion of the two excitons. The $\mathscr{T}$ matrix is defined through

$$G^{(2)}(\omega_1 + \omega_2) = \bar{G}^{(2)}(\omega_1 + \omega_2) + \bar{G}^{(2)}(\omega_1 + \omega_2)\mathscr{T}^{(g)}(\omega_1 + \omega_2)\bar{G}^{(2)}(\omega_1 + \omega_2), \tag{17.B.4}$$

where all the products are understood to be matrix products. Next we defined a new $N \times N$ Green function $\hat{G}$, which is a projection of the full noninteracting Green function onto the application space

$$\hat{G}^{(2)}_{n,n'} \equiv \bar{G}^{(2)}_{nn,n'n'}. \tag{17.B.5}$$

We find that $T^{(g)}$ is given by

$$\mathscr{T}^{(\lambda)}_{nm,n'm'} = \delta_{nm}\delta_{n'm'}g(1 + g\hat{G}^{(2)})^{-1}_{n,n'}, \tag{17.B.6}$$

where the inverse is of an $N \times N$ matrix. In the limit $g \to \infty$ we have $\mathscr{T}^{(\infty)} = \mathscr{T}$.

Notice the structure of the Green function relevant for $\mathcal{T}$: the matrix elements needed represent two excitons initially at the same site, moving over the molecules independently, both excitons end up at the same site again. We then have

$$\mathcal{T}_{nn,mm} = [\hat{G}^{(2)}]^{-1}_{nm}. \tag{17.B.7}$$

Equation (17.B.7) together with (17.B.4) yields Eq. (17.37a).

For the pure dephasing effect on the exciton population $\langle B^{\dagger}B\rangle$ variables we can proceed in a similar way. We first define the noninteracting two-exciton Green function

$$\bar{G}^{(2')}_{nm,n'm'}(\omega_1 - \omega_3) = (\omega_1 - \omega_3 + \bar{F}^{(2')})^{-1}_{nm,n'm'}. \tag{17.B.8}$$

We can then write an analogous equation to (17.B.4) with the superscript (2) replaced by (2′). Again we introduce the corresponding $\mathcal{T}^{(\hat{\Gamma})}$ matrix through

$$G^{(2')} = \bar{G}^{(2')} + \bar{G}^{(2')}\mathcal{T}^{(\hat{\Gamma})}\bar{G}^{(2')}. \tag{17.B.9}$$

In terms of the projected Green function $\hat{G}$,

$$\hat{G}^{(2')}_{n,n'} \equiv \bar{G}^{(2')}_{nn,n'n'}, \tag{17.B.10}$$

the $\mathcal{T}$ matrix is given by

$$\mathcal{T}^{(\hat{\Gamma})}_{n_1n_1,n_2n_2}(\omega_1 - \omega_3) = -i\hat{\Gamma}[1 - i\hat{\Gamma}\hat{G}^{(2')}(\omega_1 - \omega_3)]^{-1}_{n_1,n_2}. \tag{17.B.11}$$

The inverse, again, is that of an $N \times N$ matrix.

For periodic systems, $\mathcal{T}$ becomes a 1×1 matrix given by

$$\mathcal{T}(\omega_1 + \omega_2) = [\tilde{G}^{(2)}_{0,0}(\omega_1 + \omega_2)]^{-1}. \tag{17.B.12}$$

The $\mathcal{T}$ matrix is inversely proportional to the Fourier transform of the amplitude of the return of the two excitons to the origin. The noninteracting Green function for pure dephasing is

$$\tilde{G}^{(2')}_{s,s_1}(\omega_1 - \omega_3) = \delta_{ss_1}\frac{1}{\omega_1 - \omega_3 + i\gamma}. \tag{17.B.13}$$

The noninteracting two-exciton Green function is given by

$$\tilde{G}^{(2)}_{s,s_1}(\omega_1 + \omega_2) = \frac{1}{N}\sum_k e^{ik(s_1 - s)}\frac{1}{\omega_1 + \omega_2 - 2\Omega - 2J(k) + i\Gamma}, \tag{17.B.14}$$

and the scattering operator is

$$\mathcal{T}(\omega_1 + \omega_2) = N\left[\sum_k \frac{1}{\omega_1 + \omega_2 - 2\Omega - 2J(k) + i\Gamma}\right]^{-1}. \tag{17.B.15}$$

Equation (17.B.15) together with (17.B.9) thus provide a close expression for the exciton population Green function [Eq. (17.14)].

APPENDIX 17C

Green Function Solution of the Nonlinear Response

We present here the Green function solution of Eqs. (17.5) for the third-order polarization. We first solve the linearized equation for $\langle B \rangle$ to first order in the field. Next, we solve for $\langle BB \rangle$ and $\langle B^\dagger B \rangle$ to second order in the field, and finally we solve for $\langle B^\dagger BB \rangle$ and $\langle B \rangle$ to third order. The procedure is iterative; in each step we use the lower order solutions.

To first order we have

$$\langle B_n \rangle^{(1)}(t) = \frac{i}{\hbar} \sum_m \mu_m \int_{t_0}^{t} \mathrm{d}t' G_{nm}(t - t') E_m(t'). \tag{17.C.1}$$

$\langle B_n^\dagger \rangle$ are simply the complex conjugates of $\langle B_n \rangle$.

To second order we get

$$\langle B_n B_m \rangle^{(2)}(t) = \frac{i}{\hbar} \sum_{n'm'} \int_{t_0}^{t} \mathrm{d}t' G^{(2)}_{nm,n'm'}(t - t') \{ \mu_{n'} E_{n'}(t') \langle B_{m'} \rangle^{(1)}(t') + \mu_{m'} E_{m'}(t') \langle B_{n'} \rangle^{(1)}(t') \} \zeta_{n'm'}. \tag{17.C.2}$$

$$\langle B_n^+ B_m \rangle^{(2)}(t) = \frac{i}{\hbar} \sum_{n'm'} \int_{t_0}^{t} \mathrm{d}t' G^{(2)}_{nm,n'm'}(t - t') \{ \mu_{m'} E_{m'}(t') \langle B_{n'}^+ \rangle^{(1)}(t') - \mu_{n'} E_{n'}^*(t') \langle B_{m'} \rangle^{(1)}(t') \} \tag{17.C.3}$$

To third order we have

$$\langle B_n^+ B_m B_l \rangle^{(3)}(t) = \frac{i}{\hbar} \sum_{n'm'l'} \int_{t_0}^{t} \mathrm{d}t' G_{nml,n'm'l'}(t - t') \{ \mu_{m'} E_{m'}(t') \langle B_{n'}^+ B_{l'} \rangle^{(2)}(t') + \mu_{l'} E_{l'}(t') \langle B_{n'}^+ B_{m'} \rangle^{(2)}(t') - \mu_{n'} E_{n'}^*(t') \langle B_{m'} B_{l'} \rangle^{(2)}(t') \} \zeta_{m'l'} \tag{17.C.4}$$

Using these definitions and Eq. (16.37), we obtain a closed form expression for the third-order polarization:

$$P^{(3)}(t) = -\mu_n \frac{2i}{\hbar} \int_{t_0}^{t} \mathrm{d}t' \sum_m \mu_m G_{nm}(t - t') E_m(t') \langle B_m^+ B_m \rangle^{(2)}(t') + \mu_n 2i \int_{t_0}^{t} \mathrm{d}t' \sum_m \sum_{m'} J_{mm'} G_{nm}(t - t') \langle B_m^+ B_m B_{m'} \rangle^{(3)}(t') + \text{c.c.} \tag{17.C.5}$$

Substitution of Eqs. (17.C.1)–(17.C.4) into Eq. (17.C.5) yields an exact (and lengthy) closed form expression for the third-order response.

NOTES

1. T. Forster, in *Modern Quantum Chemistry, Part III: Action of Light and Organic Molecules*, O. Sinanoglu, Ed. (Academic Press, New York, 1965), p. 63.
2. D. L. Dexter, *J. Chem. Phys.* **21**, 836 (1953).
3. V. M. Agranovich and M. D. Galanin, in *Electronic Excitation Energy Transfer in Condensed Matter*, V. M. Agranovich and A. A. Maradudin, Eds. (North-Holland, Amsterdam, 1982).
4. H. J. Eichler, P. Gunter, and D. W. Pohl, *Laser-Induced Dynamic Gratings* (Springer, Berlin, 1986).
5. D. L. Dexter, *Phys. Rev.* **126**, 1962 (1962); *J. Phys. Soc. Jpn.* **18**, Suupl. II, 275 (1963); T. Miyakawa and D. L. Dexter, *Phys. Rev. B* **1**, 70 (1970); M. Altarelli and D. L. Dexter, *Opt. Commun.* **2**, 36 (1970).
6. R. H. Lehmberg, *Phys. Rev. A* **2**, 883 (1970); D. L. Andrews, D. P. Craig and T. Thirunamachandran, *Int. Rev. Phys. Chem.* **8**, 339 (1989); M. Gross and S. Haroche, *Phys. Rep.* **93**, 301 (1982).
7. G. Scheibe, *Angew. Chem.* **50**, 212 (1937); E. E. Jelly, *Nature (London)* **10**, 631 (1937); A. H. Hertz, *Adv. Colloid Interface Sci.* **8**, 237 (1977).
8. E. W. Knapp, P. O. J. Scherer, and S. F. Fischer, *Chem. Phys. Lett.* **111**, 481 (1984); E. W. Knapp, *Chem. Phys. Lett.* **85**, 73 (1984).
9. D. Möbius and H. Kuhn, *Isr. J. Chem.* **18**, 375 (1979); D. Mobius and H. Kuhn, *J. Appl. Phys.* **64**, 5138 (1988); G. Roberts, Ed. *Langmuir-Blodgett Films* (Plenum, New York, 1990).
10. J. Zyss and D. S. Chemla, in *Nonlinear Optical Properties of Organic Molecules and Crystals*, Vol. **1**, J. Zyss and D. S. Chemla, Eds. (Academic Press, Orlando, Florida, 1987), p. 23.
11. F. F. So, S. R. Forrest, Y. Q. Shi, and W. H. Steier, *Appl. Phys. Lett.* **56**, 674 (1990); F. F. So and S. R. Forrest, *Phys. Rev. Lett.* **66**, 2649 (1991).
12. F. C. Spano, J. R. Kuklinski, and S. Mukamel, *Phys. Rev. Lett.* **65**, 211 (1990); F. C. Spano, J. R. Kuklinski, and S. Mukamel, *J. Chem. Phys.* **94**, 7534 (1991).
13. H. Fidder, J. Knoester, and D. A. Wiersma, *J. Chem. Phys.* **98**, 6564 (1993); S. DeBoer and D. A. Wiersma, *Chem. Phys. Lett.* **165**, 45 (1990).
14. T. Tani, T. Suzumoto, K. Kemnitz, and K. Yoshihara, *J. Phys. Chem.* **96**, 2778 (1992); A. E. Johnson, S. Kumazaki, and K. Yoshihara, *Chem. Phys. Lett.* **211**, 511 (1993).
15. H. Haug, Ed., *Optical Nonlinearities and Instabilities in Semiconductors* (Academic Press, New York, 1988); H. Haug and S. W. Koch, *Quantum Theory of the Optical and Electronic Properties of Semiconductors* (World Scientific, Singapore, 1990); B. I. Greene and R. R. Millard, *Phys. Rev. Lett.* **55**, 1331 (1985).
16. S. Schmitt-Rink, D. S. Chemla, and D. B. Miller, *Adv. Phys.* **38**, 89 (1989); S. Schmitt-Rink, D. B. Miller, and D. S. Chemla, *Phys. Rev. B* **35**, 8113 (1987).
17. B. I. Greene, J. Orenstein, R. R. Millard, and L. R. Williams, *Phys. Rev. Lett.* **58**, 2750 (1987); B. I. Greene, J. Orenstein, and S. Schmitt-Rink, *Science* **247**, 679 (1990).
18. M. L. Steigerwald and L. E. Brus, *Acc. Chem. Res.* **23**, 183 (1990); M. Bawendi, M. L. Steigerwald, and L. E. Brus, *Annu. Rev. Phys. Chem.* **44**, 21 (1990); L. Banyai and S. W. Koch, *Semiconductor Quantum Dots* (World Scientific, Singapore, 1993).
19. J. A. Leegwater and S. Mukamel, *Phys. Rev. A* **46**, 452 (1992).
20. F. C. Spano and S. Mukamel, *Phys. Rev. A* **40**, 5783 (1989); *Phys. Rev. Lett.* **66**, 1197 (1991).

16. S. Schmitt-Rink, D. S. Chemla, and D. B. Miller, *Adv. Phys.* **38**, 89 (1989); S. Schmitt-Rink, D. B. Miller, and D. S. Chemla, *Phys. Rev. B* **35**, 8113 (1987).
17. B. I. Greene, J. Orenstein, R. R. Millard, and L. R. Williams, *Phys. Rev. Lett.* **58**, 2750 (1987); B. I. Greene, J. Orenstein, and S. Schmitt-Rink, *Science* **247**, 679 (1990).
18. M. L. Steigerwald and L. E. Brus, *Acc. Chem. Res.* **23**, 183 (1990); M. Bawendi, M. L. Steigerwald, and L. E. Brus, *Annu. Rev. Phys. Chem.* **44**, 21 (1990); L. Banyai and S. W. Koch, *Semiconductor Quantum Dots* (World Scientific, Singapore, 1993).
19. J. A. Leegwater and S. Mukamel, *Phys. Rev. A* **46**, 452 (1992).
20. F. C. Spano and S. Mukamel, *Phys. Rev. A* **40**, 5783 (1989); *Phys. Rev. Lett.* **66**, 1197 (1991).
21. H. Ishihara and K. Cho, *Phys. Rev. B* **42**, 1724 (1990); K. Cho, *J. Phys. Soc. Jpn.* **55**, 4113 (1986); *Prog. Theoret. Phys. Suppl.* **106**, 225 (1991).
22. M. Pope and C. Swenberg, *Electronic Processes in Organic Crystals* (Oxford University Press, Oxford, 1982).
23. E. I. Rashba, in *Excitons*, E. I. Rashba and M. D. Sturge, Eds. (North-Holland, Amsterdam, 1982).
24. H. Haken and G. Strobl, *Z. Phys.* **262**, 135 (1973); P. L. Bhatnagar, E. P. Gross, and M. Krook, *Phys. Rev.* **94**, 511 (1954).
25. S. Mukamel, in *Molecular Nonlinear Optics*, J. Zyss, Ed. (Academic Press, New York, 1994), p. 1.
26. V. Chernyak and S. Mukamel, *Phys. Rev. B* **48**, 2470 (1993); N. Wang, V. Chernyak, and S. Mukamel, *Phys. Rev. B* (in press); *J. Chem. Phys.* **100**, 2465 (1994).
27. M. F. Shlesinger, J. Klafter, and Y. M. Wong, *J. Stat. Phys.* **27**, 499 (1982).
28. Y. R. Kim, M. Lee, J. R. G. Thorne, R. M. Hochstrasser, and J. M. Ziegler, *Chem. Phys. Lett.* **145**, 75 (1988); J. Feldmann, G. Peter, E. O. Göbel, P. Dawson, K. Moore, C. Foxon, and R. J. Elliot, *Phys. Rev. Lett.* **59**, 2337 (1987); T. Itoh, T. Ikehara, and Y. Iwabuchi, *J. Lumin.* **45**, 29 (1990).
29. J. Knoester and S. Mukamel, *Phys. Rep.* **205**, 1 (1991).
30. T. S. Rose, R. Righini, and M. D. Fayer, *Chem. Phys. Lett.* **106**, 13 (1984); V. M. Agranovich, A. M. Ratner, and M. Salieva, *Solid State Commun.* **63**, 329 (1987).
31. U. Fano, *Phys. Rev.* **103**, 1202 (1956); U. Fano, *Rev. Mod. Phys.* **64**, 313 (1992); S. I. Pekar, *Sov. Phys. JEPT* **6**, 785 (1958); **11**, 1286 (1960).
32. G. D. Mahan, *Many Particle Physics* (Plenum, New York, 1990).
33. A. Pines, in *Proceedings of the 100th School of Physics "Enrico Fermi"* (North-Holland, Amsterdam, 1988), p. 43; R. Tycko, in *Advances in Magnetic and Optical Resonance*, Vol. 15, W. S. Warren, Ed. (Academic Press, New York, 1990), p. 1.
34. M. Orrit and P. Kottis, *Adv. Chem. Phys.* **74**, 1 (1988).
35. A. Maruani and D. S. Chemla, *Phys. Rev. B* **23**, 841 (1981).
36. O. Dubovsky and S. Mukamel, *J. Chem. Phys.* **95**, 7828 (1991); F. C. Spano, V. M. Agranovich and S. Mukamel, *J. Chem. Phys.* **95**, 1400 (1991).
37. S. Etemad and Z. G. Soos, in *Spectroscopy of Advanced Materials*, R. J. H. Clark and R. E. Hester, Eds. (Wiley, New York, 1991), p. 87; J. L. Bredas, C. Adant, P. Tacky, A. Persoons, and A. M. Pierce, *Chem. Rev.* **94**, 243 (1994).
38. C. V. Shank, R. Yen, J. Orenstein, and G. L. Banker, *Phys. Rev. B* **28**, 6095 (1983); R. D. Miller and J. Michl, *Chem. Rev.* **89**, 1359 (1989).
39. T. Kobayashi, M. Yoshizawa, U. Stamm, M. Tayi, and M. Hasegawa, *J. Opt. Soc. Am. B* **7**, 1558 (1990); S. Takeuchi, M. Yoshizawa, T. Masuda, T. Higashimura, and T. Kobayashi, *IEEE J. Quant. Electron.* **28**, 2508 (1992).
40. A. Takahashi and S. Mukamel, *J. Chem. Phys.* **100**, 2366 (1994); S. Mukamel, A. Takahashi, and H. X. Wang, *Science*, **266**, 250 (1994).
41. E. Lieb, T. Schultz, and D. Mattis, *Ann. Phys.* **16**, 407 (1961); D. B. Chestnut and A. Suna, *J. Chem. Phys.* **39**, 146 (1963); J. I. Krugler, C. G. Montgomery, and H. M. McConnell, *J. Chem. Phys.* **41**, 2421 (1964); F. C. Spano, *Phys. Rev. B* **46**, 13017 (1992).
42. B. I. Greene and R. R. Millard, *Phys. Rev. Lett.* **55**, 1331 (1985).

43. D. W. Snoke, J. P. Wolfe, and A. Mysyrowicz, *Phys. Rev. B* **41**, 11171 (1990); D. Snoke, J. P. Wolfe, and A. Mysyrowicz, *Phys. Rev. Lett.* **59**, 827 (1987); M. Hasuo, N. Nagasawa, T. Itoh, and A. Mysyrowicz, *Phys. Rev. Lett.* **70**, 1303 (1993).

44. J. J. Hopfield and D. G. Thomas, *Phys. Rev.* **132**, 563 (1963); J. J. Hopfield, *Phys. Rev.* **112**, 1555 (1958); **182**, 945 (1969); V. M. Agranovich, *Zh. Eksp. Teor. Fiz.* **37**, 430 (1959) [*Sov. Phys.—JEPT* **37**, 307 (1960)].

45. L. N. Ovander, *Usp. Fiz. Nauk* **86**, 3 (1965) [*Sov. Phys.—Usp.* **8**, 337 (1965)].

46. D. Frölich, St. Kirchhoff, P. Köhler, and W. Nieswand, *Phys. Rev. B* **40**, 1976 (1989).

47. V. N. Denisov, B. N. Mavrin, and V. B. Podobedov, *Phys. Rep.* **151**, 1 (1987).

48. C. K. Johnson and G. J. Small, in *Excited States*, Vol. 6, E. C. Lim, Ed. (Academic Press, New York, 1982); S. H. Stevenson, M. A. Connolly, and G. J. Small, *Chem. Phys.* **128**, 157 (1988).

49. G. M. Gale, F. Vallee, and C. Flytzanis, *Phys. Rev. Lett.* **57**, 1867 (1986).

BIBLIOGRAPHY

A. A. Abrikosov, L. P. Gorkov, and I. E. Dzyaloshinski, *Methods of Quantum Field Theory in Statistical Physics* (Prentice-Hall, Englewood Cliffs, New Jersey, 1963).

A. I. Akhiezer and S. V. Peletminskii, *Methods of Statistical Physics*, Sec. 6.3 (Pergamon, Oxford, 1981).

A. L. Fetter and J. D. Walecka, *Quantum Theory of Many-Particle System* (McGraw-Hill, New York, 1971).

H. Haken, *Quantum Field Theory of Solids* (North-Holland, Amsterdam, 1983).

P. Nozieres and D. Pines, *The Theory of Quantum Liquids*, Vol. 2 (Addison-Wesley, Reading, Massachusetts, 1991).

P. Ring and P. Schuck, *The Nuclear Many Body Problem* (Springer-Verlag, New York, 1980).

A. Yariv and P. Yeh, *Optical Waves in Crystals* (Wiley, New York, 1984).

Index

MX

Printed in the United States
135378LV00002B/13/P

9 780195 132915

Made in the USA
San Bernardino, CA
01 June 2017